Modern Engineering Graphics & Design

Modern Engineering Graphics & Design

Gerard G.S. Voland
Northeastern University

West Publishing Company
St. Paul New York Los Angeles San Francisco

The copyeditor for this book was Margaret Jarpey, who also prepared the index. Ernest MacMillan proofread the entire text.

The cover photographs were taken by Richard Stone. Wes Barris used MINDS (Minnesota Interactive Drafting System) to generate the computer illustration in the photograph at the top left. This is Figure 5.14 from the text and can be found on page 168.

This book was set in Kabel and Times Roman by Rolin Graphics. Rolin Graphics also prepared the interior art.

50 W. Kellogg Boulevard
P.O. Box 64526
St. Paul, MN 55164-1003

Printed in the United States of America

95 94 93 92 91 90 89 88 8 7 6 5 4 3 2 1

Library of Congress Cataloging-in-Publication Data

Voland, Gerard G. S.
Modern engineering graphics and design.

Includes index.
1. Engineering graphics. I. Title.
T353.V65 1987 604.2 86-19102
ISBN 0-314-93961-X

This book is dedicated to

my grandmother, the late Ella Kendall Burke—a loving and sweet lady,
my father, the late Norman Voland—whose last words to me have guided my life,
my mother, Eleanor Kendall Voland—who worked so exhaustively to provide me with opportunity and a happy childhood,
and my wife, Margaret Jacoby Voland—who provides her constant love, guidance, and support to me each day.

Contents in Brief

Contents

9 Size Tolerancing 243

10 Positional and Geometric Form Tolerancing 253

18 Engineering Design 415

Preface

Purpose of this Text

Engineering graphics is continually evolving as an active mode of communication within the world of technology. Its importance in this world has increased during recent years as a result of its union with the computer in computer graphics and computer aided design (CAD). Such evolution and growth has created the need for an introductory textbook that focuses upon both modern computer graphics and traditional (manual) graphics—and that also guides the student from the fundamental concepts of engineering graphics through a variety of more advanced topics. *Modern Engineering Graphics and Design* has been designed to aid the reader in his or her mastery of both manual and computer graphics.

The text attempts to achieve a balance between providing comprehensive treatment of the topics in graphics with which engineering students should be familiar and meeting the needs of the graphics instructor, who generally has limited time to cover each subject. An extensive survey of many graphics instructors in the United States (see the acknowledgments section following this preface) allowed me to identify those topics that are common to most graphics courses, and also gave me an appreciation of which topics are most emphasized in these courses. The results of this survey have influenced the design of this text.

Graphics is treated as a *language* in this work, with particular emphasis placed upon the concept of projection. Professor Gardner C. Anthony (an influential and farsighted engineering educator during the early part of the twentieth century who stressed the use of the word *graphics* to describe engineering drawing efforts) noted in the preface of his text, *Elements of Mechanical Drawing* (published by D.C. Heath & Co., Boston) in 1904:

> After a student has acquired the necessary skill in penmanship, the greatest stress should be laid on the Subject of Projection. He should be taught to regard Graphics as a language study, the grammatical construction of which is developed in

> the Theory of Projection . . . the subject should be taught as an art of expression rather than one of pictorial representation. Although most people recognize drawing as a medium for conveying thought, few appreciate the importance of teaching it as a language. But such it is in the fullest sense, possessing a well-defined grammatical construction, rich in varied forms of expression, forcible yet simple, and truly universal.

Professor Anthony's observations still ring with accuracy.

Today, computer graphics allows engineers to create drawings with both ease and accuracy. *Modern Engineering Graphics and Design* includes a careful presentation of modern computer graphics techniques.

A recent editorial in the professional journal *Engineering Design Graphics Journal* included the following passage directed toward graphics instructors and textbook authors:

> We have been criticized for writing for each other and it appears to be true. [Freshmen] can't read the graphics texts, and won't, until they are written for students.
>
> Dr. Jon Duff, Editor
> *Engineering Design Graphics Journal*
> Volume 49, Number 3, Autumn, 1985, p. 4.

This text *is* written for students, although it is rigorous in its treatment of graphic concepts and includes the latest standards of this field. Detailed explanations are included throughout the text, and the level is consistent with student abilities.

Of equal importance is the recognition that a graphics course is often the first course in engineering for a student. Unlike courses in physics, mathematics, and so forth, a graphics course should introduce the student to the engineering profession. *Modern Engineering Graphics and Design* includes numerous examples of engineering work, both past and present, that provide the student with a perspective of the profession.

Organization of this Text

As noted above, this text has been designed to aid the reader in mastering engineering graphics. This "design for success" includes both the organization of topics within the text and its content. Chapter 1 presents a brief review of engineering as a service profession within society, together with the application of graphics as a mode of communication within that profession. Chapters 2 and 3 introduce construction techniques in manual and computer graphics, respectively, together with the equipment most often used in each graphics mode.

The construction and interpretation of orthographic drawings is the focus of Chapters 4 through 10. Chapters 4 and 5 develop the student's ability to analyze data contained within a set of orthographic views; the novice is often overwhelmed by such data, seeing only lines without any logical relation to one another. With sufficient practice, such analysis can (eventually) be performed so quickly that the student may believe that he or she is immediately visualizing the object described by the drawing. Of course, such an ability to quickly read graphic descriptions of objects must be developed by the professional engineer.

Chapters 6 and 7 continue the treatment of shape/form descriptions of objects in orthographic drawings with the topics of sectioning and auxiliary views, respectively. The size description of objects in engineering drawings is then presented in Chapter 8 (dimensioning), Chapter 9 (size tolerancing), and Chapter 10 (positional and geometric form tolerancing).

Descriptive geometry forms the basis of Chapter 11 (true lengths and true-size areas of such graphic elements as lines and surfaces) and Chapter 12 (relative distances and orientations between two or more elements). This separation of descriptive geometry topics aids the student's mastery of the subtleties of this subject.

Fasteners are reviewed in Chapter 13, which is followed by two chapters devoted to pictorials. Pictorial projection theory is presented in Chapter 14 and practical construction methods for pictorial drawings in Chapter 15.

Sketching is presented in Chapter 16. Functional graphs, charts, and diagrams in which engineering data are often presented form the subject of Chapter 17. Finally, Chapter 18 reviews the engineering design process with which creative and feasible solutions to technical problems can be generated.

The instructor of the class may wish to alter the order of these topics during the course. There are numerous alternative sequences that can be very effective in the classroom.

Features and Pedagogy

The pedagogy used within the text includes a brief introductory section in each chapter (entitled "Preview"), followed by a concise set of chapter "Learning Objectives" and the main text of the chapter. Each chapter ends with a review (entitled "In Retrospect") of the major concepts that have been introduced in that chapter, together with numerous "Exercises." (In addition to the exercises given in this text, a workbook of problems is also available for use with this book.)

In addition to these features, each chapter also includes a section entitled "Engineering in Action" or "Computer Graphics in Action." Each "Engineering in Action" section presents an example of creative design in the development of an engineering solution to a challenging problem. These examples serve to introduce the student to both engineering as a profession and design as a creative activity. Each "Computer Graphics in Action" section extends the student's knowledge of the use of the computer in engineering and graphics. Chapter 3 focuses upon the use of two-dimensional computer graphics software in the creation of engineering drawings. Three-dimensional computer graphics topics (such as wire-frame, surface, and solid models and their construction) are introduced in several of the "Computer Graphics in Action" sections.

A *multistep format* in the construction (and interpretation) of engineering drawings is used in the text (particularly in Chapter 4, where the student is introduced to orthographic drawings) in order to carefully guide the student through each concept or technique. Color within the text has been judiciously used to complement this multistep format.

Special question-and-answer sections (entitled "Learning Check") have been included throughout the text in order to reinforce the student's understanding of major concepts that are sometimes misunderstood by the novice.

Historical perspective of engineering graphics (and its application in today's world) is provided in "Highlight" sections that appear throughout the text.

Exercises and Workbook

Computer graphics software allows one to store an engineering drawing on a floppy disk or within the memory of the computer system. As a result, a drawing can be developed and modified with ease, stored, and later retrieved for additional modifications as needed. In recognition of this storage capability in computer graphics systems, the exercises in this text include a large source-pool of problems (many of which are introduced in Chapters 4 through 7) that form the basis for many exercises in later chapters. For example, an exercise in Chapter 4 may require the student to develop three principal orthographic views of an object—which are then stored for later retrieval by the computer graphics system. An exercise in Chapter 7 will then refer to this earlier exercise and ask the student to develop an auxiliary view of a particular surface of the object. Similarly, an exercise in Chapter 8 will again refer to this earlier exercise and specify that the drawing should now be properly dimensioned. As a result, the student may continue to develop and complete an engineering drawing (as his or her knowledge of engineering graphics develops) by fully utilizing the storage and retrieval capability of computer graphics. Of course, such sequential modification and completion of drawings can also be performed manually by the student (without the aid of a computer).

Suggested *Term Projects* in design are included at the end of Chapter 18. These problems were developed by the National Institute for Occupational Safety and Health (NIOSH), United States Department of Health and Human Services for the 1985 American Society for Engineering Education (ASEE)—NIOSH Engineering Design Competition in Health and Safety. Each problem focuses on a *current* need in occupational safety and health. It is recommended that student design *teams* be formed to develop creative and feasible solutions to these problems. Each problem is very challenging, as one would expect of any real-life engineering project. The instructor may wish to define each problem in more specific terms so that a feasible design can be developed within the time constraints of the course. Of course, these projects are only included for possible use in those classes in which engineering design is a major course topic.

A workbook is available to accompany this text. In addition to numerous exercises which reinforce the student's understanding and ability to apply graphic techniques, blank pages are included so that additional exercises may be assigned from this text.

An instructor's manual is also available.

I trust that this text will be a useful and enjoyable tool for students of engineering graphics.

Acknowledgments

A text of this scope is truly dependent upon the skills and enthusiasm of many people. West Publishing Company has been extremely supportive of this project from its inception. I am very grateful for the guidance and sponsorship which I have received from my editor, Patrick Fitzgerald. Pat has devoted a great deal of effort to the creation of this book; thanks to his abilities, this book has been reworked and redesigned into its present (excellent) form. Nancy Hill-Whilton, development editor, has kept me on course (and nearly on schedule) during the creation of this book. Denise Bayco has been a tireless worker during the entire project. I have been most fortunate to work with a production editor who has been continually supportive and who has created the very effective design layout of this book (a difficult task for many reasons): Roslyn M. Stendahl. Finally, this text reflects the skill and devotion of Margaret Jarpey, who—as copyeditor—has immeasurably improved my original manuscript. I am grateful to each of these professionals.

The earlier and final versions of the manuscript have benefited from the thoughtful reviews of many of the foremost educators in engineering graphics and design, including:

- Yousuf Adenwala, Old Dominion University
- William G. Blakney, Auburn University
- Jack Brown, University of Alabama
- J. Harwick Butler, Middle Georgia College

- William Chalk, University of Washington
- Frank Croft, University of Louisville
- Jon Duff, Purdue University
- George Eggeman, Kansas State University
- Garland Hilliard, North Carolina State University
- Mary Jasper, Mississippi State University
- Jon Jenson, Marquette University
- Davor Juricic, University of Texas, Austin
- Robert Kelso, Louisiana Tech University
- William Kyros, University of Lowell
- B. Middleton, Ball State University
- Ed Mochel, University of Virginia
- Robert Raudebaugh, Arizona State University
- Edward Stevenson, University of Hartford
- Bruno Strack, Memphis State University
- Billy Wood, University of Texas, Austin

Robert Foster of Pennsylvania State University has been associated with the project as a reviewer since its inception and has provided extremely insightful and probing comments during its development. I warmly thank Professor Foster and each of the reviewers for their yeoman efforts.

Professor Robert Lang of Northeastern University reintroduced me to the field of engineering graphics some years ago; in addition, he has graciously served as a reviewer of this text. I thank him for his training and continuing support.

Professor Emeritus Borah Kreimer of Northeastern University has greatly affected my career through his professional support and friendship. My continuing interest in graphics and design is a direct result of his guidance and his example as a caring and extremely capable teacher.

Professor Emeritus Percy Hill and Professor William Crochetiere (former and current chairpersons, respectively) of the Department of Engineering Design at Tufts University continue to guide me through the world of engineering design. I thank each man for his kindness, patience, and expertise.

The National Institute for Occupational Safety and Health (NIOSH) kindly consented to the inclusion of several engineering design problems that served as the basis for student competition in the 1985 American Society for Engineering—NIOSH Engineering Design Competition in Health and Safety; these problems form the basis of suggested term projects in engineering design, as outlined in the exercise section of Chapter 18. Mr. John Talty, Engineer Director at NIOSH, has been continuously supportive of these efforts; I thank him for his support.

The professional publication entitled *Machine Design* has contributed many of the excellent examples of creative design which appear in the "Engineering in Action" sections near the end of many chapters. I am very grateful to the staff and publisher (Penton/IPC) of *Machine Design* for these selections.

Several industrial firms have kindly provided examples of professional engineering drawings and a wide assortment of photographs. These firms include:

- Bethlehem Steel
- Chrysler Motors
- Cincinnati Milacron
- Cleveland Twist Drill Company
- Control Data Corporation
- Deere & Company
- Digital Equipment Corporation
- General Motors Corporation
- Hughes Tool Division, Houston Texas
- Koh-I-Noor Rapidograph, Inc.
- Pratt & Whitney Machine Tool
- Texas Instruments
- Whirlpool Corporation

I happily thank each of these firms and their very helpful staffs for their contributions to this book.

I thank my wife and colleague, Margaret, for her love, understanding, and assistance, and my mother, Eleanor, for a lifetime of encouragement and support.

Finally, I thank the Lord for His guidance in my life and in the creation of this text.

As noted earlier in the preface, a survey of numerous graphics courses in the United States allowed me to identify those topics that are emphasized in most of these courses. I happily thank the following instructors who generously provided information for use in the development of this text.

Eddie R. Adams, Murray State University, Murray, KY

R. W. Adkins, University of Colorado at Denver, Denver, CO

Mallie C. Aldred, Dekalb College, Clarkston, GA

Elton Archer, LeTourneau College, Longview, TX

George Austin, Palomar College, San Marcos, CA

Bennett Lee Basore, Oklahoma State University, Stillwater, OK

Frank Blust, Edison Community College, Fort Myers, FL

Hailu Bogale, Prestonsburg Community College, Prestonsburg, KY

James P. Bosscher, Calvin College, Grand Rapids, MI

Fred Brasfield, Tarrant County Junior College, Fort Worth, TX

Sam L. Bridwell, Sr., The Pennsylvania State University, Mont Alto, PA

Daniel C. Brittigan, Virginia Military Institute, Lexington, VA

Walter N. Brown, Santa Rosa Junior College, Santa Rosa, CA

Betty J. Butler, University of Missouri at Kansas City, Independence, MO

John Campbell, Oregon State University, Corvallis, OR

James Cantrell, Caldwell Community College & Technical Institute, Hudson, NC

Les Caraway, Greenville Technical College, Greenville, SC

Charles E. Cartmill, Ricks College, Rexburg, ID

Merl Case, Central Missouri State University, Warrensburg, MO

H. A. Cazel, Georgia Institute of Technology, Atlanta, GA

Clayton Chance, Northern Arizona University. Flagstaff, AZ

Wan-Lee Cheng, University of North Dakota, Grand Forks, ND

Ginger Clark, East Central University, Ada, OK

Henry B. Cole, Jr., North Carolina A & T State University, Greensboro, NC

Alden W. Counsell, Southeastern Massachusetts University, N. Dartmouth, MA

John D. Daigh, Richland College, Dallas, TX

Wendell Deen, University of Texas at Austin. Austin, TX

Don Dekker, Rose-Hulman Institute of Technology, Terre Haute, IN

Robert Dennison, East Central University, Ada, OK

George M. Dysinger, Texas A & M University, College Station, TX
William A. Earl, New York State College of Ceramics, Alfred University, Alfred, NY
Donald L. Elfert, McNeese State University, Lake Charles, LA
William H. Eubanks, Mississippi State University, Mississippi State, MS
William L. Ferraioli, Hartford State Technical College, Hartford, CT
Alma Forman, Temple University, Philadelphia, PA
Glenn Good, Ouachita Baptist University, Arkadelphia, AR
Fryderyk E. Gorczyca, Southeastern Massachusetts University, N. Dartmouth, MA
John B. Greiner, Texas Tech University, Lubbock, TX
Bruce A. Harding, Purdue University, West Lafayette, IN
Sue Harlan, Oregon State University, Corvallis, OR
P. W. Harrison, Southern Institute of Technology, Marietta, GA
Carol L. Hoffman, University of Alabama, University, AL
William W. Holmes, University of Nebraska, Omaha, NE
Neil Hout-Cooper, Florida International University, Miami, FL
S. Louis Iannantuanu, University of Colorado at Denver, Denver, CO
Peter Ifekauche, Alabama A & M University, Normal, AL
Perry Isbell, Arkansas State University, State University, AR
Louis Jacobs, Loop College, Chicago, IL
Victor Janze, Pennsylvania State University, Dunmore, PA
William Javins, West Virginia Institute of Technology, Montgomery, WV
Robert C. Jenkins, Shoreline Community College, Seattle, WA
C. Greg Jensen, Brigham Young University, Provo, UT
Norman K. Jensen, Valparaiso University, Valparaiso, IN
Robert Jerard, Dartmouth College, Hanover, NH
Charles Kajkowski, University of Nevada, Las Vegas, NV
Daniel Kane, New York Institute of Technology, Old Westbury, NY
Donald E. Keyt, Spring Garden College, Chestnut Hill, PA and University of Pennsylvania, Philadelphia, PA
Charles O. Kishiban, University of Bridgeport, Bridgeport, CT
Richard J. Kroll, Portland Community College, Portland, OR
Thomas Krueger, Brazosport College, Lake Jackson, TX
Alexander K. Laport, Montana State University, Bozeman, MT
Keith A. Lauderbach, Millersville University, Millersville, PA
Ora Leshchinsky, Widener University, Chester, PA
Ernest Levesque, New Hampshire Vocational-Technical College, Laconia, NH
Orlyn Lockard, Longview Community College, Lee's Summit, MO
Gerald G. Lovedahl, Clemson University, Clemson, SC
Robert A. Lucas, Lehigh University, Bethlehem, PA
Robert L. Mabrey, Tennessee Technological University, Cookeville, TN
Michael D. Makinen, Baldwin Wallace College., Berea, OH
Everett Malan, Cowley County Community College, Arkansas City, KS
Joseph J. Manak, Pennsylvania State University - Beaver Campus, Monaca, PA
Douglas M. Mattick, St. Cloud State University, St. Cloud, MN
Thomas Mayerchak, Forsyth Technical Institute, Winston-Salem, NC
Barry S. McCaskill, Cleveland State Community College, Cleveland, TN
Peter W. Miller, Purdue University, West Lafayette, IN
Silver Miller, Colorado School of Mines, Golden, CO
George H. Montgomery, University of Dayton, Dayton, OH
Kenneth S. Moody, Vermont Technical College, Randolph Center, VT
H. Carlton Moore, Jr., Wentworth Institute of Technology, Boston, MA
Subir Mozumdar, Arkansas Tech University, Russellville, AR
Edward J. Nagle, Tri-State University, Angola, IN
Gabriel M. Neunzert, Colorado School of Mines, Golden, CO
Hugh Nutley, Seattle Pacific University, Seattle, WA
Wesley Pauls, McPherson College, McPherson, KS
Robert Pearce, South Plains College, Levelland, TX
Ken Perry, Indiana University-Purdue University at Fort Wayne, Fort Wayne, IN
Gordon G. Peterman, Arizona State University, Tempe, AZ
Warren F. Phillips, Utah State University, Logan, UT
Douglas L. Pickle, Southeastern Louisiana University, Hammond, LA
John W. Pierce, Chaffey College, Alta Loma, CA
Thurman Potts, Northeast Louisiana University, Monroe, LA
Roger Powell, Higheine Community College, Midway, WA
P. V. Ramakrishnaiah, University of Houston, Houston, TX
Robert Rautenstrauch, Kent State University, Kent, OH
David Rein, Central Oregon Community College, Bend, OR
Michael J. Rider, Ohio Northern University,
G. Douglas Roberts, Northern Kentucky University, Highland Heights, KY
Daniel L. Ryan, Clemson University, Clemson, SC
Gary Schrener, Southeast Missouri State University, Cape Girardeau, MO
Victor R. Serri, Rock Valley College, Rockford, IL
Frederick Sherman, University of California, Berkeley, CA
William A. Sigurdson, Indian River Community College, Fort Pierce, FL
Dale E. Simon, University of Toledo, Toledo, OH
Charles A. Slaten, Tarleton State University, Stephenville, TX
Roger Smith, Iowa State University, Ames, IA
Ron Spangler, East Carolina University, Greenville, NC
Lowell P. Stanlake, University of North Dakota, Grand Forks, ND
S. Russell Stearns, Dartmouth College, Hanover, NH
Michael Stewart, University of Arkansas, Little Rock, AR
David Stienstra, Texas A & I University, Kingsville, TX
Porter W. Stone, University of Arkansas, Fayetteville, AR
Victor Sullivan, Pittsburg State University, Pittsburg, KS
Louis M. Swiczewicz, Jr., Roger Williams College, Bristol, RI
Robert Takacs, Kings River College, Reedley, CA

Sivagnanam Thangam, Stevens Institute of Technology, Hoboken, NJ

Martin J. Thomas, Rose-Hulman Institute of Technology, Terre Haute, IN

Frank P. Tobiasz, Mohawk Valley Community College, Utica, NY

Dale M. Truitt, East Texas State University, Commerce, TX

Robert L. Turner, University of Idaho, Moscow, ID

Alok K. Verma, Old Dominion University, Norfolk, VA

Richard W. Vockroth, Corning Community College, Corning, NY

James A. Vogtsberger, Milwaukee School of Engineering, Milwaukee, WI

John C. Wagner, Northampton County Area Community College, Bethlehem, PA

Robert A. Walker, Lexington Community College, Lexington, KY

Charles W. White, Purdue University, West Lafayette, IN

Ron White, Tulsa Junior College, Tulsa, OK

E. Ray Whitfield, Richland College, Dallas, TX

Donald C. Wood, Western Carolina University, Cullowhee, NC

James P. Woughter, Alfred State College, Alfred, NY

Christopher Wyatt, Nashville State Technical Institute, Nashville, TN

Herman Yawn, York Technical College, Rock Hill, SC

Pu-Sen Yeh, Jacksonville State University, Jacksonville, AL

Lester Zimmerman, Goshen College, Goshen, IN

Lowell S. Zurbuch, Kent State University, Kent, OH

Modern Engineering Graphics & Design

1 Engineering, Graphics, and Design

Preview As you work through this text, two facts should be noted:

1. *Engineering is a service profession* devoted to solving some of the problems confronting civilization, so that people will be able to live in greater health, comfort, and safety.
2. *Engineering graphics is a visual language* that allows the engineer to communicate effectively with other professionals; and it is a language in which the engineer must be fluent. As is true of any language, fluency in graphics requires diligent effort and patience.

This text aims not only to help you develop fluency in the graphics language, but also to show you how to apply it to engineering design efforts.

Learning Objectives

Upon completion of this chapter, the reader should be able to:

- Justify the need for the professional engineer to be fluent in the graphics language.
- Describe the benefits of engineering graphics skills, including (1) the ability to communicate and (2) the ability to visualize.
- Recognize that engineering is a service profession in which science and technology are applied to the needs of civilization.
- Identify some of the areas of concern and application of various engineering disciplines (for example, civil engineering).

1.1 Engineering Graphics

Engineering graphics is a concise and accurate mode of communication; we might describe it as a **language** with its own grammar and style. It is fundamental to engineering design, architecture, manufacturing, surveying, and many other areas of engineering in which science is applied to the needs of humankind (Figures 1.1 and 1.2).

Every engineer is familiar with (at least) three specific modes of communication:

- **Verbal Languages,** which are the subject of the science known as linguistics.

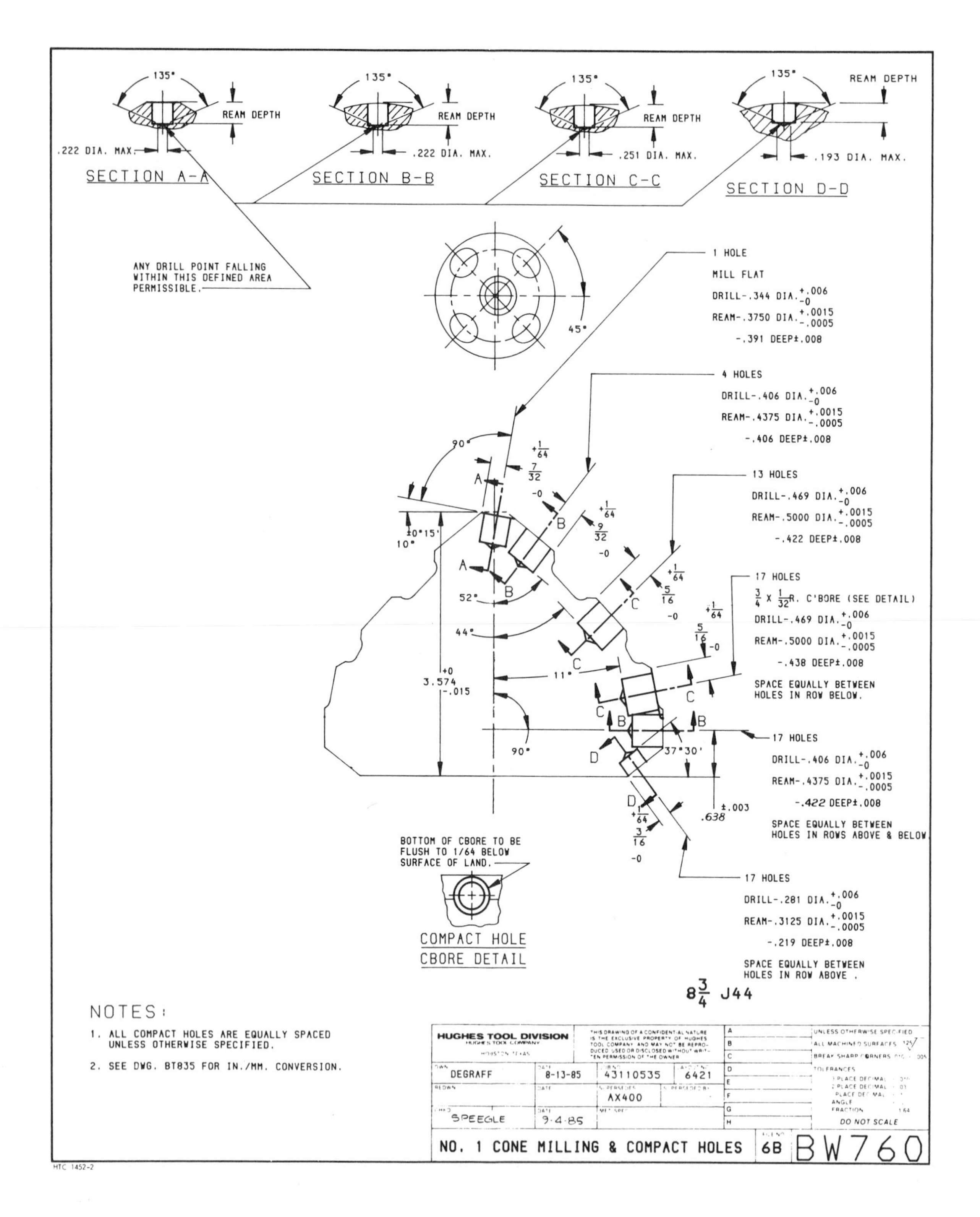

Figure 1.1 An example of a professional engineering drawing of an 8¾-inch J44 rock bit cone used for oil-drilling purposes. See photograph in Figure 1.2. (Courtesy of Hughes Tool Division, Houston, Texas.)

- **Mathematics,** in which abstract numeric symbolism is used to represent concepts in a system based upon formal logic.
- **Graphics,** in which visual symbolism is applied to the description and representation of objects and concepts.

Skill in each type of language enables the engineer to describe technical concepts accurately and clearly. A lack of skill in any of these languages inhibits communication and therefore ultimately inhibits an individual's success in his or her chosen engineering discipline.

Undoubtedly, you have already developed verbal and mathematical communication skills to a satisfactory degree, since most of your formal education has been devoted to the development of those skills. Very few engineering students, however, begin their college education with the necessary equivalent skills in graphic communication techniques. As you embark on the task of developing graphics skills, you may occasionally become frustrated with the rate at which you acquire expertise; admittedly, the learning process can be tedious. But this is true of learning *any* language. With sufficient practice and careful attention to written and oral instruction, you will gradually develop fluency in the graphics language.

Remember that graphics is perhaps the most important language with which engineers communicate. As an engineer, you must regard skill in this language as highly as you do skill in mathematics. Recognizing the importance of graphics skills in engineering will motivate you to develop these skills.

The main assets that engineering graphics provides to those entering the engineering profession are:

1. *The ability to communicate.* As noted earlier, engineers communicate through the use of three distinct types of languages: (1) verbal languages, (2) mathematics, or abstract symbolism, and (3) graphics, or visual symbolism. The professional engineer must be fluent in each type of language.

2. *The ability to visualize.* Graphics allows us to see objects, engineering components, and concepts more vividly. Engineering design focuses on the application of science and technology to develop solutions to engineering problems that need to be solved for the benefit of society. Graphics provides a way of visualizing the concepts and designs involved in that endeavor. In addition to providing the ability to communicate and to visualize, graphics also provides a structure that helps develop disciplined thinking and physical skills. French stated that graphics is the foundation upon which all designing is based.[1] Hill noted that it is the vehicle

Figure 1.2 A 7⅞-inch J22 "Tri-cone" rock bit assembly used for oil-drilling purposes. The cones used in this assembly are similar (but not identical) to that shown in Figure 1.1. Three distinct cone designs are required for a rock bit assembly. The first steel-toothed rock-drilling bit, designed and built by the Howard Hughes Tool Company, was used to sink an oil well at Goose Creek, Texas, in 1908. This bit allowed previously untapped oil deposits *below hard rock* to be drilled. Earlier wells had been sunk by applying the percussion method, that is, literally pounding soft rock repeatedly with a cutting tool. Oil-drilling practices were transformed by this bit. For the complete set of working drawings for a rock bit cone see Appendix N. (Courtesy of Hughes Tool Division, Houston, Texas.)

1. French, Thomas E., "The Educational Side of Engineering Drawing", presented at the Annual Conference of the Society for the Promotion of En-

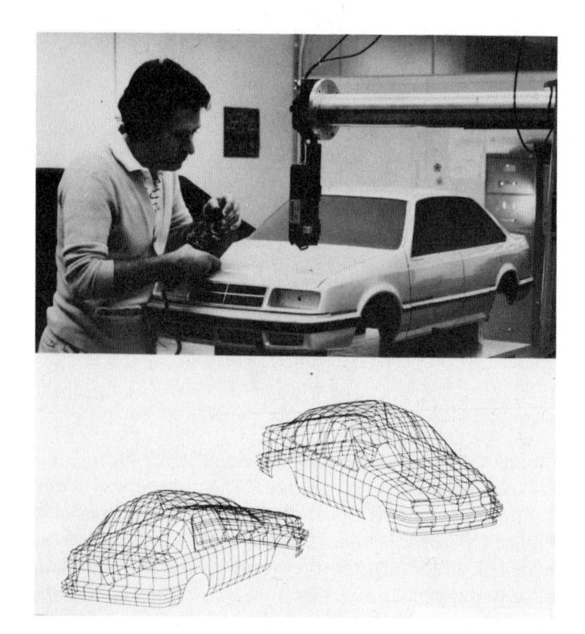

Figure 1.3 With a laser beam, Chrysler's computerized digitizer scans a three-eighths model of the 1985 Chrysler LeBaron GTS/Dodge Lancer in three dimensions. The information is relayed to Chrysler's computer-aided design (CAD) center for use in engineering design, manufacturing engineering, and production. The computer printout shown here gives two perspectives compiled from the information. (Courtesy of Chrysler Motors.)

Figure 1.4 Computer-aided design (CAD) and graphics systems allow engineers to develop solutions to challenging technical problems with both creativity and accuracy. In this example, a VAXstation 100 is shown in use. (VAXstation 100 is a trademark of Digital Equipment Corporation.)

by which design areas should be taught.[2] The significance of graphics in engineering design cannot be overestimated. However to be an asset, graphics must be mastered well enough to be applied with skill and speed in the development of engineering solutions. Such mastery comes with diligent and patient effort.

Today powerful CAD (computer-aided design) and CADD (computer-aided design drafting) systems are used to develop engineering solutions to a wide variety of problems and produce graphic descriptions of these solutions with precision and speed (Figures 1.3 and 1.4). Engineers must be proficient in the graphics language in order to fully utilize these CAD and CADD systems. Engineering graphics and engineering design have become more interlinked than ever before as a result of these systems.

1.2 Engineering—A Service Profession

Engineering is the application of science and technology to develop solutions to practical problems—problems that need to be solved for the benefit of humankind. It is a **service** profession. Each of us in this profession seeks to serve or assist others who are in need of a solution to a perplexing (and sometimes life-threatening) problem. The problem to be solved may relate to any one of the following fields:

- Food and fiber industries, in which harvesters, milk processors, feed distributors, and other farm machinery (Figure 1.5), are needed to efficiently and economically satisfy the needs of civilization **(agricultural engineering).**
- Materials production and processing industries, in which insecticides, cements, paints, plastics, fuels, and other substances need to be produced for society's needs, and in which the recycling of waste, the elimination of environmental pollution, the chemical reaction between substances, and other challenges confront the professional engineer **(chemical engineering).**
- Application of principles and knowledge of structures, hydraulics, and surveying to create buildings, dams, tunnels, bridges, and roads **(civil engineering).**
- Generation, distribution, and utilization of electrical energy, or control processes, communications, and computer technology **(electrical engineering).**

---

gineering Education, University of Minnesota, June 1913; reprinted in *Engineering Design Graphics Journal,* vol. 40, no. 3 (1976) pp. 32–35.

2. Hill, Percy H., acceptance of the Distinguished Service Award of the ASEE Division of Engineering Design Graphics, *Engineering Design Graphics Journal,* vol. 41, no. 3 (1977), p. 21.

Highlight

Leonardo da Vinci and Graphics in Engineering

Leonardo da Vinci (1452–1519) was perhaps the greatest artist and engineer of all time. He was a painter, sculptor, architect, and musician, as well as an engineer, scientist, and mathematician. His work in anatomy, hydraulics, mechanics, and machinery (for example) was both truly innovative and practical. And da Vinci appreciated the need for graphics in the development and communication of engineering designs. See his sketch of a weight-driven motor (taken from his *Codex Atlanticus*).

As we will discover in later chapters, da Vinci's work preceded the development of what we today recognize as *technical graphics;* his use of graphics in engineering design efforts showed that he valued the method of sketching ideas in order to develop and refine an engineering solution to a problem. Today, sketching remains the most valuable graphics tool available to the engineer during preliminary creative design sessions; it is a tool upon which we will focus in Chapter 16. After sketching leads to a final design solution, the rules and structure of the language known as technical (or engineering) graphics then allow that solution to be described in sufficient detail for it to be produced.

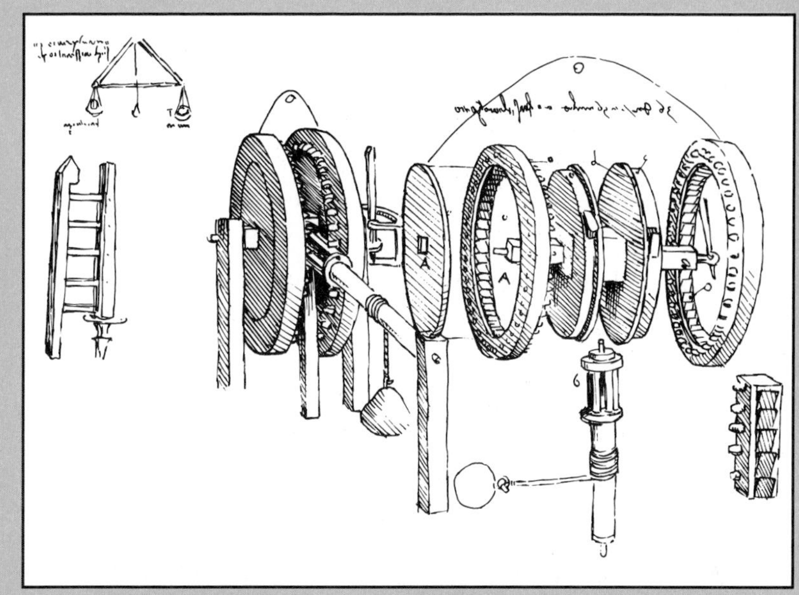

A sketch of a weight-driven motor, redrawn from Leonardo da Vinci's *Codex Atlanticus*. The assembled motor is shown on the left and an exploded assembly of the design's components on the right.

- Integrating materials, finances, equipment, and people in efficient, profitable, productive and comfortable environments **(industrial engineering).**
- Design and development of engines, automobiles, aircraft, refrigeration systems, pumps, or solar energy systems **(mechanical engineering).**
- Other engineering disciplines include **architectural, aeronautical, ceramic, environmental, mining,** and **nuclear engineering,** to name only some.

Figure 1.5 Agricultural engineers develop creative and effective machinery for use in the food and fiber industries. A Deere & Company 3430 Hay Windrower is shown in this photograph. The production, processing, and handling of food is one example of engineering as a service profession that benefits mankind. (Courtesy of Deere & Company.)

1.3 Science and Engineering

Physicists, chemists, and other scientists often direct their efforts toward analyzing and correlating experimental and theoretical discoveries; their primary goal is to develop a unified, comprehensive set of fundamental principles and

Highlight

Herbert Hoover and the Engineering Profession

Herbert C. Hoover (1874–1964), thirty-first president of the United States, chose engineering as his profession. He appreciated the service aspect of his occupation, as he demonstrated in the following statement:

> It is a great profession. There is the fascination of watching a figment of the imagination emerge through the aid of science to a plan on paper. Then it moves to realization in stone or metal or energy. Then it brings jobs and homes to men. Then it elevates the standard of living and adds to the comforts of life. That is the engineer's high privilege . . . To the engineer falls the job of clothing the bare bones of science with life, comfort and hope.

concepts that accurately describe physical behavior. In contrast, engineering is concerned with the *application* of these scientific principles and concepts to practical problems for the benefit of humankind.

Mathematicians, physicists, chemists, and other so-called pure scientists seek *unique* solutions to the problems with which they are concerned, that is, they deal with problems for which there are one-of-a-kind solutions. Engineers, on the other hand, seek the *best* of many possible solutions that could be designed and developed to a practical, real-life problem. The reason for this difference is that a practical problem may be solvable in many different ways, unlike a typical mathematical or physics problem for which there is only a single solution.

During recent years, however, the boundaries between pure science and engineering have become increasingly blurred; researchers in applied science areas have filled the region that once separated fundamental research from practical problem-solving. A continuous spectrum of activities extends from basic scientific research to the development of engineering solutions to practical problems. As a result, the engineer may be involved in both pure research and engineering applications of this research; and scientists may be involved in fundamental research that was prompted by the recognition of the need for a solution to a practical problem.

1.4 The Engineering Profession: Teamwork Among Specialists

The **engineer** is a member of a team that includes other specialists, from the **scientist** who defines the basic theoretical or experimental relationships between system variables to

Highlight

Sir Isaac Newton: Science and Mathematics

The mathematical basis for the theories of scientific research was given its greatest support by the work of Sir Isaac Newton. As a boy, he had built sundials, waterclocks, and other works; as a man, he established:

- That white light is a collection of the colors in the rainbow.
- The branch of mathematics known as the calculus (independently developed by Leibniz).
- The reflecting telescope (in which a parabolic mirror is used to reflect light, unlike refracting telescopes in which light is refracted through a lens), which advanced the study of astronomy.
- The scientific basis for the motion of heavenly bodies—and, in so doing, his three laws of motion.

Newton once noted:

> If I have seen further than other men, it is because I stood on the shoulders of giants.

Yet how many in science, mathematics, and engineering have stood upon his shoulders?

Sir Isaac Newton

the **technician** who performs laboratory experiments or supervises the production of system components. The technician serves as the human bridge between the engineer and the **crafter** who actually constructs the system. Crafters include machinists, welders, carpenters, electricians, glass blowers, and other professionals who provide skills that are not held by the engineer or the technician. Sound engineering knowledge and judgment are required in this bridging role.

Creativity in the development of engineering solutions is provided by the **designer** (or design engineer) who defines the problem and then generates multiple designs to achieve the goals required of a solution to the problem. The designer may be an engineer or a specialist from another field who has imagination and physical intuition.

Finally, the **stylist** is concerned with the appearance (as opposed to the function) of the design; he or she seeks to maximize the aesthetic beauty and the marketing potential of the engineering solution.

In addition to the professionals named thus far, **distributors, salespeople, managers, accountants, lawyers** and other workers are members of the design team; and the engineer may work with any or all of them during the development of the design solution to a problem. Effective teamwork requires precise and complete communication between all members of the team. Engineering graphics is the language that engineers can use to communicate with one another and with other members of the team accurately and clearly.

1.5 Engineering Disciplines

As you may know, an engineer often identifies himself or herself as a member of a particular engineering discipline or area of specialization. As science and technology progress, the need for specialization is creating new engineering categories in addition to the early (and now *traditional*) disciplines of engineering.

Aerospace engineering is concerned with the design and development of modern flight vehicles. These vehicles may be designed for flight in the earth's atmosphere **(aeronautical engineering),** or they may be designed to operate beyond the atmosphere **(astronautical engineering).** The design of such vehicles involves a variety of specialty component systems for life-support, propulsion, guidance, and so forth. Therefore, aerospace engineers may work in such subdisciplines as vibration control, production techniques, stress analysis, aircraft stability, instrumentation, or materials science.

Agricultural engineering involves the production, processing, and handling methods used in the food and fiber industries. As you might expect, it represents one of the earliest applications of scientific knowledge to the needs of humankind. Agricultural engineers may be striving to improve crop production through the design and fabrication of farm machinery (such as harvesters, milk processors, and feed distributors), farm structures (such as barns, silos, and granaries), or soil and water systems (such as drainage, irrigation, and soil additives). As a result, these men and women may be applying their expertise in structural mechanics, electrical circuitry, hydraulics, thermodynamics, chemistry, biology, economics, and other fields as they seek to improve food production and life for all of us.

Chemical engineering is primarily concerned with the properties, processing, and production of materials such as paints, plastics, fuels, and food. The recycling of waste products, environmental pollution control, fluid or solid material transport through pipes or conveyor systems, evaporation of liquids, chemical reactions between substances, the development of chemical compounds (such as cosmetics, dyes, insecticides, detergents, and cements), material handling, and many other activities require skilled chemical engineers.

Civil engineering applies principles and knowledge of materials, structures, hydraulics, surveying and other applicable areas to create buildings, dams, tunnels, bridges, roads, sewage and water supply systems, and so forth. Civil engineers may specialize in construction, hydraulic engineering

ENGINEERING IN ACTION

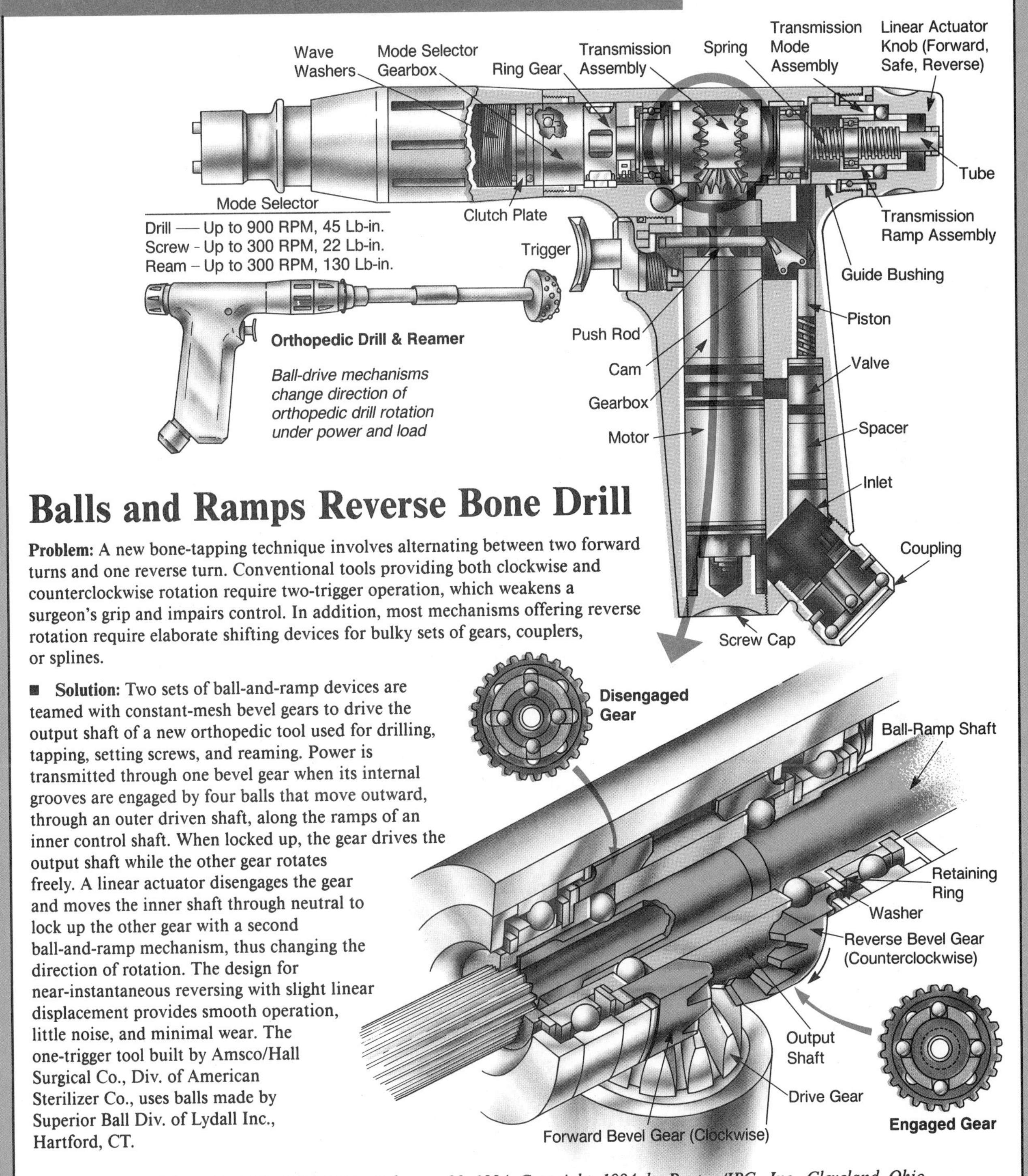

Balls and Ramps Reverse Bone Drill

Problem: A new bone-tapping technique involves alternating between two forward turns and one reverse turn. Conventional tools providing both clockwise and counterclockwise rotation require two-trigger operation, which weakens a surgeon's grip and impairs control. In addition, most mechanisms offering reverse rotation require elaborate shifting devices for bulky sets of gears, couplers, or splines.

■ **Solution:** Two sets of ball-and-ramp devices are teamed with constant-mesh bevel gears to drive the output shaft of a new orthopedic tool used for drilling, tapping, setting screws, and reaming. Power is transmitted through one bevel gear when its internal grooves are engaged by four balls that move outward, through an outer driven shaft, along the ramps of an inner control shaft. When locked up, the gear drives the output shaft while the other gear rotates freely. A linear actuator disengages the gear and moves the inner shaft through neutral to lock up the other gear with a second ball-and-ramp mechanism, thus changing the direction of rotation. The design for near-instantaneous reversing with slight linear displacement provides smooth operation, little noise, and minimal wear. The one-trigger tool built by Amsco/Hall Surgical Co., Div. of American Sterilizer Co., uses balls made by Superior Ball Div. of Lydall Inc., Hartford, CT.

Source: Reprinted from MACHINE DESIGN, February 23, 1984. Copyright, 1984, by Penton/IPC., Inc., Cleveland, Ohio.

(involving water, its usage and its transport), highway engineering (focusing upon tunnels, traffic control systems, and so forth), transportation (such as railroads and airlines), sanitation (such as water purification and pollution control), city planning (involving city growth patterns, street planning, zoning, and so on), or a variety of other subdisciplines directed towards specific needs of society.

Electrical engineering deals with the generation, distribution, and utilization of electrical energy; the design of control and communications systems; and computer designs. The major subdivisions are electronics and power: the former involves the small quantities of electrical power that are used in communications and electrical equipment, and the latter is concerned with the energy needs of large electricity users, such as cities and major industries. Transmission equipment, generators, appliances, lighting systems, communications systems, and instrumentation represent only a few of the many areas in which electrical engineering is applied.

Industrial engineering is a field that spans engineering, business management, and the social sciences. It is concerned with integrating materials, finances, equipment, and people in efficient, profitable, productive and comfortable industrial systems. Cost analysis, plant layout, quality control, plant processes, automation, performance standards and measurement, personnel management, and human factors are some of the areas in which industrial engineers are involved.

Mechanical engineering concerns include transportation devices, energy processes (including heating and cooling systems), stress analysis, manufacturing processes, and power generation. The design and development of engines, automobiles, locomotives, aircraft, fuel storage systems, refrigeration systems, pumps, and solar energy systems are examples of the practical applications of mechanical engineering.

Additional areas of engineering that are either newer, less specialized, or smaller than those already mentioned, but are of no less importance, include:

- **Computer software engineering.**
- **Biomedical engineering.**
- **Nuclear engineering.**
- **Mining and metallurgical engineering.**
- **Petroleum engineering.**

Still others exist as well, and as technology progresses and additional needs of humanity are identified, further divisions of engineering will be formed to properly direct that technology to fulfill those needs.

ENGINEERING IN ACTION

Bottle Cap Keeps Track of Drug Use

A digital module in the cap of a medicine bottle tells a patient when he last took his prescription medicine. The timepiece, recessed in the top of a cap, has two electrical contacts that fit through holes in a partition. One contact maintains constant engagement with a conductive membrane in the lower part of the cap. The other contact touches the membrane only when the cap is secure on the bottle. This causes an electrical short that disables any advancement of the cap's display device, thereby indicating the time medicine was last taken. The displayed time then catches up to the current time when the cap is removed again. The reminder cap is intended to help sometimes forgetful people take the correct amount of medication at the proper time, reports designer Bart J. Zoltan, Medical Research Div. of American Cyanamid Company's Lederle Laboratories, Wayne, N.J.

Source: Reprinted from MACHINE DESIGN, June 9, 1983. Copyright, 1983, by Penton/IPC., Inc., Cleveland, Ohio.

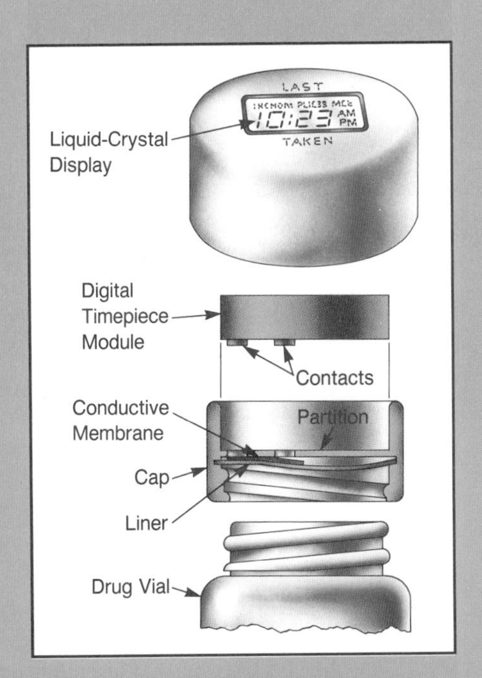

Balloon-Filled Bladder Immobilizes Injured Spine

Problem: Accident victims with injured spinal cords are often further traumatized during removal from an accident scene. The irreparable damage occurs when strapping the victim to a conventional spine board or an articulated chair. In addition, makeshift pads placed between the victim and the board or chair can create undesirable pressure points.

Solution: A conformal bladder filled with microballoons immobilizes the spine of an accident victim before any movement takes place. The flexible bladder is strapped about the head and torso of the victim, then evacuated with a vacuum pump. This causes the microballoons to expand forming a rigid mass that supports the victim's spine. The bladder is then strapped to a rescue chair using cloth wedges to fill any void between the bladder and chair. An accident victim is then transported to a hospital aboard the rigid support. A valve is opened to let air into the bladder, softening it for removal or adjusting the fit. The spine immobilizer is a design of Hubert C. Vykukal, Ames Research Center, Moffett Field, CA.

Source: Reprinted from MACHINE DESIGN, October 6, 1983. Copyright, 1983, by Penton/IPC., Inc., Cleveland, Ohio.

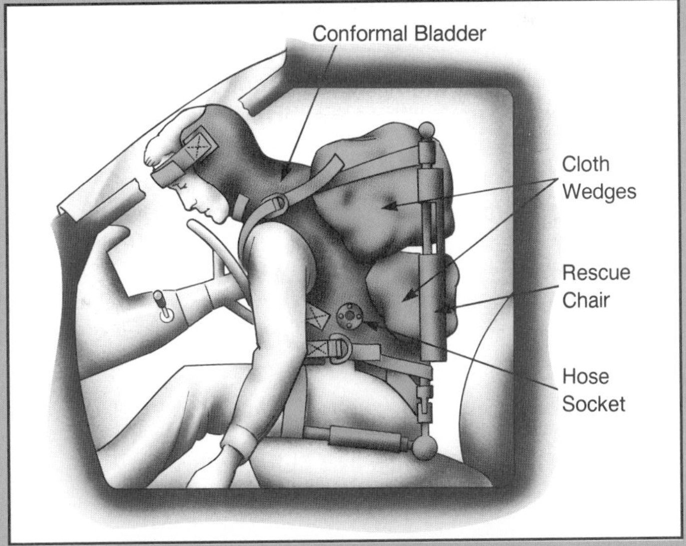

In Retrospect

- Engineering is a service profession devoted to developing solutions to problems confronting civilization, so that people will be able to live in greater health, comfort, and safety.
- The engineer is a member of a *team* that includes scientists, technicians, crafters, stylists, distributors, salespeople, managers, accountants, lawyers, the consumer (most importantly), and many others who are devoting their efforts to the development of the best design solution to an engineering problem.
- Engineering graphics is the *language* that enables the engineer to communicate effectively. It is a language in which the engineer must be fluent. Sketches and finished (instrument) drawings must be generated to describe the proposed design solution to a problem to other members of the design team. *Graphics* is the universal language (together with that of mathematics) that allows the engineer to effectively communicate with his or her colleagues.
- A broad range of engineering disciplines has developed in response to specific needs of civilization and further subdisciplines may be necessary in the future as technology progresses and additional needs are identified.

Exercises

EX1.1. Identify three engineering products that have appeared in the marketplace within the last five years. Describe the positive and negative effects on each product on individuals and on society in general. Also, identify the engineering disciplines—for example, civil, mechanical, industrial—that were involved in the design, development, and production of these three engineering solutions.

EX1.2. Throughout this text we have included examples of specific engineering design solutions to a wide variety of problems; these examples are collected under the special sections entitled "Engineering in Action" in Chapters 1, 2, 10 through 14, and 17. Select one of these designs and evaluate it in terms of the engineering disciplines that were involved in its development and its (positive and negative) impact on society.

EX1.3. Herbert Hoover and Leonardo da Vinci are only two examples of engineers who have greatly affected civilization. Identify at least ten other prominent individuals who worked as professional engineers (for example, George Washington was a civil engineer.) Indicate the particular engineering discipline of each individual.

EX1.4. List as many reasons as possible for the inclusion of graphics in engineering work.

EX1.5. Choose an engineering product (for example, a door lock, a cooking appliance, a piece of furniture, a tape dispenser, a lamp, a light switch) and sketch this product in sufficient detail so that another person would understand both the operation of the product and the function of each component.

EX1.6. List the reasons for your choice of engineering as your profession. Also list the abilities that you believe that a successful engineer must possess.

2 Manual Graphics: Equipment and Construction Techniques

Preview In this chapter we will focus upon manual graphics equipment and its use in the development of specific linework in engineering drawings. Then, in Chapter 3, we will review computer graphics equipment and its application in the development of engineering drawings.

The engineer should know how to use manual, or traditional, graphics equipment, since an equivalent set of computer graphics equipment may not always be available. Moreover, experience with manual construction techniques in graphics fosters a deeper appreciation for the precision, efficiency, and ease of computer graphics technology in developing and reproducing drawings. Another advantage of manual construction techniques is that they allow the novice in the graphics language to develop his or her skills without the additional difficulties that are sometimes associated with mastering computer graphics systems; he or she will be able to distinguish between those aspects of the *language* that require additional effort (in order to achieve an appropriate level of proficiency) and those characteristics of the *computer system* that need to be mastered.

As we will discover throughout our treatment of engineering graphics, the language is independent of the particular mode (manual or computer-based) used to construct graphic descriptions of objects. The engineer should be familiar with both construction modes.

Learning Objectives

Upon completion of this chapter, the reader should be able to:

- Identify the standardized layouts of drawing sheets in terms of both size and format specifications.
- Properly use such drawing tools as the T-square, triangles, pencils, erasers, an erasing shield, a dusting brush, a compass, dividers, and scales to construct specific types of linework in a drawing.
- Differentiate between instruments known as scales and the scale of a drawing.
- Differentiate between common types of angles (right, acute, and obtuse), sets of angles (complementary and supplementary), triangles (right, equilateral, isosceles, and scalene) and quadrilaterals (square, rectangle, rhombus, rhomboid, and so forth).
- Identify characteristics of a circle (radius, diameter, centerpoint, chord length, and so forth).
- Apply specific construction techniques to:
  - Bisect a straight line.
  - Bisect an angle.

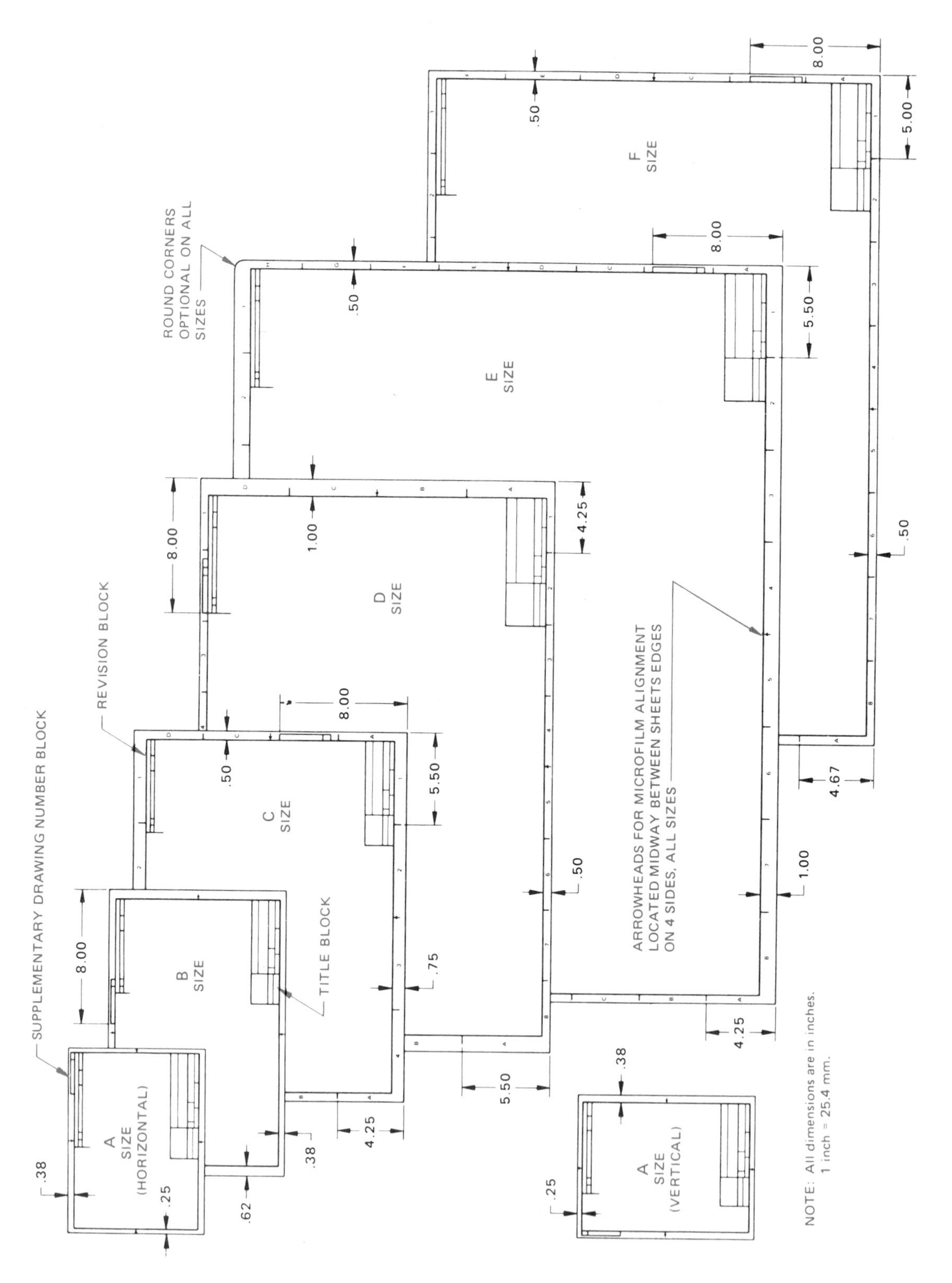

Figure 2.1 Drawing sheet (flat) sizes and formats. (ANSI Y14.1—1980, AMERICAN NATIONAL STANDARD DRAWING SHEET SIZE AND FORMAT.)

- Divide a line into more than two equal parts.
- Construct lines that are tangent to two circles of equal or unequal diameters.

- Define ellipses, parabolas, and hyperbolas as specific types of conic sections.
- Develop single-stroke gothic characters that are uniform in size and spacing for lettering in a drawing.
- Differentiate between various types of linework that may be used in an engineering drawing, including visible, hidden, section, center, phantom, extension, dimension, cutting-plane, symmetry, stitch, and chain lines.

2.1 Drawing Sheet Sizes and Formats

The paper used in manual drawing is **standardized** in both size and format. The variety of standard sheets available is presented in Figure 2.1, with the corresponding dimensions (sizes) for each standard sheet given in Table 2.1. Drawings may be stored flat or rolled, depending upon the size of the sheet.

Drawing sheets are standardized in order to ensure that:

- Drawings can be economically reproduced, bound and stored.
- Revisions to drawings can be incorporated with minimal effort.
- The available space within the drawing area is efficiently utilized.
- Communication between readers of a drawing is enhanced via the use of standardized data location.

The drawing sheet is partitioned into specific regions, or **blocks,** where standard information is inserted. A **title block** includes an identification name and address of the firm or group for whom the drawing was created, together with a drawing title, drawing number, the names of the relevant personnel involved in the production of the drawing (such as the engineer and the checker), date(s), scale(s), and any other information required by the firm. A **parts-list block,** as its name implies, includes an identification listing of the components or parts of the engineering object described in the drawing. **Supplementary blocks** can be used for dimensioning and tolerancing data, materials specification, distribution instructions, and so forth.

The title block is usually placed in the lower right corner of the drawing sheet. The parts list, beginning at the top border, can be placed directly above the title block. Supplementary blocks may then be located adjacent and to the left of the title block. With this type of layout, the title block is easily found by the reader and serves as a central location for all of the data blocks in the drawing.

Relatively thick borderlines (approximately 0.030 inches in width) are used to establish these principal blocks; thinner (approximately 0.015 inches in width) borderlines are used

Table 2.1 Standard drawing sheet sizes—flat sheets; see Figure 2.1. All dimensions are in inches. (ANSI Y14.1-1980, American National Standard Drawing Sheet Size and Format.)

Flat Sizes[a]				
Size Designation	Width[c] (Vertical)	Length (Horizontal)	Margin	
			Horizontal	Vertical
A (Horiz)	8.5	11.0	0.38	0.25
A (Vert)	11.0	8.5	0.25	0.38
B	11.0	17.0	0.38	0.62
C	17.0	22.0	0.75	0.50
D	22.0	34.0	0.50	1.00
E	34.0	44.0	1.00	0.50
F	28.0	40.0	0.50	0.50

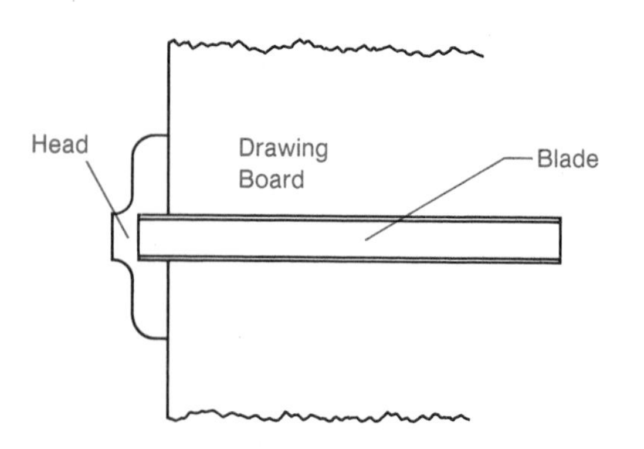

Figure 2.2 The T-Square allows the user to draw horizontal lines and define the horizontal direction on a drawing.

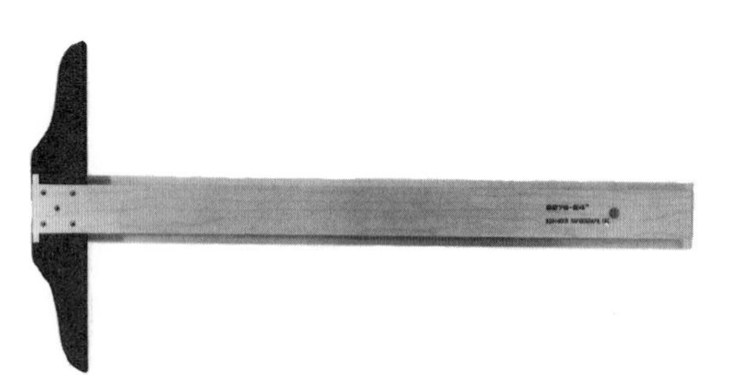

Figure 2.3 Example of a T-Square. (Courtesy of Koh-I-Noor Rapidograph, Inc., Bloomsbury, NJ.)

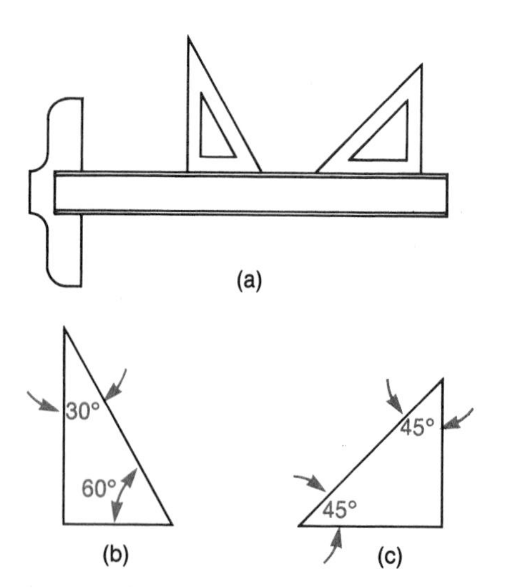

Figure 2.4 Triangles are used in conjunction with the T-Square (a) in order to draw lines that are vertical or inclined. Two basic types of triangles are used: the 30° - 60° triangle (b) and the 45° triangle (c).

to partition each principal block into smaller portions. (Examples of title blocks are shown in the figures of Appendix N.)

2.2 Triangles and the T-Square

The **T-square** consists of a rectangularly shaped blade attached (at a 90° angle) to a head (Figures 2.2 and 2.3). The blade provides a fixed horizontal direction for the drafter; it may be raised or lowered in a controlled manner by sliding the vertical edge of the T-square head along the edge of the drawing board.

Triangles are used in conjunction with the T-square to draw lines that are vertical or inclined. Two basic types of triangles are used: the 30° - 60° triangle and the 45° triangle (Figure 2.4 and 2.5). Each of these triangles includes two edges that are **orthogonal** to each other, that is, they are mutually perpendicular. A triangle may be slid along the blade of the T-square to the appropriate position at which a vertical or inclined line is to be drawn (Figure 2.6).

The 45° triangle and the 30° - 60° triangle can be used to produce inclined lines in 15° increments through 360°, as shown in Figure 2.7. Triangles are often used in conjunction with the T-square; the T-square acts as a horizontal base for the triangles.

Horizontal lines should be drawn by moving the pencil from left to right, and vertical lines by moving the pencil upward (Figure 2.8). (These recommendations apply to right-handed individuals; those who are left-handed will usually need to reverse all instrument positions.)

To construct an inclined line with a particular **orientation,** or direction, in space (for example, in Figure 2.9, the orientation is parallel to the given line $\overline{12}$, where 1 and 2 refer to the endpoints of this given line) and located at a particular **position** (for example, through the given point 3 in Figure 2.9), one may use two triangles simultaneously. The edge of one triangle is aligned with the given line ($\overline{12}$ in Figure 2.9) and braced upon the second triangle—which prevents rotation of the first triangle. The aligned triangle is then slid along the edge of the fixed triangle (which is held in place by the user's hand). If necessary, the triangles may be moved about the drawing surface until the edge of the aligned triangle passes above the point (3 in Figure 2.9) through which the line is to be drawn. In this way, we may transfer a direction, or orientation, from one (given) location to another in a drawing. The desired line may then be drawn (again, see Figure 2.9).

2.3 Drawing Pencils

Drawing pencils should be carefully chosen so that the user can produce linework with the necessary intensity and width. A particular type of clay known as Graphite Kaolin is used in pencil leads in order to product eighteen different grades of hardness. Pencil lead hardness is identified in accordance with the following scale:

SOFT	MEDIUM	HARD
7B, 6B, 5B, 4B, 3B, 2B,	B, HB, F, H, 2H, 3H,	4H,5H,6H,7H,8H,9H

Softer leads are used to produce thicker, darker linework. They wear more quickly than a harder lead. Medium grades are used for lettering (particularly the F, HB, and H grades) and general linework. The harder grades produce thinner, lighter lines. Accurate linework usually demands the thin lines achieved with these harder grades. However, the user may need to apply greater pressure to produce dark linework with them. Most construction linework is generated with the use of 4H, 5H, or 6H pencil leads, which usually are thin enough for accuracy but also dark enough with minimal exertion and without continuous sharpening of the point.

Each person has developed an individual style in writing (called "touch") that entails the application of a particular amount of pressure to the pencil lead to produce lines of desired darkness. Do not try to adjust your writing technique if you discover that the pencil lead you are using is either too hard (requiring a greater amount of pressure to produce dark lines) or too soft (requiring less pressure to avoid rapid wear of the pencil lead); instead, simply *adjust the pencil lead grade* to match your writing/drawing style. Otherwise, you

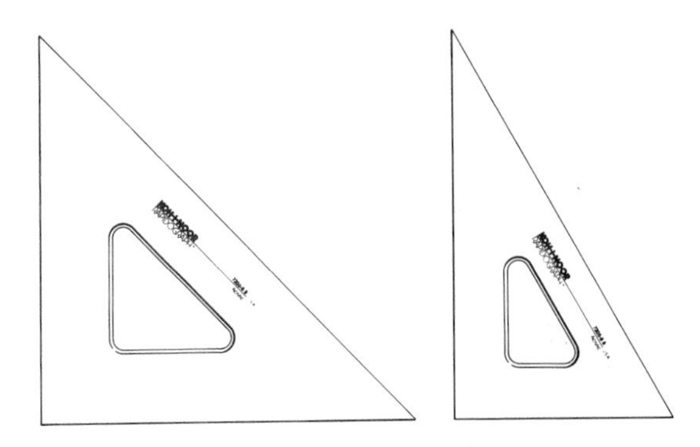

Figure 2.5 Examples of the 45° and 30° - 60° triangles. (Courtesy of Koh-I-Noor Rapidograph, Inc., Bloomsbury, NJ.)

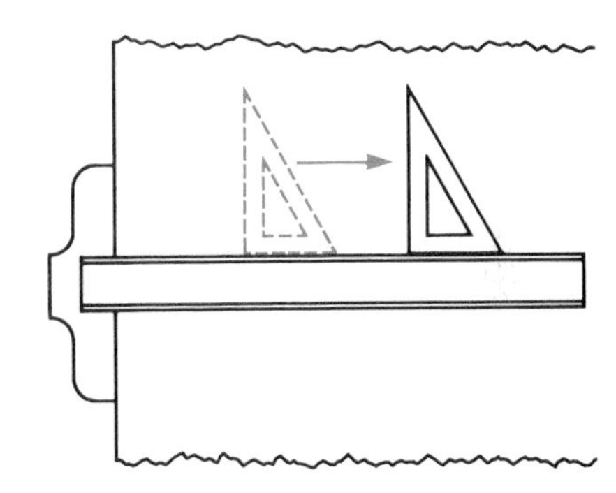

Figure 2.6 A triangle may be slid along the blade of the T-square to the appropriate position.

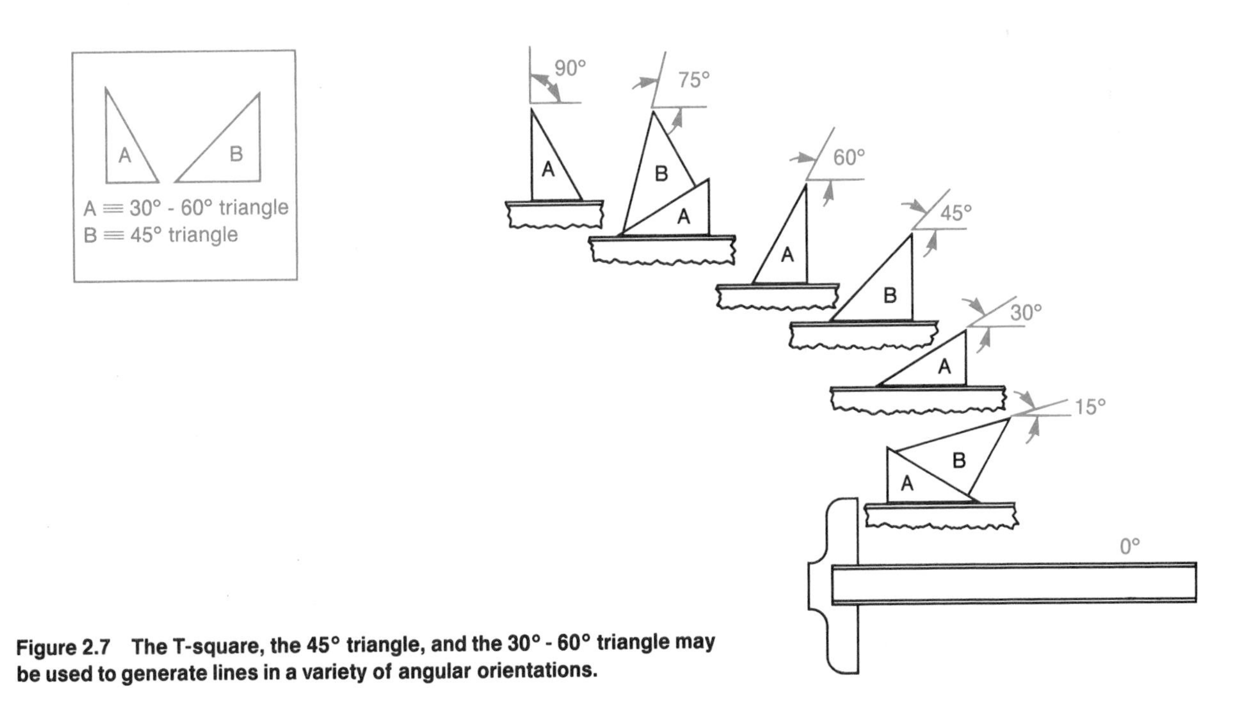

Figure 2.7 The T-square, the 45° triangle, and the 30° - 60° triangle may be used to generate lines in a variety of angular orientations.

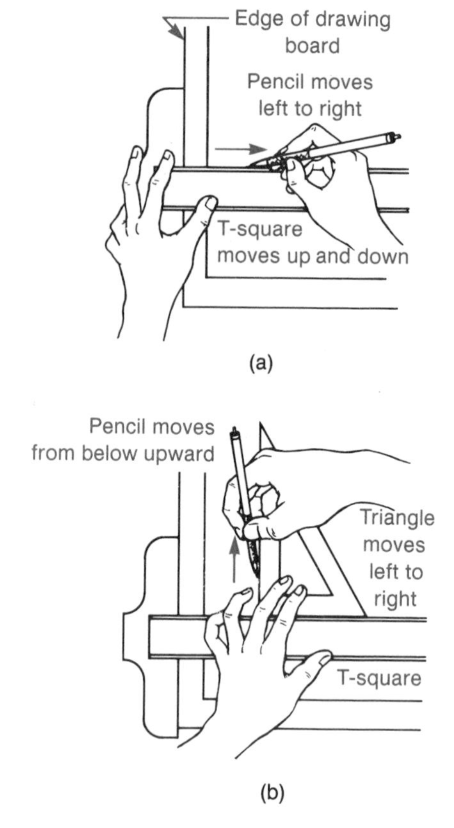

Figure 2.8 **Pencil motion for horizontal linework is left to right, (a) and for vertical linework is upward (b). (Engineering Aid 3 and 2, Volume 1, Naval Education and Training Command Rate Training Manual and Nonresident Career Course, NAVEDTRA 10634-C, published by Naval Education and Training Support Command, United States Government Printing Office, Washington, D.C.: 1976, updated 1980. This figure, and the following figures from this source were redrawn. This source will henceforth be identified as NAVEDTRA 10634-C.)**

will either need to sharpen your pencil unnecessarily often in order to maintain a sufficiently sharp point (if the pencil lead is too soft), or you may develop an ache in your drawing arm and shoulder because you are applying unusual pressure to the pencil lead (if it is too hard).

After sharpening the pencil lead, it is best to burnish the tip on a piece of scrap paper in order to produce a *slightly rounded point.* Such rounding of the pencil tip will allow you to produce consistent linework without breaking the pencil point. Sandpaper can also be used to maintain a consistent sharpness in the pencil tip.

Learning Check

Should a drafter adjust the pressure that is applied to a pencil in order to adjust the intensity (darkness) of the linework?

Answer No! He or she should change the grade of the pencil lead.

2.4 Erasure

Some products are more appropriate than others for erasing different types of linework. An Eberhard Faber Ruby eraser, for example, can be used to remove general linework from a drawing, but Artgum is a better choice for cleaning a drawing with light strokes. An electric eraser allows the engineer to quickly and effectively delete linework from a drawing. In all cases, a **dusting brush** should be used to remove erasure shavings, crumbs, and debris from the drawing; otherwise, the drawing may become smudged.

In addition, an **erasing shield** should be used to protect those portions of the drawing that are near the linework to

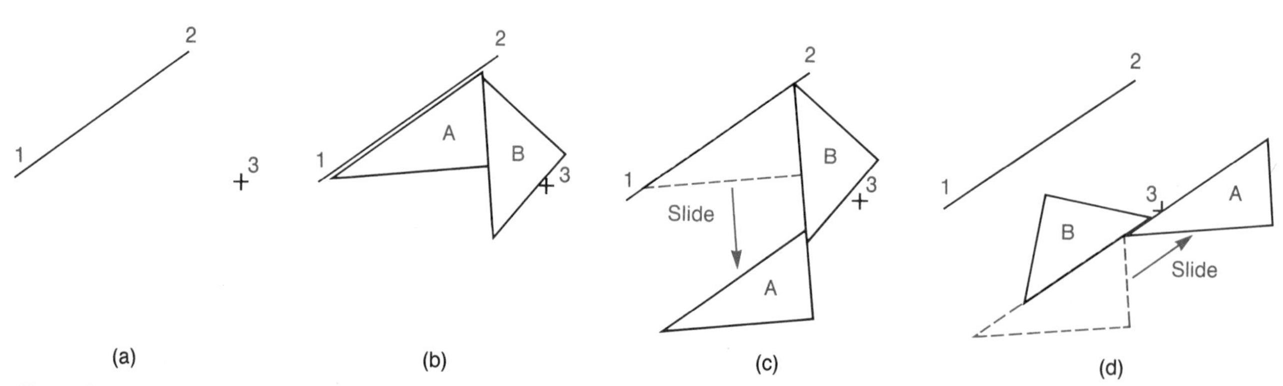

Figure 2.9 **We may construct a line with a particular *orientation*—in this case parallel to the given line $\overline{12}$—and located at a particular *position*—in this case through the given point 3) (a)—by using two triangles simultaneously (b). One triangle is held in a fixed position while the other triangle is slid along an edge of the fixed triangle (c). Repetition of this procedure allows us to transfer a direction, or orientation, from one location to another in a drawing (d). The desired line may then be drawn.**

be deleted (Figure 2.10). Such shields consist of metal strips in which small holes and other geometric cutouts have been formed.

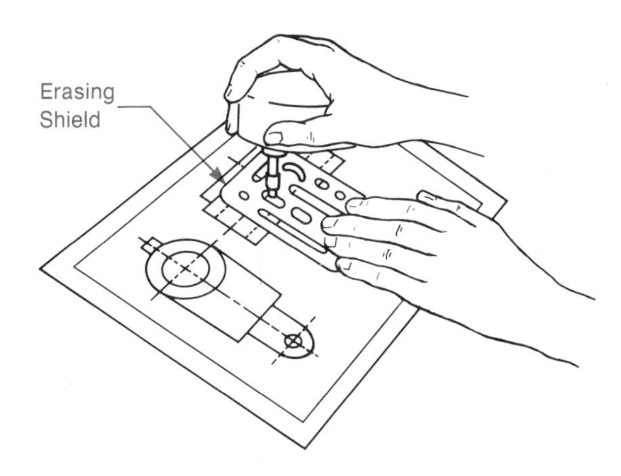

Figure 2.10 When deleting linework with an eraser, an erasing shield should be used to protect nearby portions of the drawing. (NAVEDTRA 10634-C)

2.5 Generating Arcs and Circles with a Compass

Whereas the T-square and the triangles allow us to generate straight lines (vertical, horizontal, or inclined), the **compass** allows us to create arcs and circles. A compass consists of two metal shafts connected in such a way that the relative distance between them can be controlled. Each shaft (or *leg*) of the compass has a socket in which an attachment may be inserted. One attachment is a metal needlepoint; another is a pencil lead with which linework can be generated. A centerwheel, or thumbscrew, is used to adjust the relative distance between the needlepoint and the lead.

An arc is a portion of a circle. To create an arc (or circle) of a given radius and centered about a particular location, we simply follow the procedure shown in Figure 2.11. The centerpoint is identified by two perpendicular lines (a), and the distance corresponding to the radius of the arc or circle from this centerpoint is marked along one of the perpendicular lines (b). The metal tip of the compass is placed at the location of the centerpoint (c) and the compass lead is located at the radial position that was identified earlier (d). The compass is then rotated between the user's thumb and index finger to form the desired circle or arc (e). Finally, the completed linework is analyzed to determine if the figure has been properly constructed. In particular, the diameter should be quickly measured to ensure that the size of the circle is correct (f).

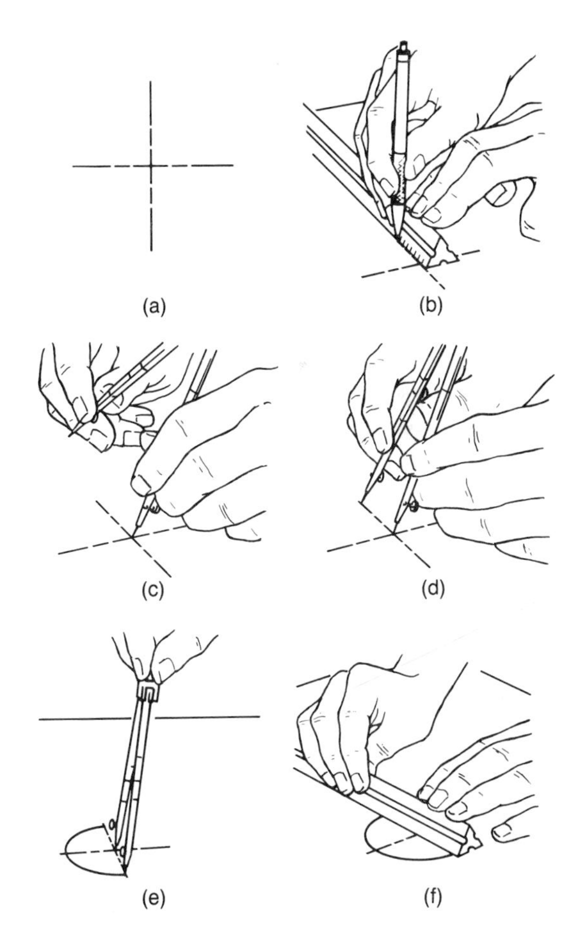

Figure 2.11 Step-by-step (a–f) procedure for using the compass to construct a circle or an arc of a specific radius and centered about a particular location. (NAVEDTRA 10634-C)

The lead tip of the compass should be sharpened by rubbing it against sandpaper to obtain an inclined surface (Figure 2.12). The lead should be adjusted in the socket joint so that the metal needlepoint extends slightly more than the pencil lead.

The proper use of the compass to create arcs and circles that are accurate and consistent in linework quality requires diligent practice. Do not be discouraged if your initial attempts are not successful. Continued practice will develop your skill.

Templates (Figure 2.13) are available for drawing circles without the use of the compass. (In addition, many other types of templates exist—for drawing common geometric figures, chemical and mathematical symbols, lettering characters, and more. Figure 2.14 shows some of these.) Although templates allow the user to quickly and accurately create circles, they also restrict the user to circles of specific diameters. Furthermore, the user may not always have access to the appropriate template. Obviously, it is wise to develop one's skills in the use of the compass.

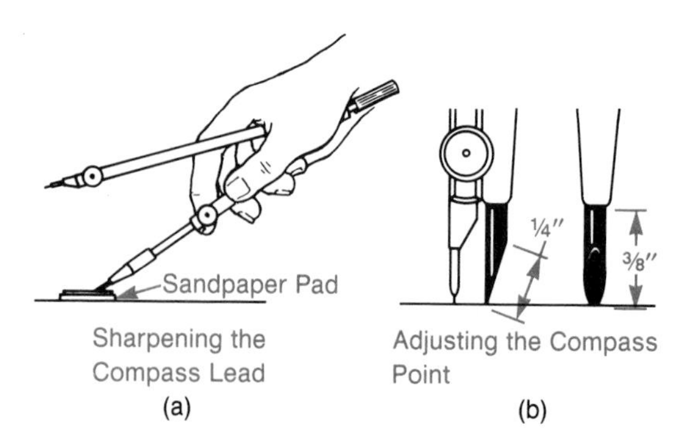

Figure 2.12 Sharpen the lead of the compass with sandpaper (a). The resultant lead point should be inclined (b). (NAVEDTRA 10634-C)

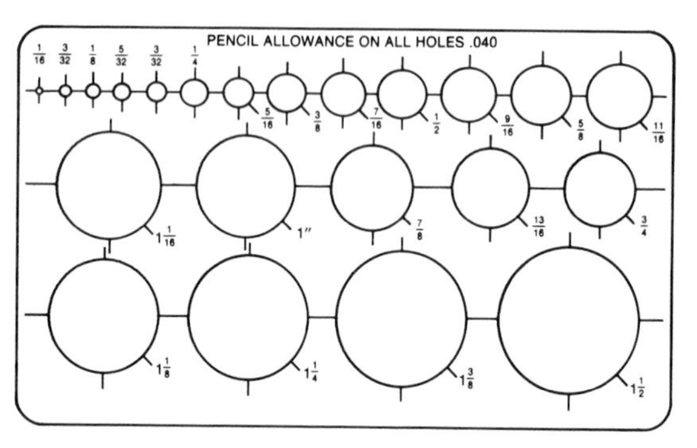

Figure 2.13 Example of a circle template. (NAVEDTRA 10634-C)

2.6 Dividers

Dividers are very similar in appearance to the compass: two metal bars are connected together in such a way that they rotate about a common point (Figure 2.15). A compass is sometimes used as a set of dividers; furthermore, a compass and dividers are often sold as a set. The difference between these two drawing instruments is that the dividers have two metal needlepoints in the sockets, whereas the compass has one needlepoint and a pencil lead.

Dividers are held between the thumb and index finger. Straight or curved lines may be divided, or sectioned, with the use of this instrument (refer again to Figure 2.15). The user simply estimates the length of the division desired, and the instrument is then rotated from one end of the line to the other. The estimation will probably be incorrect initially, so another estimate will need to be made. For example, in Figure 2.15 the distance between points X and Y is to be subdivided into four equal intervals. The initial estimate of this subdivided length is made, after which the dividers are rotated through three separate movements between point X and point Y. If the original estimate of the interval length were correct, the dividers would complete this movement at point Y; instead, as shown in this figure, the dividers complete the rotation at a point Y′. The distance between Y and Y′, corresponding to the total accumulated error in the original estimate, is then subdivided into four (approximately) equal intervals; and the length of this *error interval* is then added to the original estimate of the interval length by opening the dividers accordingly. (If the original estimate is too large—in which case point Y′ will not lie between X and Y—the error interval will need to be subtracted from the original estimate of the interval length.) Repetition of this trial-and-error procedure will eventually allow proper subdivision of a distance between two points into equal intervals.

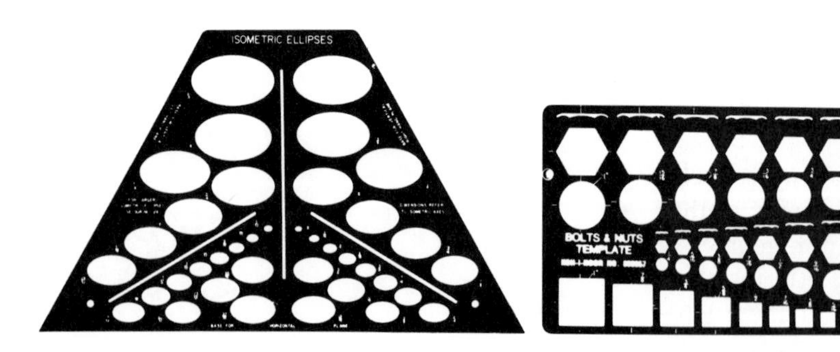

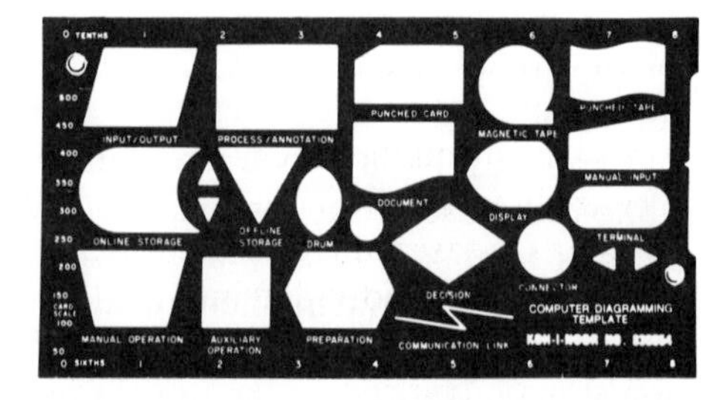

Figure 2.14 Many different types of special templates are commercially available. (Courtesy of Koh-I-Noor Rapidograph, Inc., Bloomsbury, NJ.)

Perhaps the most important use of dividers in the development of a drawing is the transference of distances from one location to another. For example, the distance between the two points X and Y in Figure 2.15(b) can be transferred from one location to another on the drawing by simply opening the dividers so that each metal needlepoint lies on one of these two points, then transferring the dividers to the new location.

Thus, dividers eliminate the task of measuring distances on a drawing in order to *transfer* a distance from one location to another or to *divide* a distance into equal intervals.

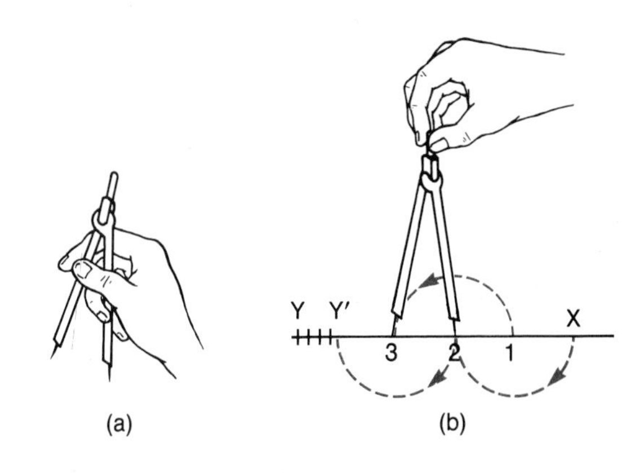

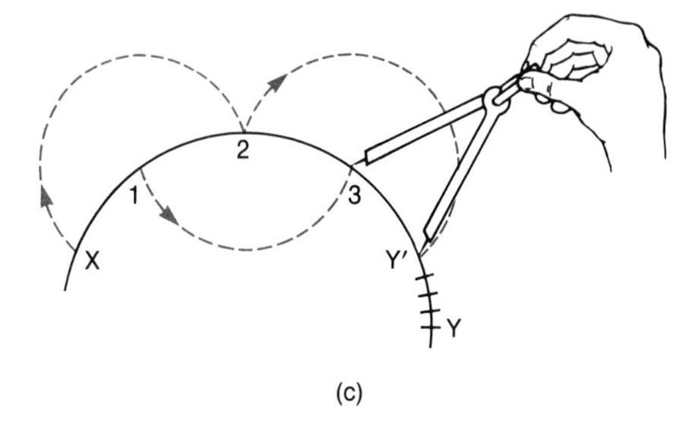

Figure 2.15 Dividers are held between the thumb and index finger (a), then rotated to produce approximately equal intervals between two points on a straight line (b) or a curved line (c). (NAVEDTRA 10634-C)

2.7 Scales

Scales are graduated measuring instruments. A ruler, a mercury thermometer, and an air pressure gauge for tires are all scales with which one can measure various quantities. The **scale** of a drawing is a different entity. It refers to the ratio between the size of the drawing and the size of the object described by the drawing. **Full-scale drawings** are identical in size to the objects they describe. **Half-scale drawings** are one-half the size of the objects they describe.

The scale of a drawing is given in the title block; it may be expressed in **words** (full), as a **ratio** (1:2, wherein the first value refers to the drawing and the second value to the relative size of the object), or as a **relation** between quantities (1 inch = 10 feet, wherein the first quantity again refers to distances on the drawing and the second value to corresponding distances on the object).

As you might expect, many objects are described by drawings considerably smaller than the objects; as a result, most scales have ratios of less than one (for example, 1:4 or 1/4). However, certain small objects need to be graphically described by drawings that are larger than the objects in order to allow details to be clearly shown, and in these cases the scales are greater than unity (for example, 5:1).

The scaling of a drawing is aided by a professional scaling instrument. The **engineer's scale** (Figure 2.16), also known as the civil engineer's scale or the chain scale, has six separate scales or rulings, each of which is subdivided decimally. The **architect's scale** (Figure 2.17) has divisions representing distances measured in feet for the actual object; these divisions are then subdivided into twelfths. In addition, the full size **mechanical engineer's scale** is included on the architect's "stick" for convenience (divided into sixteenths). For example, the ¾ scale on this instrument corresponds to a ratio or scale of ¾ inch = 1 foot between the distances on a

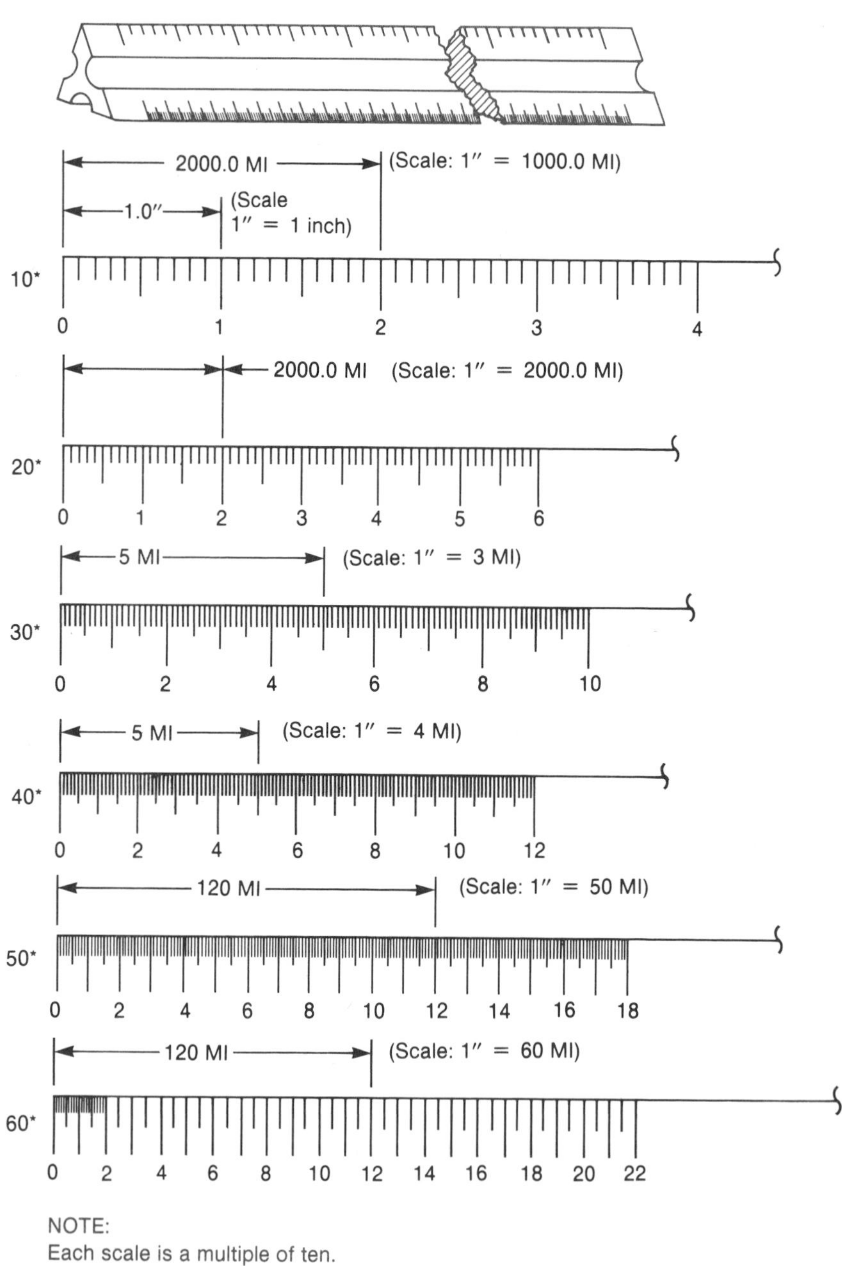

Figure 2.16 The engineer's scale has six separate rulings, each subdivided decimally. (NAVEDTRA 10634-C)

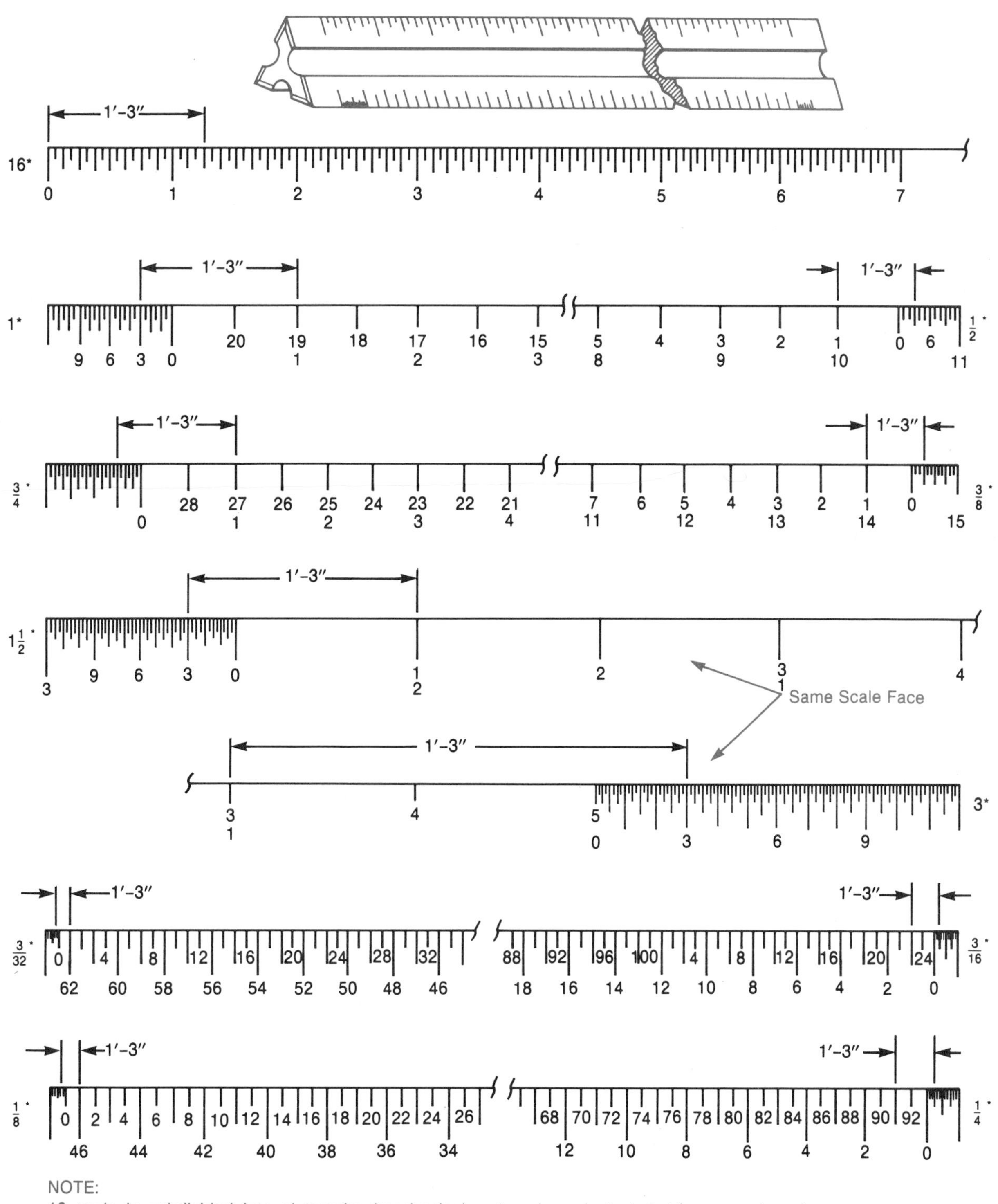

NOTE:
16 scale is subdivided into sixteenths (mechanical engineer's scale, included for convenience).
All others are divided into twelfths.
* Scale designation numbers.

Figure 2.17 The architect's scale has divisions representing distances measured in feet. (NAVEDTRA 10634-C)

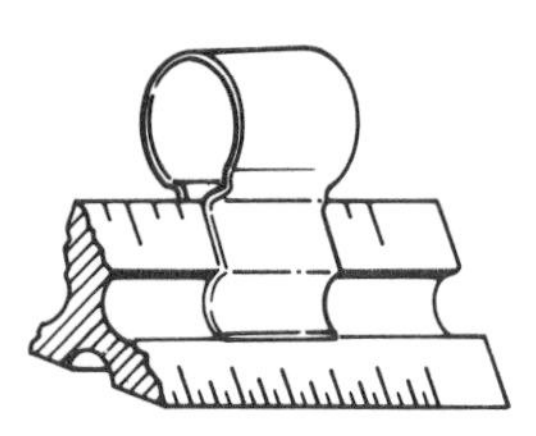

Figure 2.18 The triangular scale clip is used to ensure that the same scale is used throughout the development of a drawing. (NAVEDTRA 10634-C)

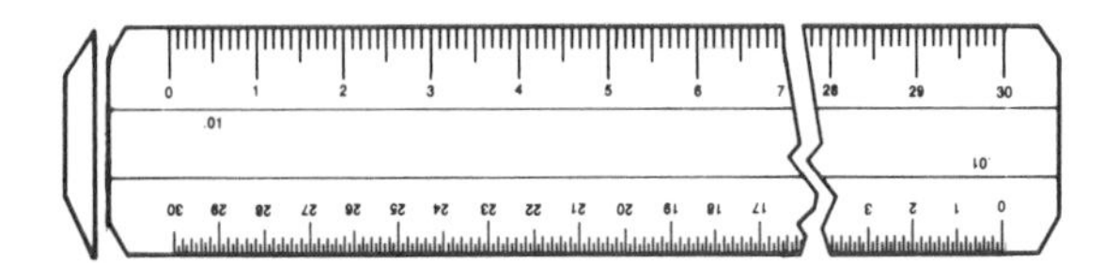

Figure 2.19 The metric scale is used to measure distances in meters, centimeters, and millimeters. (NAVEDTRA 10634-C)

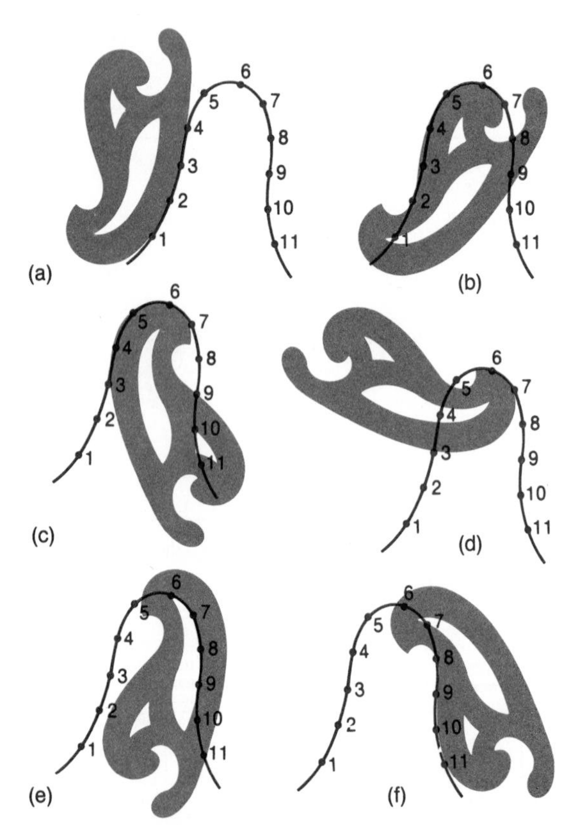

Figure 2.20 Using the French, or irregular, curve to construct a smooth (irregular) line through a set of specific points (a–f). The instrument is aligned with only three or four points at a time so that a very smooth curve through all points is produced. (NAVEDTRA 10634-C)

drawing and the actual object being described, that is, a distance of 1 foot on the object has been reduced by a factor of $^1/_{16}$ to a distance of $^3/_4$ inch of the drawing.

A simple triangular scale clip (Figure 2.18) can be used to ensure that the same scale is used throughout the development of a drawing.

Metric scales (Figure 2.19) allow the user to measure distances in terms of meters, centimeters, and millimeters. Some useful conversions are:

1 inch = 2.54 centimeters
= 25.4 millimeters
1 meter = 100 centimeters
= 1000 millimeters
1 centimeter = 10 millimeters

Distances may be transferred from a scale to the drawing through the use of dividers (as described in Section 2.6).

2.8 The Irregular Curve

Having discussed the construction of both straight lines and arcs, we now direct our attention to irregular curves, that is, lines that curve through space in an irregular, or nonarcing, path. Such irregular curves are defined by the specific points that lie on the curves. Engineers are often confronted with the task of "fitting" a smooth but irregular curve through a set of specific points that represent data describing a process, a surface in space, or some other quantity or relationship. For example, consider the eleven points shown in Figure 2.20(a) through which a smooth curve is to be constructed. An instrument known as a **French curve** or an **irregular curve** is used to construct this smooth line as shown in the figure.

There are many varieties of irregular curves, each consisting of a series of curving edges of different forms with which one may "fit" data points. As shown in Figure 2.20, the irregular curve is used to fit three (or more) points at a time by aligning an edge of the curve with these points; a line is then drawn along the path defined by the curve through these fitted points. The line is drawn through all points *except* the last point in the set aligned with the curve. For example, if points 1, 2, and 3 are aligned with the curve in Figure 2.20(a), a line is drawn from point 1 through point 2 toward point 3—but the line is not drawn to point 3; instead, the line ends at some arbitrarily chosen location between points 2 and 3. Points 2, 3, 4—and perhaps a fourth point, 5, as shown in Figure 2.20(b)—are then aligned with the irregular curve. The line is continued from the location at which it ended between points 2 and 3, through point 3, to a position between points 3 and 4 (or through point 4 to a position between points 4 and 5, if point 5 was included in the set of points used to align the irregular curve). This procedure is continued until a line has been drawn through the entire set of points. To obtain the best fit, through all points, one should never try to fit more than three or four points simultaneously. Although the procedure is somewhat tedious when limited to three or four points at a time, the resulting smooth curve through all points justifies the extra effort.

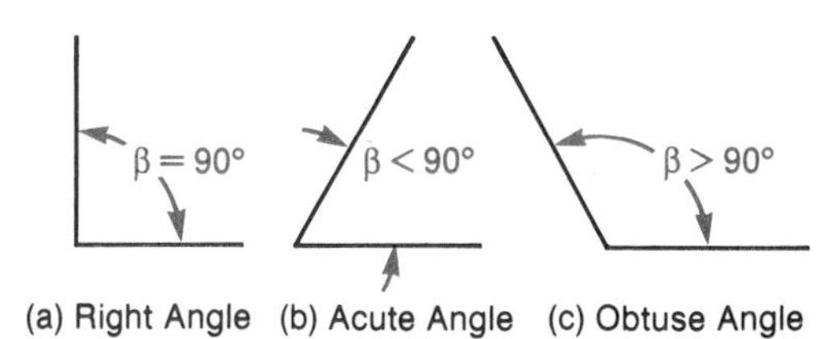

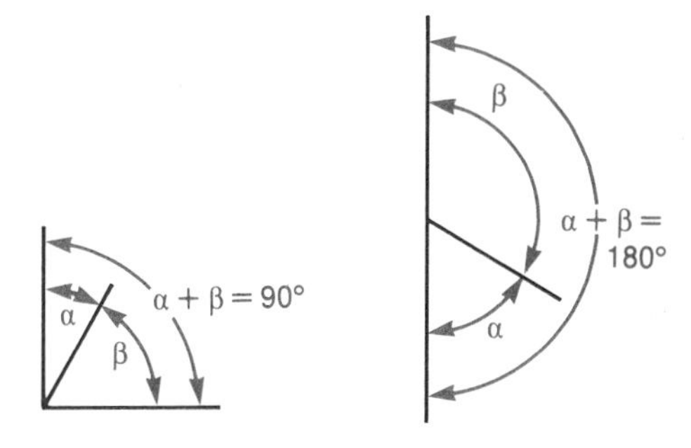

Figure 2.21 Common types of angles are the right angle which is 90° (a), acute angle, which is less than 90° (b), and obtuse angle, which is greater than 90° (c). Complementary angles together equal 90° (d). Supplementary angles together equals 180° (e).

2.9 Angles

Figure 2.21 presents the following common types of angles between straight lines:

- **Right angle,** in which the two lines are mutually perpendicular to one another.
- **Acute angle,** in which the lines are oriented in such a way that the angle between them is less than 90°.
- **Obtuse angle,** in which the lines are oriented in such a way that the angle between them is greater than 90°.
- **Complementary angles,** in which two angles (α and β in the figure) are oriented in such a way that their sum is equal to 90°.
- **Supplementary angles,** in which two angles (α and β in the figure) are oriented in such a way that their sum is equal to 180°.

2.10 Triangles

Triangles are three-sided geometric figures. The sum of the internal angles of a triangle is equal to 180°. Figure 2.22 presents the following common types of triangles:

- **Right triangle,** in which one internal angle is equal to 90°.
- **Equilateral triangle,** in which all sides are equal in length and all internal angles are equal to 60°.
- **Isosceles triangle,** in which two sides are equal in length and two internal angles are equal.
- **Scalene triangle,** in which all three sides are of unequal length and all three internal angles are unequal.

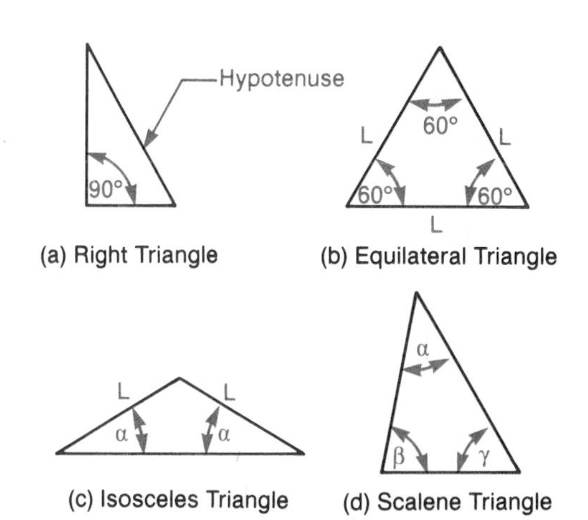

Figure 2.22 Common types of triangles are the right triangle, in which one angle equals 90° (a); equilateral triangle, which has sides of equal length (*L*) and equal internal angles equal to 60° (b); isosceles triangle, in which two sides are of equal length (*L*) and two internal angles are equal (c); and scalene triangle, in which all three sides are of unequal length and all three internal angles are unequal (d).

2.11 Quadrilaterals

Quadrilaterals are four-sided geometric figures (as triangles are three-sided figures). Figure 2.23 presents the following common types of quadrilaterals:

- **Square,** in which all four sides are equal in length and all four internal angles are equal to 90°.
- **Rectangle,** in which opposite sides of the figure are equal in length and parallel to each other, and all four internal angles are equal to 90°.
- **Rhombus,** in which all four sides are equal in length, opposite sides are parallel, and the internal angles are not equal to 90°; a slanted square.

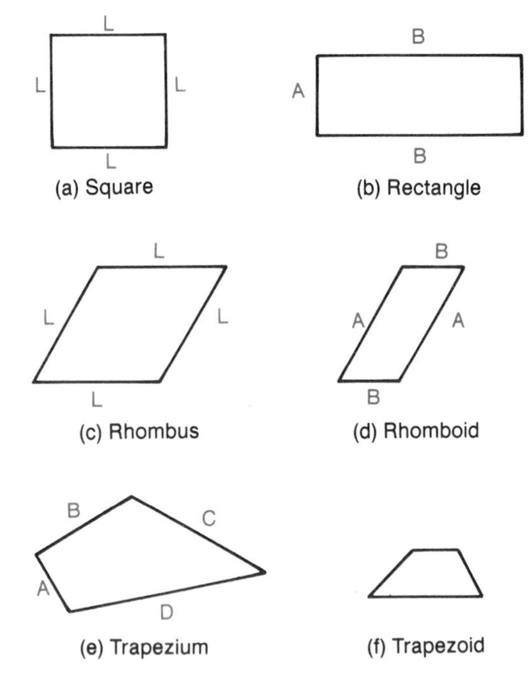

Figure 2.23 **Common types of quadrilaterals are the square, in which all sides are equal (a); rectangle, in which opposite sides are parallel and equal (b); rhombus, in which opposite sides are parallel, all sides are equal, and internal angles are not equal to 90° (c); rhomboid, in which opposite sides are parallel and equal, and adjacent sides are not equal (d); trapezium, in which all sides are nonparallel (e); and trapezoid, in which two opposite sides only are parallel (f).**

- **Rhomboid,** in which opposite sides of the figure are equal in length and parallel; however, adjacent sides are not equal; a slanted rectangle.
- **Trapezium,** in which all sides are nonparallel to one another.
- **Trapezoid,** in which two opposite sides only are parallel to each other.

2.12 Circles

The following significant characteristics of circles are presented in Figure 2.24:

- **Centerpoint.** Denoted in a drawing by a set of perpendicular lines (known as **centerlines** and discussed in detail in Section 2.25).
- **Radius.** The distance between all points on the circle and the centerpoint.
- **Diameter.** The distance between the extreme points of intersection on a line drawn through the centerpoint of the circle; it is equal in length to twice the value of the radius.
- **Arc.** A portion of the circle; specifically, an arc is a collection (or locus) of points that are equally distant from a (center)point.
- **Chord.** The linear (straight-line) distance between two points on a circle.
- **Concentric circles.** Two or more circles that share a common centerpoint.
- **Inscribed circle.** A circle enclosed by a polygon and tangent to it on each side.
- **Circumscribed circle.** A circle that encloses a polygon and passes through each vertex.

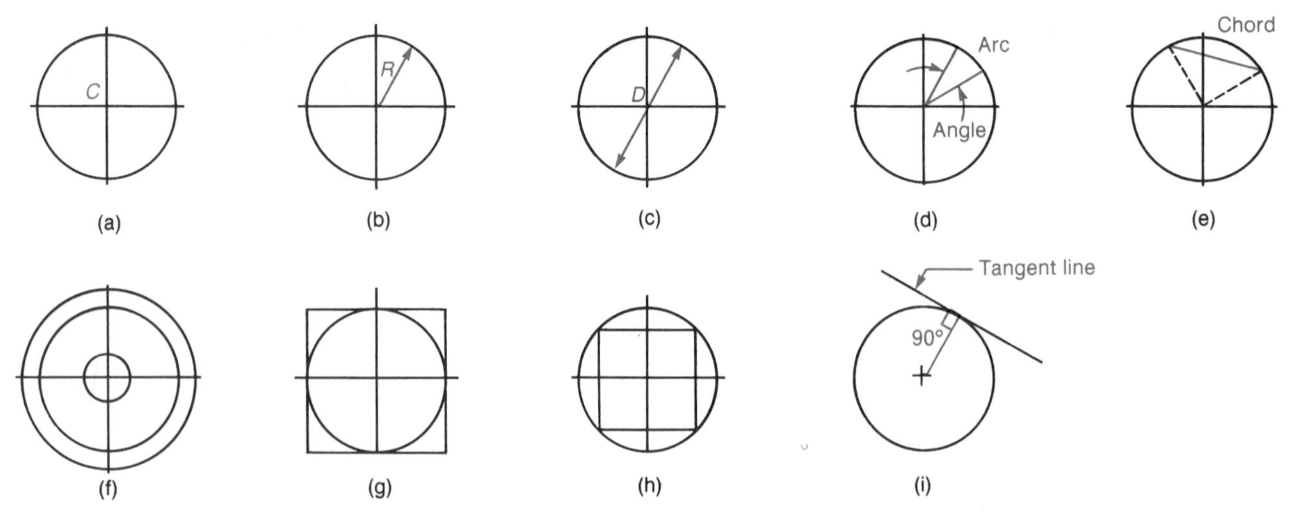

Figure 2.24 **Important characteristics of circles are the centerpoint, *C*, which is denoted by the intersection of centerlines (a); radius, *R*, of circle (b); and diameter, *D*, of circle (c). An arc (d) is a portion of a circle. A chord (e) is the linear distance between two points on a circle. Concentric circles (f) share a common centerpoint. An inscribed circle (g) is enclosed by a square. A circumscribed circle (h) encloses a square. A line that is tangent to a circle is perpendicular to a second line passing through the point of tangency and the centerpoint (i).**

- **Tangent line.** A line that is tangent to a circle and is also perpendicular to a second line passing through the point of tangency and through the centerpoint of the circle.

2.13 Solid Figures

Figure 2.25 presents some common types of solid figures with which all engineers should be familiar. Other types of solids will appear in forthcoming chapters.

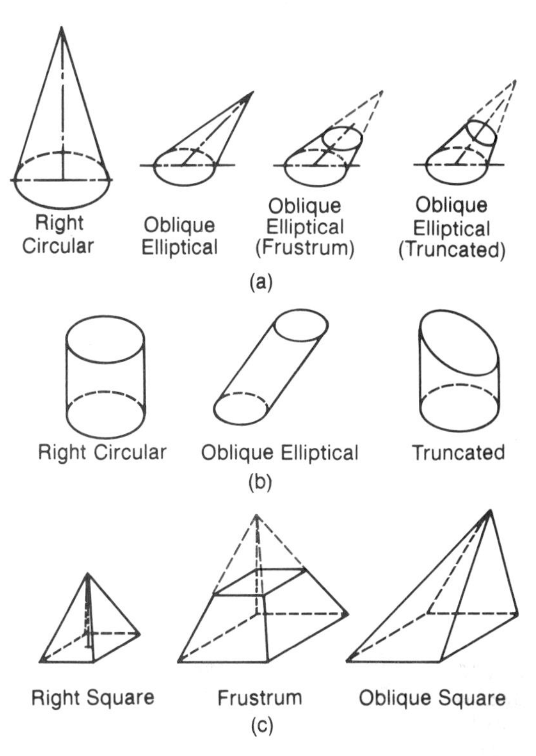

Figure 2.25 Common types of solid figures are cones (a), cylinders (b), and pyramids (c).

2.14 Bisecting a Straight Line

For *quick* and *accurate* development of a drawing, we need to be able to bisect, or divide, a line without taking time to measure it with a scale. Dividers are one tool for quick bisection. A compass is another.

Figure 2.26 shows how a straight line between the two points 1 and 2 (a) is bisected, or divided, into two equal portions with the aid of a compass. First, two small construction arcs of equal radius are drawn, each centered at point 1 (b). The radius of these arcs is arbitrarily chosen with the constraint that it should be greater than one-half the length of the line ($\overline{12}$) that is to be bisected. Two more arcs are next drawn, each centered at point 2 (c). All four construction arcs share the same radius. The intersection points 3 and 4 between these arcs (d) define a straight line, which contains the midpoint, 5, of the line $\overline{12}$. The bisection of the line $\overline{12}$ is then completed.

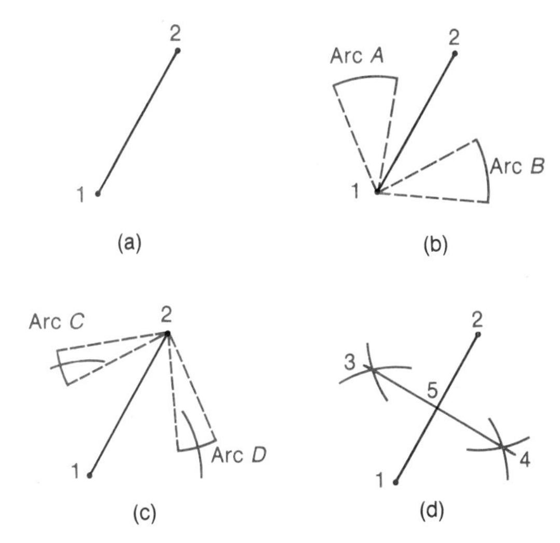

Figure 2.26 To bisect the straight line $\overline{12}$ (a), two arcs (*A* and *B*), centered at point 1, are drawn (b). The radius of these arcs is arbitrarily chosen; however, it must be greater than one-half of the distance between points 1 and 2. Next, arcs *C* and *D*, centered at point 2, are drawn (c). The arcs *A*, *B*, *C*, and *D*, share the same radius. The straight line between the intersection points 3 and 4 of arcs *A*, *B*, *C*, and *D* intersects line $\overline{12}$ at its midpoint, 5 (d).

2.15 Bisecting an Angle

In order to bisect an angle without taking measurements, we again use the compass, as shown in Figure 2.27. In this figure the angle β, between the two lines $\overline{12}$ and $\overline{13}$, is to be bisected (divided into two equal portions). An arc is drawn with the compass in such a way that it is centered at point 1 with the radius R_1 (b). The intersection points 4 and 5 of this arc, with the lines $\overline{12}$ and $\overline{13}$, are then identified (c). Points 4 and 5 then act as centerpoints for two additional construction arcs of radius R_2, which intersect at point 6 (d). The line $\overline{16}$ then bisects the angle β, as required (e). Note that the accuracy is improved as R_1 increases.

2.16 Dividing a Line into More Than Two Segments of Equal Length

We can divide a line into more than two segments of equal length by using the dividers, as described in Section 2.6, or by using a compass, as shown in Figure 2.28. The line $\overline{12}$ in the figure (a) is to be divided into four segments of equal length. We first construct a line of arbitrarily chosen length and direction, which has point 1 or point 2 as one of its end-

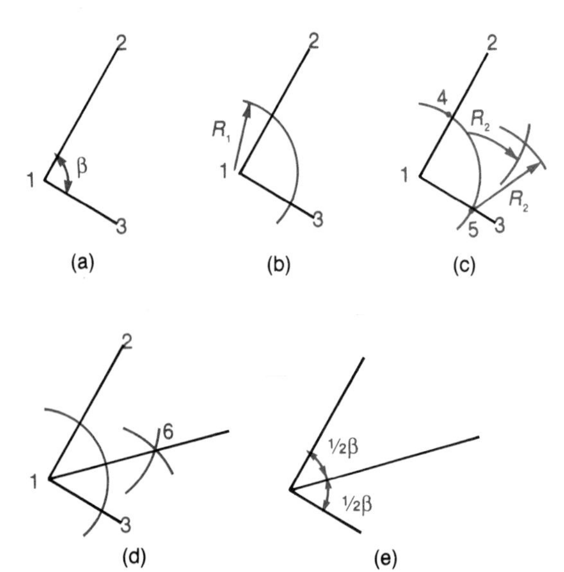

Figure 2.27 To bisect the angle β (a), an arc, centered at point 1 with an arbitrarily chosen radius R_1, is drawn (b). The intersection points 4 and 5 then act as centerpoints for two additional arcs of radius R_2 (c). The intersection point 6, together with point 1, then defines a line that bisects the angle β (d). The final result is that the angle has been bisected (e).

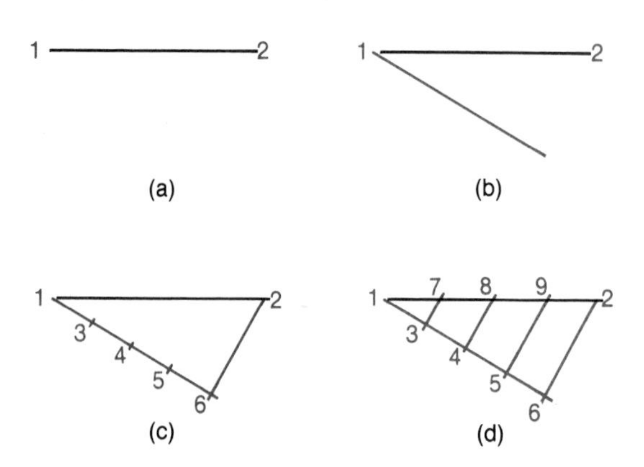

Figure 2.28 To divide line $\overline{12}$ (a) into four equal segments, first draw a line of arbitrarily chosen length and direction, with point 1 as its endpoint (b). The construction line is then divided into four equal segments (c). Parallel lines are drawn from points 3, 4, 5, and 6 to line $\overline{12}$, thereby dividing $\overline{12}$ into segments of equal length.

points (b). The compass is used to mark off segments of equal length along this construction line (c). (Points 3, 4, 5, and 6 in the figure are identified on the construction line where segments $\overline{13}$, $\overline{34}$, $\overline{45}$, and $\overline{56}$ are all of equal length.) A series of parallel lines are then drawn between points 3, 4, 5, and 6 and the line $\overline{12}$, in such a way that each of these lines is parallel to the line $\overline{26}$ (d). The intersection points 7, 8, 9, and 2 between these parallel lines and line $\overline{12}$ then define segments $\overline{17}$, $\overline{78}$, $\overline{89}$, and $\overline{92}$ of equal length on line $\overline{12}$, as required.

Learning Check

Why not measure the length of a line or the angle between two lines in order to bisect or otherwise divide it?

Answer We need to be able to *quickly* yet *accurately* divide lines or angles during the construction of a drawing, which means eliminating the need to take measurements. In addition, we do not want to be dependent upon scales or protractors, since such equipment may not always be readily available to us.

2.17 Constructing Lines Tangent to Two Circles of Equal Diameter

Figure 2.29 shows how to construct straight lines tangent to two circles of equal diameter. Between the centerpoints of the circles, identified as 1 and 2 (a), draw the line $\overline{12}$ (b). Then draw two lines ($\overline{34}$ and $\overline{56}$) perpendicular to line $\overline{12}$ and containing the centerpoints 1 and 2 (b). The points 3, 4, 5, and 6 are **points of tangency** between the two given circles, and the two straight lines $\overline{35}$ and $\overline{46}$ can then be drawn (c), thereby completing our construction effort (d). Note that these tangent lines, $\overline{35}$ and $\overline{46}$, are parallel to line $\overline{12}$ between the centerpoints.

2.18 Constructing Lines Tangent to Two Circles of Unequal Diameter

Engineers may need to construct lines tangent to two circles of unequal diameter in a drawing—for example, to represent a set of belts and pulleys. Figure 2.30 shows how to construct a **crossed-belt** configuration, in which two straight lines are tangent to two circles of unequal diameter and also intersect. Figure 2.31 shows how to construct an **open-belt** configuration, in which the lines do not intersect.

The given circles, centered at points 1 and 2, are shown with radii R_1 and R_2 in Figure 2.30(a). A construction circle is then formed with a radius equal to R_1+R_2 and centered about point 1 (b). In addition, the midpoint 3 of the line $\overline{12}$ is identified.

Next, an arc of radius, R, is drawn, centered at point 3, where R is equal to one-half of the length $\overline{12}$ (that is, R is equal to $\overline{13}$ or $\overline{23}$ in length). This arc intersects the construction circle at the points 4 and 5 (c). Two lines are then drawn from point 1 through points 4 and 5. These lines intersect the original (given) circle (centered at point 1) at the points 6 and 7 (d), which are the points of tangency between the straight lines under construction and this given circle. The tangent lines may then be drawn so that they are perpendicular to the radial lines $\overline{16}$ and $\overline{17}$ (e).

One may determine the points 8 and 9 (which are the points of tangency on the circle centered at point 2) by drawing the two additional construction lines $\overline{28}$ and $\overline{29}$, which are parallel to lines $\overline{17}$ and $\overline{16}$, respectively; the lines $\overline{69}$ and $\overline{78}$ can be drawn directly between the appropriate points. Figure 2.30(f) shows the final set of tangent straight lines, together with the given circles.

The open-belt configuration (in which the straight lines tangent to the given circles do not intersect each other) can be constructed in a similar manner, as shown in Figure 2.31.

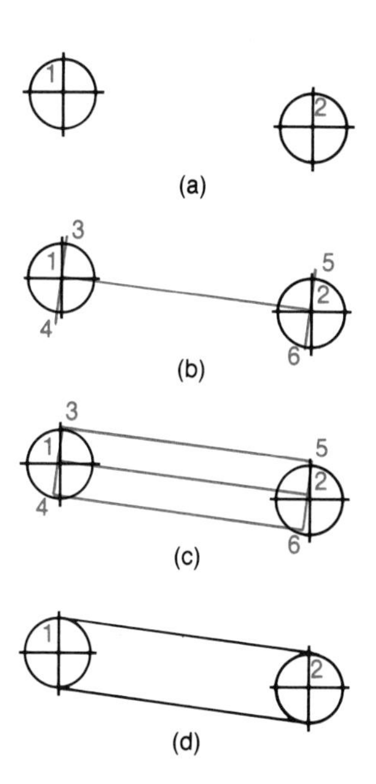

Figure 2.29 To construct tangent lines between circles of equal diameters (a), first draw line $\overline{12}$ between the centerpoints (b), then lines $\overline{34}$ and $\overline{56}$ (b), then lines $\overline{35}$ and $\overline{46}$ (c) to complete the task (d).

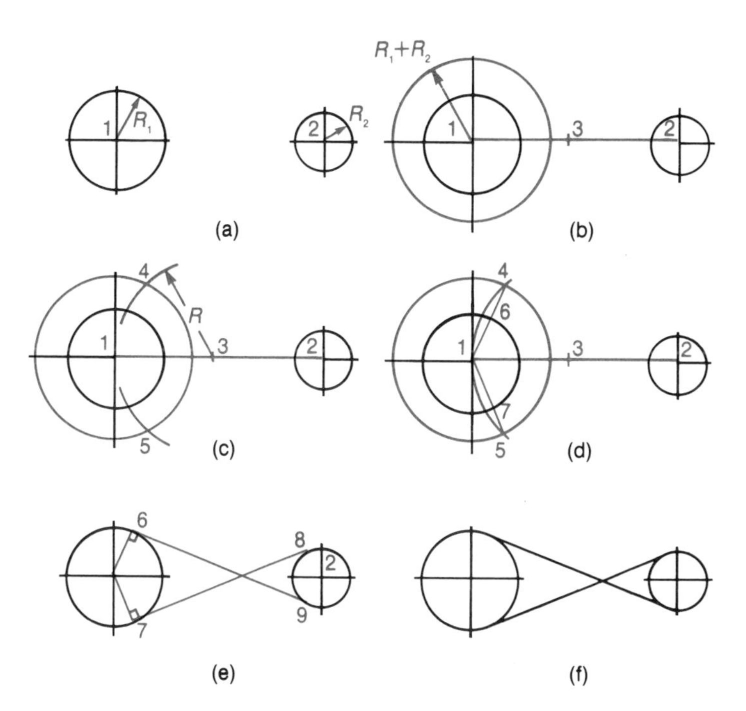

Figure 2.30 To construct a crossed-belt configuration, in which lines tangent to circles of unequal diameters (a) intersect each other, first form a construction circle with a radius equal to the sum of the radii of the two circles, and center it on point 1 (b). Then draw an arc centered at point 3 of line $\overline{12}$ of a radius equal to half the length of line $\overline{12}$ (c). Draw two lines through points 4 and 5 (d). Points 6 and 7 are the points of tangency on the first circle and points 8 and 9 on the second (e). A crossed-belt configuration is the result (f).

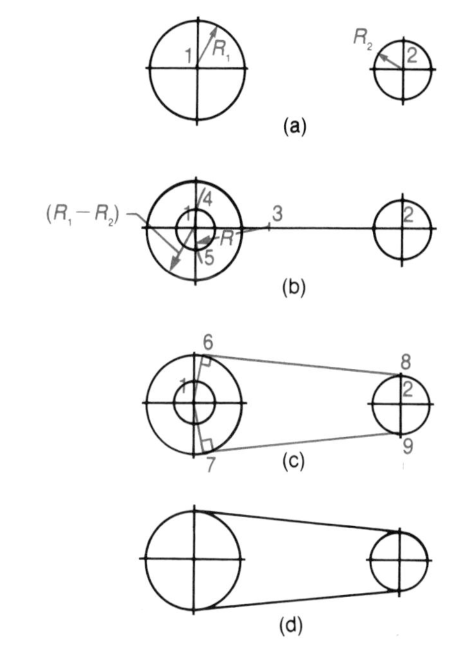

Figure 2.31 To construct an open-belt configuration, in which lines tangent to circles of unequal diameters do not intersect each other, follow the same procedure as for lines that do intersect (in Figure 2.30) with one exception: for an open-belt configuration the construction circle is drawn with a radius equal to R_1-R_2 instead of R_1+R_2.

The difference between the two configurations is that the construction circle is drawn with a radius equal to $R_1 - R_2$ for the open-belt case, instead of $R_1 + R_2$ as in the crossed-belt procedure.

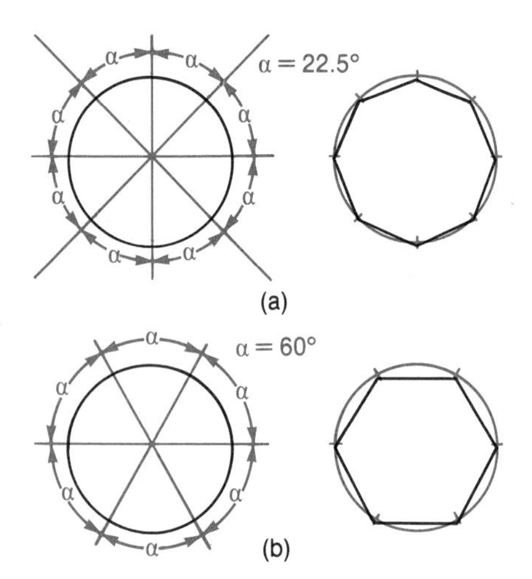

Figure 2.32 To construct an *n*-sided polygon, divide a circle into *n* equal segments; each segment is defined by an angle, α, where $\alpha = 360°/n$. Construction of an octagon ($n = 8$) (a) and a hexagon ($n = 6$) (b) is shown.

2.19 Constructing a Regular Polygon

A **regular polygon** is an *n*-sided, ($n > 2$) two-dimensional geometric figure in which all sides are of equal length. A **pentagon** corresponds to $n = 5$, a **hexagon** corresponds to $n = 6$, a **heptagon** to $n = 7$, an **octagon** to $n = 8$, and so forth.

An *n*-sided regular polygon may be constructed as shown in Figure 2.32. A construction circle is partitioned into *n* equal arc segments by dividing 360° by *n*, thereby producing angle, α, for each segment (where $\alpha = 360°/n$). Chords are then drawn connecting the *n* equally spaced points that lie on the circle between these segments, thereby forming the desired polygon. Figure 2.32 shows how an octagon (a) and a hexagon (b) are developed.

Other construction techniques exist as well for developing a regular polygon, and we will review these techniques where appropriate in succeeding chapters.

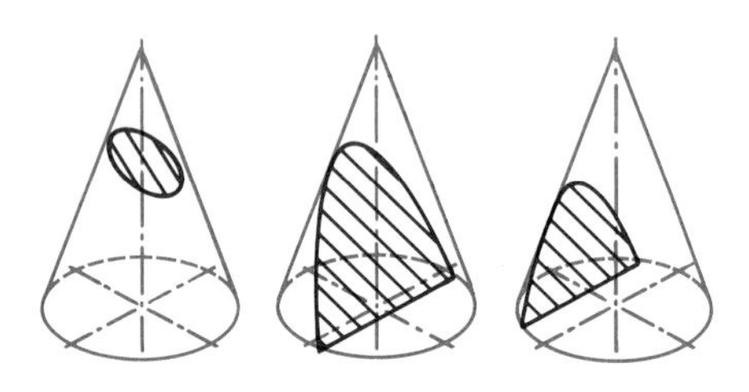

Figure 2.33 Three types of conic sections, depending on where the cone intersects with the plane, are the ellipse (a), parabola (b), and hyperbola (c) (NAVEDTRA 10634-C).

2.20 Conic Sections

The intersection of a cone with a plane can produce a variety of geometric curves, as shown in Figure 2.33. An **ellipse** is formed from the intersection of the cone with an inclined plane that passes through the conic surface (a). (A **planar** surface is one which is contained entirely within a single, flat, two-dimensional plane in space.) A **parabola** is formed from the intersection of the cone with an inclined plane that passes through the planar base of the cone (b). A **hyperbola** is formed from the intersection of the cone with a vertical plane passing through the planar base of the cone (c). Each of these curves can be represented by a corresponding mathematical equation, given in Sections 2.21 and 2.22.

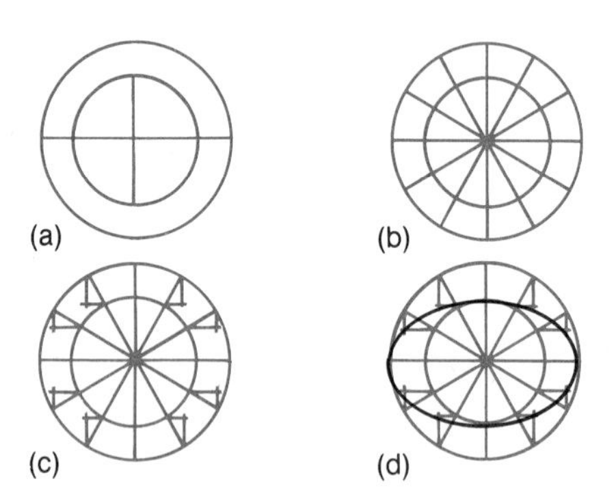

Figure 2.34 To construct an ellipse via the concentric circle method, draw two circles equal to the major and minor diameters of the ellipse (a), draw a series of inclined lines through the centerpoint (b), draw horizontal and vertical lines through the points of intersection as shown (c), and construct the ellipse through the appropriate points (d). (NAVEDTRA 10634-C).

2.21 Constructing an Ellipse

One particular type of conic section deserves special attention: the ellipse. As we will discover in forthcoming chapters, various techniques for constructing ellipses can be used during the development of certain engineering drawings. One method, among the many that exist, for constructing an ellipse is based upon concentric circles (Figure 2.34). In this method, one must know the major and minor diameters of the ellipse to be constructed. The major diameter is the largest distance between opposite points on the ellipse, and the minor diameter is the smallest distance.

Highlight

Archimedes and His Use of Regular Polygons and Circles to Determine the Value of Pi (π)

Archimedes (287?–212 B.C.) was one of the greatest mathematicians and engineers of ancient times. His contributions include:

- The principle of buoyancy, which states that the volume of liquid displaced by a submerged object is identical to the object's volume.
- The principle of the lever, with which great weights can be moved, and which therefore lay the foundation for the field of statics in engineering.
- The *screw of Archimedes,* which was a water pump in the form of a rotating, hollow, helical cylinder.

In addition, Archimedes determined the most accurate value for **pi** (which is the ratio of the circumference, *C*, of a circle to its diameter, *D*, that is, $\pi \equiv \frac{C}{D} = 3.14159$) in the ancient world. He obtained this value by (1) inscribing a circle within a polygon, (2) circumscribing a second polygon about this circle, and (3) increasing the number of sides of each polygon so that their total circumferences each approached that of the circle. These circumferences, composed of many small straight lines, could then be easily determined. As a result, Archimedes concluded that the value of pi was between $\frac{223}{71}$ and $\frac{220}{70}$. Today circles are approximated by many-sided polygons in computer graphics (plotting) systems.

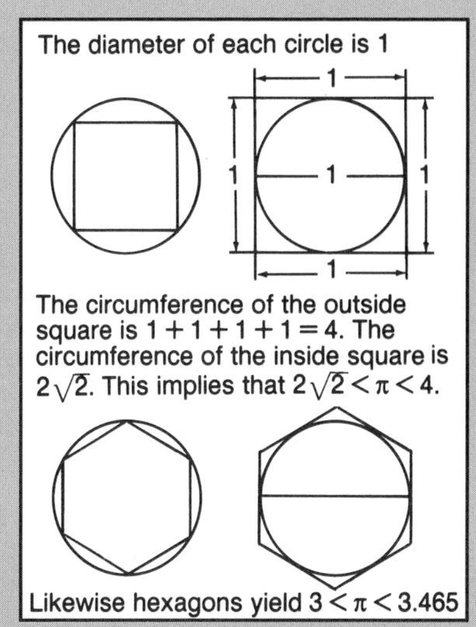

NOTE: Further details can be found in Biographical Encyclopedia of Science and Technology, *by I. Asimov, Doubleday and Company, Garden City, NY, 1964.*

Two concentric circles are drawn with diameters equal to the major and minor diameters of the ellipse under development (a). The 30° - 60° triangle is then used to draw a series of inclined lines (at 30° increments) through the common centerpoint of these circles (b). Vertical and horizontal lines are then drawn as shown in Figure 2.34: Horizontal lines are drawn through the points of intersection between the inclined straight lines and the smaller circle, and vertical lines are drawn through the points of intersection between the inclined lines and the larger circle (c). The intersections between these vertical and horizontal lines then form a set of points that lie on the ellipse to be formed. A French, or irregular, curve is used to construct the ellipse through these points (d). Figure 2.35 describes the ellipse in terms of its mathematical and geometric characteristics.

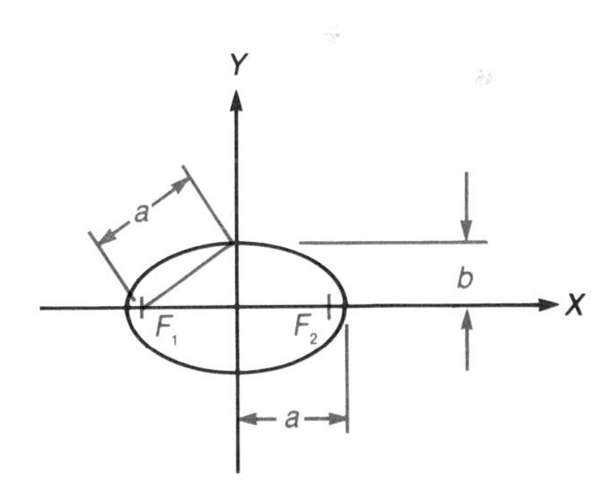

Figure 2.35 An ellipse is mathematically defined by the equation:

$$\frac{x^2}{a^2} + \frac{y^2}{b^2} = 1 \quad (\text{where } a > b > 0)$$

where *a* and *b* represent the radii of the major and minor axes, respectively. The geometric definition for an ellipse is that the sum of the distances between any point on the ellipse and the foci (F_1 and F_2) is constant (that is, the value of this sum is identical for all points on the ellipse).

2.22 Constructing Parabolas and Hyperbolas

Figure 2.36 and 2.37 present popular methods for constructing a parabola if one knows certain information about the parabola to be formed. To use the method in Figure 2.36, one must know the **directrix** and the **focus** of the parabola.

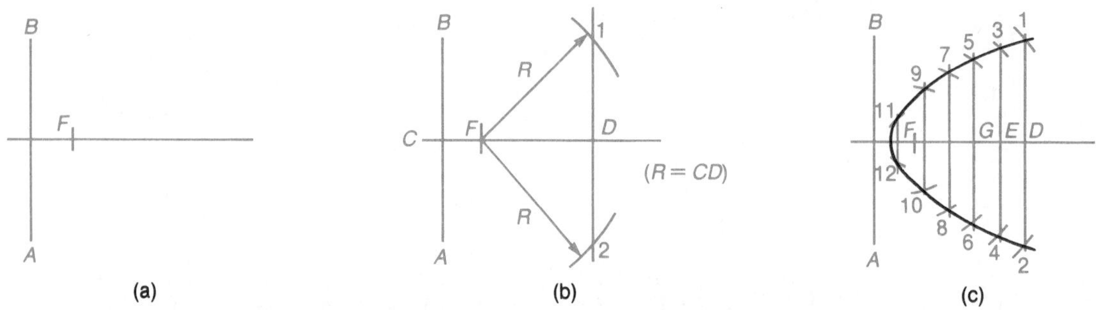

Figure 2.36 To construct a parabola when one knows both the directrix, $\overline{AB}$, and the focus, *F* (a), an arbitrary point, *D*, is chosen, and a line is drawn through *D* (the line is perpendicularly oriented with respect to the axis containing the focus, *F*) (b). Two arcs or radius, *R*, are drawn, centered at *F* and with *R* equal to the distance $\overline{CD}$, between the directrix and point *D*. The intersection points 1 and 2 lie on the parabola. Additional points are constructed, after which the parabola is drawn with an irregular curve (c).

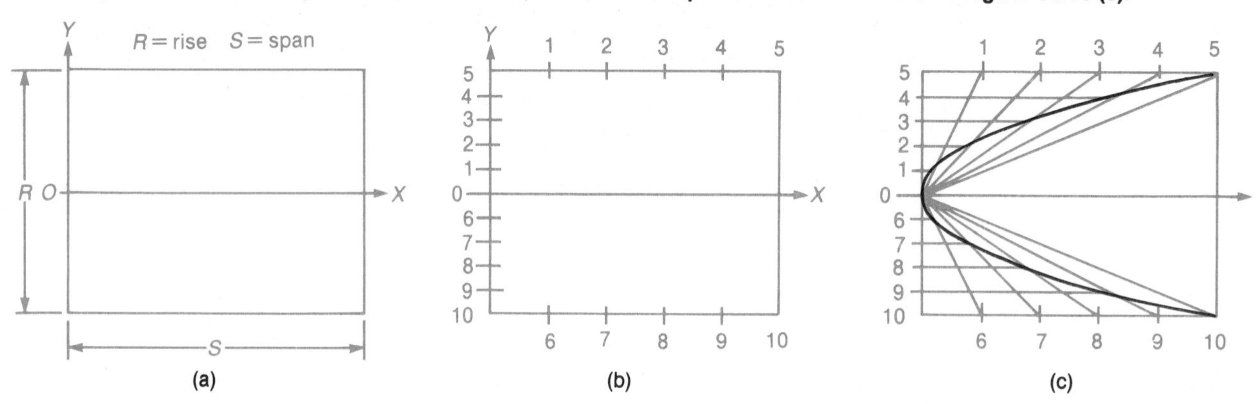

Figure 2.37 To construct a parabola when one knows both the span (width) and rise (depth) of the curve to be formed (a), a set of arbitrarily selected points are identified that divide the rise and span into an equal number of sections, in this example, five sections (b). Lines are drawn between the origin, *O*, and the points along the direction of the span (c). Lines parallel to the span are then drawn between the (inclined) lines and the point along the direction of the rise. The intersections between *corresponding* construction lines (lines with similarly numbered endpoints 1, 2, and so on) are points that lie on the parabola. The parabola is then constructed with the use of an irregular curve.

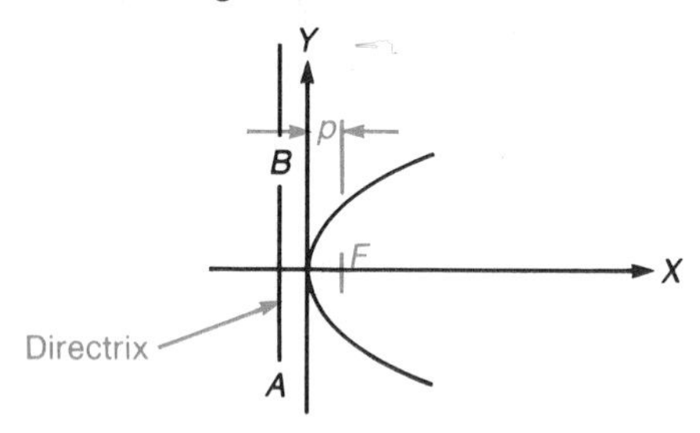

Figure 2.38 The parabola is mathematically defined by the equation:

$$y^2 = 4px$$

where *p* is the distance between the focus of the parabola and the origin of the rectilinear coordinate system. Geometrically, all points on the parabolic curve lie at positions equally distant (*equidistant*) from both the focus *F*, of the parabola and the straight line $\overline{AB}$ known as the parabola's directrix.

The directrix is a straight line, and the focus is a fixed point, the two arranged in such a way that the shortest distance from any point on the parabola to the directrix is equal to the distance between that point and the focus.

To use the method in Figure 2.37, one must know the **span** (width) and **rise** (depth). Figure 2.38 describes the mathematical and geometric characteristics of a parabola.

Figures 2.39 and 2.40 present methods of constructing a hyperbola. To use the method in Figure 2.39, one must know the location of its foci and its transverse axis. To use the method in Figure 2.40 to construct an equilateral, or rectangular, hyperbola, with perpendicular **asymptotes,** one must know the location of a point on the hyperbola with respect to the positions of the asymptotes. (Asymptotes are lines to which the curve becomes tangent at an infinite distance from the origin, *O*.) Figure 2.41 describes the hyperbola in terms of its mathematical and geometric characteristics.

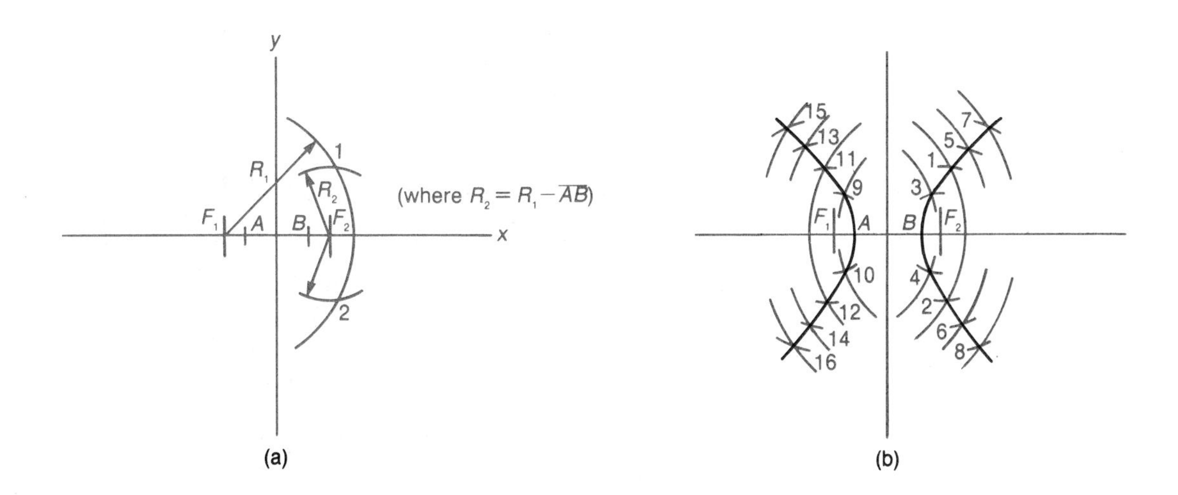

Figure 2.39 To construct a hyperbola when one knows the location of its foci, F_1 and F_2, and its transverse axis, $\overline{AB}$, where A and B are the points at which the hyperbola intersects this axis, two arcs are drawn, centered at F_1 and F_2 and with radii R_1 and R_2, respectively, where $R_2 = R_1 - \overline{AB}$. The points 1 and 2 at which these arcs intersect lie on the hyperbola (a). The procedure is repeated to obtain a sufficient number of points on the hyperbola for its construction.

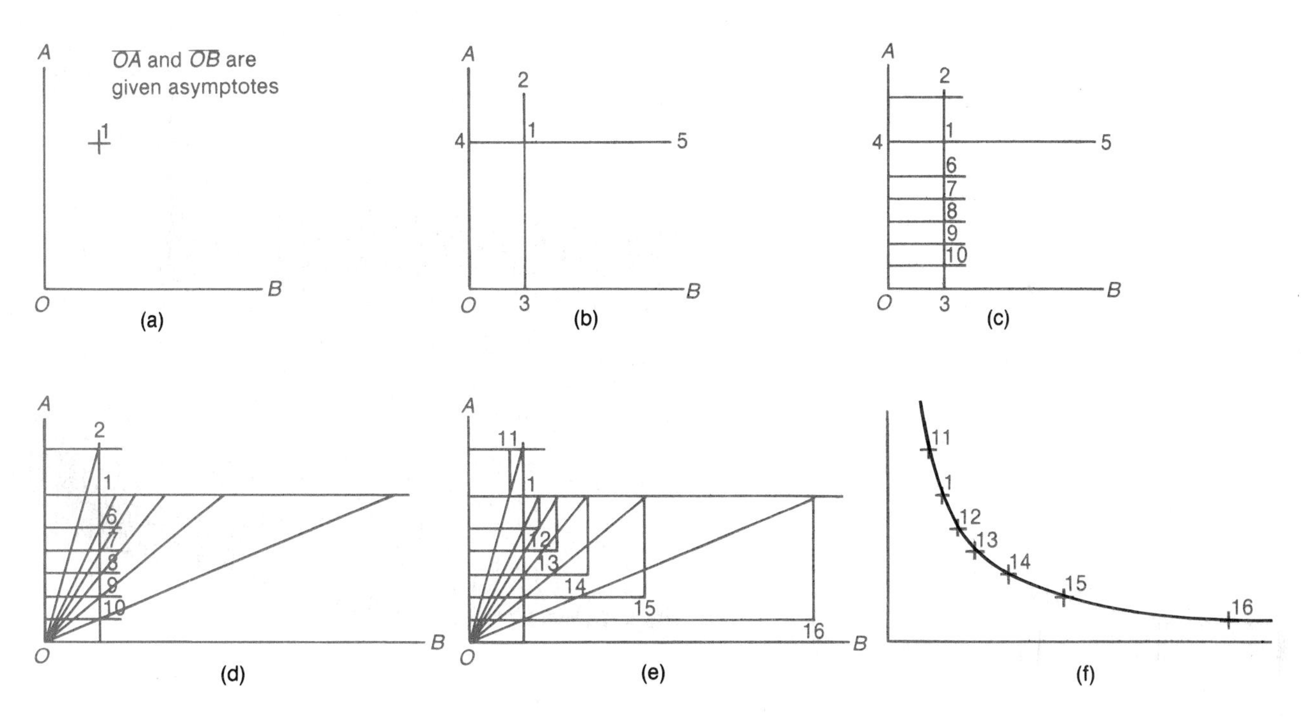

Figure 2.40 To construct an equilateral hyperbola (or rectangular hyperbola) with perpendicular asymptotes when one is given a point, 1, on the hyperbola and the asymptotes, $\overline{OA}$ and $\overline{OB}$ (a). Two lines are drawn through point 1, each perpendicular to an asymptote (b). Lines perpendicular to one asymptote are drawn through line $\overline{23}$ at arbitrarily selected points 6, 7, 8, 9, and 10 (c). Lines through these selected points are drawn, intersecting at the origin, O (d). At those points of intersection between these lines passing through the origin and the line containing points 1 and 4, one draws another set of lines perpendicular to $\overline{14}$. The points of intersection between these lines and those containing points 6 through 10 (perpendicular to $\overline{23}$) are the points 11 through 16 that lie on the hyperbola (e). The final curve is then constructed (f).

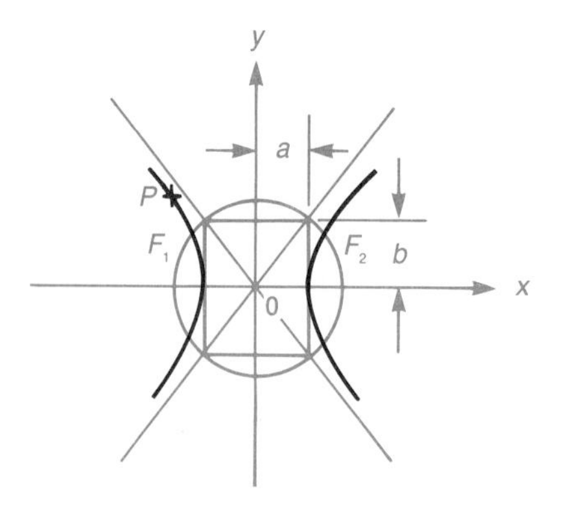

Figure 2.41 A hyperbola is mathematically defined by the equation:

$$\frac{y^2}{a^2} - \frac{x^2}{b^2} = 1$$

where *a* and *b* are the distances between the origin of the rectilinear coordinate system and the asymptotes of the hyperbola, as shown. Geometrically, the difference between the distances $\overline{PF_1}$ and $\overline{PF_2}$ for *any* point *P* and the foci F_1 and F_2 is identical for *all* points on the curve.

2.23 Additional Construction Techniques of Value

Figures 2.42 through 2.47 present specific construction techniques for the following combinations of drawing elements:

- Construction of an arc of a given radius tangent to two given perpendicular lines (Figure 2.42).
- Construction of an arc of a given radius tangent to a given arc and a given straight line (Figure 2.43).
- Construction of an arc of a given radius tangent to two given arcs (Figure 2.44).
- Construction of a hexagon, given the distance across its corners (Figures 2.45 and 2.46).
- Construction of a hexagon, given the distance across its flats (Figure 2.47).

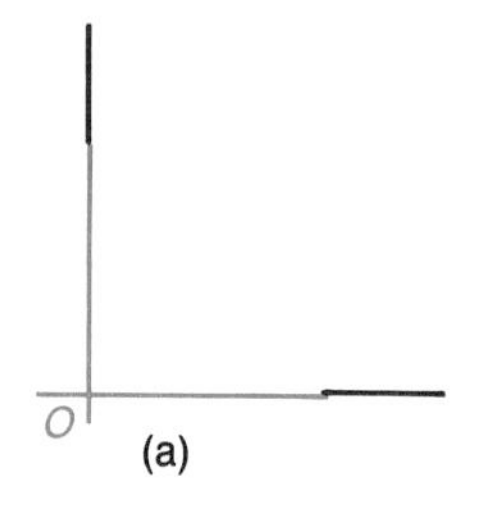

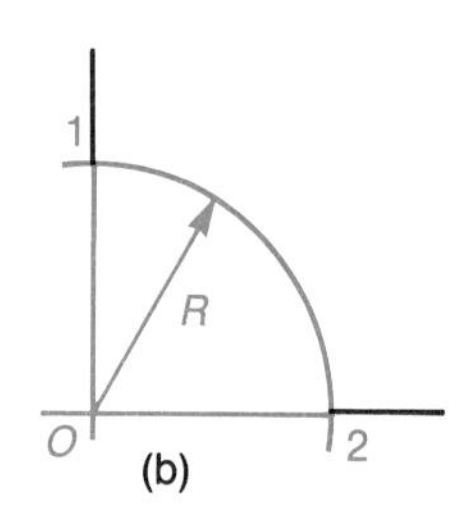

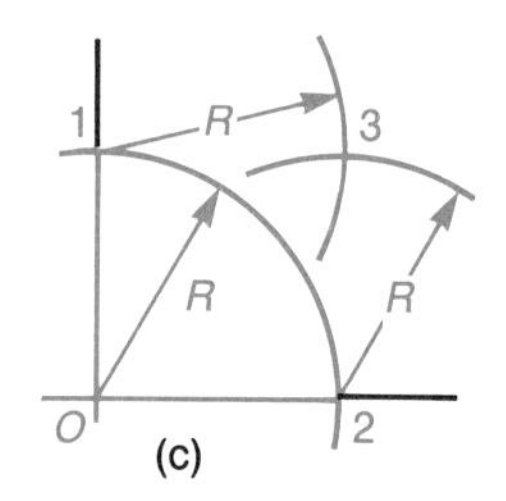

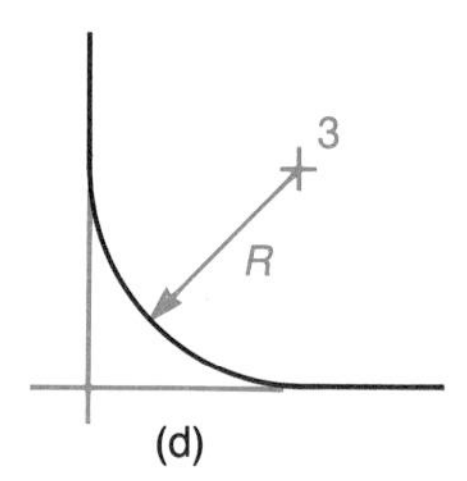

Figure 2.42 To construct an arc of given radius, *R*, which is tangent to two given perpendicular lines (a), an arc of radius *R*, is drawn, centered at the intersection, *O*, of the perpendicular lines; the given straight lines and this arc then intersect at points 1 and 2 (b). Two additional arcs are drawn, centered at points 1 and 2; the intersection of these arcs is point 3, which is the center of the arc under construction (c). Finally, the finished arc is drawn between points 1 and 2, centered at point 3 and with a radius equal to *R* (d).

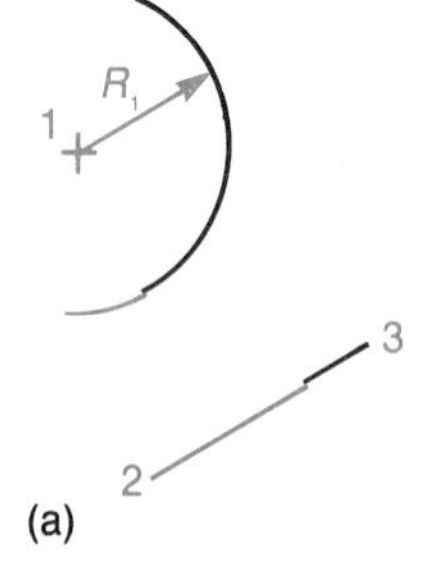

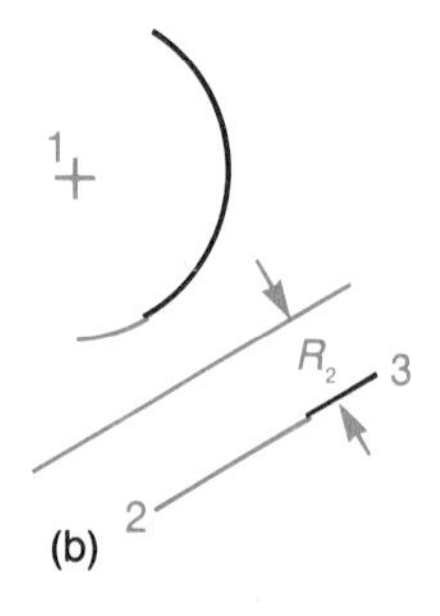

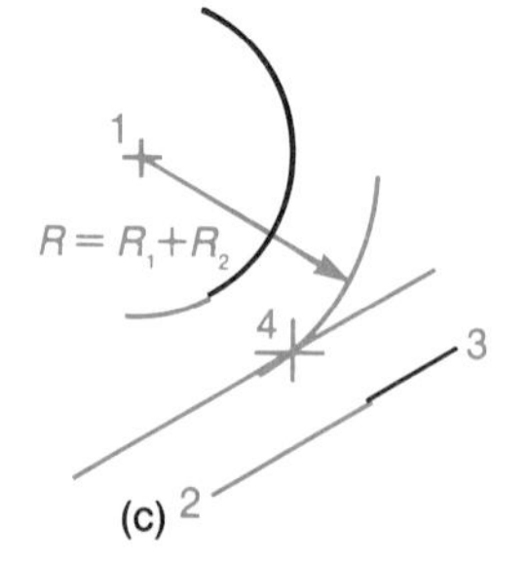

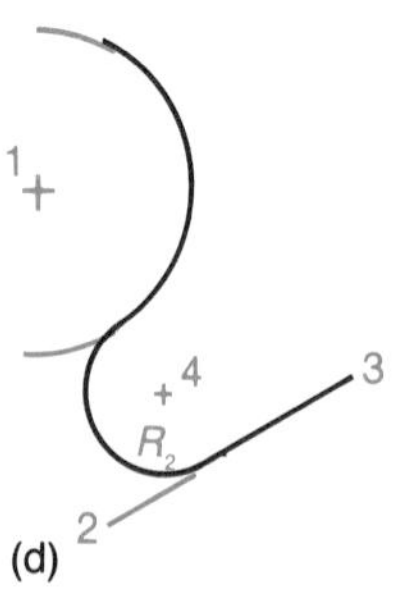

Figure 2.43 To construct an arc tangent both to a straight line and to another arc, given the straight line $\overline{23}$ and the given arc with radius R_1, centered at point 1 (a), a construction line is drawn parallel to line $\overline{23}$ and at a distance equal to R_2, the radius of the arc to be constructed (b). A construction arc of radius $R = R_1 + R_2$, centered at point 1, is drawn; this arc intersects the construction line at point 4, which is the center of the arc that is sought (c). The finished arc, with radius R_2 and centered at point 4, is then drawn (d).

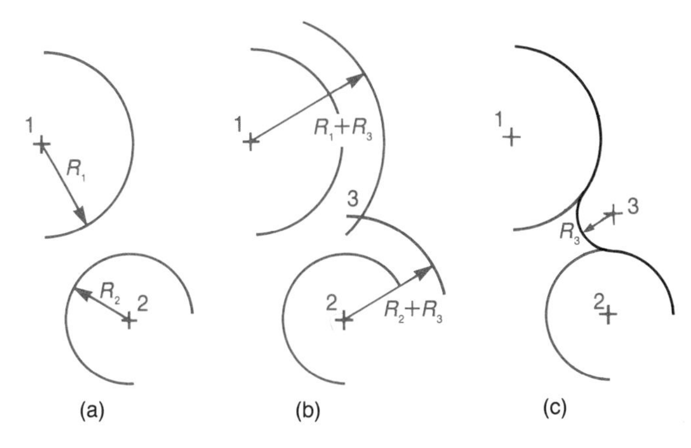

Figure 2.44 To construct an arc tangent to two given arcs with radii R_1 and R_2 (a), two construction arcs are drawn, centered at points 1 and 2 and with radii equal to R_1+R_3 and R_2+R_3, respectively, where R_2 is the radius of the arc (b). The intersection point 3 between these construction arcs is the center of the arc to be formed (c). The finished arc, of radius R_3 and centered at point 3, is then drawn (d).

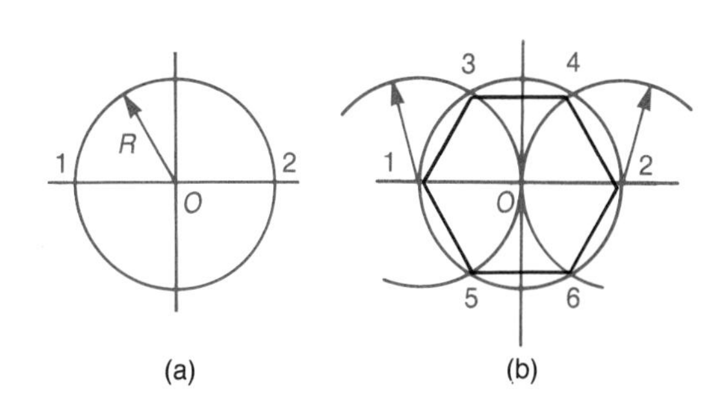

Figure 2.45 To construct a hexagon, given the distance 2*R* across corners, draw a circle of radius *R* (a). Draw two additional arcs, centered at points 1 and 2 and of radius *R*; the points 3, 4, 5, and 6 at which these arc intersect the first construction arc, together with 1 and 2, are corners of the hexagon, which is then finished (b).

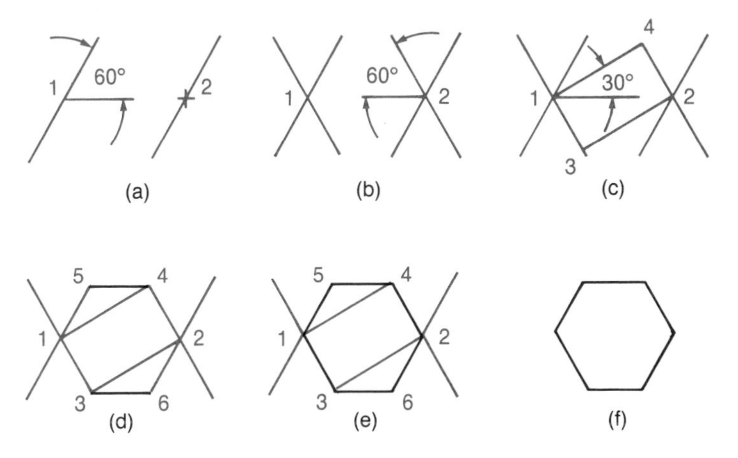

Figure 2.46 An alternative method for constructing a hexagon, given the distance $\overline{12}$ across corners, involves the use of the 30° - 60° triangle, as shown (a–f).

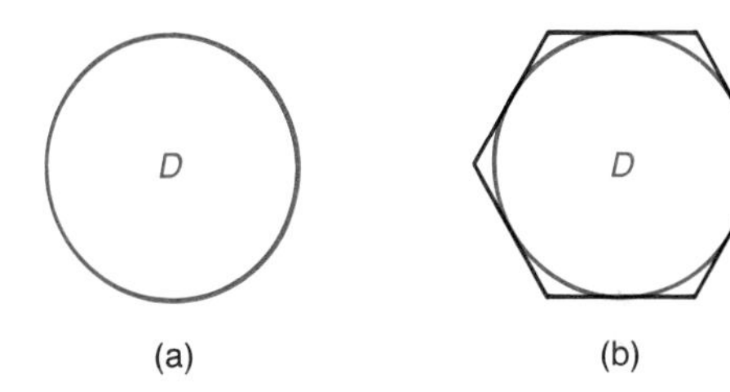

Figure 2.47 Given the distance, *D*, across the flats of a hexagon (a), one can form the hexagon by simply drawing a circle of diameter *D* and then drawing lines (with the 30° - 60° triangle) tangent to this circle, as shown (b).

2.24 Lettering

Lettering is used for notes, dimensioning, and other information on engineering drawings; it must be neat, correct, and consistent, or the entire drawing will appear poorly produced. **Single-stroke gothic characters** are recommended for engineering drawings. Such characters can be quickly and legibly constructed with a series of simple pencil strokes in either of two formats: vertically oriented or inclined (Figure 2.48).

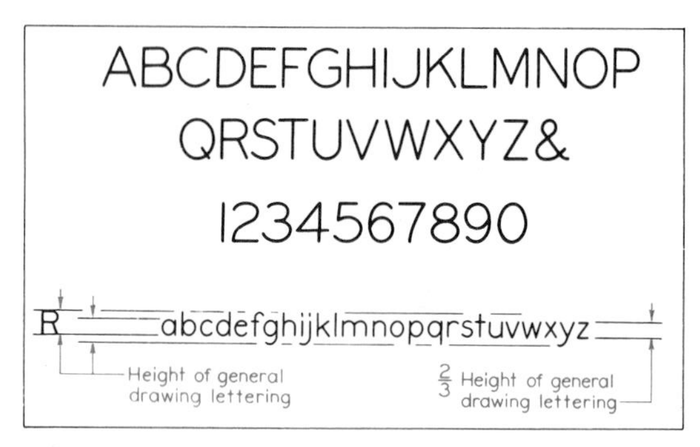

Figure 2.48 Gothic lettering can be rendered in inclined characters (a), or vertical characters (b). (ANSI Y14.2M-1979, American National Standard Line Conventions and Lettering.)

ABCDEFGHIJKLMNO
PQRSTUVWXYZ
1234567890

Figure 2.49 Microfont characters for modified gothic style of lettering (generated with the aid of digital equipment). (ANSI Y14.2M-1979, American National Standard Line Conventions and Lettering.)

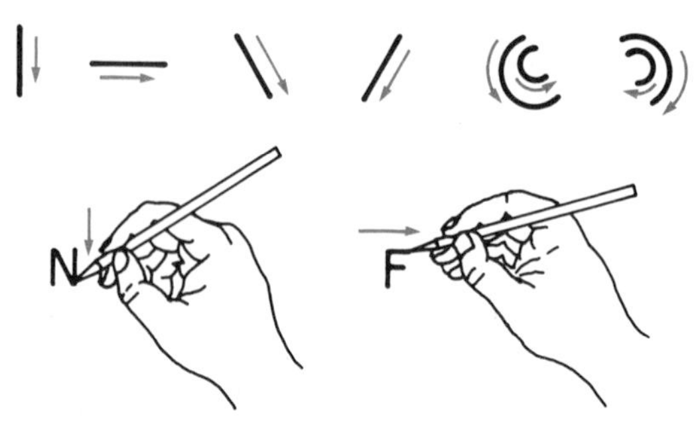

Figure 2.50 Only six fundamental lettering strokes are used in gothic lettering. (NAVEDTRA 10634-C).

The preferred slope for inclined lettering is 2/5, as indicated in Figure 2.48(a); the corresponding slope angle is 68°. Uppercase characters should be used unless lowercase characters are specified for a particular drawing or application.

The American National Standard Institute (ANSI) also recommends a modified gothic style (developed by the National Micrographics Association) for reproduction purposes (see Figure 2.49). This style, known as **Microfont,** is generated by digital equipment. Whichever style is chosen, it should be used throughout a drawing.

Figure 2.50 presents the fundamental lettering strokes for gothic characters. Notice that there are only six strokes, proof of the simplicity of this lettering style.

Very light construction lines, known as **guidelines,** should always be used in the development of notes, dimensions, and other lettered data in a drawing. As indicated in Figure 2.51, characters should extend to, but not beyond, the guidelines so that these characters will be consistent in height (also see Figure 2.52). In addition, spacing between characters should be chosen so that the background (white space) areas between characters in a word appear to approximately uniform; in other words, the distances between letters in a word are not necessarily identical. Letters spaced too closely or too loosely are difficult to read and very unappealing in appearance (Figure 2.53).

Vertical spacing (D in Figure 2.54) between lines should be related to the height (H in the figure) of the lettering in accordance with the following inequality:

$$\frac{1}{2}H < D < H$$

Furthermore, notes should be aligned or *left-justified,* as shown in this figure. The height, H, of most notes and dimensions in a drawing is usually about 3 mm. or 0.125 in.

Additional, vertical guidelines should be arbitrarily spaced throughout the construction region on a drawing for lettering to help in the construction of vertical characters (Figure 2.55).

Spacing between words appearing on a single line should be between $\frac{1}{2}H$ and H (where H represents the height of the characters). Spacing between sentences should be equal to (or slightly greater than) H. Finally, spacing between characters separated by a decimal point should be equal to (or slightly greater than) $\frac{2}{3}H$ (see Figure 2.56).

Fractions should be drawn with a height equal to twice that of a single character (Figure 2.57). In addition, the

N

Correct

N

Incorrect

N

Incorrect

Figure 2.51 **Characters should extend to but not beyond, the guidelines. *Always use guidelines.***

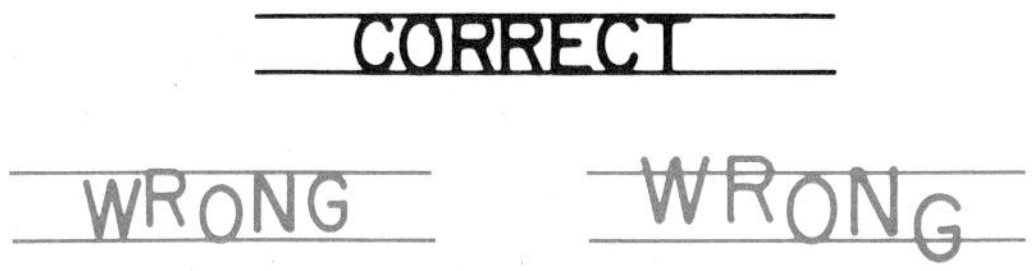

CORRECT

WRONG WRONG

Figure 2.52 **Guidelines keep characters consistent in height. *Always use guidelines.***

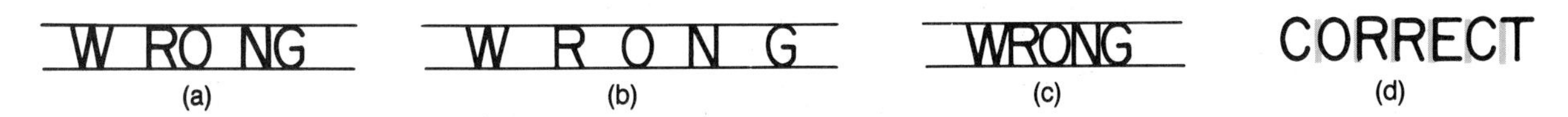

Figure 2.53 **Spacing between characters in a word should be chosen so that the *background* (i.e., white space) between letters appears to be approximately uniform.**
(a) Spacing between letters is inconsistent in both background areas and distance between letters. This is incorrect.
(b) Spacing between letters is consistent in distance between letters but inconsistent in background areas. This is incorrect.
(c) Spacing between letters is too small. This is incorrect.
(d) Spacing between letters is consistent in terms of background areas between characters. This is correct.
Spacing which is too small or too large is difficult to read and very unappealing to the eye. *Always use guidelines.*

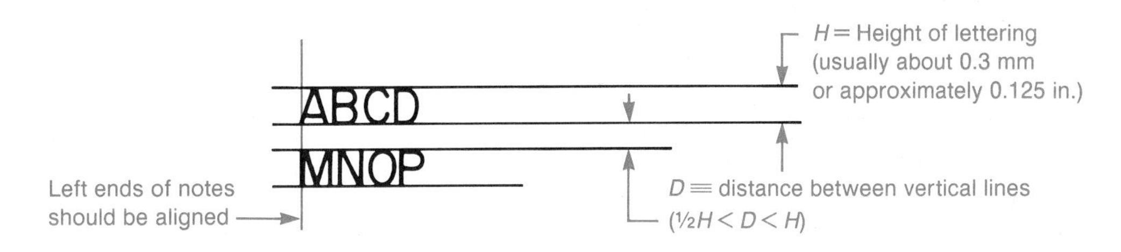

Figure 2.54 **Vertical spacing between lines, *D*, should be related to the height of the lettering, *H*, as shown. Notes should be aligned on the left.**

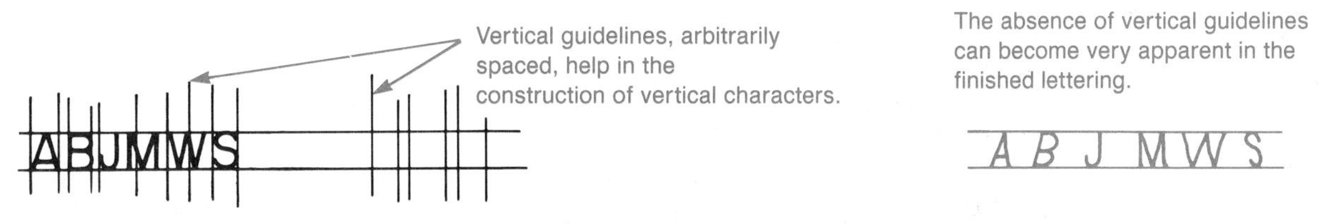

Figure 2.55 **Vertical as well as horizontal guidelines should be used to properly construct lettering.**

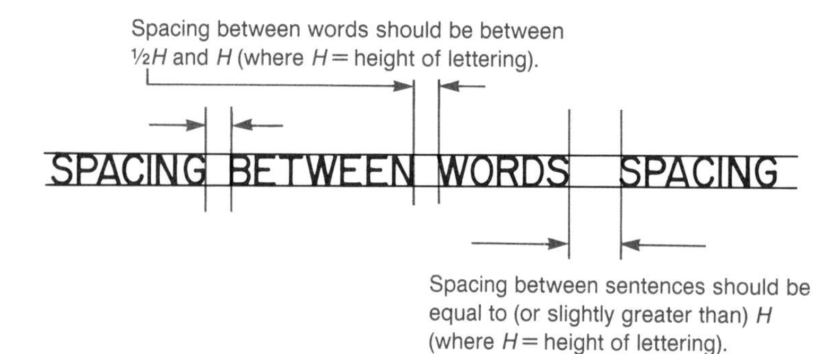

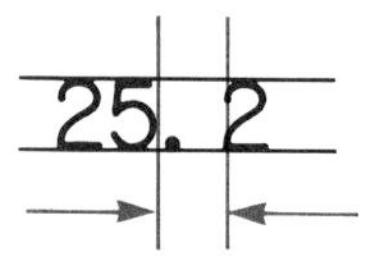

Figure 2.56 Spacing between words, sentences, and characters separated by a decimal point should be related to the height of the lettering, *H*, as shown.

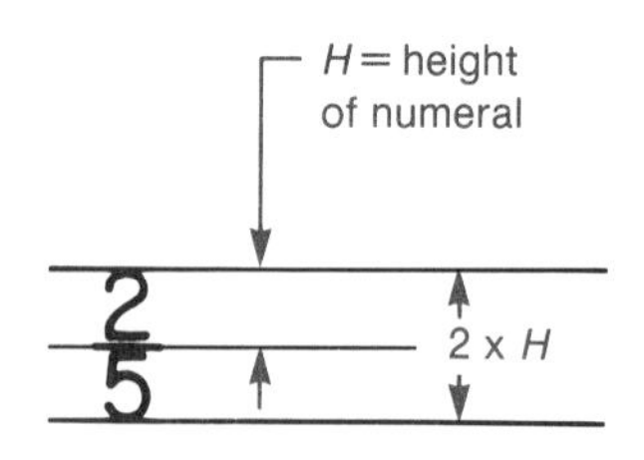

Figure 2.57 Fractions should be drawn with a height equal to twice that of a single numeral. Notice that the width of the line separating the numerals in the fraction is equal to the width of the numerals.

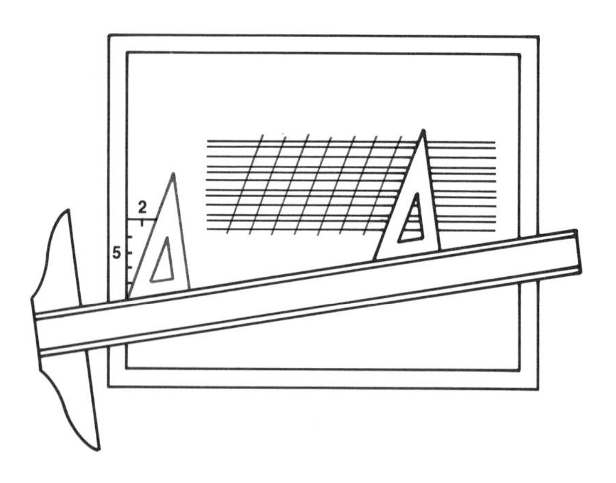

Figure 2.58 To develop inclined guidelines with the T-square and a triangle, orient the T-square and triangle so that the hypotenuse of the triangle corresponds to a slope of 2/5. Simply measure five units in the vertical direction and two units in the horizontal direction (as shown) to obtain this orientation. The triangle may then be slid along the stationary T-square while one draws the inclined guidelines parallel to the hypotenuse. (NAVEDTRA 10634-C).

width of the line separating the numerals in a fraction should be equal to the width of the numerals.

Figure 2.58 describes an arrangement of the T-square and a triangle that can be used to construct inclined guidelines at the appropriate slope of 2/5, or 68°.

2.25 Standard Types of Lines

Figure 2.59 presents examples of standard types of lines used in engineering drawings. The particular application of each type of line will be developed in subsequent chapters. Lines of the same type in a drawing should be identical in intensity and width; such consistency is needed if the drawing is to be easily read and interpreted.

Lines in drawings that have been **manually produced** may be of two distinct widths: **thin** (corresponding to widths of about 0.35 mm. or 0.016 in.) and **thick** (0.7 mm. or 0.032 in.). Drawings generated by computer can be produced with lines of identical width (the width is determined by the equipment in use, the size and purpose of the drawing, and other factors).

Line types may be summarized as follows:

- **Visible lines** are solid and relatively thick. These lines represent visible surface boundaries of objects described in engineering drawings (see Chapter 4).
- **Hidden lines** consist of a series of short, thin dashes separated by spaces with widths of approximately ¼ *W*, where *W* is the width of each dash. These lines represent surface boundaries that are hidden by solid material in a drawing of an object.
- **Section lines** are solid, thin lines used to indicate surfaces that have been cut, or sectioned, in a so-called section view (see Chapter 6).
- **Center and symmetry lines** are composed of alternating short and long dashes to indicate centerpoints and axes of symmetry of features in a drawing.

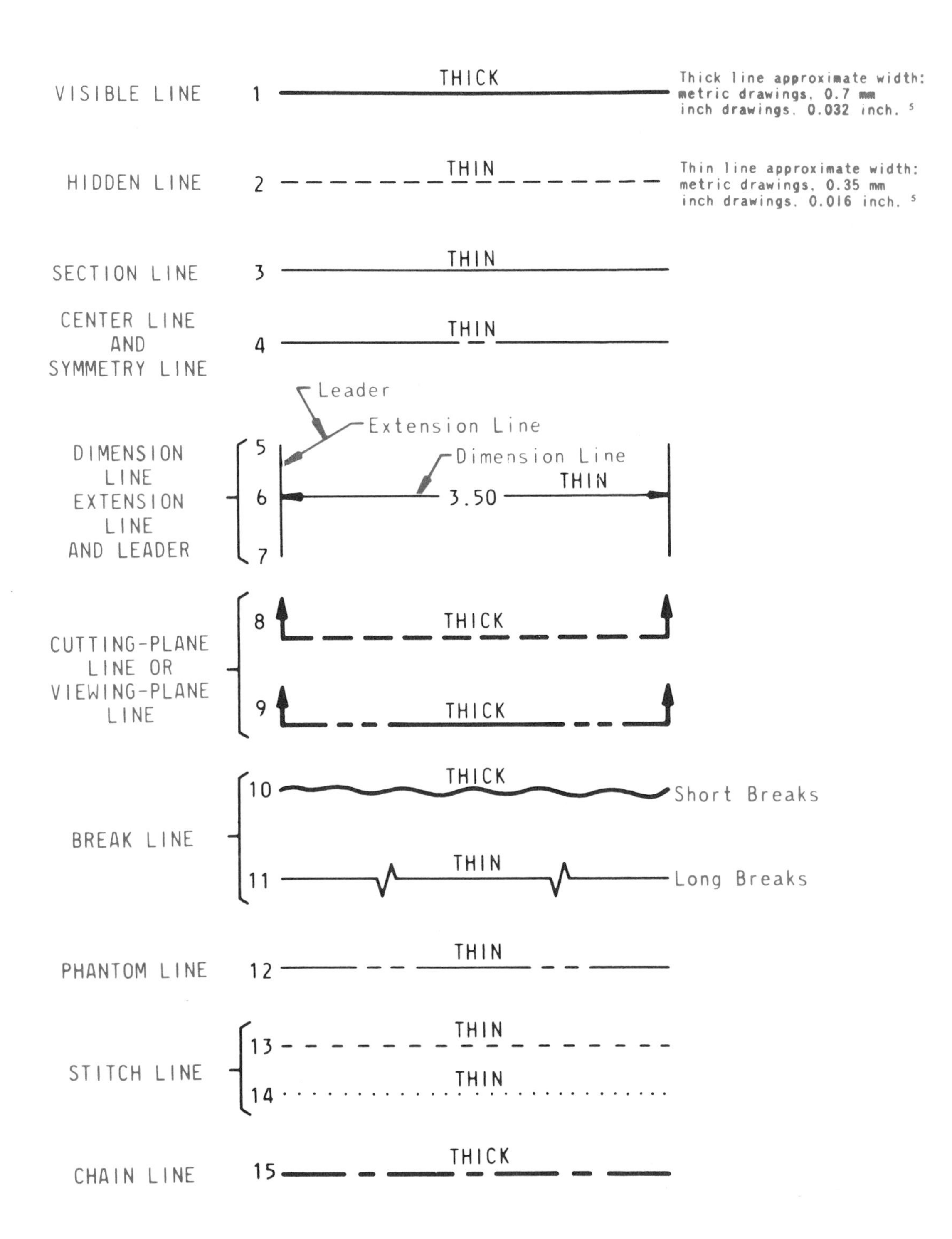

Figure 2.59 Standard types of lines used in engineering drawings. (ANSI Y14.2M-1979, American National Standard Line Conventions and Lettering.)

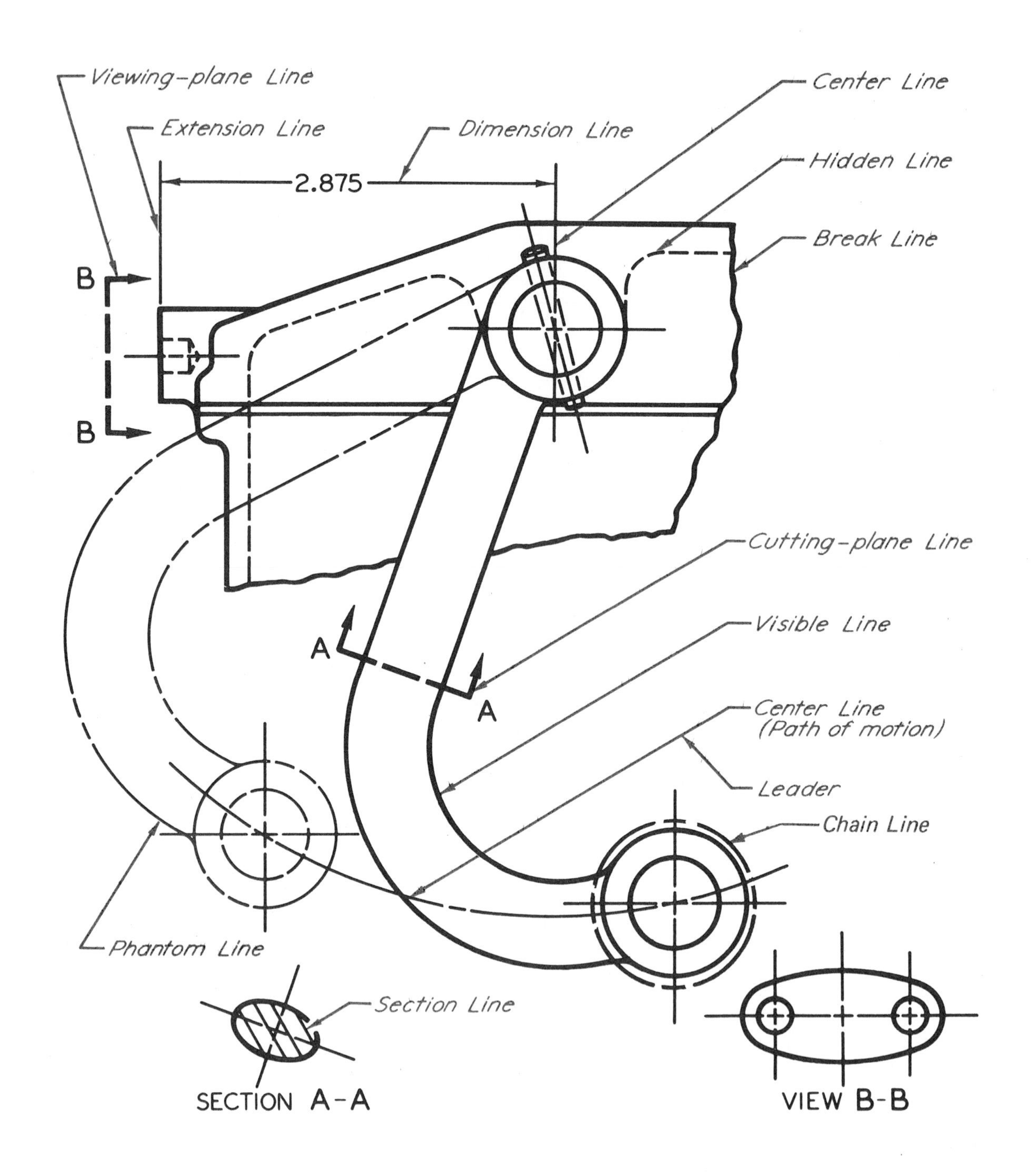

Figure 2.60 Examples of lines in an engineering drawing. (ANSI Y14.2M-1979, American National Standard Line Conventions and Lettering.)

- **Dimension lines, extension lines, and leaders** are solid, thin lines used to specify the size of a feature on a drawing and other significant information (see Chapter 8).
- **Cutting-plane or viewing-plane lines** may be composed of short dashes of identical length or long dashes that alternate with pairs of short dashes. The ends of such lines are short, perpendicular lines with arrowheads. These arrowheads indicate the direction of sight with which the observer views the sectioned view in an engineering drawing (see Chapter 6).
- **Break lines** may be of two separate forms: (1) thick lines drawn without the aid of instruments or (2) thin lines joined by jagged or zigzag lines. These lines indicate that a portion of an object is not shown in a graphic description.
- **Phantom lines** consist of long, thin dashes separated by pairs of short dashes. Such lines are used to describe alternative locations for the components of an object or the relative positions of parts in a working configuration.
- **Stitch lines** are composed of short, thin dashes and spaces of *equal* length, or a series of dots approximately 0.35 mm. (0.016 in.) in diameter and spaced at intervals of approximately 3 mm. (0.12 in.). Stitch lines are used to indicate a stitch or sewing pattern in an object.
- **Chain lines** consist of alternating long and short dashes of relatively thick width. Chain lines are used to identify a specific region or surface on a drawing for additional consideration.

Figure 2.60 presents an example of the applications of these lines in an engineering drawing. The application of phantom lines is shown in Figure 2.61.

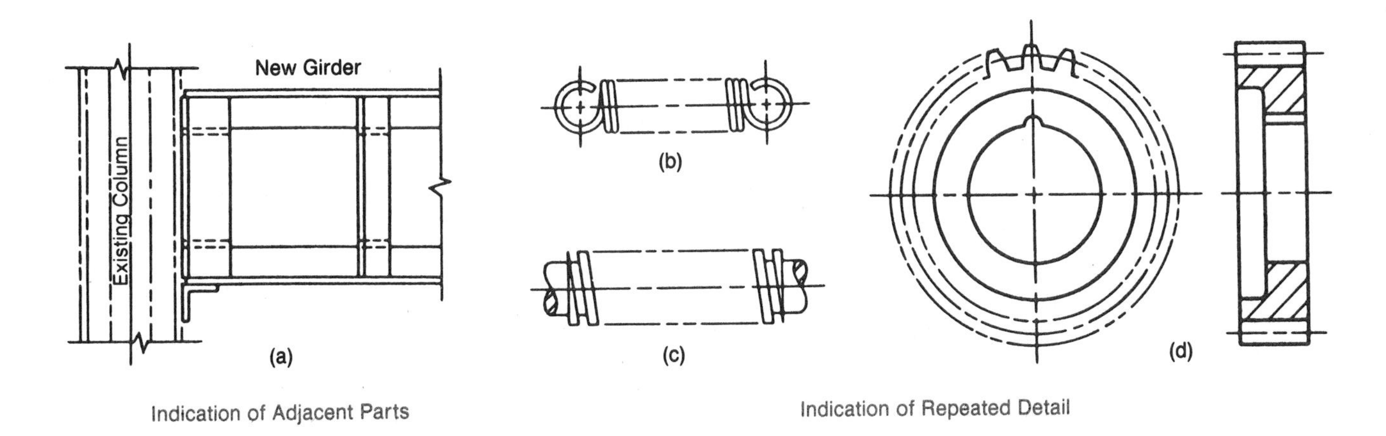

Figure 2.61 Applications of phantom linework. (ANSI Y14.2M-1979, American National Standard Line Conventions and Lettering.)

Figure 2.62 To construct an arrowhead, follow steps a through c or a through d. (NAVEDTRA 10634-C).

Finally, dimension lines and cutting-plane lines include **arrowheads** that must be carefully constructed. Figure 2.62 presents a construction method for arrowheads. Such arrowheads should have widths equal to approximately one-third their lengths. Arrowheads may be *filled in* as shown in Figure 2.62(d) or not. In any event, arrowheads should be *uniform* in both size and style within a drawing. (Further details may be found in Chapters 8, 9, and 10.)

Construction lines are not finished lines and are therefore not included in the preceding list of types of lines. Construction linework must be very, very light so that it will not be easily seen on a drawing. It is often not erased. If it is to be erased, such erasure must be accomplished without affecting the finished linework on the drawing. A rule of thumb for determining whether construction lines are too dark in a drawing is that they should not be seen if one views a drawing from arm's length.

Alternatively, **visible lines,** the most apparent type of linework in a drawing, should be thin yet dark (intense) enough to be clearly visible to an observer at a distance of approximately 5 feet from the drawing. Such linework will then be reproducible via various printing processes. Other types of finished linework will be less apparent due to spacing within the lines or lesser intensity. We will return to the subject of finished and construction linework in engineering drawings in Chapter 4.

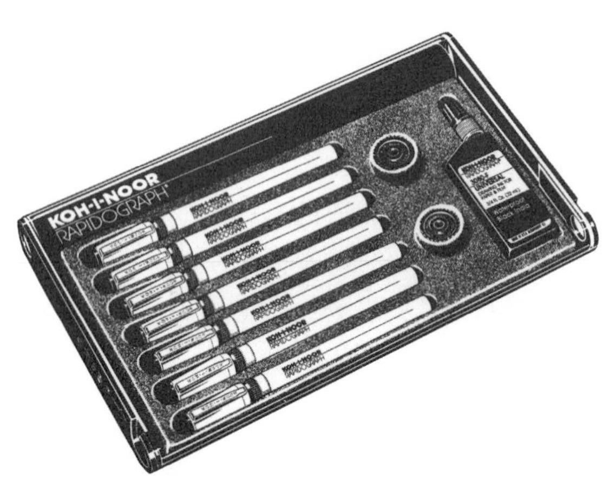

Figure 2.63 There is a wide variety of pens available for working in ink. (Courtesy of Koh-I-Noor Rapidograph, Inc., Bloomsbury, NJ.)

2.26 Ink Graphics

Finally, a word about drawing in ink is in order. Although we have focused our attention upon pencilled drawings, an accomplished drafter will finish a drawing in ink. There is a wide variety of graphics aids for working in ink, including pens that can produce a broad range of line widths (Figures 2.63 through 2.65) and lettering guides or scribers (Figures 2.66 and 2.67). As your skill in graphics increases, you may want to develop your ability to work in ink as well as pencil.

Figure 2.64 An example of a technical pen drawing point. (Courtesy of Koh-I-Noor Rapidograph, Inc., Bloomsbury, NJ.)

6x0	4x0	3x0	00	0	1	2	2½	3	3½	4	6	7
.13	.18	.25	.30	.35	.50	.60	.70	.80	1.00	1.20	1.40	2.00
.005 in.	.007 in.	.010 in.	.012 in.	.014 in.	.020 in.	.024 in.	.028 in.	.031 in.	.039 in.	.047 in.	.055 in.	.079 in.
.13 mm	.18 mm	.25 mm	.30 mm	.35 mm	.50 mm	.60 mm	.70 mm	.80 mm	1.00 mm	1.20 mm	1.40 mm	2.00 mm

Figure 2.65 A broad range of line widths can be produced with ink pens. (Courtesy of Koh-I-Noor Rapidograph, Inc., Bloomsbury, NJ.)

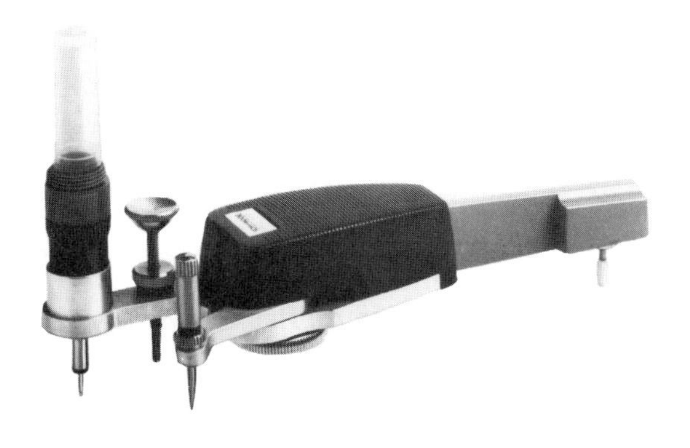

Figure 2.66 An ink scriber is an aid for working in ink. (Courtesy of Koh-I-Noor Rapidograph, Inc., Bloomsbury, NJ.)

Figure 2.67 A scriber set is another aid for working in ink. (Courtesy of Koh-I-Noor Rapidograph, Inc., Bloomsbury, NJ.)

In Retrospect

- Drawing sheets are standardized in both size and format.
- A T-square allows the user to draw horizontal lines and define a particular direction as *horizontal* in the development of a drawing.
- Triangles are used in conjunction with the T-square to draw vertical and inclined lines.
- Drawing pencils must be carefully selected so that the user can produce linework of the appropriate intensity and width. Pencil leads are produced in eighteen separate grades.
- To delete unwanted linework from a drawing, erasers, an erasing shield and a dusting brush should be used.
- A compass allows the user to produce arcs and circles of arbitrary radii, and templates may be used to easily construct circles of specific diameters.
- Dividers are used to quickly partition a given line into equal parts or to transfer the distance between two points on a drawing without taking time to measure.
- Scales are graduated measuring instruments, whereas the *scale of a drawing* refers to the ratio between the size of the drawing and the size of the object described by the drawing.
- The irregular, or French, curve can be used to construct smooth curves through a given set of points in space.
- Common types of angles are right, acute, obtuse, complementary, and supplementary angles. Common types of triangles are right, equilateral, isosceles and scalene.
- Quadrilaterals are four-sided geometric figures that include the square, rectangle, rhombus, rhomboid, trapezium and trapezoid.
- Significant characteristics of circles include the centerpoint, radius, diameter, arc, chord, and concentricity of multiple circles.
- Construction methods exist for:
  - Bisecting a straight line.
  - Bisecting an angle.
  - Dividing a line into equal segments.
  - Constructing lines tangent to two circles of equal or unequal diameters.
- Conic sections include ellipses, parabolas and hyperbolas. Various construction techniques for these conic sections exist, depending on what is known about the figure to be constructed.
- Lettering is used for notes, dimensioning, and other data in a drawing. Single-stroke gothic characters allow one to produce lettering quickly and legibly with a series of basic strokes. Lettering may be inclined (with a slope of 2/5, or 68°) or vertical.
- Standard types of lines used in engineering drawings include visible, hidden, section, center, symmetry, dimension, extension, cutting-plane, break, phantom, stitch and chain lines.

ENGINEERING IN ACTION

Hook-and-Loop Fasteners Speed Boardwork

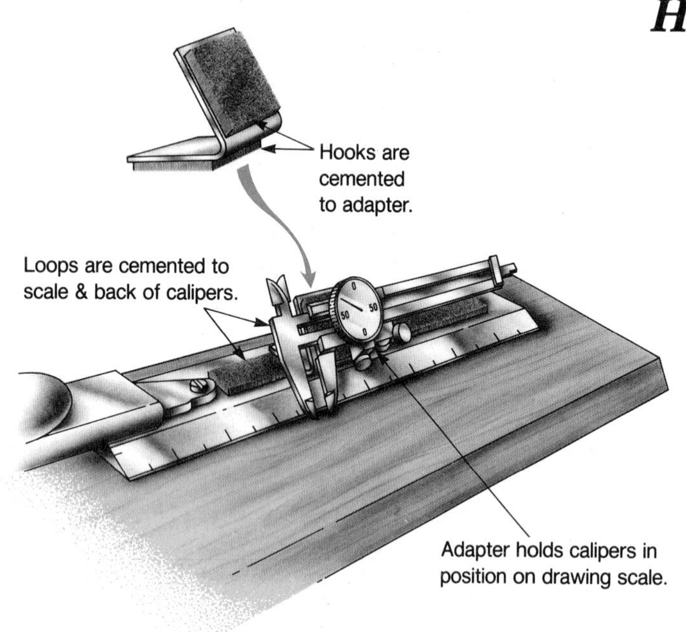

■ Draftsmen often use vernier calipers to transfer measurements directly to a drawing. The process can be simplified by using an adapter that holds the calipers on the drafting machine scale. The technique allows the calipers to be positioned for easy viewing and frees both of the draftsman's hands for drawing. A mounting flange and hook-and-loop fasteners are used to hold the calipers on the scale. Hooks are cemented to the mounting flange while loops are cemented on both the calipers and scale. In operation, a measurement is made with the calipers which is then simply attached to the drafting scale with the mounting flange. The technique is the idea of W. E. Pulley, Reuter Stokes, Cleveland, Ohio.

Source: Reprinted from MACHINE DESIGN, June 10, 1982. Copyright, 1982, by Penton/IPC., Inc., Cleveland, Ohio, p. 40.

Dashpot Cushions Plotting Pen

■ **Problem:** Pens used with precision plotters must descend rapidly onto the paper, leave a trace, and retract almost instantly. The pen must contact the paper with just enough force to produce a clean trace, but not hard enough to be destroyed by repeated impact.
■ **Solution:** This pen plotter drive uses an air dashpot to cushion impact of the pen against the plotter paper. Although the pen cycles at high speed (20 times per second), the dashpot snubs the pen just before impact and lowers it gently onto the paper. Airpot Corp. makes the dashpot used in the Hewlett-Packard design. The device dissipates 0.12 in.-lb. of energy produced by the descending motion of the pen, protecting the pen nib from high impact forces that would cause deformation.

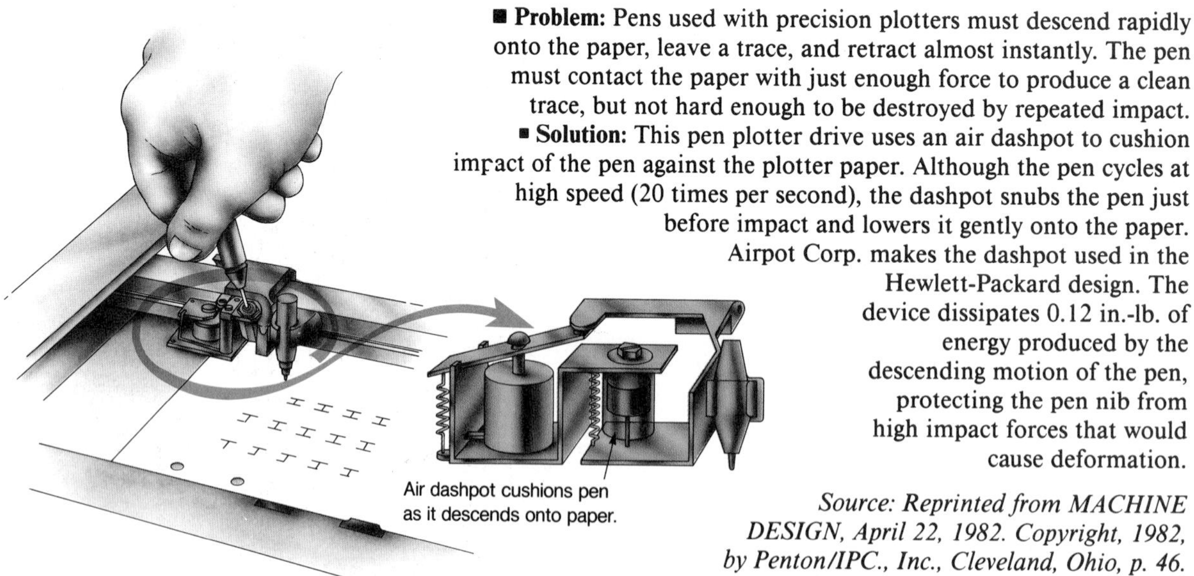

Source: Reprinted from MACHINE DESIGN, April 22, 1982. Copyright, 1982, by Penton/IPC., Inc., Cleveland, Ohio, p. 46.

Exercises

Your instructor will specify the appropriate drawing sheet formats, lettering style, and scale(s) for the following exercises.

EX2.1. Complete a general title block; include your name, your institution, the date and additional information as specified by your instructor.

EX2.2. Letter, in heights of 0.125 in., 0.25 in., and 0.50 in., the following information: your name, the english alphabet, and the numerals 0 through 9. Use guidelines. Repeat this effort. Center your work within the drawing area on your paper.

EX2.3. Divide your drawing sheet into four equal areas (your title block will be contained within one of these areas, thereby reducing the available drawing space within that region).

a. In the upper left region draw five visible lines, five hidden lines, five centerlines, and five construction lines. All lines should be equally spaced and of identical length. Recall that the dashes in a hidden line should be approximately equal to 4 W, where W is the width of the space between dashes. (Spacing and dashes should be consistent.) In addition, the length of the long dashes in a centerline should be equal to $4L$, where L is the length of the short (central) dash. Spacing between dashes in a centerline should be equal to $\frac{1}{4}L$. As a result, you will need to consider the total length of the hidden and center lines to determine the appropriate dash lengths and spacing for these lines.

b. In the upper right region draw five concentric circles with diameters equal to 0.5 in., 0.75 in., 1.0 in., 1.25 in., and 1.5 in.

c. In the lower left region draw a set of visible lines with angles of orientation of 0°, 15°, 30°, 45°, 60°, 75°, 90°, 105°, 120°, 135°, 150°, and 165° with respect to the horizontal direction. All lines should share a common endpoint.

d. In the lower right region draw a circle with a radius equal to 1.5 in., inscribed within a square.

EX2.4. Draw concentric circles with radii equal to 5 mm., 8 mm., 1 cm. (centimeter), 1.5 cm., 2.0 cm., 3.7 cm., and 5 cm.

EX2.5. Draw a scalene triangle with sides equal to 5 cm., 7 cm., and 3 cm.

EX2.6. Draw a scalene triangle with sides equal to 2 in., 4 in., and 5.5 in.

EX2.7. Draw an equilateral triangle with each side equal to 4.4 cm.

EX2.8. Draw an equilateral triangle with each side equal to 3.8 cm.

EX2.9. Draw an isosceles triangle in which the sides of equal length are 2.8 cm.

EX2.10. Draw an isosceles triangle in which the sides of equal length are 3.5 in.

EX2.11. Draw a square with each side equal to 2.75 cm.

EX2.12. Draw a square with each side equal to 2.75 in.

EX2.13. Draw a set of squares sharing a common centerpoint and with sides equal to 2 cm., 3.5 cm., and 5.5 cm.

EX2.14. Draw a set of squares sharing a common centerpoint and with sides equal to 2 in., 3.5 in., and 4.75 in.

EX2.15. Draw a rectangle with sides equal to 2.5 cm. and 3.8 cm.

EX2.16. Draw a rectangle with sides equal to 3.5 in. and 1.75 in.

EX2.17. Draw a rhombus with sides equal to 3.7 cm.

EX2.18. Draw a rhombus with sides equal to $2\frac{1}{4}$ in.

EX2.19. Draw a rhomboid with sides equal to 2.54 cm. and 4.2 cm.

EX2.20. Draw a rhomboid with sides equal to 1.0 in. and $3\frac{3}{8}$ in.

EX2.21. Draw a trapezium with sides equal to 1 cm., 2 cm., 3 cm., and 4 cm. One side should be oriented along the horizontal direction.

EX2.22. Draw a trapezium with sides equal to 1 in., 2 in., 2.5 in., and 3.5 in. One side should be oriented along the vertical direction.

EX2.23. Draw a trapezoid in which the parallel sides are equal to 2.5 cm. and 4 cm.; a third side should be equal to 3.4 cm.

EX2.24. Draw a trapezoid in which the parallel sides are equal to 2.5 in. and 3.25 in.; a third side should be equal to 2.75 in.

EX2.25. Draw an ellipse with minor and major diameters equal to 2 cm. and 3.5 cm., respectively.

EX2.26. Draw an ellipse with minor and major diameters equal to 1.5 in. and 2.5 in., respectively.

EX2.27. Draw a polygon with n sides of 2.0 cm. length, where $n = 5$ (a pentagon).

EX2.28. Draw a polygon with n sides of 2.0 cm. length, where $n = 5$ (a pentagon).

EX2.29. Draw a polygon with n sides of 2.75 in. length, where $n = 6$ (a hexagon).

EX2.30. Draw a polygon with n sides of 2.9 cm. length, where $n = 6$ (a hexagon).

EX2.31. Draw a polygon with n sides of 1.8 cm. length, where $n = 7$ (a heptagon).

EX2.32. Draw a polygon with n sides of 2.25 in. length, where $n = 7$ (a heptagon).

EX2.33. Draw a polygon with n sides of 1.7 cm. length, where $n = 8$ (an octagon).

EX2.34. Draw a polygon with n sides of 1.5 in. length, where $n = 8$ (a octagon).

EX2.35. Subdivide the line $\overline{12}$ shown in Figure EX2.35 into five equal parts; do not use a scale.

EX2.36. Divide the line $\overline{12}$ shown in Figure EX2.36 into three equal parts; do not use a scale.

EX2.37. Bisect the angle between lines $\overline{12}$ and $\overline{23}$ in Figure EX2.37; do not measure the angle.

EX2.38. Bisect the angle between lines $\overline{12}$ and $\overline{23}$ in Figure EX2.38; do not measure the angle.

EX2.39. Bisect the angle between lines $\overline{12}$ and $\overline{23}$ in Figure EX2.39; do not measure the angle.

EX2.40. Bisect the line $\overline{12}$ in Figure EX 2.35; do not measure the length of the line.

EX2.41. Bisect the line $\overline{12}$ in Figure EX2.36; do not measure the length of the line.

EX2.42. Draw the linework shown in Figure EX2.42; use the scale specified by your instructor.

EX2.43. Draw the linework shown in Figure EX2.43; use the scale specified by your instructor.

EX2.44. Draw two complete figures as follows: construct a set of tangent lines between the two circles shown in Figure EX2.44 in:
a. An open-belt configuration.
b. A crossed-belt configuration.

EX2.45. Draw Figure EX2.45; use a scale as specified by your instructor.

EX2.46. Draw Figure EX2.46; use a scale as specified by your instructor.

EX2.47. Draw Figure EX2.47; use a scale as specified by your instructor.

EX2.48. Draw Figure EX2.48; use a scale as specified by your instructor.

EX2.49. Draw Figure EX2.49; use a scale as specified by your instructor.

EX2.50. Draw Figure EX2.50; use a scale as specified by your instructor.

EX2.51. Draw Figure EX2.51; use a scale as specified by your instructor.

EX2.52. Draw a smooth curve through the twenty-three points given in Figure EX2.52; use the irregular, or French, curve and a scale specified by your instructor.

EX2.53. Draw a set of (six) tangent lines between the three circles shown in Figure EX2.53. Use an open-belt configuration and a scale specified by your instructor.

EX2.54. Draw a set of (six) tangent lines between the three circles shown in Figure EX2.53. Use a closed-belt configuration and a scale specified by your instructor.

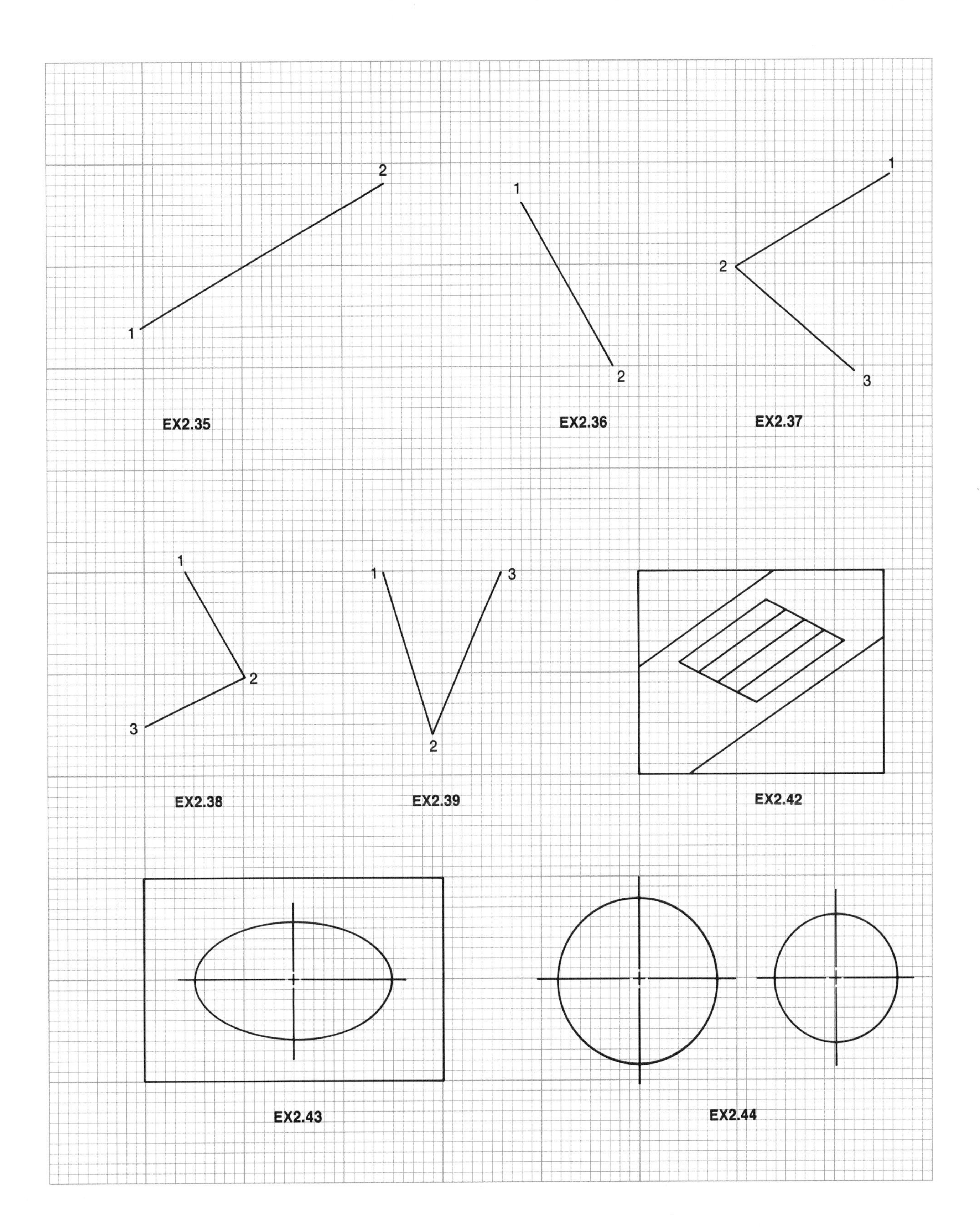
2
1
EX2.35
1
2
EX2.36
1
2
3
EX2.37
1
2
3
EX2.38
1
3
2
EX2.39
EX2.42
EX2.43
EX2.44

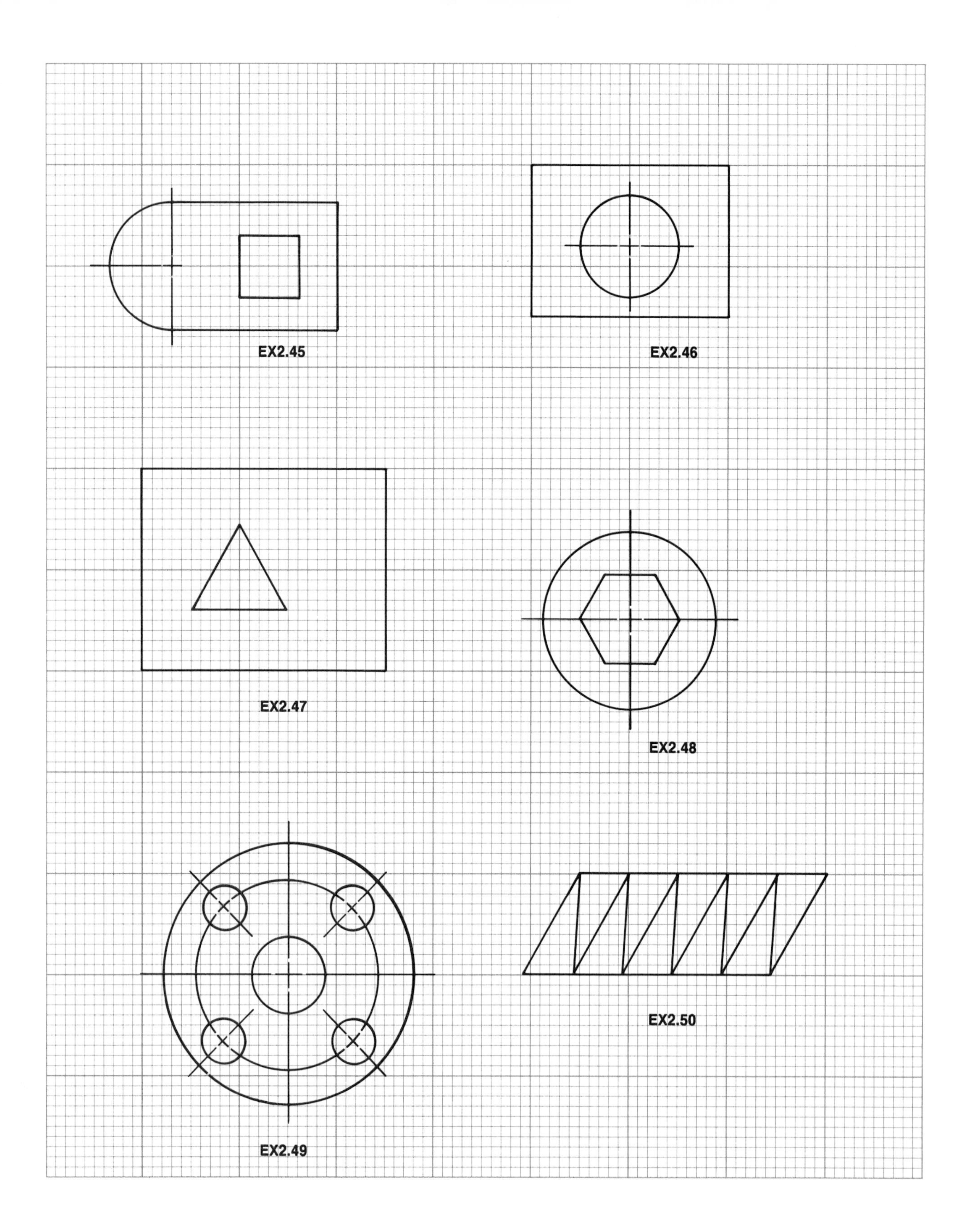
EX2.45
EX2.46
EX2.47
EX2.48
EX2.49
EX2.50

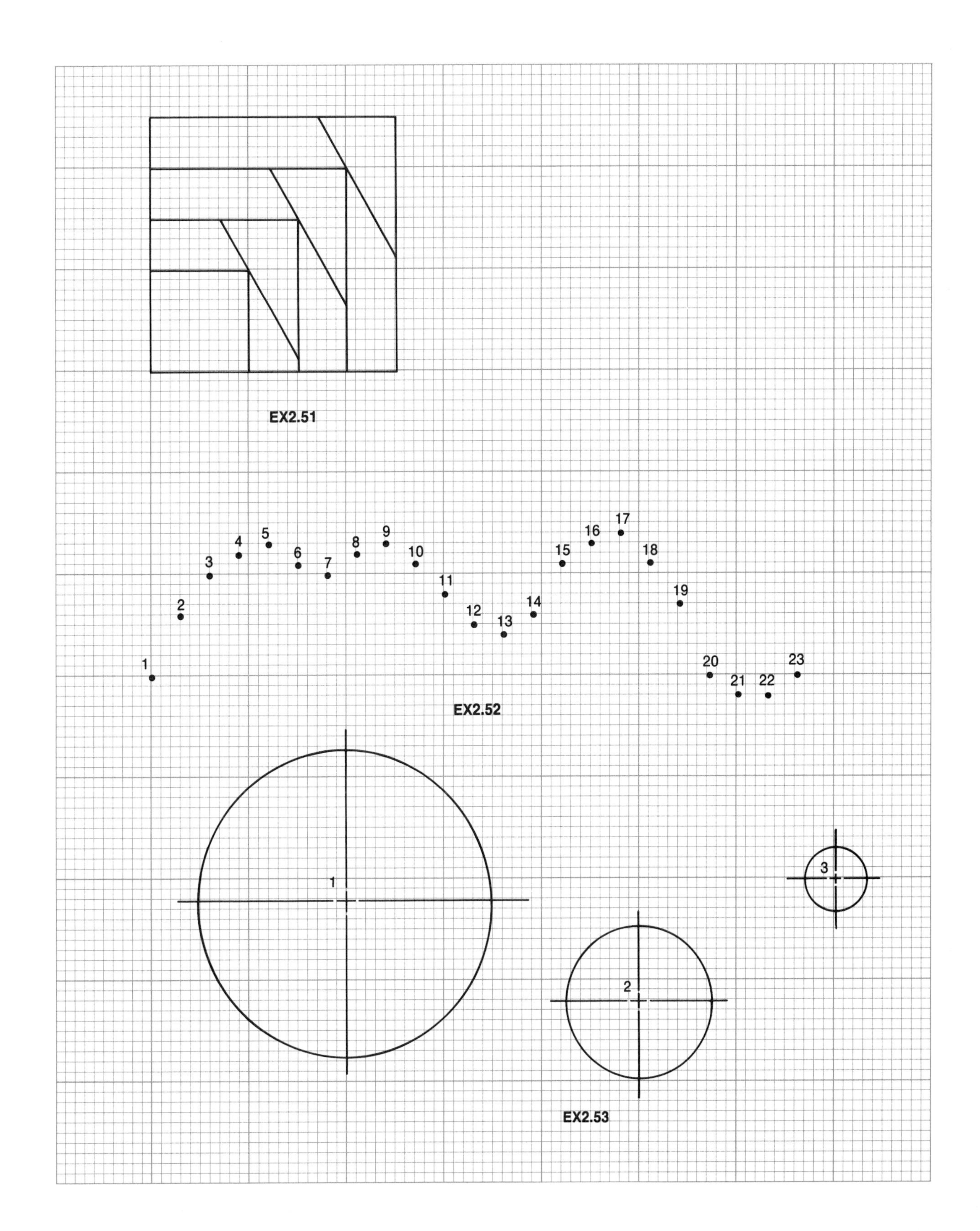
EX2.51
1
2
3
4
5
6
7
8
9
10
11
12
13
14
15
16
17
18
19
20
21
22
23
EX2.52
1
2
3
EX2.53

3 Two-Dimensional Computer Graphics

Preview We reviewed manual construction techniques for engineering graphics in Chapter 2. We now focus upon graphic construction techniques in which a computer is used as the primary drawing tool. Each year computers are used more extensively in engineering applications. In manufacturing and production facilities they are used to control and coordinate machinery, such as cutting tools, robots, and conveyor systems. Such applications form part of the field known as **computer-aided manufacturing** (CAM). Computers are also used in the development of engineering components and products. Alternative designs may be generated and compared by means of **computer-aided design** (CAD) principles and equipment (discussed in Chapter 18).

Computer-aided design drafting (CADD), as its name implies, allows the engineer/designer to develop final working drawings and other graphic descriptions of a part or product that can be produced in a variety of formats. Drawings can be plotted in hard-copy (paper) form with the use of a plotter, or they can be displayed on a CRT (cathode-ray tube) monitor (which is similar to a television screen). Three-dimensional (3D) objects can be represented by a series of two-dimensional (2D) elements, as we will discover. An object may be described by geometric data that are stored within the computer; these data may then be used to **rotate, translate,** and/or **scale** the object with respect to a fixed coordinate system. A set of engineering drawings corresponding to various orientations of the object in space may then be generated via computer graphics.

In this chapter we will focus upon computer graphics tools that allow the engineer/designer to construct technical drawings with ease and accuracy. In particular, we will focus upon **2D computer graphics** applications and methods in which engineering **drawings** of a **part** or **product** are created by the user with the aid of a computer. In the "Computer Graphics in Action" sections near the end of chapters 3 through 7, we will focus upon **3D geometric modeling computer graphics** in which the part or product under consideration is described in the form of geometric and physical data; these data are then stored within the computer's memory and used to generate drawings of the object as needed.

Learning Objectives

Upon completion of this chapter, the reader should be able to:

- Identify the necessary hardware (equipment) components for computer graphics applications.
- Differentiate between 3D geometric modeling computer graphics and 2D computer graphics in terms of the type of information that must be supplied by the user to the computer in each case.
- State the functions that must be performed during the creation of a drawing via 2D computer graphics.
- Apply menu-driven 2D computer graphics software packages in the generation of engineering drawings.
- Differentiate between several common functions in 2D menu-driven software packages with which one may create drawings with both speed and accuracy.

3.1 Computer Hardware

Computer **hardware** is the machinery and equipment used in computer applications; computer **software** is the instructions (or *programs*) that direct the activities of the hardware. Computer graphics applications require several hardware components in order to produce engineering drawings with ease, speed, and accuracy, including (1) a central processing unit (CPU), (2) a graphics display terminal, (3) input devices, and (4) output devices.

3.1.1 Central Processing Unit

The **central processing unit** (CPU) is the core of a computer system. It consists of three distinct subsystems: a **control unit** that receives and transmits signals in order to process software instructions within the CPU; an **arithmetic-logic unit** that performs arithmetic calculations and manipulates data in an orderly (logical) manner; and a **primary storage unit** (or primary memory unit) in which data is stored until needed for processing. In simple terms, the control unit is a *traffic director* of signals within the system; the arithmetic-logic unit performs the appropriate processing of data; and the primary storage unit maintains an inventory of data for the system.

3.1.2 Graphics Display Terminal

The **graphics display terminal** is a visual display monitor that is similar to a television screen. A **cathode-ray tube** (CRT) is an integral part of the display terminal in which a beam of electrons is generated by a heated cathode (see Figure 3.1). This beam is directed toward the CRT screen by metal deflection plates between which an electrical voltage potential difference is maintained; that is, electric and magnetic fields direct and focus the electron beam toward the screen. The electrons strike the atoms and molecules of a phosphor coating on the interior surface of the screen, thereby exciting the electrons of this coating to higher energy

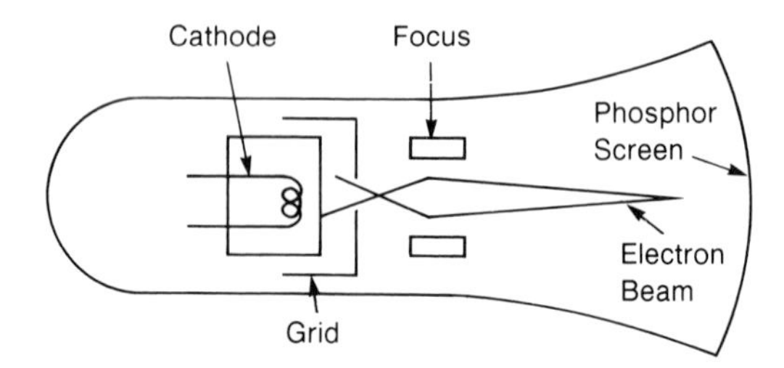

Figure 3.1 Structure of a cathode-ray tube.

states. As the excited electrons subsequently return to their original energy levels, *energy is emitted in the form of visible light.* By varying the strength and the direction of the focusing electromagnetic fields, one can move the electron beam so that it strikes specific locations of the phosphor-coated screen, thus changing the position of the visible light emitted from the screen. Table 3.1 summarizes some of the characteristics associated with various types of computer graphics display terminals. The screen image is generated via one of two approaches: stroke-writing and raster-scan. Terminals that use stroke-writing may be the directed beam refresh type or the direct-view storage tube type.

Stroke-writing Some types of terminals use stroke-writing, in which the image is etched on the screen with the electron beam moving from one position to the next (that is, linear or straight-line motion). Curved paths may then be approximated by a collection of very short, straight lines.

Raster-Scan Another type of terminal uses raster-scan, in which the screen is divided into many discrete elements known as **pixels,** which together constitute the **raster,** or the matrix of elements forming the screen area. A square (x by x) matrix of pixels is thereby created on the screen. Depending on the pixel density, x may equal, for example, 256 or 1024. As x increases, the necessary computer memory increases; however, resolution of the screen image is also increased.

Each pixel is energized by the electron beam as it sweeps across the screen in a horizontal direction. At the end of each sweep, the beam moves downward to begin the next sweep. The entire screen is thereby scanned as many as sixty times each second. Pixels may be energized to glow with *varying intensity;* and color screens also allow the pixels to generate visible light of *varying color.*

Table 3.1: Some Characteristics of Computer Graphics Display Terminals.

	TERMINAL TYPE		
	Directed Beam Refresh	Direct-View Storage Tube (DVST)	Raster-Scan
Method of Image Generation	Stroke-writing	Stroke-writing	Raster-scan
Picture Quality	Excellent	Excellent	Moderate to good
Selective Erasure	Yes	No	Yes
Color	Moderate	No	Yes

Source: M.P. Groover and E.W. Zimmers, Jr., **CAD/CAM: Computer-Aided Design and Manufacturing** *(Englewood Cliffs, N.J.: Prentice-Hall, 1984).*

Directed Beam Refresh Terminals One type of terminal in which stroke-writing is used to generate the screen image is the directed beam refresh terminal. Selective erasure of the elements that form the image is possible with this type of terminal.

Direct-View Storage Tube (DVST) Terminals Another terminal in which stroke-writing is used to create the screen image is the DVST terminal. An **electron flood gun** is used in this type of terminal to maintain the energy excitation of the screen's phosphor elements, thereby allowing an image on the screen to be retained following its creation. Since the entire screen image is retained with the aid of the electron flood gun, selective erasure of particular lines is not possible.

Raster-Scan Terminals In terminals where the image is generated by raster-scan, selective erasure is possible, since each pixel forming the raster is energized separately by the electron beam.

3.1.3 Input Devices

Input devices allow instructions and other data to be transmitted to the CPU. Examples of input devices include:

- **Alphanumeric keyboards,** which are similar to typewriter keyboards and are used to type data.
- **Cursor control devices,** which the user manipulates to control the position of a **cursor** (a bright spot on the CRT screen, usually denoted by the intersection of a vertical line and a horizontal line, known as **crosshairs**). These devices include:
  - **Thumbwheels,** which the user rotates to move the cursor about the CRT screen (vertically and horizontally).
  - A hand-held device known as a **mouse,** which is moved about a surface (such as a desk). It consists of a **graphic tablet,** (either electrosensitive, electromagnetic, or acoustic) and a cursor **puck,** or pad, that is moved over the tablet. Such movement is monitored by the computer system and directly transferred to corresponding movement of the cursor.
  - A **joystick** (a toggle stick contained within a box), which is pushed by the user in the desired direction of motion for the cursor.
  - A **light pen,** which the user may apply as a pointer to detect a particular location (an existing vector or line) on the screen. (A light pen does not project light; rather, it is a *detector* of light on the monitor screen.)
  - A **digitizer electronic tablet** and a **pen stylus** with which the user may identify specific locations on the tablet. Digitizers also may be used to trace linework from an existing paper drawing and to insert this data into the

computer system for monitor screen display and manipulation.

3.1.4 Output Devices

Computer-generated drawings can be created with the aid of a hard-copy (paper) **output device,** such as:

- A **drum plotter,** consisting of a slide-mounted pen that moves over the surface of paper rolled about a drum.
- A **flat-bed plotter,** which also features slide-mounted pens moving over a paper surface; however, the paper is attached to a horizontal or vertical flat bed.

The reader should become familiar with the hardware configuration at his or her institution.

3.2 Interactive Image Generation

Interactive computer graphics refers to an equipment and software configuration that allows the user to *modify a drawing during its construction* in order to eliminate errors or create alternative designs. In other words, the user may interact directly and immediately with the computer through the picture on the screen during its generation of the drawing.

Interactive computer graphics capability, together with the availability of relatively low-cost microcomputers with significant primary memories and high processing speed, has produced dramatic improvements in engineering graphics. Accuracy, speed, versatility, productivity, cost minimization, ease of operation, and ease of communication between engineering professionals have all significantly increased in recent years as a result of interactive computer graphics hardware and software developments.

3.3 2D and 3D Computer Graphics

Microcomputers, used in conjunction with high-resolution color monitors, color printers, and plotters, allow the user to create sophisticated engineering drawings with ease and accuracy. In **three-dimensional (3D), geometric modeling computer graphics,** the *object* under consideration is described in the form of geometric data (with respect to a rectangular x, y, z coordinate system) that is stored within the computer's memory. The object—in the form of this stored data— may then be translated, rotated, scaled, or otherwise manipulated in order to produce the desired engineering drawings. The key point in 3D computer graphics is that a *description of the object is stored* within the computer memory from which one may then create any desired graphic representations of the object.

In two-dimensional (2D) computer graphics work (in which a two-dimensional x, y coordinate system is used), the desired drawings of the object are created and stored within

Highlight

Charles Babbage and Herman Hollerith

Hollerith's tabulating machine

Charles Babbage has been called the father of computers because of his design for a mechanical *analytical engine* that could add, subtract, divide, and multiply values in accordance with instructions given on a coded input card. His original design of 1833 was eventually constructed by his son (in 1871), who demonstrated that the analytical engine did indeed work.

A significant problem arose with the 1880 census of the United States: it required seven and one-half years to tabulate the results! It was anticipated that with the expected rise in the size of the population, subsequent census studies would require more than the ten years apiece to tabulate results. Herman Hollerith (a statistician) experimented with the concepts of edge-marked cards and hole-punched paper tape in order to speed the tallying process; neither concept proved fruitful.

While traveling by train to St. Louis, Hollerith noticed that the train conductors created a record of each passenger's distinguishing characteristics (hair color, eye color, weight, and so forth) by punching holes in the passenger's ticket in specific locations according to a particular code. (This practice had been instituted in an attempt to decrease the danger of train robbery by thieves traveling as passengers.) Hollerith adapted the concept to create the *Hollerith card,* together with a keyboard punching machine. The Hollerith card allowed data to be stored and tabulated in the form of prepunched holes. It allowed the next census to be tabulated in only two and one-half years. Hollerith then established his own firm (the Tabulating Machine Company) for the manufacture of punched-card equipment. In 1911 this firm merged with other companies to form International Business Machines Corporation (IBM).

Sources: Stephan Konz, Work Design: Industrial Ergonomics, *2d ed. (Columbus, Ohio, Grid Publishing, Inc., 1983) and Steven Mandell,* Computers and Data Processing: Concepts and Applications, *2d ed. (St. Paul, MN: West Publishing Co., 1982).*

the computer memory, after which hard-copy (paper) plots of these drawings can be generated as needed. The object itself—in the form of geometric and other structural data—is not described by the user; instead, *2D drawings of the object are created and stored* for later modification and hard-copy production.

We will return to the subject of 3D computer graphics and geometric modeling of engineering objects after we explore the subject of 2D computer graphics more thoroughly.

Learning Check

What is the basic distinction between 2D and 3D computer graphics representations of an object?

Answer In 2D computer graphics, a *drawing* of the object is defined and constructed. In 3D computer graphics, data that describes the *object* is entered and stored within the computer memory; that data can then be manipulated to generate graphic descriptions of the object as needed.

3.4 Necessary Functions of Computer Graphics Ware

In using 2D computer graphics hardware and software instead of the traditional manual graphics equipment to generate engineering drawings, we must be able to:

1. **Create** the variety of linework (such as solid lines, dashed lines, and construction lines) that may be required in the development of a finished engineering drawing.
2. **Erase/edit/modify** linework in a drawing under construction, just as is done in a manually constructed drawing.
3. **Save and retrieve** drawings that are either under construction or have been completed. The ability to store numerous drawings in a computer memory file that can then be easily accessed and retrieved by the user is a significant advantage of computer graphics.
4. **Generate hard copy** of any engineering drawings that have been created with the use of the computer.

As we will discover in the following sections, certain types of 2D computer graphics software and hardware are currently available for performing each of these functions.

3.5 Menu-Driven Software

In the "Computer Graphics in Action" sections of Chapters 3, 4, 5, 6, and 7, we will focus upon 3D computer graphics, which can be used to (1) define a 3D object in terms of specific geometric data, and (2) instruct a computer to rotate, translate, scale, or otherwise manipulate the object under consideration in preparation for the construction of whatever engineering drawings are needed. Such manipulations greatly increase the engineer's ability to design and analyze objects.

However, before we focus upon 3D computer graphics and the principles of **solid modeling,** we must first master 2D computer graphics and its application in the creation of en-

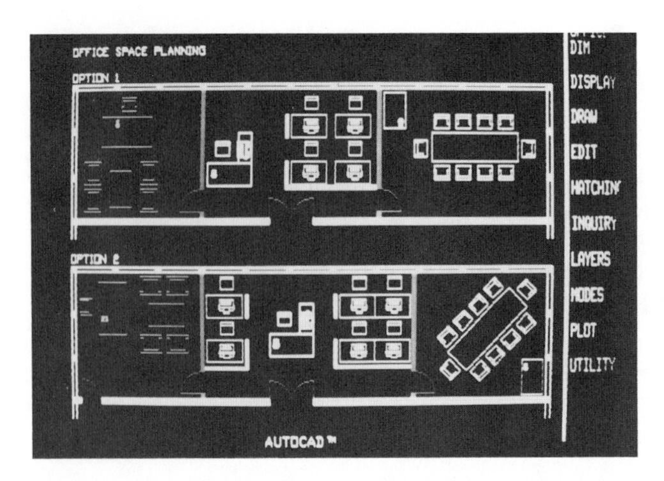

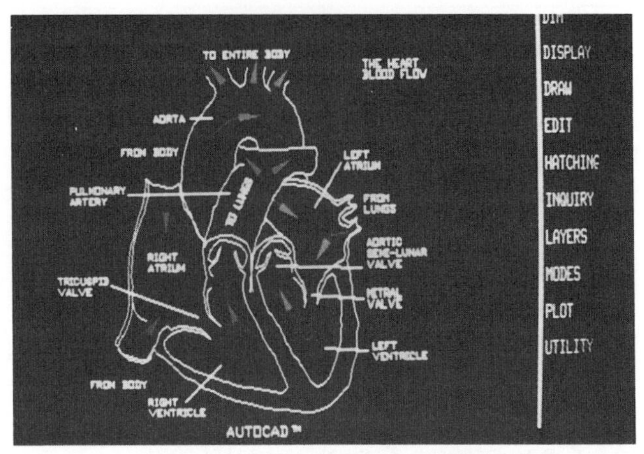

Figure 3.2 **Menu-driven computer graphics hardware allows engineers to construct detailed engineering drawings with ease and accuracy. The photograph on the left shows office-space planning layouts that can be quickly modified by the designer; the one on the right shows a drawing of the blood flow through the human heart. Both drawings were constructed with the use of AutoCAD™ software. (Courtesy of Texas Instruments.)**

gineering drawings. Many 2D computer graphics software packages, such as AutoCAD™ and FIRSTDRAW 2.0™, are **menu-driven** (Figure 3.2). Such software does not require that the user be proficient in a particular programming language, such as FORTRAN or BASIC; instead, the user simply selects the type of graphics function that he or she desires from a menu of possible selections (Figure 3.3). Common functions in such a menu selection include:[1]

- **LINE,** which allows the user to add a particular line to a drawing.
- **CIRCLE,** which allows the user to create a circle of a desired size and centered at a specific location within the drawing.
- **SAVE,** which stores a drawing in computer memory.
- **TEXT,** which allows the user to add textual information (alphanumeric characters) to a drawing.
- **ERASE** or **DELETE,** which allows the user to remove linework from a drawing.
- **PLOT,** which generates hard copy of a drawing.
- **GRID,** which creates a grid (usually in the form of equally spaced dots) on the CRT display screen to be used as a scale or measuring aid in the construction of a drawing.
- **SNAP,** which allows the user to move about the screen in discrete amounts from one GRID point to another.

1. Many of the examples given in this text have been generated with the use of the graphics software package known as AutoCAD™.

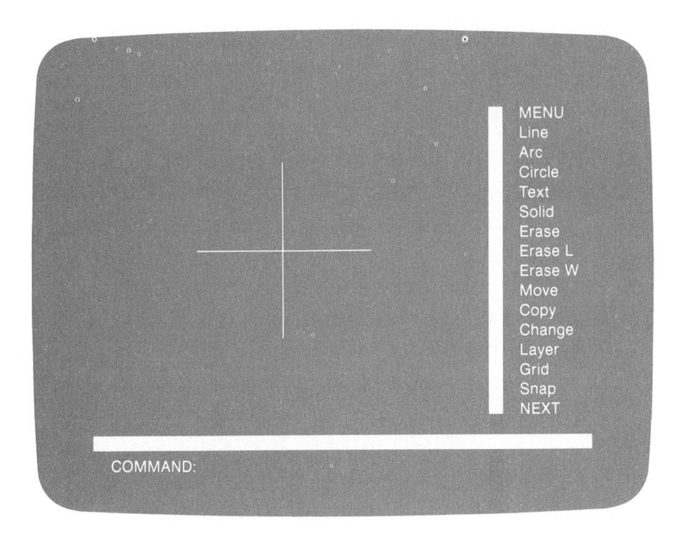

Figure 3.3 An example of a CRT screen in which a menu of 2D graphics functions is shown. Crosshairs represent the screen cursor with which the user may select a particular function. The user may also use the keyboard to type the needed selection. COMMAND indicates that the system is prepared to receive a new function specification.

The names of these (and other) functions may differ in different software packages. In addition, the specific procedures that must be used for each of these functions may differ from one software package to another. The user must become familiar with the specific requirements of the software with which he or she is working. In this (and subsequent) chapters, we will focus upon the *general* characteristics in 2D computer graphics software.

3.6 Menu Selection Modes

The user of a menu-driven software package may select a command from the menu via **keyboard** selection or **cursor control,** depending upon the hardware to which he or she has access. The keyboard mode requires that the user, when instructed by the computer to select a command from the menu, type that selection on the computer keyboard. For example, the menu selection entitled LINE can be selected by typing the characters L, I, N, and E on the keyboard, followed by striking the carriage return key (which informs the computer that the entire command specification—in this case, LINE—has been completed by the user).

As noted in Section 3.1, a **mouse** is a hand-held device used in conjunction with an electrosensitive (or electromagnetic or acoustic) pad. The relative position of the mouse on the pad is monitored by the computer system and displayed on the CRT screen by a set of crosshairs (vertical and horizontal lines) that represent the cursor (refer again to Figure

3.3). The user may control the position of the cursor on the CRT display screen by moving the mouse on the electrosensitive pad. Menu-driven software displays a list of possible command selections on the CRT screen (as shown in Figures 3.3 and 3.4); any of these selections may be chosen by:

1. Manipulating the computer software so that the menu is displayed on the CRT screen with the *software in the command mode,* that is, the software should be ready to accept a command selection from the menu.

2. *Locating the cursor* (crosshairs) so that it lies on the desired menu command.

3. *Pushing an appropriate button on the cursor control,* thereby initiating the use of that command by the software.

(If the user has a digitizing tablet, the user may select menu items by moving the tablet's cursor to the appropriate position and striking the *pick* button of the tablet.)

Cursor movement may be restricted to only specific directions, or it may be unrestricted, depending upon the particular type of software in use. The speed and the amount of cursor movement also varies from one software package to another.

With practice, a user can become very proficient in quickly moving the cursor about a menu and selecting the appropriate commands. Once such proficiency is attained, the use of a mouse, light pen or other selection device becomes much more efficient in the creation of a drawing than menu selection performed via the keyboard. However, all selection modes (keyboard, mouse, light pen, and so forth)

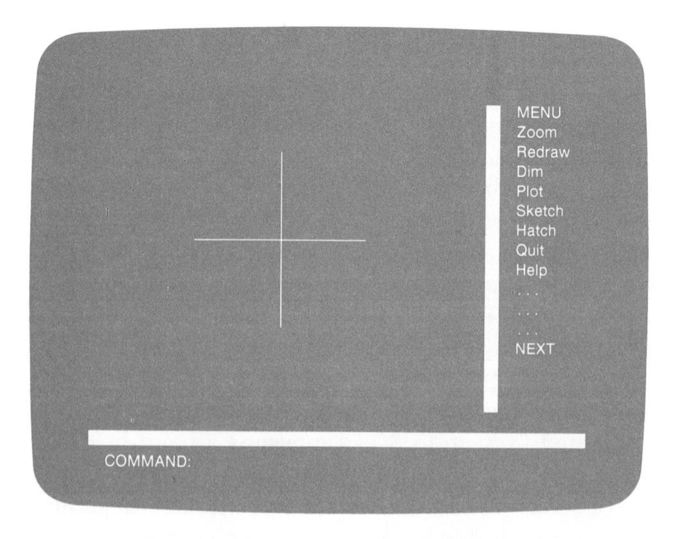

Figure 3.4 Another example of menu selections shown on a CRT screen. NEXT indicates that the user may request the system to display still another column of menu selections.

provide the user with access to 2D computer graphics software with which engineering drawings can be generated with both ease and accuracy.

3.7 GRID and SNAP Commands

Most types of 2D menu-driven computer graphics software include a GRID command that allows the user to display a **grid** (a collection of equally spaced dots) on the CRT screen in order to aid in the development of a drawing (Figure 3.5). The user can usually specify the spacing of the dots that form the grid so that the grid can be used as a scaling or measuring guide in specific directions (note the different spacing patterns in Figure 3.5). A choice of rectangular and nonrectangular (such as isometric) gridwork is available in some software packages.

The SNAP command, used in conjunction with the GRID command, allows one to move the cursor discretely from one grid **dot** (point) to another grid dot. The cursor can be moved from dot to dot along the grid as lines, arcs, and other elements of a drawing are created. Some software packages allow *continuous* motion about the grid region so that the cursor can be fixed on a non-grid point (a location between dots). Other packages require *discrete* movements of the cursor from one grid point to another. Still others, such as AutoCAD™, allow the user to specify either continuous or discrete motion about a grid.

Grids are not (usually) part of the drawing under construction. That is, gridwork is usually not stored in the computer memory with the lines, arcs, and other elements forming a drawing.

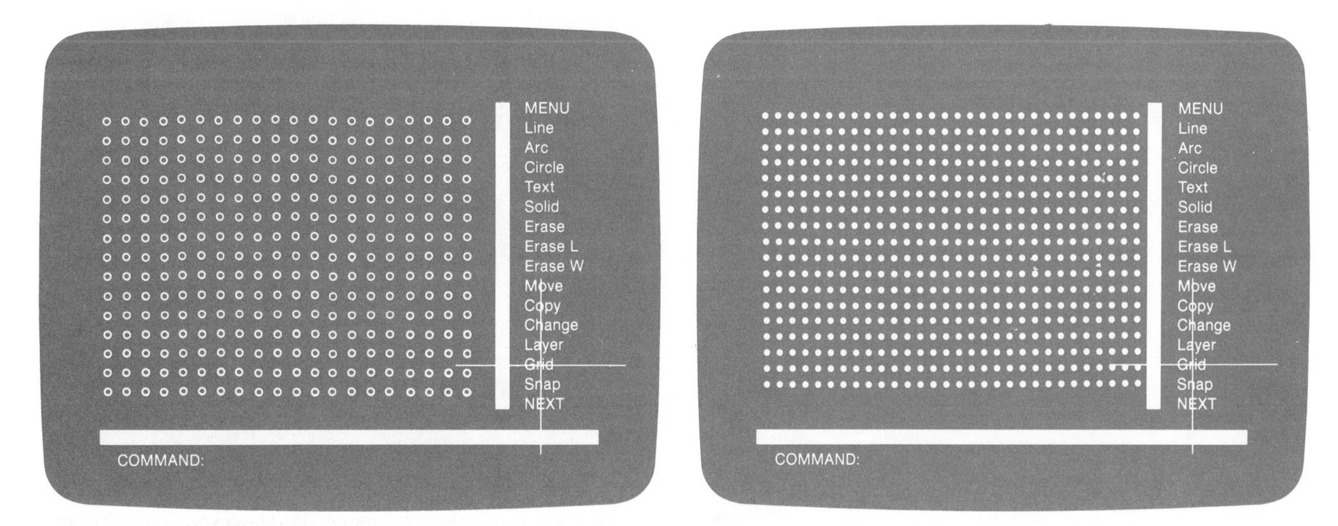

Figure 3.5 The display of a grid—a collection of dots—via the GRID command can be used to visually measure distances between positions on the CRT screen. The spacing of grid points can be specified by the user. Two different spacing patterns are shown here.

3.8 LINE Command

During the construction of any drawing by 2D computer graphics, we must be able to specify distinct types of linework. Some lines need to be solid, or unbroken, with a specific width; other lines need to be dashed, or otherwise broken, with a range of segment lengths. The length and the direction of each line must also be specified. Consequently, most software packages allow the user to specify the *type, length,* and *direction* of lines within a drawing.

The basic command to draw a line is usually LINE. The user must specify the location of *two endpoints* of a straight line by (1) moving the cursor to the appropriate position for each point and pushing a button on the cursor control, which then initiates the storage of this location information in the computer memory, or by (2) identifying the horizontal and vertical distances (the coordinates) of each point with respect to a given reference location (the origin)—see Figure 3.6. The user defines these coordinate values by striking the appropriate numeric keys on the keyboard.

For example, AutoCAD™ will ask the user to specify the point *from* which the line is to begin and the point *to* which the line is to be drawn in the following form:

FROM:. . . .

TO:. . . .

For the first endpoint, or starting point of the line *(FROM),* the user should give the coordinate values or, alternatively, locate the cursor at the appropriate location and push a button on the cursor control. For the second endpoint, the user should again give the coordinate values or move the cursor to this endpoint's location and push a button on the cursor control.

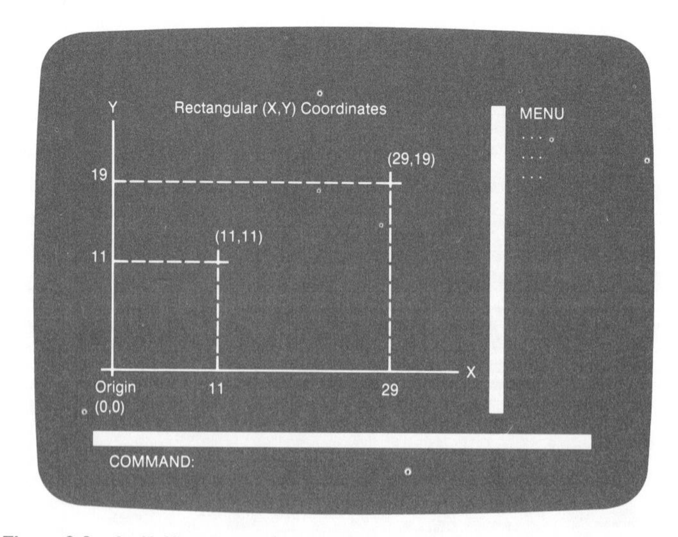

Figure 3.6 An X, Y rectangular coordinate system can be used to specify the locations of specific points on the CRT screen with respect to an origin.

Most software packages will continue to draw straight lines from one point to the next until the user specifies that the last line in the sequence has been drawn.

Figure 3.7 shows the selection of the LINE command and identification of two endpoints with a cursor (a) and the finished line drawn between these two endpoints (b).

Learning Check

Why are two points in space needed to define a line?

Answer A line is completely specified by its *location* in space and its orientation, or *direction.* The position of one point (on the line) fixes the location of the line, whereas the position of a second point uniquely specifies the direction of that line in space.

3.9 Construction Linework and LAYER Command

Construction linework can sometimes be designated as such by using the command of TEMPORARY, meaning it is to be displayed on the CRT screen but is not to be transferred to computer memory with finished (non-temporary) linework. An alternative approach for generating construction linework (available with AutoCAD™ software) involves the use of a LAYER command. With this command, linework may be defined to lie within a particular 2D *layer,* and the user may reserve one layer for construction linework while using

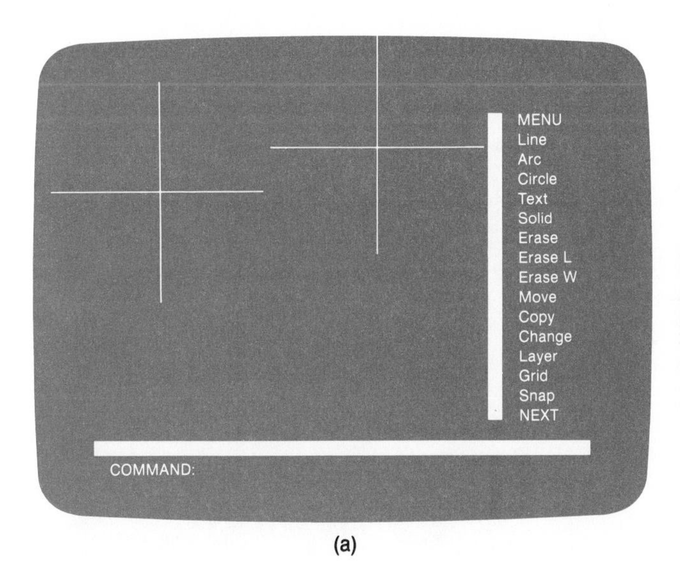

(a)

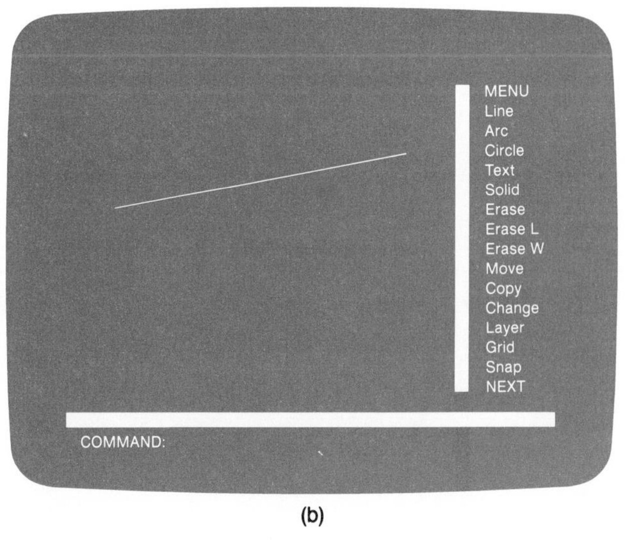

(b)

Figure 3.7 The LINE command (or its equivalent) asks the user to specify the endpoints of the straight line to be drawn (a). The desired line is then drawn between the given endpoints (b).

another layer for finished linework. Each layer of linework may be stored separately within the computer memory or in various combinations. Different layers of linework may then be shown simultaneously on the CRT screen during the development of a drawing. On a CRT color monitor the layers can be displayed in different colors. The LAYER command expands the user's choice of construction methods for the development of 2D computer graphics drawings.

3.10 ERASE/DELETE Command

We need to be able to erase, or delete, linework that has been displayed on a CRT screen or stored in computer memory if we are to achieve truly *interactive* computer graphics capability (explained in Section 3.2). Most menu-driven software includes the command ERASE or DELETE, with which specific lines can be removed from a drawing (Figure 3.8). A keyboard or a cursor control device can be used to specify the endpoints of the line to be erased.

A command available in some software packages is ERASE-LAST (or its equivalent), with which the user can specify that the latest addition to a drawing (for example, a line or certain text characters) is to be deleted. We often recognize a mistake immediately after we include it in a drawing; the ERASE-LAST command allows us to respond to such immediate recognition with both speed and ease.

Certain software packages, specifically, AutoCAD™, allow the user to specify a 2D range (a **window**, or 2D *box*) on the CRT screen within which all linework is to be erased. A window option allows the user to quickly modify and/or correct a drawing with minimal effort. Figure 3.9 shows how a window option (ERASE W) is used to selectively erase one

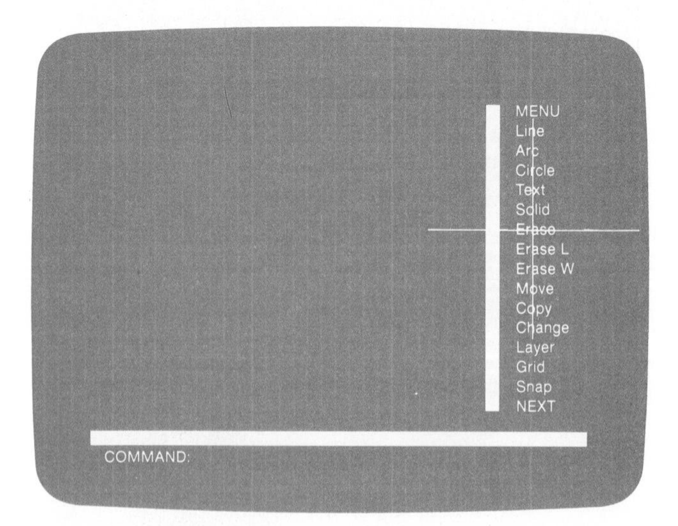

Figure 3.8 The ERASE command can be selected from the menu in order to delete lines or other graphic elements from a drawing.

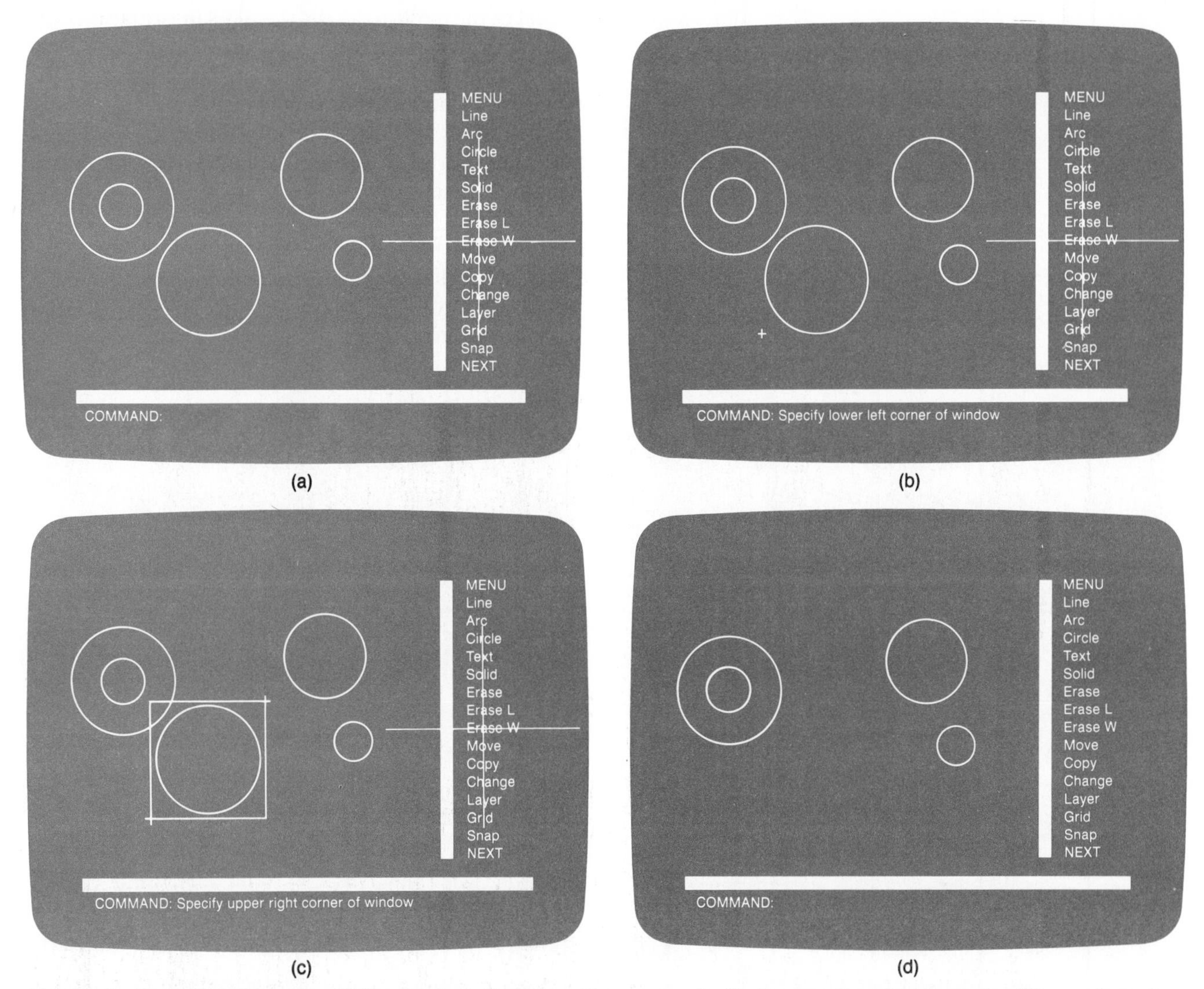

Figure 3.9 The ERASE W command (*W* meaning *window*) allows the user to selectively erase all graphic elements within a rectangular portion of the drawing. To erase one of the circles in this drawing (a), the lower left corner of the window is first specified by the user (b). Next, the upper right corner of the window is specified, thereby fully locating the window (c). (The lower left and upper right corners pinpoint the desired size and position of the window.) All elements within the window are then erased from the drawing (d).

circle from a drawing of several circles. Two opposite corners (lower left and upper right) of the window are first identified by the user, after which the linework within the window is erased from the screen.

3.11 CIRCLE Command

Many 2D computer graphics software programs include such menu selections as CIRCLE, ARC, RECTANGLE, and POLYGON, with which the user may generate the corresponding geometric entity. In response to a CIRCLE command, the user will be requested to specify both the *size* and the *location* of the circle to be drawn. This information must be provided in the specific form required by the software. For example, the user may need to provide any one of the following sets of information (Figure 3.10):

- The locations of the *centerpoint* and a point that lies on the circle itself, thereby establishing the *radius* of the circle as the distance between these two points.
- The *centerpoint* location and the *diameter* of the circle.
- The location of *two points* that lie at opposite positions on the circle; the distance between these points is then equal to the diameter of the circle with its centerpoint located midway between these points.

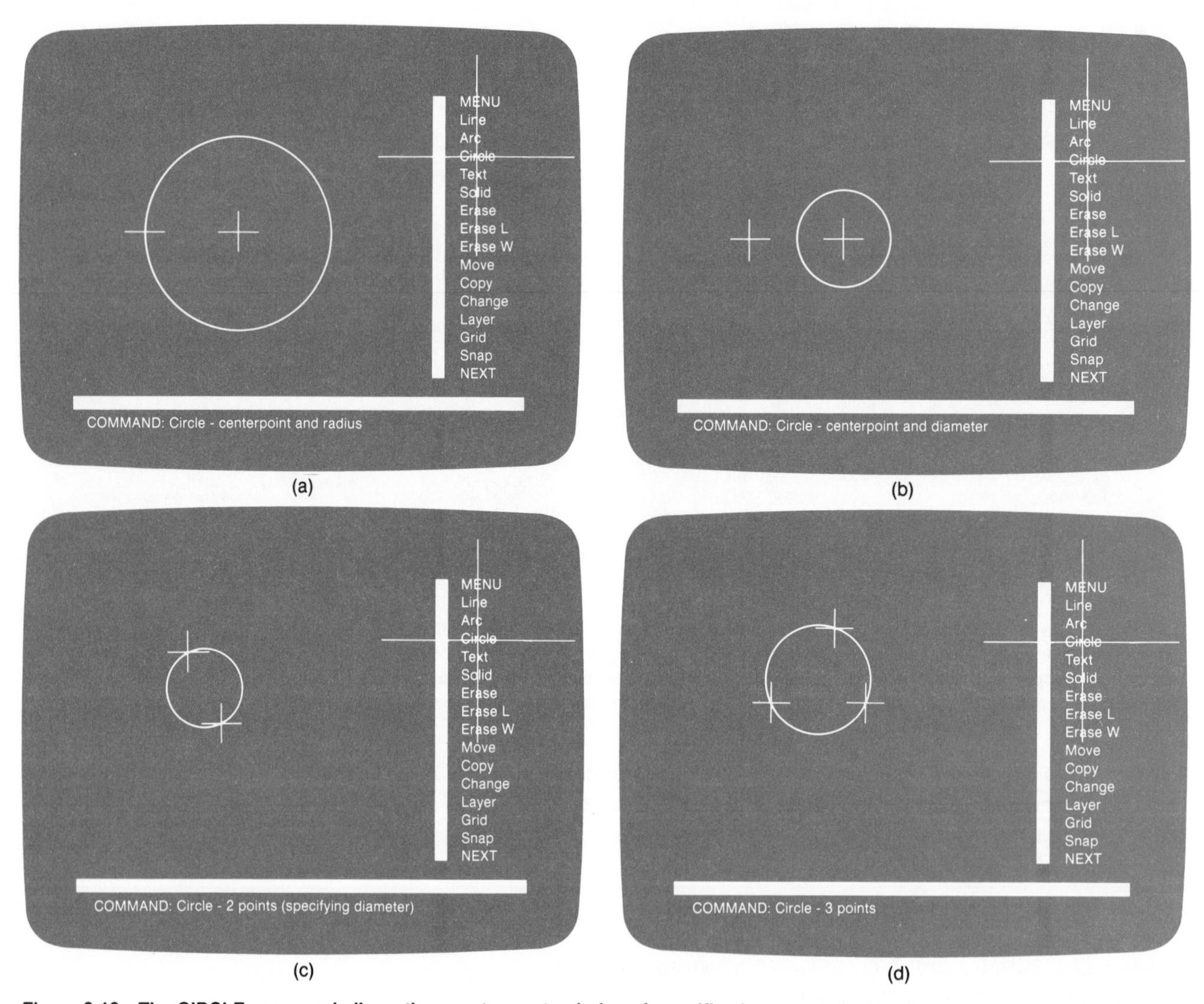

Figure 3.10 The CIRCLE command allows the user to create circles of specific sizes at particular locations in a drawing. The size and the location may be specified in terms of the centerpoint and the radius (a), the centerpoint and a second point that is located at a distance from the centerpoint equal to the diameter of the circle (b), two points that lie on the circle (the centerpoint is located midway between these two points) (c), or three points that lie on the circle (d).

- The location of *three points* that lie on the circle, thereby uniquely establishing the circle's diameter and location.

The first two sets of information represent the most direct approaches in specifying a circle, since the user most often knows both the centerpoint location of the circle and the size of the circle (in terms of the required radius or diameter).

3.12 ARC Command

An arc is formed by a locus (or collection) of points that are equidistant from a particular location in a plane (a plane is a two-dimensional, flat surface in space). In other words, an arc is a particular portion of a circle with a specific diameter and centerpoint location. Software for 2D computer graphics may request the user to define the arc which is to be drawn in terms of any one of the following sets of information (Figure 3.11):

- The location of *three specific points* that lie on the arc, where two of these points are the endpoints of the arc.
- The location of the arc's *centerpoint,* together with the positions of its *two endpoints.*
- The locations of the arc's *centerpoint* and *one endpoint,* together with the *included angle* (the angle of the arc).
- The location of the arc's *centerpoint* and *one endpoint,* together with the *chord length* for the arc (the linear or straight-line distance between the arc's endpoints).
- The location of the *two endpoints* and the value of the arc's *radius.*

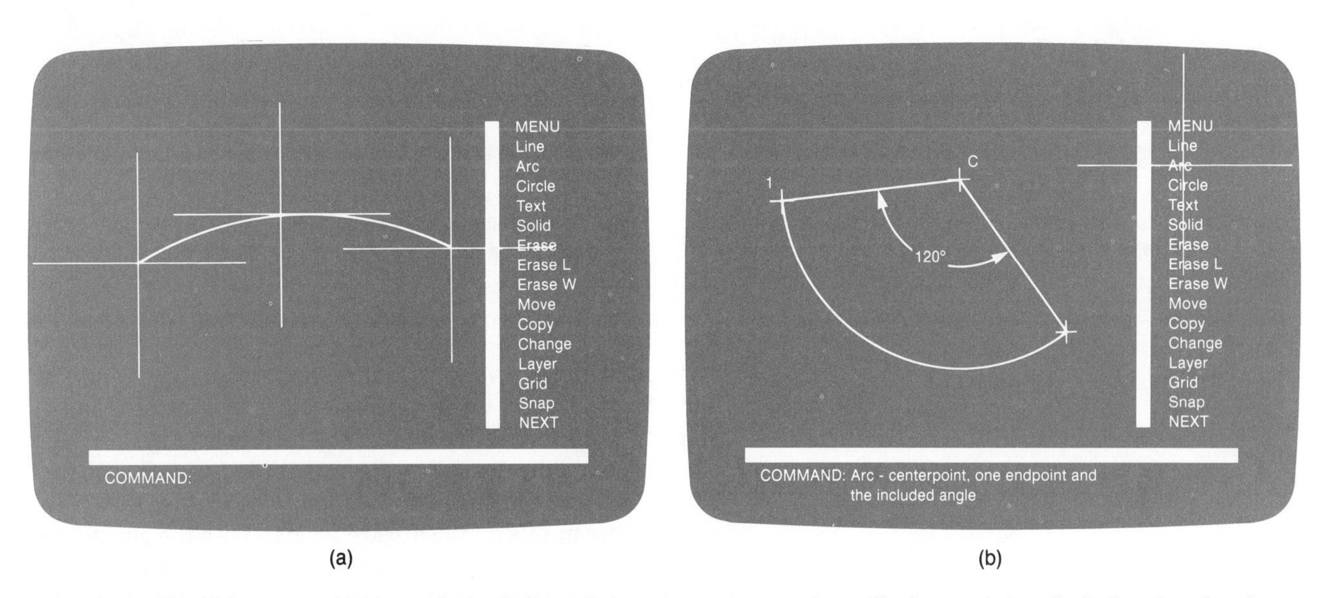

Figure 3.11 The ARC command (or its equivalent) allows the user to create arcs of specific sizes and at particular locations in a drawing. The size and location may be specified in terms of three points that lie on it (a), its centerpoint, one endpoint, and the included angle (b), its centerpoint, one endpoint, and the chord length (c), or its two endpoints and its radius (d). ***Continued on next page.***

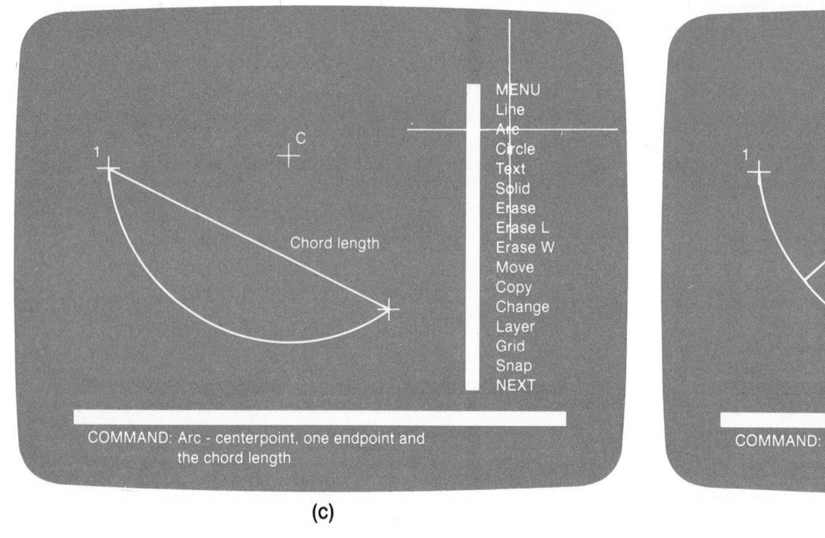

(c)

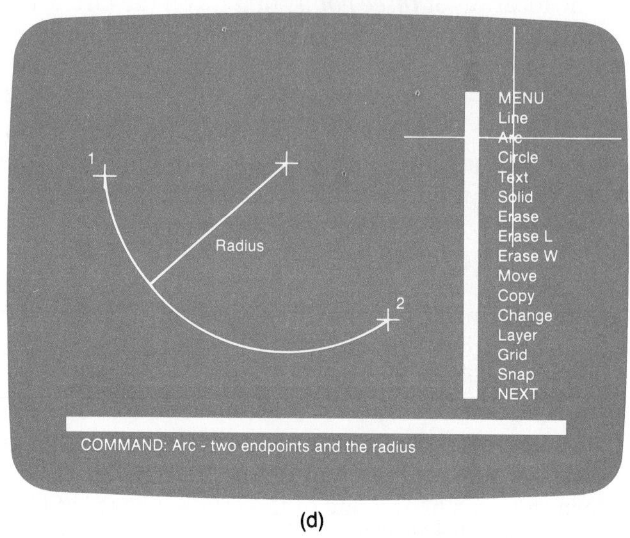

(d)

Figure 3.11—*Continued*

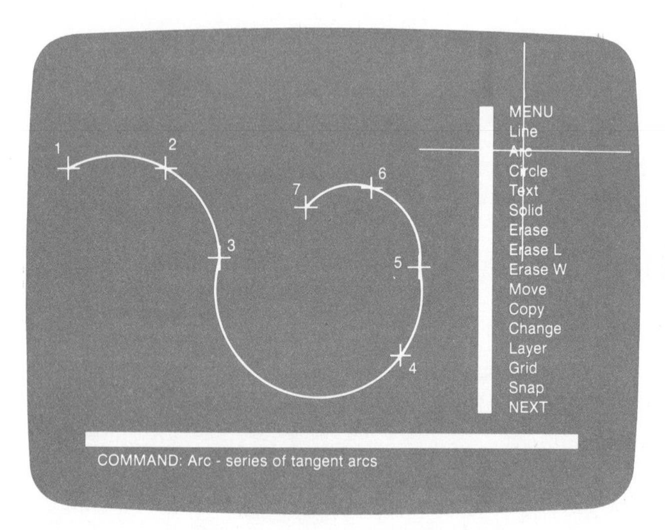

Figure 3.12 The ARC command (or its equivalent) can be used to generate a series of arcs that share common endpoints.

Figure 3.13 Points used to construct an arc may be defined with respect to a rectangular (X, Y) coordinate system.

In addition, a series of arcs that share common endpoints can be generated by using the final endpoint of one arc as the starting endpoint for the next arc to be drawn (Figure 3.12). The point and distance data that are needed to define an arc can be inserted via cursor movement (see Figure 3.11(a) in which three points on the arc are located with the cursor) or by specifying the horizontal (X) and vertical (Y) distances of the points from a given coordinate origin (Figure 3.13). The arc is then generated by the computer, as shown in Figure 3.14.

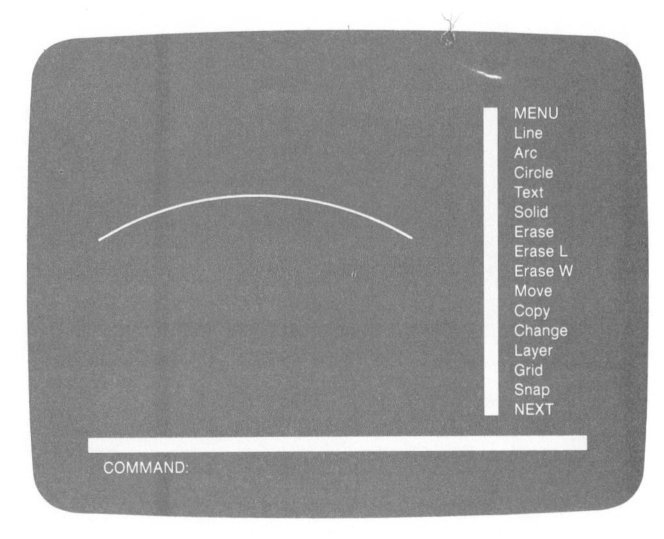

Figure 3.14 The final result of the ARC command (or its equivalent) is the desired arc—of a specific size at a particular location.

3.13 POLYGON Command

A polygon (an *n*-sided figure) can be constructed by drawing *n* lines to form the sides of the figure. Some software packages include menu options with which one may draw specific types of polygons (such as rectangles and triangles) or a general *n*-sided figure.

3.14 MOVE, COPY, and BLOCK Commands

AutoCAD™ and other software packages allow the user to relocate, or *move,* a figure from one location to another on the display screen. This capability is a truly convenient aspect of computer graphics. The MOVE option requires the user to specify the *current location* of a point on the geometric entity (which could be a straight line, a circle, an arc, a polygon, and so forth) to be moved with the cursor or via the keyboard, followed by the *new location* of this point in the drawing (Figure 3.15). The figure is then moved to the new position and simultaneously erased from the drawing at the original location. Some software packages allow the user to enclose a portion of the drawing space within a **window,** after which all of the linework contained within this window is moved to a new location in the drawing.

The COPY command (or its equivalent in the software package in use) is similar to the MOVE command in that it allows the user to relocate linework to different positions on the display screen. The difference between the COPY and

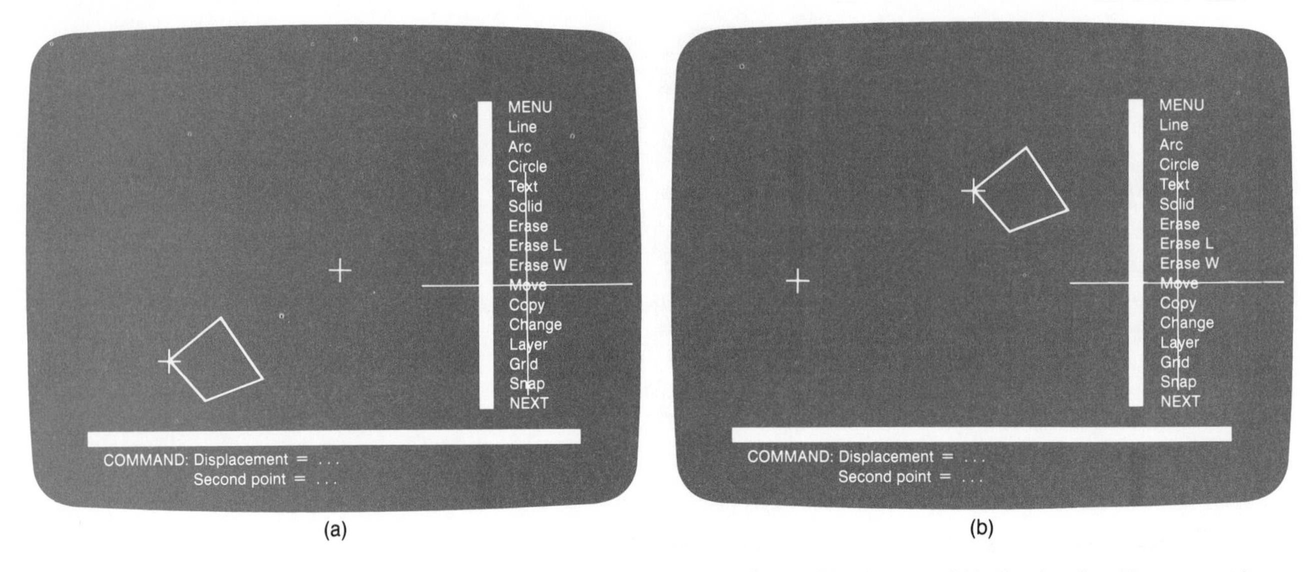

Figure 3.15 The MOVE command (or its equivalent) can be used to relocate a graphic element within the drawing. The present location (a point on the element that is to be moved) is first identified (a). Then the new location of this second point is identified, resulting in the movement of the element.

the MOVE commands is that COPY does not include the erasure of the linework at its original position (Figure 3.16). Obviously, the COPY command is very useful in those instances where a particular feature (such as a circle of a specific diameter) must be duplicated at several locations in an engineering drawing.

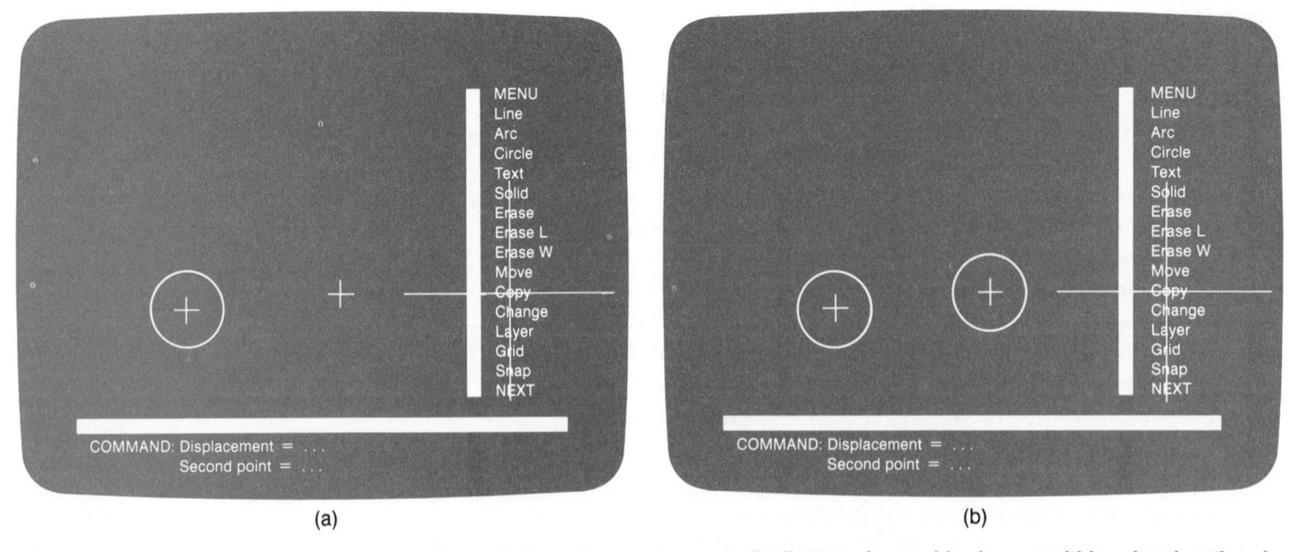

Figure 3.16 The COPY command (or its equivalent) allows the user to create duplicates of a graphic element within a drawing, thereby saving time and minimizing effort in construction. In this example, a circle (a) is duplicated once (b), then many times (c). ***Continued on next page.***

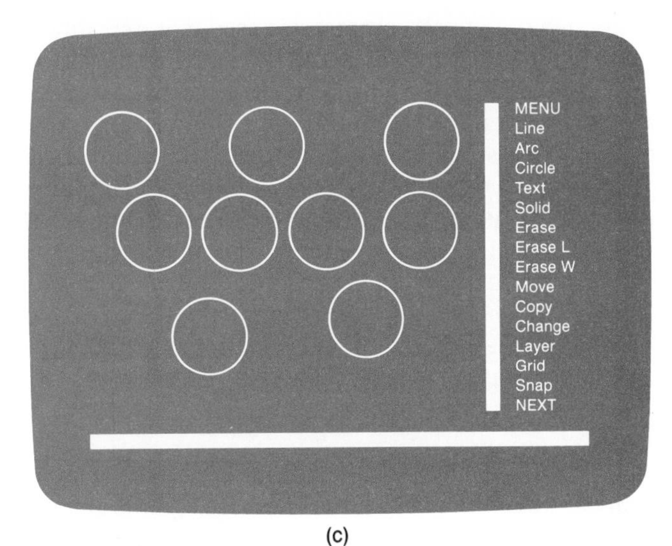

(c)

Figure 3.16—*Continued*

The COPY command is used in Figure 3.17 to create six octagons within a single drawing. The first octagon is constructed with eight straight lines (using the LINE command or its equivalent). After this initial octagon is created, it can be duplicated at various locations in the drawing by following the steps shown in the figure.

Step 1 in the figure presents the preliminary drawing in which the octagon is to be inserted at six distinct locations. Step 2 shows the insertion of the initial octagon. Step 3 presents the drawing after the COPY command has been used to create two duplicates of this octagon. In Step 4, the COPY command has been used *once* again to duplicate the entire set of three octagons shown in Step 3.

Learning Check

Six octagons are developed in Figure 3.17. If these octagons were constructed without the aid of the COPY command, forty-eight separate LINE commands would be needed (eight straight lines per octagon). With the use of the COPY command, how many distinct commands (COPY and LINE commands) are needed to create these six octagons?

Answer Only eleven commands are needed to create the six octagons: eight LINE commands to create the initial octagon, followed by three COPY commands. (Two COPY commands are used to create two duplicates of the initial octagon; a third and final COPY command is then used to create another set of three octagons, thereby completing the entire group of six octagons.)

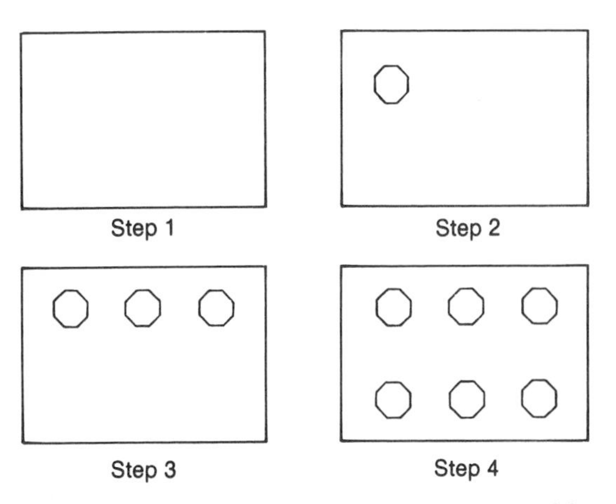

Figure 3.17 In this example, the LINE command is used eight consecutive times to construct the initial graphic element—an octagon. Then the COPY command is used to duplicate it.

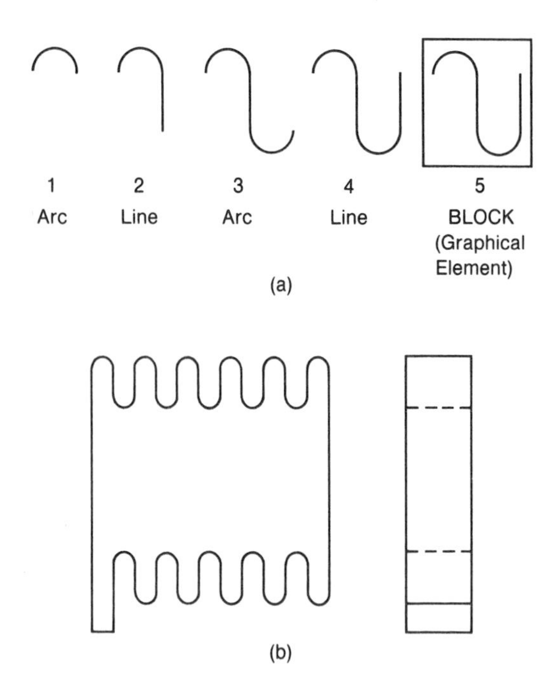

Figure 3.18 The BLOCK command can be used to *define* a particular graphic element and then *store* this element within the computer system's memory. In this example the ARC and LINE commands are used to create a graphic element that is then stored via the BLOCK command. This stored element is used to construct an engineering drawing with speed and accuracy. (Additional straight lines are created through the use of the LINE command.)

Although the COPY command is used to duplicate a feature (such as a circle or an octagon) in a drawing, the BLOCK command (or its equivalent) can also be used to duplicate features. For example, the LINE and ARC commands (or their equivalents) can be used to construct a portion of a surface boundary that is then **defined** as a specific graphic element with the aid of the BLOCK command (Figure 3.18). The user may then insert (repeatedly, if necessary) this BLOCK element in the drawing. In addition, *the BLOCK element will be stored* within the computer system's memory for future use. In contrast, the COPY command does not result in such storage of the duplicated graphic element. An entire drawing may be stored as a single BLOCK element if necessary.

The BLOCK command allows the user to **translate** (move in a linear direction) and/or **rotate** a stored graphic element to a new location in a drawing. As a result, we can create dynamic and useful graphic descriptions of engineering parts or modify the relative positions and angular orientations of the components within a design quickly and easily. (In Figure 3.19, an engineering drawing is rotated to three distinct relative positions through the use of the BLOCK command.)

3.15 CHANGE Command

The CHANGE command (or its equivalent), if it is available among the menu selections of a software package, allows the user to modify a feature in a drawing under construction without redefining the entire feature via a specific command such as LINE, CIRCLE, or ARC. For example, the endpoint of a line can be changed, after which the line is redrawn—or the radius of a circle can be redefined by the use of this command.

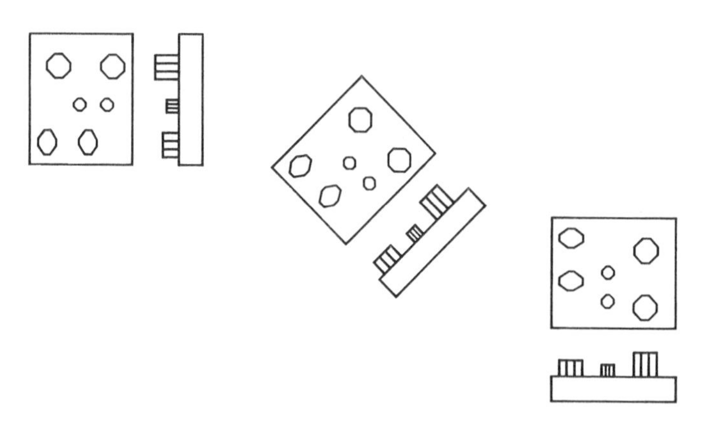

Figure 3.19 An engineering drawing is rotated to different orientations through the use of the BLOCK command.

3.16 TEXT Command

In addition to linework, engineering drawings must include textual information (such as fabrication specifications, part names, dates, and scales). The TEXT command (or its equivalent) allows the user to insert this information within a drawing by specifying its *location, height,* rotation angle or *orientation,* and *content.* As Figure 3.20 indicates, alphanumeric characters can be inserted into a drawing with this command; in addition, special characters (such as !, @, #, $, %, ¢, &, *, ?, +, /) can be inserted. The angle of orientation can be specified for any text (in Figure 3.20b, textual information is given at 0°, 20°, and 180° orientations with respect to the horizontal direction), together with its relative size (c) and uppercase/lowercase characters (d).

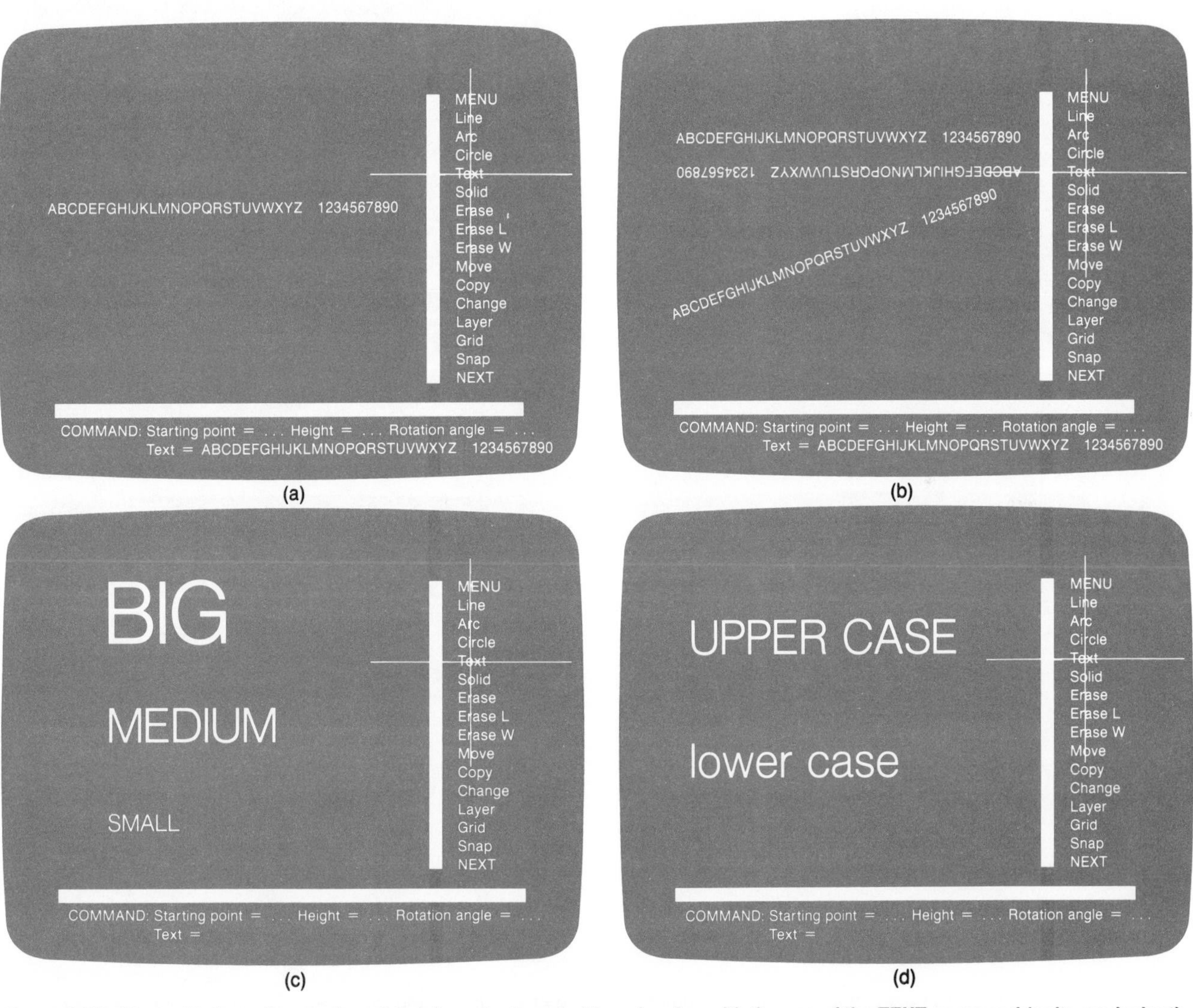

Figure 3.20 Textual information such as this (a) can be inserted in a drawing with the use of the TEXT command (or its equivalent). The user specifies the angle of orientation of the textual material with respect to the horizontal direction (b). The textual information may be created in a variety of sizes and heights (c) and in both uppercase and lowercase letters (d).

3.17 ZOOM Command

The ZOOM command (or its equivalent) allows the user to adjust the relative distance between the drawing and the observer so that the *image* of the drawing on the CRT screen, but not the actual drawing itself, changes in size as the viewing distance increases or decreases in accordance with the ZOOM command. One simply specifies the **magnification factor** to be used in the ZOOM command and the portion of the drawing to be affected. Figure 3.21 presents an engineering drawing undergoing three different types of magnification via the ZOOM command. With this command, the user

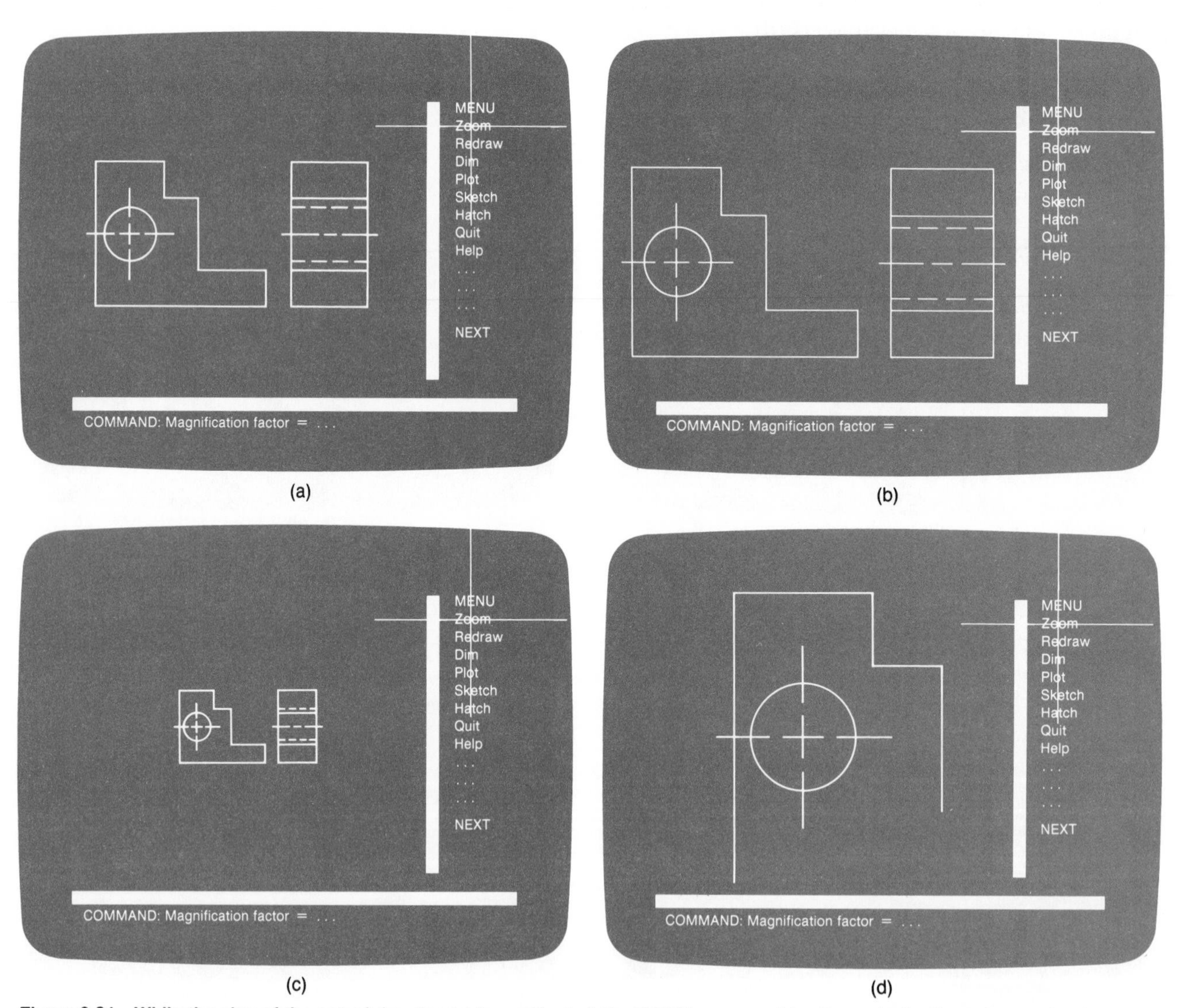

Figure 3.21 While the size of the actual drawing (a) is unaffected, the ZOOM command (or its equivalent) can be used to increase (b) or decrease (c) the *image* of that drawing on the CRT screen. Also, one portion of the image on the CRT screen can be increased (d) or decreased via this command.

may include extremely detailed information within a drawing that may be viewed by another user of a software package that includes a ZOOM command.

3.18 HATCH Command

The HATCH command (or its equivalent) allows the user to fill a region of a drawing with a particular **(hatchwork)** pattern of linework. This pattern may be one of several standard formats included in the software package. For example, in Figure 3.22 six standard patterns in the AutoCAD™ software are presented. It may also be a special pattern designed by the user. Hatchwork patterns are used in engineering drawings to differentiate between different components of an object and/or to indicate the specific materials to be used for these components.

3.19 DIMENSIONING Command

As we will discover in forthcoming chapters, engineering drawings in which three-dimensional objects are described must include size and location (or dimensioning) data if the objects are to be fabricated. The DIMENSIONING command (or its equivalent) allows the user to insert such data by specifying the type of dimensioning information to be included, together with its location and other characteristics. Figure 3.23 presents examples of drawings in which the following specific types of dimensioning data have been included: horizontal, vertical, aligned, angular, diametral, and radial. Each type of dimension will be fully discussed in Chapter 8.

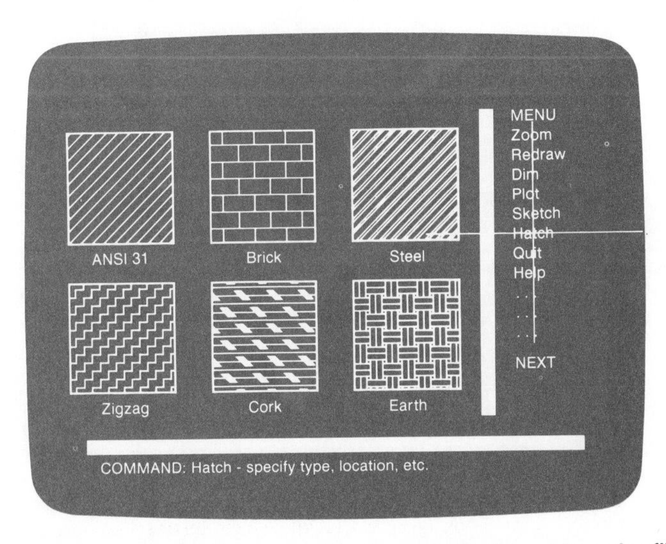

Figure 3.22 The HATCH command (or its equivalent) can be used to fill an area of a drawing with standard or special patterns.

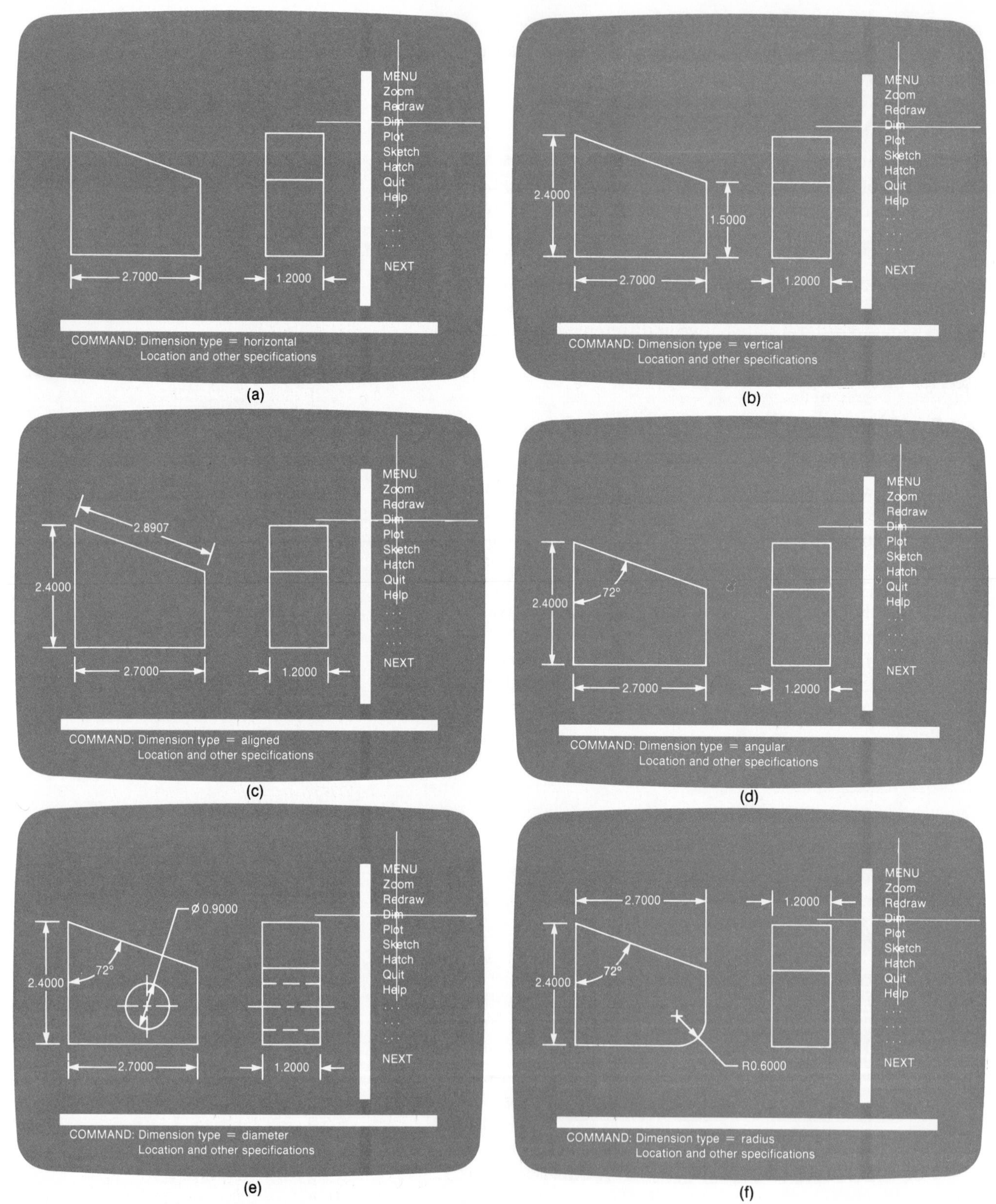

Figure 3.23 The DIMENSIONING command (or its equivalent) can be used to insert horizontal (a), vertical (b), aligned (c), angular (d), diametric (e), and radial (f) dimensions.

3.20 PLOT Command

The PLOT command (or its equivalent) allows the user to specify that a hard-copy (paper) plot of a particular drawing is to be created with the aid of a plotter attached to the computer. A plotter may include several pens of different colors so that a multicolored drawing can be created.

3.21 SAVE/END/FILE and RETRIEVE Commands

A computer-generated drawing may be stored (or saved) within the computer memory or on a floppy disk by using a command such as SAVE, END, or FILE (or the equivalent for the software in use). A command such as RETRIEVE (or its equivalent) allows the user to recall a previously stored drawing from memory.

3.22 Special Commands

Software for 2D computer graphics may also include special commands for selecting specific color schemes for a drawing, defining particular objects or portions of a drawing which can then be stored and retrieved as needed, or otherwise modifying a drawing under construction. It is important to be familiar with all the commands and other operational details of the software.

Learning Check

Some of the necessary functions of computer graphics ware are the ability to *create, erase, edit, modify, save, retrieve,* and *generate hard copy.* Identify the menu commands that correspond to these functions.

Answer

Function	Menu Command(s)
CREATE	LINE, CIRCLE, ARC, TEXT, HATCH, etc.
ERASE/EDIT/MODIFY	ERASE, DELETE, CHANGE, MOVE, COPY.
SAVE and RETRIEVE	SAVE, STORAGE, RETRIEVE, BLOCK.
GENERATE HARD COPY	PLOT.

COMPUTER GRAPHICS IN ACTION

3D Computer Graphics: Wire-Frame Models

In this chapter, we have focused upon two-dimensional computer graphics in which the engineer constructs the desired drawing(s) of the object under consideration. However, three-dimensional computer graphics software packages are now available and in popular use within many industries. Three-dimensional (geometric) computer graphics models of engineering objects allow one to quickly and accurately construct virtually any graphic description of an object with greater ease than that provided by two-dimensional software systems because the computer develops the drawing(s) *based upon the data used to describe the object* (that is, the engineer describes the object in terms of geometric and other physical data; any desired drawing can then be generated via three-dimensional computer graphics from this data). Three distinct forms of three-dimensional computer graphics models—**wire-frame, surface,** and **solid**—will be briefly reviewed in the "Computer Graphics in Action" sections near the end of Chapters 3 through 7.

Wire-frame models can be very confusing to the reader of an engineering drawing (and may be misinterpreted) because *all* linework is generally shown in such a drawing—including surface boundaries that would be hidden from the observer of the object by the solid material forming the object. The figure shown here presents a simple wire-frame model of an object. Surface boundaries are represented by a set of interconnected lines between key points on the object. The advantages of wire-frame models include:

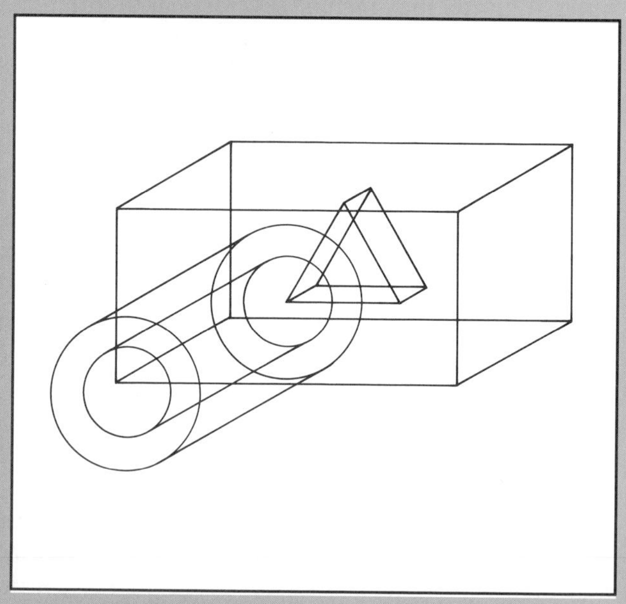

A wire-frame model of an object.

- Simplicity in the development of computer programs (software) for the manipulation of geometric data.
- Relatively low costs for development and minimal need for computer memory usage.

However, the disadvantages associated with such models, which are significant, are:

- Exterior and interior surfaces are not distinguished, creating difficulty in the interpretation of a drawing.
- Part surfaces are poorly defined in terms of boundary edges.

More useful and informative than the wire-frame model is the **surface** model, which we will focus upon in the "Computer Graphics in Action" section near the end of Chapter 4.

In Retrospect

- Computer-aided design drafting (CADD) allows the designer to develop final working drawings and other graphic descriptions of a part or a product.
- Drawings can be plotted in hard-copy form or displayed on a CRT (cathode-ray tube) screen.
- Interactive computer graphics refers to an equipment (hardware) and software configuration that allows the user to modify a drawing during its construction by interacting with the computer through the picture on the screen.
- In 3D geometric modeling computer graphics, the object under consideration is described in the form of geometric and physical data that is stored within the computer's memory. Drawings of the object may then be generated from this data.
- In 2D computer graphics, the desired drawings of the object are created by the user and then stored within the computer memory; hard-copy plots of these drawings can then be generated as needed. The object itself is not defined, *only specific images or graphic representations* of the object.
- Necessary functions that must be performed during the creation of a drawing via 2D computer graphics include:
  - The creation of linework.
  - The erasure, editing, and modification of linework.
  - The storage and subsequent retrieval of drawings.
  - The generation of hard copies of drawings.
- Many 2D computer graphics software packages are menu-driven, which means the user can select the desired graphics function from a group (or menu) of possibilities. Selection may be accomplished via the keyboard, the controlled movement of a cursor about the CRT screen, or through interaction with a digitizer pad.
- Common functions of 2D menu-driven software packages include the following selections (or their equivalent):
  - GRID, a collection of points/dots on the screen that are used as visual reference points for discrete movements on the drawing with the aid of the SNAP command.
  - LINE, which allows the user to draw straight lines of specific types, lengths, and directions.
  - LAYERING/TEMPORARY, which allows the user to draw lines on the CRT screen that will not be stored in memory with other (finished) linework for a drawing. Thus, the user may generate construction lines with this type of command.
  - ERASE/DELETE, with which linework may be removed from a drawing.
  - CIRCLE, which allows the user to generate circles of a particular size at a given location in a drawing.
  - ARC, with which an arc of a given size, centered at a particular location, may be created.
  - POLYGON, with which n-sided 2D figures may be created.
  - MOVE and COPY, with which an existing figure can be moved about the screen or duplicated as many times as necessary.
  - CHANGE, with which an existing element (such as a line, circle, or arc) in a drawing can be quickly modified.
  - TEXT, which allows the user to insert textual information at a given location at a specified height and angle of orientation.
  - ZOOM, which allows the user to reduce or magnify the image of the drawing shown on the CRT screen.
  - HATCH, with which an area of a drawing may be filled with a standard or special pattern.
  - DIMENSIONING, with which dimensioning information can be inserted into an engineering drawing.
  - PLOT, with which hard-copy (paper) plots of a drawing can be produced.
  - SAVE/STORAGE, which creates a file within the computer's memory (or on a portable floppy disk) corresponding to a particular drawing.
  - RETRIEVE, with which a stored file can be recalled from memory for further modification and development.

Other functions are also available in most 2D software packages. The reader is strongly encouraged to become familiar with all the functions (and their identifying names) that are available for 2D computer graphics work at his or her institution.

Use a computer graphics system to complete the following exercises. First, determine the specific characteristics of the computer graphics system to be used. The scale(s) for many of the exercises may be specified by the instructor or if the instructor so desires, chosen by the student. Hard-copy plots should be generated as required. All work should be centered within the available drawing space, unless otherwise specified by the instructor.

EX3.1. Create a standard title block that can be added to each of your drawings. Include such data as your name and institutional affiliation, together with spaces for the drawing number, date, drawing title, scale, and so forth. Store this title block (if possible) as a distinct graphical element that can be retrieved later for use in other drawings.

EX3.2. A rectangular coordinate (x,y) system is commonly used to define point locations in the two-dimensional drawing space in computer graphics systems. The origin of this coordinate system is at the lower left corner of the drawing space, with coordinate values of 0, 0.

Draw lines that intersect at point 5, 5 and that have the following angular orientations with respect to the horizontal direction: 10°, 20°, 30°, 40°, 50°, 60°, 70°, 80°, and 90°. The length of these lines is chosen by the reader (or the instructor). (Note that you must become familiar with the units of length measurement—for example, inches—used in the coordinate system at your institution. *Many systems use the inch as the basic unit,* in which case 5, 5 would denote a location 5 in. to the right and 5 in. above the origin 0, 0 at the lower left corner of the drawing space.)

EX3.3. Draw concentric circles of radii 1 cm., 2 cm., 3 cm., 4 cm., and 5 cm. Center these circles within the drawing space.

EX3.4. Draw concentric circles of radii 1 in., 2 in., 3 in., 3.25 in., and 3.75 in. Center these circles within the drawing space.

EX3.5. Draw an arc centered at point 4, 5 in the drawing space. The arc radius should equal 3 in. The endpoints of the arc may be arbitrarily selected.

EX3.6. Draw a line at 38° with respect to the horizontal direction. Use the COPY command (or its equivalent) to then create ten additional lines that are parallel to this line. The length and relative spacing of these lines may be arbitrarily selected; however, the spacing between lines should be identical.

EX3.7. Draw two circles, one of radius equal to 2 in. centered at point 2.25, 2.25 and one of radius equal to 1.25 in. centered at point 7, 5. Then draw lines that are tangent to each circle, first in an open-belt configuration and then in a crossed-belt configuration. (See Section 2.18.)

EX3.8. Draw a square with sides 2 in. in length. Next, copy this square to a location 4 in. to the right and 4 in. above the original position of the square (this original position should be carefully chosen by the reader). Change this second square into a rhombus in which two sides are horizontal and two sides are oriented at 30° to the horizontal direction. NOTE: the final figure should then consist of the original square and the rhombus.

EX3.9. Draw Figure 3.1 (in which the structure of a cathode-ray tube is shown). Use a scale of your choice (or your instructor's choice).

Use a computer graphics system to complete each exercise. EX3.10 through EX3.50 refer to specific exercises in Chapter 2. For example, for EX3.10, use 2D computer graphics to draw the scalene triangle described in EX2.5 from Chapter 2. Properly center the finished drawing within the drawing region, and include a title block in accordance with the directions of your instructor.

EX3.10. See EX2.5.

EX3.11. See EX2.6.

EX3.12. See EX2.7.

EX3.13. See EX2.8.

EX3.14. See EX2.9.

EX3.15. See EX2.10.

EX3.16. See EX2.11.

EX3.17. See EX2.12.

EX3.18. See EX2.13.

EX3.19. See EX2.14.

EX3.20. See EX2.15.

EX3.21. See EX2.16.

EX3.22. See EX2.17.

EX3.23. See EX2.18.

EX3.24. See EX2.19.

EX3.25. See EX2.20.

EX3.26. See EX2.21.

EX3.27. See EX2.22.

EX3.28. See EX2.23.

EX3.29. See EX2.24.

EX3.30. See EX2.25.

EX3.31. See EX2.26.

EX3.32. See EX2.27.

EX3.33. See EX2.28.

EX3.34. See EX2.29.

EX3.35. See EX2.30.

EX3.36. See EX2.31.

EX3.37. See EX2.32.

EX3.38. See EX2.33.

EX3.39. See EX2.34.

EX3.40. See EX2.42.

EX3.41. See EX2.43.

EX3.42. See EX2.45.

EX3.43. See EX2.46.

EX3.44. See EX2.47.

EX3.45. See EX2.48.

EX3.46. See EX2.49.

EX3.47. See EX2.50.

EX3.48. See EX2.51.

EX3.49. See EX2.53.

EX3.50. See EX2.54.

4 Orthographic Projection, Part 1: Foundations

Preview **Multiview** drawings are used in industry to describe objects that are to be machined or otherwise fabricated. Such drawings can be created by using Monge's **planes of projection,** which will be explained in this chapter. **Orthographic drawings** are images created by theoretical projections of the object onto perpendicular reference planes. (Two Greek words form the basis for the word 'orthographic': **orthos,** which refers to right angles, and **graphikus,** which means "to write" or "to draw".) Each orthographic projection is a particular two-dimensional view of the object under consideration; a set of multiple views (multiviews) of an object can be used to completely describe the object. Figure 4.1 presents three orthographic views of an object, together with a **pictorial** (three-dimensional) drawing.

Orthographic views are usually constructed with drawing instruments. As a result, such graphical descriptions of an object are also known as **mechanical drawings.** Orthographic drawings with complete dimensions and notes are known as **working drawings.** Engineers, technicians, drafters, and others who are involved in the design, manufacture, and application of engineering products communicate through the use of orthographic drawings.

In this chapter, we will explore the foundations of orthographic projection, beginning with Monge's planes of projection (circa 1795). We will also introduce the rules for properly constructing orthographic views. Orthographic drawings form a special type of graphical language that is similar to written languages: rules must be obeyed if the given information is to be properly correlated and understood by all concerned.

Throughout this chapter, we encourage you to *analyze* sets of orthographic views rather than attempt to immediately visualize the objects described by these views. A novice often believes that he or she should be able to simply "see" the object described in a set of views because other, more experienced people seem to be able to "see" the object almost immediately upon viewing the drawings. Actually, those with more experience simply *analyze and correlate* the given information in the drawings more quickly—so quickly, in fact, that they may not realize that they are *not* simply visualizing the object through an innate or intuitive ability but instead are using graphical reading skills that they have developed through practice.

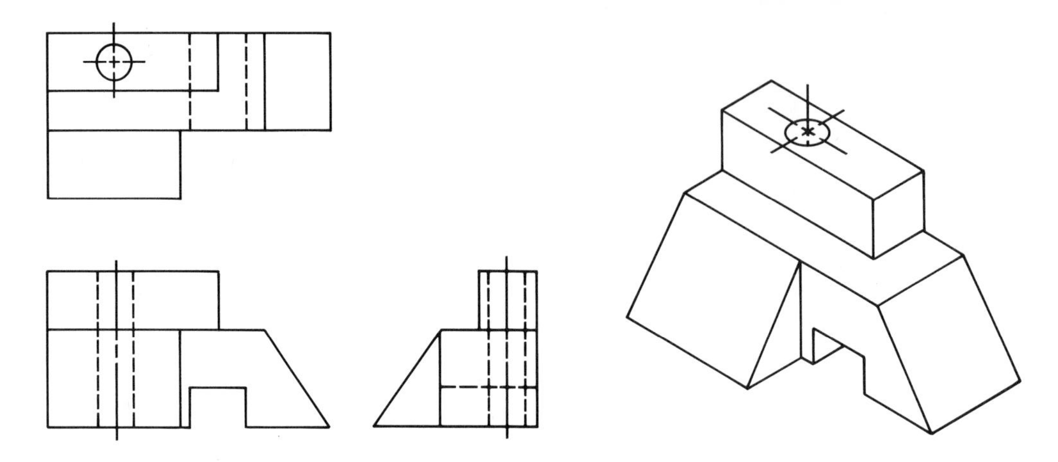

Figure 4.1 Three orthographic views and a pictorial of an example object.

Do not be discouraged if you cannot immediately visualize the object described by a set of orthographic views. You will eventually be able to visualize it if you carefully and patiently analyze the information contained in the drawings, in accordance with the rules and techniques given in this chapter. Remember, you did not learn to read a written language (such as English) immediately, but only after sufficient practice. So, too, will you learn to read and interpret orthographic drawings after sufficient practice.

Learning Objectives

Upon completion of this chapter, the reader should be able to:

- Identify the planes of projection and apply these planes in the development of orthographic views.
- Relate the **glass box** concept to the six principal orthographic views of an object.
- Identify the two dimensions associated with each principal view.
- Apply the three rules for orthographic views:
  - Rule 1–common dimensions.
  - Rule 2–mutual alignment.
  - Rule 3–adjacent placement.
- Select the *best and necessary* orthographic views for an object.
- Identify three distinct categories of planar surfaces—parallel to a projection plane, inclined, and oblique—together with their resulting projections onto the planes of projection.
- Apply the configuration rule during an analysis of a set of orthographic views.
- Interpret given orthographic views of an object and develop any necessary missing views via application of the surface-by-surface, point-by-point method.
- Apply the cut-by-cut approach in the interpretation of orthographic views.
- Explain the reasons for the priority of linework used in orthographic views.
- Properly intersect visible and hidden lines in a drawing.
- Select views for two-view drawings and one-view drawings.
- Differentiate between first-angle projection and third-angle projection.
- Recognize the need for clarity and ease of interpretation in orthographic drawings.

4.1 Monge's Planes of Projection

Multiview projection—a subset of orthographic projection—is the method of depicting three-dimensional objects or concepts with sets of two-dimensional drawings. (Since only two dimensions can be given in a single orthographic drawing, the third dimension must be supplied in an additional (second) drawing, hence the name *multiview.*)

Gaspard Monge has been called the father of both descriptive geometry and orthographic projection. His book *Geometrie descriptive* presented many principles of technical drawing involving the use of points, lines, and planes as references from which distances can be measured. (A plane is simply a flat surface.) The use of Monge's planes of projection (frontal, horizontal, and profile) to describe three-dimensional objects on two-dimensional surfaces is the basis of orthographic projection.

Consider the object shown in Figure 4.2; we will develop a set of orthographic multiviews that describe this object.

In developing orthographic views, we begin with two basic assumptions:

1. Your sight line or viewing direction is perpendicular to the principal surface onto which *projection* is to occur.
2. The object is viewed from a long distance.

The first assumption explains the name *orthographic.* We will draw the object as we see it in a given direction, projecting the object's edge-lines onto a reference plane that is perpendicular, or at right angles (which is what the Greek word *orthos* means) to the viewing direction. The second assumption emphasizes the fact that we will not draw objects with true perspective but rather with dimensions of true size; at long distances, parallel edges of an object appear to be identical in length. The apparent differences in length caused by perspective effects at shorter viewing distances are eliminated. In other words, the object will be described in orthographic views as we wish it to be built or manufactured, not as we would see the object with true perspective (Figure 4.3).

After choosing a particular surface of the object as its front (a choice we will discuss in detail shortly), one may imagine a vertical reference plane (Monge's frontal projection plane) onto which **key points** of the object may be projected, as shown in Figure 4.4. Key points are those deemed to be important in defining the shape of the object as seen from a direction perpendicular to the projection plane. In Figure 4.4 seven points have been selected as key points projected onto the **frontal reference plane.** The viewing direction perpendicular to the frontal reference plane in Figure 4.4 is called the frontal viewing direction; the frontal viewing direction is, of course, also parallel to the projection lines, or projectors, from the object to the FRP.

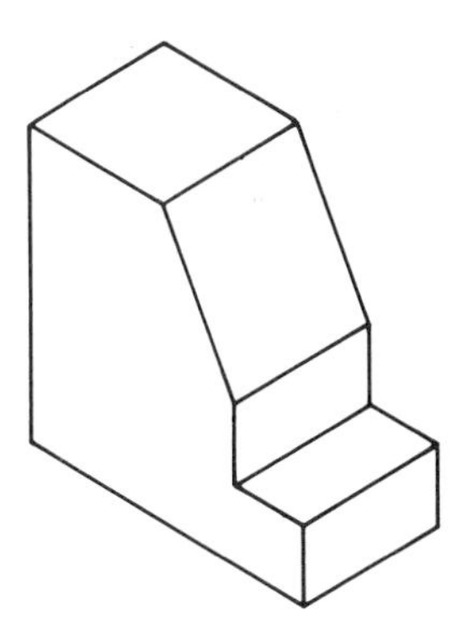

Figure 4.2 An example object to be the subject of successive figures in this chapter.

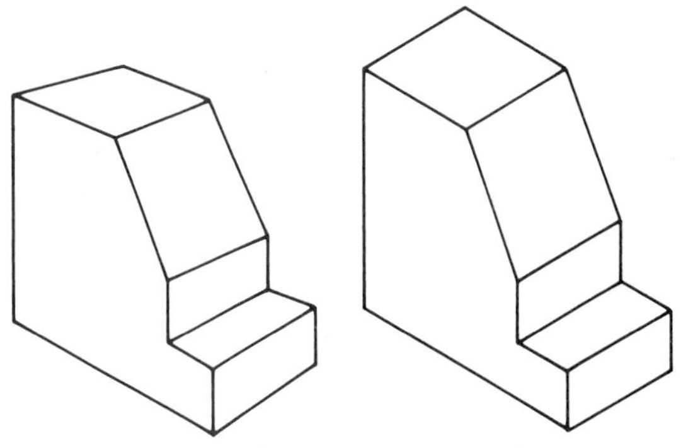

Figure 4.3 Perspective in engineering drawing is not the same as true perspective.

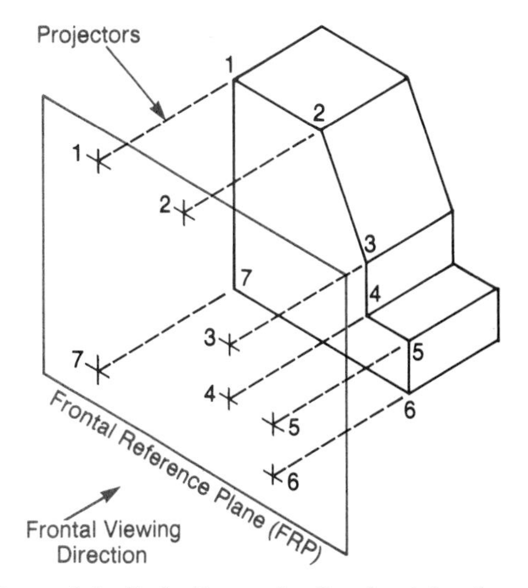

Figure 4.4 Projection onto the frontal reference plane.

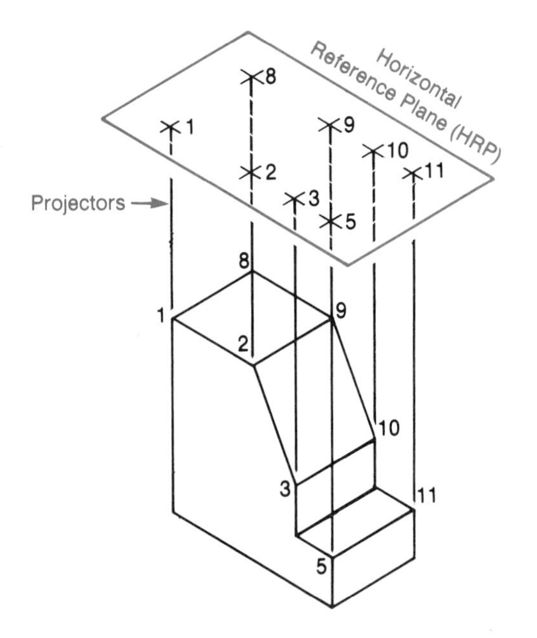

Figure 4.5 Projection onto the horizontal reference plane.

Next, we may imagine a horizontal reference plane (Monge's horizontal projection plane) onto which key points of the object may be projected, as shown in Figure 4.5. The **horizontal reference plane** (HRP) is, by definition, perpendicular to the FRP. Once again, projectors are perpendicular to the plane of projection. Eight points have been projected from the object to the HRP in Figure 4.5

Finally, we may imagine a second vertical reference plane (a profile[1] projection plane) that is mutually perpendicular to both the FRP and the HRP. Once again, key points are projected from the object onto this **profile reference plane** (PRP) along projectors that are perpendicular to the PRP. Eight points have been projected onto the PRP in Figure 4.6.

The key points that have been projected onto the three principal planes of projection in Figures 4.4, 4.5, and 4.6 are the endpoints of surface boundaries on the object. Each surface boundary appears as a line between two key points in Figure 4.2; we now wish to connect the projected points on the three reference planes with lines corresponding to the surface boundaries of the original object. However, before

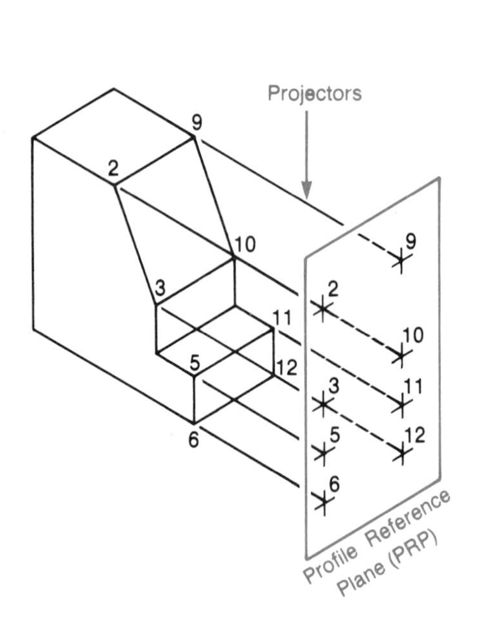

Figure 4.6 Projection onto the profile reference plane.

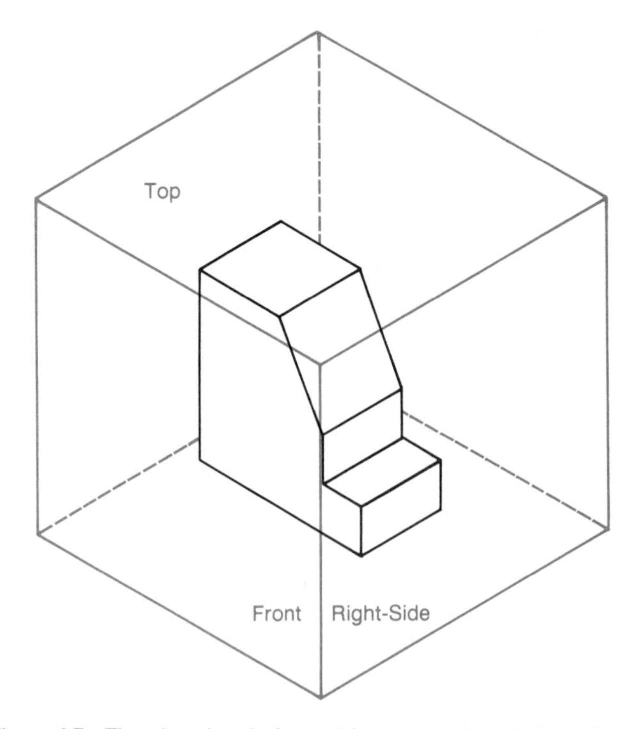

Figure 4.7 The glass box is formed from sets of vertical, horizontal, and profile planes.

1. Monge actually generated only two planes of projection for his work; however, the three planes (FRP, HRP, PRP) presented in this section are an extension of Monge's work. Three planes of projection are now commonly used for the development of orthographic views.

we "connect the dots" on the projection planes, let us consider the **glass box** shown in Figure 4.7, which surrounds the object of our consideration. This glass box is formed from sets of vertical, horizontal, and profile planes; it will aid us in understanding the final orthographic views that we will obtain from Monge's planes of projection.

Figure 4.8 shows the glass box with the projected key points of the object on three of its walls; these projected points have then been connected with lines corresponding to the surface boundaries of the object. Indeed, we have simply combined the results of our point projections onto the FRP, HRP, and PRP (in Figures 4.4, 4.5, and 4.6) with the concept of the glass box. Figure 4.9 shows the result of such point projection onto the three walls of the glass box under consideration. (The glass box actually has six walls, but we have ignored the three walls hidden from view by the visible walls nearest to the observer.) One can see that three two-dimensional drawings of the object, as seen from three different viewing directions, have been formed on the walls of the glass box.

In Figure 4.10, three walls of the glass box are opened so that they are parallel to one another; Figure 4.11 shows the three walls (corresponding to the three principal planes of projection—the FRP, the HRP, and the PRP) lying flat on a single plane (the drawing paper). The three projected views of the object shown in Figure 4.11 are the **front view** (corresponding to projection onto the FRP), the **top view** (corresponding to projection onto the HRP) and the **right-side view** (corresponding to projection onto the PRP). The edges of the glass box form the boundaries of these three views; such boundaries are known as **fold lines.** (In finished orthographic views, one does not always show fold lines; however,

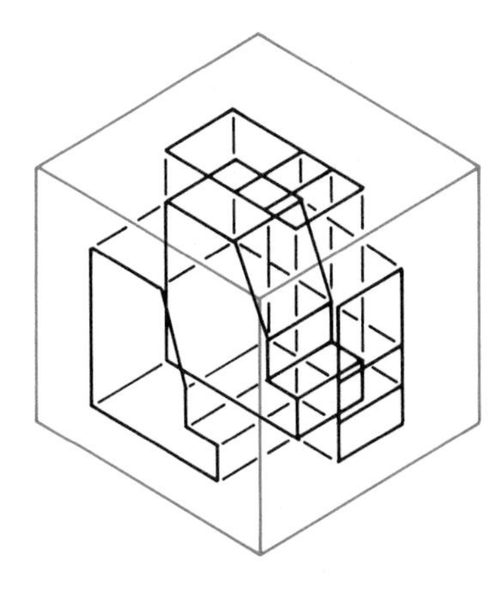

Figure 4.8 Projection onto the glass box.

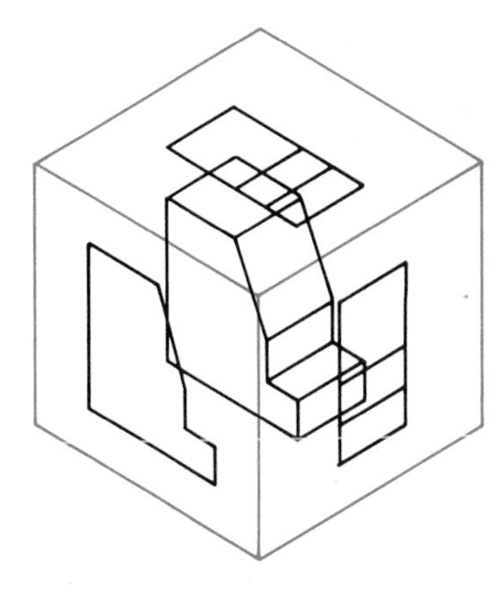

Figure 4.9 Result of projection onto the glass box.

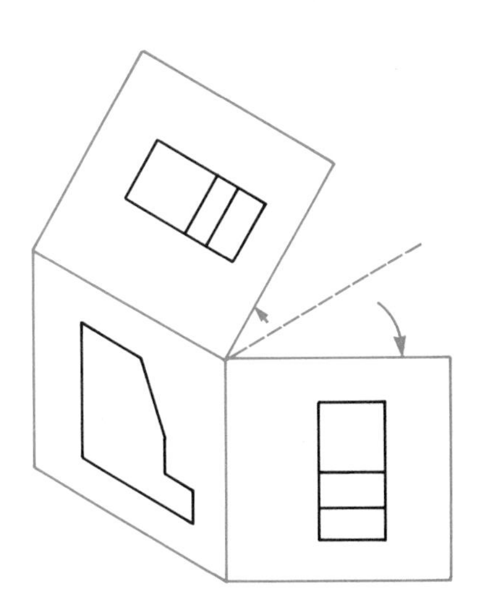

Figure 4.10 Opening the glass box.

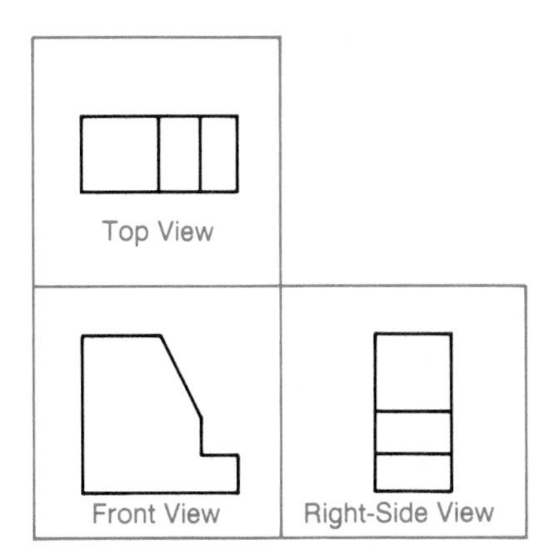

Figure 4.11 Three walls of the glass box correspond to three orthographic views.

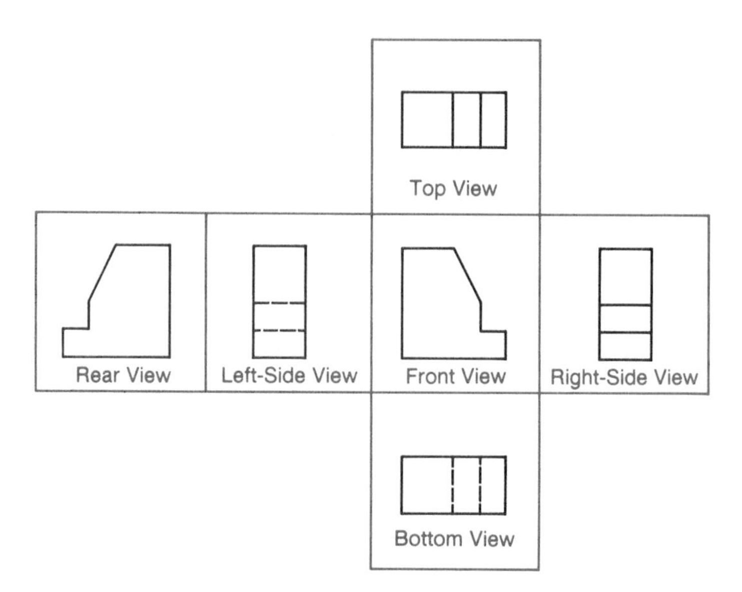

Figure 4.12 The six walls of the glass box correspond to the six principal orthographic views.

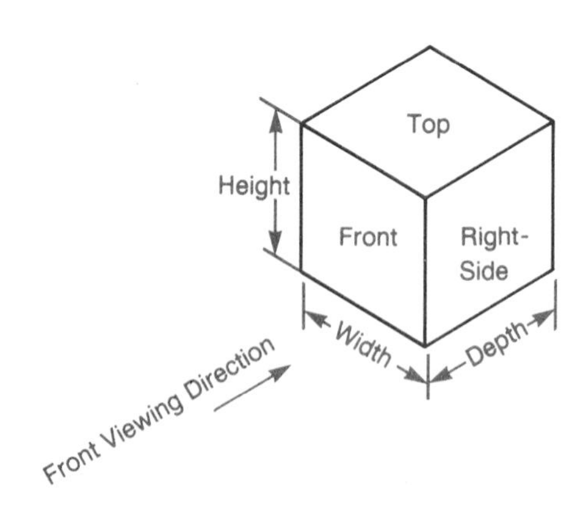

Figure 4.13 Once the front surface of this rectangular prism is identified, all other surfaces may then be readily identified.

the concept of a fold line—representing the edge of intersection between two walls of the glass box—will be very helpful in our discussion of auxiliary orthographic views in Chapter 7.)

We have ignored the fact that the glass box has six, not three, walls and that one may project views of the object onto all six walls of the box. Such projection will produce six views of the object, as shown in Figure 4.12: the front view, the top view, and the right-side view are again obtained; and, in addition, the **bottom view,** the **rear view,** and the **left side view** of the object. These six views (front, top, right-side, bottom, left-side, and rear) are known as the six principal views of an object; any views in addition to these six principal views are known as auxiliary views (see Chapter 7).

One should note that the right-side view shows the right side of the object relative to the position of the observer, the top view clearly shows the top of the object and so forth. Indeed, once one has identified the front of an object, all other directions (right, top, rear, and so on) are automatically defined.

Consider the rectangular prism shown in Figure 4.13: once we have identified the front surface of this prism, all six surfaces may then be readily identified. Figure 4.14 shows the front view of the prism, in which (by definition) the front surface of the prism is shown as an area. The four lines representing the boundaries, or edges, of this front surface are the edges of intersection between it and the four perpendicular surfaces in contact with it, that is, the top, right-side, bottom, and left-side surfaces of the prism. From the perspective of the viewer, names are given to each of these surfaces: the right-side surface (denoted as RSS in Figure 4.14) is represented *on edge* as the line on the right of the front view, the top surface (denoted as TS) is represented on edge as the line at the top of the front view, and so on.

The two dimensions associated with the front view are the **height** *(H)* and the **width** *(W)* (Figure 4.15). The third dimension, **depth** *(D)*, must be given in a second orthographic view—for example, the right-side view.

In summary, then:

1. The multiview-drawing form of orthographic projection is the method of depicting three-dimensional objects or concepts with sets of two-dimensional drawings.
2. Monge's planes of projection (frontal, horizontal, and profile) provide the basis for modern orthographic projection.
3. The viewing direction, or sight line, of the observer is always perpendicular to the principal plane onto which projection is to occur.
4. True perspective is not shown in orthographic views; the object or part is described in the views as it is to be fabricated.
5. The planes (FRP, HRP, and PRP) of projection may be used to form an imaginary glass box that may then be unfolded in order to reveal the relationship between the projected views.
6. The dimensions associated with each orthographic view are as follows:

- Front view, rear view: height *(H)*, width *(W)*.
- Right-side view, left-side view: Height *(H)*, depth *(D)*.
- Top view, bottom view: width *(W)*, depth *(D)*.

Having reviewed the six principal orthographic views based upon Monge's planes of projection and the concept of the glass box, we must go on to develop an appreciation for the rules of orthographic projection, together with a deeper understanding of the development and interpretation of orthographic views.

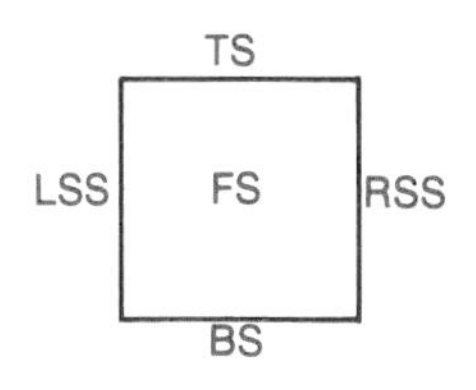

FS = Front Surface
RSS = Right-Side Surface (Edge View)
LSS = Left-Side Surface (Edge View)
TS = Top Surface (Edge View)
BS = Bottom Surface (Edge View)
REARS = Rear Surface (Not Visible)

Figure 4.14 In the front view of this rectangular prism, the four lines representing the edges of the front surface are the edges of intersection between it and the four perpendicular surfaces in contact with it.

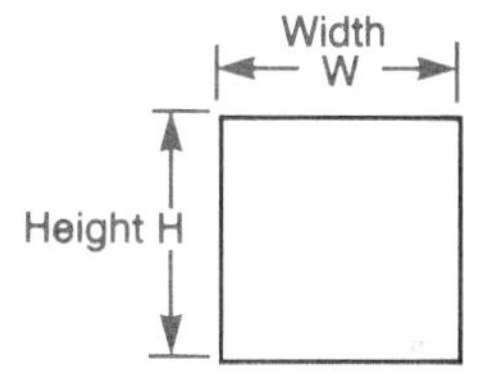

Figure 4.15 The two dimensions associated with the front view of this rectangular prism are the height and width.

4.2 Three Rules for Orthographic Views

The development of the six principal views of the object shown in Figure 4.12 implies the following three rules for orthographic projection:

- Rule 1. Orthographic views must be drawn with consistent (common) dimensions.
- Rule 2. Orthographic views must be mutually aligned.
- Rule 3. Orthographic views must obey the rule of adjacency.

Highlight

Gaspard Monge

Gaspard Monge had a far-reaching effect upon both engineering drawing and technology through his innovative drawing methods—based upon geometry and the concept of projection planes—and his development of a training school (École Normale) for teachers, established in 1794. He realized that the development of industry and manufacturing was dependent upon the extent and the quality of technical education.

Monge was born in Beaune (near Dijon), France in 1746. His skill in mathematics and science led to a position as student draughtsman (draftsman) in the fortification design office of the school at Mézières for French army officers. At Mézières, he developed an entirely new drawing method which required less time and produced results of greater accuracy than the traditional empirical approximation methods used at that time. He then was promoted to the position of assistant to a professor of physics at the school; during this period, he developed the first principles for his master work in descriptive geometry.

Stone-cutters at Mézières quickly adopted Monge's methods, but carpenters were reluctant to abandon their techniques that had been passed from father to son for generations. Twenty years passed before the Monge methods were adopted by the carpenters at Mézières.

Monge's work in descriptive geometry was classified as a military secret by the authorities at Méziéres, preventing the publication of Monge's book on the subject.

Monge eventually became Naval Examiner and Minister for Marine; then the French Revolution exploded. The Executive Council was a group of ministers who acted as the government of France during this period. Monge became a minister and a member of the Executive Council immediately following the outbreak of revolution.

With Lavoisier, Carnot, and others, Monge sought the establishment of the École Normale. In 1794 the school was formed. In addition, the ban on Monge's work in descriptive geometry was removed, leading to the publication of his book entitled *Geometrie descriptive* (1795). Descriptive geometry became one of the fundamental topics in French technical education. Monge also wrote extensively on the subjects of gunpowder, iron and steelmaking, and other topics of military and industrial significance.

Gaspard Monge

Monge was later requested to journey to Italy, Syria, and Egypt for scientific work. After Napoleon's defeat and the reestablishment of the monarchy, Monge was out of favor with the authorities. He died in 1818.

Monge's work led, both directly and indirectly, to improved engineering methods involving mathematics and science. In addition, his efforts produced recognition that education must be supported if a society is to progress.

Source: Peter Booker, A History of Engineering Drawing *(London: Northgate Publishing Company, Ltd.,1979).*

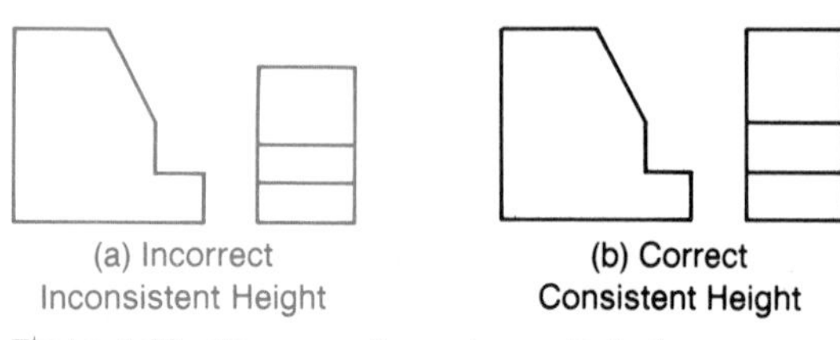

Figure 4.16 Common dimensions—Rule 1.

Figure 4.16 shows two versions of two views (front and right-side) of the object shown initially in Figure 4.2. Notice that the height of the object is a common dimension in these two views, that is, a dimension shared by both views. Inconsistency in the height of the object shown in one version of the two views (a) violates Rule 1. The height must be shown by identical lengths in these two views, as in the correct version (b).

Height *(H)* is a common dimension, and thus must be shown with consistent lengths in the following four views:

front, right-side, left-side, and rear views. Similarly, the dimension of width *(W)* must be shown with consistent lengths in those views in which it is a common dimension, namely, the front, top, bottom, and rear views. Finally, the dimension of depth *(D)* is a common dimension in the right-side, left-side, top, and bottom views. Return to Figure 4.12 and consider the common dimensions that are associated with the given six orthographic views of the object.

Like Figure 4.16, Figure 4.17 also shows two versions of the front view and the right-side view of the object shown in Figure 4.2. One version (a) is incorrect because these two views are not aligned with one another in accordance with Rule 2; in the other version (b) these two views are properly aligned. The shared common dimension of height between these two views is clearly indicated by proper alignment. As another example of alignment, consider the words on this page: each word on a given line is aligned with the other words appearing on that line. Such alignment simplifies our effort to read and relate these words. In fact, we expect to find these words in alignment. Furthermore, we expect to find these words printed in a particular order—from left to right—and each line printed from top to bottom on the page, because this is the accepted order for words and lines in English text. We would be quite dismayed to discover that the words were arranged in some other order. The subject of ordering textual information on the printed page leads us to the subject of Rule 3—the rule of adjacency. Just as the reader of the English language expects to find words printed in a particular order, so does the reader of an orthographic drawing expect to find views presented in a particular order. Thus, certain views must be placed *adjacent* to other views. In other words, the rule of adjacency defines the orientation of the object in a set of orthographic views. For example, the right-side view of the object in Figure 4.12 is placed adjacent (nearest) to the right-side surface edge of the front view. Furthermore, the edge of the right-side view that is adjacent to the front view represents the front surface seen on edge as a line.

In order to see this more clearly, let us return to the concept of a frontal reference plane (FRP) and a profile reference plane (PRP). We will allow these mutually perpendicular planes to *contact* the object under consideration, as shown in Figure 4.18. The dashed lines in the figure represent the FRP and PRP seen on edge as lines in the right-side and front views, respectively (b). Notice that the FRP contains the front surface of the object, seen on edge in the right-side view, whereas the PRP contains the right-side surface of the object, seen on edge in the front view. The FRP must be adjacent to the front view, whereas the PRP must be adjacent to the right-side view, as shown in the correct version

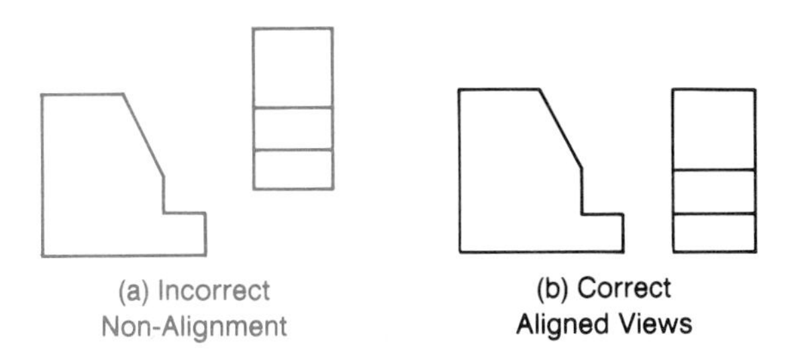

Figure 4.17 Alignment of views—Rule 2.

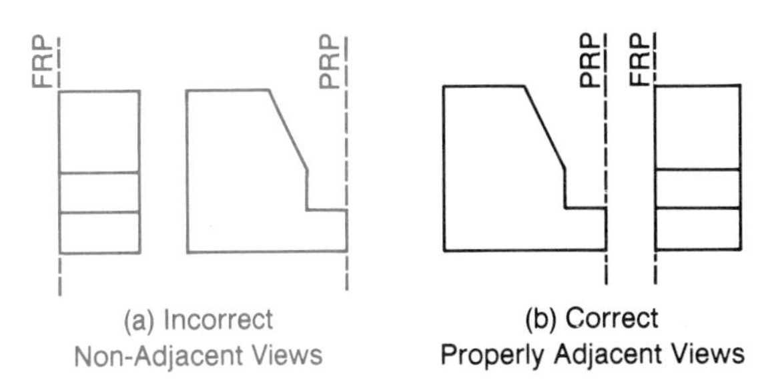

Figure 4.18 Adjacent placement of views—Rule 3 (rule of adjacency).

(b). In the incorrect version (a) these views are presented in wrong relationship to each other.

A simplification of this rule is simply that the right-side view must be drawn to the right of the front view; similarly, the top view must be drawn above the front view, the left-side view must be drawn to the left of the front view, and so on. However, the concept of the FRP containing the front surface of the object and appearing on edge as a line in the right-side view (adjacent to the front view) allows us to define the depth *(D)* of an object in terms of its perpendicular (horizontal) distance from the FRP to the rear of the object. Similarly, the dimension of width *(W)* can be defined as the distance between the PRP (which contains the right-side surface of the object) and the left-side surface of the object. The concept of reference planes, such as the FRP and the PRP, obviously facilitates the development of orthographic views.

Figure 4.19 shows a cube described in terms of the six principal orthographic views. Notice that the top view could be placed above the front, right-side, left-side, or rear views and still be in accordance with the rule of adjacency; however, when one has a choice, the top view should be placed above the front view. Similarly, the bottom view should be placed below the front view. This is standard practice, since there will always be a front view—it is the first view defined in a set of orthographic views.

In summary, then:

1. Three rules for orthographic projection must be obeyed:

- Rule 1—common dimensions.
- Rule 2—alignment of views.
- Rule 3—adjacent placement of views.

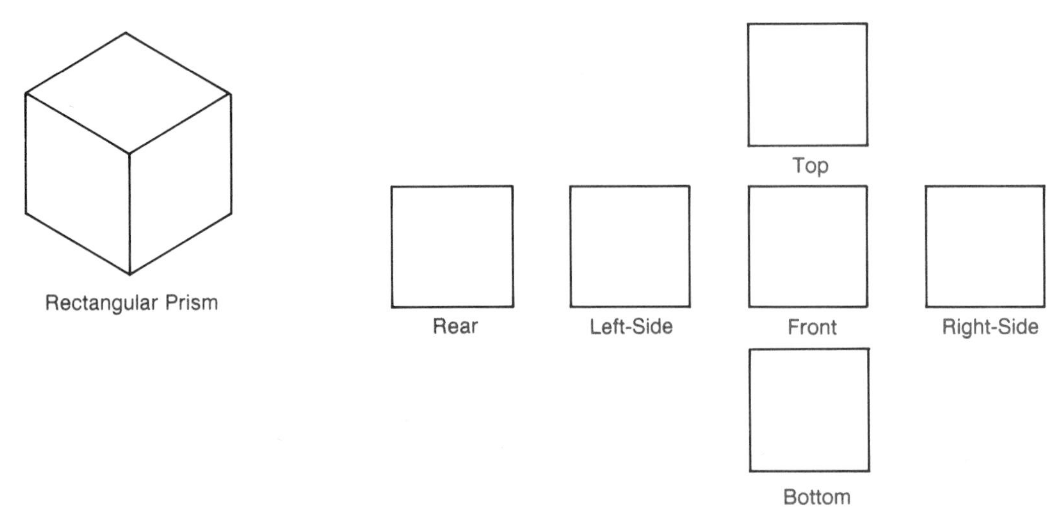

Figure 4.19 Six principal views of a rectangular prism.

Highlight

Rotation

Orthographic projection can be interpreted in terms of the observer moving with respect to the object (for example, the observer moves from the front viewing position to the right-side viewing position) or in terms of the object moving with respect to the observer. If interpreted the second way, orthographic views may be better understood if the reader attempts to visualize rotation of the object. In the sketches shown here an object is moved from the front view to the top view and also to the right-side view. Sketches such as these are helpful in developing an intuitive understanding of orthographic views; however, the reader must recognize that the rule of adjacency (and the other rules previously described) must be followed in the process of rotation. Thus, if you imagine rotating an object to view its right side, remember that only rotation in the proper direction (according to the rule of adjacency) will allow you to 'see' the right side of the object and not some other surface (for example, the left side). The rules for orthographic projection must be obeyed even in the imaginary rotation of an object from one position to another to aid interpretation of orthographic views.

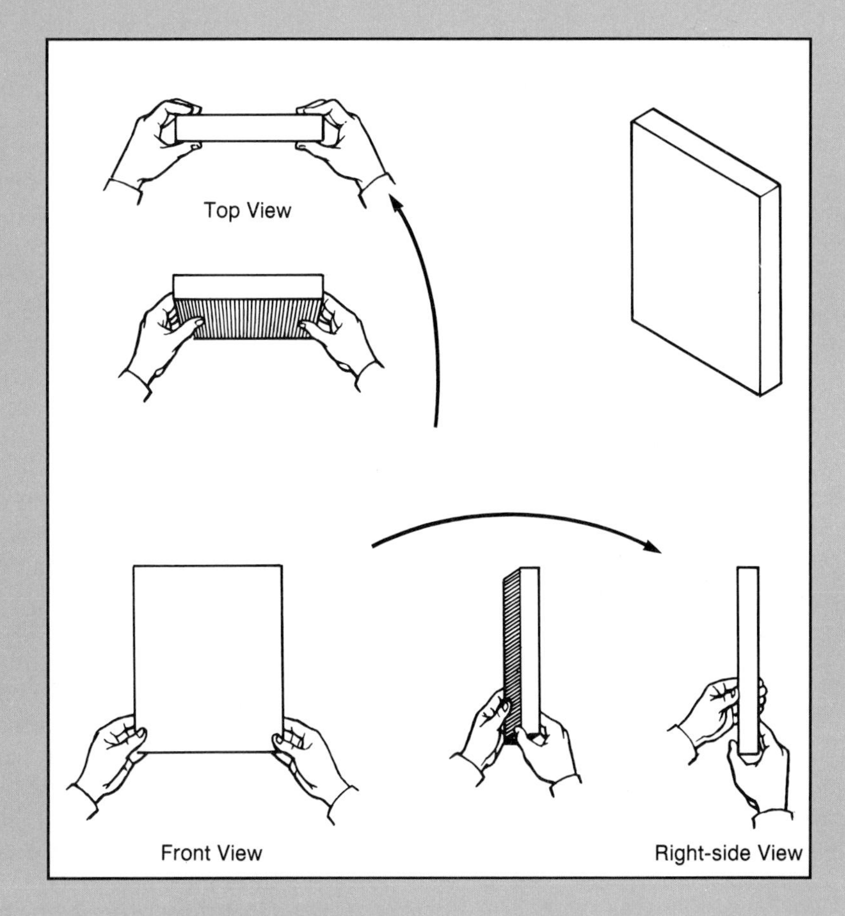

Source: Anthony Zipprich, Freehand Drafting for Technical Sketching *(New York, D. Van Nostrand Company, Inc., 1954).*

2. In the placement of orthographic views in a drawing, *a front view is always included;* as a result, the front view should be considered the *common view*, that is, the view that shares a common dimension with any view adjacent to it. The front view is the one about which other views may be placed in accordance with the rule of adjacency (Rule 3). (In the next section, we will discuss guidelines for choosing the front view of an object.)

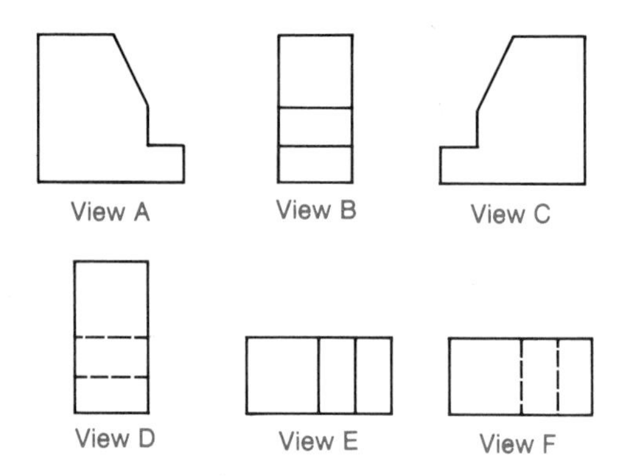

Figure 4.20 Six views of the object in Figure 4.2. Which view should be selected as the front view?

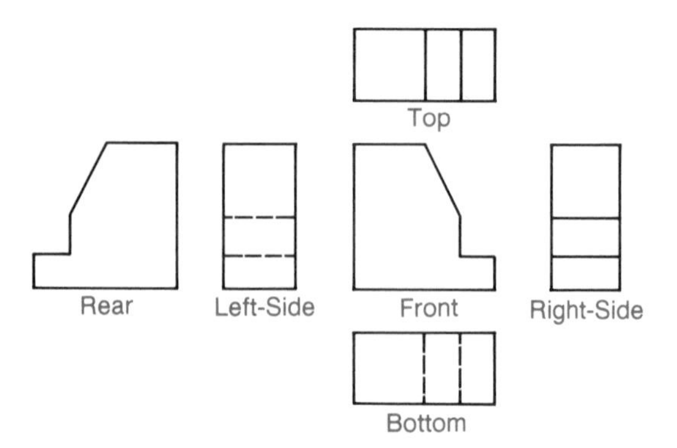

Figure 4.21 The resulting six principal views if view A in Figure 4.20 is selected as the front view.

4.3 Selecting the Best and Necessary Views

Figure 4.12 presents six orthographic views of the object shown in Figure 4.2; each view in Figure 4.12 is labeled in accordance with the rule of adjacency and with the aid of the glass box concept. (Once the front view of an object is chosen, all other views are automatically defined.) Figure 4.20 presents the same six orthographic views of this object; however, these six views are randomly shown in Figure 4.20—without any consideration for the rule of adjacency and without the label *front view* attached to any of the views. In this section we will discuss how one view is selected as the front view of an object. Of course, in the case of Figure 4.21, the reader knows which view will be ultimately selected, since the object is the same as the one in Figure 4.12; but now this choice will be justified as the most appropriate one, according to the following guidelines:

- The front view of the object should be the one that is commonly recognized as the front view if such a recognized view exists (for example, a television set has a commonly recognized front surface).
- The front view of the object should be the one that provides the most unambiguous (least confusing) description of the shape of the object. If two or more views provide such accurate information about the shape of the object, select the one that has the fewest hidden lines.

Applying these guidelines to the multiple views of the object shown in Figure 4.20, we note that there is no commonly recognized front view of this object. Most objects do not have such recognized views (and, of course, no object has the label *front surface* attached to one of its surfaces). Next we note that four of the views shown in Figure 4.20 (B, D, E, and F) are rectangular in shape, whereas two views (A and C) are seven-sided figures. Our second guideline therefore suggests that either view A or view C should be chosen as the front view of this object, since each of these views provides detailed information about the shape of the object (unlike views B, D, E, and F). But which view should we choose—view A or view C?

In order to choose between views A and C, we must consider the other views that will appear in our final set of orthographic views used to describe this object. There are similar guidelines for choosing secondary views (right-side, top, rear, and so forth) after one has selected the front view; it is usually not necessary to provide the reader with all six principal views of the object. Instead, one simply provides as many views of the object as is necessary to describe the object. The left-side and right-side views of an object often contain similar information about the shape of the object; if the information contained in these two views is *identical,* the right-side view is traditionally chosen. Similarly, if the top view and the bottom view of an object provide identical information about the shape of the object, then the top view is traditionally chosen. The rear view is included only if it provides additional information that is not clearly indicated in other views.

The word *identical,* as used in referring to orthographic views, means that each view provides the same information about the shape of the object with identical types of linework; if one view contains more hidden lines than the other view, these views are not identical. If one must choose between two views that provide the same amount of information about the shape of the object, but one view contains more hidden lines than the other view, one should select the view with the fewest hidden lines. As we will discover shortly, it is much more difficult for the reader to interpret the meaning of hidden lines in a drawing than to interpret visible linework; therefore, we wish to minimize hidden linework in our final set of orthographic views.

We may now state the guidelines for the selection of orthographic views in addition to the front view:

- Select those views (in addition to the front view) that provide the most unambiguous information about the shape of the object.
- Do not use more views than are necessary to describe the object.
- If two views are identical in terms of information and linework, select the *right-side view* instead of the left-side view and the *top view* instead of the bottom view—in accordance with tradition.
- If two views are identical in terms of information, but one view contains more hidden linework than the other, select the view with fewer hidden lines (in other words, *minimize hidden linework*).

We may now return to our earlier problem of selecting either view A or view C in Figure 4.20 as the front view of the object. When confronted with two views that appear to serve equally well as the front view, consider the additional views that will be associated with the selected front view. For example, if view C is chosen as the front view, then view B or view D will be selected as the corresponding side view. However, according to the guidelines for selecting additional views, view B is preferable to view D as a side view of the object because it contains fewer hidden lines. View B corresponds to the left-side view of the object if view C is the front view. Since we prefer to give the right-side view of the object instead of the left-side view—in accordance with tradition—and since the information contained in the selected views is identical (whether we choose view A or view C as the

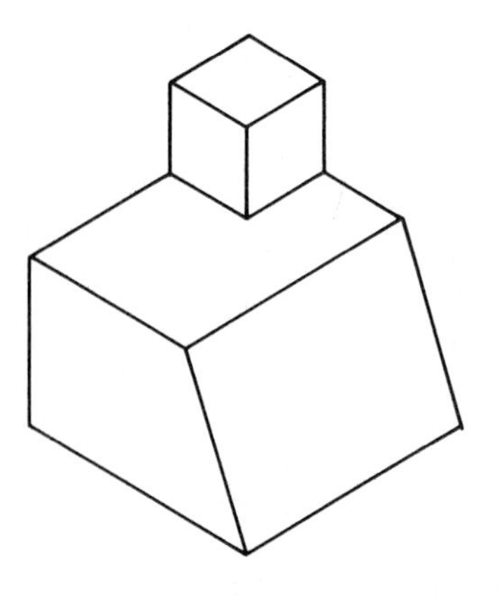

Figure 4.22 An example object to be the subject of successive figures in this chapter.

front view), we therefore choose view A as the front view, allowing the inclusion of view B as the right-side view.

Finally, view E will be the top view of the object in accordance with the rule of adjacency, and the concept of the glass box. Figure 4.21 shows the resulting six principal views of the object as defined in accordance with the choice of view A as the front view.

Learning Check

Figure 4.22 shows an object that will serve as the subject of this "Learning Check" and the next. Its corresponding six principal views are given *in random order* in Figure 4.23. Which of these six views should be selected as the front view of the object?

Answer Views E and F do not provide as much information about the shape of the object as the other four views, so they should not be selected as the front view. Views B and D are not as complex in shape as views A and C; the outlines of views B and D are composed of mutually perpendicular and parallel lines, unlike those of views A and C, which indicate there is an inclined surface on the object. View A contains less hidden linework than view C; as a result, view A should be selected as the front view of this object. Figure 4.24 identifies the resulting six principal views of the object, if view A is the front view.

Of the six principal views in Figure 4.24, which ones should be selected as necessary to describe this object and thus included in the final set of orthographic views?

Answer The front view must always be included among any final set of orthographic views of an object. Each of the side views in Figure 4.24 provides similar information about the shape of the object through outer boundary lines, but the right-side view contains less hidden linework than the left-side view, so the right-side view should be included in the final set of views.

A third view, in addition to the front and right-side views, should be included in order to unambiguously describe this object. The rear view provides no additional information to that contained in the front view, so it is eliminated. The top view and the bottom view do provide information about the shape of this object that will aid the reader in the interpretation of the front and right-side views; however, the information is the same in both top and bottom views, but the top view contains less hidden linework than the bottom view. Therefore, we choose the top view of the object as the third

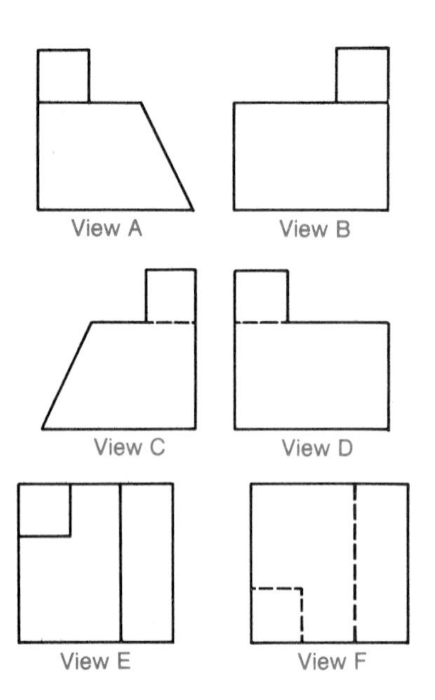

Figure 4.23 Six views of the object in Figure 4.22. Which view should be selected as the front view?

necessary view. Three views (front, top, and right-side) provide a complete description of the shape of this object. These are the three necessary (and best) views (Figure 4.25). Other views (rear, bottom, or left-side) are not necessary, since they will not provide any additional information about the object's form. (Some objects require more than three views in order to completely and unambiguously describe the shape of such an object, and others require less than three views, as will be described later in this chapter.)

In summary, then:

1. Guidelines for selecting a front view are:

- The front view of the object should be the one that is commonly recognized as the front view if such a recognized front view exists.
- The front view of the object should be the one that provides the *least confusing* description of the shape of the object. If two or more views provide such detailed information about the shape of the object, select the one that has the *fewest hidden lines.*

2. Guidelines for selecting views in addition to the front view are:

- Additional views should be the ones that provide the least confusing information about the shape of the object.
- If two views are identical in terms of information but one contains more hidden linework than the other, select the one with the fewest hidden lines (to minimize hidden linework).

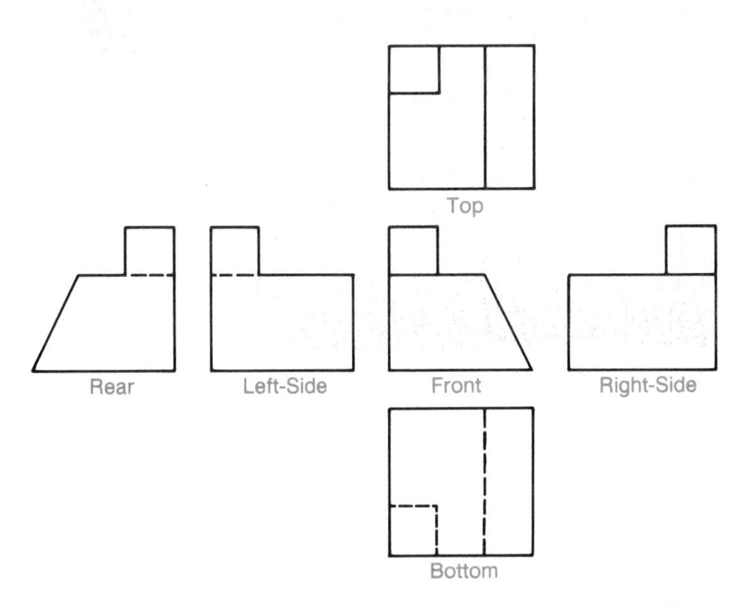

Figure 4.24 The resulting six principal views if view A in Figure 4.23 is selected as the front view.

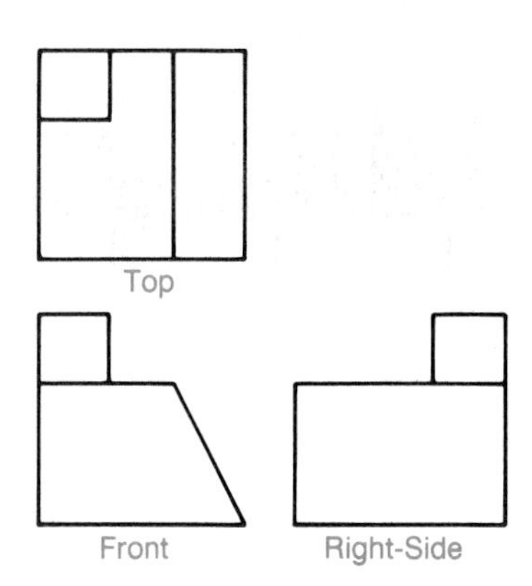

Figure 4.25 Necessary and best views for the object in Figure 4.24.

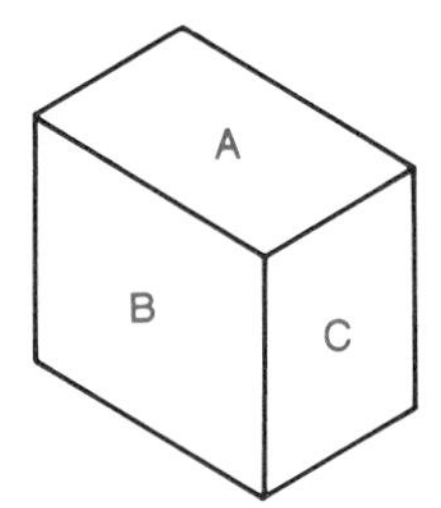

Figure 4.26 Planar surface parallel to a projection plane.

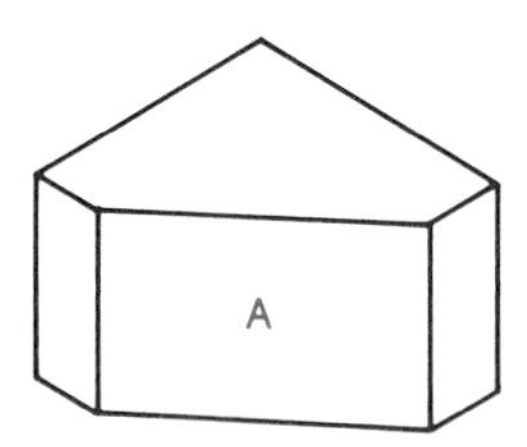

Figure 4.27 Inclined planar surface.

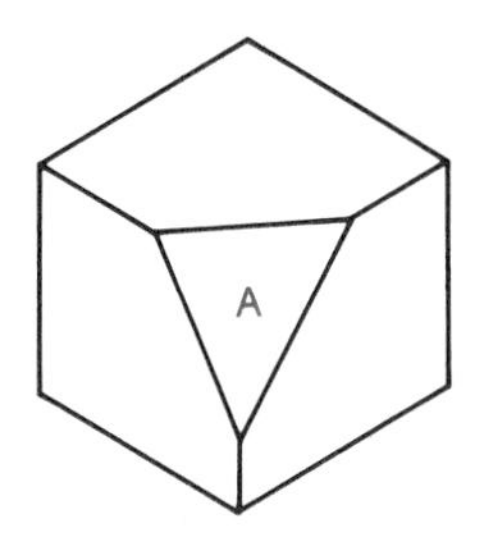

Figure 4.28 Oblique planar surface.

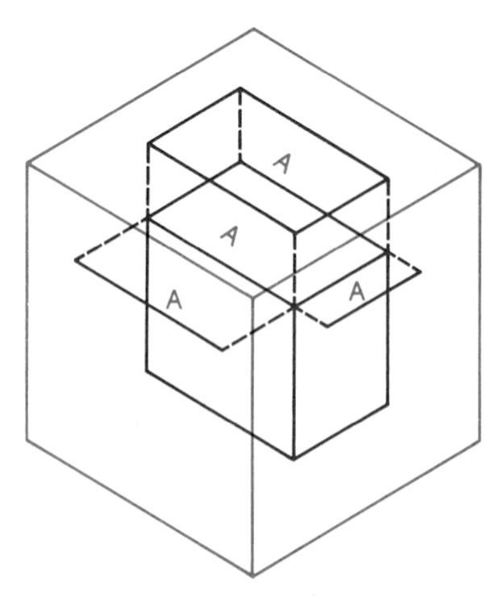

Figure 4.29 Projection of a surface that is parallel to a projection plane.

- If two views are identical in terms of information and linework, select the right-side view instead of the left-side view and the top view instead of the bottom view—in accordance with tradition.

3. Do not use more views than are necessary to describe a part.

4.4 Planar Surfaces

We can extend our understanding of orthographic projection and our ability to interpret orthographic views by dividing surfaces of an object into two categories: **planar** and **nonplanar.** Planar surfaces, as the name implies, are those contained within a single plane in space. In other words, planar surfaces are flat. Nonplanar surfaces curve or bend through space so that a single plane cannot be used to contain the entire surface. Planar surfaces will be discussed in this section and nonplanar surfaces in the next.

Planar surfaces can be classified into three subcategories: surfaces parallel to a projection plane, inclined surfaces, and oblique surfaces. Each type of planar surface results in a unique type of projection onto a set of reference planes.

Figure 4.26 shows a planar surface, A, that is **parallel to a projection plane,** in this case to the horizontal reference plane (HRP). Similarly, surface B in Figure 4.26 is parallel to the frontal reference plane (FRP), and surface C to the profile reference plane (PRP).

Figure 4.27 presents an object with a vertical surface, A, that is not parallel to any projection plane; however, surface A is perpendicular to the HRP. Surfaces that are inclined to two projection planes but perpendicular to a third projection plane are known as **inclined** surfaces. Thus, surface A in Figure 4.27 is an inclined surface.

Figure 4.28 shows an object with a surface, A, that is neither parallel nor perpendicular to any of the three projection planes; this type of surface is called **oblique.**

4.4.1 Surfaces Parallel to a Projection Plane

A planar surface that is parallel to one of the projection planes (such as surface A in Figure 4.26) will project onto that reference plane without distortion; that is, it will project as an area in true size. For example, Figure 4.29 shows the projection of surface A in Figure 4.26 onto the walls of the glass box; such projection results in one view (the top view corresponding to projection onto the HRP that is parallel to surface A) in which the projected surface A appears as an area identical in size to the original surface A. Projection onto the FRP and the PRP—projection planes that are perpendicular to surface A—results in surface A appearing *on edge*, as a line, in both the front view and the right-side view of the object. Figure 4.30 shows the results of projection of surface A onto the FRP, HRP, and PRP. Figure 4.31 presents the complete orthographic views that correspond to projection of the entire object onto the walls of the glass box.

We may summarize these results as follows:

Any surface that is parallel to a projection plane will appear as a *true-size* (and true-shape) area in the orthographic view corresponding to this projection plane; furthermore, this surface will appear on edge, as a line, in the views corresponding to those projection planes that are perpendicular to the surface.

Figure 4.30 Results for the surface projection in Figure 4.29.

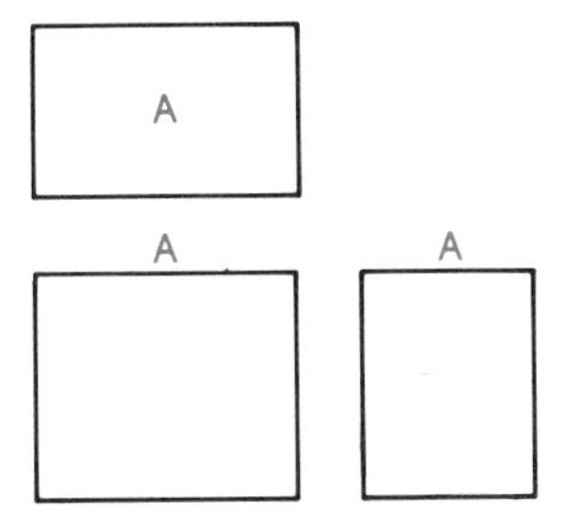

Figure 4.31 Complete orthographic views for the object in Figure 4.29.

4.4.2 Inclined Surfaces

A planar surface that is inclined to two projection planes and is perpendicular to the third projection plane (such as surface A in Figure 4.27) will project as a line onto that plane perpendicular to the surface. Furthermore, it will project as a **foreshortened** area (an area smaller in size than the true area of the surface) onto the planes to which it is inclined. Figure 4.32 shows the projection of an inclined surface onto the walls of the glass box; Figure 4.33 presents the results of such projection. Notice that surface A appears *on edge,* as a line, in the top view, corresponding to projection onto the HRP that is perpendicular to surface A. Surface A appears as foreshortened, or reduced, areas in both the front view and the right-side view, corresponding to projection onto the FRP and the PRP—both of which are inclined to surface A. Figure 4.34 presents the complete front, top, and right-side orthographic views for this object.

We may summarize these results as follows:

An inclined surface will project as a line onto that projection plane to which it is perpendicular. Furthermore, it will appear as a foreshortened, or reduced, area in those orthographic views corresponding to projection onto reference planes that are inclined to the surface.

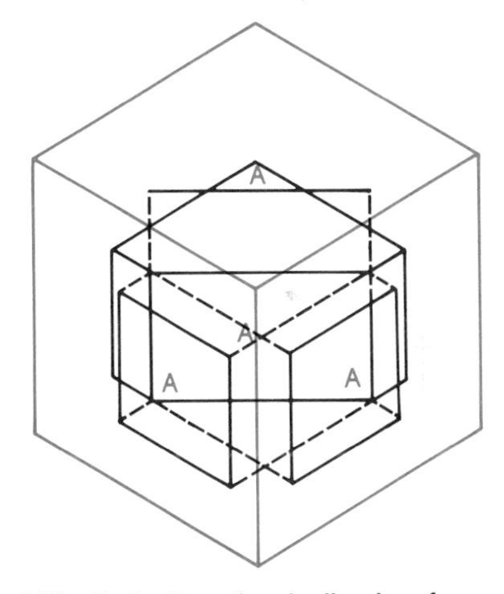

Figure 4.32 Projection of an inclined surface.

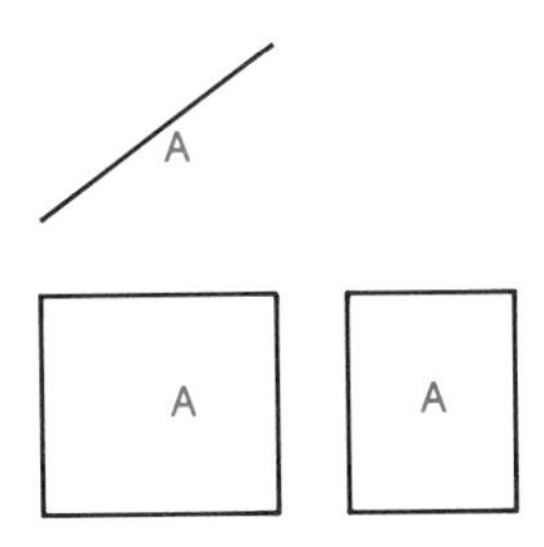

Figure 4.33 Results for the surface projection in Figure 4.32.

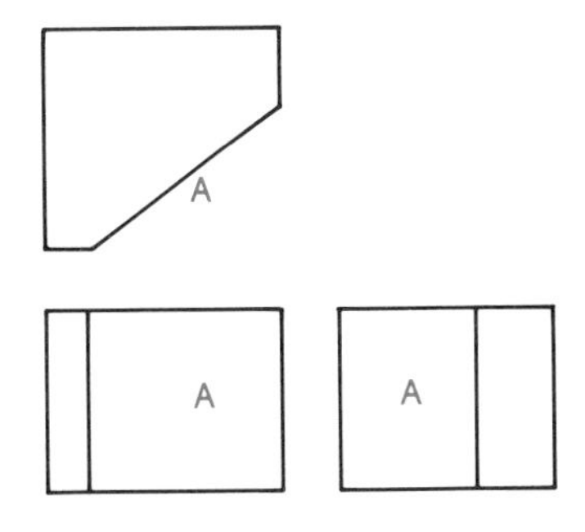

Figure 4.34 Complete orthographic views for the object in Figure 4.32.

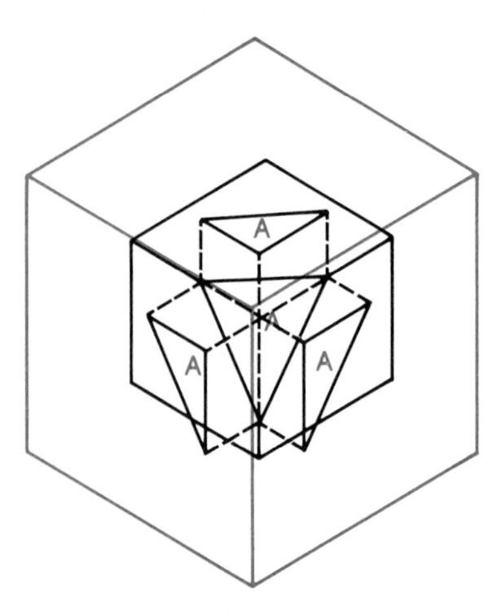

Figure 4.35 Projection of an oblique surface.

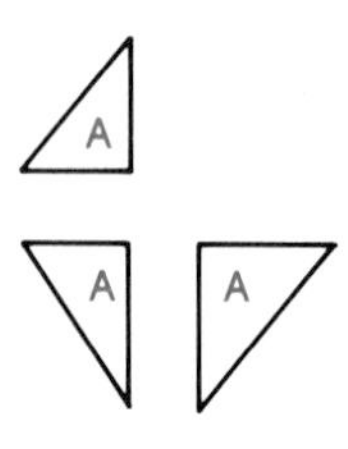

Figure 4.36 Results for the surface projection in Figure 4.35.

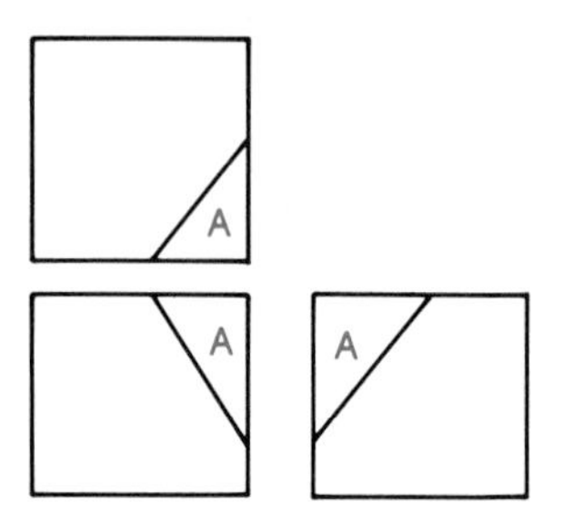

Figure 4.37 Complete orthographic views for the object in Figure 4.35.

4.4.3 Oblique Surfaces

A planar surface that is oblique (inclined to all projection planes) will be projected as foreshortened, or reduced, areas in all principal orthographic views. Surface A of the object shown in Figure 4.28 is an oblique surface; the projection of this surface onto the walls of the glass box is shown in Figure 4.35. As can be seen in Figure 4.36, surface A appears as an area in all (top, front, and right-side) views. Notice that each projected area in Figure 4.36 is smaller than surface A; in other words, surface A is distorted in the orthographic views because it is inclined to all three projection planes. Such distortion results in surface A appearing to be smaller in the resultant orthographic views than its actual size. Figure 4.37 presents the complete front, top, and right-side orthographic views for this object.

We may summarize these results as follows:

> An oblique surface will project as a foreshortened area in all principal orthographic views.

Learning Check

Match the appropriate type of planar surface with the geometrical form(s) that it may produce in a principal orthographic view:

___ Inclined surface	A. Will appear as a true-size area or as a line.
___ Surface parallel to a projection plane	B. Will appear as a line or as a foreshortened area.
___ Oblique surface	C. Will appear as a foreshortened area.

Answer Inclined surfaces may appear as either foreshortened areas or on edge, as lines. Planar surfaces that are parallel to a projection plane may appear as true-size areas or on edge, as lines. Finally, oblique surfaces appear as foreshortened areas in principal orthographic views. Therefore, the correct set of matching answers are:

_B_ Inclined surface
_A_ Surface parallel to a projection plane
_C_ Oblique surface

(If you did not match the descriptions with the appropriate types of planar surfaces, you should review the material in this section before continuing with this chapter.)

In summary, then:

1. Planar surfaces are those surfaces that are contained within a single plane in space.

2. Planar surfaces that are parallel to a projection plane will appear as true-size areas in the orthographic view that corresponds to that projection plane; these surfaces will appear on edge, as lines, in the views corresponding to the projection planes that are perpendicular to the surfaces.

3. Surfaces that are inclined to two projection planes but perpendicular to a third projection plane are known as inclined surfaces. An inclined surface will project as a line onto that projection plane to which it is perpendicular. Furthermore, it will appear as a foreshortened, or reduced, area in those orthographic views corresponding to projection onto reference planes that are inclined to the surface.

4. A surface that is neither parallel nor perpendicular to any of the three projection planes is called oblique. An oblique surface will project as a foreshortened area in all principal orthographic views.

4.5 Nonplanar Surfaces

Nonplanar surfaces curve, or bend, through space so that a single plane cannot be used to contain the entire surface. Figure 4.38 presents an object that contains nonplanar surfaces (surfaces A and B). As a result of such curving through space, a nonplanar surface may or may not be entirely visible to the observer from a single viewing direction. For example, if the object shown in Figure 4.38 is viewed from the top, one will see only a portion of surface A and a portion of surface B—the bottom portions of these nonplanar surfaces are hidden from the top-viewing observer. Similarly, the observer will see only portions of these surfaces when viewing the object from the right-side viewing direction.

Figure 4.39 shows two orthographic (front and right-side) views of this object. Surfaces A and B are labeled in each of these views. Notice that both surfaces A and B appear as circles in the front view but as rectangles in the right-side view. Only a portion of each surface is shown in the right-side view as a visible area because each surface curves through space and continues on the left side of the object.

Horizon lines (also known as **surface limit lines**) are identified in Figure 4.39; such lines represent *apparent* edges or boundaries of the nonplanar surfaces A and B. However, these lines are simply the viewing limits of these surfaces—beyond which we cannot see the remaining (left-side) portions of the surfaces.

Nonplanar surfaces may appear on edge, as lines, in orthographic views (such as the circles representing surfaces A and B in the front view of Figure 4.39); they may also appear as foreshortened areas or as *incomplete* areas where a portion of the surface is hidden from the viewer.

Nonplanar surfaces may be difficult to interpret in a given set of orthographic views because sometimes such sur-

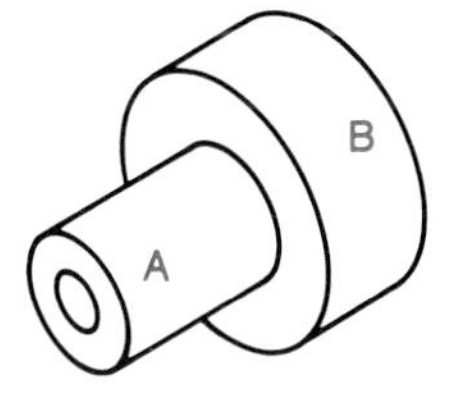

Figure 4.38 An example object containing nonplanar surfaces.

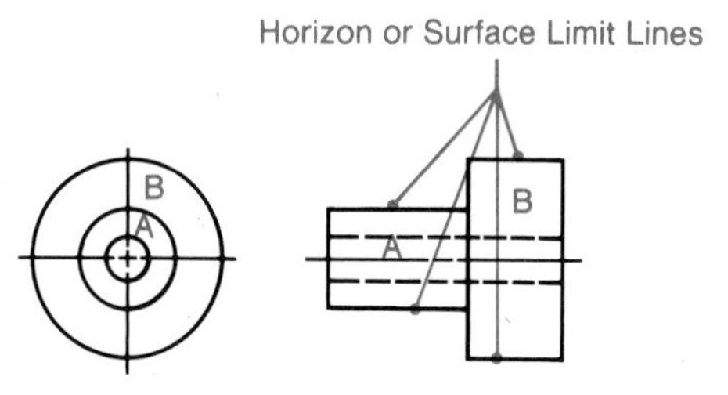

Figure 4.39 Orthographic views of the object in Figure 4.38.

faces extend in space beyond our viewing limit (beyond the horizon of the viewing direction for a given orthographic view). As a result, the reader of such orthographic views should initially focus on any planar surfaces of the object under consideration before considering the nonplanar surfaces. (The next section covers the topic of reading orthographic views.)

Figure 4.40 shows another example object that contains nonplanar surfaces (surfaces A and B). Figure 4.41 shows a set of orthographic views (front, top, and right-side views) for this object in which the nonplanar surfaces have been identified.

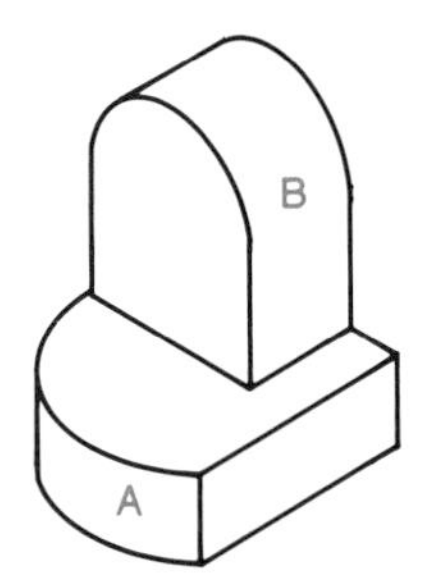

Figure 4.40 An example object containing nonplanar surfaces.

4.6 Reading Orthographic Views

Graphics is a language—it allows us to communicate ideas and concepts to one another. Like any language, graphics includes certain rules that must be obeyed. These rules help the reader interpret and assimilate all the bits of information contained within a graphic presentation (a set of orthographic drawings).

Reading orthographic drawings requires care and patience of the novice. When you first began to read English, you discovered that each letter and each word had to be considered *individually* and *then as a set* through which an idea could be communicated. The same is true of reading orthographic drawings. Furthermore, as you read this page of type, you do not anticipate which concepts or ideas will be presented before the page is read (except in the most general way); only after you have read the entire page do you know for certain which concepts have been discussed.

Similarly, we do not know what type of object is described within a set of orthographic views until we have analyzed and correlated all of the information contained in the views. It is a mistake to assume that the object is cylindrical or rectangular, for example, until one has correlated all of that information.

People have a natural inclination to "read" orthographic drawings before all of the information contained within the drawings has been analyzed—that is, to prematurely attempt to visualize the object. Such visualization, based on incomplete information, and thus possibly incorrect assumptions about the shape of the object, may delay the correct interpretation of the drawings and only slow down the reading process. After one develops skill in reading orthographic drawings through time and practice, of course, one can more quickly interpret the drawings and visualize the object.

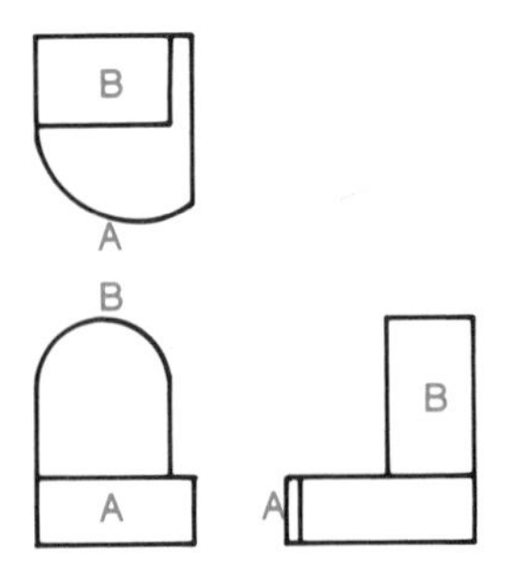

Figure 4.41 Orthographic views of the object in Figure 4.40.

Two guidelines for reading orthographic views are:

1. Do not visualize the object that is described in a set of orthographic views before you have analyzed and correlated the information contained in these views. (With sufficient practice, you will be able to perform such analysis and correlation so quickly that it may seem that you are "seeing" the object almost immediately.)
2. Remember that orthographic views of an object must be mutually consistent.

The second guideline means that since a set of orthographic views must be developed in obedience to the rules of common dimension, adjacency, and alignment that were discussed earlier in this chapter, they should be read with those rules in mind. Furthermore, the information contained in the drawings about the object should be consistent, so that one can identify each surface of an object in each orthographic view of that object. In some views a given surface may appear as a visible area, while in others it appears as a hidden area, a visible line, or a hidden line. In all views, however, one should be able to identify each surface of an object as an area (visible or hidden) or as a line (visible or hidden). All of these identifications must then be correlated into a consistent whole.

Consider Figure 4.42, in which a given front view (a simple rectangle) is shown, together with a set of additional orthographic views (A through H). Each of the views A through D are *possible* right-side views that are consistent with the given front view, whereas views E through H are inconsistent with the front view. (Of course, each of the consistent views A through D correspond to separate and distinct objects.)

Figure 4.43 presents the given front view together with the consistent right-side views A through D. There are many other possible right-side views that would be consistent with the given front view; in fact, we are limited only by our imagination in the number of possible right-side views that can be formulated for this given front view.

Figure 4.44 presents the original given front view of Figure 4.42, together with the inconsistent right-side views E through H. In addition, Figure 4.44 shows the objects that correspond to the inconsistent right-side views, so that one can see the front views that do correspond to (are consistent with) these right-side views.

Compare each of these front views with the original front view. Notice that each of the new front views contains linework not contained in the original front view. Each of the

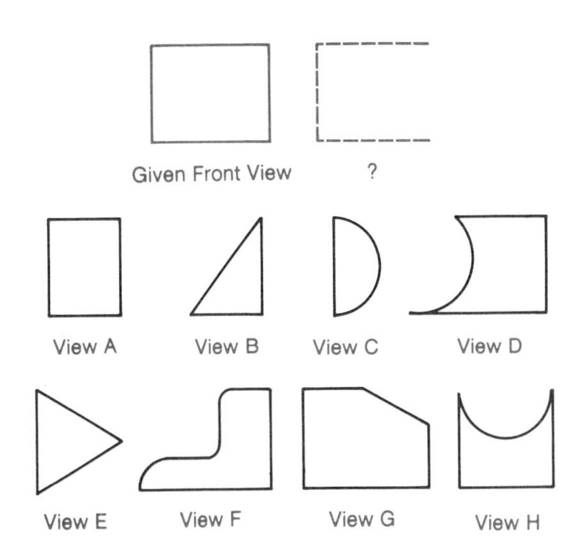

Figure 4.42 Consistent and inconsistent right-side views for a given front view.

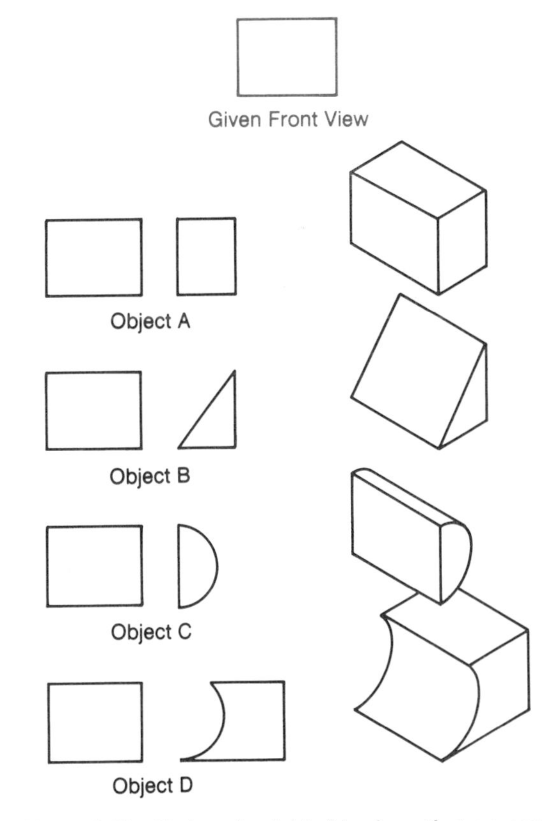

Figure 4.43 Choices for right-side views that are consistent with the given front view.

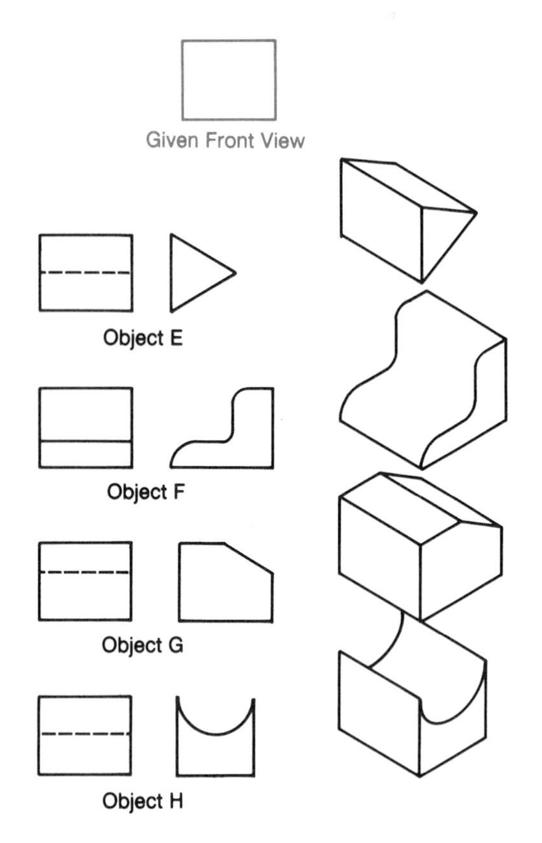

Figure 4.44 Choices for right-side views that are inconsistent with the given front view.

views E through H indicate that (1) surfaces are intersecting at a height location between the top and the bottom of the object or that (2) a nonplanar surface changes direction at a height location between the top and the bottom of the object. In either case, linework must appear within the corresponding front view if the two views are to be mutually consistent, since such linework represents the common boundary edge between intersecting surfaces, or the horizon line of curvature for the nonplanar surface. Such interior linework **(in-lines)** is absent in the original front view of Figure 4.42; thus, this original front view and the views E through H of Figure 4.42 are inconsistent.

Unlike outer boundary lines—which clearly delineate the extent of the object in a given direction—in-lines can be very difficult to interpret. One of the difficulties associated with in-lines is that they indicate the separation of surfaces in an orthographic view, but without indicating which surface is nearer to the reader. For example, the given front view in Figure 4.45 (consisting of a square in which a circle is drawn) can have four different interpretations. Object A (the first interpretation) is a cube in which a hole exists; the circle contained in the front view represents the boundary of the hole. The circle of the front view represents a cylindrical extension in objects B and C and the boundary of a hole in object D. The circle and outer boundary lines of the front view hide the rear portion of object D, which includes a cylindrical joint and a triangular prism.

Figure 4.45 reminds us that we must read all of the given orthographic views in order to determine the shape of an object under consideration. Obviously, we could develop an endless variety of possible (consistent) right-side views to correspond to the given front view in Figure 4.45.

Figure 4.46 presents a set of given front and top views; the right-side view is not shown. Surfaces A, B, C, and D are identified in this figure. We know that surfaces A and B are different surfaces because they are separated by interior linework in the given front view. Furthermore, we know that surfaces C and D are separate and distinct by the same type of reasoning. However, we do not know whether surfaces A and C are distinct surfaces or whether they are, in fact, the same surface. The width dimension (the extent of the surface from the right side of the object toward the left) is identical for both surface A and surface C. Furthermore, both surface A and surface C are rectangular in shape. Since surfaces A and C have the same dimension and the same general shape, they may or may not be the same surface seen as an area in both the top and front views. For the same reasons, we do not know if surfaces B and D are different surfaces or a single surface.

Learning Check

Why have we considered only the width of surfaces A, B, C, and D in the preceding discussion and ignored the dimensions of height and depth?

Answer We focused upon a comparison of the extent of each surface in the width direction because the width is the dimension that is *common* to the front and top orthographic views. It is therefore noteworthy that a surface that appears as an area in both the front and top views has the same width in these views.

Figure 4.47 shows four possible interpretations of the given front and top views in Figure 4.46. Figures 4.48 through 4.51 present sets of orthographic views that correspond to the four different interpretations of the given top and front views in Figure 4.47. Carefully review each of these interpretations and try to identify each surface of the particular object in each orthographic view.

It bears repeating that an incomplete set of orthographic views cannot completely identify an object under consideration. The front and top views shown in Figure 4.49 do not define a unique object, as proved by the various possible interpretations in Figures 4.48 through 4.51. Similarly, incomplete consideration of the information contained in a set of orthographic views may lead to an incorrect interpretation; one must both assimilate and correlate *all* of the information contained within a complete set of views for accurate interpretation.

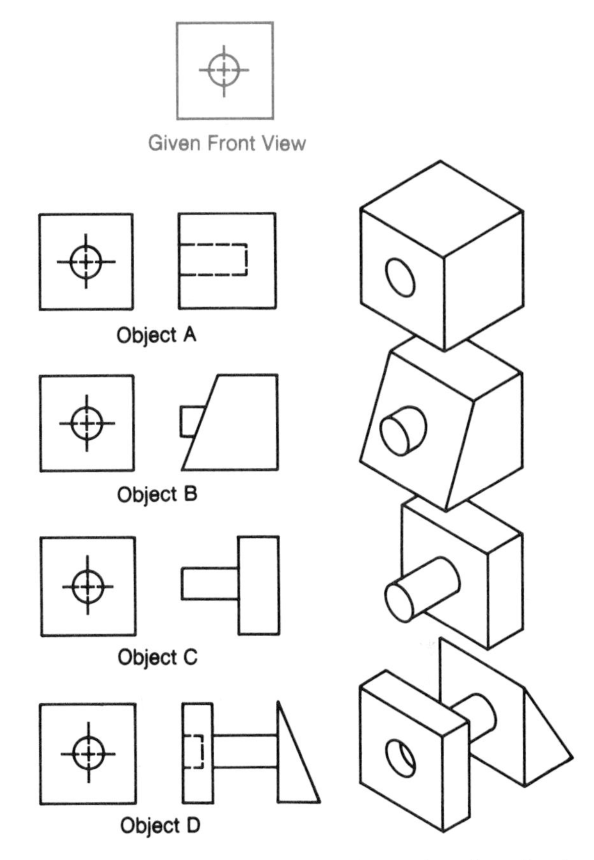

Figure 4.45 Multiple interpretations of a given front view.

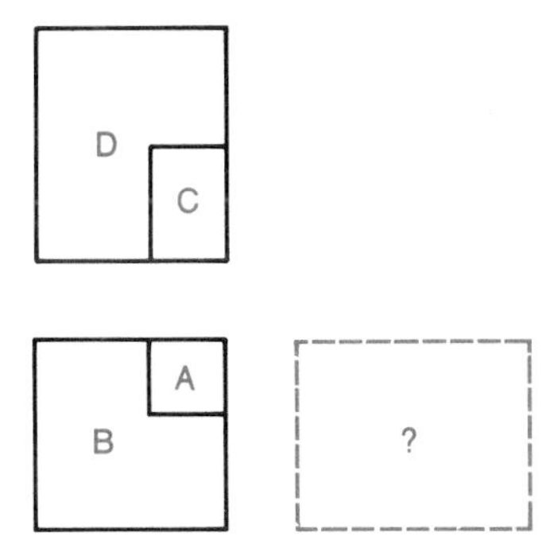

Figure 4.46 Two orthographic views (front and top). The right-side view is not shown. These views are the subject of successive figures in this chapter.

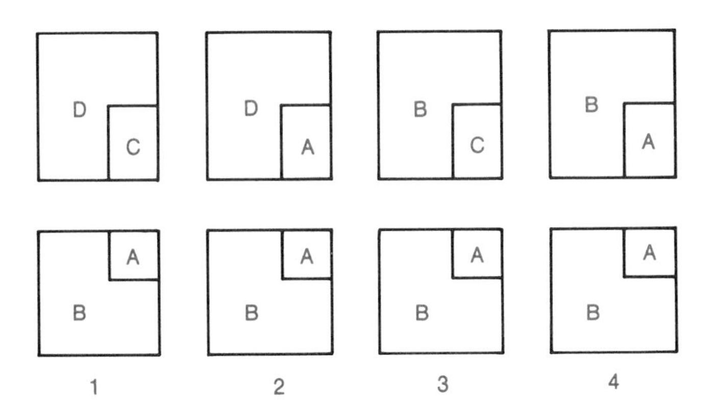

Figure 4.47 Multiple interpretations of the views in Figure 4.46.

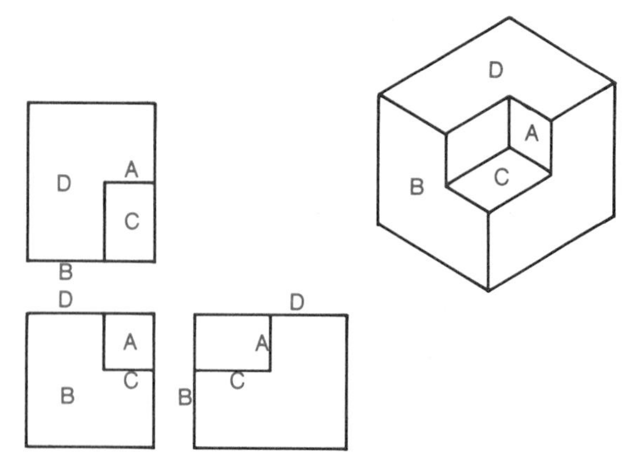

Figure 4.48 Interpretation 1 from Figure 4.47.

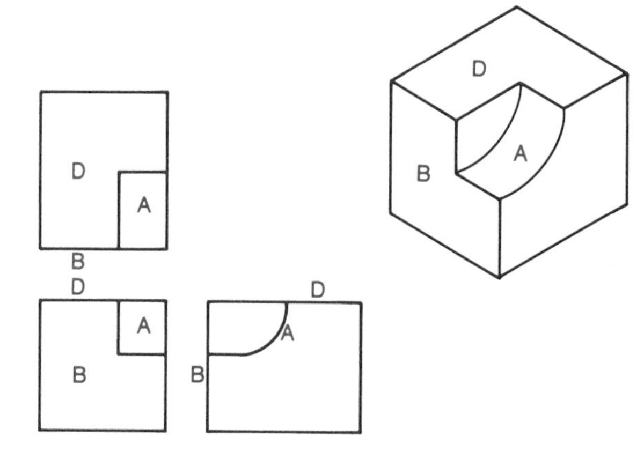

Figure 4.49 Interpretation 2 from Figure 4.47.

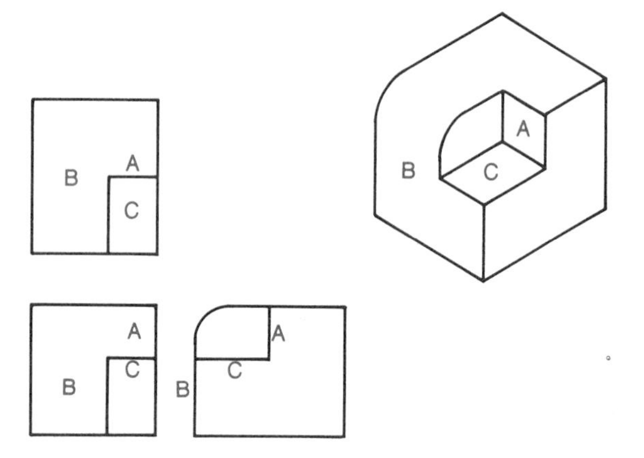

Figure 4.50 Interpretation 3 from Figure 4.47.

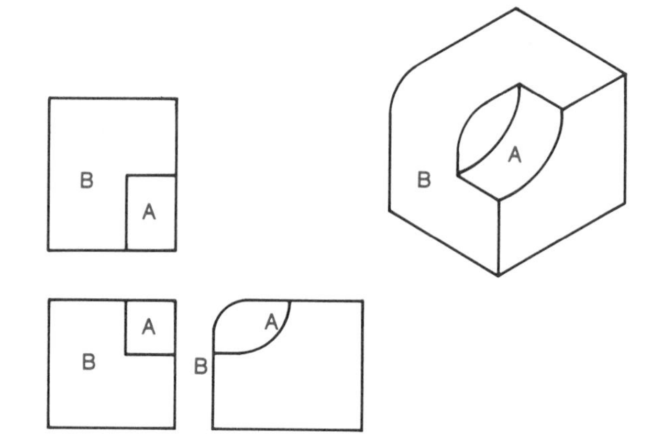

Figure 4.51 Interpretation 4 from Figure 4.47.

Learning Check

Is it possible for the areas A and B in Figure 4.47 to represent the same surface?

Answer Areas A and B are separated by in-lines, or interior linework; if these areas represented a single, unique surface of the object, there would be no boundary lines separating the areas in any view. Therefore, areas A and B must represent different surfaces.

In summary, then:

1. Graphics is a language that includes certain rules for proper communication between professionals.

2. The reader of a set of orthographic views should analyze and correlate all of the information contained in the views. The reader should resist the temptation to visualize the object described in a set of orthographic views until all information has been analyzed and correlated.

3. Orthographic views of an object must be mutually consistent.

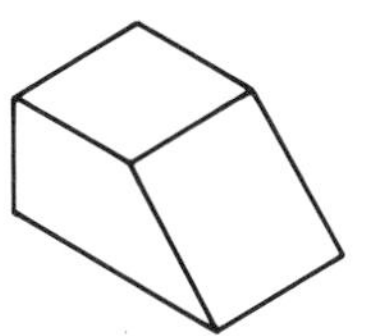

Figure 4.52 An example object to be the subject of successive figures in this chapter.

4.7 Surface-by-Surface, Point-by-Point Technique and Cut-by-Cut Technique

Several techniques can be used to help us identify and correlate the information contained in a set of orthographic views and thus develop the correct interpretation. The **surface-by-surface, point-by-point** technique requires one to identify *each* surface of the object described by a set of orthographic views in *each* view, one surface at a time. As a surface is identified in each view, key points on that surface are also identified in all views. This careful approach minimizes errors in interpretation. Application of this approach is greatly aided by the **configuration rule.**

> *The configuration rule:* A planar surface (a surface contained within a single plane in space) that appears as an area with *n* sides or boundaries in one orthographic view will appear either as an area with *n* sides or on edge, as a line, in all other orthographic views.

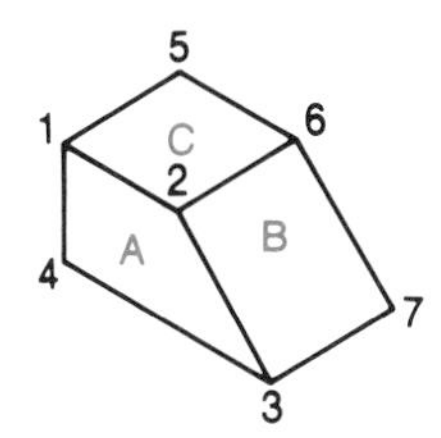

Figure 4.53 Identification of surfaces and points.

The configuration rule means, for example, that a triangular surface will always appear as a triangle (or on edge, as a line) —it will never be seen as a rectangle, hexagon, or any area of a type that is not triangular. Consider slowly rotating a rectangular sheet of paper away from you; it will slowly change its apparent size, but it will always appear as a quadrangle—until it is eventually seen on edge, as a line. The configuration rule is extremely useful when one is attempting to identify surfaces in various orthographic views.

A continuation of the configuration rule states that if two edges of a surface are parallel in one orthographic view, then these edges will also be parallel in all views in which the surface appears as an area.

Let us reconsider the process of projection. Figure 4.52 shows an object that we will describe in a set of orthographic views. In order to develop these views, identify three surfaces (A, B, C) of the object, together with some key points, as shown in Figure 4.53. Projection of the object onto the frontal reference plane (FRP) then proceeds as shown in Figure 4.54. Initially, points 1, 2, 3, and 4 are projected onto the

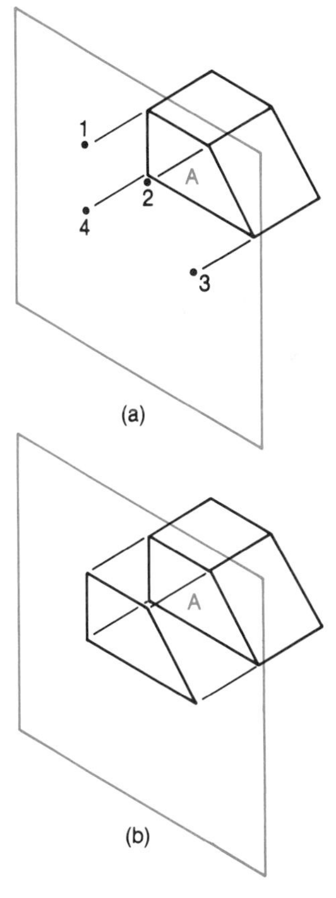

Figure 4.54 Projection of surface A onto the frontal reference plane (FRP). First points are projected (a), then connected by lines corresponding to the boundary edges of surface A.

FRP (a in the figure), after which the projections of these points are connected by lines corresponding to the boundary edges of surface A (b). Surface A then appears in the projection as an area. Surfaces B and C (not labeled in this figure) appear in this projection as lines (line $\overline{12}$ and line $\overline{23}$, respectively); that is, they appear on edge in the projected front view because they are perpendicular to the FRP.

Figure 4.55 presents the projection of the object onto the horizontal reference plane (HRP). Points 1, 2, 3, 5, 6, and 7 are projected onto the HRP, then connected by linework that corresponds to the boundaries of surfaces B and C on the object. Surface A (which has key corner points 1, 2, 3, and 4) appears on edge, as a line, in the projected view since it is perpendicular to the HRP.

Projection of the object onto the profile reference plane (PRP) is shown in Figure 4.56. Points 2, 6, 3, and 7 are projected onto the PRP, then connected by linework corresponding to the boundary edges of surface B. Surfaces A and C, both of which are perpendicular to the PRP, appear on edge, as lines, in the projected view.

Learning Check

How does one determine which key points should be projected onto a given reference plane? (Consider Figures 4.54 through 4.56 and identify the characteristic that is shared by the points chosen for projection onto a given reference plane.)

Answer One projects the key points of the object that are nearest to the reference plane under consideration and that are necessary in order to fully describe the object (and all of its surfaces) in the projected view. Some points that are distant from the reference plane may need to be projected in

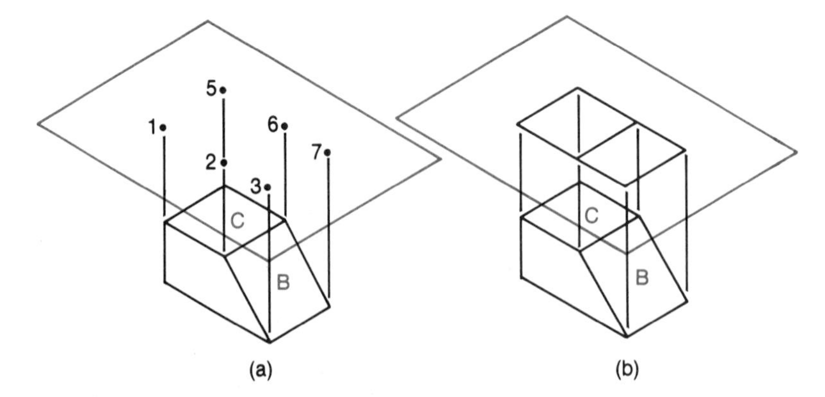

Figure 4.55 Projection of surface B and surface C onto the horizontal reference plane (HRP). First points are projected (a), then connected by lines corresponding to the boundary edges of surfaces B and C.

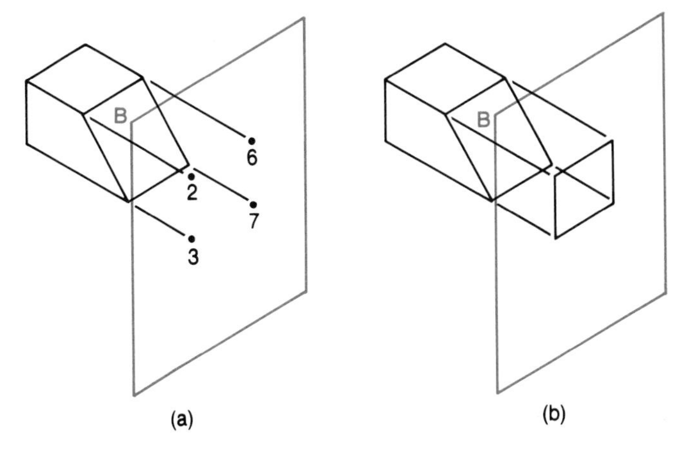

Figure 4.56 Projection of surface B onto the profile reference plane (PRP). First points are projected (a), then connected by lines corresponding to the boundary edges of surface B.

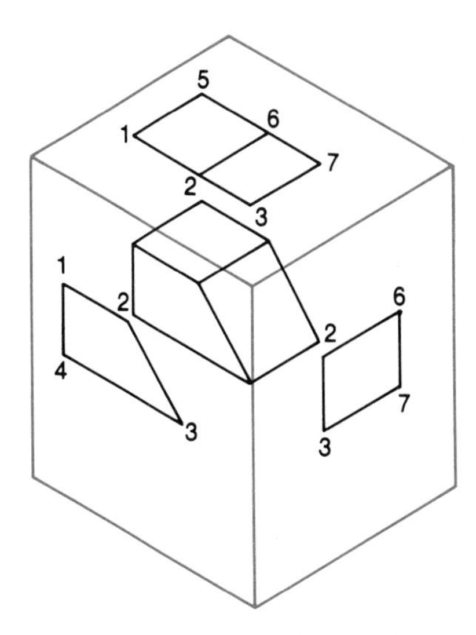

Figure 4.57 Glass box formed from projection planes.

order to describe hidden surfaces of an object in a projected view (such distant points were not needed in the projection work shown in Figures 4.54 through 4.56).

Figure 4.57 presents the glass box projections that correspond to the work developed in Figures 4.54 through 4.56. Notice that the line $\overline{23}$, connecting points 2 and 3, appears in all three (front, top, and right-side) projected views. Line $\overline{23}$ corresponds to the front, top, right-side boundary of the object; hence, its appearance in all three projections.

Figure 4.58 shows the opening of the glass box, and Figure 4.59 shows the final result of such opening with the three projected views of the object contained within a common plane (the plane of the paper), together with the **fold lines,** or boundary edges, of the glass box. (The inclusion of fold lines, as noted earlier, will aid us in our development of auxiliary orthographic views.)

Figure 4.60 shows the three projected views without fold lines. Figure 4.61 presents these views without the identification of key points. (Such point identification is only needed for the *development* of the views; it should not be included in the final drawings.)

> To enhance your understanding, consider *reversing* the order of Figures 4.53 through 4.61. Beginning with the set of orthographic views, identify key points and develop an interpretation that coincides with the form of the object described.

In order to better explain the surface-by-surface, point-by-point technique, we will start with an incomplete set of orthographic views and use this technique to develop the

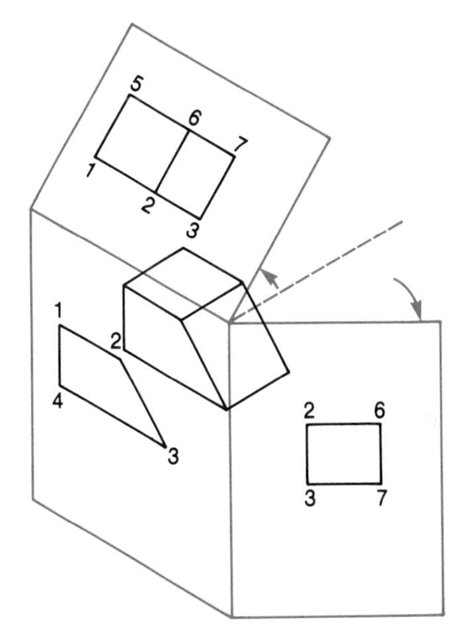

Figure 4.58 Opening the glass box.

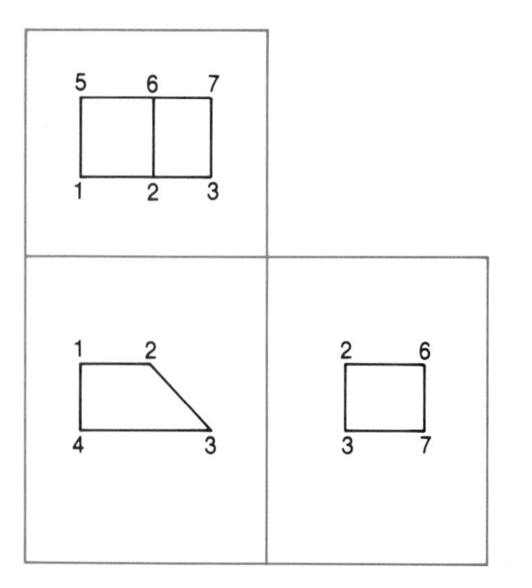

Figure 4.59 Projection results of opening the glass box in Figure 4.58, including fold lines.

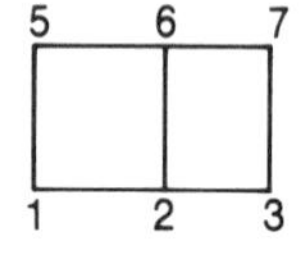

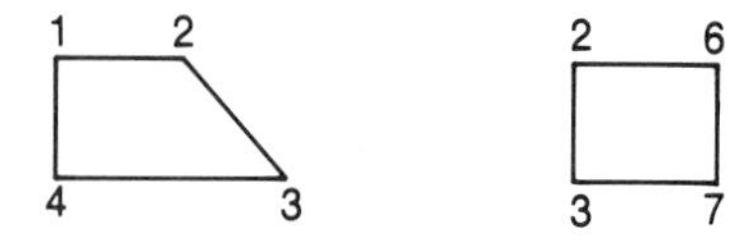

Figure 4.60 Projection results of opening glass box in Figure 4.58 without fold lines.

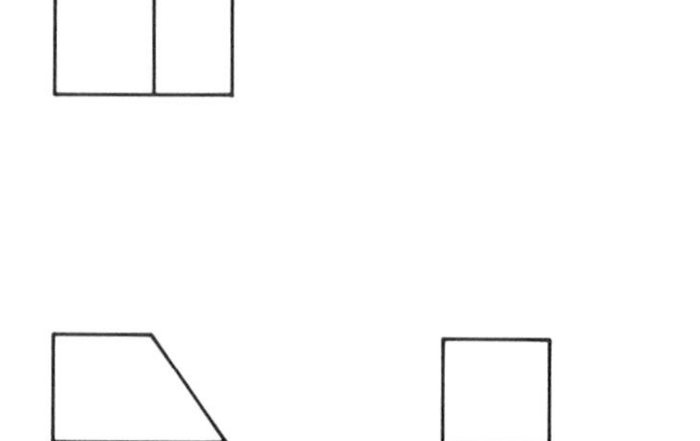

Figure 4.61 Final set of orthographic views from Figure 4.58 without key points.

missing view. In Figure 4.62 a top view is developed by the surface-by-surface, point-by-point technique to be consistent with a given front view and a given right-side view of an object (a). The initial step is to identify the surfaces shown in the given views (b). Surface A appears as a triangular area in the given front view, whereas surface B appears as a four-sided area in the given right-side view. (We have also included the FRP—which appears on edge, as a line—in the right-side view in order to aid our *construction* of the missing top view; notice that the FRP is vertical and contains the portion of the right-side view that is nearest to the given front view—again, in order to aid our construction effort.)

We must now locate surface A in the right-side view and surface B in the front view. The configuration rule states that A and B must be distinct (different) surfaces if we assume that they are planar. Surface A appears as a three-sided area in the front view; if it is assumed to be planar, it cannot appear as a four-sided area (labeled B) in the right-side view. That is, A and B must be different surfaces, as we stated earlier.

One assumes that the surface(s) under consideration is (are) planar in order to apply the configuration rule. If we are *then* confronted with an inconsistency in our development, we may need to reevaluate our initial assumptions regarding the planar characteristics of the object.

Returning to Figure 4.62, we must now locate surface A in the given right-side view. Application of the rules of common dimension, alignment, and adjacency (Rules 1, 2, and 3 discussed earlier in this chapter) leads us to the assumption that surface A appears as the inclined straight line nearest to the FRP in the right-side view (c). Furthermore, similar application of Rules 1, 2, and 3 lead us to the assumption that surface B appears in the front view as the inclined line that is nearest to the right side of this view. (These identifications must be recognized as assumptions, not conclusions, until after we have interpreted all of the information contained in the given views in a complete and consistent manner.)

The identification of surfaces A and B as *straight* lines in the given views is consistent with the initial assumption that these surfaces are *planar;* if either surface were identified in a given view as a curved line, then that surface would need to be labeled nonplanar (or one would need to reevaluate the initial assumptions that led to such an identification).

Figure 4.62(c) shows the FRP as a horizontal construction line in that area of the drawing in which the top view is to be drawn. The FRP indicates the portion of the top view that will be nearest to the given front view, in accordance with the rule of adjacency and the identification of the FRP in the given right-side view. We have placed the FRP in the

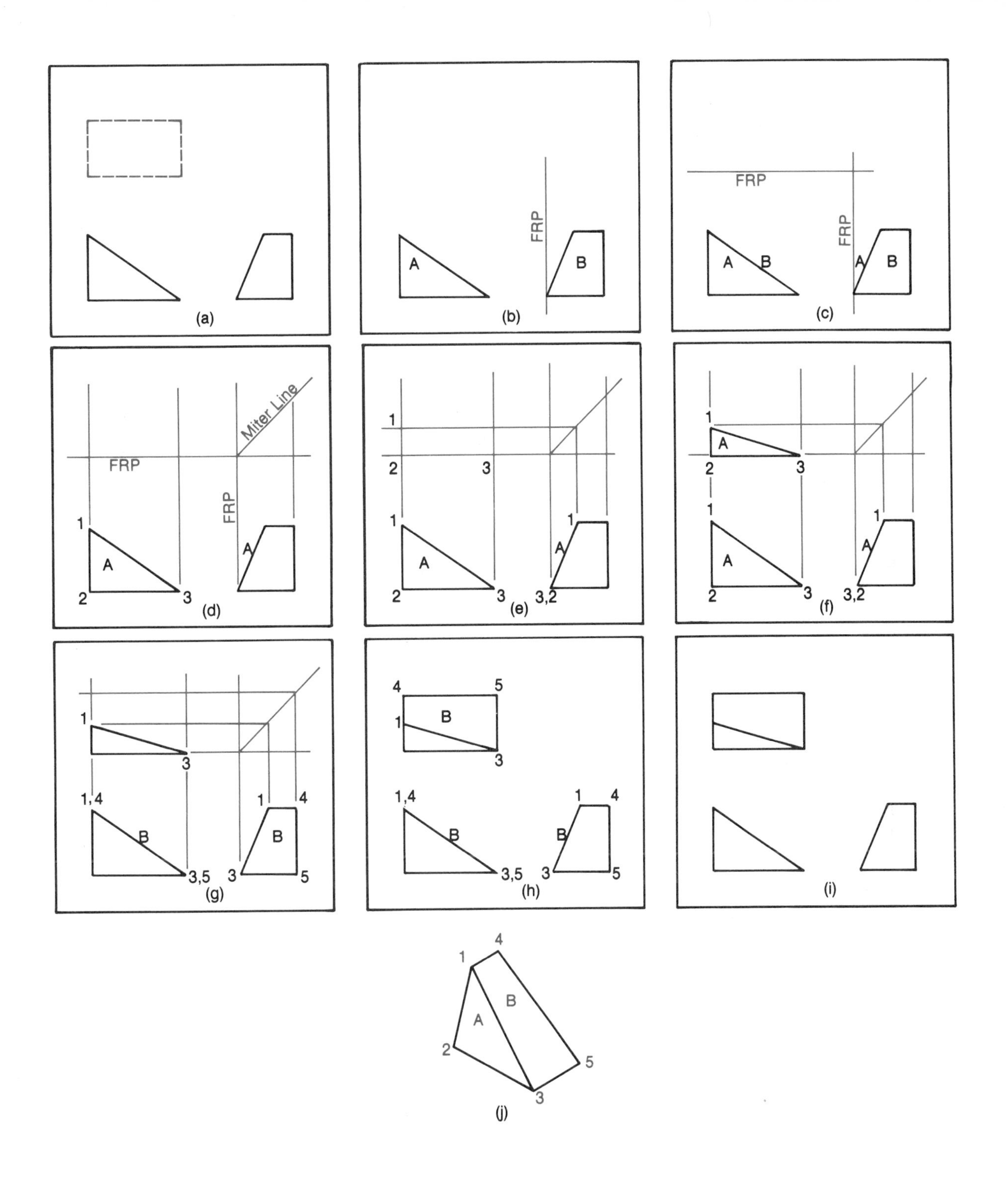

Figure 4.62 Surface-by-surface, point-by-point development of a top view consistent with given front and right-side views (a). This technique (b–h) leads to a complete set of orthographic drawings (i) for the object under consideration (j).

top-view construction area in such a way that it lies at a (vertical) distance above the front view, which is identical to the (horizontal) distance separating the FRP in the right-side view from the given front view. Simply stated, we locate the FRP in the top-view construction area so that the distance between the top and front views will be identical to the distance that separates the given front and right-side views. This type of symmetrical placement of the orthographic views will produce a pleasing balance in the final drawings.

Next, we identify three key points (1, 2, 3) defining surface A (d). If we can now locate these three key points in the given right-side view, we will know the height, width, and depth of each point. Points 1, 2, and 3 can then be properly located in the top-view construction area (in which the dimensions of width and depth are used). Finally, these three points can be connected by linework so that surface A will be completely specified in the developing top view.

In order to develop surface A in the top view, we introduce a **miter line** (d). A miter line passes through the intersection of the two construction lines that represent the FRP in the (given) right-side view and the (to-be-constructed) top view, as shown in the figure. Furthermore, a miter line is located at a 45° angle relative to the (vertical and horizontal) FRP construction lines. The miter line then allows one to transfer depth information from the right-side view to the top view and vice versa.

We do not recommend that the reader use the miter-line technique for transferring depth information from one view to another. It is better to use a set of dividers or a compass to transfer depth data, since they require less time and produce less construction work (which must eventually be erased from the drawing). We have chosen to use the miter-line technique in this section only because this technique illustrates the concept of depth transference with greater clarity than the alternative approaches in which a compass or a set of dividers is used. However, the use of dividers and the compass for this task will be reviewed shortly; in addition, the reader should refer to Sections 2.5 and 2.6 for a general discussion of this equipment.

The use of the miter line is shown in part e of Figure 4.62. Points 1, 2, and 3 have been located in the given right-side view. Each of these three points, by definition, must lie on surface A; therefore, they must lie on the line that represents A in the right-side view. Furthermore, the rule of common dimensions states that since point 1 is shown to lie on the top of the object in the front view, it must also be located at the top of the object in the right-side view. Point 1 must therefore be located at the top of the line that represents surface A in the right-side view, as shown in the figure. Similar analysis results in the location of points 2 and 3 at the bottom of the line that represents surface A in the right-side view.

Learning Check

What is the common, or shared, dimension of the front and right-side views?

Answer Height is the common dimension that is shared by the front and right-side views. We have used this common dimension in these two views to locate points 1, 2, and 3 in Figure 4.62(e); point 1, for example, cannot change its height position, so it must lie at the top of the object in the right-side view as it does in the given front view.

Returning to Figure 4.62(e), note that both points 2 and 3 are located at the same height (the bottom of the object); as a result, these points *appear* to be located at the same position in the right-side view on the line representing surface A. Point 2 is hidden in the right-side view by point 3; that is, one sees point 3 when viewing the object from the right-side viewing direction, but point 2 cannot be seen since it lies on the left side of the object. In order to signify this fact, we have introduced the notation 3, 2, which specifies the location of these two points in the right-side view as well as the relative distance from the right-side view observer of each point. This notation indicates that point 3 is visible in the right-side view and that point 2 lies at a greater distance from the observer (point 3 hides point 2 from view).

The notation a, b, c indicates that point a is the visible or nearest point of the object at a particular location in an orthographic view, followed by the hidden point b, which is located at a greater distance from the viewer than point a; and finally, that point c lies at the same location as points a and b in the view but is farthest from the viewer. This notation aids the development of the surface-by-surface, point-by-point technique.

The next step, in Figure 4.62(e), is to locate points 1, 2, and 3 in the top-view construction area. The *width* positions of these three points are projected vertically from the front view; points 1 and 2 both lie on the left side of the front view, so they must also lie on the left of the top view. Similarly, point 3 lies on the right side of the object, so it must be located on the right side of both the front and top views.

The *depth* positions of these three points are then obtained from the right-side view. The miter line allows us to first project the depth information vertically from the right-side view, followed by horizontal projection from the miter line to the top-view construction area. (Once again, a set of dividers can be used to quickly transfer distances, such as the depth of a point from the FRP, about the drawing.) Points 2 and 3 are located on the FRP in the right-side view; these points must also lie on the FRP in the top view, as shown in the figure. Point 1 lies at a depth distance from the FRP in the right-side view; it must be located at the same depth distance from the FRP in the top view. Figure 4.62(e) shows the proper placement of these points in the top-view construction area.

Points 1, 2, and 3 are connected by straight lines in the front view to form the triangular surface A. Surface A has been assumed to be planar, so the edges of this surface must in actuality be straight lines if they appear as straight lines in an orthographic view showing the surface as an area (as is the case for surface A in the front view). Points 1, 2, and 3 should then be connected by straight lines in order to form a triangular area representing surface A in the top view, as shown in part f of Figure 4.62. Surface A has now been completely determined in all three views, including the partially constructed top view. We now turn our attention to surface B.

Part g in Figure 4.62 shows that surface B can be defined in the given right-side view by straight lines connecting the four points labeled 1, 3, 4, and 5. Surface B has been assumed to be represented in the front view by the inclined straight line lying near the right side of this view, in accordance with the rule of adjacency and the configuration rule. Points 1 and 4 are located at the highest position of the right-side view; these points must also be located at the top of the line representing surface B in the front view. The notation 1, 4 in the front view indicates that point 1 is nearer to the front-view observer than point 4. Similarly, points 3 and 5 have been located at the bottom of the line representing surface B in the front view, with the notation 3, 5 indicating that point 3 lies nearer the front-view observer than point 5.

Points 1 and 3 have been located in the top-view construction area as a result of our earlier work in the development of surface A. We must now locate points 4 and 5 in the top-view construction area. The miter line allows us to transfer the depth locations of these two points from the given right-side view to the top view construction region. Both point 4 and point 5 lie at the rear of the right-side view, so they must also lie at the rear of the top view. Part h of Figure 4.62 shows the location of points 1, 3, 4, and 5 in the top view; these points are then connected by linework in order to form the completed four-sided area representing surface B in the top view.

Part i of Figure 4.62 presents the completed set of three (front, right-side, and top) orthographic views. The corresponding object is shown in part j of the figure. The surfaces A and B, together with points 1 through 5, are also identified in part j.

4.7.1 Cut-by-Cut Technique

Another helpful technique in the interpretation of orthographic views is known as the **cut-by-cut** technique. It is used as an adjunct to or quick substitute for the more complete surface-by-surface, point-by-point analysis. One begins by visualizing (or drawing) a simple block of material from which the object described in the set of orthographic views can be "formed"; Figure 4.63 shows a drawing of a rectangular prism corresponding to such a block of material (a). If one wishes to produce an instrument drawing of such a rectangular prism, one may draw an isometric pictorial (a), in which two axes (the width and the depth) are located at 30° inclinations relative to the horizontal direction, and the third axis (height) is drawn in a vertical direction. We will develop a complete description of isometric pictorials in Chapters 14 and 15. The pictorial is drawn with dimensions along the principal directions—height, width, and depth—equal in distance to those shown in the orthographic views under consideration. Thus, the height of the rectangular prism is equal to the height of the object shown in the orthographic views.

After the rectangular block is constructed, one "cuts" material from this block in accordance with the outer boundary lines of the orthographic views. For example, in Figure 4.62(a) the front view indicates that the entire *top, right* portion of an original block of material would need to be removed by an inclined cut, denoted as cut 1 in Figure 4.63 (b and c). The right-side view in Figure 4.62(a) indicates that a second cut, cut 2 in Figure 4.63(d and e) must be made in the "raw material" from which we are forming the object. Cut 2 should remove the *front, top* portion of the object in accordance with the right-side view in Figure 4.62(a). The placement of cut 2 is determined by the dimensions given in the right-side view in Figure 4.62(a). Comparison in Figure 4.63(f) with the final object in Figure 4.62(j) indicates that we have obtained a complete and correct interpretation of the object described in Figure 4.62 by means of the cut-by-cut technique.

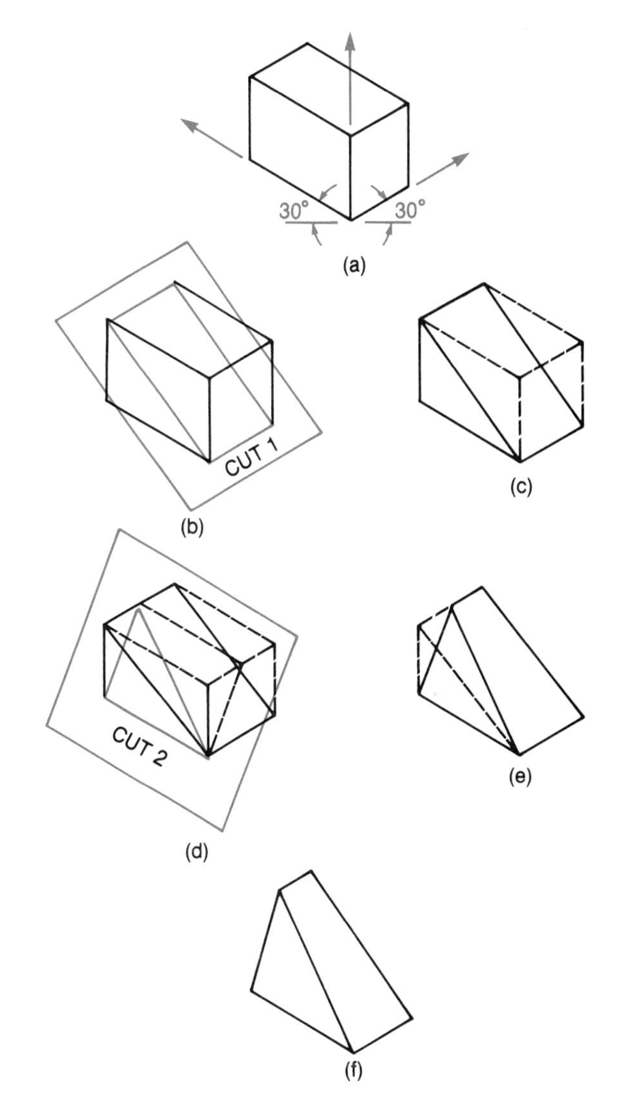

Figure 4.63 Cut-by-cut development gives the same result as that obtained in Figure 4.62. This would not be true of all objects, however, since interior linework may require the surface-by-surface, point-by-point technique in Figure 4.62.

The cut-by-cut approach requires only the available information contained in the outer boundary linework of orthographic views to develop a pictorial representation. This approach also helps one develop an understanding of the construction of an object. However, this approach *does not guarantee* an accurate and complete interpretation of the object, since interior linework may need to be interpreted via the surface-by-surface, point-by-point technique.

4.7.2 Developing Skill in These Techniques

Once again, it is important to recognize that as skill increases in the interpretation of orthographic views, the time needed for analysis will correspondingly decrease. The surface-by-surface, point-by-point technique and the cut-by-cut technique help *develop* skill in reading orthographic drawings. Eventually, it may seem that you are visualizing the object immediately upon viewing the orthographic views used to describe it; in fact, your mind is simply analyzing the information (areas, line, and points) contained within the given views very quickly so that a complete image of the object forms in your mind. This ability to quickly "read" a set of orthographic drawings can be developed through diligent practice.

4.7.3 Summary of These Techniques

Let us review the procedure for the surface-by-surface, point-by-point technique:

Step 1. Identify areas that appear in each view.

Step 2. Analyze each surface, one surface at a time, by identifying its representation (area or line) in all views. (The configuration rule should be used.)

Step 3. Identify key points that can be used to define each surface and determine the location of these points in all views.

Step 4. If a missing orthographic view needs to be constructed, each surface of the object should be developed by locating its key points in the construction region. These points can then be connected by linework to form the appropriate representation (area or line) of the surface under consideration.

The cut-by-cut technique can be used to develop a pictorial representation of the object, if necessary, in accordance with the outer boundary linework given in a set of orthographic views. This pictorial may be incomplete, since any interior linework will need to be interpreted through additional analysis, but it can be helpful as one seeks to "read" the given orthographic views.

4.7.4 Applying These Techniques

Figure 4.64 presents a set of orthographic views (front and right-side) (a) for which we will develop the corresponding top view by using the surface-by-surface, point-by-point procedure outlined in Steps 1 through 4 in Section 4.7.3.

Step 1. Four distinct areas (A, B, C, D) are identified (b).

Step 2. Surface A is identified in the given views (c) by using the configuration rule under the assumption that surface A is planar.

Step 3. Four key points (1, 2, 3, 4) are identified in the given front view; these points can be used to define the four-sided area labeled A (c). Points 1, 2, 3, and 4 are then located in the given right-side view (d and e).

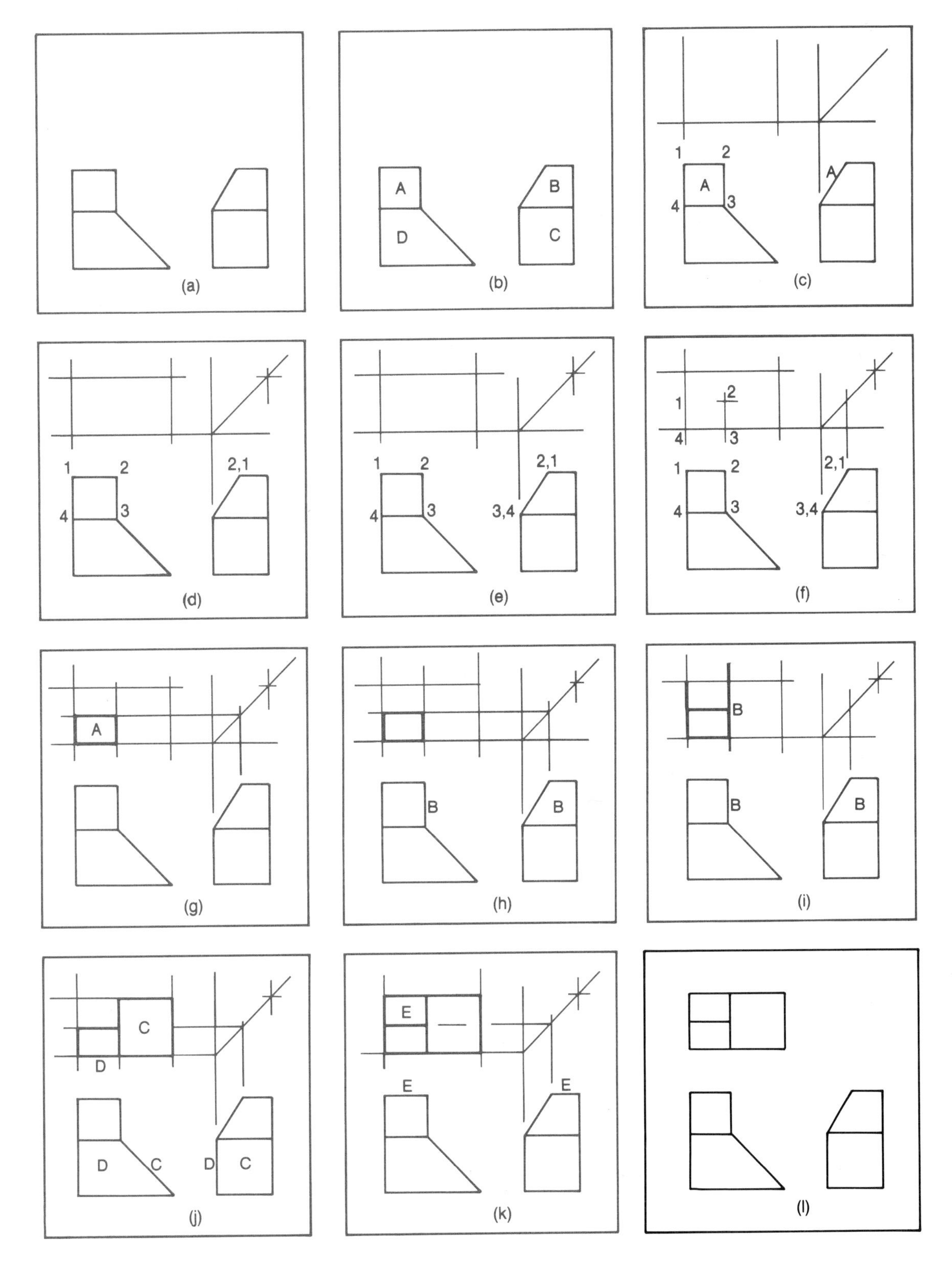

Figure 4.64 Surface-by-surface, point-by-point development of the missing top view for two orthographic views (a) starts with identification of areas (b) and proceeds through analysis of each surface and each point (c–k), resulting in a final set of views (l).

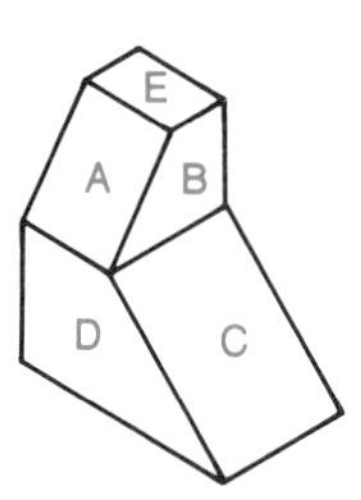

Figure 4.65 Object described by the orthographic views in Figure 4.64.

Step 4. The key points 1, 2, 3, and 4 are then located in the missing-view construction area through the projection of width and depth information for each point from the front and right-side views, respectively (f). These points are then connected by linework to form the appropriate representation (a four-sided area) of surface A in the top view (g).

Steps 2 through 4 are then repeated in order to produce representations of surfaces B, C, and D (h through j). Finally, a fifth surface, E, is identified and developed (k). (Surface E is necessary in order to explain all of the linework appearing in the given views and in order to complete the missing top view.) In the final set of three orthographic views (l), construction work has been erased. Figure 4.65 presents a pictorial of the object described by the orthographic views in Figure 4.64.

Figure 4.64 includes the use of the **miter line** for the transference of depth information between views (top, right-side). Alternative approaches are illustrated in Figures 4.66 and 4.67. In Figure 4.66 a **compass** is used to transfer necessary data from the right-side view to the top view through rotation about the point of intersection between the FRP construction lines. Figure 4.67 shows the transference of depth information through the use of **dividers** (or, alternatively, a *compass in which the data is transferred directly* to the construction region of the missing view). The use of dividers or a compass is recommended for greater efficiency and less construction work.

Figure 4.68 shows the cut-by-cut technique applied to the given orthographic views in Figure 4.64(a). Three separate cuts (a–c) are made in a rectangular prism (block of material), in accordance with the outer boundary linework of the given views. The final pictorial of Figure 4.65 is produced as a result of these three cuts (d), just as it was produced by the surface-by-surface, point-by-point method (although such is not always the case for all objects).

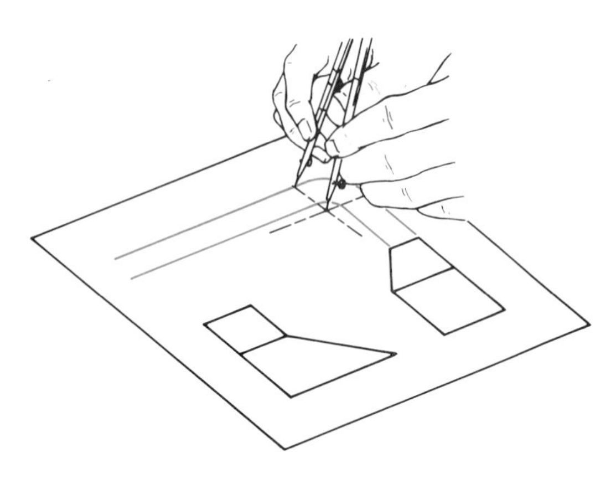

Figure 4.66 Transfer of depth information via compass.

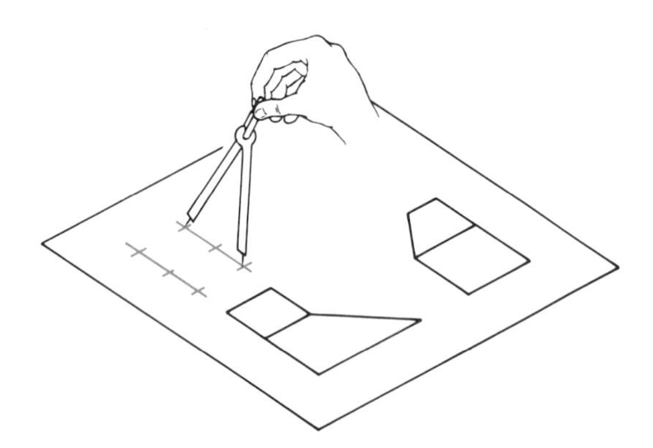

Figure 4.67 Transfer of depth information via dividers.

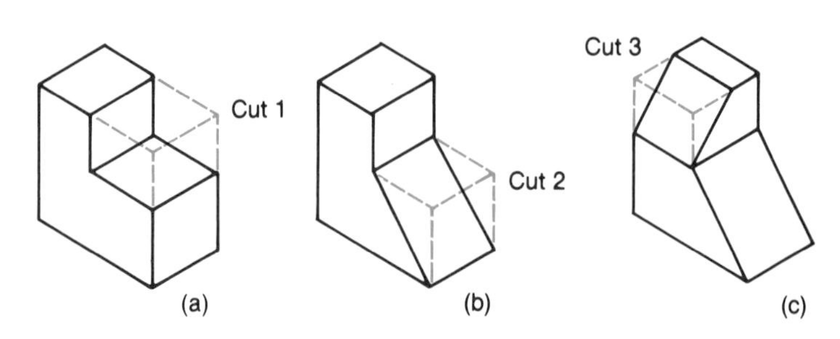

Figure 4.68 Cut-by-cut technique (a-c) applied to the orthographic views in Figure 4.64 produces the same pictorial as the surface-by-surface, point-by-point technique does (Figure 4.65).

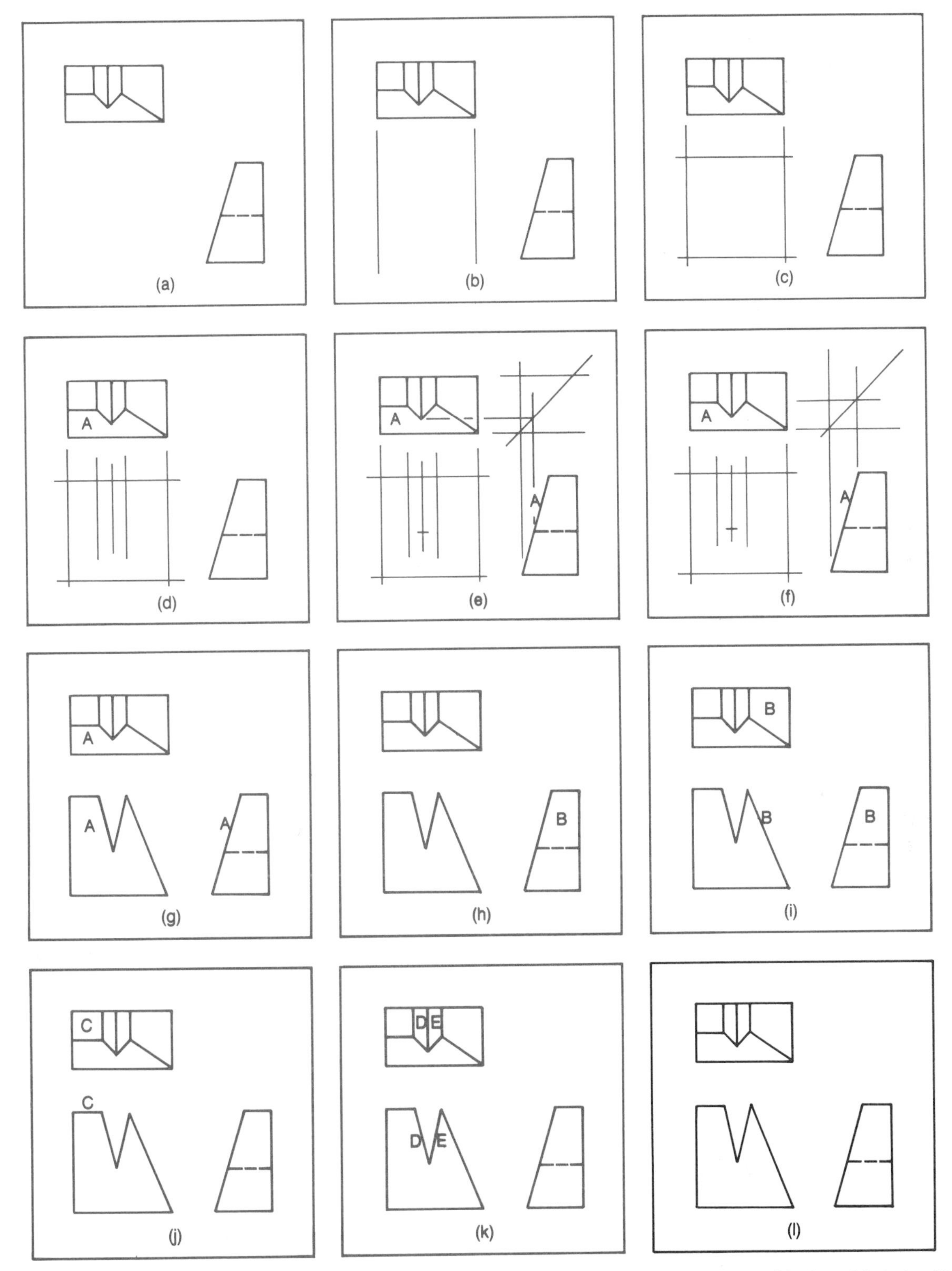

Figure 4.69 Surface-by-surface, point-by-point development of the missing front view for two orthographic views (a) starts with projection of width and height (b and c) and proceeds through analysis of each surface and each point (d–k) resulting in a final set of views (l).

In another application of the surface-by-surface, point-by-point procedure, consider the two orthographic views shown in Figure 4.69(a). We seek to develop a front view consistent with these two given (top and right-side) views. First, the *width* of the missing front view is projected from the given top view (b). The *height* of the front view is then projected from the given right-side view (c).

We then identify all areas appearing in the two given views (this step is not shown in the figure). Beginning with area A (d), we then apply the surface-by-surface, point-by-point technique to develop the front view. Surface A is identified in both of the given views (e–g), after which significant distances are projected from the top view to the front-view construction area; in addition, distances of *depth* are transferred from the top view to the right-side view (via a miter line) so that relevant height data can then be projected from the right-side view to the front-view region.

Key points should also be identified during the construction of surface A's representation in the missing front view. We have purposely not shown the identification of key points in Figure 4.69 in order to (1) enhance the clarity of the development and (2) explicitly demonstrate that, with skill developed through practice, one can perform most of the "cataloging" of points in one's mind rather than upon the drawing paper. The reader should consider which key points should be identified during the development of surface A in the front view of Figure 4.69. Hint: surface A is shown as a six-sided area in the given top view.

Surfaces B, C, D, and E are then found to each appear on edge as a line in the front view (h–k). Furthermore, each of these lines representing surfaces B, C, D, and E are boundary lines of the area representing surface A in the front view.

In Figure 4.70 the cut-by-cut technique is applied to the given views of Figure 4.69(a). Three cuts are needed (a–c) to complete the pictorial of the object (d) in accordance with the outer boundary lines of the two given views.

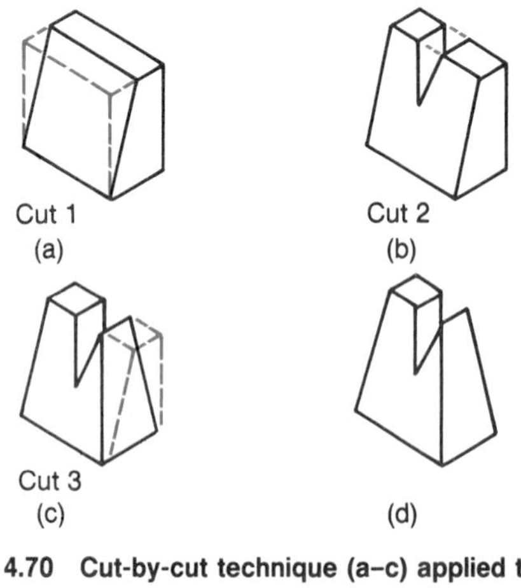

Cut 1
(a)

Cut 2
(b)

Cut 3
(c)

(d)

Figure 4.70 Cut-by-cut technique (a–c) applied to the orthographic views in Figure 4.69 produces a pictorial representation (d) in three cuts.

In summary, then:

1. The configuration rule states that a planar surface that appears as an area with n sides, or boundaries, in one orthographic view will appear as an area with n sides (or on edge as a line) in all other orthographic views.

2. The surface-by-surface, point-by-point approach may be used in the interpretation of a set of orthographic views. Each surface of an object is identified, one surface at a time, in all given views; key points for each surface are then located in all views. This approach results in careful analysis of the information contained within the set of views, which should ensure a correct interpretation.

3. The cut-by-cut approach, in which the final object is produced from a series of imaginary cuts from a rectangular block of material, may be used as an additional aid in the de-

Highlight

The Learning Curve

Educators are familiar with the concept of the learning curve (shown in the accompanying sketch), which traces the systematic development of a person's understanding of a concept. In fact, this growth curve can be applied to the physical development of a person from conception through adolescence to maturity. It also can be used to describe the development and application of technology, the growth of a business, the honing of a physical skill, and so on. In each case there is a relatively slow initial period of growth, followed by an interval of rapid development and ending with marginal growth leading to a final plateau. The learning curve reminds us that time is required to develop skill in graphics. We should expect an initial period of slow development, followed by relatively rapid growth nurtured by diligent practice. If you find that you require more time and effort to interpret a given set of orthographic views than you expected, remember the learning curve and the slow period of growth at the beginning of any learning process. With practice, your skill will inevitably grow with increasing speed until you have achieved the desired plateau of ability.

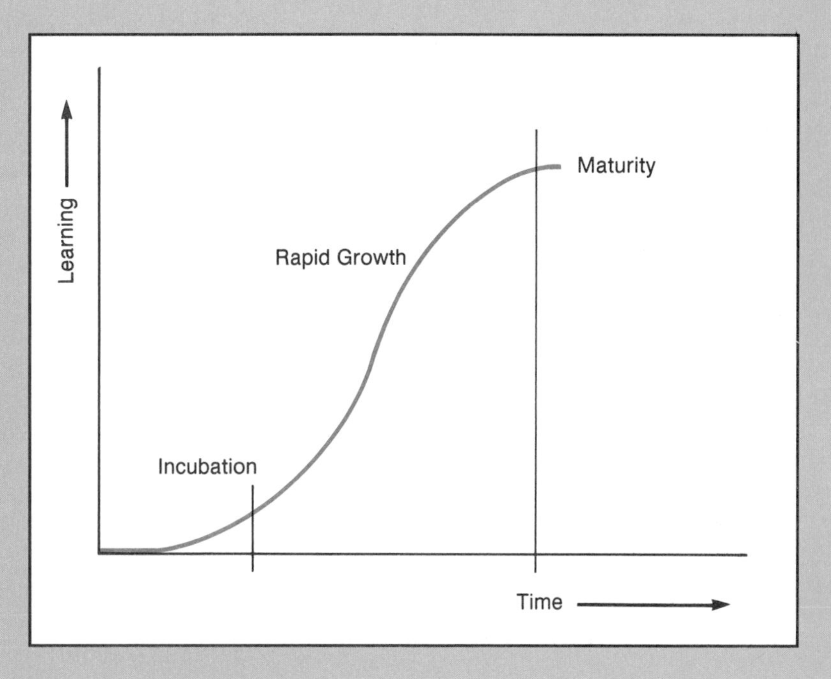

velopment of a correct interpretation of a set of orthographic views or as a faster method of obtaining a preliminary pictorial representation. Remember, however, that the pictorial produced by this approach may not always be accurate or complete. Consequently, the cut-by-cut method should never be used alone.

4.8 Developing Planar and Nonplanar Surfaces

The surface-by-surface, point-by-point technique can be applied to both planar and nonplanar surfaces that appear in orthographic views. The configuration rule, however, applies only to planar surfaces; consequently, if one can determine which surfaces are planar in a set of views, one should use these surfaces for the initial analysis before considering any nonplanar surfaces contained in the drawings.

Figure 4.71 presents two (front, right-side) orthographic views of an object (a), for which two surfaces can be identified (b). Surface A appears as a four-sided area in the given front view, and surface B appears as a four-sided area (three of which sides are straight and one of which curves through space) in the right-side view. Application of the rules of adjacency and common dimensions then leads us to the following conclusion: surface A is represented by the curving line appearing in the given right-side view, whereas surface B appears as a straight line in the front view (b). Surface A, seen on edge as a curving line in the right-side view, must therefore be a nonplanar surface (curving through space). Since we want to begin with development of planar rather than nonplanar surfaces, we will construct surface B's representation in this missing top view first (c–e). To accomplish this, we identify four key points (points 1, 2, 3, and 4) in the given views and project these points for the missing view by transferring the width data from the front view and the depth data from the right-side view.

The results (c) might tempt one to connect points 1 through 4 with either four straight lines or three straight lines and a curved line. But the configuration rule tells us that the planar surface B cannot change its shape from one set (three straight, one curved) of boundary lines to another set (four straight), so we must reject the interpretation of four straight boundary lines for surface B's representation in the top view. It must appear as an area with three straight and one curving boundary line.

In order to draw the curving line between points 1 and 4 in the top view, we need to locate several additional intermediate points (d). Each of these intermediate points (5 through 8) are first arbitrarily defined along the curving line appearing in the right-side view, then found in the front view by projecting the height of each point from the right-side view to the front view. In this manner, we obtain the width (in the front view) and the depth (in the right-side view) of each of the points 5, 6, 7, and 8. Finally, this width and depth data is projected from the two given views to the top-view construction region (d). The resulting points are then connected to form the representation of surface B in the top view; as expected, this representation is a four-sided (three straight lines, one curving line) area (e).

The nonplanar surface A can now be developed in the construction region. Four key points (1, 4, 9, and 10) are identified in the front view; these points are the four corners of the four-sided area representing surface A in this view. The same four points are then found in the given right-side view. All four must lie along the line representing surface A in the side view, since they have been defined as lying along surface A. Once all four points have been located in both of

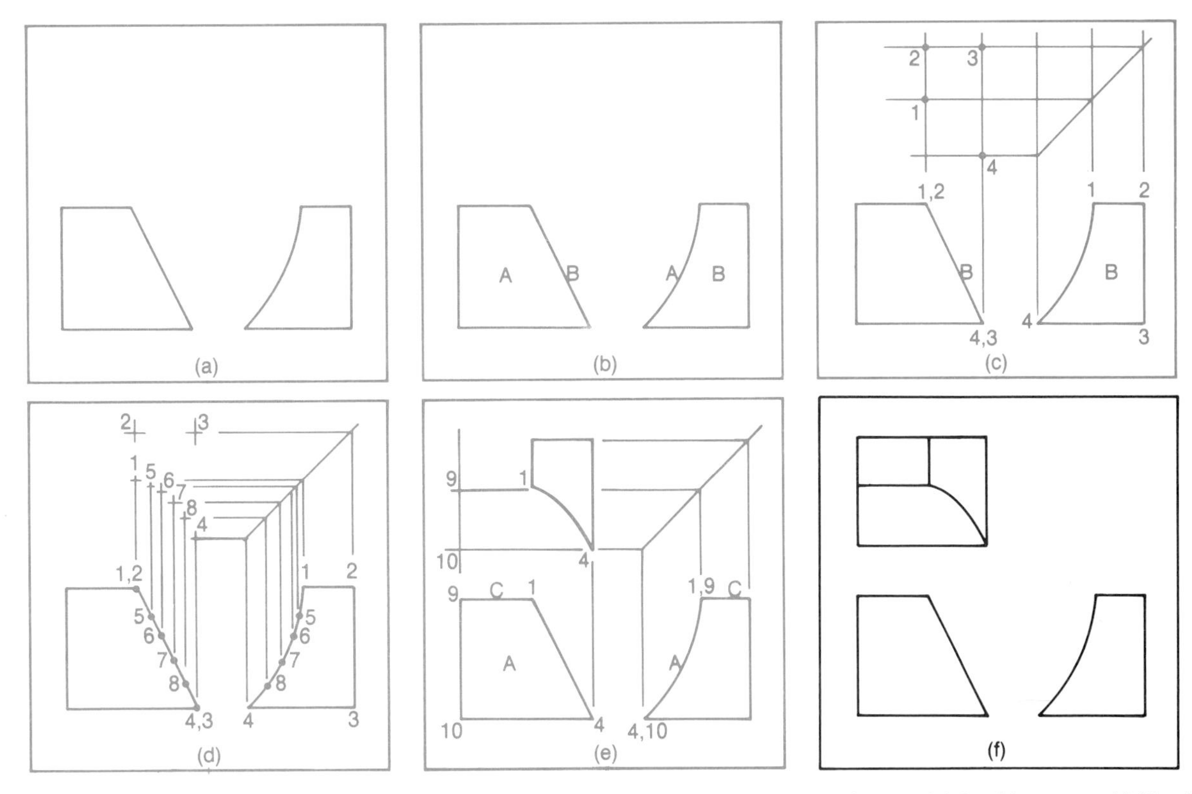

Figure 4.71 Development of planar and nonplanar surfaces from two orthographic views (a), front and right-side, starts with identification of any planar surfaces (b)—for example, surface B—and construction of same (c and d) before construction of any nonplanar surfaces (e)—surface A in this example—leading to development of the missing third view (f).

the given views, the width and depth of each point is then projected from these views to the top-view construction region (e). Finally, linework is used to connect these points to form the representation of surface A in the top view. In the complete set of three views for this object, a third surface, denoted as C in part e, in which it appears as a horizontal line in the two given views, appears in the final top view as a rectangular area. Surface C is the horizontal top surface of the object.

Learning Check

In the final top view shown in Figure 4.71(f), surface A appears as a four-sided area—an area with three straight lines and one curving line as its boundaries. In the given front view, surface A appears as a four-sided area—an area with four straight lines as its boundaries. In other words, surface A changes its shape in these two views. Is this a violation of the configuration rule?

Answer It is not a violation of the configuration rule, since this rule applies only to planar surfaces. Surface A has been shown to be nonplanar (curving through space and not contained within a single plane). As a result, surface A can "change its shape" in different orthographic views. (Of course, it is only our view of the surface, not the surface itself, that changes from one view to another. In some views, only part of the nonplanar surface may be visible; in others its shape may appear distorted due to perspective.)

Figure 4.72 presents another example of developing a missing third view, given two complete views of an object. In this case, the given views are the front and top views (a), and the right-side view must be constructed. Two distinct surfaces (A and B) are identified in the two given views (b) through the application of the various rules (rule of adjacency, configuration rule, rule of common dimensions). In addition, the FRP is identified in the given top view and also in the construction region of the right-side view (so that the distances between views will be identical). The missing right-side view is then developed by following the proper

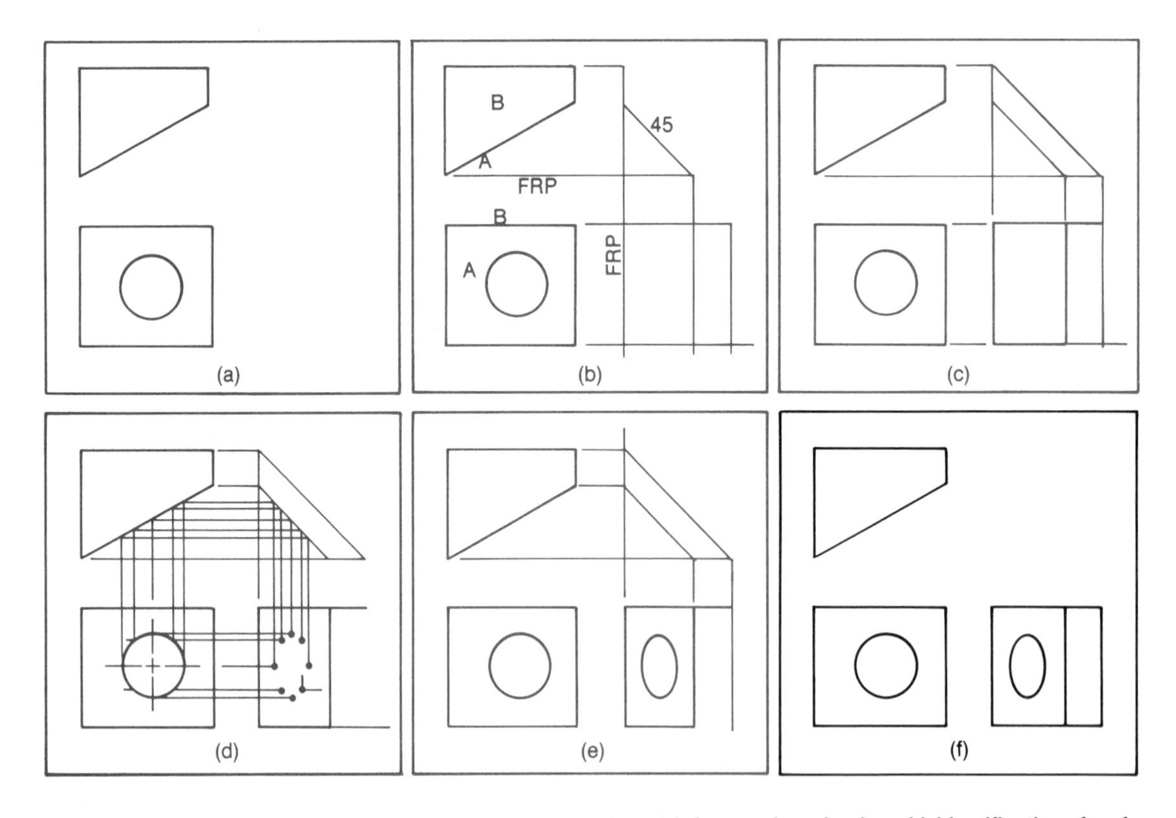

Figure 4.72 Development of a missing view from two orthographic views (a), front and top, begins with identification of surfaces A and B and the FRP (b). Using projection techniques (c–e), the missing right-side view is developed (f).

projection techniques (c–e), resulting in a final set of three complete orthographic views (f).

4.9 Developing Hidden Linework

Figure 4.73 shows a set of two given (front, right-side) views (a). The top view that must be constructed will contain **hidden linework.** Surface-by-surface, point-by-point development is the procedure used to develop an orthographic view in which hidden linework appears. Surfaces A and B in Figure 4.74 are identified in the given views through the application of the rule of adjacency, the rule of common dimensions, and the configuration rule (b). Key points are then identified for surface A (points 1, 2, 3, and 4) in the given front view and the given right-side view (c). The width and depth dimensions of each point are then projected to the top-view construction region. All four points lie in the FRP, as does the entire surface A. Surface A appears on edge in the top view as a line (c). Notice that point 3 is denoted in the top view by the following notation: (3). This notation, in which a point number is given in parentheses, indicates that the point is hidden by another (visible) point not identified by number or name. For example, point 3 is hidden in the top

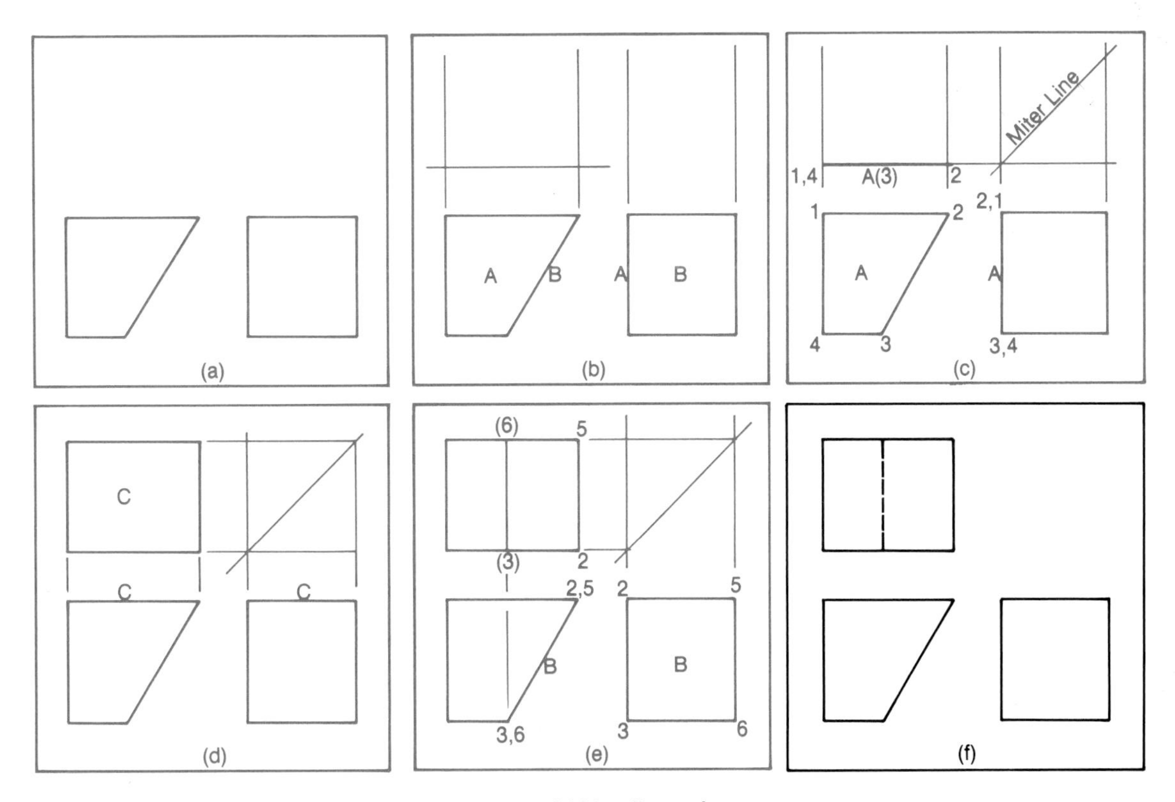

Figure 4.73 Developing a missing view that includes hidden linework.

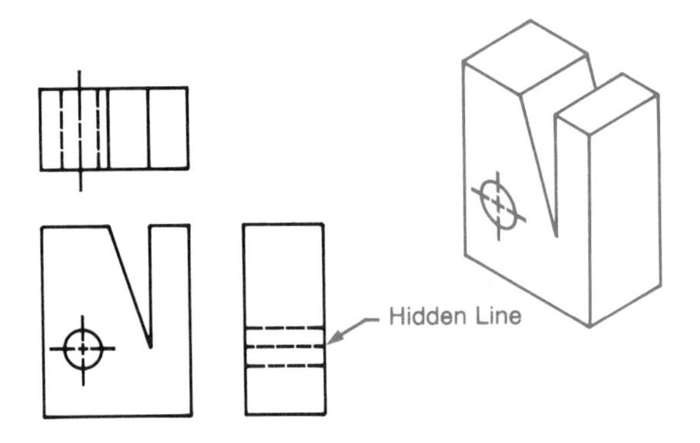

Figure 4.74 Linework priority—hidden line/centerline.

view of Figure 4.73 by the point lying on the line $\overline{12}$ at the same width position as point 3; this visible point on line $\overline{12}$ has not been (and will not be) identified by number because it is not a key point (meaning it is not needed for the development of the missing top view).

The top horizontal surface, C, of the object is developed by the rules of projection (d). Surface B is analyzed by first identifying its key points (2, 3, 5, and 6) in the given views and then locating these points in the construction region via projection (e). Two points (3 and 6) are hidden in the top view; as a result, the line connecting these two points in the top view, which represents a boundary of the surface B, is hidden. In fact, the entire surface B is hidden in the top view, as indicated in the final set of three orthographic views (f).

If a point is hidden in an orthographic view, then the portion of the line that contains that point must be hidden in that view (the entire line may even be hidden). Furthermore, the portion of the surface that contains the hidden part of this line must also be hidden (again, the entire surface may even be hidden). One must carefully analyze points on the line in order to determine which portions of the line are hidden and which portions, if any, are visible.

Remember that all surfaces of an object should be identifiable as visible areas, hidden areas, visible lines, or hidden lines in every orthographic view of an object; otherwise, the orthographic view is incomplete.

In summary, then:

1. Hidden linework in a missing orthographic view should be developed through the same surface-by-surface, point-by-point procedure that is applied to visible linework.

2. If a point is hidden in an orthographic view, then the portion of the line (the surface boundary) that contains that point must be hidden in that view. The entire line, and, indeed, the entire surface, may be hidden, or only a portion of the line and the surface may be hidden.

3. *All surfaces of an object should be identifiable as areas (visible or hidden) or as lines (visible or hidden) in every orthographic view; otherwise the view is incomplete.*

4.10 Priority of Linework

Lines may occasionally overlap in an orthographic view. When they do, linework priority is:

1. Visible linework.
2. Hidden linework.
3. Center linework.

Figure 4.74 presents three orthographic views of an object, together with a pictorial. The right-side view contains three hidden lines. One of these hidden lines has the same height location as the center line (for the hole in the object), which would appear in the right-side view if the hidden line were absent. However, in accordance with the standard system for prioritizing linework, the hidden line is shown in this view whereas the center line is not shown.

Figure 4.75 presents three orthographic views of another object, together with a pictorial. A visible line and a hidden line overlap in the top view. In this case, priority is given to the visible line. (Since visible linework always has priority over all other linework in a view, it dominates any orthographic view.)

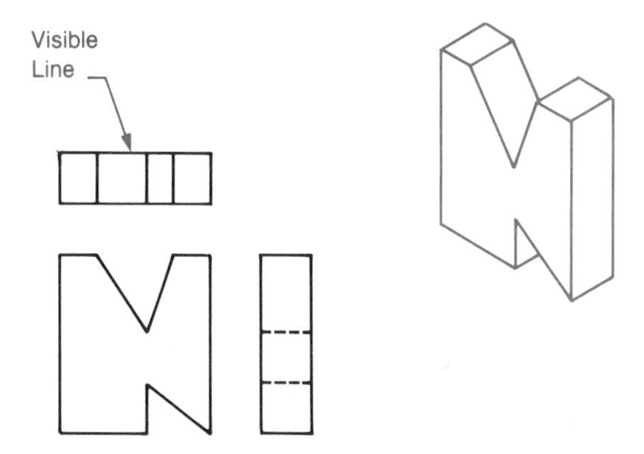

Figure 4.75 Linework priority—visible line/hidden line.

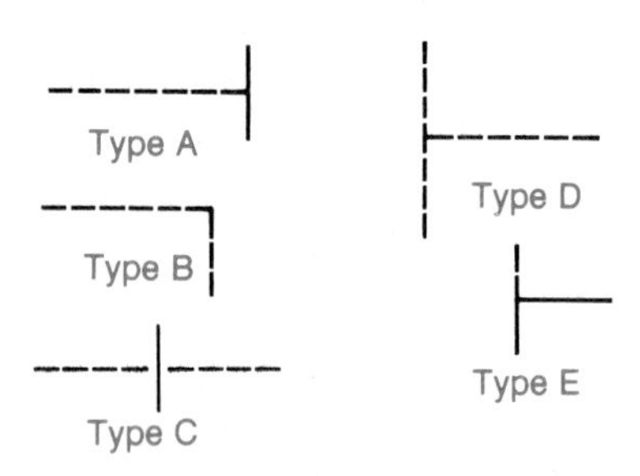

Figure 4.76 Clarity in linework intersection demands specific handling of the five types of intersections, which are shown here and further illustrated in successive figures.

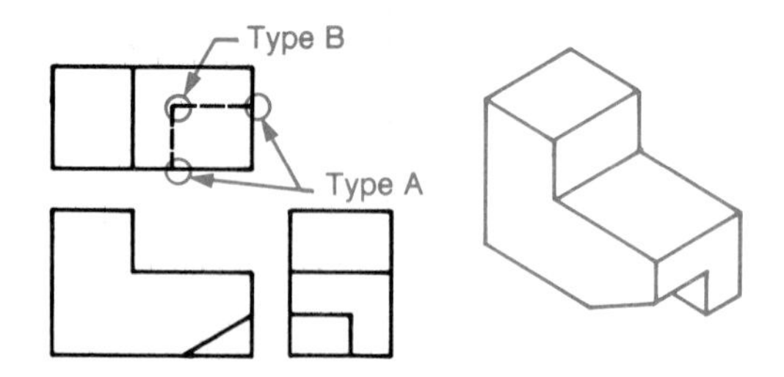

Figure 4.77 Example of types A and B intersections.

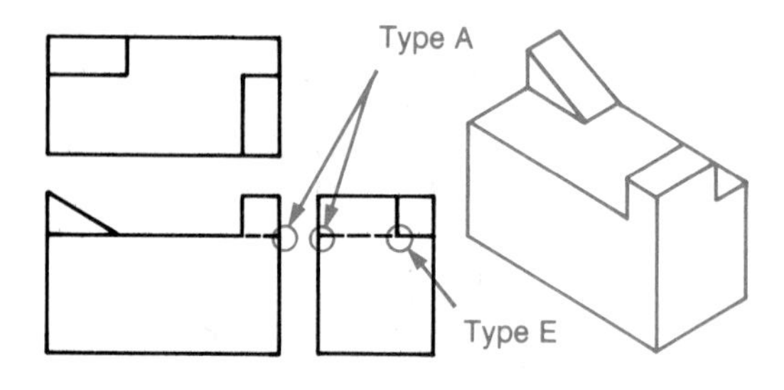

Figure 4.78 Example of types A and E intersections.

NOTE: Centerlines not included for clarity.

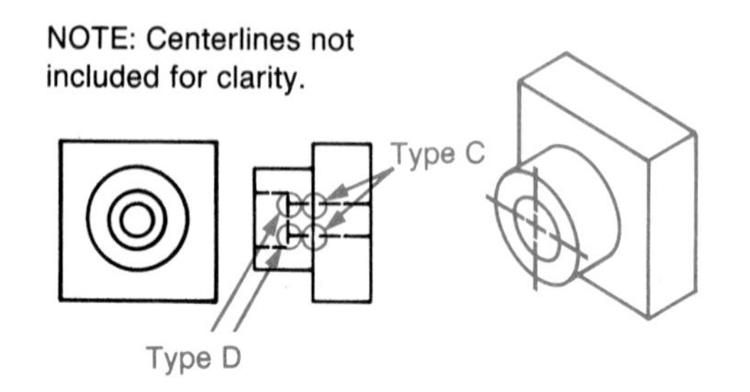

Figure 4.79 Example of types C and D intersections.

Learning Check

Explain why visible linework must dominate hidden linework in the same location within an orthographic view.

Answer Visible linework represents surface boundaries that can be seen by the observer in a given view, whereas hidden linework represents surface boundaries that cannot be seen because solid material lies between the observer and the hidden boundary. In order to properly represent the appearance of the object in a given view, therefore, one must show its visible boundaries. Hidden linework representing boundaries lying "underneath" the visible boundaries in a view cannot be shown unless the visible linework is eliminated—and visible linework must never be eliminated.

Why is center linework given lower priority than both visible and hidden linework?

Answer Visible and hidden linework both represent true physical boundaries of surfaces on the object. Center linework is included in orthographic drawings only to provide information about the *location* of centerpoints for circles and arcs. It is used in dimensioning the drawings so that the machinist may construct the object in the appropriate size from the orthographic views. But *centerlines do not represent physical characteristics (boundaries)* of the object; therefore, they do not have as much significance as visible and hidden linework in a drawing.

In summary, then:

1. Linework occasionally overlaps in orthographic drawings. The standard system for prioritizing overlapping linework is (1) visible linework, (2) hidden linework, (3) center linework.
2. The priority system given is justified as follows: Both visible and hidden lines represent real surface boundaries, so these lines have priority over center lines, which are used only for dimensioning purposes. Visible lines must be shown in a view in order to properly represent the appearance of an object; hence, these lines have priority over hidden lines.

4.11 Intersection of Linework

Linework in an orthographic drawing may intersect in a variety of ways; Figure 4.76 presents five different types of intersections that frequently occur in orthographic views. The major goal in linework intersection is *clarity*. The reader must be able to discern the *extent* of each line—its beginning and end—in a drawing so that he or she will be able to properly interpret the graphical description of an object.

Intersections of type A (in Figure 4.77) involve nonparallel visible and hidden lines. The dash of the hidden line *contacts* the visible line but it does not extend beyond it. Type B intersections involve nonparallel hidden lines only. Once again, the point of intersection between these lines is clearly defined by the intersection of dashes at the end of the lines.

Type C intersections involve the overlapping of nonparallel visible and hidden lines. They differ from type A in that a dash of the hidden line is drawn so that it does not intersect the visible line. As a result, the reader will understand that the hidden line extends beyond the position of the visible line. Type D intersections again involve nonparallel hidden lines; one line extends to the point of intersection but not beyond it. Two dashes (one dash in each hidden line) are used to explicitly identify the point of intersection by forming a T-shaped intersection between these dashes.

Finally, type E intersections involve three lines (two visible lines and one hidden line). All three lines end at the point of intersection. In order to illustrate this fact most clearly, a space (rather than a dash) in the hidden line is used at the point of intersection. (If one used a dash at the end of the hidden line, the reader might believe that the visible line, which is parallel to the hidden line, extends beyond the point of intersection.)

Figures 4.77, 4.78, and 4.79 illustrate the use of each type of intersection in orthographic drawings. Each figure includes a set of three orthographic views and a pictorial of an object. Carefully review (that is, "read") each set of views.

4.12 Additional Practice in Reading and Developing Views

For additional practice in interpreting orthographic drawings, consider Figures 4.80 (object A) and 4.81 (object B). In each case, two orthographic views of an object are given; we must develop a third view for each set.

Figure 4.82 presents an analysis of the two given views of object A. Two surfaces (A and B) are identified in Figure 4.82(a), after which surface-by-surface, point-by-point development of a third view is applied. Possible representations of surfaces A and B are identified (b) simply by following the rules of adjacency and common dimensions and the configuration rule. This set of representations is denoted as Interpretation 1 (b). Continuing with Interpretation 1, other surfaces (C, D, E) are identified (c and d), but then we notice that surface E, though identified in one given view, cannot be identified in the other view (d). Our initial identification of surface representations has led us to an *inconsistent* and *incorrect* interpretation! We must begin again.

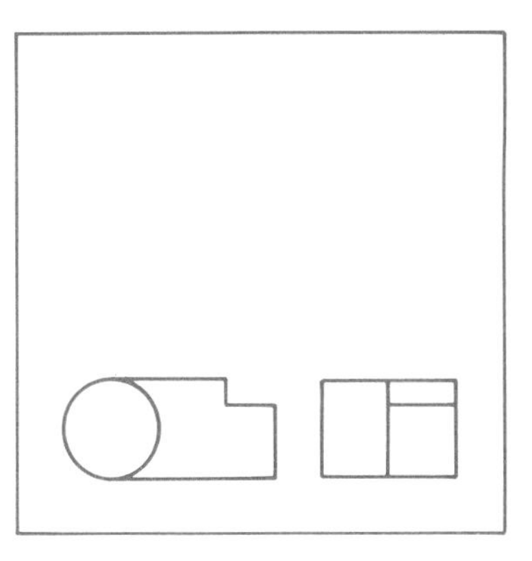

Figure 4.80 Example object A, to be the subject of successive figures.

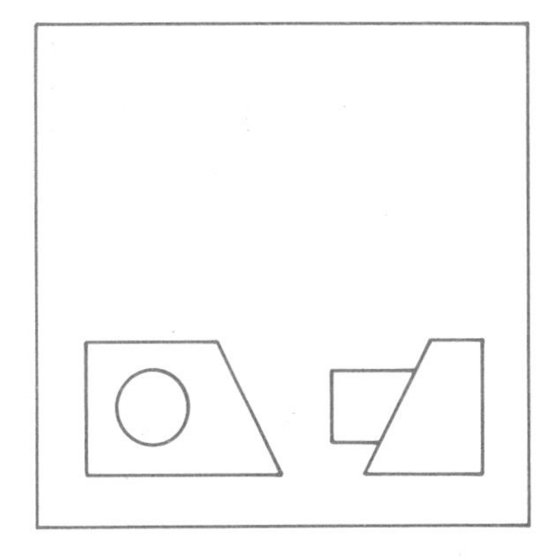

Figure 4.81 Example object B, to be the subject of successive figures.

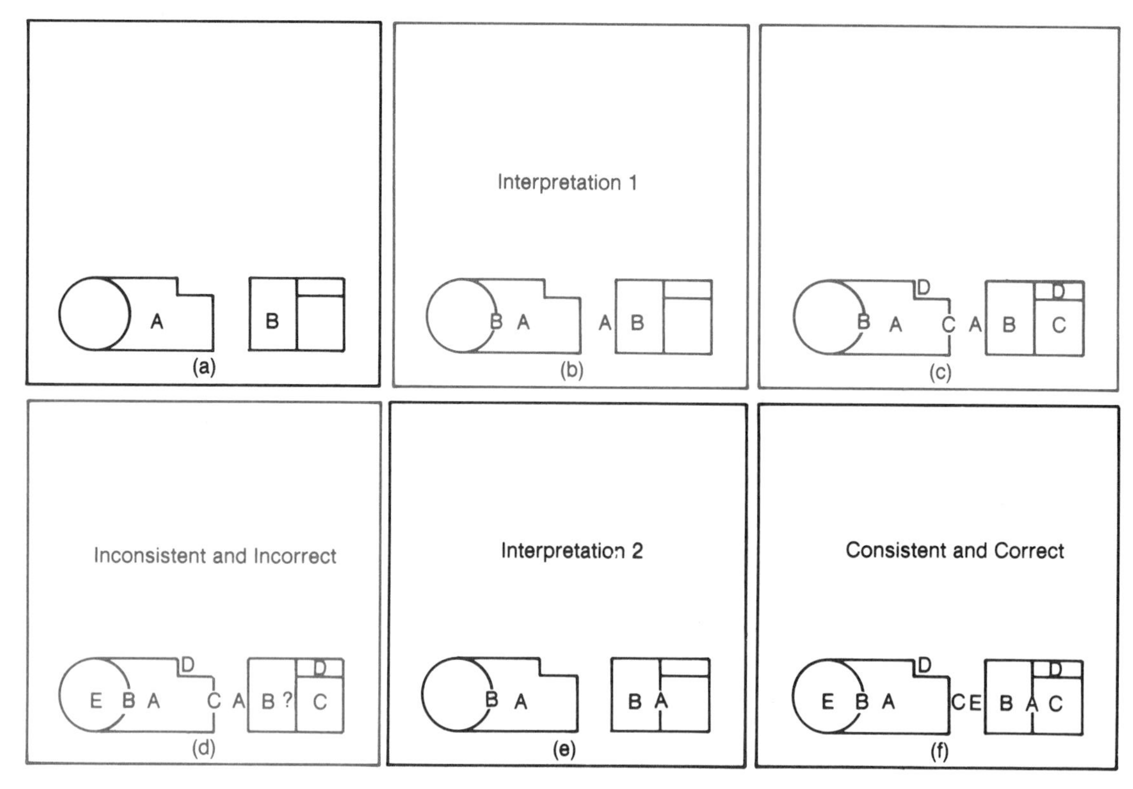

Figure 4.82 Interpreting orthographic views of object A in Figure 4.80.

Carefully analyze parts c and d in Figure 4.82. Why is Interpretation 1 inconsistent and incorrect? Why have we concluded that surface E cannot be identified in both views, based upon this interpretation?

Answer Surface E has been identified as the circular area in one (front) of the given views and surface A as the six-sided area that forms the remainder of this view (d). Surface A has been identified as a line in the other (right-side) given view, located at the FRP—at the front of the object in this side view. Application of the configuration rule leads us to conclude that surface E must also be represented by a line in the side view since there are no circular areas shown in this view. The rule of common dimensions tells us that the line representing surface E in the side view must have the same height as the area representing E in the given front view; in other words, the line representing E in the side view must extend from the bottom of the object to the top of the object. Only two vertical lines remain as possibilities for E's representation in the side view now that the line in the FRP has been assumed to represent surface A. One of these two lines lies at the rear of the object. If such a line represented surface E in the side view, then surface E would be infinitesimally thin (of exceedingly small depth). As a result, we assumed that the second possibility (an *inside line* denoted by a question mark in part d of the figure) must represent surface E in the side view. However, this assumption leads to an inconsistency in the interpretation: surface E would then lie nearer to the rear of the object than surface A, and it would lie to the left of the object with surface A near the right side of the object (as shown in the front view). As a result, surface E would have to be represented by a *hidden* line in the side view; of course, there is no hidden line in the given view! We must then conclude that our original interpretation is incorrect.

Figure 4.82 goes on to present a second interpretation (Interpretation 2) in which surface A is assumed to be represented by an inside line in the right-side view (e). Application of the rules for interpretation then lead to a consistent and correct identification of all surfaces in all views (f).

As Figure 4.82 proves, the identification of all surfaces as areas or lines must be completely consistent throughout all views. If an incorrect assumption leads to an inconsistent interpretation, one simply begins analysis again.

Figure 4.83 presents the systematic development of a third (top) view for object A (the subject of Figures 4.80 and 4.82), and Figure 4.84 does the same for object B (the subject of Figure 4.81). The surface-by-surface, point-by-point technique is used, together with a miter line. Points and surfaces are not explicitly identified in these constructions. Carefully review the development of these missing views. Figure 4.85 presents pictorials of objects A and B.

In summary, then:

1. A completely consistent interpretation of all orthographic views must be developed if one is to understand the object under description.
2. If a particular interpretation leads one to an inconsistency, analysis should begin again, with a different set of underlying assumptions. One must be patient during the analysis of a drawing to obtain a correct interpretation.

4.13 One-View and Two-View Descriptions of Objects

One should never use more views than are necessary to describe an object since additional views will require more time for the reader to interpret all of the given information.

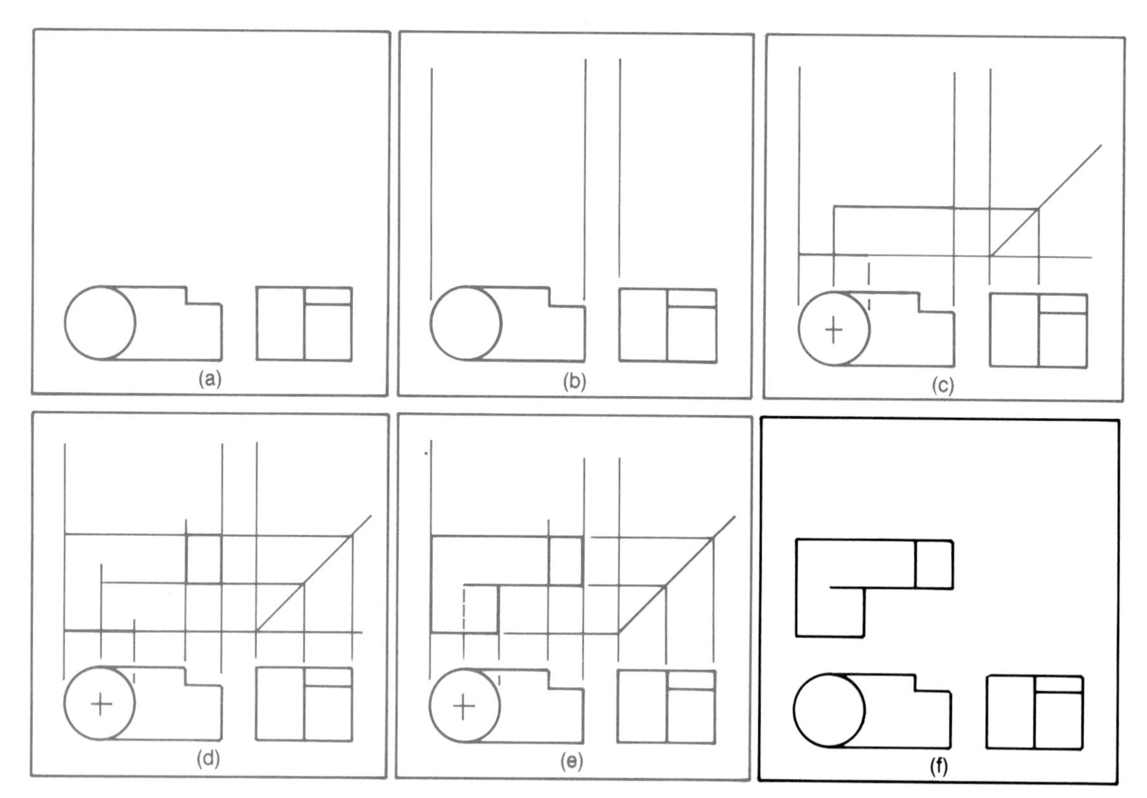

Figure 4.83 Developing the missing top view for object A in Figure 4.80.

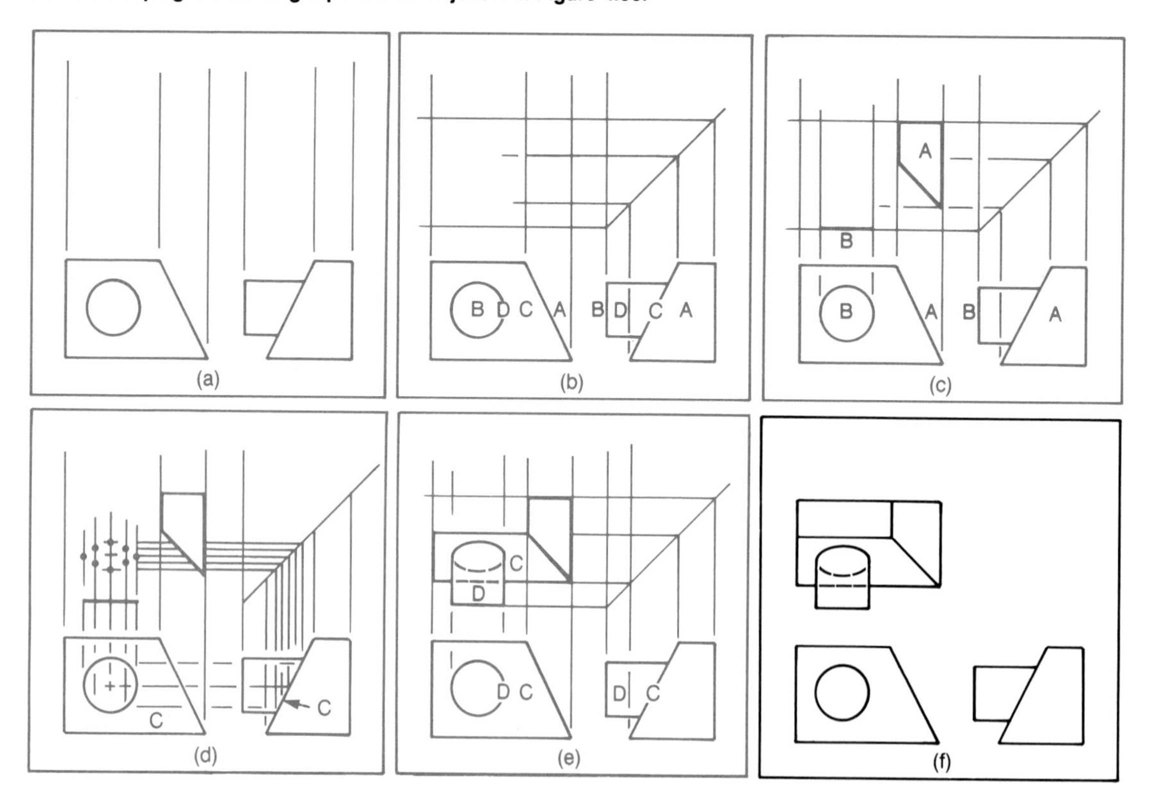

Figure 4.84 Developing the missing top view for object B in Figure 4.81.

Machinists and others who work with orthographic drawings must be able to correctly and quickly interpret a set of views; by minimizing the number of views in a set, we can save both the reader and ourselves time and effort. (Of course, the object must be *fully* described in a set of views.)

Cylindrical and disc-shaped objects can usually be fully defined by two views, as shown in Figure 4.86. Simple objects, such as that shown in Figure 4.87, can also be described by two views. A few types of objects can even be described by only one orthographic view if one provides the third dimension (depth) in a note. Such objects must be of uniform thickness so that a simple note is sufficient to provide the machinist with enough information to fabricate the object. Figures 4.88 and 4.89 show examples of such objects.

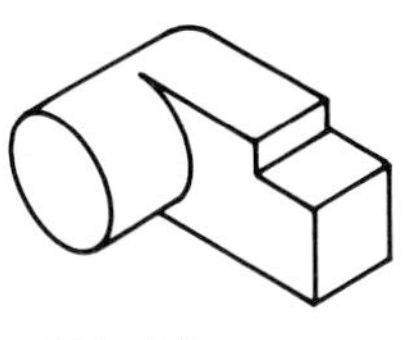

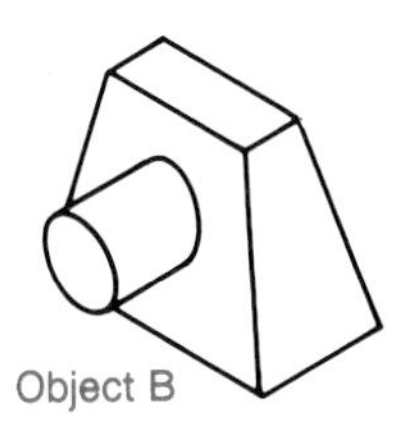

Figure 4.85 Pictorials of objects A and B from Figures 4.80 and 4.81.

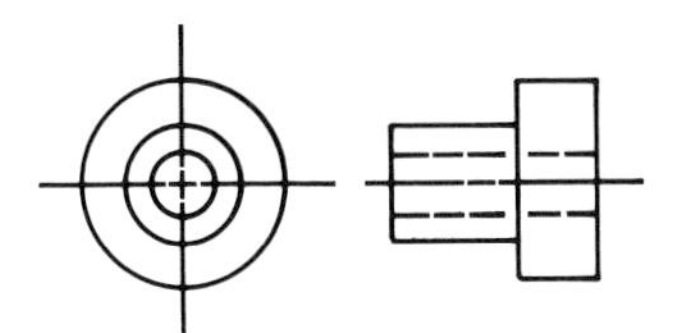

Figure 4.86 Two orthographic views are sufficient for most cylindrical and disc-shaped objects such as the one shown here.

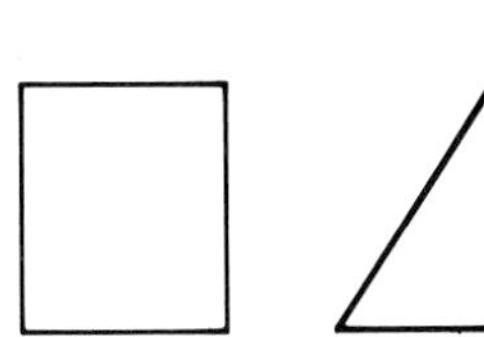

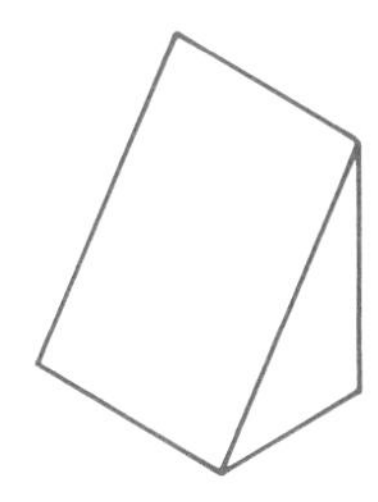

Figure 4.87 Two orthographic views are also sufficient for most simple objects such as the one shown here.

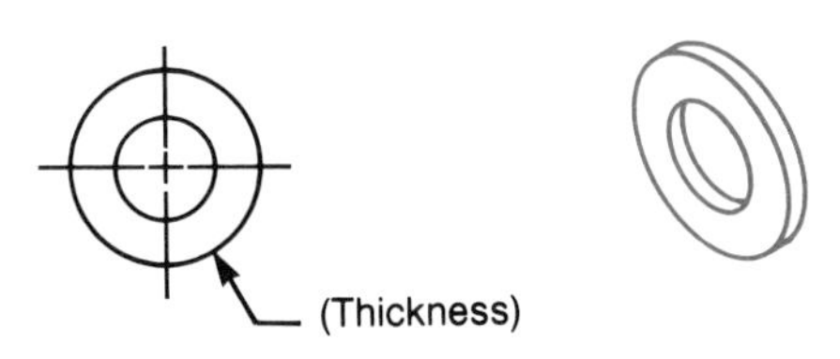

Figure 4.88 One orthographic view is sufficient for an object such as this if the third dimension (depth) is provided in a note.

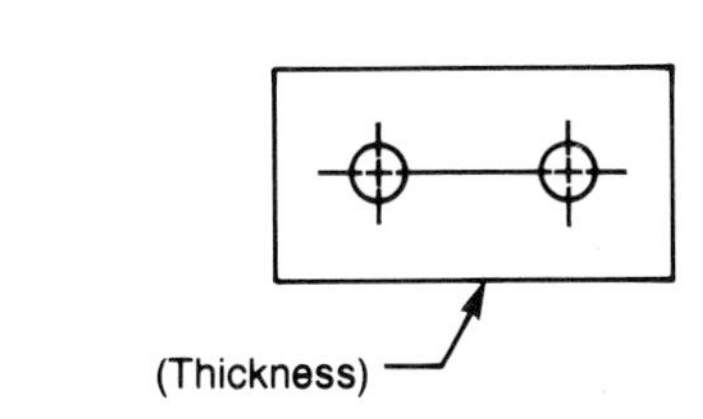

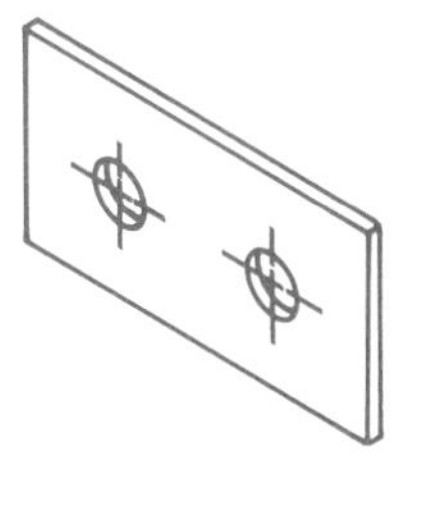

Figure 4.89 One orthographic view is also sufficient for an object such as this if the third dimension (depth) is provided in a note.

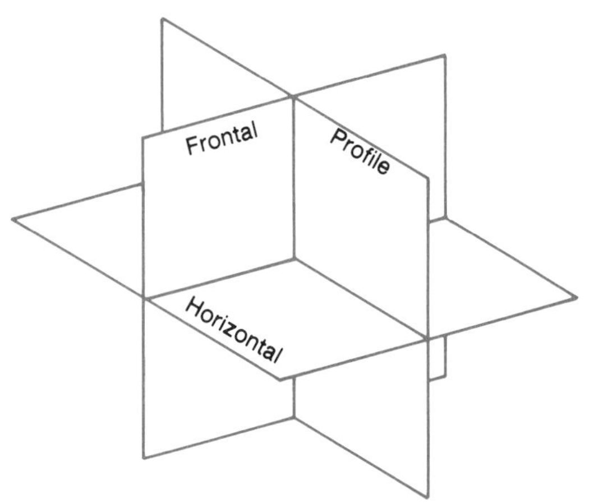

Figure 4.90 Intersection of Monge's planes of projection.

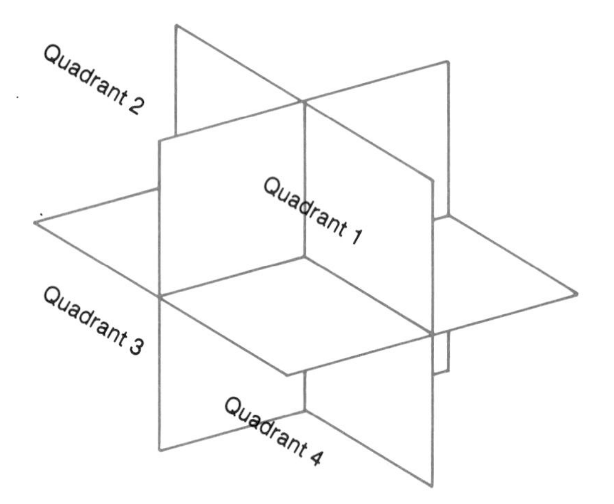

Figure 4.91 The first and third quadrants created by the intersection of Monge's planes of projection correspond to first-angle and third-angle projection, respectively.

4.14 First-Angle Projection and Third-Angle Projection

Thus far we have focused upon a particular form of orthographic projection known as **third-angle projection,** which is the popular format in the United States, Canada, and (to a lesser extent) Great Britain. **First-angle projection** is used in Europe and other areas of the world. Both types of projection can be defined with respect to Monge's planes of projection. The intersection of the mutually perpendicular Monge planes (Figure 4.90) creates quadrants, or sections, in space, as shown in Figure 4.91. Placement of an object (to be "projected" onto the Monge planes) in the first quadrant corresponds to first-angle projection, whereas placement in the third quadrant corresponds to third-angle projection.

In third-angle projection, the object is viewed from *behind* the Monge planes onto which projection is then performed (Figure 4.92). The resulting set of views (Figure 4.93) conforms to those previously discussed in this text. In first-angle projection, the object is viewed in *front* of the Monge planes onto which projection is then performed (Figure 4.94). The resulting set of views (Figure 4.95) is different from those previously discussed: the left-side view appears to the right of the front view, and the top view is placed below the front view.

The unusual placement of views in first-angle projection has limited the adoption of this format in the United States. Third-angle projection views are believed to be easier for the reader to interpret. Information is presented in a more intuitive form in third-angle projection: the right-side view of the object is located to the right of the front view, the top view is placed above the front view, and so on. The glass box technique, with its subsequent unfolding into a set of orthographic views for visualizing results, is based upon third-angle projection. We will continue to focus upon third-angle projection throughout the rest of this text because of its greater clarity and its popularity in the United States.

Symbols should be included near the title block on a drawing to indicate whether third-angle projection or first-angle projection has been used to develop the set of orthographic views. Figure 4.96 and 4.97 present the symbols for third-angle projection and first-angle projection, respectively, recommended by the International Standards Organization (ISO).

In summary, then:

1. Orthographic views may be presented in two equivalent formats: first-angle projection (used extensively in Europe) and third-angle projection (popular in the United States of America).

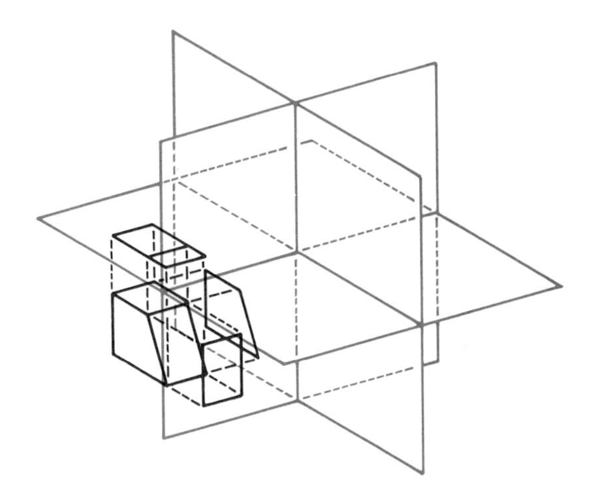

Figure 4.92 In third-angle projection, the object is viewed from behind the Monge planes.

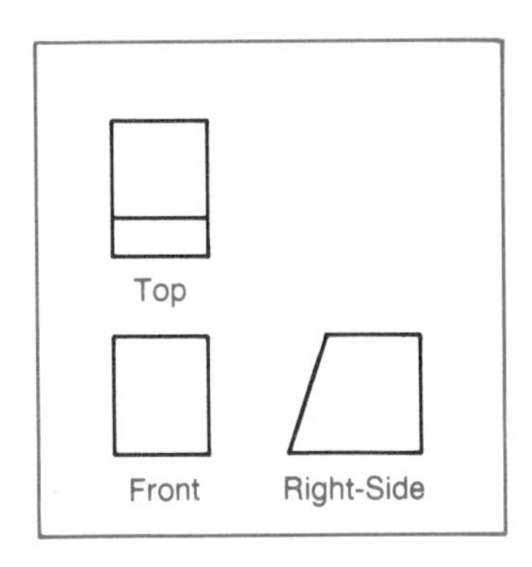

Figure 4.93 Resulting views from third-angle projection.

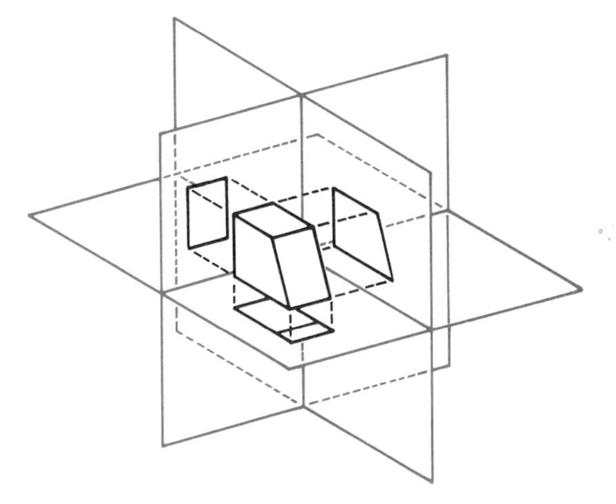

Figure 4.94 In first-angle projection, the object is viewed in front of Monge's planes.

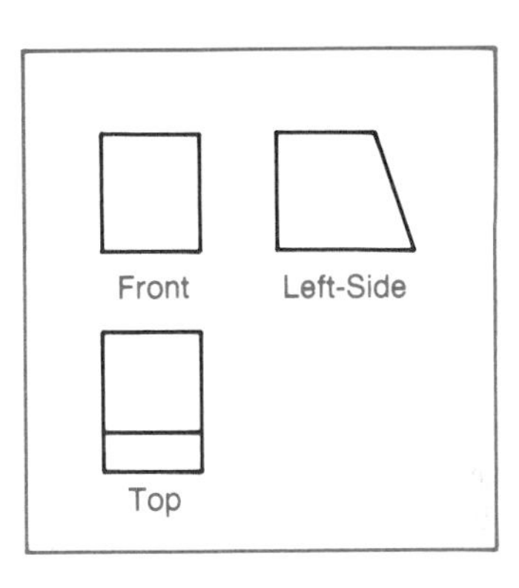

Figure 4.95 Resulting views from first-angle projection.

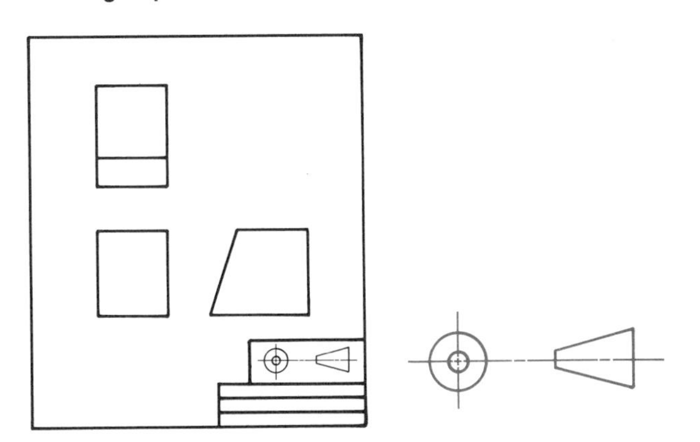

Figure 4.96 Symbol for third-angle projection (included near the title block of a drawing).

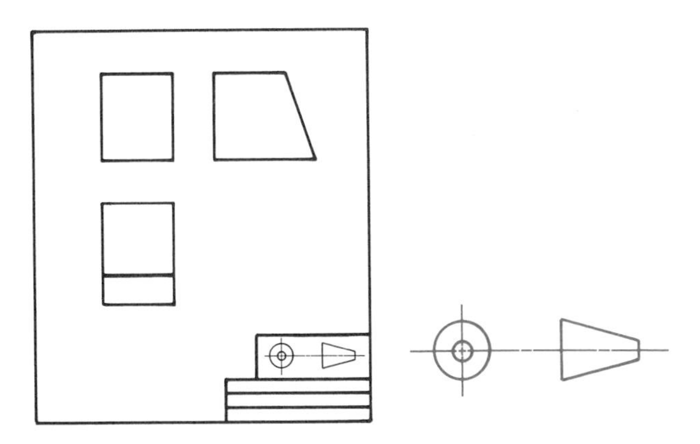

Figure 4.97 Symbol for first-angle projection (included near the title block of a drawing).

Highlight

First-Angle Projection and Third-Angle Projection

Monge's descriptive geometry was introduced in the United States by Claude Crozet at the U.S. Military Academy at West Point. Crozet was a graduate of the École Polytechnique and had served as an artillery officer for Napoleon. He arrived at West Point in 1816; in 1821 his text entitled *A Treatise on Descriptive Geometry* was published for use at the military academy. This work may have been the first treatment in English on the topic of descriptive geometry. Crozet has become known as the father of descriptive geometry in the United States.

Monge defined only two planes of projection in his original work, thereby allowing one to create only two views. First-angle projection developed in Europe from the Monge planes and descriptive geometry. In the early part of the nineteenth century, Crozet and others promoted first-angle projection in America. However, others arranged orthographic views in non-traditional ways for *ease of execution* in the drawing effort as opposed to clarity in interpretation and communication. During the industrial development of America between 1850 and 1900, college-trained draftsmen joined the ranks of those who had served as apprentices to seasoned professionals in order to learn the drafting trade. College teachers and textbook writers focused upon the arrangement of orthographic views that we now call third-angle projection because of its clarity and simplicity in layout. Third-angle projection became popular in America through this effort.

Points, lines, and planes were developed in terms of first-angle projection, but solid geometry was treated in terms of third-angle projection in textbooks. By the time of World War I, it was recognized that a consistent development was needed for both descriptive geometry and third-angle projection. A physical model was proposed, the *glass box* model, which provided the theoretical foundation for third-angle projection and its particular arrangement of orthographic views. Three planes of projection and a comprehensive treatment of both first-angle and third-angle projection could then be formulated.

During World War II, communication via orthographic drawings between Allied forces necessitated that draftsmen on both sides of the Atlantic be familiar with both first-angle projection and third-angle projection. Today British industry is adopting third-angle projection while maintaining its traditional use of first-angle projection in order to serve its European counterparts.

The first American Drawing Standard was approved in 1935; it included third-angle projection for the arrangement of orthographic views.

Source: Peter Booker, A History of Engineering Drawing *(London: Northgate Publishing Company, Ltd., 1979.)*

2. In first-angle projection, the object is viewed in *front* of the Monge planes onto which projection is then performed. In third-angle projection the object is viewed from *behind* the Monge planes (in accordance with presentations throughout this text).
3. A symbol should be included near the title block on a drawing in order to indicate the type of projection used.

In Retrospect

- The three planes of projection—frontal reference plane (FRP), horizontal reference plane (HRP), and profile reference plane (PRP), which are based upon Monge's two original planes of projection, allow us to develop orthographic, two-dimensional views of an object. A set of such views can be used to completely describe an object.
- The planes of projection generated the concept of the glass box, from which the six principal orthographic views resulted.
- Two specific dimensions are associated with each principal view:
  - Front view, rear view—height, width
  - Right-side view, left-side view—height, depth
  - Top view, bottom view—width, depth.
- Three rules must be obeyed in the development and arrangement of orthographic views:
  - Rule 1—Orthographic views must be drawn with consistent (common) dimensions.
  - Rule 2—Orthographic views must be mutually aligned.
  - Rule 3—Orthographic views must obey the rule of adjacency in their relative arrangement in a drawing.
- The selection of the best and necessary views should be based upon the major goal of achieving clarity in the description of the object under consideration. As a result, we seek to minimize the number of views to only those necessary to describe the object. Views that provide the most unambiguous information about the shape of an object should be selected. Finally, if two views provide identical information about the object, that view which contains less hidden linework should be selected.
- There are three categories of planar surfaces: surfaces parallel to a projection plane, inclined surfaces, and oblique surfaces. Each type of planar surface produces a distinct set of projection results in the six principle orthographic views.
- The configuration rule states that a planar surface that appears as an area with *n* sides or boundaries in one orthographic view will appear as an area with *n* sides or will appear on edge as a line in all other orthographic views. The configuration rule can be very helpful as one develops an interpretation of a set of orthographic views (that is, as one "reads" a set of views).
- The surface-by-surface, point-by-point technique requires one to identify each surface of an object (which is described by a set of orthographic views) in *each* view, one surface at a time. As each surface is identified in each view, key points of that surface are also identified in each view. This careful approach allows one to analyze a given set of views with minimal danger of committing an error in interpretation.
- Linework priority in orthographic views is (1) visible linework, (2) hidden linework, and (3) center linework. (Additional types of linework will be introduced in forthcoming chapters.)
- Intersections between visible and hidden lines must be carefully drawn to ensure clarity.
- Certain objects can be adequately described by only two orthographic views. A few types of objects can even be described by one view if the third dimension (depth) is provided in a note on the view.
- Third-angle projection is the popular format for the arrangement of orthographic views in the United States, Canada, and (to a lesser extent) Great Britain. First-angle projection is used elsewhere. In this text, we use third-angle projection because of its general acceptance in North America.

COMPUTER GRAPHICS IN ACTION

3D Computer Graphics: Surface Models

In the "Computer Graphics in Action" section of Chapter 3, we mentioned three types of three-dimensional computer graphics models: wire-frame, surface, and solid. We then focused briefly upon the advantages and the disadvantages of wire-frame models. A more useful and informative computer graphics representation of a three-dimensional object is the **surface model,** in which the user constructs the part by using **predefined surface elements** (planar and nonplanar). As a result, surface boundaries are more well-determined; this can be an important characteristic in such industrial applications as **NC (numerical control) machining operations,** in which a cutting tool is controlled by a computer to produce the desired part—based upon the geometric model of the object within the computer data base. However, as in wire-frame modeling, interior surfaces are not well-defined. Surface definitions are restricted to the exterior of the object, and the interior is not specified in any form.

In order to fully and clearly describe an object via three-dimensional computer graphics, we may use a third form of representation, solid modeling, which will be examined in the "Computer Graphics in Action" sections of Chapters 5, 6, and 7.

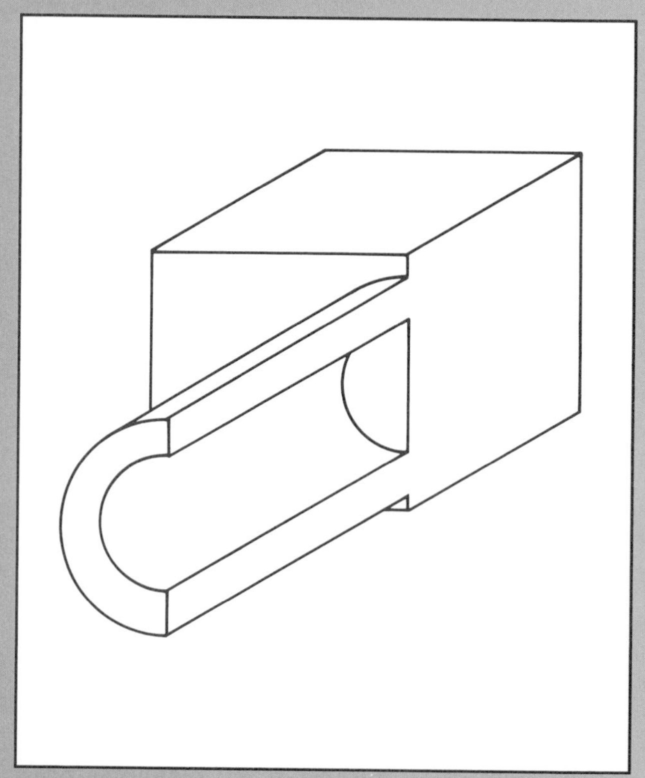

A surface model of an object.

Exercises

Your instructor will identify the appropriate scale for each exercise.

EX4.1 and EX4.2 Two surfaces, A and B, have been labeled in the given views. Identify these two surfaces in all views.

EX4.3 through EX4.35 Two complete orthographic views are given in each exercise. Draw the most appropriate missing view (or that missing view indicated by the small corner identification mark shown in some exercises).

EX4.36 through EX4.50 A pictorial sketch of an object is given in each exercise. Draw the best and necessary views for each object.

EX4.51 through EX4.59 Draw the best and necessary views of the objects described in these figures.

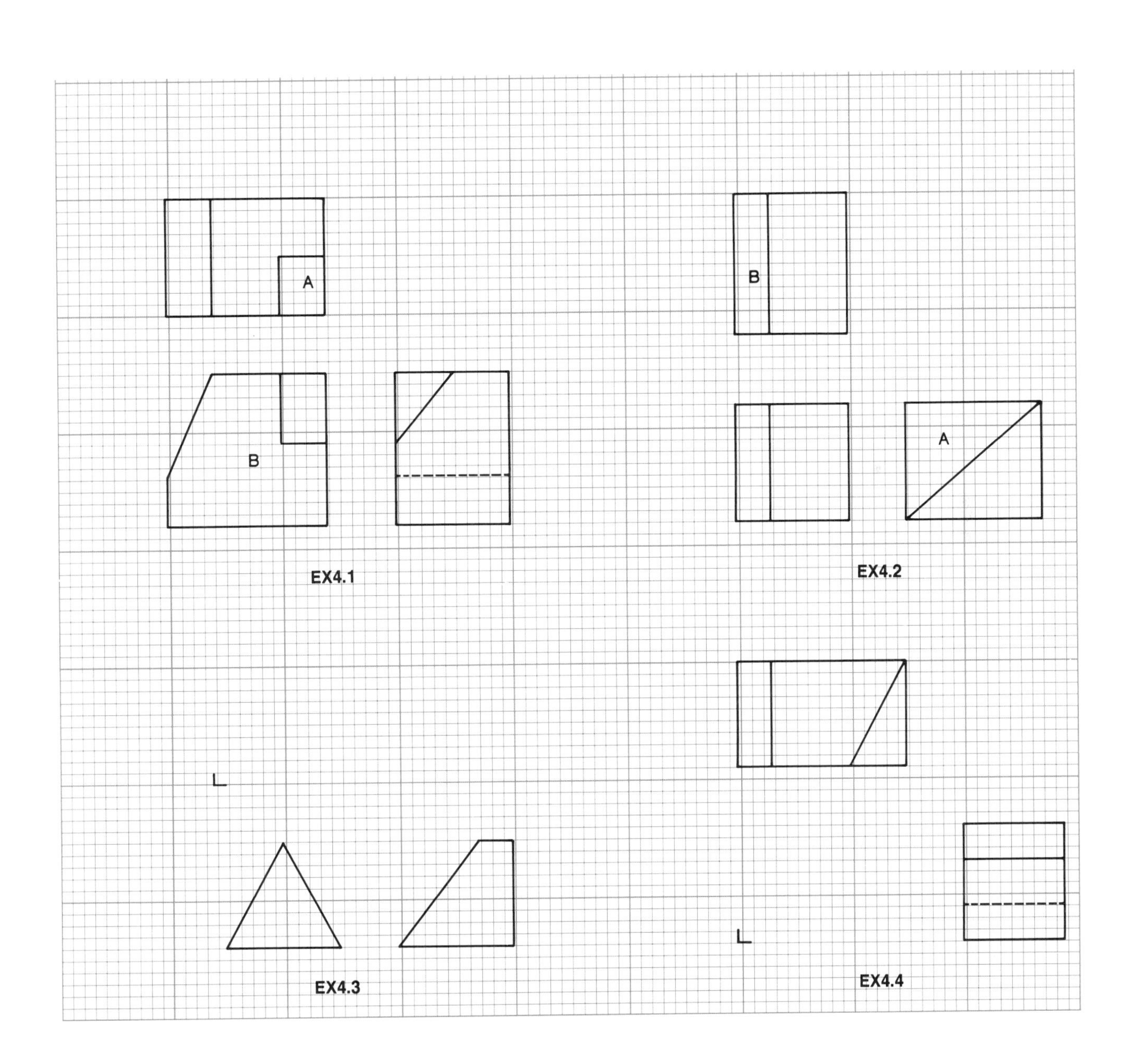

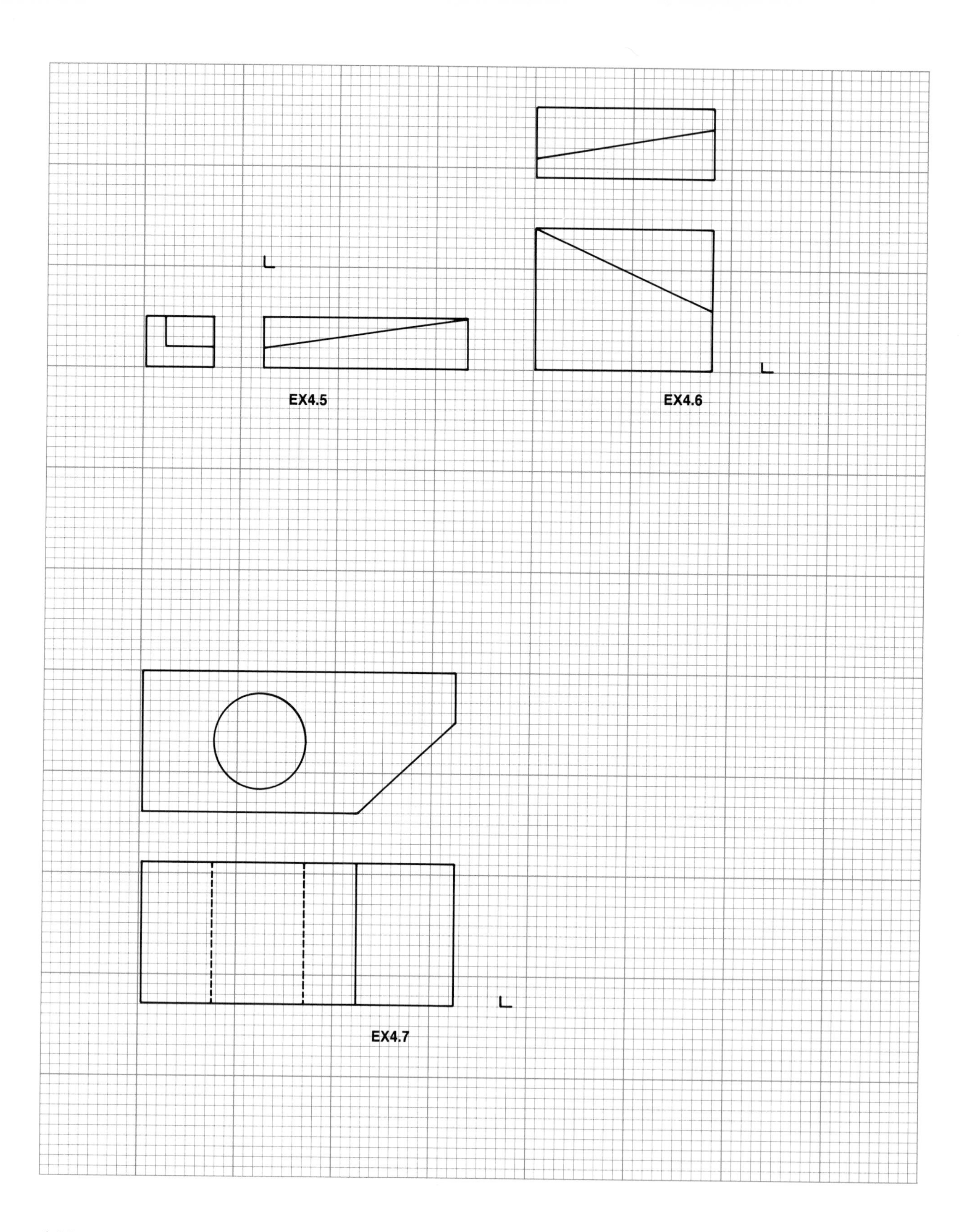
EX4.5
EX4.6
EX4.7

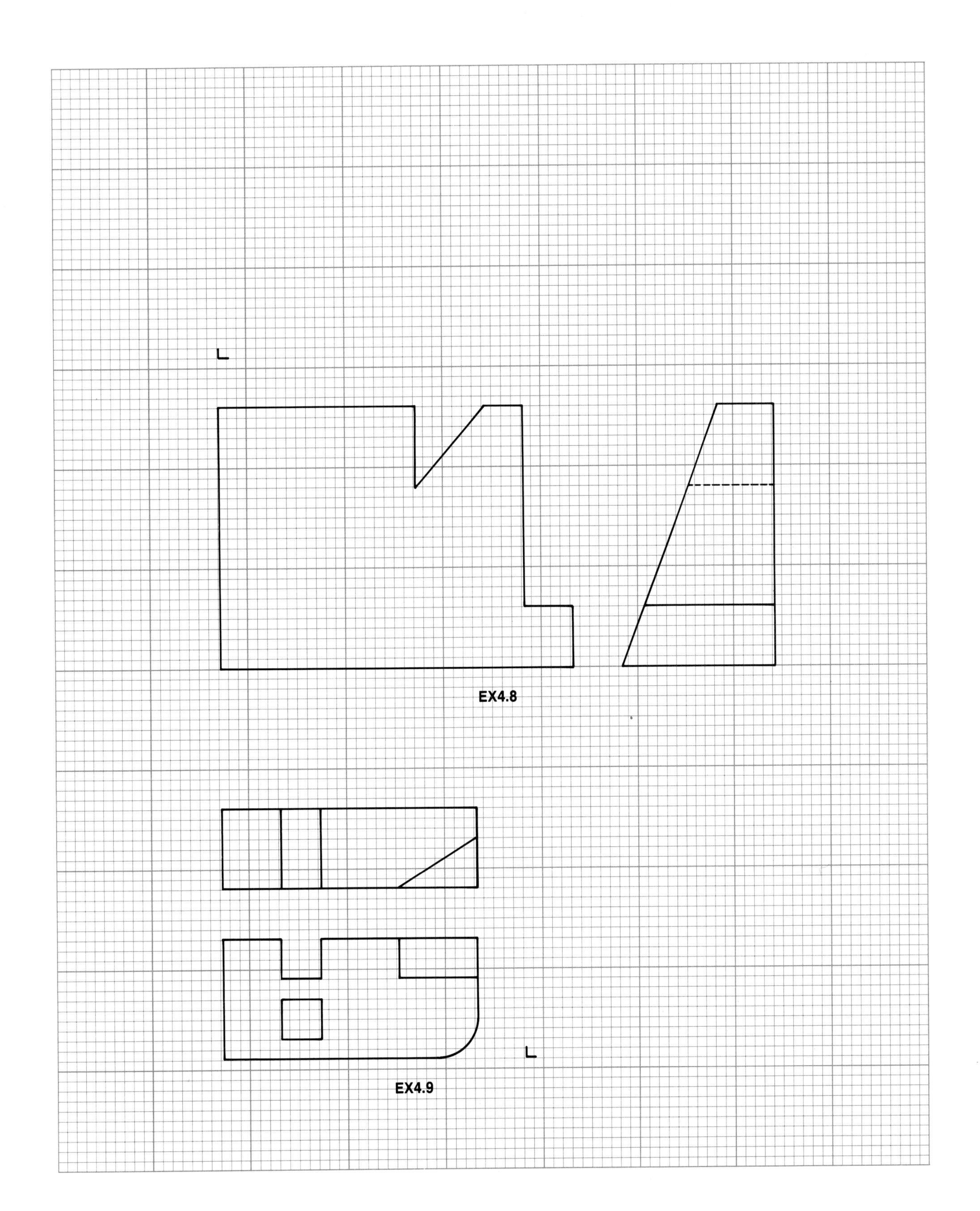
EX4.8
EX4.9

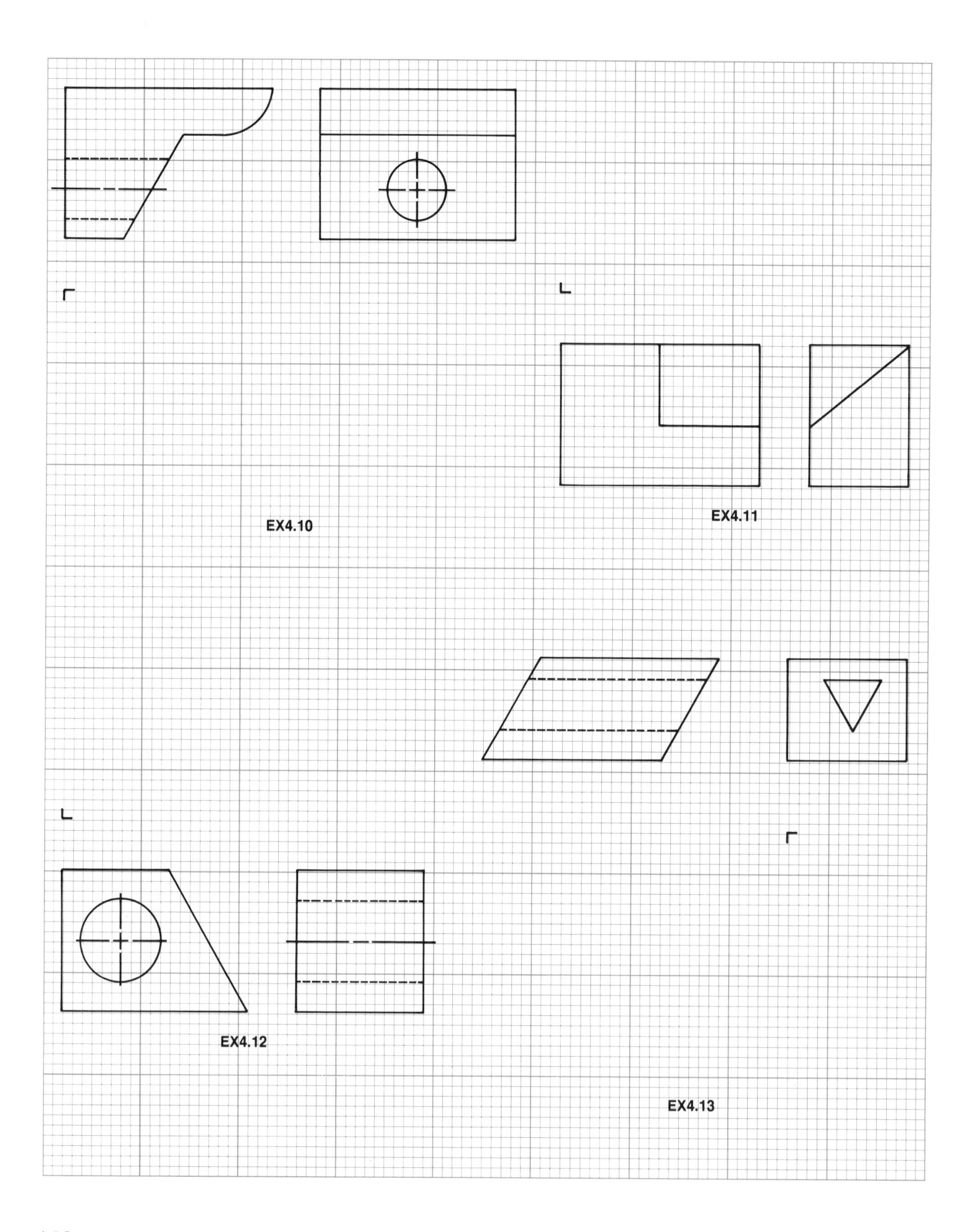
EX4.10
EX4.11
EX4.12
EX4.13

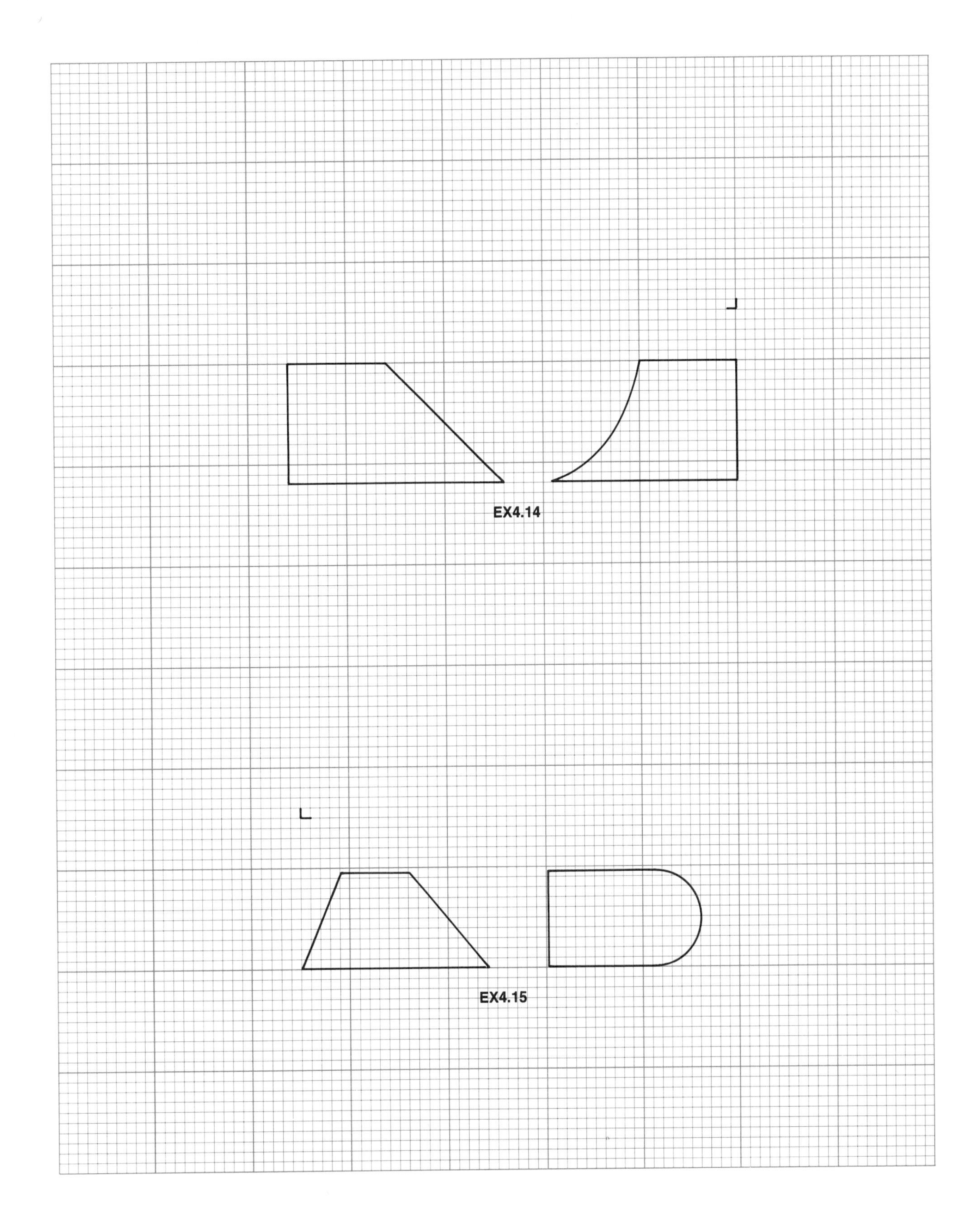
EX4.14
EX4.15

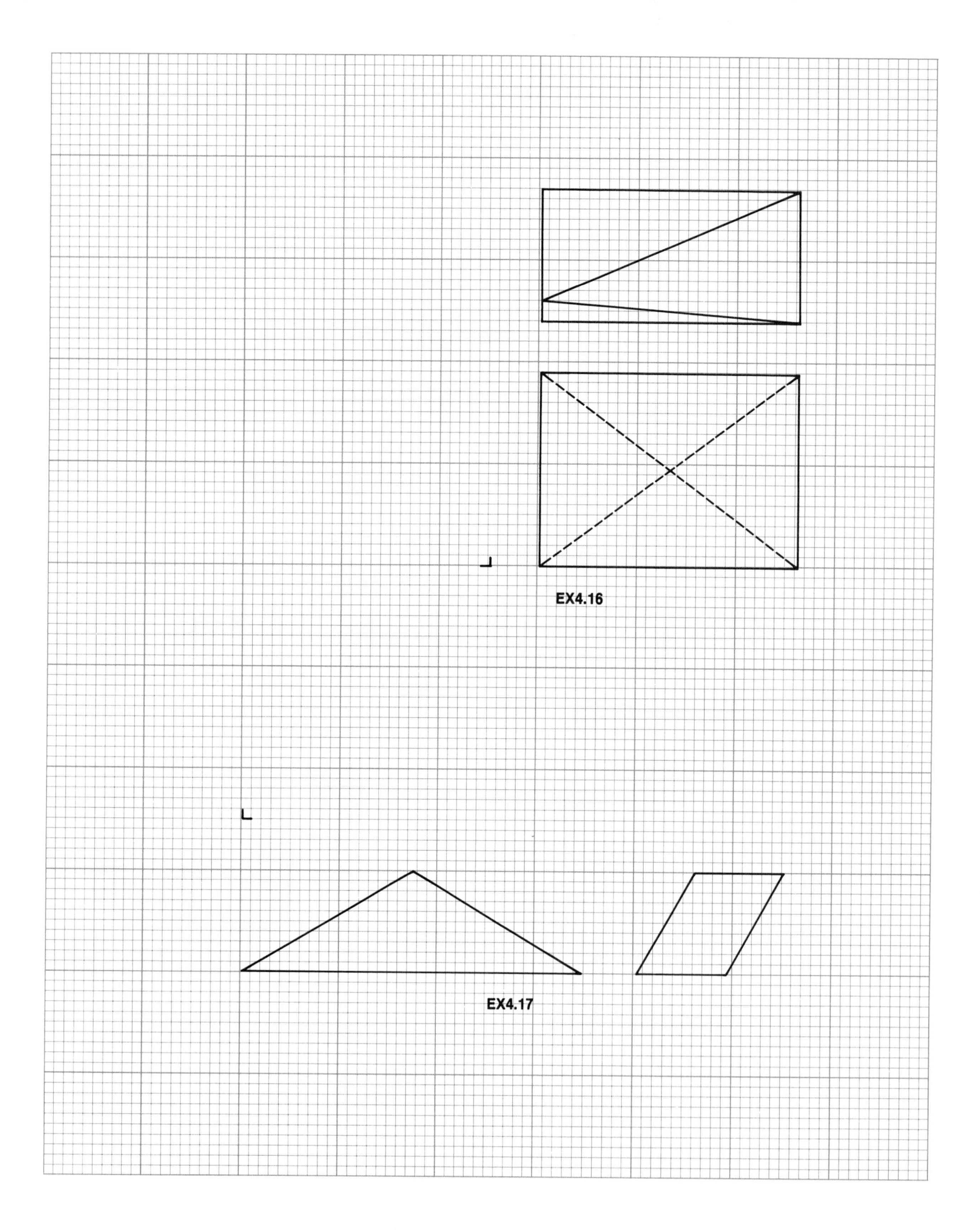
EX4.16
EX4.17

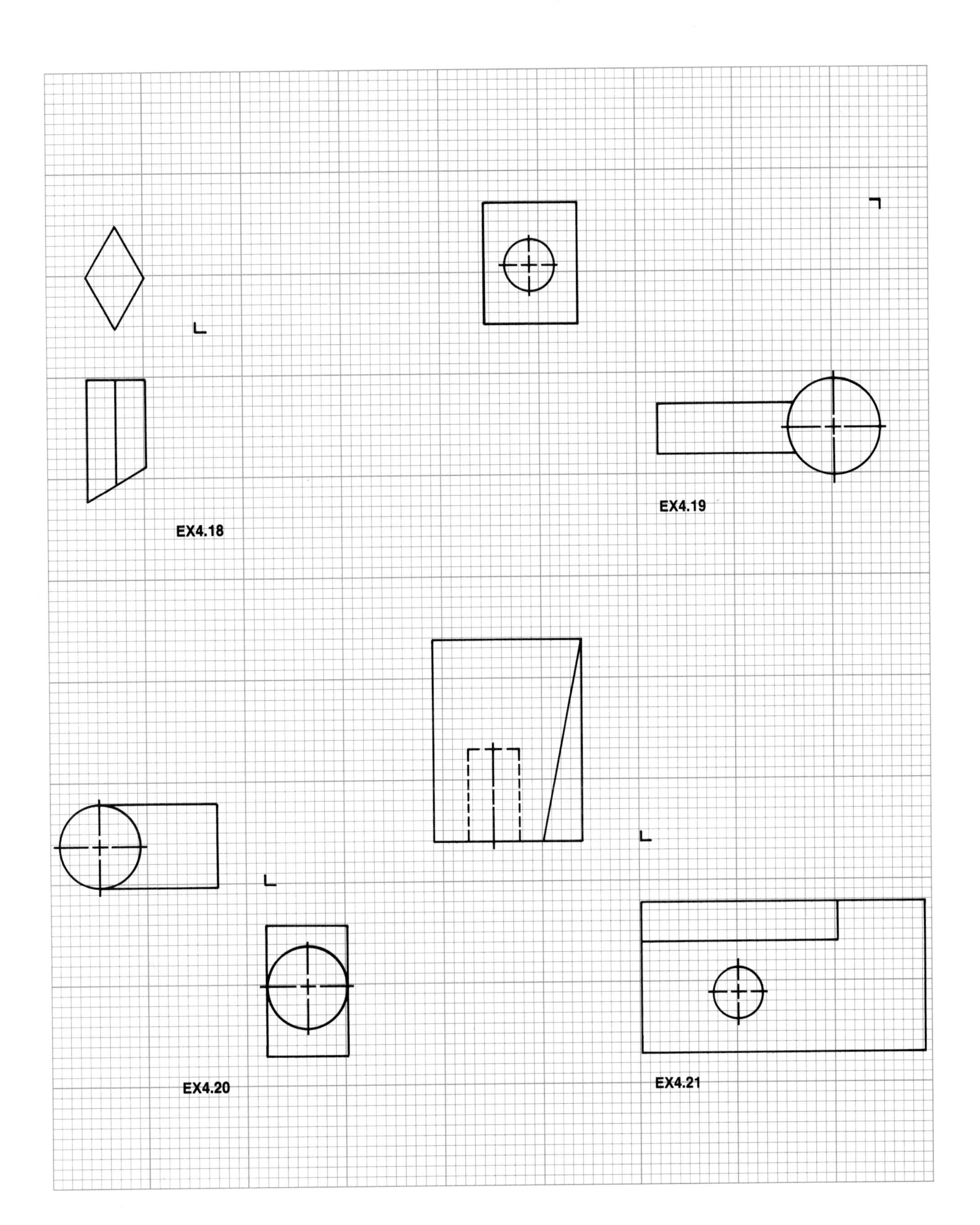

EX4.18
EX4.19
EX4.20
EX4.21

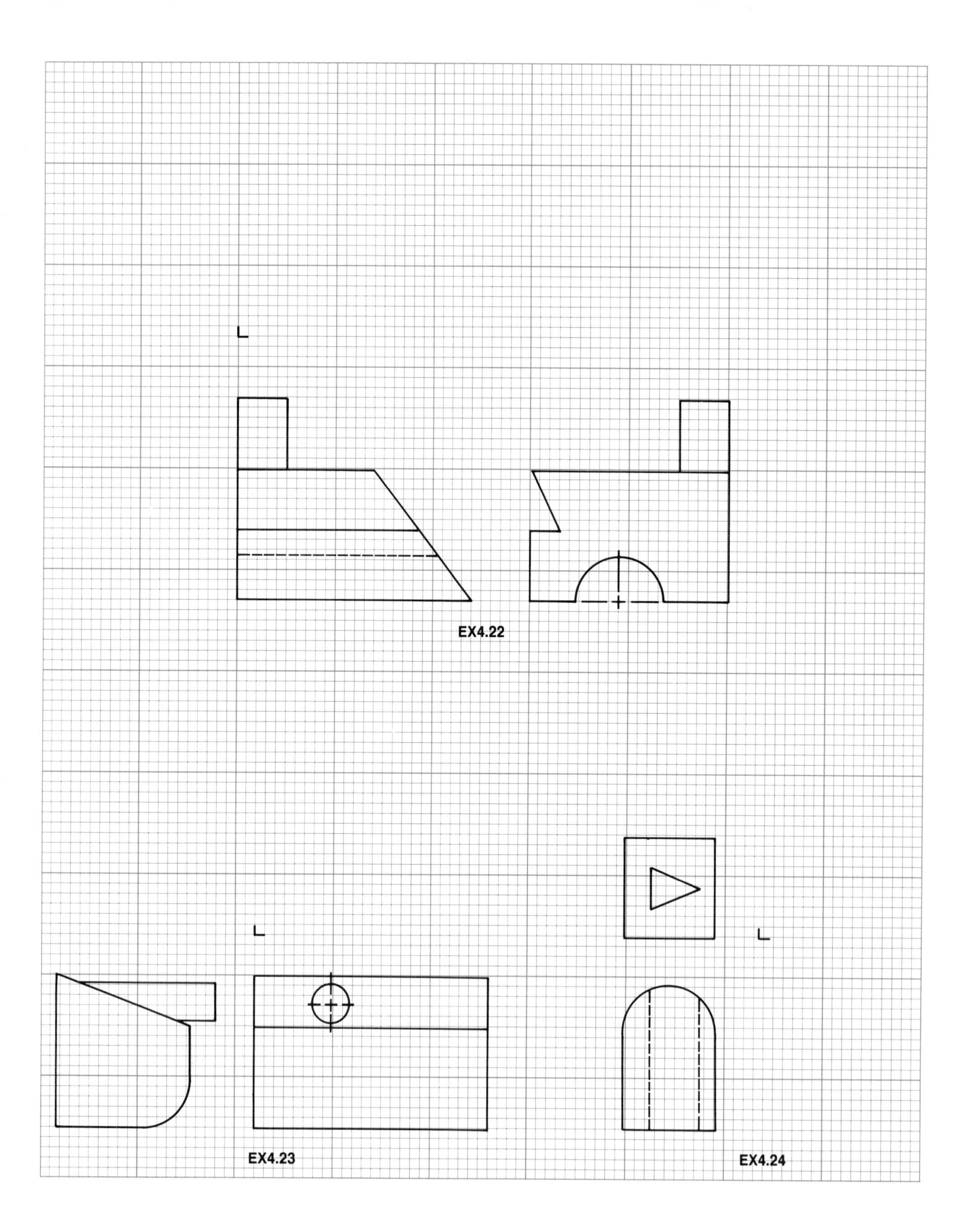
EX4.22
EX4.23
EX4.24

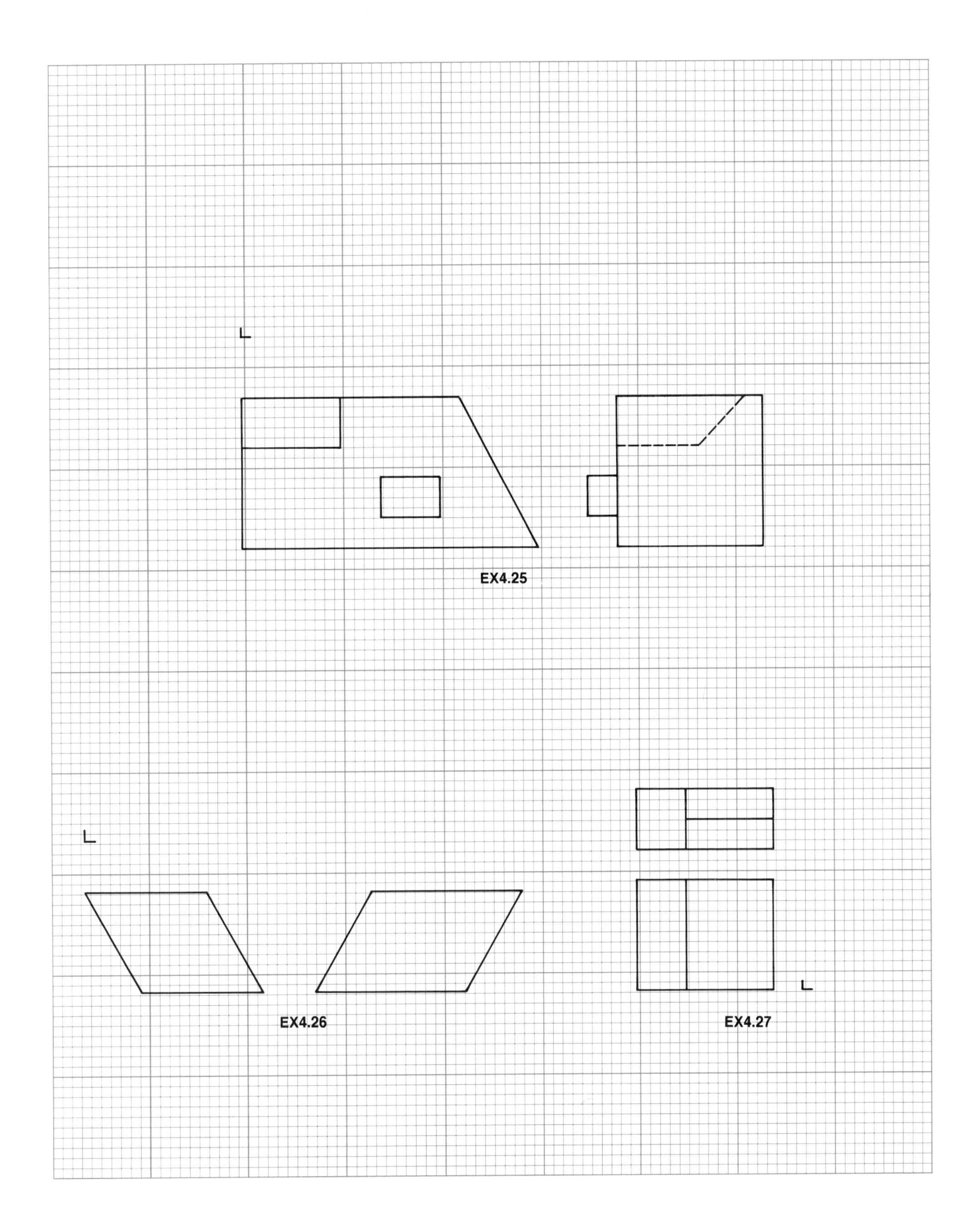
EX4.25
EX4.26
EX4.27

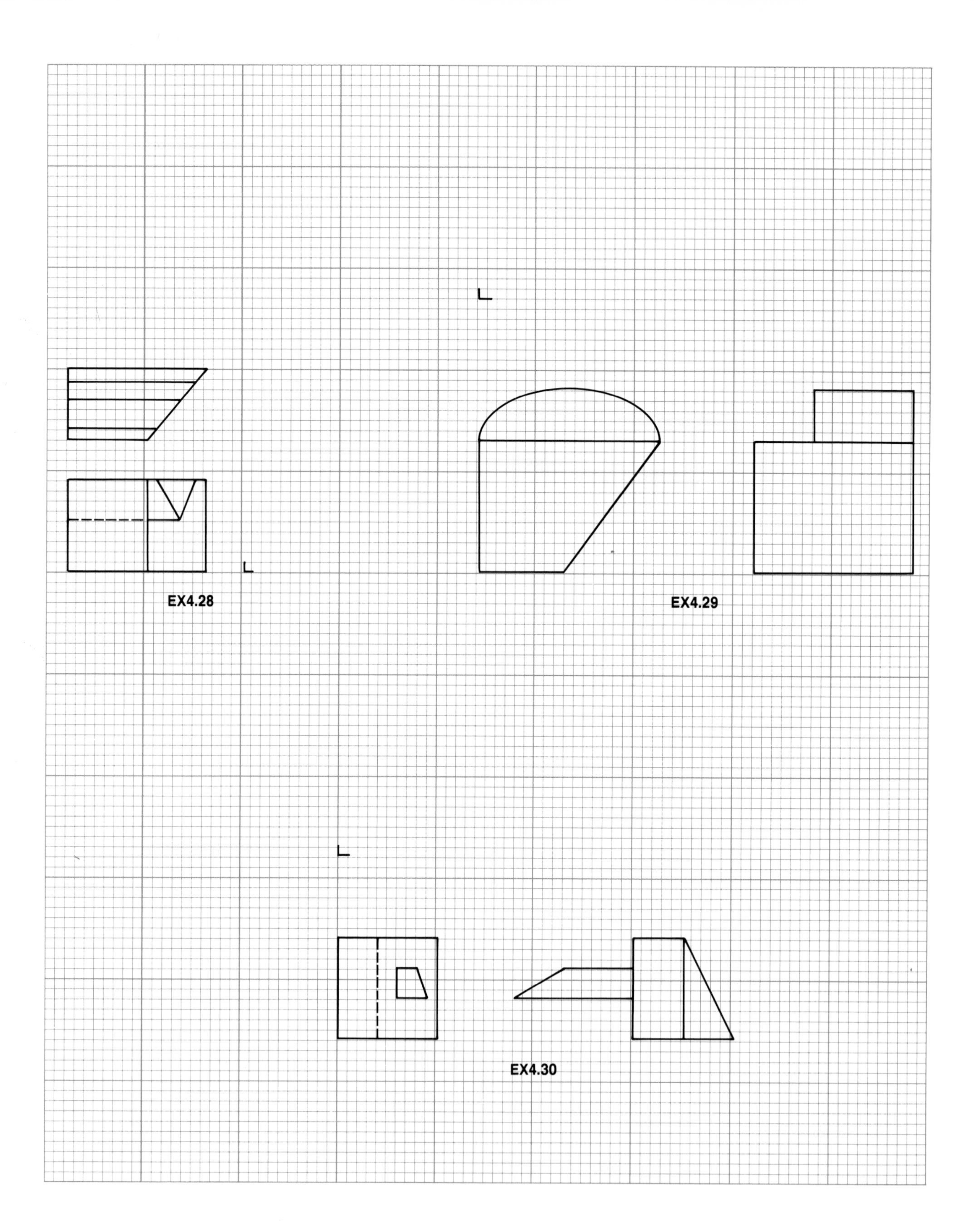
EX4.28
EX4.29
EX4.30

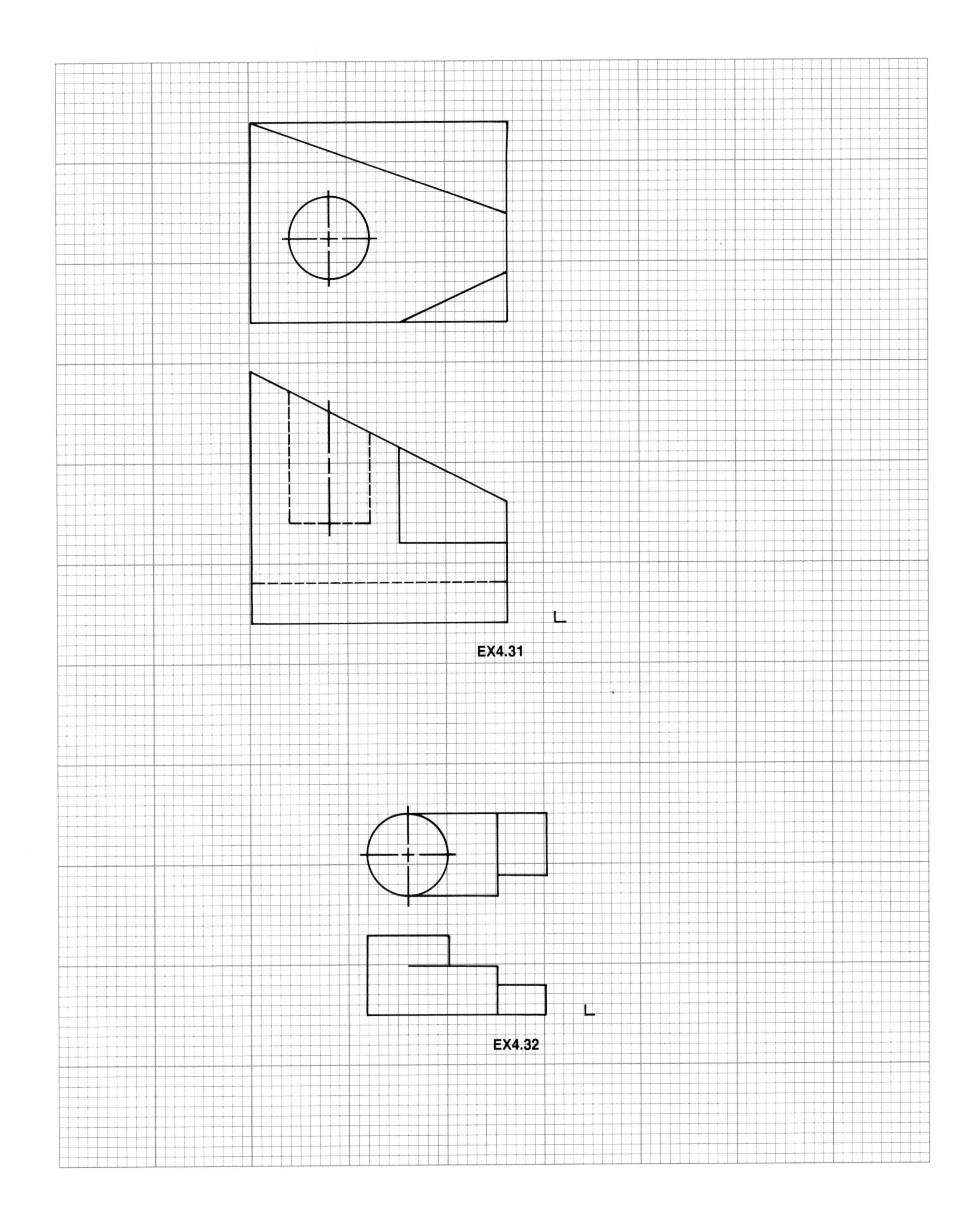

L
EX4.31
L
EX4.32

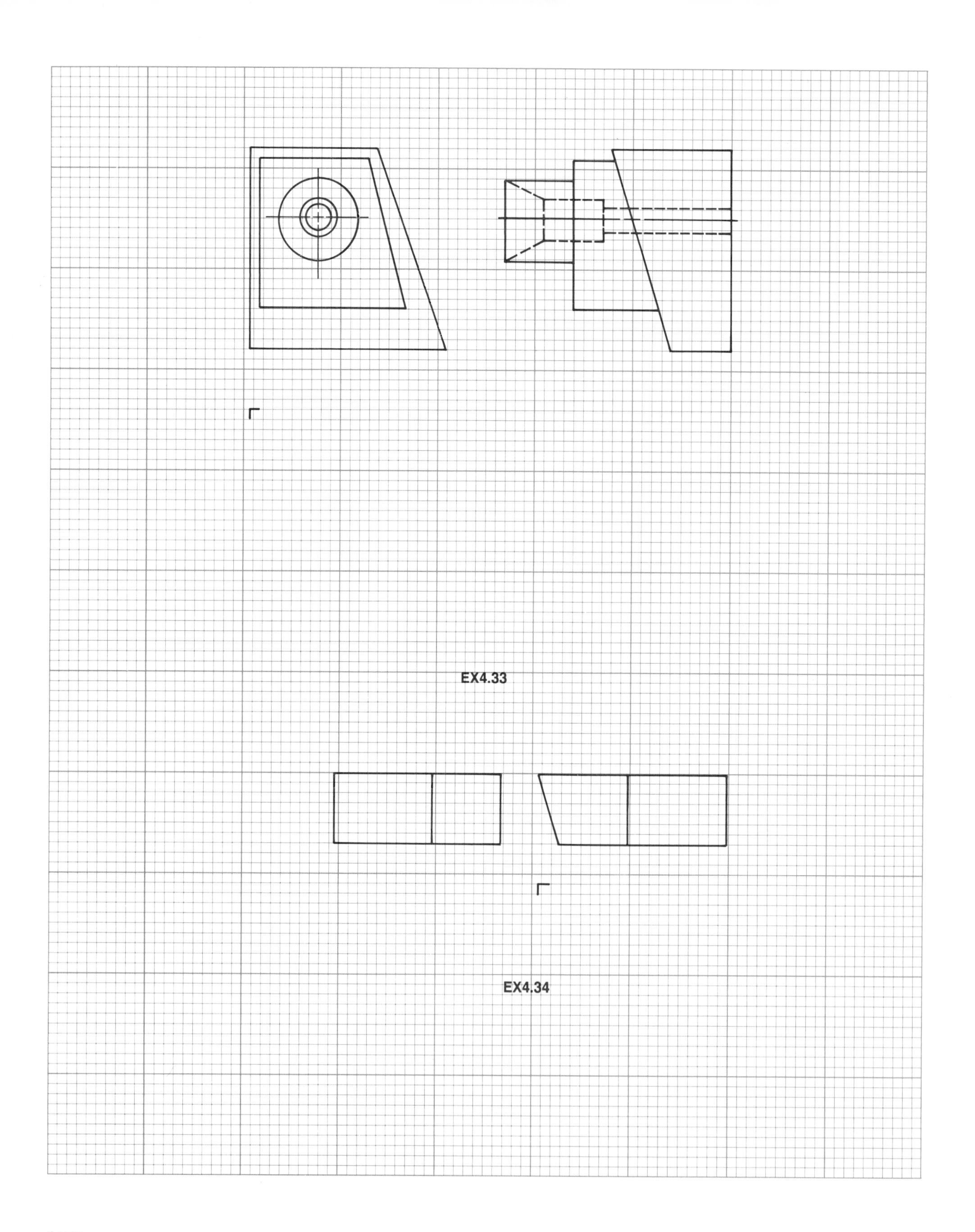
EX4.33
EX4.34

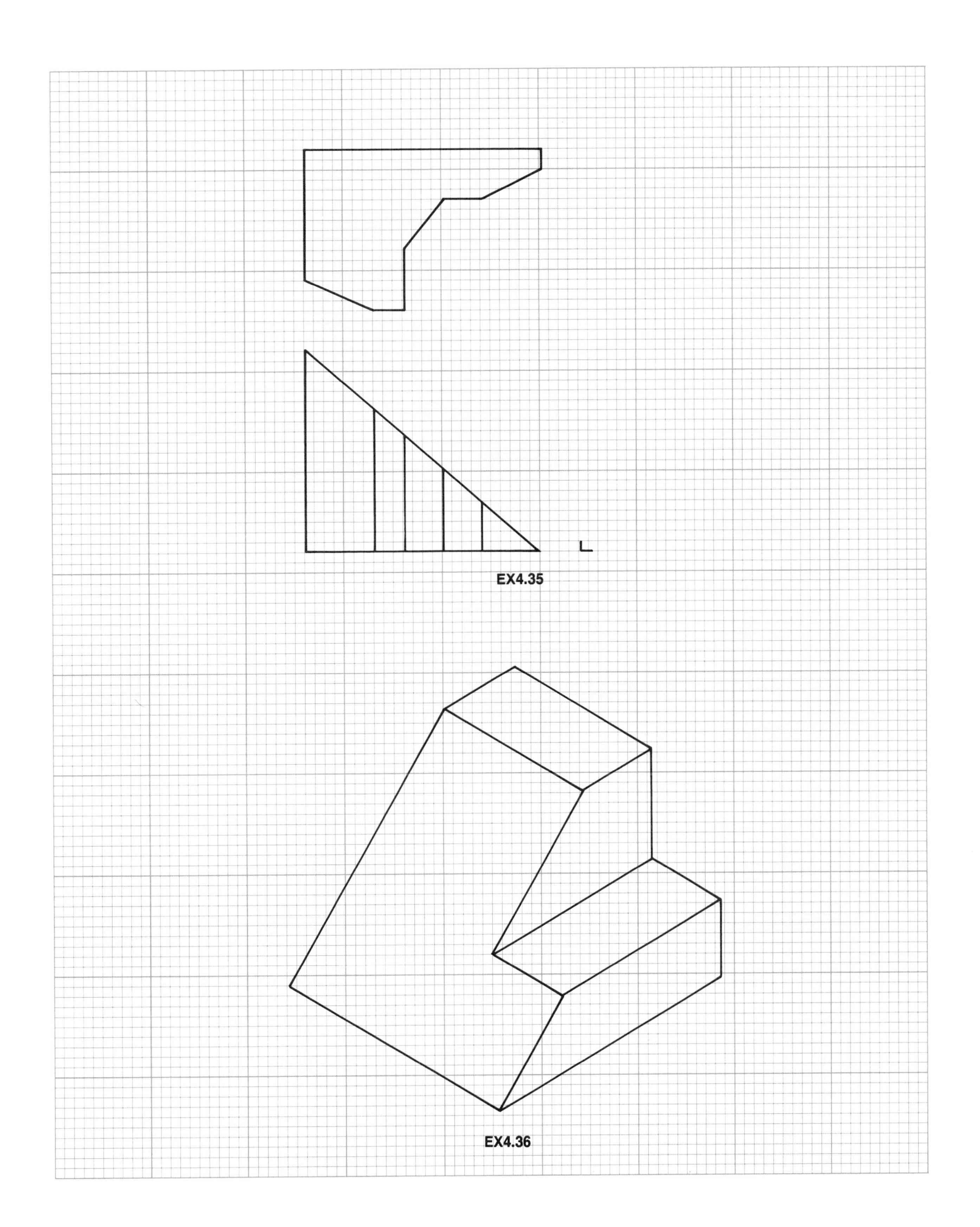
EX4.35
EX4.36

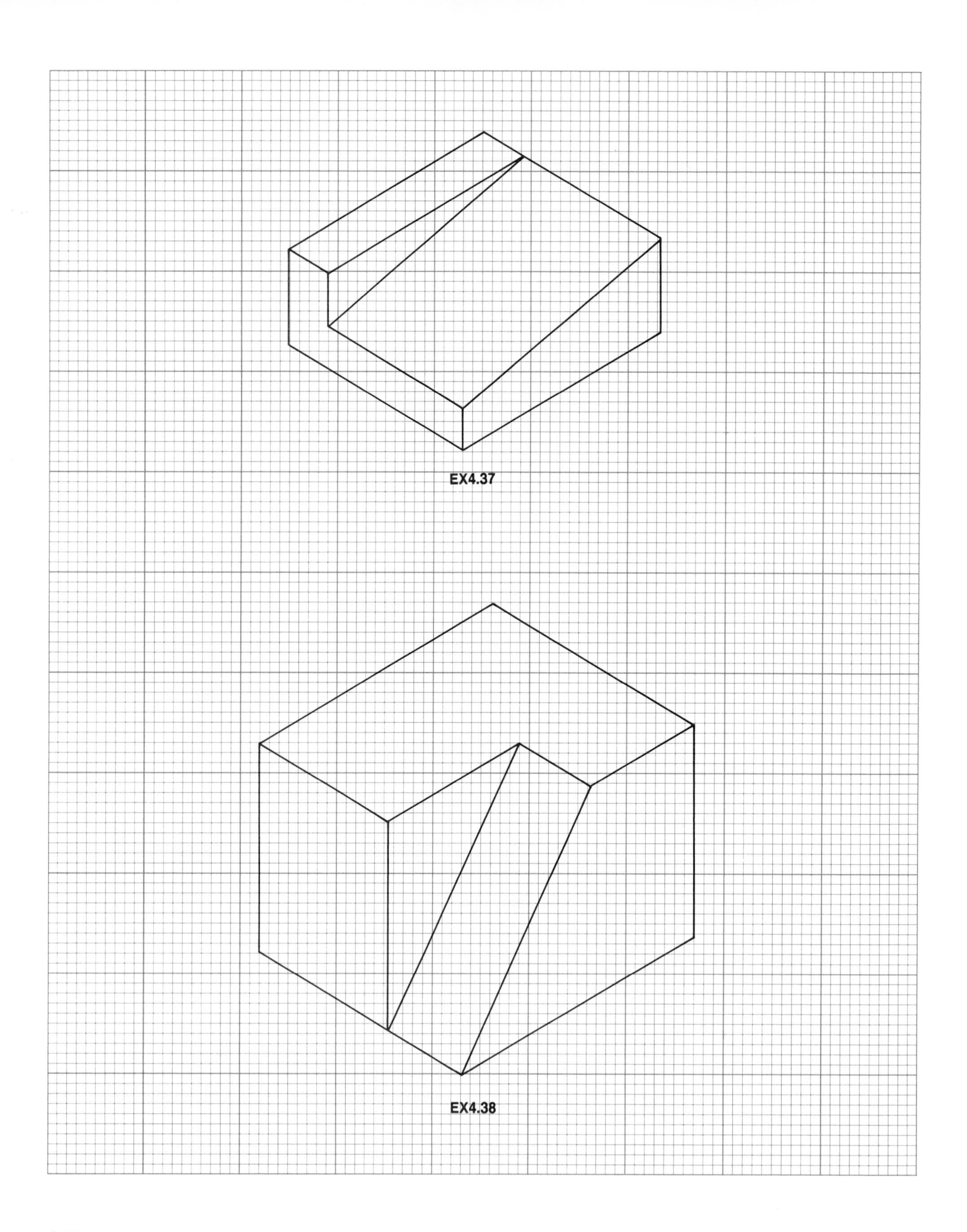
EX4.37
EX4.38

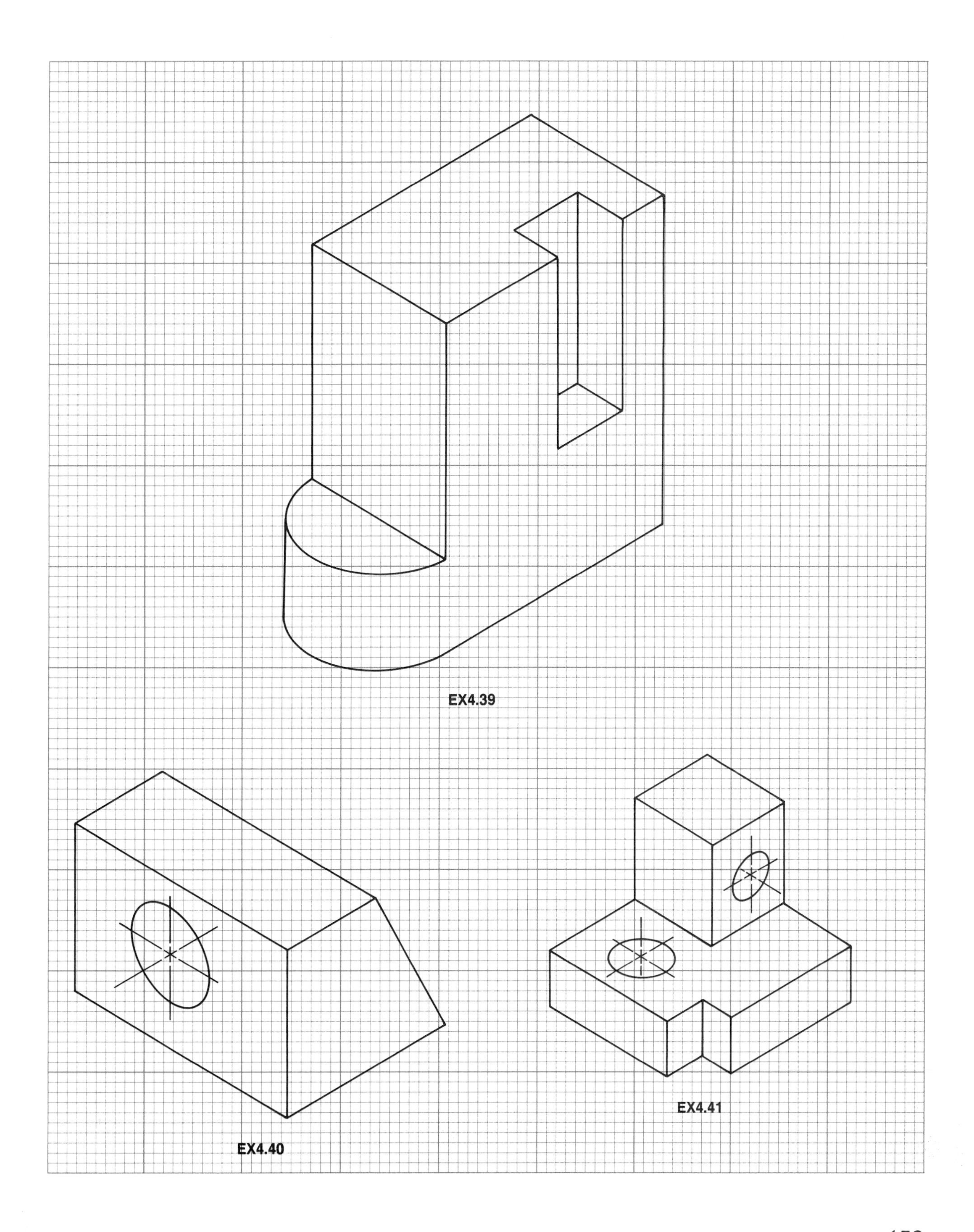
EX4.39
EX4.40
EX4.41

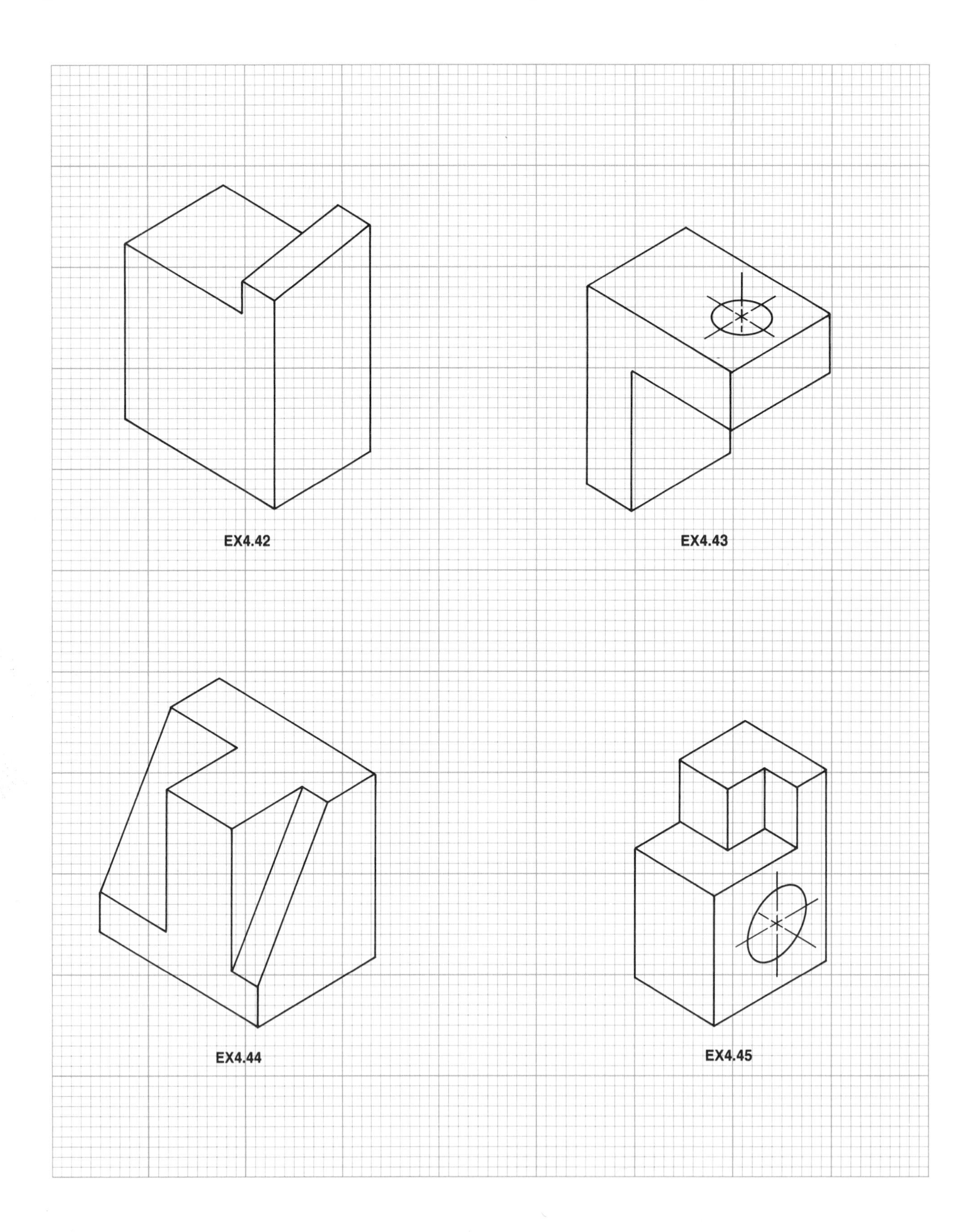
EX4.42
EX4.43
EX4.44
EX4.45

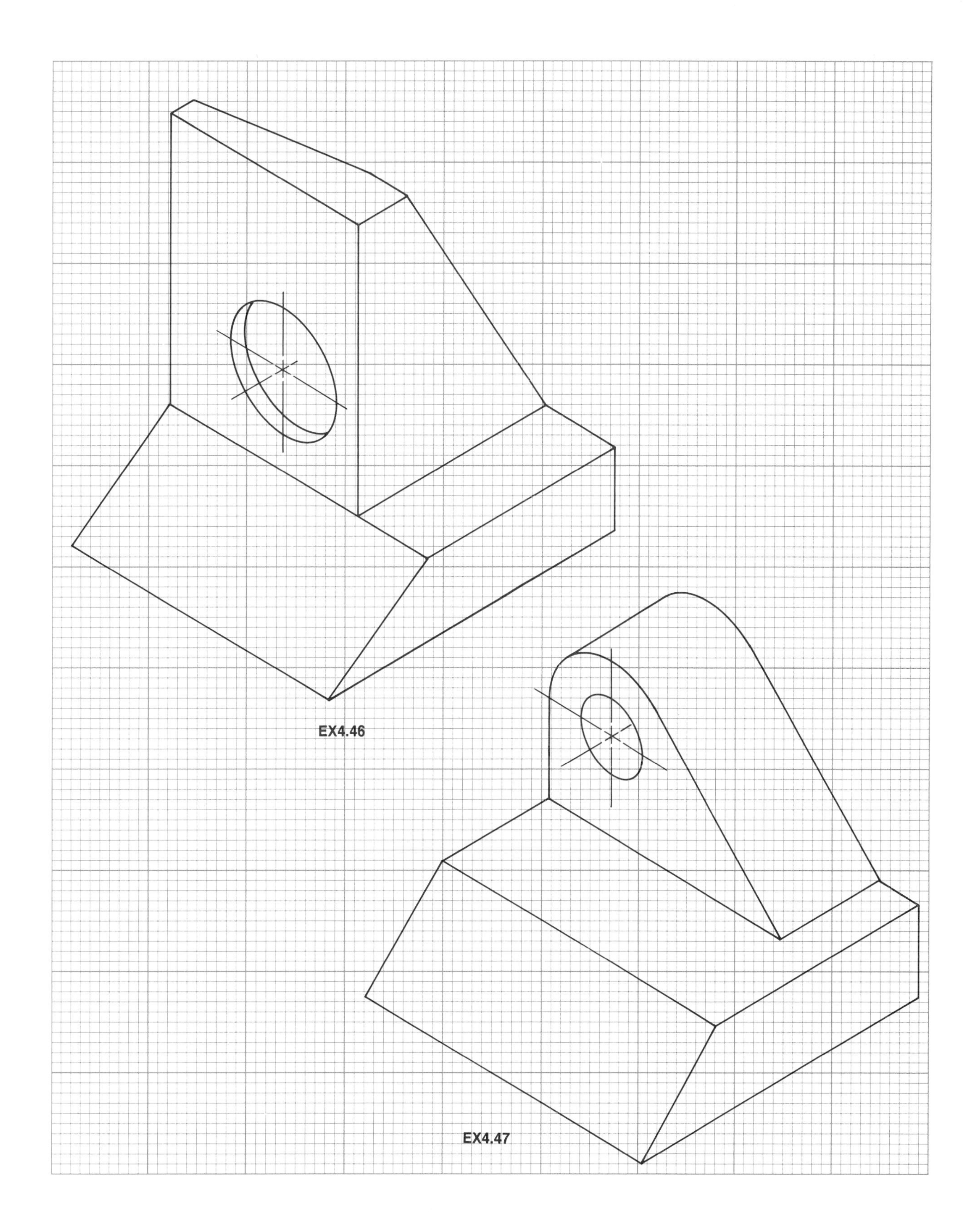
EX4.46
EX4.47

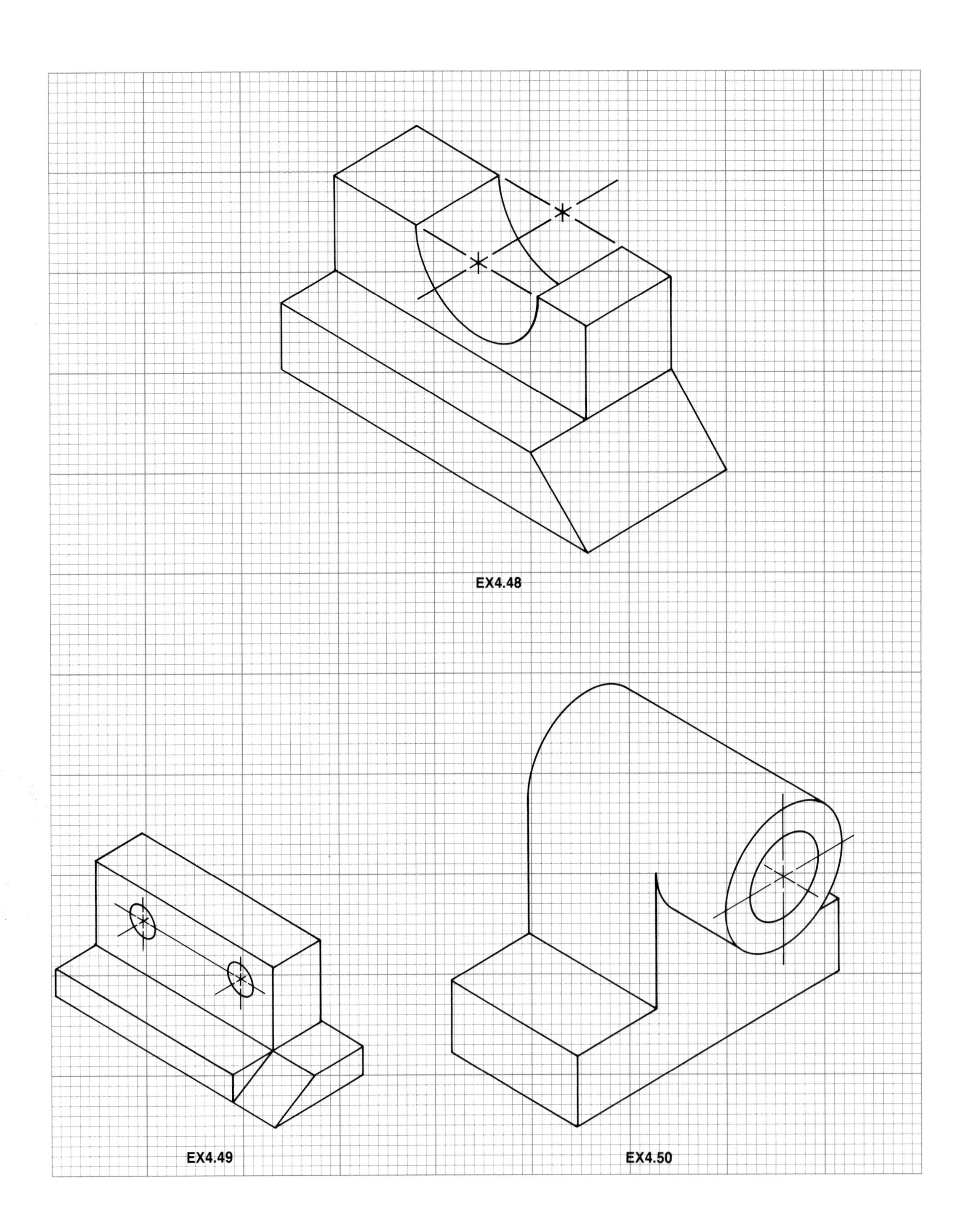
EX4.48
EX4.49
EX4.50

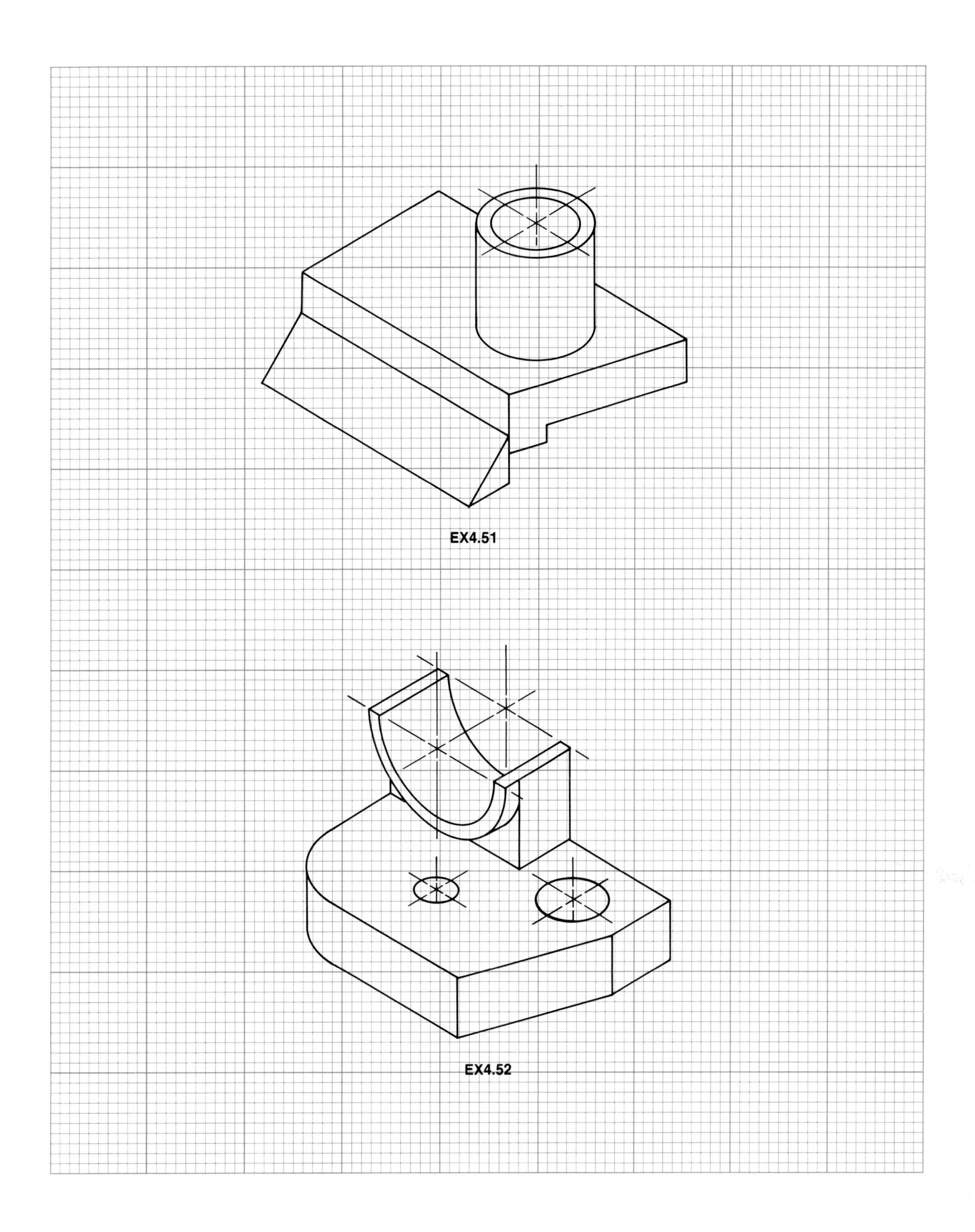
EX4.51
EX4.52

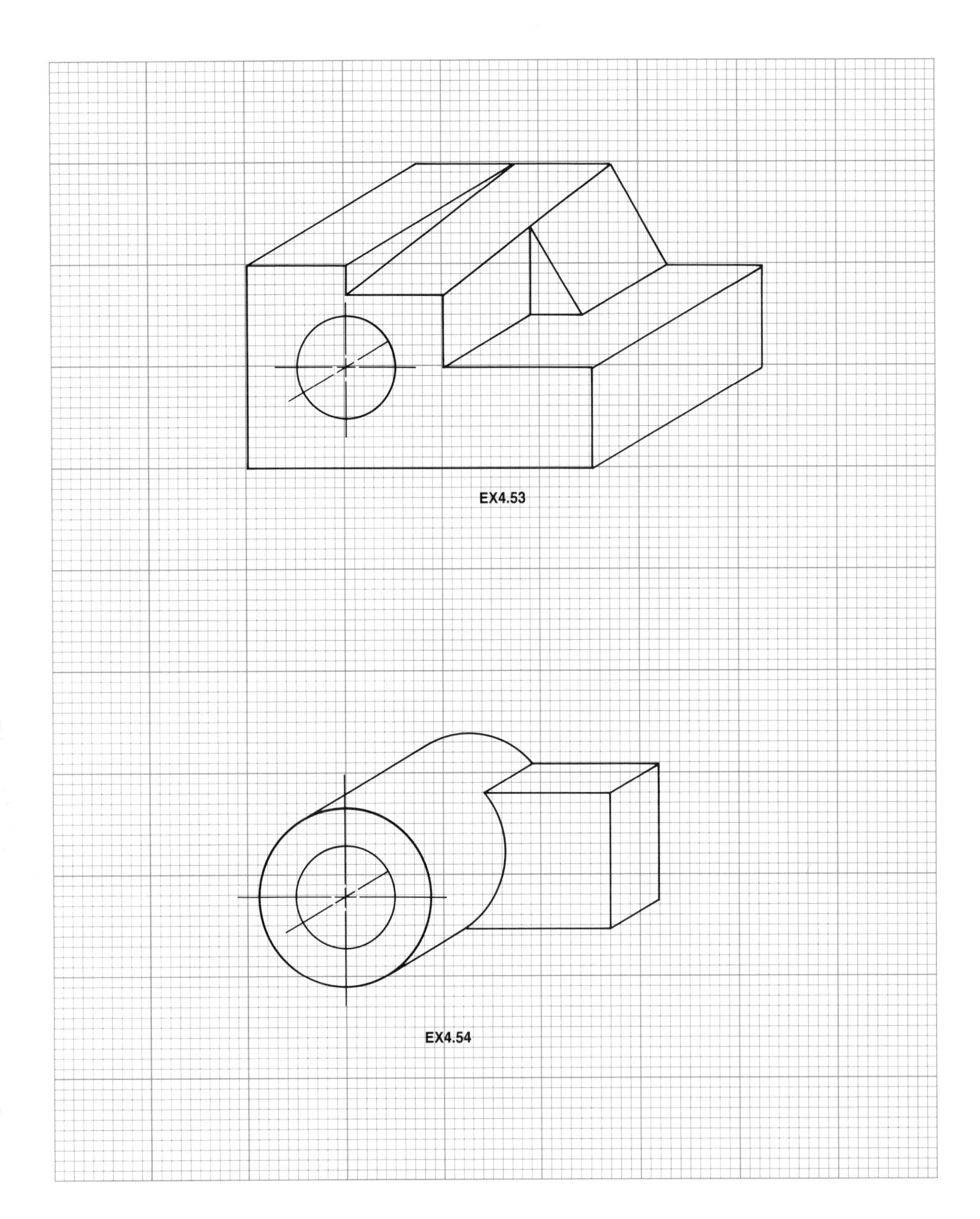

EX4.53

EX4.54

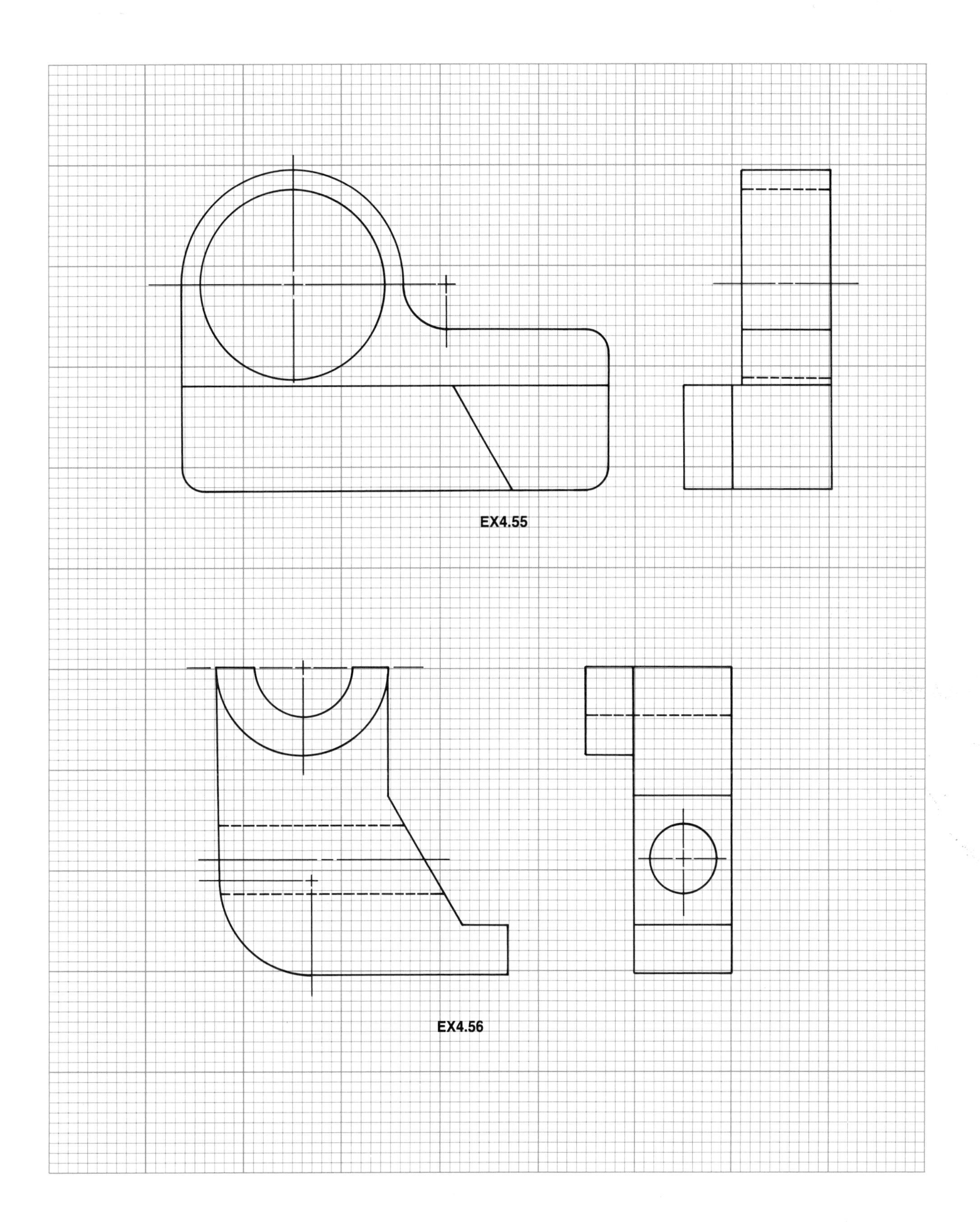
EX4.55
EX4.56

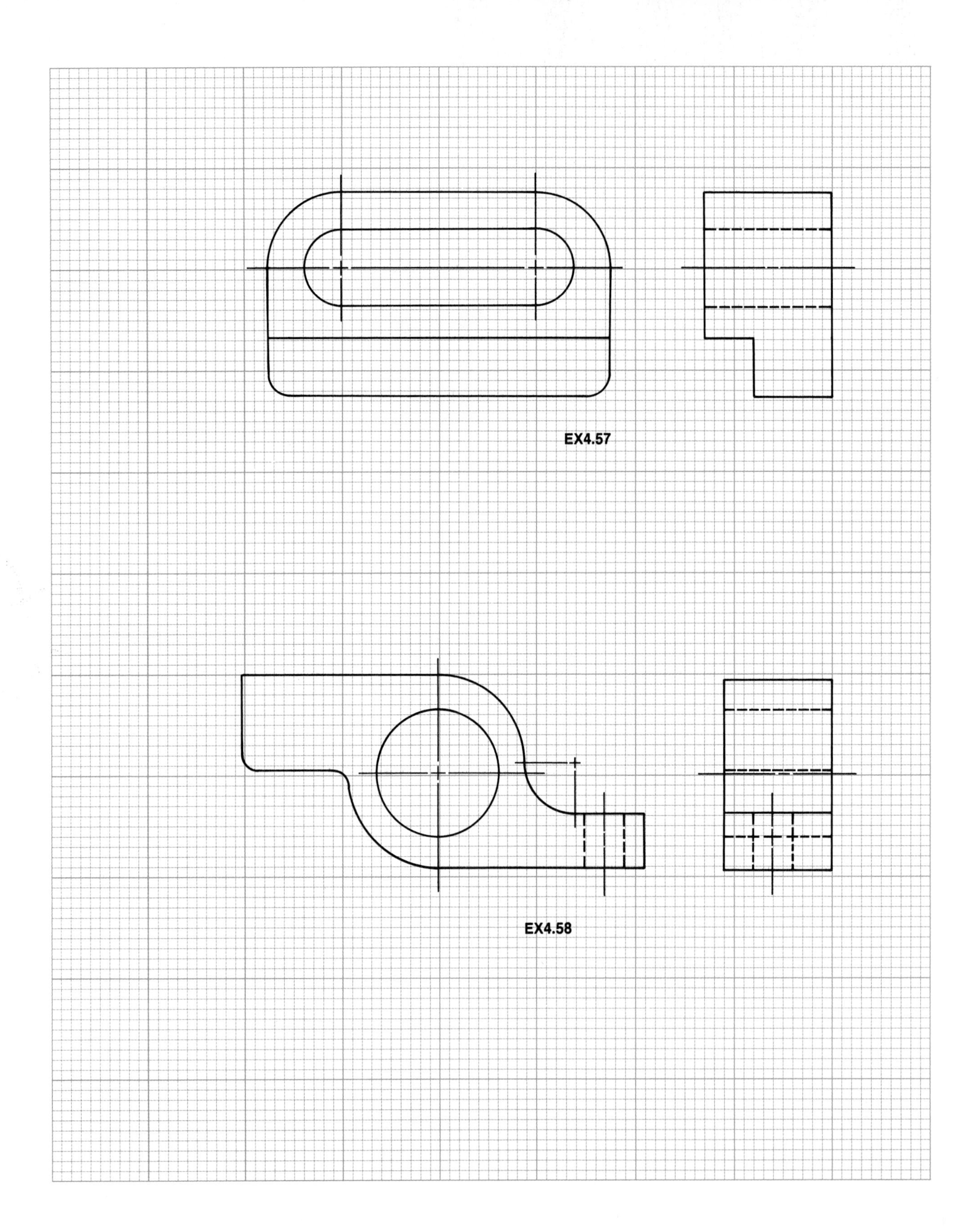
EX4.57
EX4.58

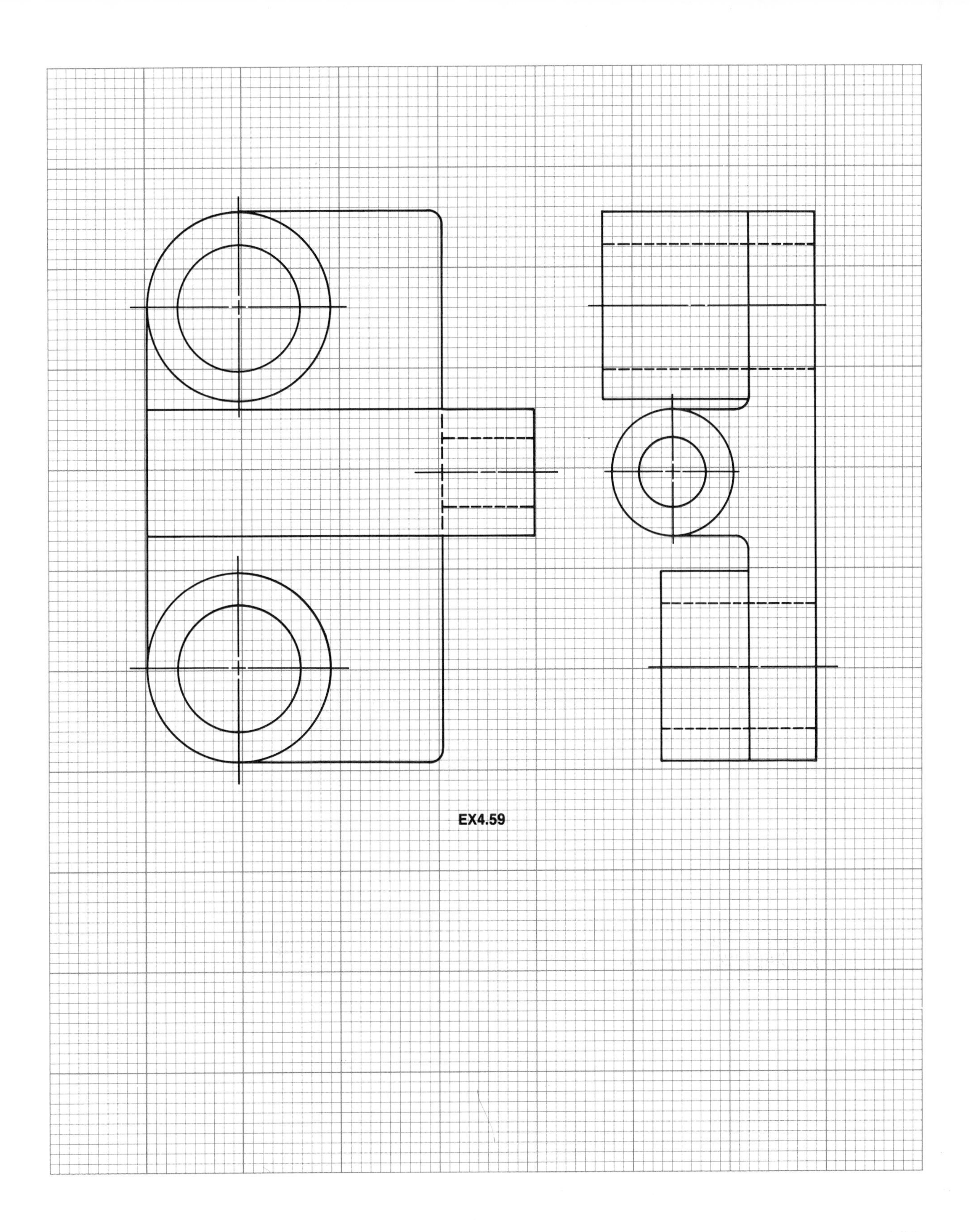
EX4.59

5 Orthographic Projection, Part 2: Maximizing Clarity in a Drawing

Preview In Chapter 4, we reviewed the fundamental characteristics of orthographic projection and its application in the description of engineering devices. In this chapter we will consider the advanced practices used in orthographic drawings to describe objects (or **parts**) with maximum clarity.

The rules for orthographic projection must sometimes be violated in order to enhance the clarity of a drawing. If the true projection of an object would create confusion and ambiguity in its graphic description, one may use **partial views, removed views, incomplete views,** and **conventional revolutions** that are not in strict conformance with true projection principles in order to ease the difficulty encountered in graphic description.

In addition, **machined holes** (blind, through, countersunk, and counterbored) will be discussed in this chapter. Since finished (bored) holes in a part will increase its fabrication cost, the designer of the part must justify the need for a finished hole.

Finally, the topics of left-hand and right hand parts, intersections between cylindrical surfaces, rounds, and fillets will be developed in this chapter.

Learning Objectives

Upon completion of this chapter, the reader should be able to:

- Recognize that the rules for orthographic projection must at times be violated in order to enhance the clarity of a drawing.
- Develop partial views to describe symmetrical or cylindrical objects.
- Develop removed views to represent features that are difficult to represent with clarity in full orthographic views because of conflicting linework for other features.
- Develop (and judiciously use) incomplete views to aid the reader's interpretation of a set of orthographic views by omitting confusing linework that would otherwise appear in the corresponding complete view.
- Develop conventional revolutions to emphasize the radial symmetry of such features as holes and ribs around the center of a circular plane.
- Represent left-hand and right-hand parts with a single set of orthographic views and appropriate notations.
- Recognize the need to justify the use of machined holes that will increase the fabrication cost of a part.

- Describe blind, through, countersunk, and counterbored holes in orthographic drawings.
- Select and construct the appropriate representations of intersections between cylindrical surfaces in a set of orthographic views.
- Identify the benefits associated with rounded corners (rounds and fillets) in a part.
- Describe rounds and fillets in orthographic drawings.

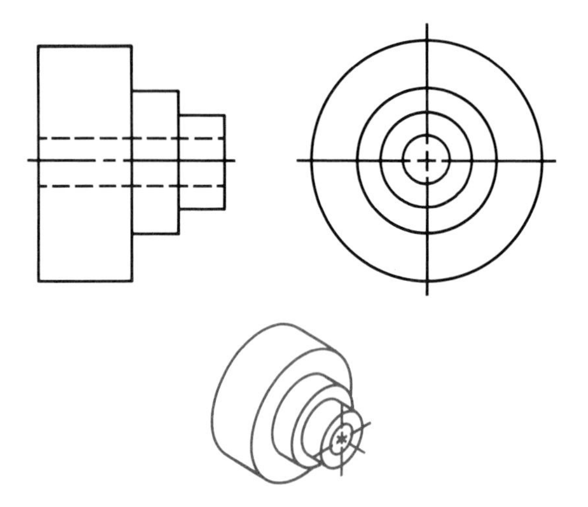

Figure 5.1 Symmetrical part with complete orthographic views.

5.1 Partial Views

Parts are often symmetrical or cylindrical in form and may be adequately described by a set of **partial views.** For example, the part shown in Figure 5.1 with a full set of orthographic views could be represented adequately by the views shown in Figure 5.2. This **equivalent** set of views includes a partial right-side view in which the centerline indicates that the unseen half of the object is symmetrical to the given portion. (Note that this partial right-side view presents the portion of the object that is nearest to the given front view.)

The full right-side view in Figure 5.1 does not provide any additional information to the partial view, since the object is symmetrical about the vertical centerline of the side view. We may save both drawing time and space by using partial views without any loss in the orthographic description of an object.

To help the reader recognize a partial view in a set of orthographic drawings, use a **partial break,** which is simply a jagged line indicting that the given (partial) view is not complete (Figure 5.3). Partial breaks should never coincide with surface boundary lines (visible or hidden lines).

Figure 5.4 presents another symmetrical object described in both orthographic and pictorial form. The preferred set of orthographic views (in which a partial view is used) is given in Figure 5.5. (Notice that a partial break has been used in Figure 5.5.)

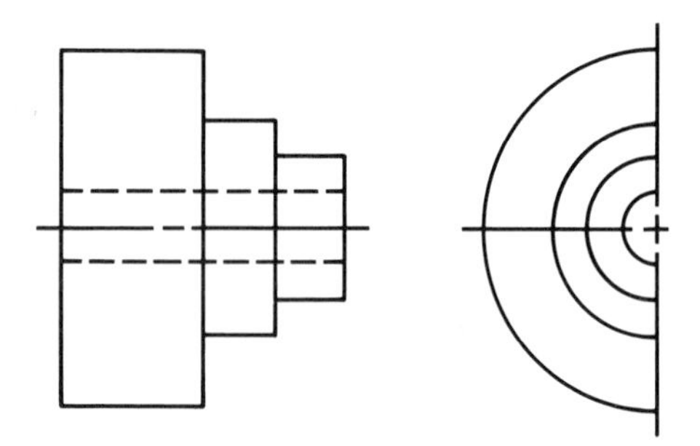

Figure 5.2 Partial view of symmetrical part in Figure 5.1 is sufficient.

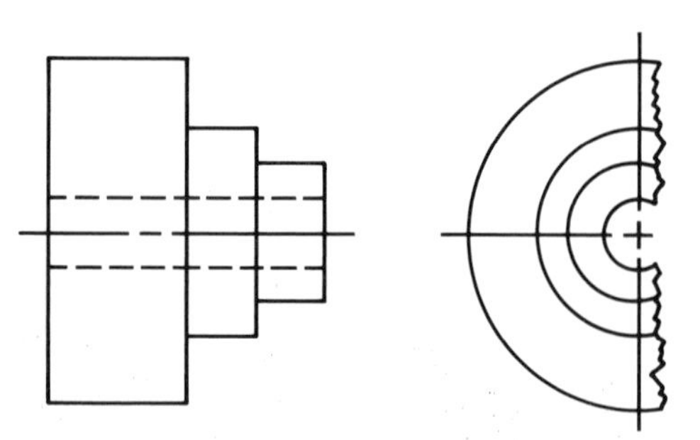

Figure 5.3 Partial break in the partial view (of the part in Figure 5.1) indicates to the reader that the view is not complete.

5.2 Removed Views

A part may include a feature that is difficult to represent with clarity in full orthographic views because of conflicting linework from other features. In such cases, one should use a **removed view,** which concerns itself only with the particular feature under consideration. Figure 5.6 presents a set of three full orthographic views of an object, together with a removed view labeled A-A. Notice that the viewing direction for view A-A is shown in the given front view by a **viewing plane line** of the following form:

The directional arrows indicate the viewing direction for the removed view relative to the full view. (A second removed view in Figure 5.6 would be labeled B-B, a third C-C, and so forth.)

If necessary, a removed view may be shown with a larger scale than that used for the given full views in order to present the details of the feature shown in the removed view. In addition, removed views may be placed in any convenient location on the drawing; see, for example, Figures 1.1 and 1.2, in which enlarged and removed views are used.

In summary, then:

1. Removed views should be used to represent features that are difficult to represent with clarity in full orthographic views because of conflicting linework from other features.

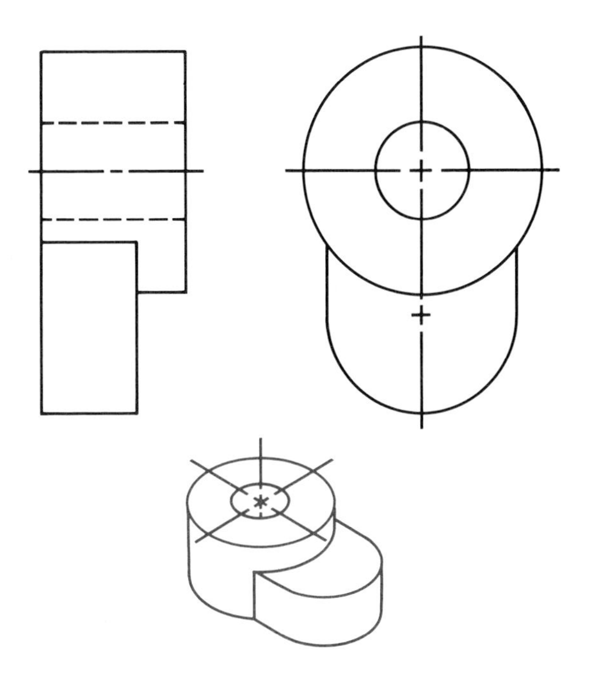

Figure 5.4 Symmetrical part with complete orthographic views.

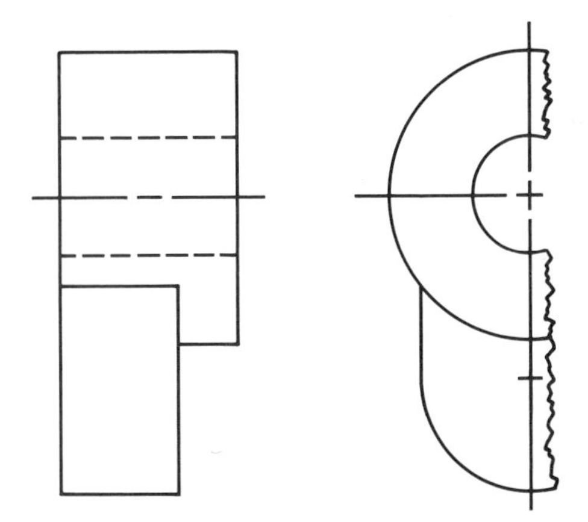

Figure 5.5 Partial view (with partial break) of symmetrical part in Figure 5.4 is preferable to the complete view in that figure.

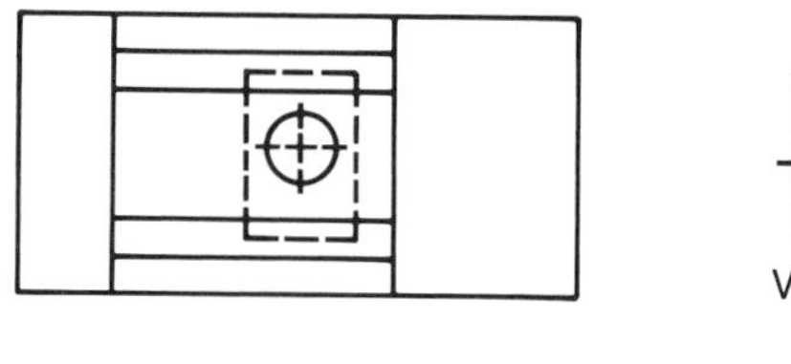

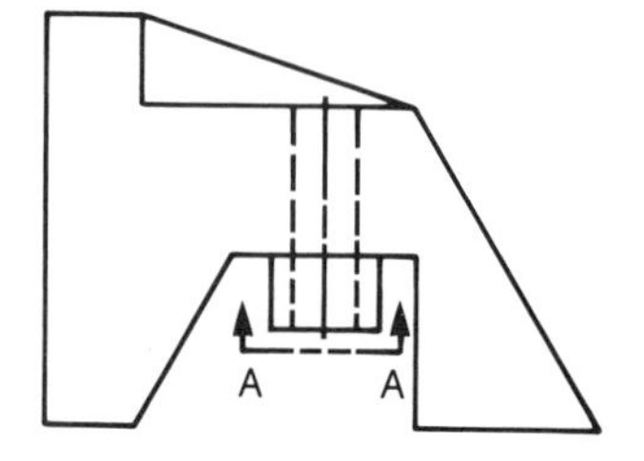

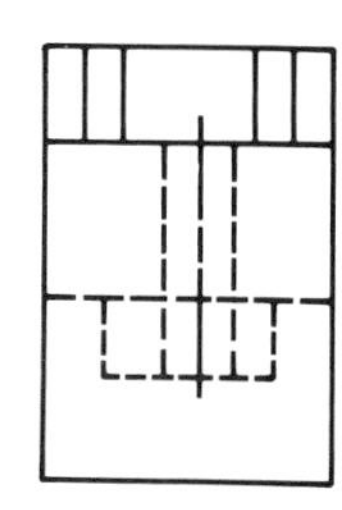

Figure 5.6 Orthographic drawing, including a removed view. Note the viewing direction for the removed view is given in the front view.

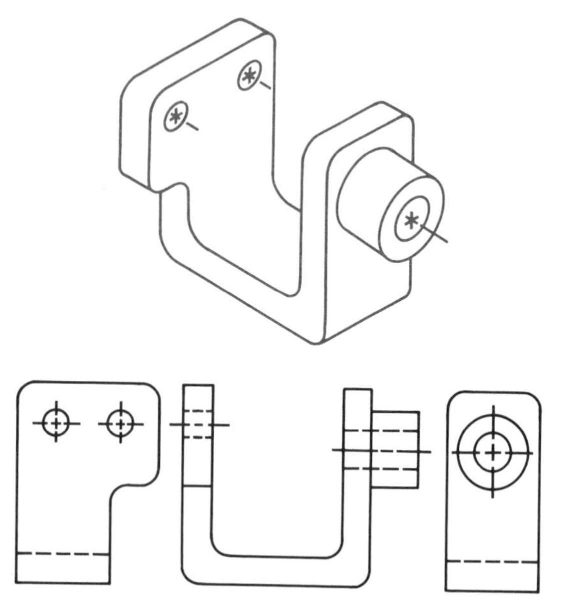

Figure 5.7 Full views for this asymmetrical part would include overlapping linework confusing to the reader. Incomplete side views make interpretation easier.

2. Removed views allow us to represent such features without drawing additional complete views of the object.
3. A removed view shows only the particular feature that is not clearly shown in the given full views.
4. A viewing plane line is used to indicate the viewing direction for the removed view with respect to one of the given full orthographic views.
5. A removed view may be shown with a different scale than that used for the full views.
6. A removed view may be placed in any convenient location on the drawing.

5.3 Incomplete Views

If complete orthographic views of an object might confuse the reader because of overlapping linework or numerous hidden lines, some views may be drawn intentionally **incomplete.**

Consider the part in Figure 5.7. A set of complete orthographic views of this part would include overlapping linework that could confuse the reader's interpretation of the part's shape. In the set of views presented in the figure (front, left-side, and right-side), the side views are incomplete making it easier for the reader to interpret the description of the object.

Of course, it is important that incomplete views be judiciously used. Readers expect—with good reason—to be presented with complete orthographic views of an object. An incomplete view should be used only for those cases where the corresponding complete view of the object would be unnecessarily confusing.

Learning Check

What is the difference between an incomplete view, as discussed in this section, and a partial view?

Answer Both types of drawings are, in fact, incomplete orthographic views. However, partial views present (essentially) one-half of the object under description where this object is *symmetrical* about a plane or an axis; as a result, equivalent full views could not contain any more information about the object than that given in the partial view. In contrast, incomplete views describe an *asymmetrical* object where overlapping linework in an equivalent full view would simply confuse the reader. Both types of views must be judiciously chosen so that the clarity of a drawing is enhanced, not diminished!

5.4 Conventional Revolutions

Conventional revolutions are another instance in which the rules for orthographic projection are violated in order to increase clarity. Figure 5.8 presents a part in which four holes appear; three small holes are symmetrically located about the outer rim of the disc-shaped base of the part, whereas one larger hole is located at the center of the object. The given right-side view is a true orthographic projection consistent with the given front view of the part. However, notice that the hidden lines in the right-side view that correspond to the small holes in the object appear in asymmetrical positions with respect to the principal axis of the object; the lower hole appears to be closer to this axis than does the upper hole because of the relative height positions of these holes in the given front view.

In order to increase the clarity of such views, one should "rotate" features—such as holes, ribs, and lugs—that are symmetrically located within a circular plane about the axis of symmetry so that these features will be shown at their *true radial distance* from this axis. For example, the small holes of the object shown in Figure 5.8 can be revolved about the principal axis of the object in the front view so that the hidden lines representing these holes will appear at their true radial distance from the principal centerline in the right-side view (Figure 5.9). Such a revolution is *imaginary;* it is only performed in order to emphasize the radial symmetry of these holes in the object to the reader. The finished orthographic views for this object are given in Figure 5.10.

Figure 5.11 presents the true orthographic views for an object in which three symmetrically placed ribs are included about the perimeter of the circular disc base. The ribs appear in the right-side view as if they are not symmetrically located about the principal axis of the object. Figure 5.12 shows the conventional revolution that should be performed in order to show these ribs at their true radial distances from the principal centerline in the right-side view. The finished views are shown in Figure 5.13. Once again, this conventional (imaginary) revolution emphasizes the radial symmetry of the ribs with respect to the principal axis of the object.

As a final example, Figure 5.14 presents the true orthographic views for an object in which both holes and ribs are symmetrically located about the principal axis. Conventional revolutions are applied to these features, as indicated in Figure 5.15, in order to emphasize its radial symmetry in the right-side view. The finished views are shown in Figure 5.16.

In the case of sectional orthographic views (which we will discuss in the next chapter), a conventional revolution (or **revolved view**) is called an *aligned view.*

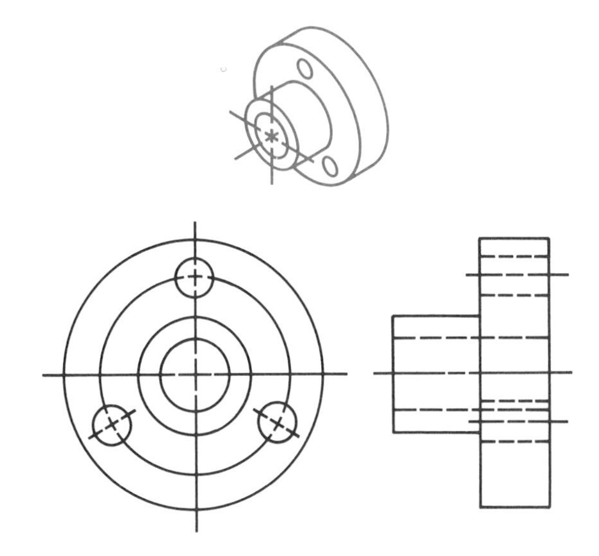

Figure 5.8 True orthographic views for a part with three symmetrically located holes. Note that the hidden lines in the right-side view corresponding to the symmetrically located holes appear in asymmetrical positions. This problem can be solved by conventional revolution (see Figure 5.9).

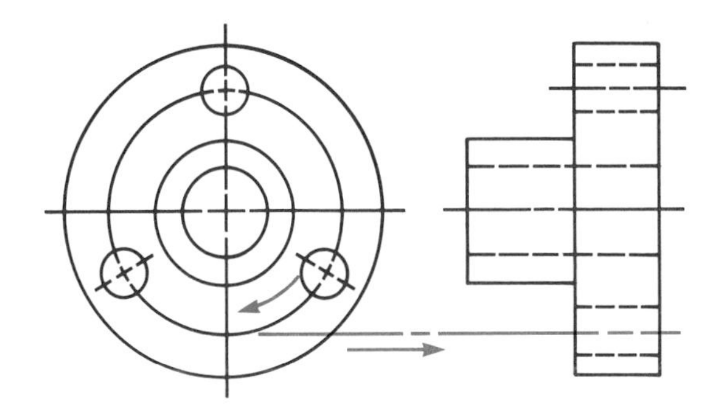

Figure 5.9 Construction of a conventional revolution for the part in Figure 5.8.

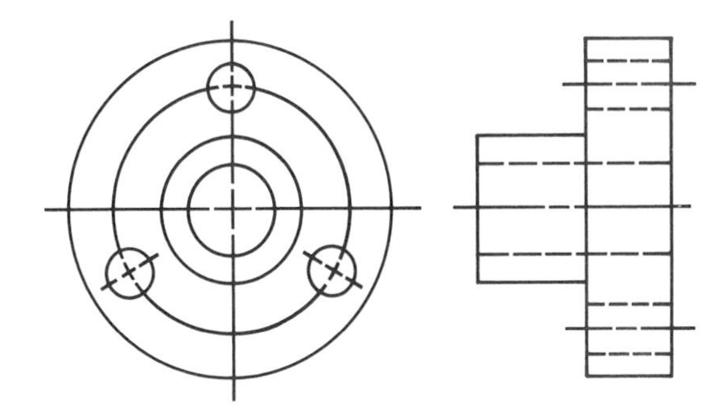

Figure 5.10 Finished set of orthographic views for the part in Figure 5.8 with conventional revolution.

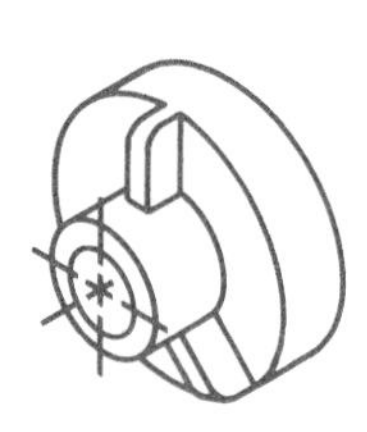

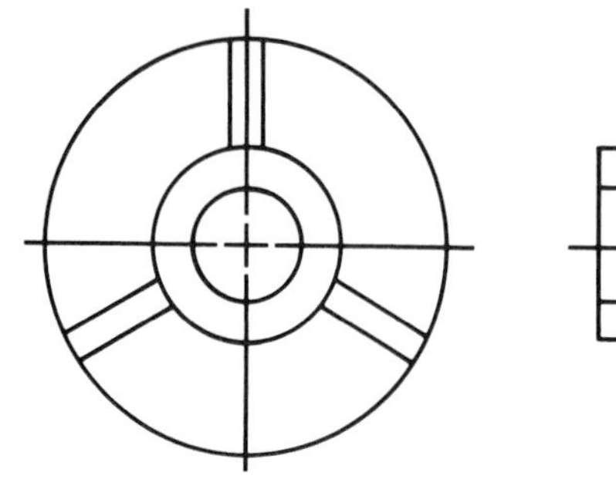

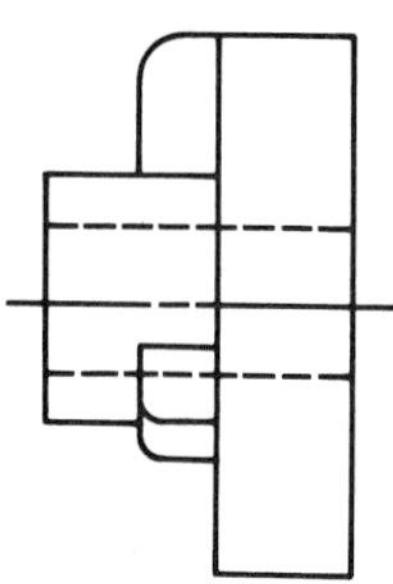

Figure 5.11 True orthographic views for a part with three symmetrically placed ribs. Note that the ribs appear asymmetrically located in the right-side view. This problem can be solved by conventional revolution. See Figure 5.12.

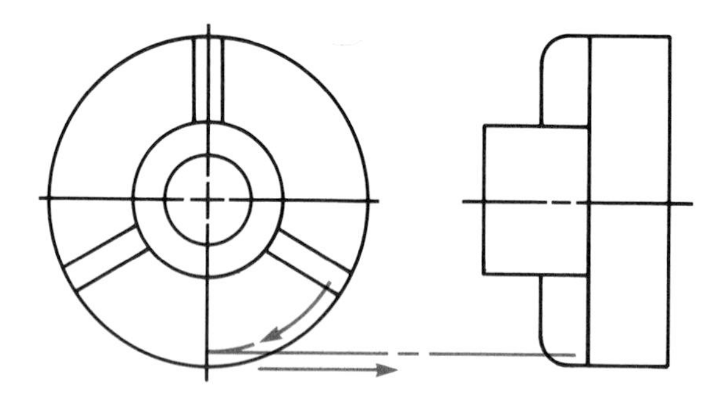

Figure 5.12 Construction of a conventional revolution for the part in Figure 5.11.

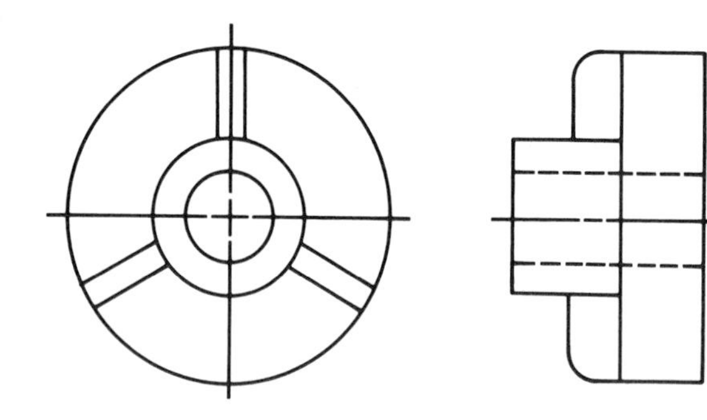

Figure 5.13 Finished set of orthographic views for the part in Figure 5.11 with conventional revolution.

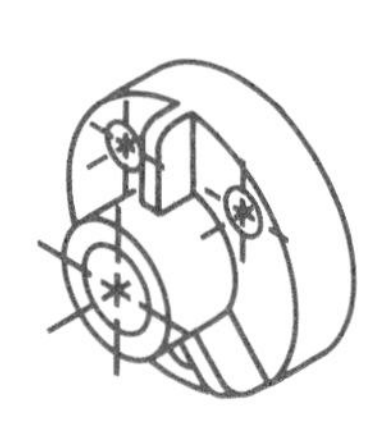

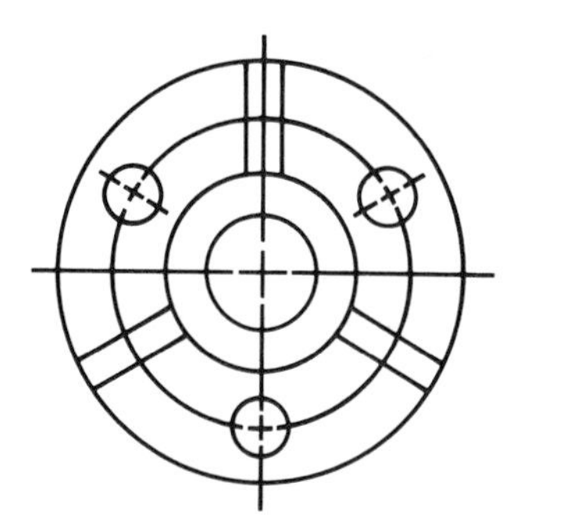

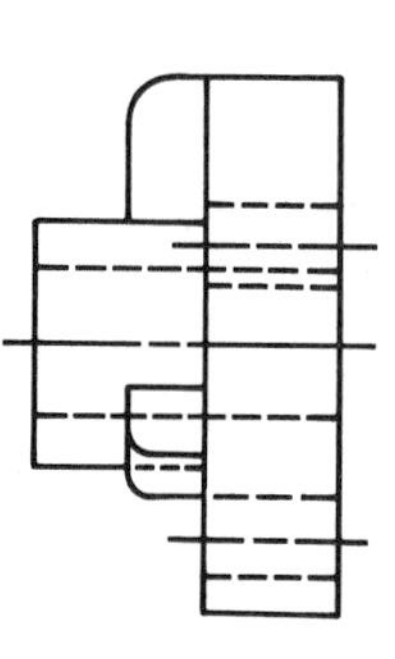

Figure 5.14 True orthographic views for a part with symmetrically placed holes and ribs. See Figure 5.15 for conventional revolutions required.

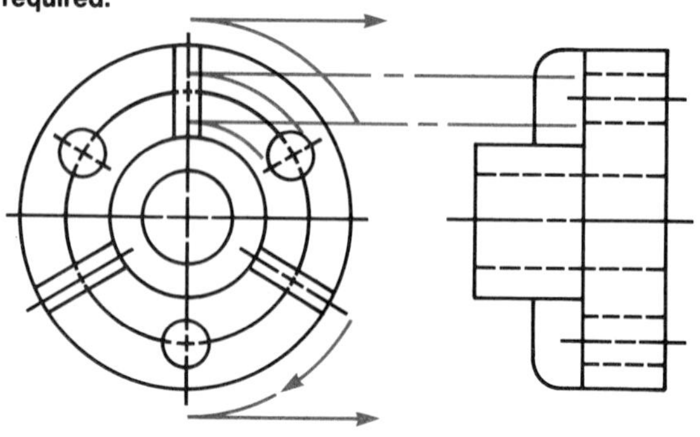

Figure 5.15 Construction of conventional revolutions for the part in Figure 5.14.

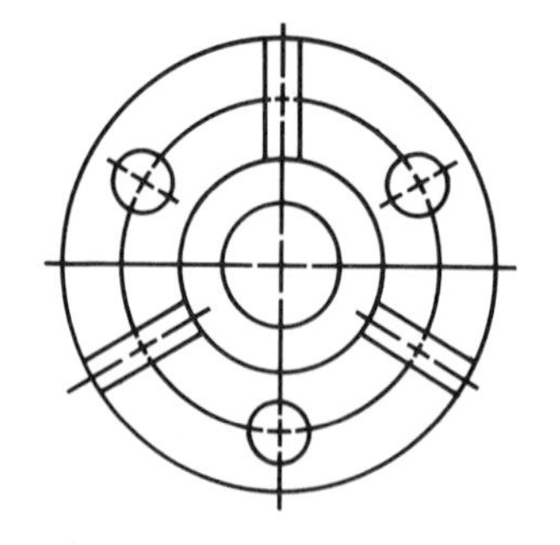

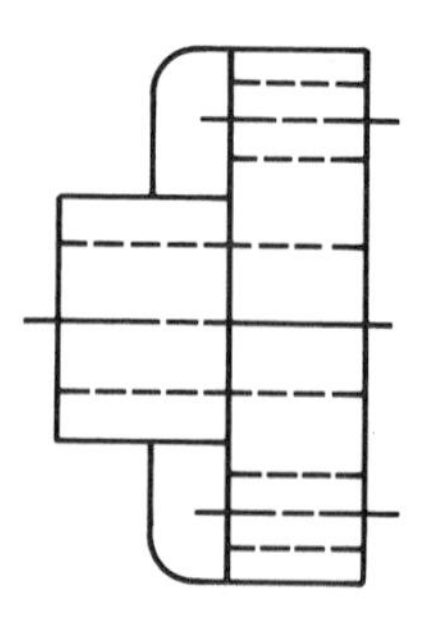

Figure 5.16 Finished set of orthographic views for the part in Figure 5.14 with conventional revolutions.

In summary, then:

1. Conventional revolutions are used to emphasize the radial symmetry of such features as holes and ribs located about a principal axis of an object.

2. Conventional revolutions are imaginary. They do not represent actual modifications in the object; such revolutions are used in an orthographic representation only for the purpose of greater clarity.

3. A conventional revolution is an instance in which the rules for orthographic projection are violated in order to achieve greater clarity.

5.5 Left-hand and Right-hand Parts

Some parts are reversed images of one another, forming a set in which one part may be called the **right-hand** part and the second part the **left-hand** part (Figure 5.17). Only one of these parts need be drawn if one includes a note on the drawing indicating that the other part is to be produced as a reversed image. Figure 5.17 presents a set of views in which a right-hand part is described and a note indicates that the left-hand part must also be fabricated. Such a note replaces an entire set of views showing the left-hand part, saving both drawing time and space.

In summary, then:

1. Left-hand and right-hand parts are reversed (mirror) images of one another.

2. Left-hand and right-hand parts may be adequately described by a set of views in which only one part is shown, together with an appropriate note indicating that the reversed-image part must also be produced.

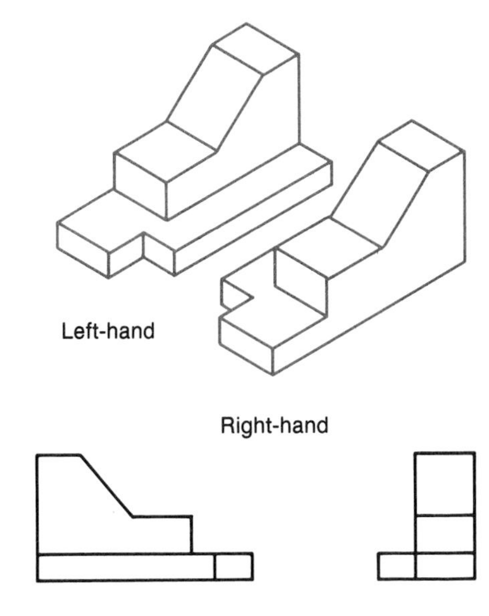

Figure 5.17 Left-hand and right-hand parts.

5.6 Machined Holes

Machined holes (holes that are drilled, reamed, counterbored, and so forth) must be carefully described in orthographic drawings, since the designer must consider the need for a particular type of hole, together with the cost associated with its fabrication. For example, a drilled but unfinished (not bored) hole requires only one machining operation (drilling), whereas a bored, finished hole requires two operations (drilling and reaming); so the finished hole requires more time and labor than the unfinished hole. The designer must be able to justify the need for the more expensive finished hole in a part; otherwise, an unfinished hole should be specified in the drawing.

Figure 5.18 presents two orthographic views each for four types of holes drilled into a simple rectangular part. For the **blind hole** (a), which is a hole that does not extend entirely through an object, the triangle in the drawing is constructed so that its sides lie at 30° relative to the hypotenuse (hidden

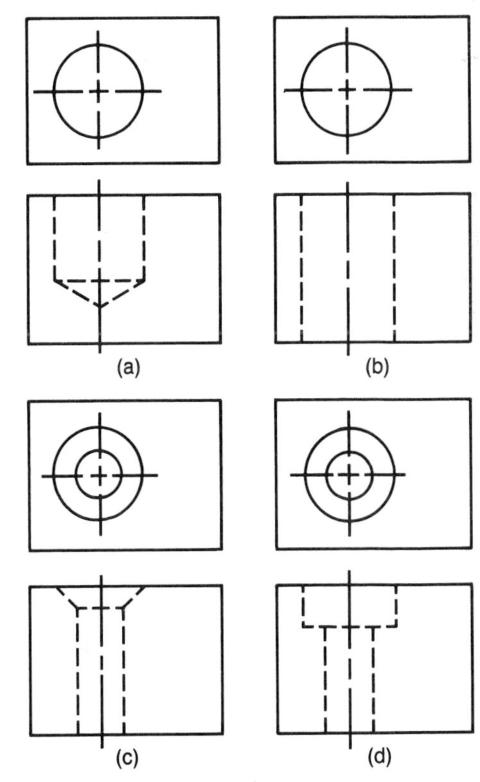

Figure 5.18 Four types of holes that must be carefully described in orthographic drawings are the blind hole (a), through hole (b), countersunk hole (c), and counterbored hole (d).

linework). The hole is to be drilled to a specified depth that corresponds to the hypotenuse of this triangle formed with hidden linework. (The depth of a hole refers to the extent of the cylindrical portion of the hole, not the conical shape associated with the most interior portion of an unfinished hole.) The triangle in this drawing indicates to the machinist that the hole is to remain unfinished, that is, that it is not necessary to bore the hole in order to produce a finished (flat) interior surface.

The **through hole** (b) is a hole that does extend entirely through an object. The **countersunk hole** (c) is to be first drilled, then enlarged in its outer portion in accordance with the given conical specifications (countersunk angle and countersunk diameter). The **counterbored hole** (d) is first drilled, then enlarged in its outer portion in accordance with the given cylindrical specifications (counterbored depth and counterbored diameter).

Other types of machining operations are reviewed elsewhere throughout this text (please refer to the index for specific operations).

In summary, then:

1. The designer of a part must be able to justify the need for particular types of machined holes that increase the cost for fabrication of the part.
2. The depth of a hole refers to the extent of the cylindrical portion of the hole.
3. A triangle (with two interior angles equal to 30°) is used to indicate an unfinished (drilled but not bored) hole.
4. A blind hole does not extend through the part as do through holes.
5. Countersunk holes are first drilled, then finished by enlarging the outer portion of the hole in accordance with given conical specifications (countersunk angle and countersunk diameter).
6. Counterbored holes are first drilled, then finished by enlarging the outer portion of the hole in accordance with given cylindrical specifications (counterbored depth and counterbored diameter).

5.7 Cylindrical Intersections

The primary considerations for intersections between cylindrical surfaces are

- The *accuracy* with which such intersections must be shown in an orthographic view.
- The *ease* with which such intersections may be constructed in an orthographic view.
- The *relative size* of the cylindrical radii.

Figures 5.19 through 5.22 present intersections between cylinders of various sizes. In Figure 5.19 two cylinders of

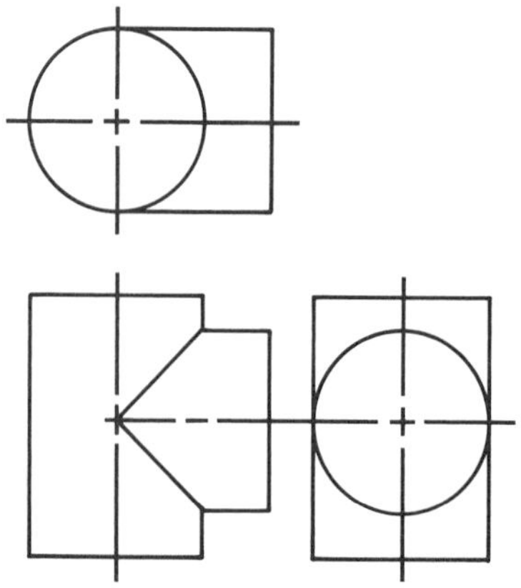

Figure 5.19 Intersection of cylinders of equal diameters.

equal diameter intersect, resulting in intersection lines that appear to be straight lines in the given front view. (Of course, these intersection lines actually curve through space along a semi-elliptical path. The straight-line appearance in the front view is the result of the orthographic projection of the intersection between cylinders onto the projection plane.

In Figure 5.20 two cylinders with different diameters intersect, resulting in an intersection line in the front view that must be constructed by transferring the width and the height of arbitrarily chosen points along the intersection path from the top and right-side views, respectively. An accurate representation of the intersection is thereby obtained in the front view.

In Figure 5.21 two cylinders of significantly different diameters intersect. One could plot the intersection in the front view through the use of arbitrarily chosen points, as shown in Figure 5.20; however, an alternative approach—which will save drawing time and yet produce an adequate approximation of the intersection in the front view—is shown in Figure 5.21. The radius, *R*, of the larger cylinder is transferred to the front view (as shown) in order to locate the point along the axis of the small cylinder that lies at the distance, *R*, from the highest (and lowest) points of intersection between the cylinders. An arc of radius *R*, centered at this point along the axis of the small cylinder, is then drawn as an approximation of the true line of intersection between the cylinders in the front view.

Intersecting cylinders of extremely different diameters are shown in Figure 5.22. The line of intersection between these cylinders is so small—relative to the other dimensions of the object—that it is not shown in the front view.

In summary, then:

1. The intersections between cylindrical surfaces of different sizes may be shown in a variety of forms:

- With accuracy—by carefully locating points along the intersection line (Figure 5.20).
- In an approximate form—if the cylinders are of significantly different sizes—by using the radius, *R*, of the larger cylinder in order to form an arc representation of the intersection (Figure 5.21).
- Not shown whatsoever if the cylinders are of extremely different diameters (Figure 5.22).

The last two choices allow one to save both drawing time and effort, although only an approximation of the intersection is generated.

2. The intersection between cylindrical surfaces of equal diameter (where the axes for the cylinders lie at 90° relative to one another) appear as a pair of straight lines in an orthographic projection (Figure 5.19).

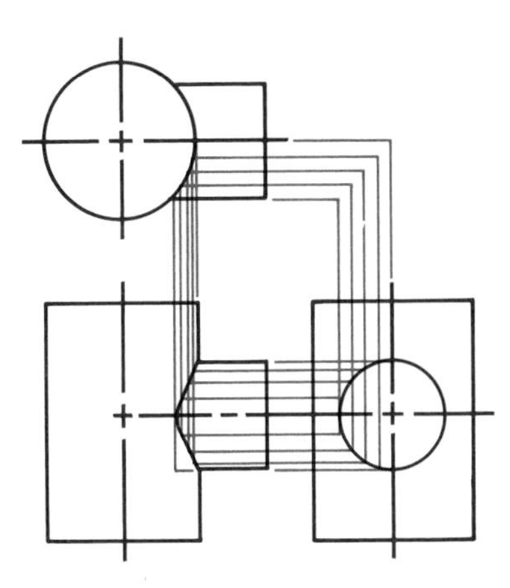

Figure 5.20 Intersection of cylinders of unequal diameters.

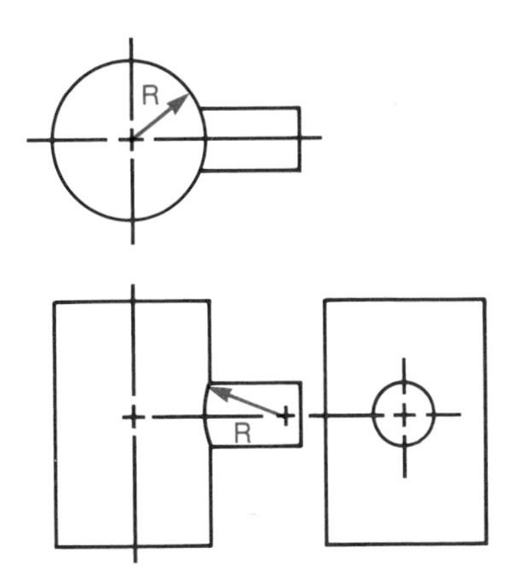

Figure 5.21 Intersection of cylinders of significantly different diameters.

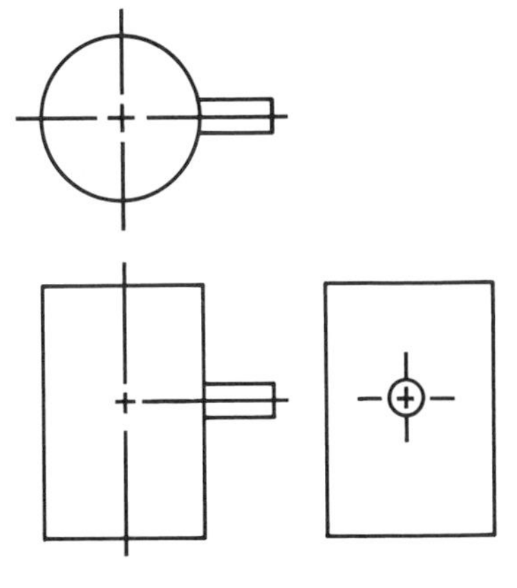

Figure 5.22 Intersection of cylinders of extremely different diameters.

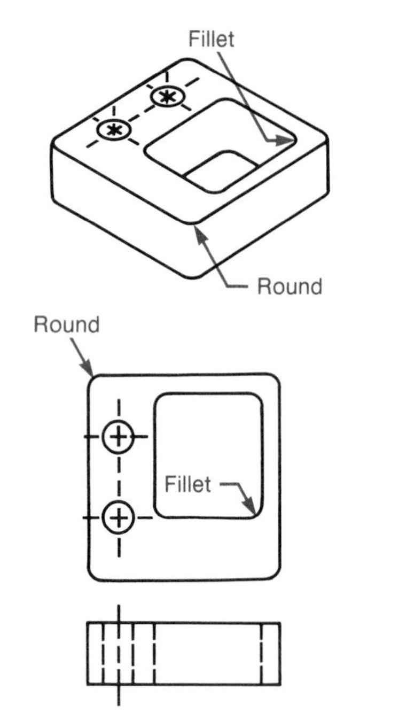

Figure 5.23 Orthographic drawing for a part with rounds and fillets.

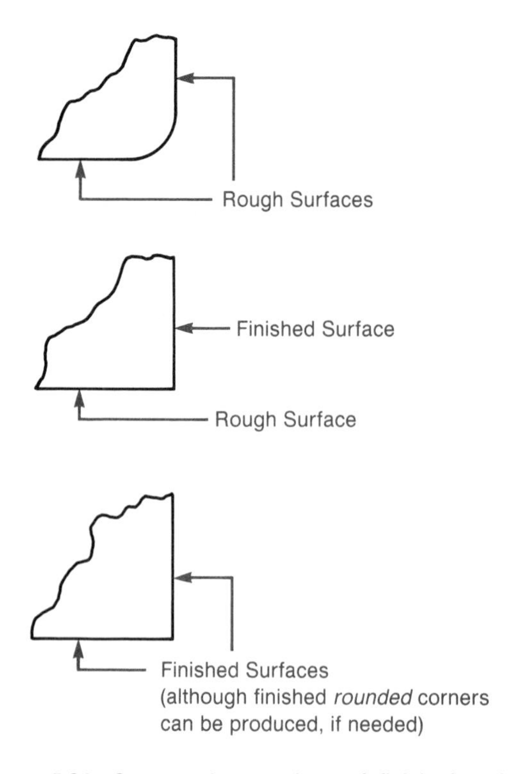

Figure 5.24 Common intersections of finished and rough surfaces.

5.8 Fillets, Rounds, and Runouts

Parts are often fabricated (through forging or casting) with rounded corners in order to:

- Increase the strength of the part (in the case of interior corners).
- Decrease the difficulty associated with the production of sharp corners.
- Enhance the appearance of the part.

A **round** is a rounded exterior corner, whereas a **fillet** denotes a rounded interior corner (Figure 5.23).

Rounded corners occur if the two intersecting surfaces are both rough, or unfinished. A finished (machined) surface *usually* intersects with another surface (finished or unfinished) sharply, as shown in Figure 5.24 (although machined, rounded corners are possible).

A round or a fillet should be drawn accurately with a bow pencil or pen or with the aid of a template. The common radius for a round or fillet in a drawing is about 6 mm. (or about 0.25 in.).

A **runout** is the curve produced by a fillet located at the point of tangency between a planar surface and a cylindrical surface. A runout is represented by a ⅛-circular (45°) arc with a radius equal to that of the fillet. Figures 5.25 through 5.27 present objects in which runouts appear, compared with the same objects without runouts. Figure 5.25 shows a part in which no fillet is to be formed (a) and a similar part in which a fillet is included (b). Similarly, Figures 5.26 and 5.27 show parts in which sharp intersections occur between a planar surface and a cylindrical surface (a in both figures) and parts in which rounded intersections are indicated by runouts (b in both figures). (Notice that, as shown in Figure 5.27 one must determine the *exact point of tangency* between intersecting planes in order to properly locate the intersection line (if no fillet is to be formed) or the runout (in the case of a fillet).

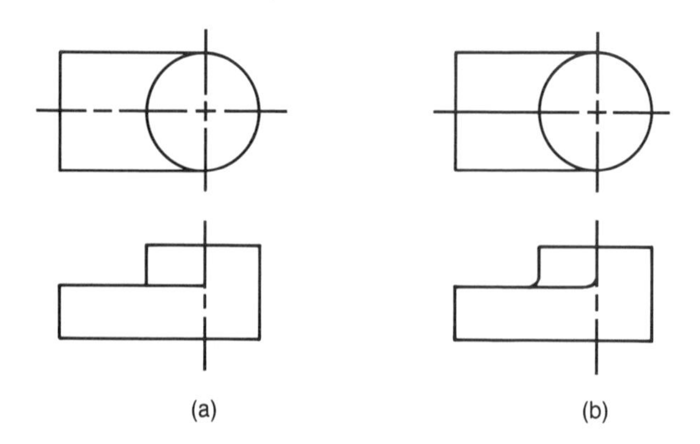

Figure 5.25 A part without runout (a) and similar part with runout (b).

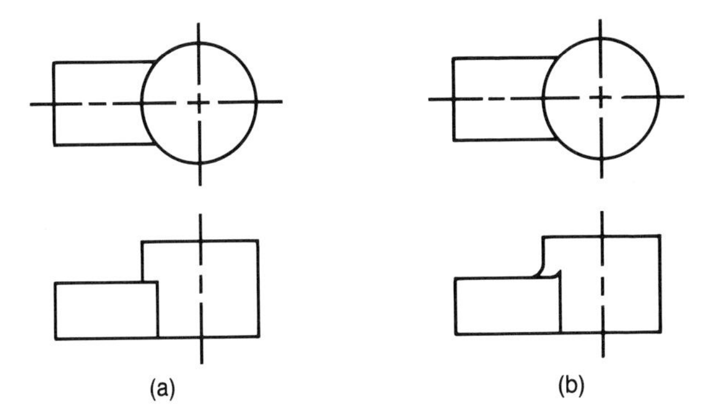

Figure 5.26 A part without runout (a) and similar part with runout (b).

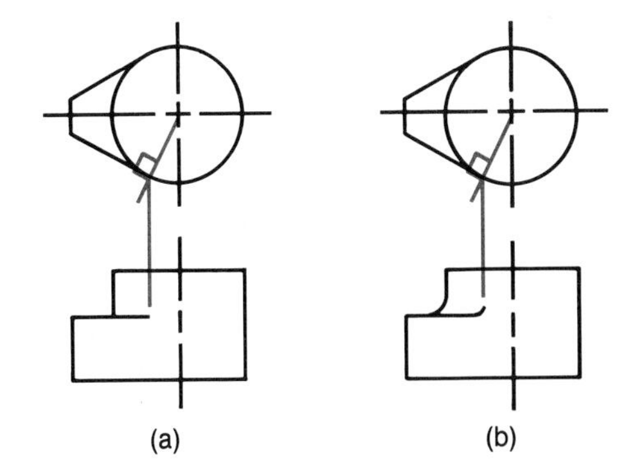

Figure 5.27 A part without runout (a) and similar part with runout (b).

Figure 5.28 presents a part in which a rib intersects other surfaces with rounded corners, as indicated by the fillets shown in the given orthographic views. Figure 5.29 describes a similar object in which the rib itself has rounded edges. Note the subtle but significant differences between the orthographic views for this part and the views for the part shown in Figure 5.28; these differences identify the form of the rib that is to be produced by the machinist.

In summary, then:

1. Parts are fabricated with rounded corners in order to:

- Increase the strength of the part (in the case of interior corners).
- Decrease difficulties in the fabrication of the part.
- Enhance the appearance of the part.

2. A round is a rounded exterior corner.

3. A fillet is a rounded interior corner.

4. Fillets and rounds occur if both intersecting surfaces are unfinished; a finished (machined) surface will usually result in sharp intersections.

5. A runout is the curve produced by a fillet at the point of tangency between a planar surface and a cylindrical surface.

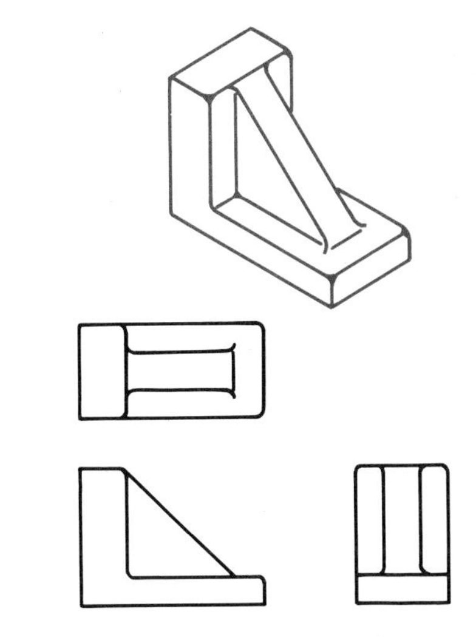

Figure 5.28 A part with a sharp-edged rib.

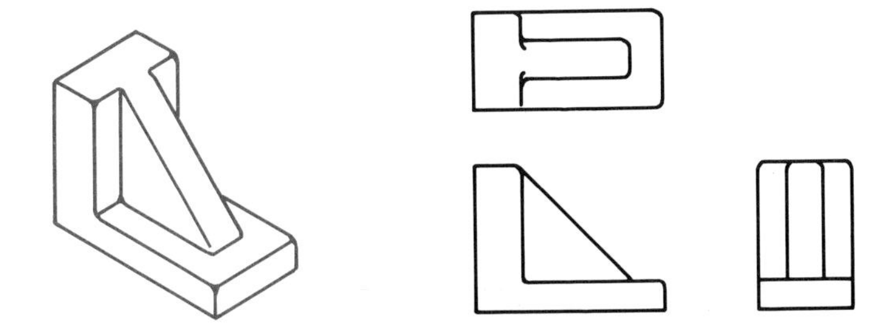

Figure 5.29 A part with a rounded-edged rib.

COMPUTER GRAPHICS IN ACTION

3D Computer Graphics: Solid Models

Solid models are used to define three-dimensional objects in a complete mathematical form. Both geometric and mass data are used in the specification of a particular part. As a result, solid models offer several advantages not provided by wire-frame or surface models:

- Each point in an object can be specified as an exterior or interior point, enabling well-defined interior and exterior surface specifications and appropriate visibility representations of surfaces in graphical descriptions of the object.
- Mass properties (such as weight, center of gravity, and moment of inertia) of the object can be easily determined since both surfaces (and volumes) and mass characteristics of the part are used to develop the solid model.
- The relative positions of parts in an assembly can be investigated in solid modeling efforts; this capability allows us to develop **kinematic** analysis of the relative motion between moving parts in a system. One can imagine what a powerful tool for the analysis of alternative engineering designs (see Chapter 18) is thereby provided by solid models.
- Solid models are used to create **finite-element models** of systems in which a part can be subdivided into numerous small (but finite) three-dimensional elements for detailed analysis of its properties and behavior in a variety of situations.
- Solid models can be used to create very realistic graphical descriptions of an object in which surfaces (and masses) are well-defined.

Surface models are the most useful form in which three-dimensional objects may be represented with the aid of a computer. Although the computer programs that are required for solid modeling are more complex (and require more computer memory storage and processing time) than that used for wire-frame and surface modeling efforts, the advantages of solid models justify this additional complexity. Furthermore, many commercial solid modeling software programs are available. As a result, the engineer does not need to develop such software, but instead simply familiarize himself or herself with the particular program that is to be used in a given modeling effort.

Solid models are developed in one of two ways: through the use of **primitives** or **boundary definition.** We will review each of these techniques in the "Computer Graphics in Action" sections of Chapters 6 and 7.

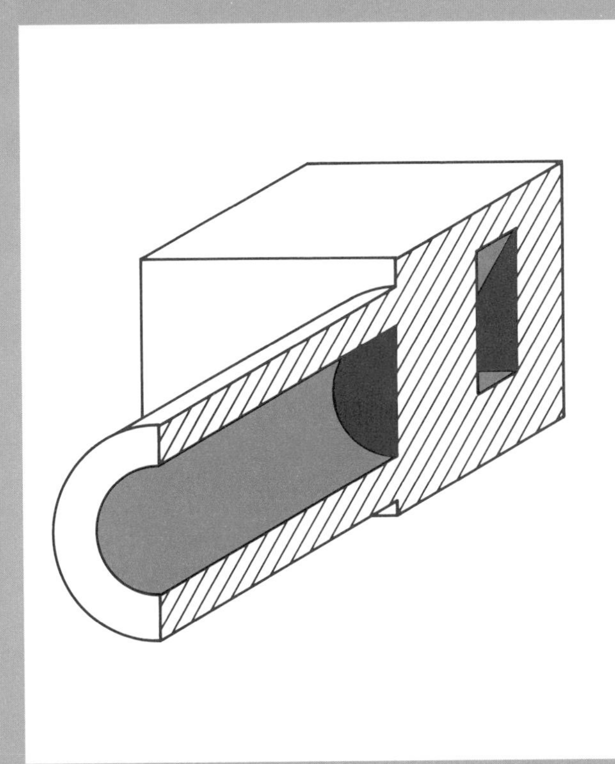

A solid model of an object in which interior detail and mass properties are included in the description.

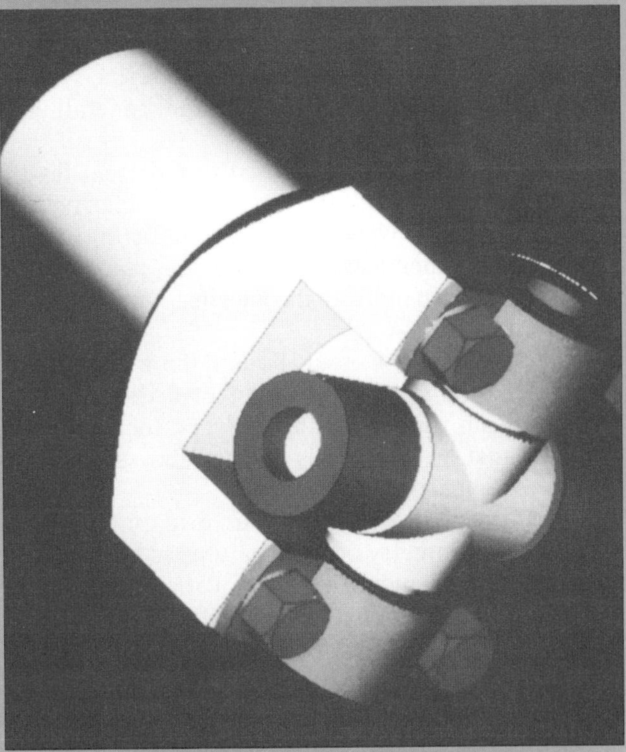

Using Computervision's Solidesign™ software, solid model geometry can be converted into a realistic three-dimensional model. This model of a universal joint is an example of that capability. (Courtesy of Computervision Corporation.)

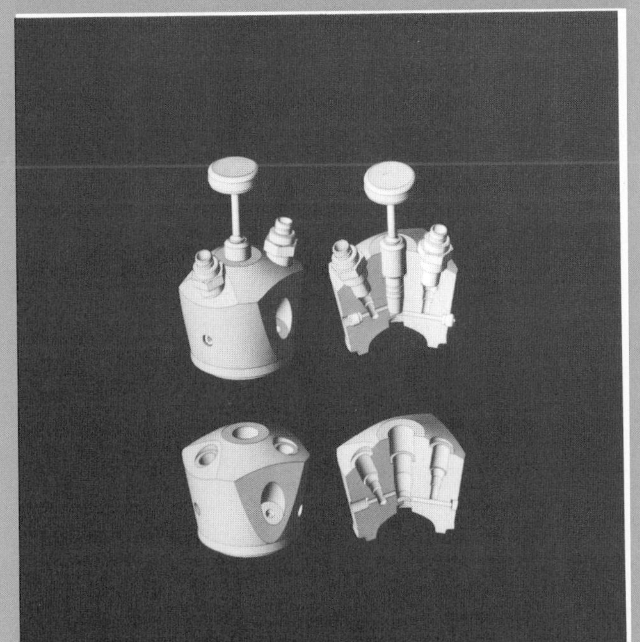

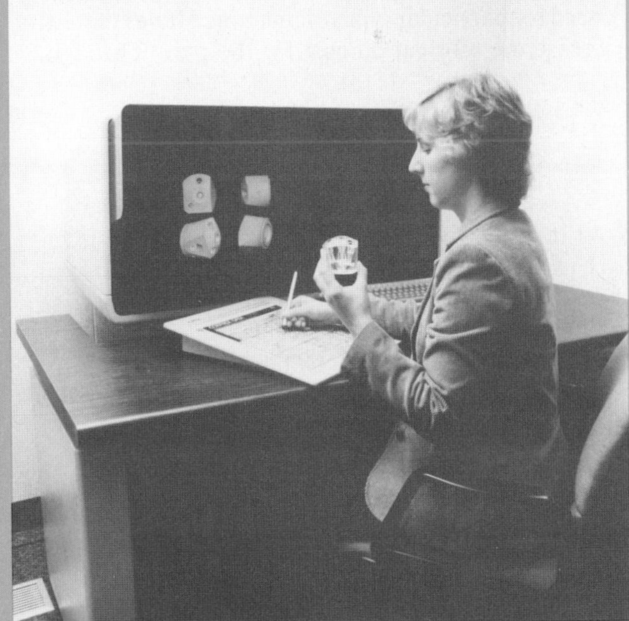

An example of a solid model in which interior detail and mass properties are included in the description of the object (which, in this case, is a valve housing). Also shown is a physical model of the housing, which is compared to the solid model seen on a Control Data ECEM Ergonomic Workstation screen. (Photos courtesy of Control Data Corporation.)

In Retrospect

- Partial views are often adequate to describe symmetrical or cylindrical objects without generating a full set of complete views.
- Removed views should be used to represent features that are difficult to represent with clarity in full orthographic views because of conflicting linework from other features. A viewing plane line must be used to indicate the viewing direction for the removed view.
- An incomplete view should be used if it will aid the reader's interpretation of a set of orthographic views by omitting confusing linework. Incomplete views must be judiciously chosen; they are only used to enhance clarity.
- Conventional revolutions are used to emphasize the radial symmetry of such features as holes and ribs about the center of a circular plane. Such revolutions do not represent actual modifications in a part; they are only used for greater clarity in an orthographic drawing.
- Left-hand and right-hand parts may be adequately described by a set of views in which only one part is shown, together with an appropriate note indicating that the reversed-image part must also be produced.
- The designer of a part must be able to justify the need for particular types of machined holes that increase the fabrication cost for the part. Through, blind, countersunk, and counterbored holes are four types of holes that must be clearly identified.
- The primary considerations for intersections between cylindrical surfaces are:
  - The accuracy that is required.
  - The ease with which such intersections may be constructed.
  - The relative size of the intersecting cylindrical surfaces.
- Rounded corners are often included in parts so that:
  - The strength of the part is increased (in the case of interior corners).
  - The difficulty associated with the production of sharp corners (which are associated with finished or machined surfaces) may be avoided.
  - The appearance of the part may be enhanced.
- Rounds and fillets are round exterior and interior corners, respectively. A runout is the curve produced by a fillet located at the point of tangency between a planar surface and a cylindrical surface.
- The rules for orthographic projection may be violated only if such a violation (such as incomplete views and conventional revolutions) will increase the clarity of an orthographic drawing.

Exercises

Your instructor will specify the appropriate scale for each exercise. All holes are *through holes* (they extend through the entire object) unless otherwise indicated.

EX5.1 through EX5.4. Draw the best and necessary views of each object in the figures.

In the drawings for EX5.5 through EX5.9 clarity in presentation has not been maximized. You will be asked to redraw these exercises.

EX5.5. Draw a more appropriate orthographic representation of this object.

EX5.6. Draw a correct orthographic representation of this object.

EX5.7 through EX5.9. Draw more appropriate orthographic representations of the objects in these figures.

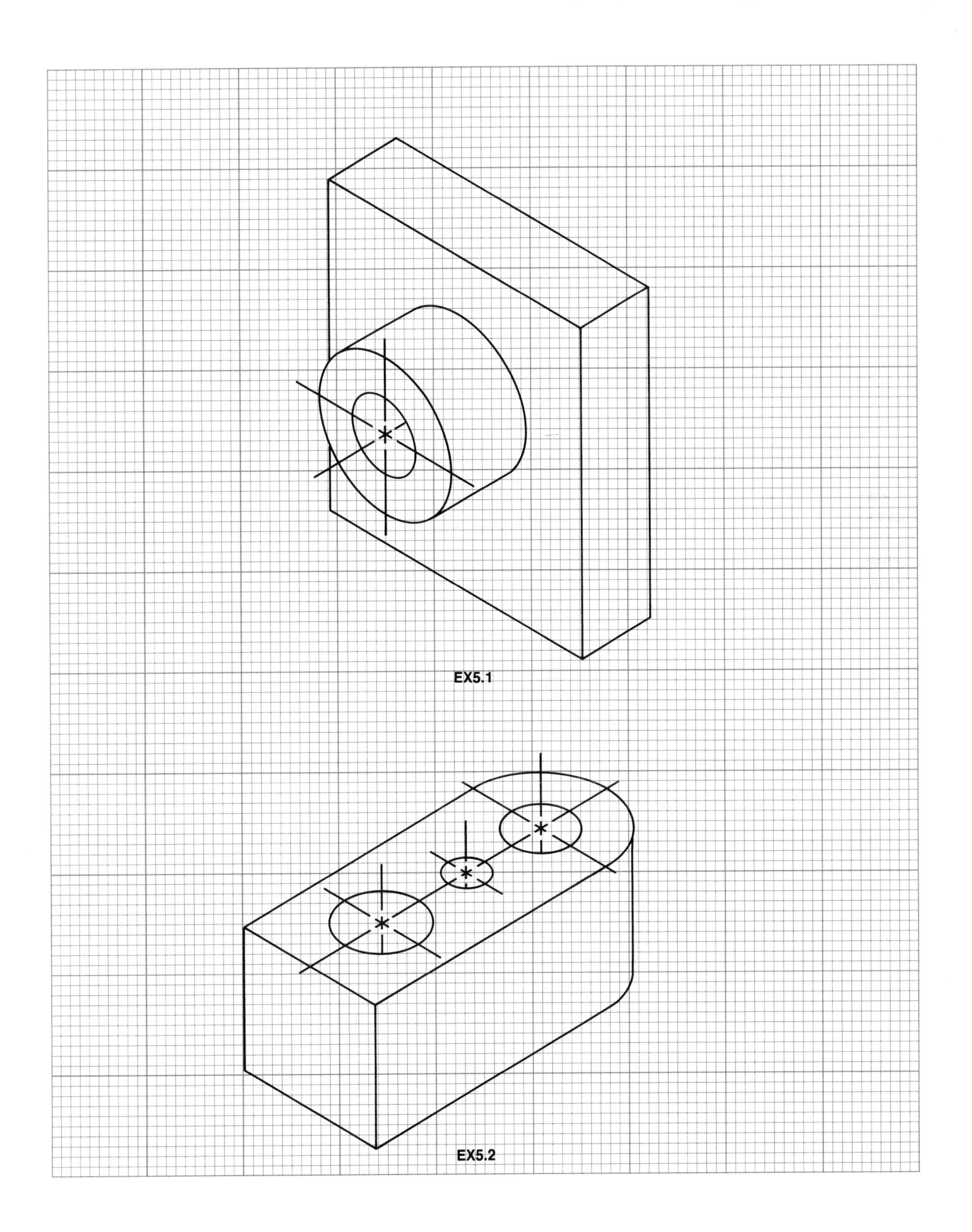
EX5.1
EX5.2

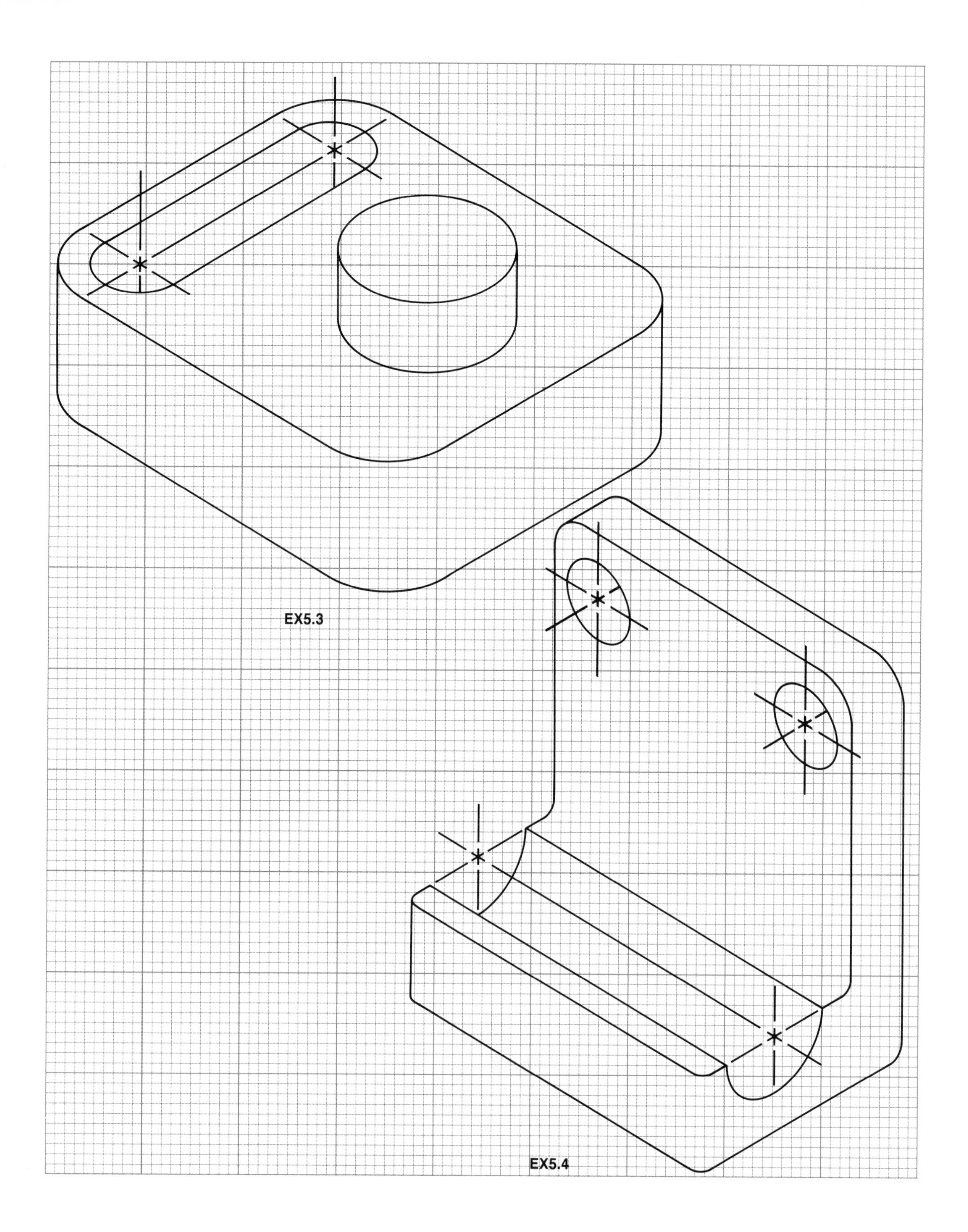
EX5.3
EX5.4

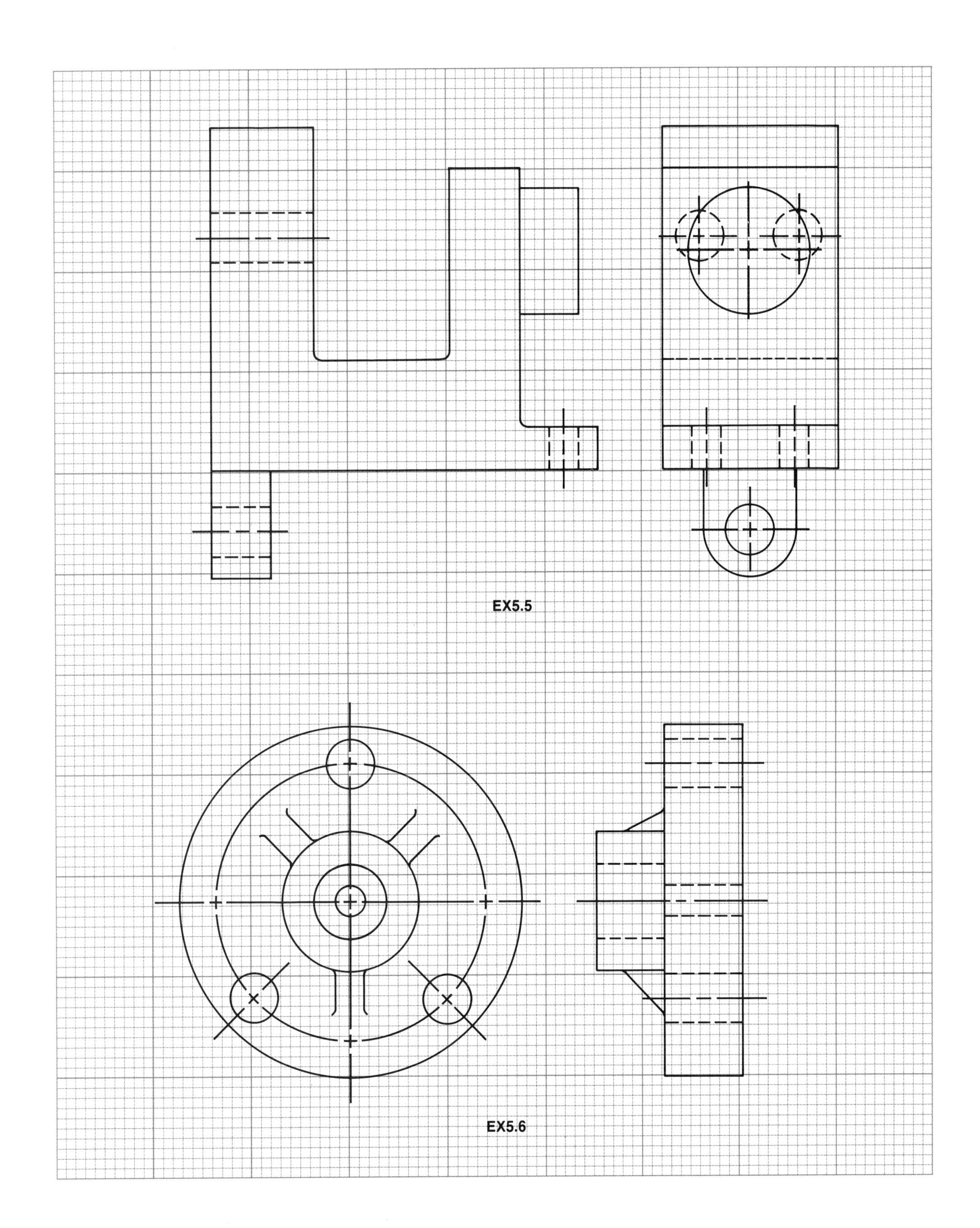
EX5.5
EX5.6

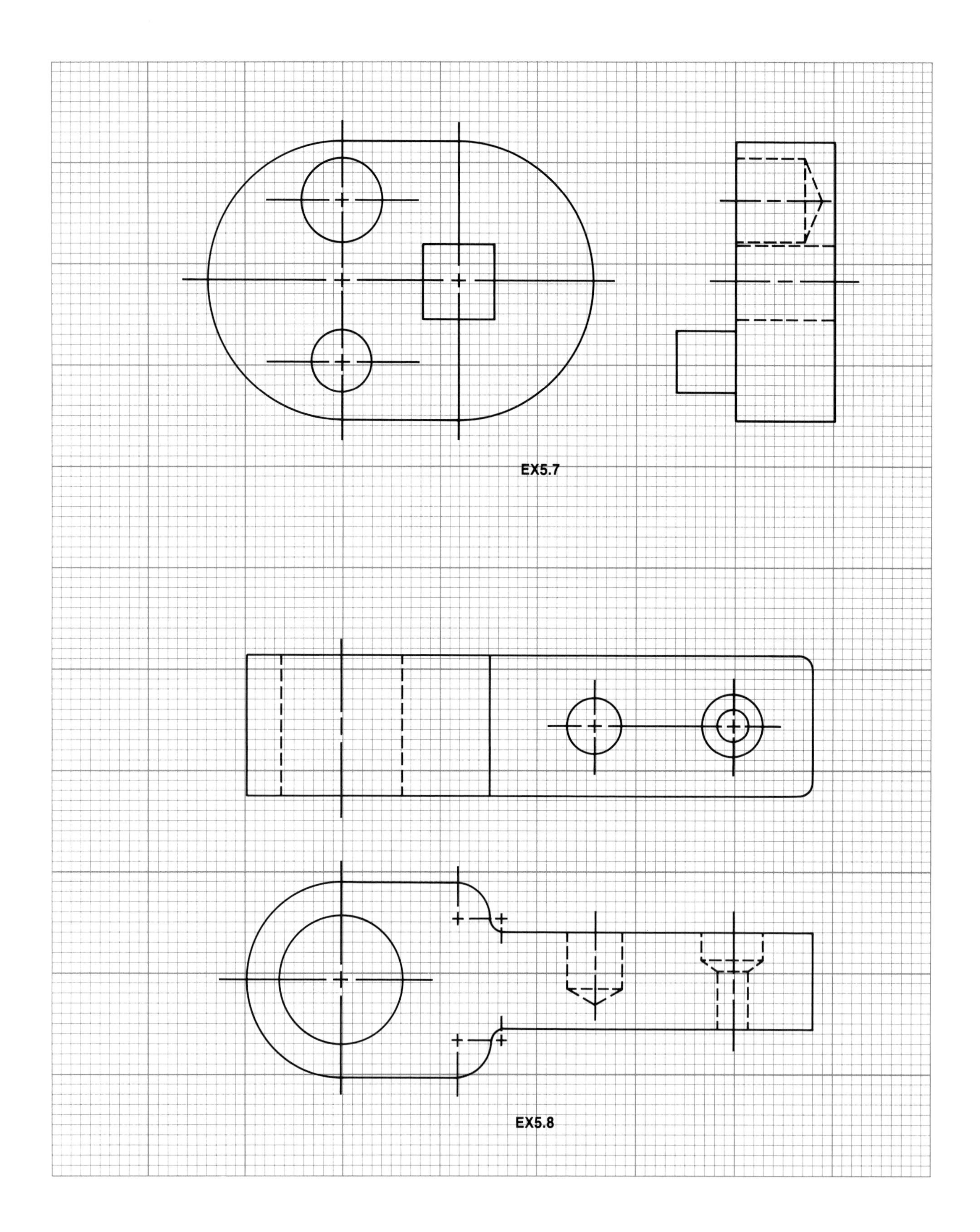
EX5.7
EX5.8

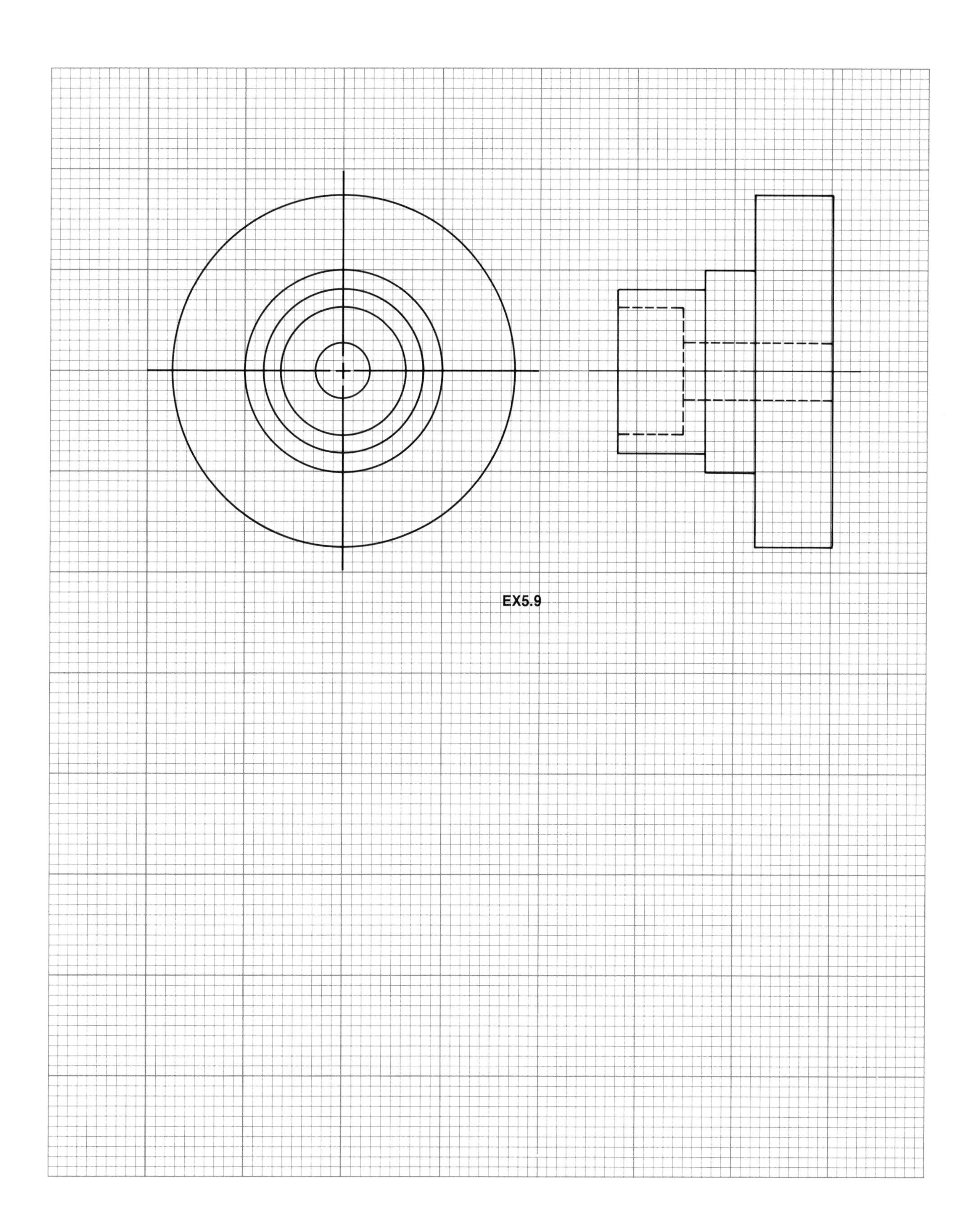
EX5.9

6 Sectioning

Preview As we have seen repeatedly in our review of orthographic projection (Chapters 4 and 5), *clarity must always be our major goal in engineering drawings.* The rules for orthographic projection may even be violated in certain instances in order to enhance clarity, as in the case of conventional revolutions and incomplete views. In addition, special orthographic drawings known as **sectional views** should be added when they can describe the internal features of a part with greater clarity.

Hidden linework in orthographic views often creates confusion for the reader of a drawing. A sectional view shows the part as if a portion of it had been removed (or cut away), thereby allowing the reader to see its internal features without the overlapping linework of the external features.

Learning Objectives

Upon completion of this chapter, the reader should be able to:

- Interpret sectional views to understand the interior features of a part.
- Denote a sectional view by an appropriate cutting-plane line in an orthographic drawing.
- Include the appropriate section linework in a sectional view.
- Differentiate between full, half, revolved, removed, and broken-out sectional views and generate such views for various parts.
- Interpret assembly sections of an object composed of multiple components.
- Apply conventional breaks in orthographic drawings.

6.1 Sectional Views (Sections) and Cutting Planes

Sectional views (or **sections**) allow one to represent the internal features of a part with clarity in an orthographic drawing. An imaginary **cutting plane** is used to section the object in accordance with the given sectional view. For example, the object shown in Figure 6.1 includes a cylindrical boss and two holes that must be clearly described in a set of orthographic views. An imaginary cutting plane is used in Figures 6.2 and 6.3 to "remove" a portion of the object so that we may clearly see that both holes extend through the entire object (that is, they are through holes). A second imaginary cutting plane is used in Figures 6.4 and 6.5 to remove a portion of the object so that we may clearly see the intersection of the

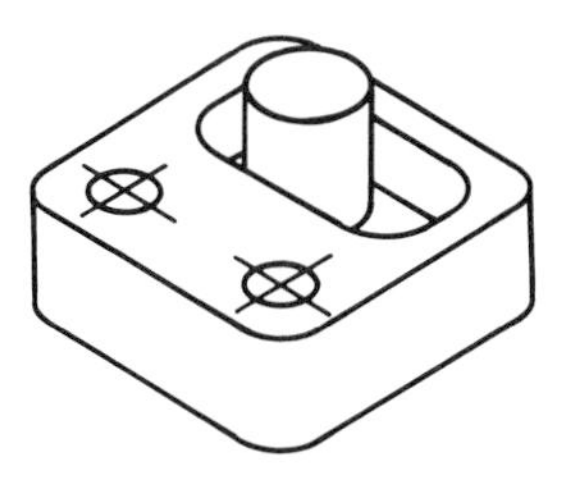

Figure 6.1 Example object to be the subject of successive figures in this chapter. Note the cylindrical boss and two holes.

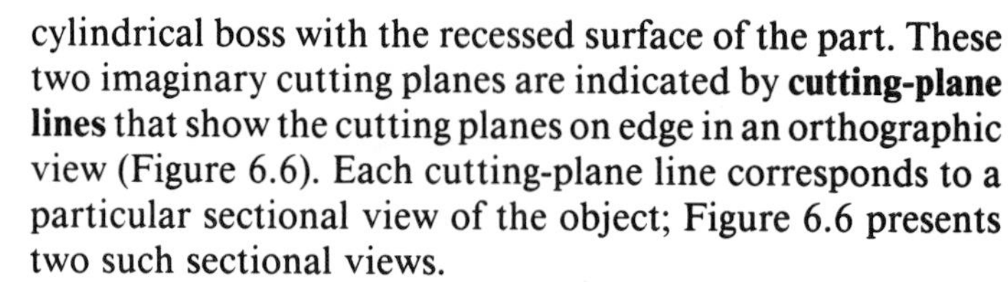

cylindrical boss with the recessed surface of the part. These two imaginary cutting planes are indicated by **cutting-plane lines** that show the cutting planes on edge in an orthographic view (Figure 6.6). Each cutting-plane line corresponds to a particular sectional view of the object; Figure 6.6 presents two such sectional views.

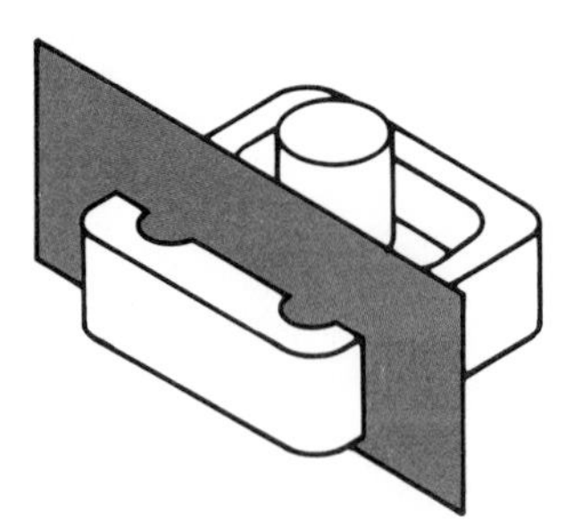

Figure 6.2 Imaginary cutting plane passing through the object in Figure 6.1.

Cutting-plane lines have the following features:

- They are composed of (1) two short dashes, each about 3 mm. or 1/8 in. in length in most cases, that alternate with a long dash about 25 mm. or 1 in. in length or (2) a series of dashes, each about 6 mm. or 1/2 in. in length.
- They are drawn with the intensity of visible linework (very dark) and should be thick.
- They have priority over centerlines for circles and arcs.
- They include arrows indicating the viewing direction that corresponds to a particular sectional view.
- They use *identification callout letters* (A-A, B-B, and so on) to label multiple sectional views in an orthographic drawing.

Cutting-plane lines inform the reader that a sectional view of a part is given in an orthographic drawing. In order to further emphasize that the view is a sectional representation of a part and not a true orthographic projection, one must include **section lines** across those interior surfaces that contact the imaginary cutting plane. General section linework (also known as *crosshatching* or *hatchwork*) is composed of parallel, equally spaced, thin lines, as shown in Figure 6.7. Notice the examples of poor section linework shown in this figure, and avoid similar mistakes.

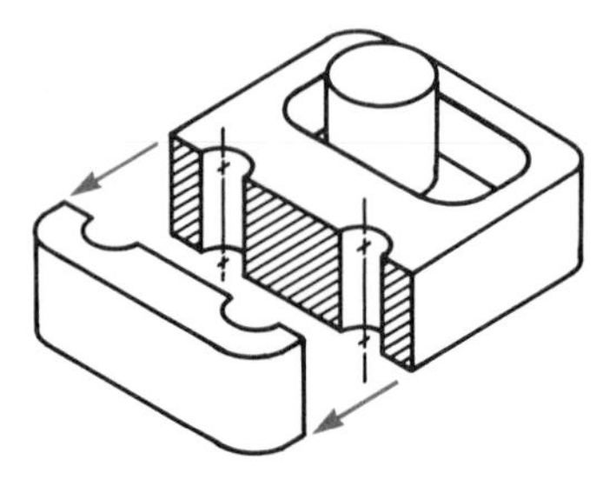

Figure 6.3 A portion of the object in Figure 6.1 is removed in order to show interior detail.

Section lines are usually drawn at an angle of 45° relative to the horizontal direction unless such an orientation will result in a conflict or an overlapping with other linework. Special symbolic section linework, as shown in Figure 6.8, is used to specify particular metals or other materials in the fabrication of a part. Hidden linework is not included in a

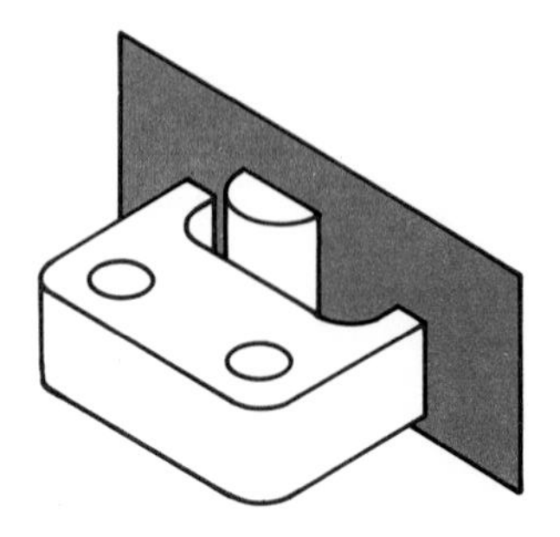

Figure 6.4 Second imaginary cutting plane passing through the object in Figure 6.1.

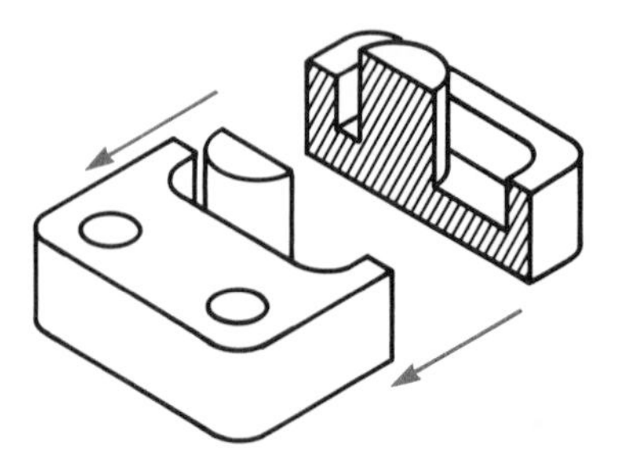

Figure 6.5 A second sectioned view of the object in Figure 6.1 is produced.

sectional view unless it is absolutely necessary to describe the true form of the part.

In summary, then:

1. Sectional views are used to describe the internal features of a part with greater clarity.

2. An imaginary cutting plane passing through the part is indicated by a cutting-plane line in an orthographic drawing. Each cutting-plane line corresponds to a particular sectional view.

3. Identification callout letters are used to label multiple sectional views in a drawing (such callout letters are not necessary in the case of a single cutting-plane line and its corresponding sectional view in a drawing).

4. Section linework is used to identify all interior surfaces that contact the imaginary cutting plane. Symbolic section

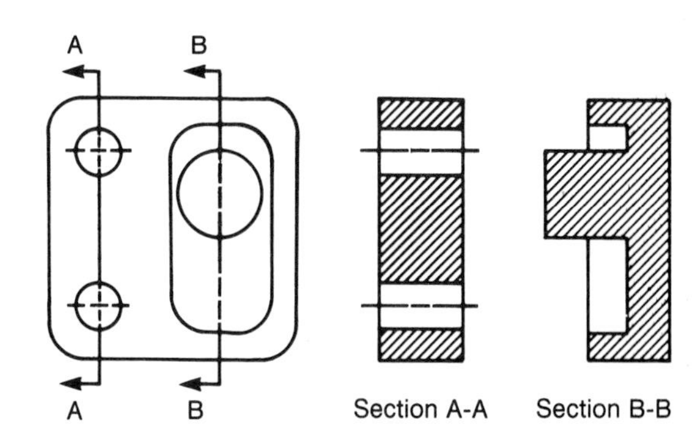

Figure 6.6 Orthographic views of the example object in Figure 6.1, including two sectional views.

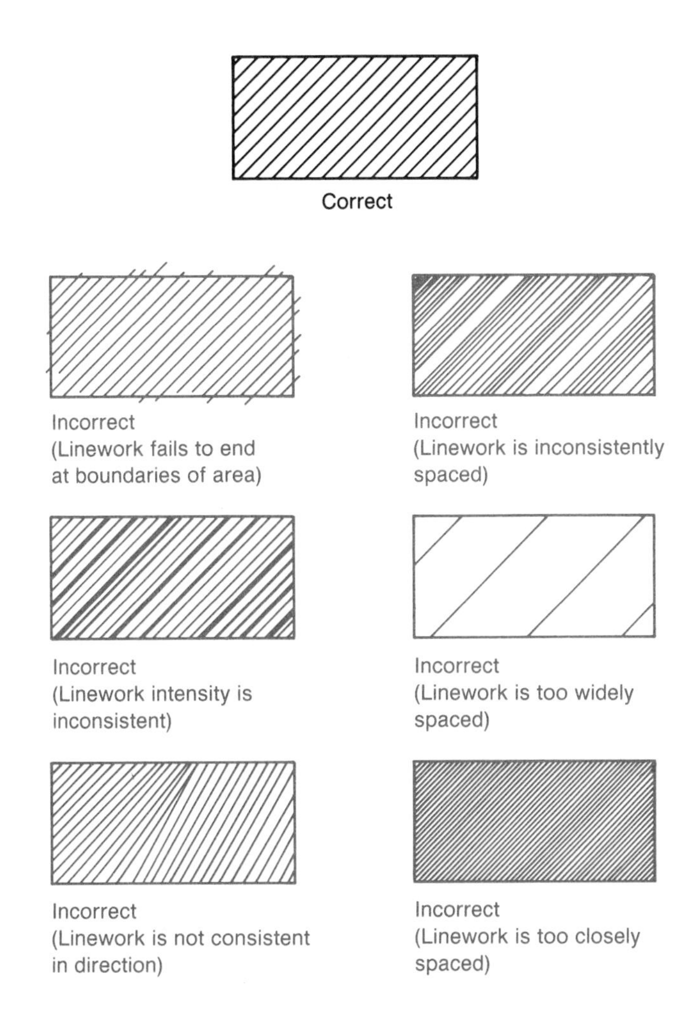

Figure 6.7 Section linework must be carefully drawn, with parallel, equally spaced, thin lines.

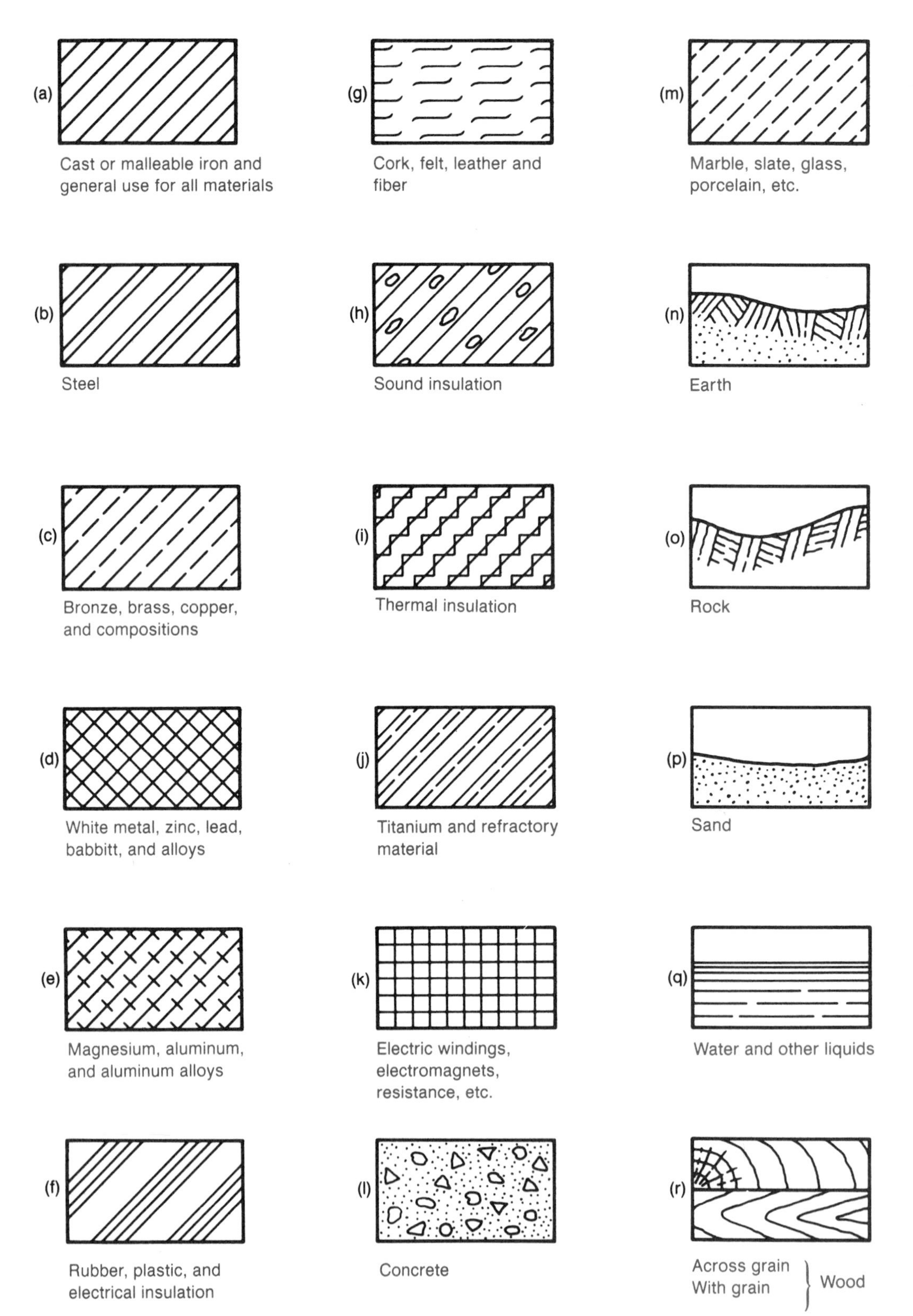

Figure 6.8 Symbolic section linework is used to specify particular materials for the fabrication of a part. (ANSI Y14.2M-1979, American National Standard Line Conventions and Lettering.)

linework also may be used to indicate the type of material to be used in the fabrication of a part.

5. Hidden linework is not included in sectional views unless it is absolutely necessary in order to properly describe the shape of the part (or for dimensioning purposes).

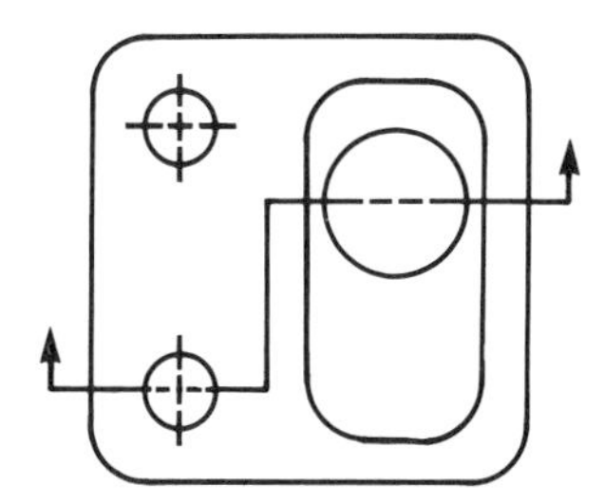

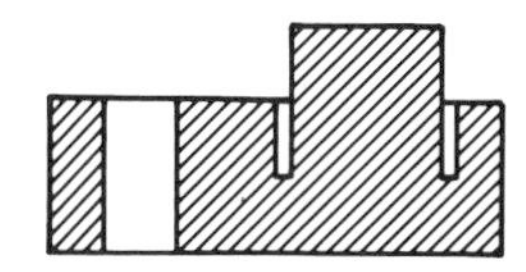

Figure 6.9 An offset sectional view of the object in Figure 6.1.

6.2 Offset Sections

The example object of Figure 6.1, described by multiple sectional views in Figure 6.6, can be more efficiently represented by a single **offset sectional view** (or **offset section**), shown in Figure 6.9. In such a view, the cutting plane is offset, or broken, so that it passes through both of the interior portions of the object that require a sectional view for clarity of interpretation. Multiple sectional views can often be replaced this way by a single equivalent offset view. The breaks, or discontinuities, in offset cutting-plane lines are not shown in the corresponding sectional view.

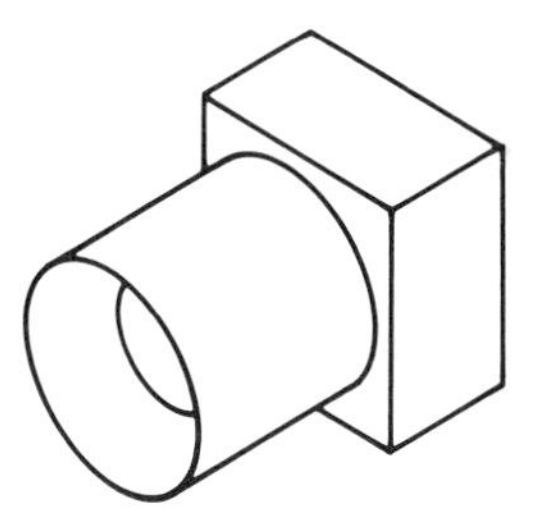

Figure 6.10 Example object to be the subject of successive figures in this chapter.

6.3 Full Sections

A **full sectional view** (or **full section**) of an object is produced if the cutting plane is passed through the entire part. For example, the object in Figure 6.10 can be clearly described by a sectional view. In Figure 6.11 an imaginary cutting plane has passed through the entire part, after which one-half of the part has been removed. Now we can see that the hole in the part does not extend entirely through the object. Figure 6.12 presents the corresponding orthographic drawing for this part, which includes the full sectional view. (Notice that identification callout letters are not necessary since only one cutting-plane line and one corresponding sectional view appear in the drawing.) An offset view, in which the cutting-plane line passes through the entire object (see Figure 6.9), is a full sectional view.

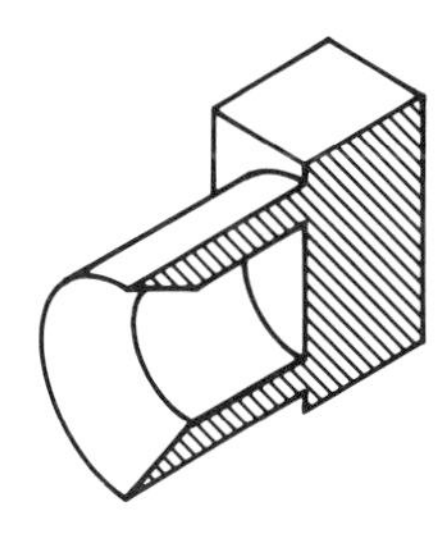

Figure 6.11 One-half of the object in Figure 6.10 remains after a full sectional cut has been applied.

6.4 Half Sections

The object shown in Figure 6.10 is symmetrical about a central axis; such parts can be clearly described with either full sectional views (Figure 6.12) or **half sectional views (half sections)** in which cutting-planes pass through one-half of the object, after which one-quarter of the object is removed (Figures 6.13 through 6.15).

A half-sectional view (Figure 6.15) should be used if a full section will remove a portion of the part that cannot be otherwise described. No hidden lines are included in a half-sectional view unless such linework is absolutely necessary to describe the shape or the size (via dimensioning) of the part.

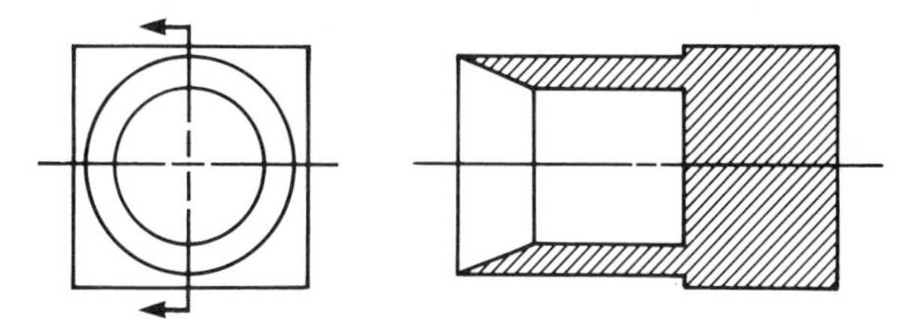

Figure 6.12 The set of orthographic views for the object in Figure 6.10 includes a full sectional view.

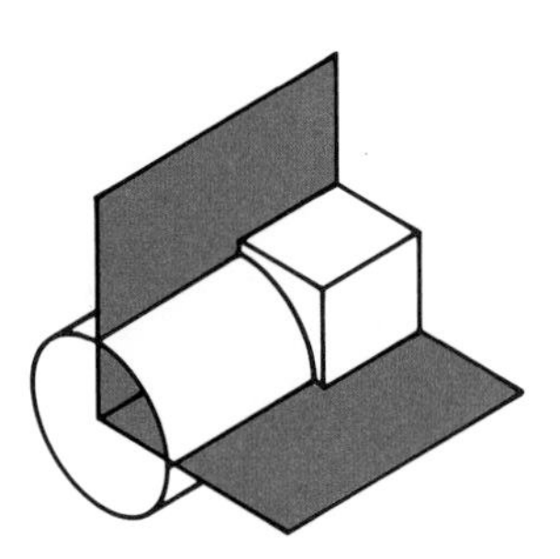

Figure 6.13 Two cutting planes are applied to the object in Figure 6.10 in order to form a half-sectional view.

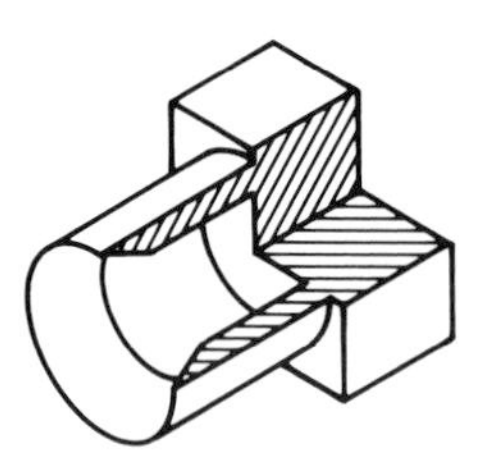

Figure 6.14 One-quarter of the object in Figure 6.10 is removed for a half-sectional view. Nonparallel section linework is used on nonparallel interior surfaces that contact the cutting planes.

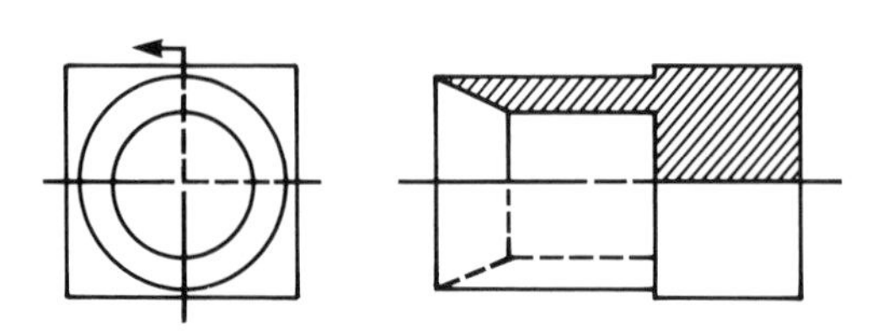

Figure 6.15 The set of orthographic views for the object in Figure 6.10 includes a half-sectional view.

Notice that an arrowhead is attached to only one end of the cutting-plane line in Figure 6.15. This is standard practice to clearly indicate the viewing direction for the section view. If the two section views were given—for example, top and right-side views—then two arrowheads would be needed. A centerline is used to separate the sectioned portion of a half sectional view from the unsectioned portion.

Learning Check

Why are orthographic views sectioned within a drawing? Are not the unsectioned views sufficient to fully describe the shape of an object?

Answer A view may need to be sectioned in order to describe the internal features of a part with greater clarity. *A drawing should describe a part with maximum clarity.*

Why is section linework added to a sectional view?

Answer Section linework is used to clearly identify the internal surfaces of a part in a sectional view.

6.5 Revolved Sections and Removed Sections

Partial sectional views can be used to describe the internal features of some parts instead of complete sectional views. In one type of partial sectional view known as a **revolved section,** a cutting plane is passed through the part and then revolved 90°. The resulting cross-sectional cut is shown in the given orthographic view (Figure 6.16).

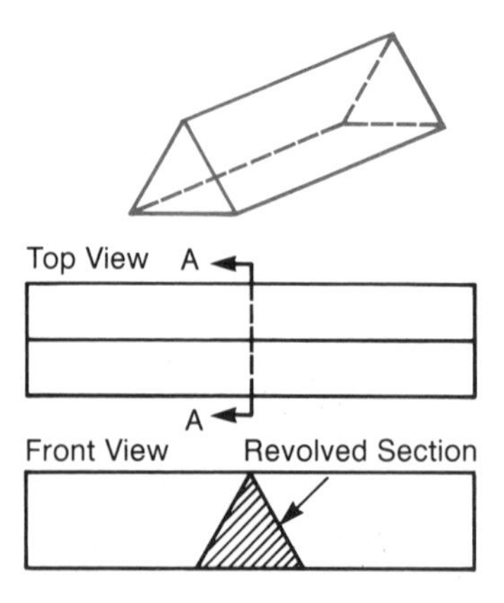

Figure 6.16 In a revolved section, the cutting plane is passed through the object and then revolved 90°. (NAVEDTRA 10634-C.)

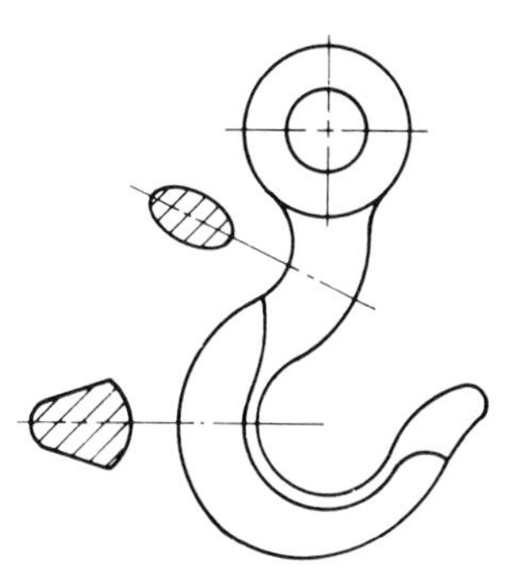

Figure 6.17 Removed sections are shown at a separate location on the drawing from the view in which the cutting-plane line appears. (ANSI Y14.3-1975, American National Standard Multi and Sectional View Drawings.)

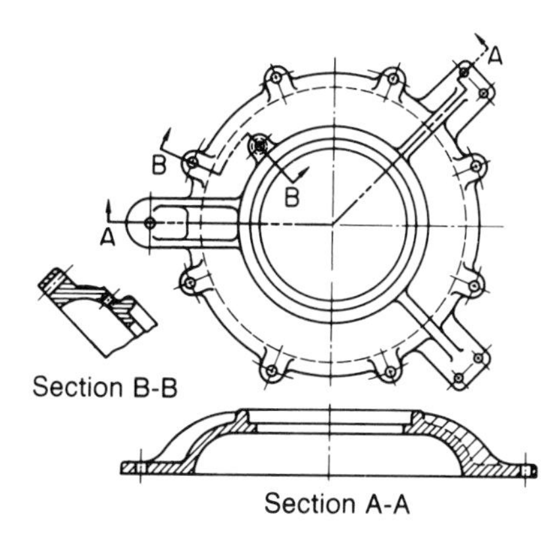

Figure 6.18 Section B-B is the removed section in this example. (ANSI Y14.3-1975, American National Standard Multi and Sectional View Drawings.)

A revolved section may be used to save both drawing time and space. It focuses the reader's attention upon the internal feature that is of particular importance and that can be properly described by a sectional view—yet, the revolved section takes the place of a complete sectional view.

If there is insufficient space within an orthographic view to include a revolved section, one may either omit unnecessary linework or use a type of revolved section known as a **removed view.** A removed (revolved) sectional view is shown at a separate location on the drawing from the view in which the cutting-plane line is used to indicate the sectional revolution (see Figures 6.17 and 6.18).

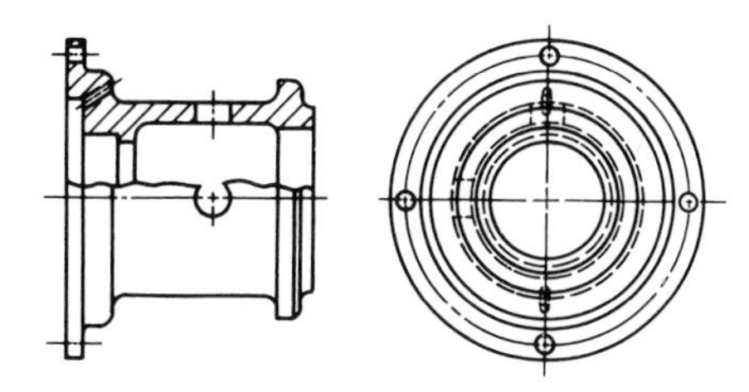

Figure 6.19 A broken-out section shows a portion of the object. An irregular break line is used. (ANSI Y14.3-1975, American National Standard Multi and Sectional View Drawings.)

6.6 Broken-Out Sections

In another type of partial sectional view known as a **broken-out section** a portion of the orthographic view is shown (Figure 6.19). A broken-out section is used when a full or half-sectional view would eliminate necessary visible linework or when a complete sectional view is not necessary and a revolved section is not appropriate due to the shape of the part. An irregular **break line** is used to clearly show the separation between the sectioned and unsectioned portions of a broken-out section (Figure 6.19).

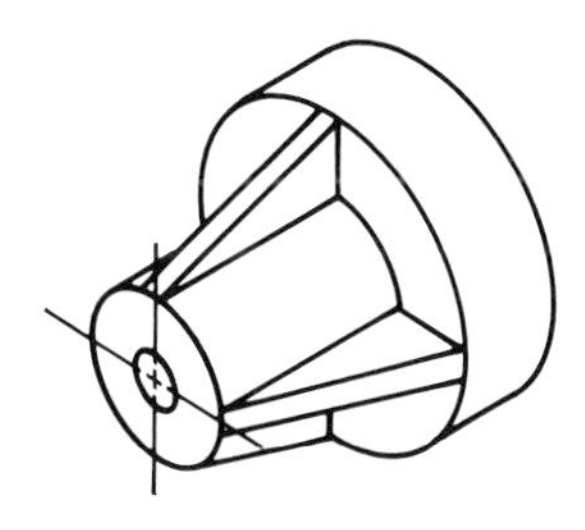

Figure 6.20 Example object to be the subject of successive figures in this chapter.

6.7 Ribs, Webs and Spokes in Section

Certain parts of an object should never be shown in section even though the cutting plane may pass through them. For example, the ribs of the object shown in Figure 6.20 should *not* be shown in section, that is, with section lining (see Figures 6.21 and 6.22) since such section linework, as shown in the incorrect drawing in Figure 6.22, may lead the reader to conclude that the part has a solid, tapered, conical form instead of its true shape.

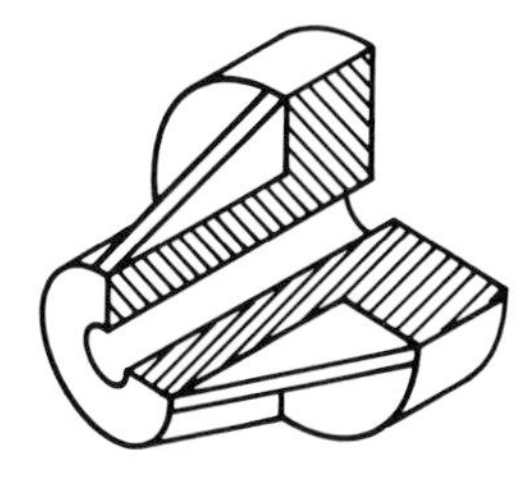

Figure 6.21 Half-sectional of the object in Figure 6.20. Note that ribs are not section-lined.

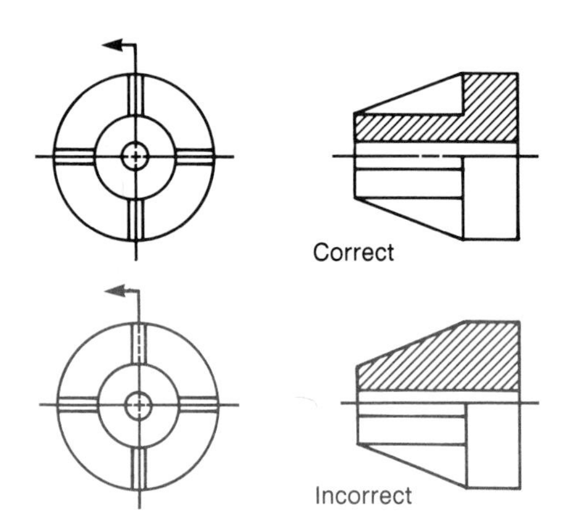

Figure 6.22 Correctly drawn and incorrectly drawn sectional views of a part that includes a thin rib.

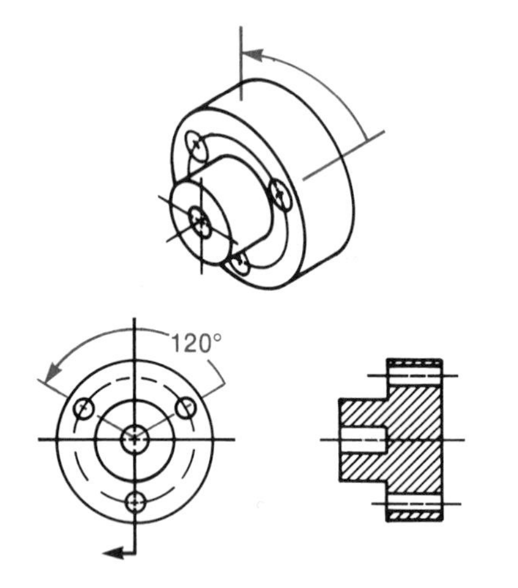

Figure 6.23 Aligned view of an object with symmetrically located holes.

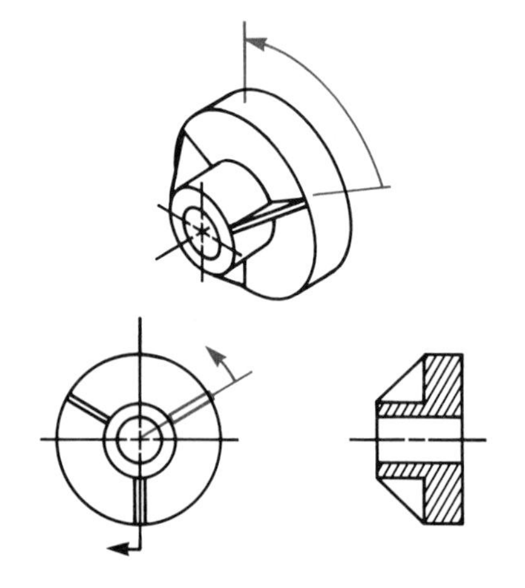

Figure 6.24 Aligned view of an object with symmetrically located ribs.

Webs and spokes, in addition to ribs, are usually not shown in section in order to clearly specify to the reader that these features are thin portions of the part. Alternate section linework can be used for webs only if the web outline coincides with other outlines. Nuts, rivets, shafts, bearings, and machine screws are also not section-lined in order to enhance clarity in a drawing (see Section 6.9, "Assembly Sections").

6.8 Aligned Sections

In Section 5.4 we reviewed the practice of conventional revolutions, in which the radial symmetry about a circular plane of certain features is emphasized by rotating these features about the center of a circular plane. Such conventional revolutions are, of course, imaginary; they are included only in order to enhance clarity in an orthographic drawing.

In sectional views one may also use conventional revolutions to enhance the clarity of the drawing. In such drawings, called **aligned sections,** the (symmetrically located) features are aligned with one another in the sectional view if these features are symmetrically arranged around the center. In asymmetrical objects, the choice of cutting-plane can produce a drawing that resembles an aligned section.

Figure 6.23 presents an object in which three holes are symmetrically located about the perimeter of the disc-shaped base of the part. In the orthographic views for this object, the cutting-plane line is used to indicate both the location of the cutting plane as it passes through the object and the direction of alignment, or rotation, for the holes in the sectional view. The aligned holes in the sectional view are shown at their true radial distance from the axis of symmetry for the part.

Figure 6.24 presents an object with symmetrically located ribs similarly described in an aligned sectional view. Still another example of an aligned section is section A-A back in Figure 6.18.

6.9 Assembly Sections

Assembly drawings describe the mating of the components that form a part. Such drawings are often sectioned in order to enhance the clarity of the description; in these cases, the drawings are known as **assembly sections** (Figure 6.25).

Components in an assembly section that are adjacent are shown with different section-lining in order to emphasize each component in the drawing. Common parts such as nuts, bolts, and machine screws are not shown in section in order to aid the reader's interpretation of the drawing. *Thin parts* (such as shims, sheet metal, gaskets, and plates) can be shown as darkened solid areas (Figure 6.26).

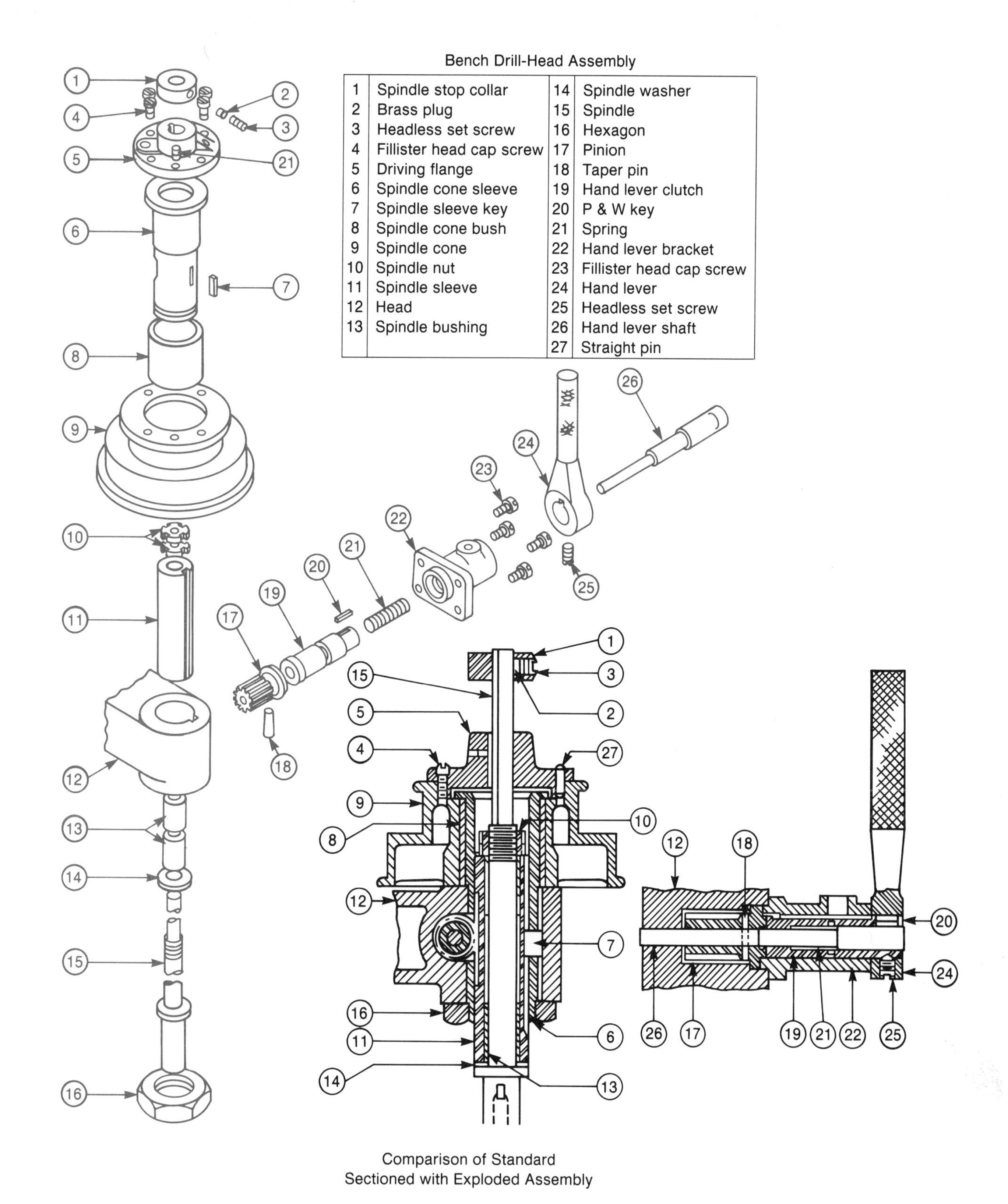

Bench Drill-Head Assembly

1	Spindle stop collar	14	Spindle washer
2	Brass plug	15	Spindle
3	Headless set screw	16	Hexagon
4	Fillister head cap screw	17	Pinion
5	Driving flange	18	Taper pin
6	Spindle cone sleeve	19	Hand lever clutch
7	Spindle sleeve key	20	P & W key
8	Spindle cone bush	21	Spring
9	Spindle cone	22	Hand lever bracket
10	Spindle nut	23	Fillister head cap screw
11	Spindle sleeve	24	Hand lever
12	Head	25	Headless set screw
13	Spindle bushing	26	Hand lever shaft
		27	Straight pin

Figure 6.25 Example of an assembly section used in conjunction with an exploded pictorial. (ASA Y14.4-1957, Pictorial Drawing.)

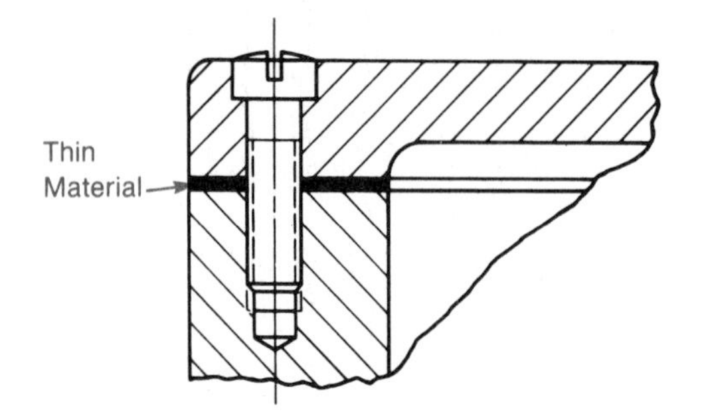

Figure 6.26 Thin material may be shown as a darkened solid area in an assembly section. (ANSI Y14.2M-1979, American National Standard Line Conventions and Lettering.)

6.10 Conventional Breaks in Sections

Conventional breaks are used to shorten long portions (such as shafts and arms) of an object in an orthographic view. Such breaks indicate that a portion of the part has been removed. In addition, they may be used to indicate the geometric shape of a part or that a revolved section is shown in the view. Figure 6.27 presents several conventional breaks.

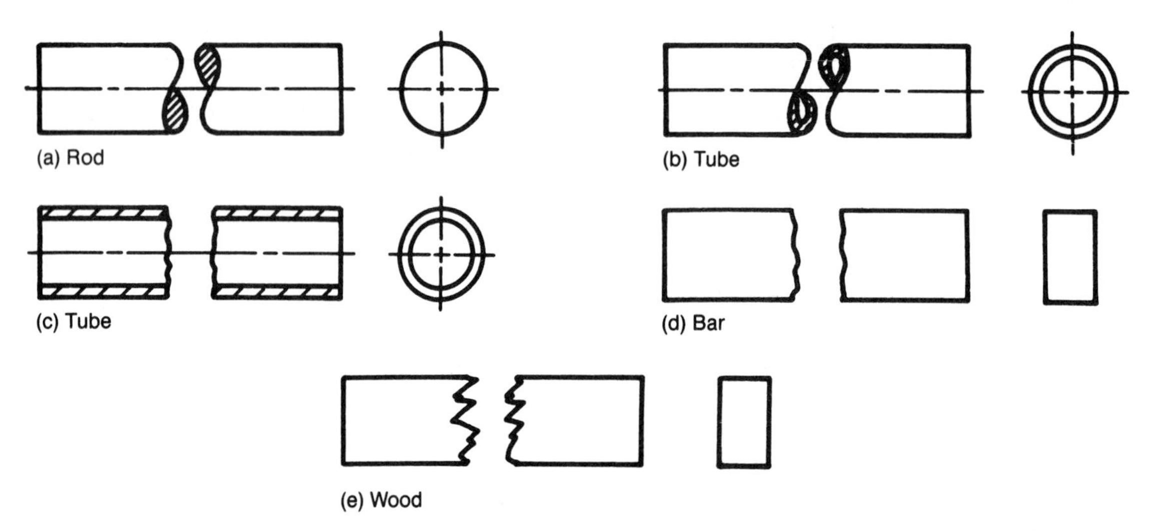

Figure 6.27 Conventional representation of breaks in elongated parts. (ANSI Y14.3-1975, American National Standard Multi and Sectional Drawings.)

In Retrospect

- A sectional view depicts an object as if a portion of the object had been removed.
- An imaginary cutting plane which passes through the object is indicated by a cutting-plane line in an orthographic drawing. Such lines indicate both the position of the cutting plane as it passes through the part and the viewing direction for the corresponding sectional view.
- Section linework is used to identify all interior surfaces in a sectional view that contact the imaginary cutting plane. Symbolic section linework may be used to indicate the type of material to be used in the fabrication of the part.
- Hidden linework is not included in sectional views unless it is absolutely necessary in order to properly describe the shape or the size of a part.
- Multiple sectional views can be replaced by an equivalent offset view that corresponds to a broken or offset cutting plane passing through the object. Discontinuities in the cutting-plane line are not shown in the offset view.
- A full sectional view is produced by a cutting-plane

3D Computer Graphics: Constructing Solid Models with Primitives

Primitives are simple geometric shapes (such as cylinders, squares, and spheres) that are defined and stored with the computer system's memory for use in the construction of solid models (discussed in "Computer Graphics in Action" in Chapter 5). These primitives are combined in various ways to form the desired model. For example, the illustration presents two primitives that have been combined in three different ways. The first is **intersection,** resulting in a solid model composed of the points common to both primitives when they are combined in a specific relative geometric position. The second is **union,** in which the model is composed of both primitives joined in a specific orientation. The third is **difference,** in which the volume and form of one primitive is subtracted or removed from the other to form the desired model. The user of a solid modeling software package can translate primitives to any location with respect to one another as needed, after which an intersection, union, or difference operation is specified to be performed between the given primitives. Since this method of solid modeling is dependent upon a set of common geometric shapes (primitives), the objects to be modeled should be of a form that can be constructed through the manipulation of such primitives. This may seem restrictive; however, most machined parts may be represented as a combination of primitives since machining processes (drilling, cutting, pressing, and so forth) correspond to the creation of basic geometric shapes.

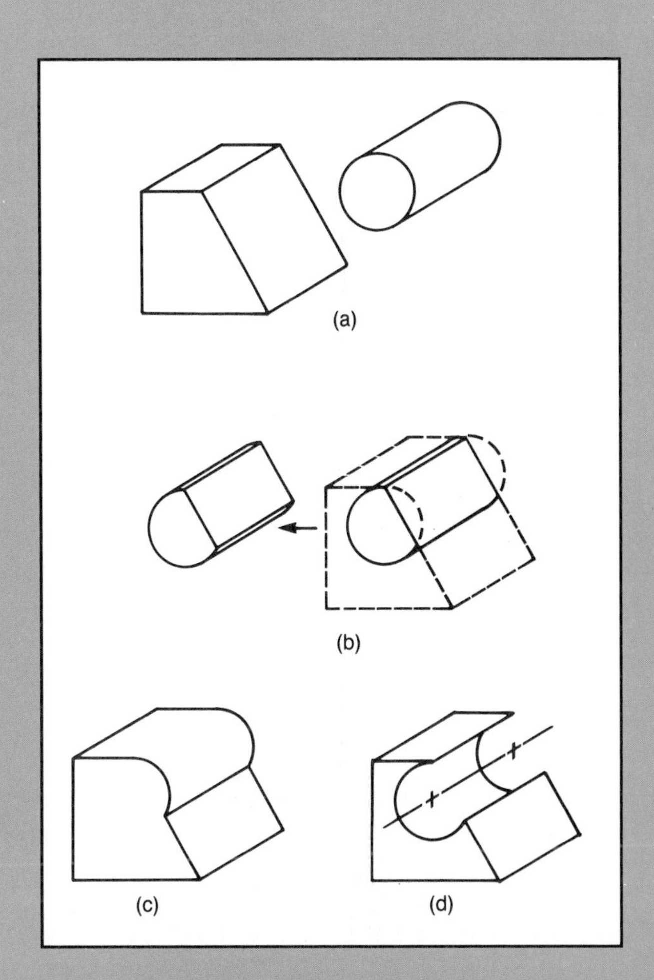

Two primitives (a) can be developed into a solid model through intersection (b), union (c), or difference (d).

Solid models may also be developed through the method known as **boundary definition,** which we will review in the "Computer Graphics in Action" section in Chapter 7.

that passes through the entire object. A half-sectional view corresponds to the imaginary removal of one-quarter of the object.

- Half-sectional views allow one to show both the internal and external features of a part in a single view, separated by a centerline (although standard practices allow the centerline to be replaced with a solid line).
- Section linework on nonparallel surfaces are drawn with nonparallel linework; however, section linework must be identical (parallel, thin, dark lines with equal spacing) on parallel surfaces of the same part.
- A revolved section is a partial sectional view in which a cross-sectional representation of a part is shown in an orthographic view. Revolved sections are generated by rotating the cutting plane 90° so that the cross-sectional representation may be projected onto the given orthographic view.
- A removed section is a revolved sectional view that has been removed, or separated, from the orthographic view in which the cutting-plane line is shown; removed sections are used if there is insufficient space for a revolved section in the complete orthographic view or if multiple revolved sections must be used to properly describe a part.
- Another type of partial sectional view is known as a broken-out section; such views are used if a full or a half-sectional view would eliminate necessary visible linework, or if a revolved section (or complete sectional view) is not appropriate because of the shape of the part. Irregular break lines are used to indicate the extent of the sectioned portions in a broken-out section.
- Webs, ribs, and spokes are not shown in section; alternate section linework may be used if the web outline coincides with other outlines.
- Bolts, nuts, rivets, machine screws, shafts, and other standard parts are not shown in section since such sectioning will not aid the reader in the interpretation of the drawing.
- Aligned sections are produced by conventional revolutions of a part in order to emphasize the symmetrical location of features about the center of a circular plane. Aligned sections present these features (holes, ribs, and so forth) at their true radial distance from the axis of symmetry. Cutting planes used for asymmetrical parts can produce views that resemble aligned sections.
- Assembly sections describe the way in which the components of an object are to be joined.
- Conventional breaks may be used to shorten particularly long portions of a part in an orthographic view, thereby saving space in a drawing.

Exercises

Your instructor will specify the appropriate scale for each exercise.

EX6.1. Draw the given views of this object, including the sectional views as indicated.

EX6.2. Draw the appropriate set of orthographic views of this object, including the three sectional views as indicated.

EX6.3. Draw the appropriate set of orthographic views, including the offset sectional view as indicated.

EX6.4. Draw the appropriate set of orthographic views, including the offset sectional view as indicated.

EX6.5 through EX6.22 refer to specific exercises in Chapters 4 and 5. In each case draw an appropriate set of orthographic views of the particular object, and include a section (full, partial, offset, and so on) that clearly describes the interior structure of the object or part.

EX6.5. See EX4.33.

EX6.6. See EX4.41.

EX6.7. See EX4.49.

EX6.8. See EX4.50.

EX6.9. See EX4.52.

EX6.10. See EX4.55.

EX6.11. See EX4.56.

EX6.12. See EX4.57.

EX6.13. See EX4.58.

EX6.14. See EX4.59.

EX6.15. See EX5.1.

EX6.16. See EX5.2.

EX6.17. See EX5.3.

EX6.18. See EX5.5.

EX6.19. See EX5.6.

EX6.20. See EX5.7.

EX6.21. See EX5.8.

EX6.22. See EX5.9.

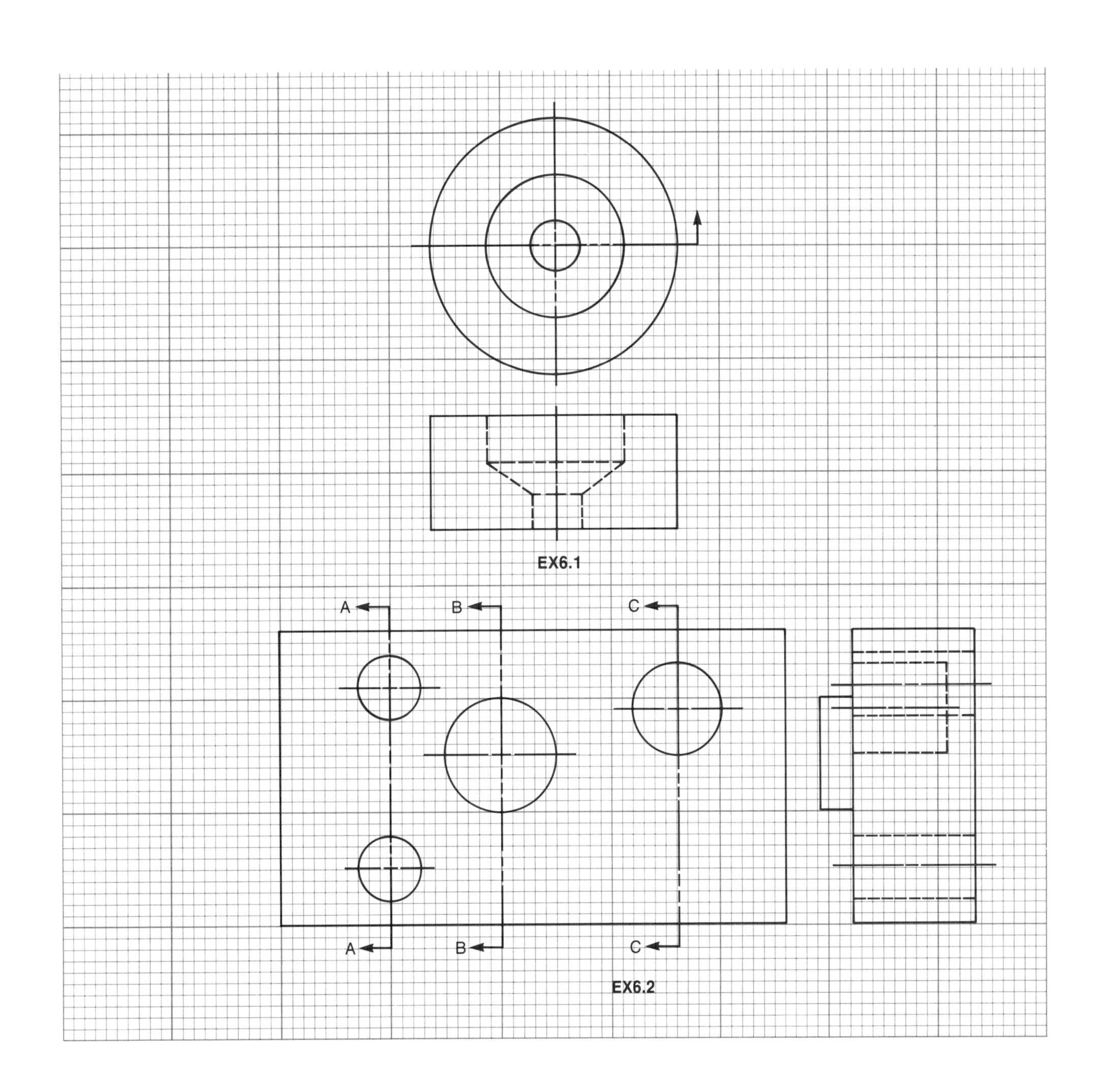

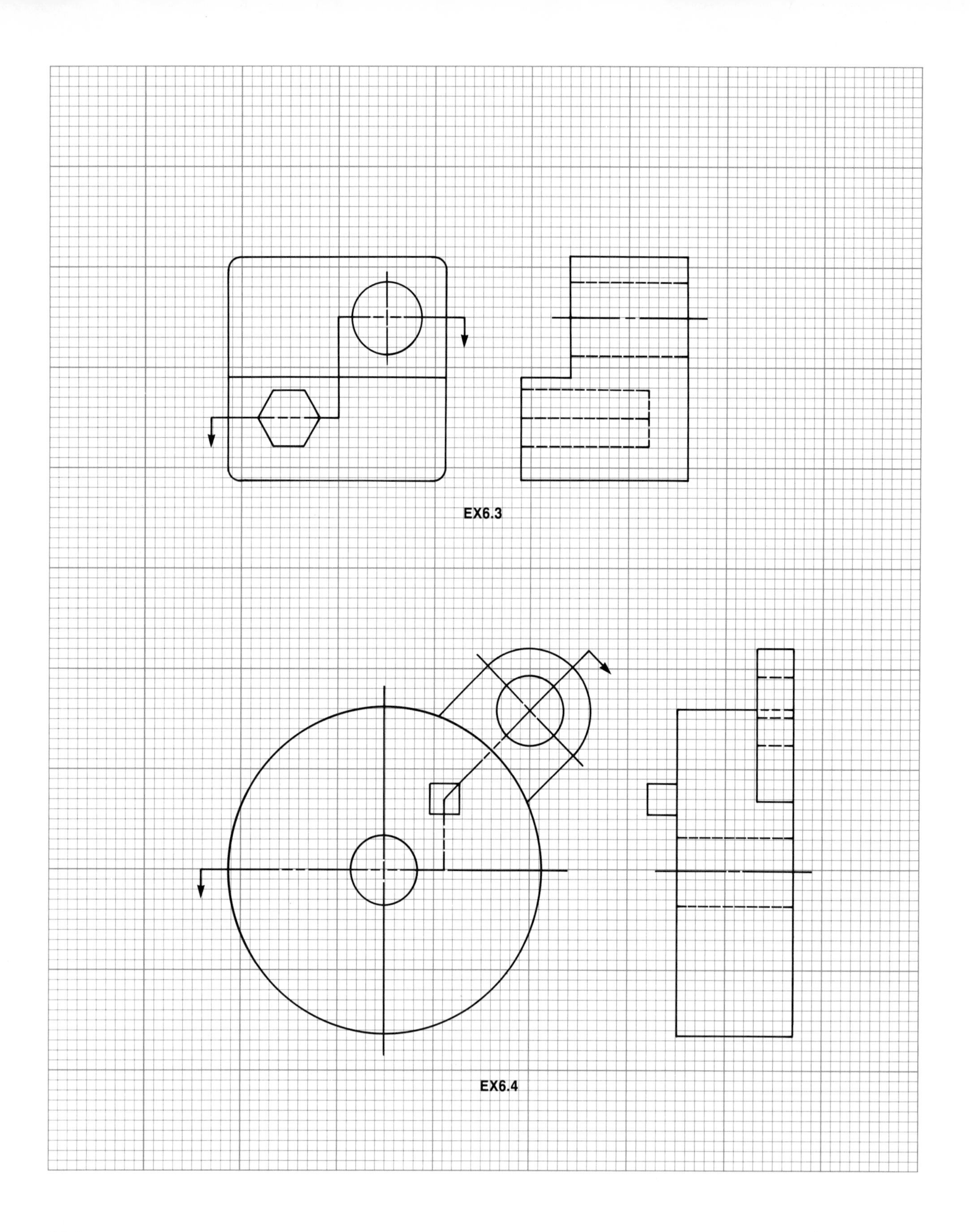
EX6.3
EX6.4

7 Auxiliary Views

Preview The orthographic projection views discussed thus far are the six **principal views:** front, rear, right-side, left-side, top, and bottom. Inclined surfaces of an object appear on edge (as lines) or as foreshortened (reduced) areas in principal views; oblique surfaces also appear as foreshortened areas in principal views. Such surfaces can be shown as true-size, undistorted areas only in views where the direction of sight is perpendicular to the surface. When an undistorted representation of an inclined or oblique surface is essential for an accurate and complete graphic description of an object, **auxiliary views** should be developed.

In this chapter, we will review construction techniques for the development of both **partial** and **complete** auxiliary views. In addition, we will carefully consider **primary** and **secondary** auxiliary views. Partial and secondary auxiliary views will be essential in our application of descriptive geometry techniques to develop accurate orthographic descriptions of points, lines, and planes used to define an engineering device (see Chapters 11 and 12).

Learning Objectives

Upon completion of this chapter, the reader should be able to:

- Construct auxiliary orthographic views of an object in which inclined and oblique surfaces are shown as true-size areas—without distortion.
- Differentiate between partial and complete auxiliary views.
- Differentiate between primary and secondary auxiliary views.
- Select the appropriate basic view of an object in order to generate the particular auxiliary view that is desired.

7.1 The Glass Box and the Auxiliary Viewing Direction

Figure 7.1 presents an object with an inclined surface, A, which is enclosed within a glass box. (An **inclined surface** is, by definition, perpendicular to one of the principal projection planes but not parallel to any of these planes.) Projection along three principal (front, top, and right-side) viewing directions onto the walls of the glass box results in three distinct orthographic views of the object (Figures 7.2 through 7.4). In these views the inclined surface A is seen on edge, as a line (in the front view), or as a foreshortened, or reduced,

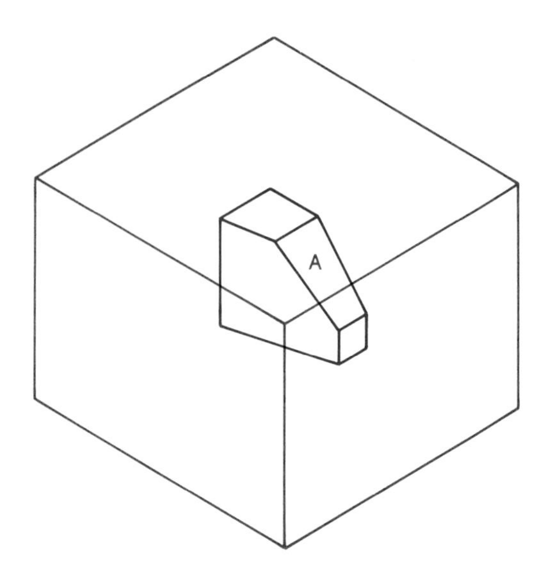

Figure 7.1 Example object, to be the subject of successive figures in this chapter, enclosed by a glass box. Note that surface A is inclined.

area (in the top and right-side views). In order to generate an orthographic view in which surface A is shown in a true-size (TS) area representation, the observer must view the object along a sight-line direction that is perpendicular to the plane containing surface A (see Figure 7.5). Projection along this **auxiliary viewing direction** onto an auxiliary projection plane (that is, a plane parallel to surface A) then results in an orthographic image of A that is a true-size area representation (Figures 7.6 and 7.7).

7.2 Development of a Primary Auxiliary View of an Inclined Surface

Before we review the procedure for developing an auxiliary orthographic view of an inclined surface, recall the strategy used to develop a missing principal orthographic view. As shown in Figure 7.8, a missing principal view is developed through the use of the following:

1. A *viewing direction* for the missing principal view. This is chosen with respect to a given view (for example, the missing right-side view in Figure 7.8(a) is to be "drawn off," or drawn with respect to the given front view.

2. A *reference plane* with respect to which distances may be transferred from a given view to the missing-view construction region. For example, the frontal reference plane (FRP) is used in Figure 7.8 to transfer distances of depth from the given top view to the (missing) right-side view (b).

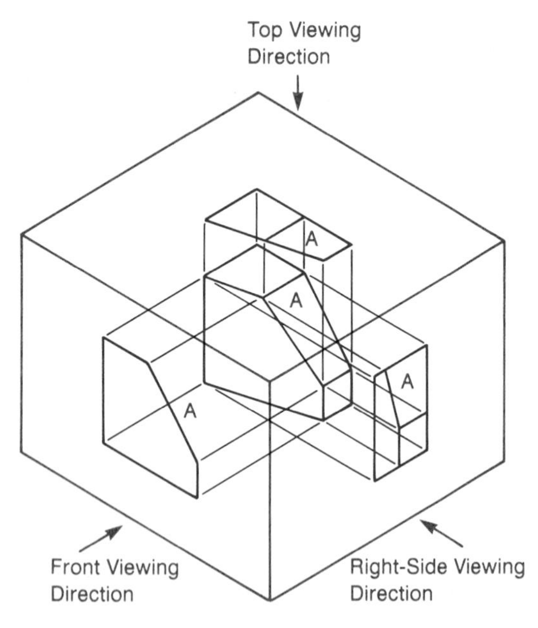

Figure 7.2 Projection of the object in Figure 7.1 along three principal viewing directions onto the walls of the glass box.

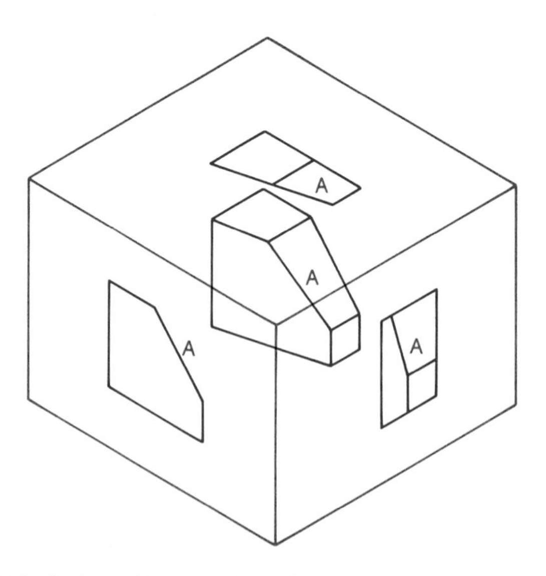

Figure 7.3 Projected images of the object in Figure 7.1 on the walls of the glass box. Inclined surface A appears as either a foreshortened area (top and right-side images) or on edge, as a line (front image).

3. *Key points,* which are used to define the object and its surface boundaries. Each point is located in at least two of the given views, after which the location of the point in space is completely defined. The point is then located in the missing-view construction region. Surfaces of the object are drawn in the construction region through the use of these key points (c), thereby leading to the completion of the missing view (d).

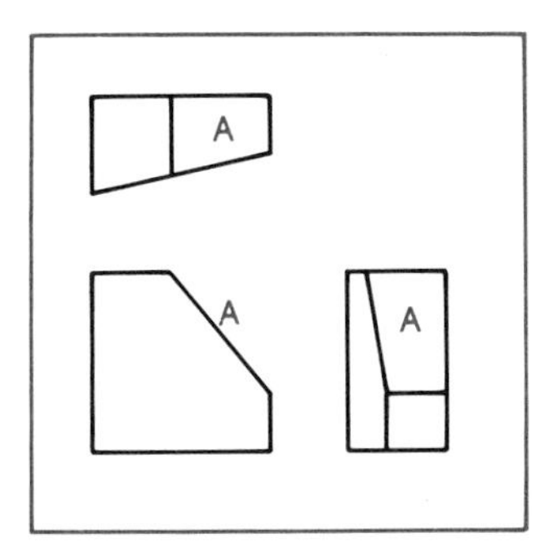

Figure 7.4 Set of principal orthographic views of the object in Figure 7.1. Surface A is seen on edge, as a line, and as foreshortened areas.

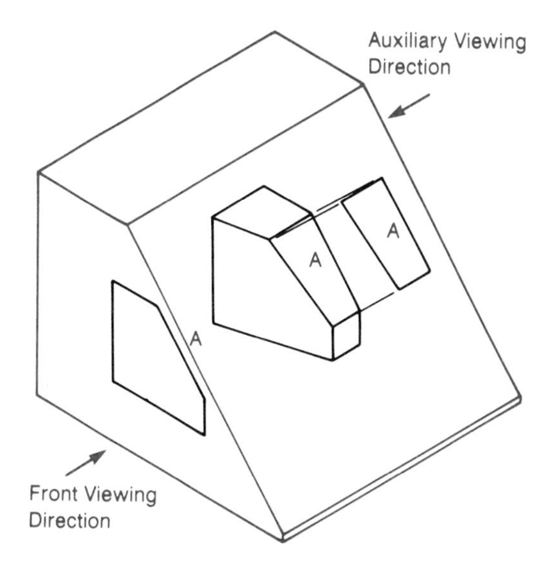

Figure 7.5 Projection of surface A (of the object in Figure 7.1) onto a plane perpendicular to the auxiliary viewing direction (that is, a plane parallel to the plane containing surface A) results in an image representation of A which is the true size.

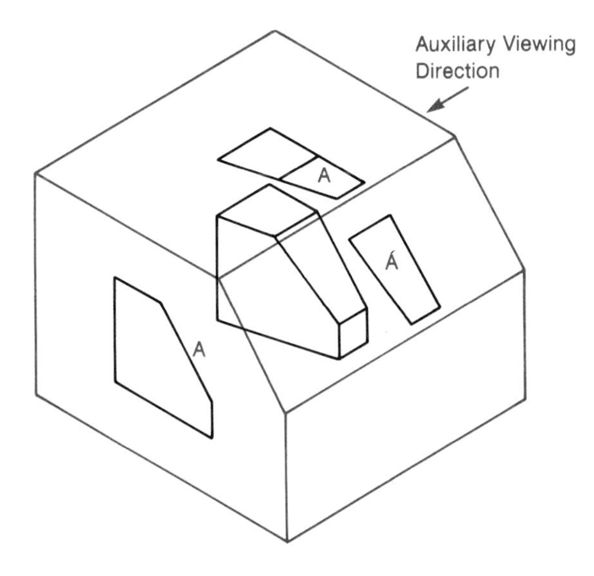

Figure 7.6 A glass box of appropriate dimensions allows one to project multiple principal views of the object enclosed within the box, together with an auxiliary view of surface A. Surface A appears as a true-size area in the auxiliary view shown herein; it is shown on edge, as a line, in the front-view projection and as a foreshortened, reduced, area in the top-view projection.

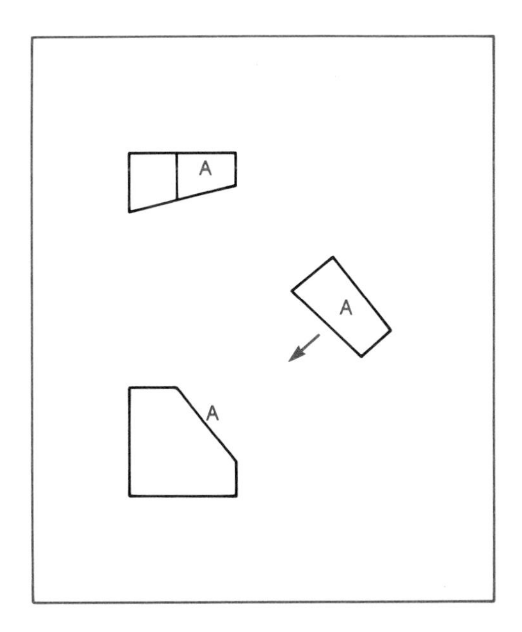

Figure 7.7 Two principal (front and top) views and an auxiliary view of the object in Figure 7.1. The auxiliary view presents a true-size area representation of the inclined surface A.

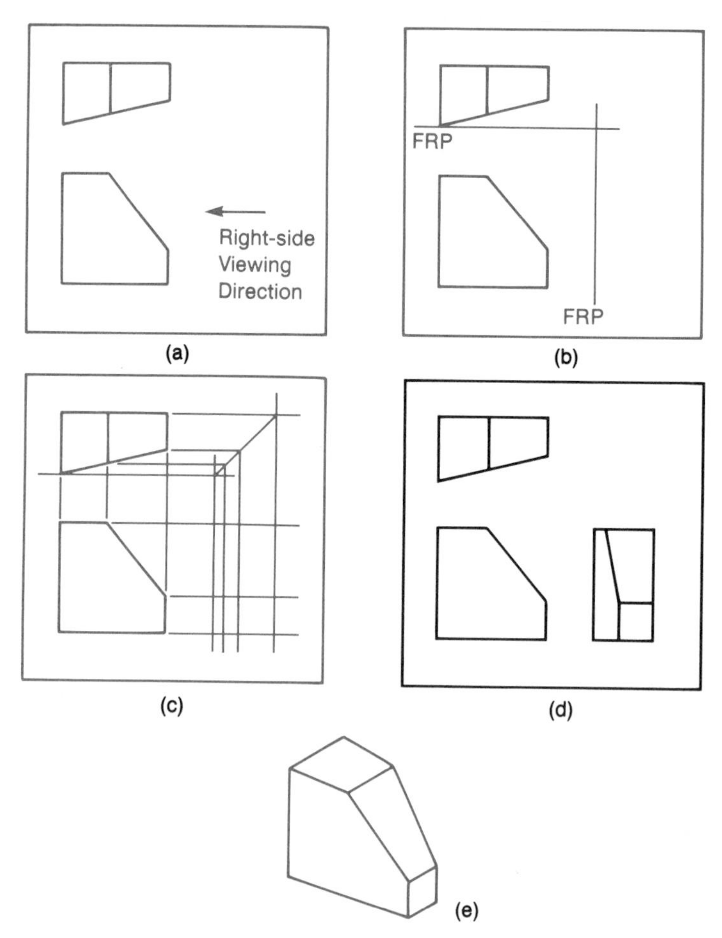

Figure 7.8 Review of the procedure (described in Chapter 4) for development of a missing principal orthographic view through the use of a reference projection plane, the FRP in this case. First, the viewing direction of the missing (right-side) view is identified with respect to a given (front) view (a). Distances are transferred to the missing-view area from the given views (b and c), resulting in a final set of views (d) of the object (e).

The procedure for developing a missing auxiliary view of an inclined surface is similar to the one for developing a principal view: an appropriate viewing direction is chosen, a reference plane is identified, and key points are located in the construction region until the view can be completed. Figure 7.9 shows the following procedure:

1. The auxiliary viewing direction is chosen (a). This direction must be perpendicular to the plane containing the surface that is to appear as a *true-size area* in the resulting auxiliary view. As a result, one must first identify the given **basic view** in which the surface appears on edge, as a line. (The basic view is also known as the *central view* because it is adjacent to two views that are related.) If the basic view is a principal view, the resulting auxiliary view is called **primary.** Thus, the front view in Figure 7.9(a) is the basic view of the primary auxiliary view under construction.

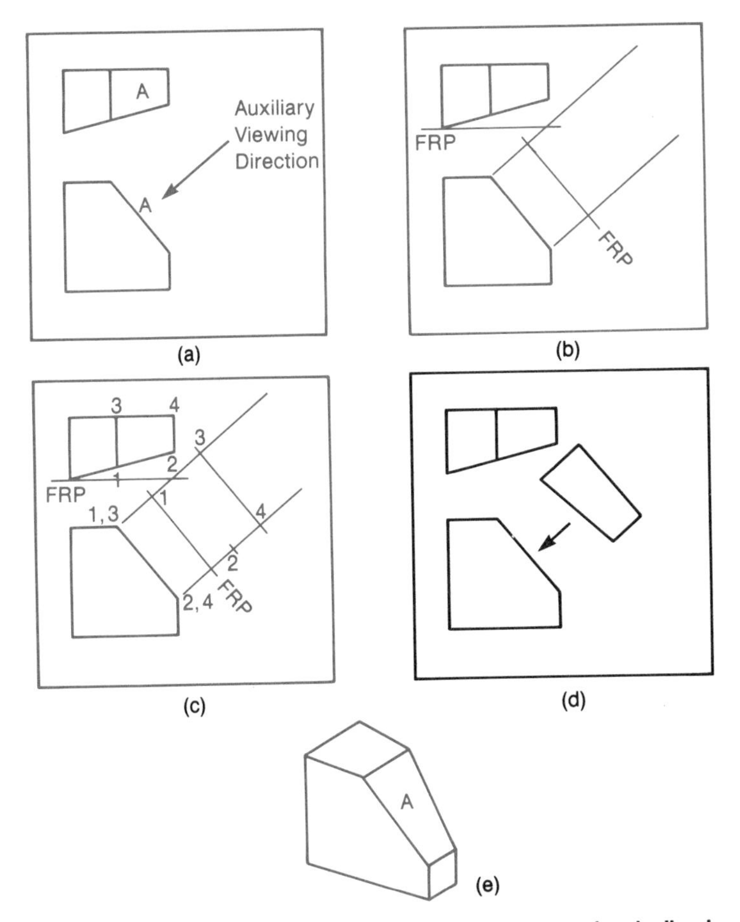

Figure 7.9 Developing a missing partial auxiliary view of an inclined surface through the use of a reference projection plane, FRP in this case. First, the auxiliary viewing direction is chosen (a), then a reference plane identified (b), and key points located (c) to complete the missing view (d) of the object (e). Note that the auxiliary view (d) only shows surface A; such a view is known as a partial auxiliary view.

The desired auxiliary viewing direction is then perpendicular to the line that represents the surface in the basic view. Projection or construction lines are drawn—parallel to this auxiliary viewing direction—from the basic view to the construction region.

2. The position of the auxiliary view relative to the given principal views is chosen by specifying the location of an appropriate reference plane. In this case the frontal reference plane (FRP) is indicated in the construction region and in the given top view (b).

3. Key points, which may be used to define the surface (or surfaces) under consideration are then identified in the given views. Each point is located in the construction region by transferring appropriate distances from the given views (c). Note that the rule of adjacency (Chapter 4) must be obeyed in the development of a missing auxiliary view. Surfaces are then completed to form the auxiliary view (d).

An auxiliary view in which the entire object is not shown is called a **partial** view. Figure 7.9(d) includes two complete principal views of the object and a partial (that is, incomplete) auxiliary view of the inclined surface A. Partial views are often sufficient since the function of an auxiliary view is to show—without distortion—a particular surface that is not parallel to any of the principal projection planes. However, a **complete** auxiliary view can be developed by transferring additional point locations from the given principal views until the entire object has been constructed, as shown in Figure 7.10, which presents two principal views and a complete auxiliary view of the object in Figure 7.9.

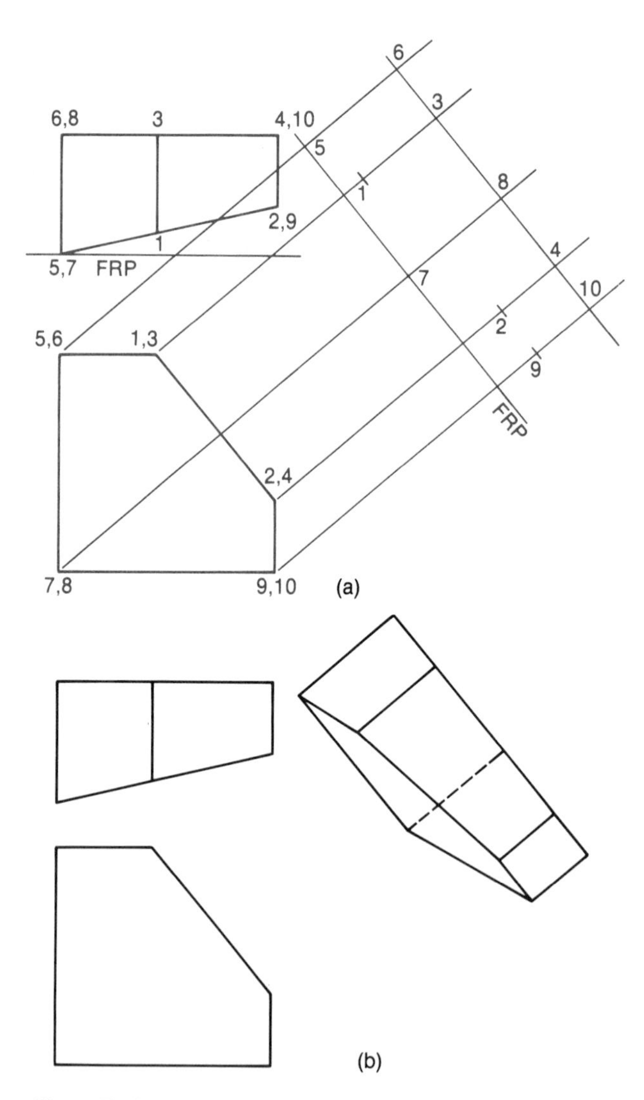

Figure 7.10 Developing a missing complete auxiliary view instead of the partial view shown in Figure 7.9 requires transference of additional points from the principal given views (a–b).

As another example of the development of a primary auxiliary view, consider Figure 7.11, in which the given top view (a) is the basic view of inclined surface A. Identification of an appropriate reference plane (b) is followed by construction of the primary auxiliary view, partial (c and d) and then complete (e and f).

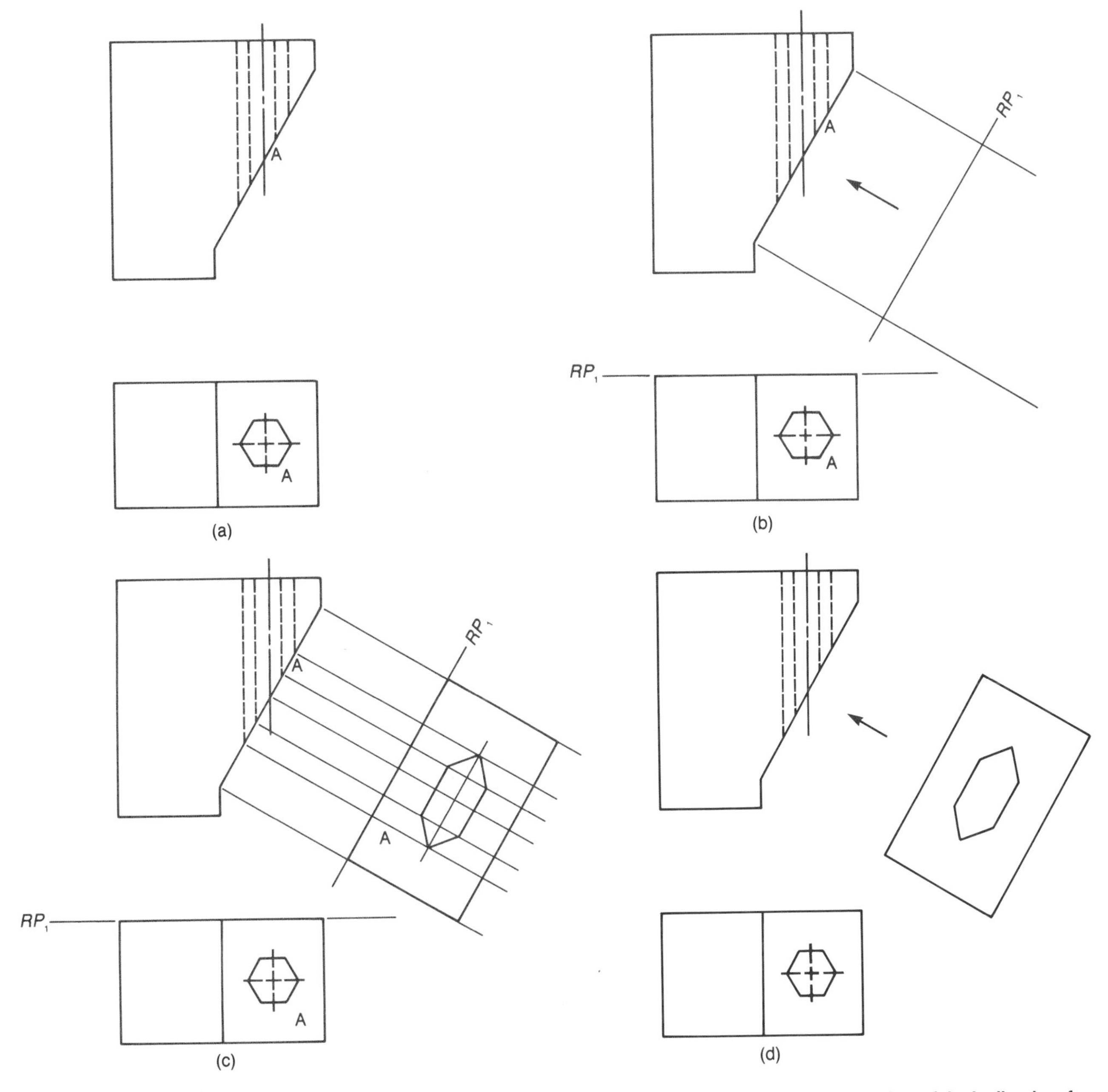

Figure 7.11 **Developing a partial and a complete primary auxiliary view. The given top view is the basic view of the inclined surface (a). Reference plane RP_1 is identified (b), followed by construction of the primary auxiliary view of surface A (c), resulting in the final set of given principal views and partial auxiliary view (d). Next, construction of a complete auxiliary view (e) results in final set of given principal views and complete auxiliary view (f).** ***Continued on next page.***

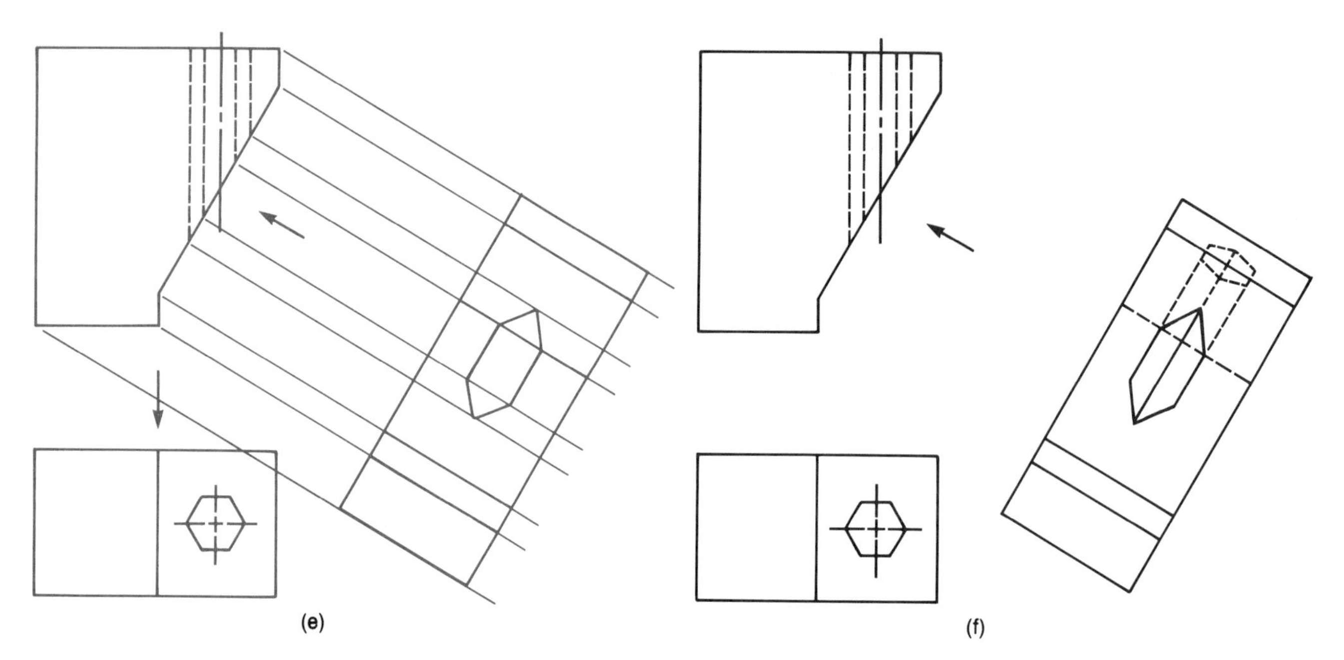

Figure 7.11—***Continued.***

Why is it necessary to sometimes use an auxiliary view of a surface (or of the entire object) in a drawing?

Answer An auxiliary view is sometimes needed in a drawing in order to show an inclined or oblique surface *without distortion* for an accurate and complete description of a part.

7.3 Development of a Primary Auxiliary View of an Inclined Surface with a Curving Boundary

An auxiliary view of an inclined surface with a curving boundary can be developed as shown in Figure 7.12. In this example, the basic view is the given right-side view (a). The appropriate viewing direction is identified (perpendicular to the edge representation of the surface under consideration in the basic view), together with a convenient reference plane (b). Key points are then identified in the given views and located in the construction region (c). The desired primary auxiliary view of the inclined surface is a partial view (d). For a complete auxiliary view, additional key points are located for all surfaces of the object (e and f). Once again, the rules for the development of orthographic views must be obeyed in the construction of complete primary auxiliary views.

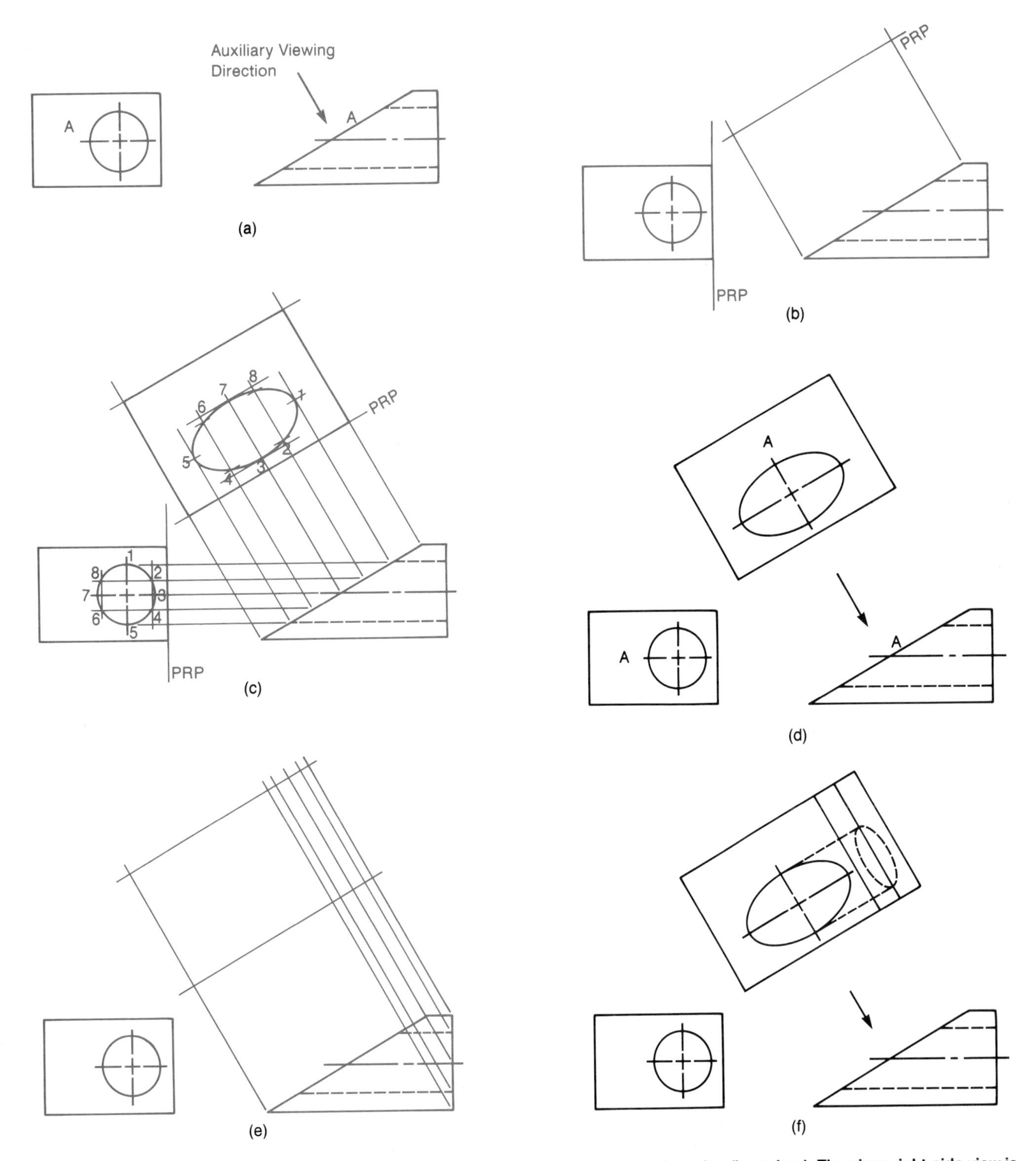

Figure 7.12 Developing an auxiliary view of an inclined surface that includes a curving edge (boundary). The given right-side view is the basic view (a). Identification of the appropriate viewing direction and reference plane (b) is followed by locating arbitrarily chosen points in the auxiliary-view construction region via the transference of (width) distances from the front view to the auxiliary view (c). Distances are measured with respect to the PRP. The result is an incomplete auxiliary view of surface A (d). Distances are projected from the right-side view to the auxiliary-view construction region (e) in preparation for a complete auxiliary view of the object (f).

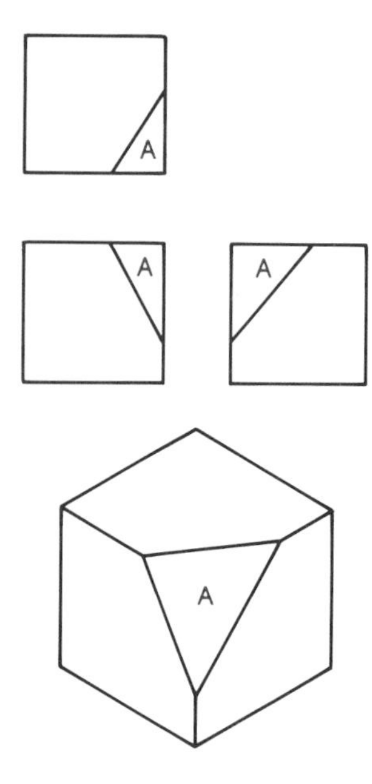

Figure 7.13 Example object to be the subject of successive figures in this chapter. Note the oblique planar surface A. Three principal views (top, front, and right-side) are given.

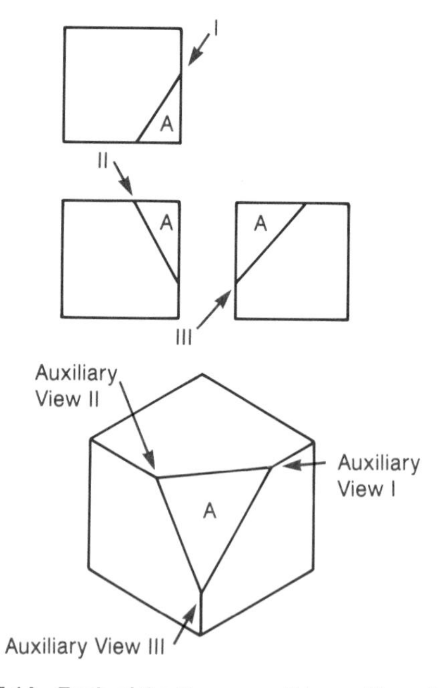

Figure 7.14 Each of the three possible auxiliary viewing directions will produce edge views of the oblique surface A (that is, auxiliary views in which A appears on edge, as a line).

7.4 Development of a Secondary Auxiliary View of an Oblique Surface

A primary auxiliary view of an inclined surface is drawn with respect to a basic view in which the surface appears on edge, as a line. In the object shown in Figure 7.13, an oblique surface, A, is identified. Recall that an **oblique surface** is inclined to all principal projection planes and, as a result, appears as a foreshortened area in all principal orthographic views. None of the six principal views can be used as a basic view for an auxiliary drawing in which an oblique surface appears as a true-size area. Such a basic view will need to be constructed, followed by the construction of the desired auxiliary view of the oblique surface. This type of auxiliary view—developed by a two-step process involving (1) a preliminary construction of an auxiliary (basic) view and (2) construction of the final view—is known as a **secondary auxiliary view.**

A basic view may be developed if one identifies the appropriate (basic) viewing direction with respect to the object and the given principal orthographic views. Each of these viewing directions is parallel to a straight edge, or boundary, of the oblique surface, as shown in Figure 7.14. Such a boundary (line) will appear as a point in the basic view, whereas the oblique surface itself will appear on edge, as a line.

Figure 7.15 presents the development of a basic view and a secondary auxiliary view of the oblique surface A. A basic (primary) viewing direction is chosen so that the basic view will be drawn with respect to the given top view (a). A reference plane (denoted as RP_1) is defined in both the basic-view construction region (so that it is perpendicular to the basic viewing direction) and the given principal views (b). Key points of the oblique surface are then located in the basic-view construction region so that the partial view (of the surface) can be formed (c). (We will not need any other portion of the basic view of the object if we intend to develop only a partial secondary auxiliary view in which only the oblique surface is shown.)

The secondary auxiliary view is developed from the basic view and from the principal view that was used to construct the basic view. A secondary viewing direction is identified (d), and a second reference plane (denoted RP_2) is used (e) to develop a partial secondary auxiliary view in which the oblique surface A appears as a true-size area (f). The final set of orthographic views includes a partial primary (basic) auxiliary view and a partial secondary auxiliary view of the oblique surface A (g). The development of complete primary (basic) and complete secondary auxiliary views of the object in Figure 7.13 is shown in Figure 7.16 (a–c).

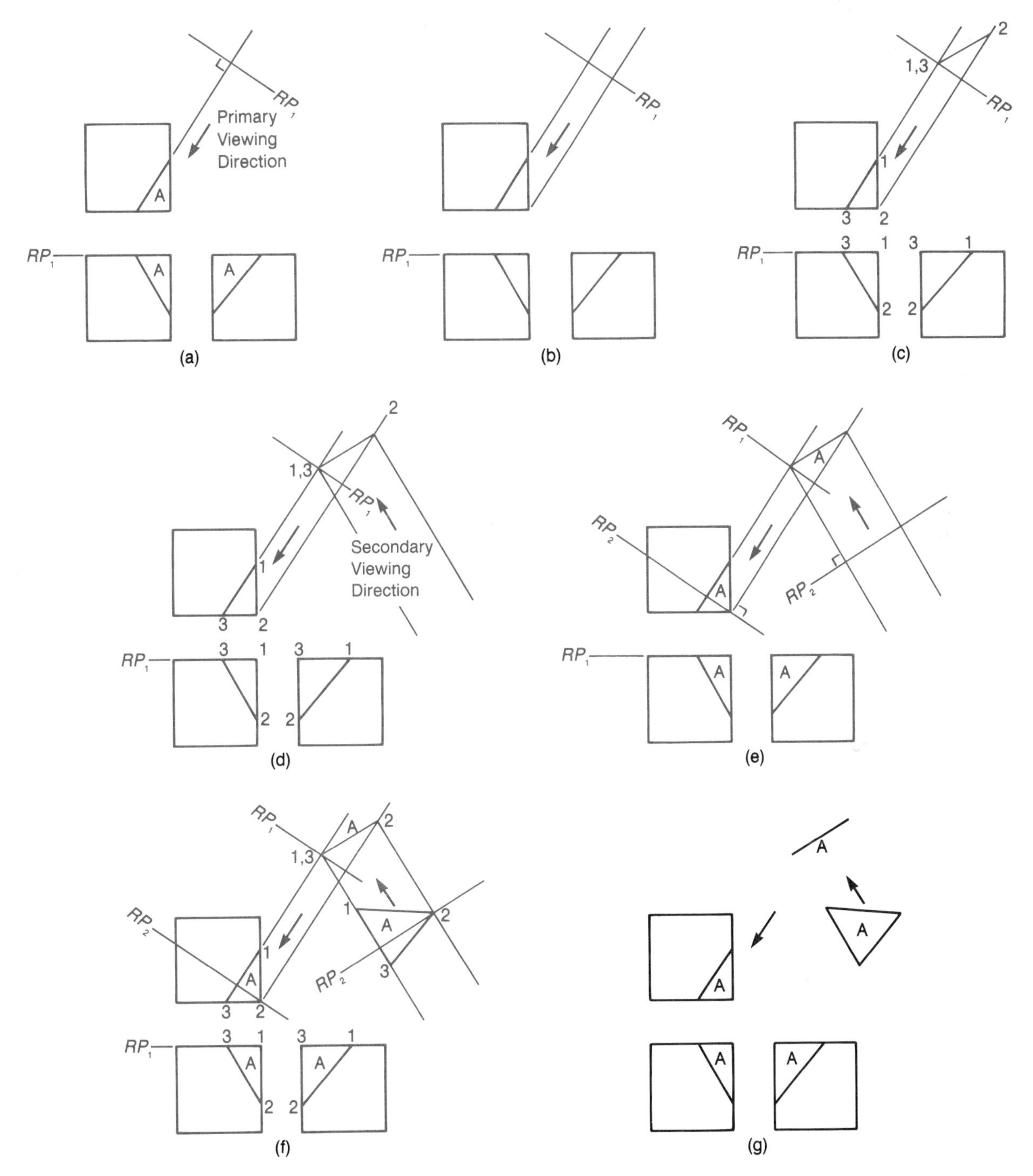

Figure 7.15 Developing a primary and a secondary auxiliary view of an oblique surface (for the object in Figure 7.14). The primary auxiliary viewing direction is chosen so that surface A will be seen on edge, as a line (a). Projection of distances into the auxiliary-view construction region (b) is followed by construction of a primary auxiliary view of surface A, which appears on edge, as a line (c). Identification of a secondary auxiliary viewing direction that is perpendicular to the plane containing surface A (d) is followed by identification of a second reference plane (RP_2) in preparation for the development of the secondary auxiliary view of surface A (f). In the final set of orthographic views, three principal views are shown, together with an incomplete (edge) primary auxiliary view and an incomplete (area) secondary auxiliary view of surface A. Surface A is shown as a true-size area in the secondary auxiliary view.

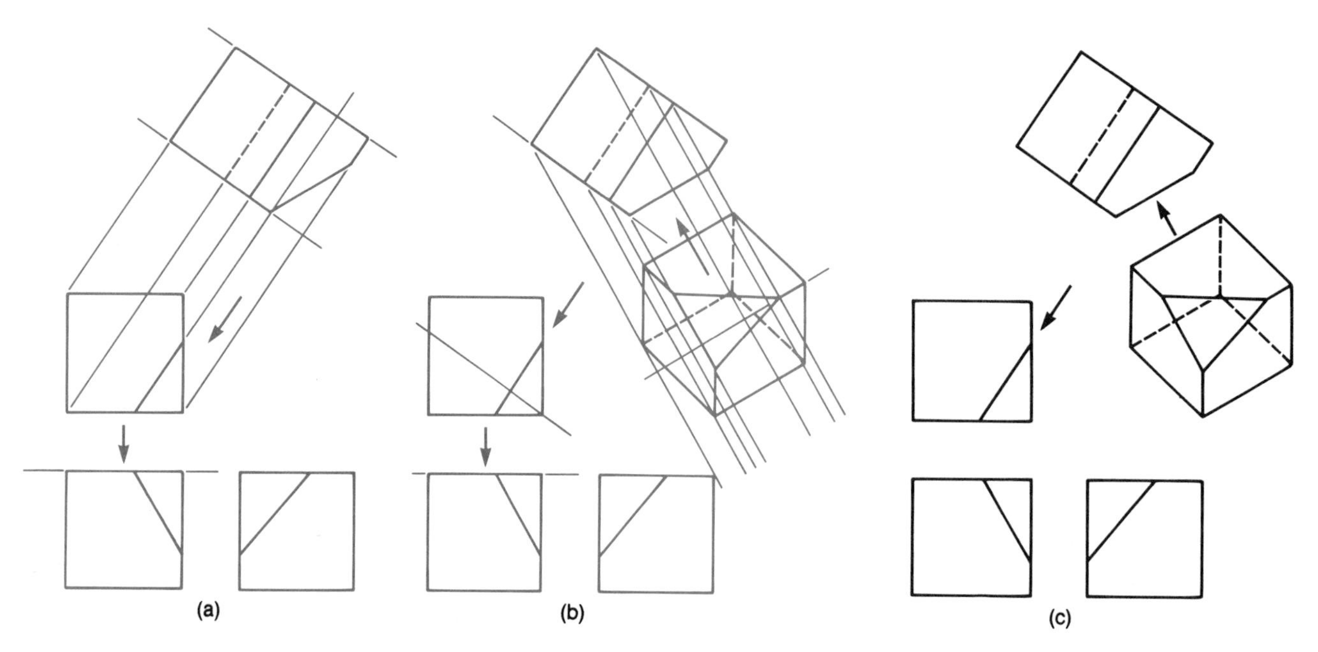

Figure 7.16 Developing a complete primary (basic) view and complete secondary auxiliary view (a–c) for the object in Figure 7.14.

What is a primary auxiliary view?

Answer A primary auxiliary view is one in which a principal orthographic view acts as its basic view.

In Retrospect

- Auxiliary orthographic views are needed to show inclined and oblique surfaces of an object as undistorted, true-size areas. Such views may be complete (in which the entire object is shown) or partial (in which only a particular surface is shown).
- An auxiliary view of an inclined surface may be developed with respect to a basic view, in which the surface appears on edge, as a line. The auxiliary viewing direction is perpendicular to the line representing the surface in the basic view.
- A primary auxiliary view of a surface or an entire object is developed from a (basic) principal view. A secondary auxiliary view is developed from a preliminary auxiliary basic view that must first be constructed. True-size area representations of oblique surfaces are shown only in secondary auxiliary views.

3D Computer Graphics: Constructing a Solid Model through Boundary Definition

In the "Computer Graphics in Action" section of Chapter 6, we reviewed the use of primitives in the construction of a solid model. Solid models may also be developed through the method known as **boundary definition.** In this method, the surfaces of the object are generated via one (or a combination) of three techniques:

- **Sweeping,** in which a two-dimensional surface is "swept" through space to form the desired volume.
- **Gluing,** in which predefined solid models are combined ("glued") to form a new shape.
- **Tweaking,** in which a predefined solid model is modified by adjusting the relative location of key points, that is, "tweaking" the object, or pulling it in particular directions as if it were a soft or elastic mass.

Solid models represent the current state of the art in computer graphics efforts. Any opportunity to use a solid modeling software system should be utilized to its fullest extent.

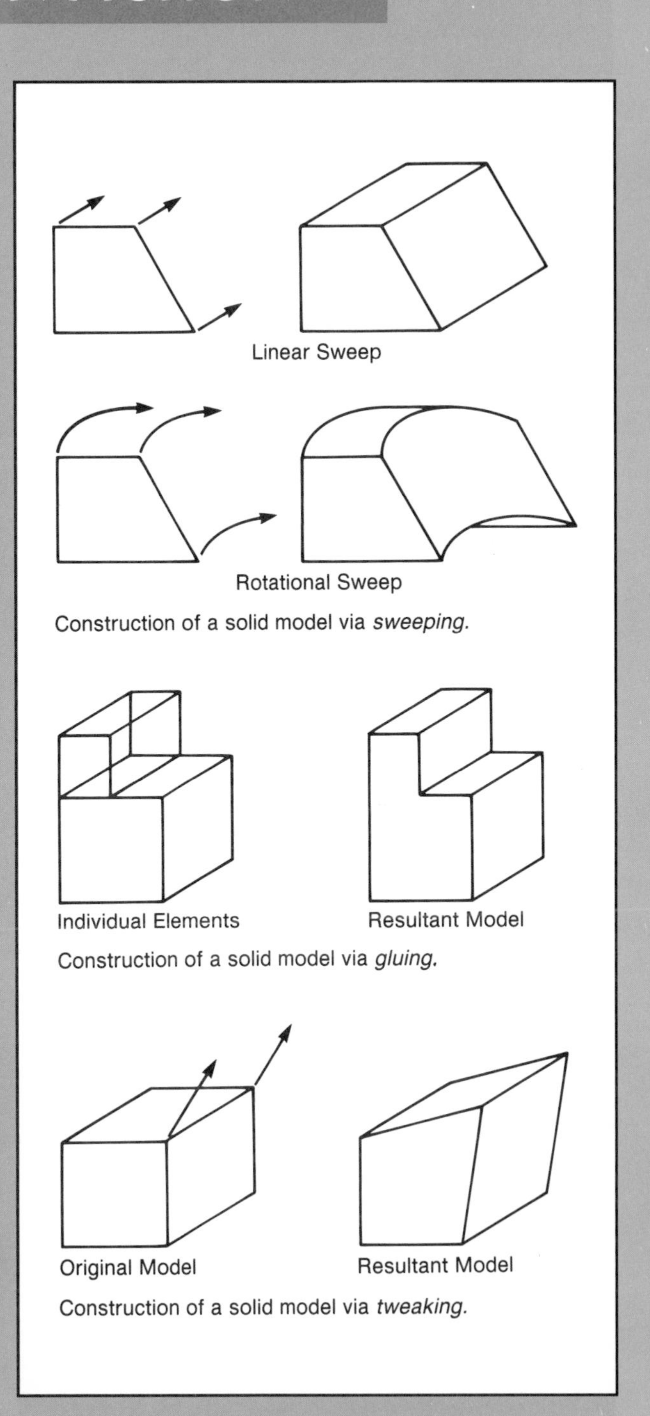

Construction of a solid model via *sweeping.*

Construction of a solid model via *gluing.*

Construction of a solid model via *tweaking.*

Exercises

Your instructor will specify the appropriate scale for each exercise. In addition, your instructor will specify if the auxiliary view to be developed should be a *partial* or a *complete* view. Each exercise refers to a specific exercise in Chapter 4. In each case, draw a set of the best and necessary views of the object under consideration, including *primary* and/or *secondary auxiliary views* in which the true shape and size of each surface is clearly shown. In addition, you may need to convert one or more of these auxiliary views into appropriate *section* views in order to describe the internal features of certain objects. (Your instructor may specify additional requirements or modifications for these exercises.)

EX7.1. See EX4.1.

EX7.2. See EX4.2.

EX7.3. See EX4.3.

EX7.4. See EX4.4.

EX7.5. See EX4.5.

EX7.6. See EX4.6.

EX7.7. See EX4.7.

EX7.8. See EX4.8.

EX7.9. See EX4.9.

EX7.10. See EX4.10.

EX7.11. See EX4.11.

EX7.12. See EX4.12.

EX7.13. See EX4.13.

EX7.14. See EX4.14.

EX7.15. See EX4.15.

EX7.16. See EX4.16.

EX7.17. See EX4.17.

EX7.18. See EX4.18.

EX7.19. See EX4.21.

EX7.20. See EX4.22.

EX7.21. See EX4.26.

EX7.22. See EX4.28.

EX7.23. See EX4.30.

EX7.24. See EX4.31.

EX7.25. See EX4.33.

EX7.26. See EX4.34.

EX7.27. See EX4.35.

EX7.28. See EX4.36.

EX7.29. See EX4.37.

EX7.30. See EX4.38.

EX7.31. See EX4.40.

EX7.32. See EX4.44.

EX7.33. See EX4.46.

EX7.34. See EX4.48.

EX7.35. See EX4.51.

EX7.36. See EX4.52.

EX7.37. See EX4.53.

EX7.38. See EX4.54.

EX7.39. See EX4.55.

EX7.40. See EX4.56.

8 Dimensioning

Preview In order to fabricate a part, machinists and other production workers must know both the *shape* of the part and its *size*. Earlier chapters were devoted to the description of the shape of a part in a set of orthographic views; now we begin our discussion of size specifications with the topic of dimensioning.

Dimensioning information must be included in a drawing in order to specify both the size of the part (and its features) and the *location* of the part's features relative to particular reference positions (points, axes, or planes). However, in adding the dimensioning data, the designer must avoid any reduction in the clarity of the description of the part's shape. Consequently, specific rules for dimensioning a set of orthographic views for a part must be obeyed.

A mistake in the size of a part (either in its graphic description or in its fabrication) can be catastrophic in a manufacturing or production effort. Dimensioning of an orthographic drawing must be performed with care so that all those in the industrial network will produce a part of the appropriate shape and size.

Learning Objectives

Upon completion of this chapter, the reader should be able to:

- Specify the uses of dimensions in engineering drawings.
- Recognize that dimensions describe the size of the OBJECT, not of the drawing.
- Recognize that every dimension needed for the fabrication of a part must be shown in a drawing.
- Follow a step-by-step procedure for dimensioning a set of orthographic views that facilitates a proper perspective of the dimensioning data as well as greater consistency in linework and lettering.
- Dimension a set of orthographic views for a part in accordance with the basic rules for dimensioning, which are given in this chapter.
- Define such terms as tolerance, datum, and scale.
- Identify extension lines and dimension lines in a set of orthographic views.
- Differentiate between basic dimensions and reference dimensions.
- Use both SI units (*Le Systeme International d'Unites,* or the International System of Units) and English units in engineering drawings as appropriate.

- Use both rectangular coordinates and polar coordinates to properly dimension a part.
- Present dimensioning data in tabular form in order to eliminate (or minimize) confusing linework from a drawing.
- Apply a finishing symbol to a drawing in order to specify those surfaces to be machined (finished).
- Dimension special, but not uncommon, features such as parts with rounded ends, slotted holes, fillets, chamfers, keyseats, knurling, spherical surfaces, angles, arcs, chords, cones, pyramids, and conical tapers.

8.1 Terminology

Dimensions are used in a graphic description of a part in order to specify both (1) the *size* of the part and its features and (2) the *location* of the part's features relative to particular reference positions (points, axes, planes). Such specifications are given in appropriate numerical units. These units of measure (such as millimeters and inches) must be specified on a drawing.

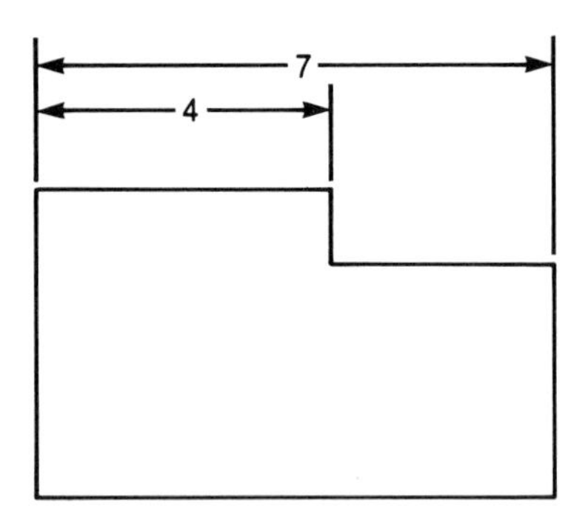

Figure 8.1 Basic dimensions specify theoretically exact values for size, location, and orientation.

Terms used in dimensioning include:

- **Basic dimensions** specify the theoretically exact value for the size, location (or "true position"), and orientation of a part's features (Figure 8.1).
- **Reference dimensions** are used to provide information to the reader. They either *repeat* a basic dimension or are *derived* from the basic dimensions. Reference dimensions should be enclosed within parentheses or given the superscript REF in order to emphasize their use as reference values (Figure 8.2).
- **Tolerances** specify the total *range of acceptable variation* in the corresponding actual size of the fabricated part. We will focus upon the use of tolerances in Chapters 9 and 10.
- **Datum,** the exact *reference point, axis, or plane* from which one may specify the location of a feature.
- **MAX/MIN,** abbreviations that indicate the **maximum** and **minimum** values, respectively, for the size of a feature.
- **LMC,** an abbreviation for the **least material condition** that is acceptable for a feature. The LMC is a specification of the size that corresponds to the use of minimal material. Examples include a maximum diameter for a hole or the minimum value for the diameter of a shaft.
- **MMC,** an abbreviation for the **maximum material condition** that is acceptable for a feature. The MMC is a specification of the size that corresponds to the use of maximum material. Examples include a minimal diameter for a hole or the maximum value for the diameter of a shaft.

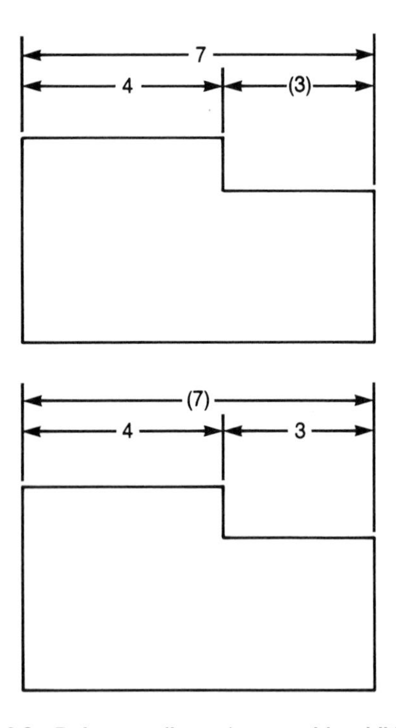

Figure 8.2 Reference dimensions provide additional information and are enclosed in parentheses.

Please note that the size of the *object,* not of the drawing, is described by the dimensions on a drawing.

A *scale* should be included in a drawing to indicate the ratio of the length of a line on the drawing to the length of the corresponding feature of the object. For example, a one-half or half-size scale indicates that each distance shown on the drawing is one-half of the corresponding distance of the object.

In summary, then:

1. Dimensions are used to specify both the size of a feature and its location.
2. Basic dimensions specify the theoretically exact values for the size, orientation, and location of a feature.
3. Reference dimensions provide additional information to the reader. Since they are either repetitions of given basic dimensions or can be derived from these basic values, reference dimensions should always be enclosed within parentheses to distinguish them from basic dimensions.
4. The tolerance associated with a dimension for a part specifies the total range of acceptable variation in the actual size of the fabricated part.
5. A datum is the exact reference point, axis, or plane from which one may specify the location of a feature.
6. Abbreviations such as MAX, MIN, LMC, and MMC indicate maximum and minimum values for the sizes of particular features.
7. Dimensions describe the size of the object, not of the drawing.
8. A scale should be included on a drawing in order to indicate the ratio between the length of a line on the drawing and the size of the corresponding feature of the part.

8.2 Units of Measurement

Units of measurement must be specified on a drawing with a note such as:

UNLESS OTHERWISE SPECIFIED, ALL
DIMENSIONS ARE IN MILLIMETERS

The preceding example, of course, would apply only if all dimensions are indeed given in millimeters. Otherwise, the fundamental (metric) unit is 1 meter.

Units may be given in the *Le Systéme International d'Unites* (or the International System of Units, abbreviated as **SI units**) with the fundamental unit used on drawings of the millimeter. Alternatively, units may be given in **customary linear** units **(English units)** with the fundamental unit of the (decimal or fractional) inch.

Angular units may be expressed in degrees (°), minutes (′) and seconds (″). A complete circular arc is composed of 360°, 1 degree contains 60 minutes, and 1 minute contains 60 seconds. Angles may be specified in these units or purely in degrees with decimal fractions (Figure 8.3).

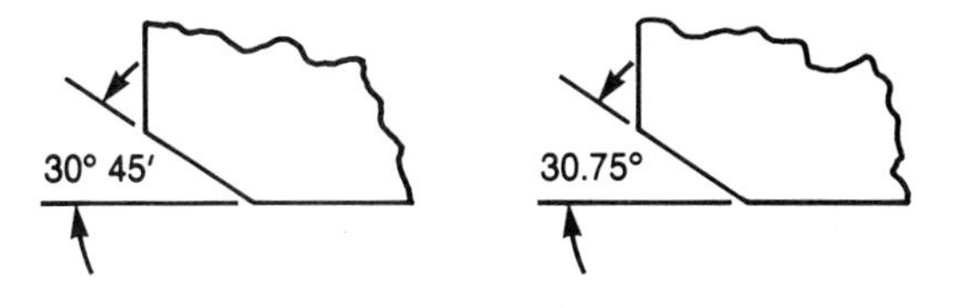

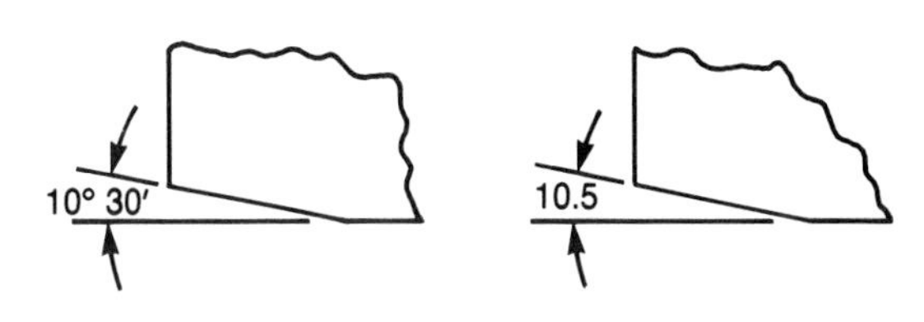

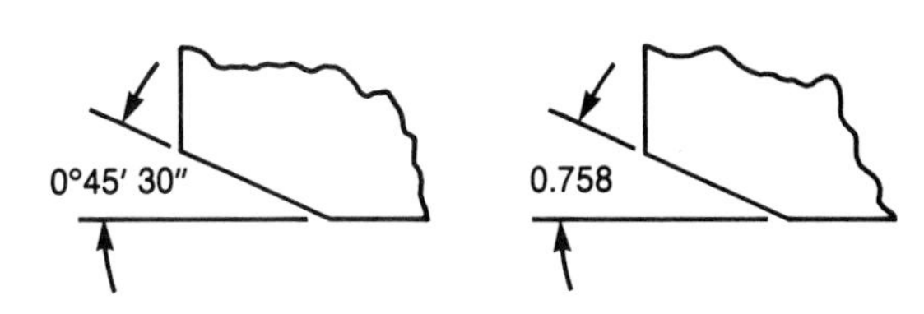

Figure 8.3 Angular units are expressed in degrees (°), minutes, (′), or seconds (″).

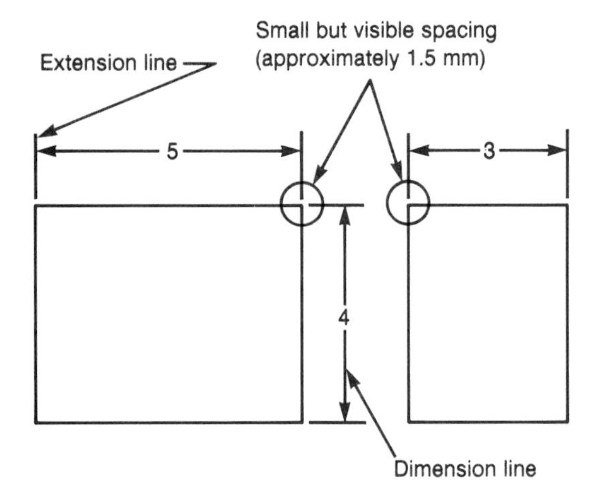

Figure 8.4 Extension lines indicate the extent of a feature or surface. Dimension lines run between the extension lines and (usually) perpendicular to them.

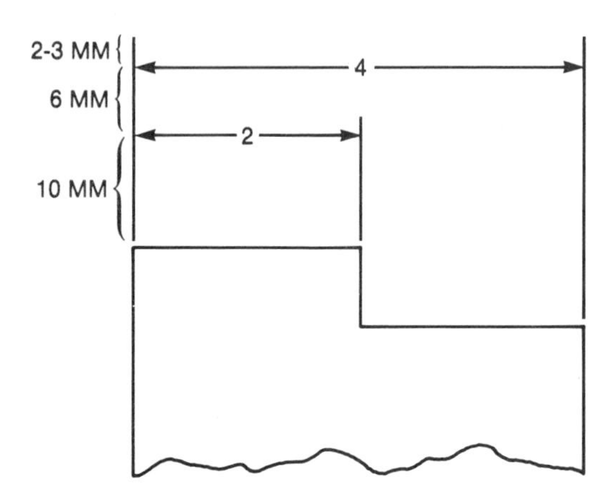

Figure 8.5 Dimension lines should be placed at least 10 mm. from the orthographic views. Parallel dimension lines should be spaced 6 mm. apart. Extension lines should extend 2 to 3 mm. from the dimension line.

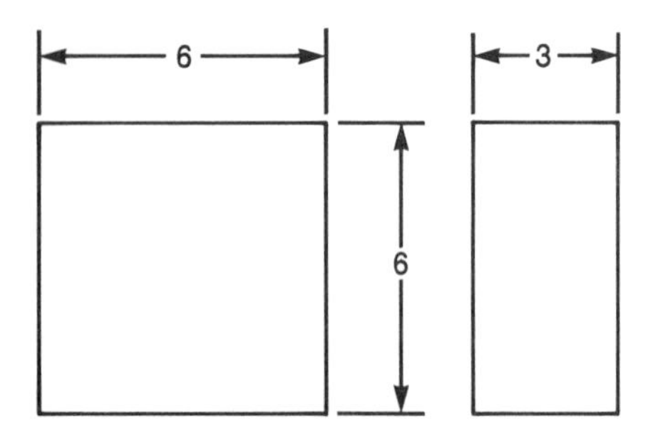

Figure 8.6 Principal (total) dimensions for a rectangular prism or block.

8.3 Extension and Dimension Lines

Extension lines, as the name implies, indicate the extent of a feature or surface in a given (linear or angular) direction. A corresponding **dimension line** runs between these extension lines, usually perpendicular to them. The numerical value of the dimension is given in an open, or broken, space in the dimension line. Figure 8.4 presents two orthographic views of a rectangular prism in which extension and dimension linework are identified.

Arrowheads are drawn at the ends of the dimension lines. The width of the arrowhead should be approximately one-third of its length, and its length should be approximately 3 mm. Arrowheads should be uniform in size on a single drawing. They should touch, but not extend beyond, the extension lines.

The distance between an extension line and the boundary linework for the corresponding feature should be small (approximately 1.5 mm.) but visible (refer again to Figure 8.4).

The distance between the dimension line and the orthographic view that is being dimensioned should be sufficiently large (at least 10 mm.) to avoid crowding of information, yet not be so large that it becomes difficult for the reader of the drawing to relate a dimension to the appropriate feature (Figure 8.5). Parallel dimension lines should be placed at least 6 mm. from each other in order to avoid crowding (Figure 8.5). Extension lines should extend approximately 2 to 3 mm. beyond the dimension line that lies farthest from the orthographic view (Figure 8.5).

One should lay out dimensions for a part on a **rough sketch** of the necessary orthographic views for the object before transferring these dimensions to the finished drawing. Such an approach ensures that sufficient space will be available for the required dimensioning information in the finished orthographic drawing.

8.4 Basic Dimensioning Principles

We will now gradually develop the procedure for applying dimensions to orthographic views in which a variety of different features are described. Figure 8.6 presents a set of views in which a rectangular prism or block is described. The only dimensions necessary to describe this object are its principal (or total) dimensions of width, height, and depth. Each dimension is given only once in the drawing. Two primary rules for dimensioning are:

- Rule 1. Show every necessary dimension in a drawing.
- Rule 2. Never repeat a basic dimension in a drawing.

NOTE: Throughout this chapter, we will develop such rules for dimensioning; please note that *some computer graphics systems "change" or disobey these rules to allow automatic dimensioning of a drawing.* These systems and the accepted dimensioning standards (rules) will eventually become mutually consistent, but we should recognize that minor inconsistencies do sometimes exist.

Figure 8.7 presents front and right-side views of a rectangular prism in which a hole is to be drilled. In addition to the principal dimensions for the prism, the location for the centerpoint of the hole must be specified (and its diameter and depth). Centerlines are used as extension lines in order to locate a centerpoint.

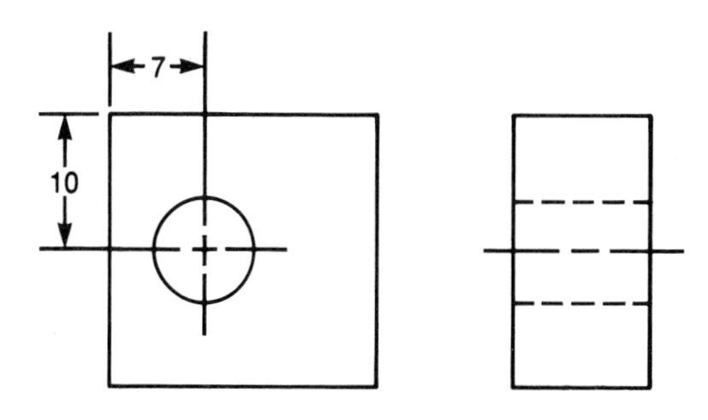

Figure 8.7 Centerlines are used as extension lines in order to locate the centerpoint of a hole in this rectangular prism.

Notice that only those portions of the centerlines that must reach beyond the boundaries of the orthographic view for dimensioning purposes are extended; the long dashes of the centerlines that remain in the view are not enlarged. Furthermore, notice that the same reference point (top left corner of the front view) is used for both the relative width and height dimensions that locate the position of the centerpoint. Such *linked* dimensions should be located with respect to the same datum, if possible.

We may state this practice in the form of a rule:

- Rule 3. Locate dimensions that are linked to one another (such as dimensions for the location of a centerpoint) with respect to the same datum if possible.

Figure 8.8 gives the fully dimensioned set of views for the rectangular prism with a through hole. The larger (principal) dimensions for the prism are located at a greater distance from the front view than the smaller dimensions that locate the position for the centerpoint of the hole. This practice is followed because the machinist will *initially* need the principal (total) dimensions for the part before working inward toward individual (smaller) features of the part (such as a hole).

- Rule 4. Place larger dimensions farther from an orthographic view than smaller dimensions.

All of the dimensions given in Figure 8.8 are shown outside the orthographic views in order to avoid confusing the reader's interpretation of the object's shape with additional interior linework in the views. On occasion, however, some dimensional information may need to be shown within an orthographic view because of space limitations or for greater clarity.

- Rule 5. Place dimensioning information outside the orthographic views if possible.

The diameter of the hole, instead of its radius, is specified in Figure 8.8 for two reasons:

1. Drill bits are identified according to diametral values.
2. Inspection of the finished part requires measurement of the diameter, not the radius, of the hole in order to determine if it has been properly machined.

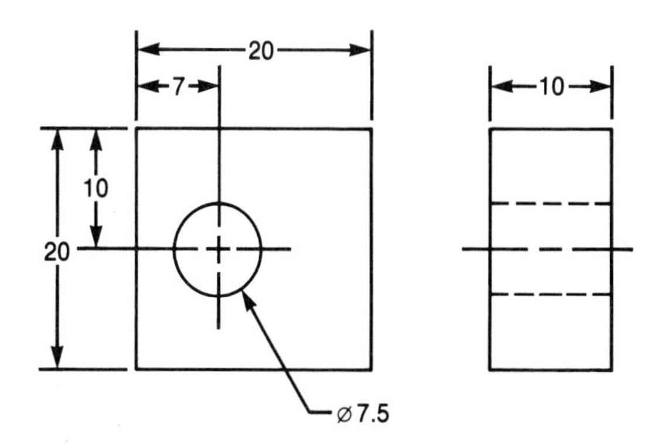

Figure 8.8 A fully dimensioned set of orthographic views for a rectangular prism with a through hole.

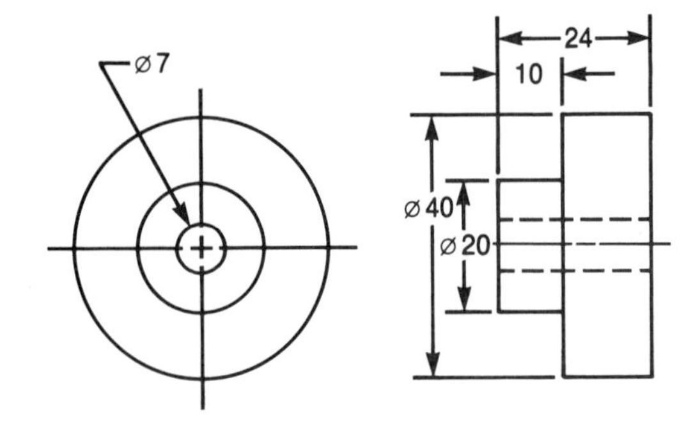

Figure 8.9 The dimension for the through hole in this cylindrically shaped object is preceded by the symbol ⌀.

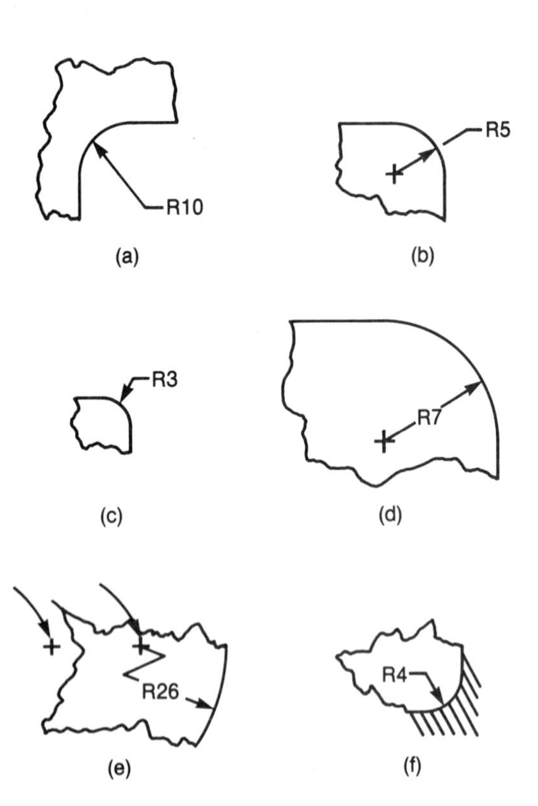

Figure 8.10 A variety of formats are acceptable in dimensioning arcs. Leaders may be used (a, b, c, and f) or dimension lines (d and e).

The dimension for the hole diameter is identified with the symbol ⌀, which should precede the diametral value (Figure 8.9), although older drawings sometimes show this symbol following the value. It is associated with a special type of dimension line known as a **leader.** Leaders are directed toward the center of the circle, which represents a hole in a drawing. In other words, leaders are **radially directed.** The arrowhead of a leader is directed toward this centerpoint; it touches, but does not extend beyond, the circle.

Furthermore, a **shoulder** should be used with a leader. A shoulder is a horizontal line approximately 6 mm. in length that runs from the leader to the numerical information. This shoulder should be centered between the guidelines used in the construction of the numerical data; it is always located either at the *beginning* or at the *end* of the informational note. Notice that a leader is used to specify the diameter of the through hole of the cylindrical object in Figure 8.9. Radial leaders such as this are used only in conjunction with circles that represent holes, not circles that represent shafts or solid cylinders. For solid cylinders, rectangular dimensions are used to indicate the diameter. As a result, the reader's interpretation of a circle as a hole or as a shaft is aided by this distinctive use of leaders.

Arcs are sometimes dimensioned with the use of leaders and sometimes with dimension lines. Both situations are shown in Figure 8.10. The numerical value for the radius of the arc should be preceded by the symbol **R.** The variety of acceptable formats shown in Figure 8.10 allows one to maximize clarity by avoiding any overlapping or crowding of dimensions in limited drawing space.

Of particular interest is the case shown in Figure 8.10(e) where the true centerpoint of the arc is not located within the given view (because of the very large size of the arc's radius). A **jagged** dimensioning line is used to indicate that the centerpoint shown in the view is not the true centerpoint. The portion of the leader that includes the arrowhead should be directed toward the location of the true centerpoint.

- Rule 6. Leaders should be used to dimension arcs (radii) and holes (diameters) but not cylindrical solids.
- Rule 7. The direction of leaders used to dimension circles and arcs should be radial.
- Rule 8. Specify the diameter (not the radius) of a hole, together with the symbol ⌀, which should precede the diametral value. The symbol R follows the radial value of an arc.

In summary, then:

1. Every dimension that is needed for the fabrication of a part must be shown in a drawing.
2. A basic dimension should be given only once in a drawing (never repeated).

Highlight

Engineering Drawings and Mass Production

Precision products were mass-produced in America during the early part of the nineteenth century through the distribution of so-called perfect specimens to manufacturing centers. Muskets, for example, were produced in this manner: perfect specimens were fabricated from **master** gauges and jigs and then distributed to armories where **manufacturing** gauges and jigs were produced from the part specimens. These manufacturing gauges and jigs were then used to produce muskets in large quantities. (The concept of **interchangeable parts** has been credited to Eli Whitney, although others were more successful in achieving such interchangeability; we will return to this concept in the next chapter.)

Such mass production was achieved without the use of engineering drawings because graphic descriptions were not yet capable of fully specifying a product—and all of its components—to multiple craftsmen. A drawing could be used to describe the needed product to only one craftsman, who would then fabricate the entire object. Sizes could be obtained from the drawing with proportional dividers so that universal units of measurement were not required. The craftsman would fabricate the components of the object so that they would fit together properly. If several people were involved in the fabrication process, a master set of gauges were first produced, as noted in the example of musket manufacturing.

As methods for graphics reproduction improved, drawings were gradually recognized as a preferred means of communication between those involved in the manufacture of products and those who design such products. Standardization in shape and size descriptions, together with the development of precision measuring equipment and improved fabrication processes, has led to the current status of engineering drawings as the primary means of communication in technology.

Source: Peter Booker, A History of Engineering Drawing *(London: Northgate Publishing Company, Ltd., 1979).*

3. Locate linked dimensions with respect to the same datum if possible.
4. Place larger dimensions farther from an orthographic view than smaller dimensions.
5. Place dimensioning information outside the orthographic views in a drawing if possible.
6. Leaders should be used to dimension arcs and circles, but only if the circle represents a hole, not a cylindrical solid.
7. Leaders should be radially directed.
8. The diameter of a hole, not its radius, should be specified. The symbol ∅ should precede the diametral value. The symbol R follows the radial value for an arc.

8.5 Dimensioning Layout—Further Details

Figure 8.11 presents a set of orthographic views in which the placement of extension and dimension linework is shown (numerical values have been omitted). Notice that extension lines may cross each other, but *dimension lines should not cross one another* unless it is impossible to avoid. Furthermore, an extension line should not be broken unless it passes

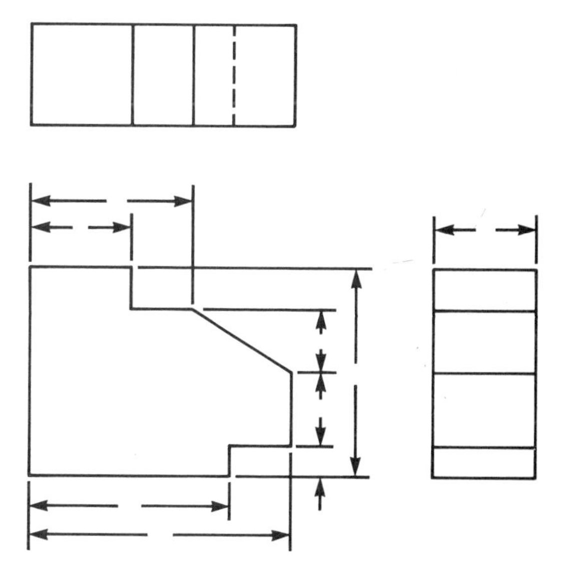

Figure 8.11 In a proper dimensioning layout, dimension lines should not cross one another, but extension lines may cross one another. Furthermore, a dimension line may cross an extension line.

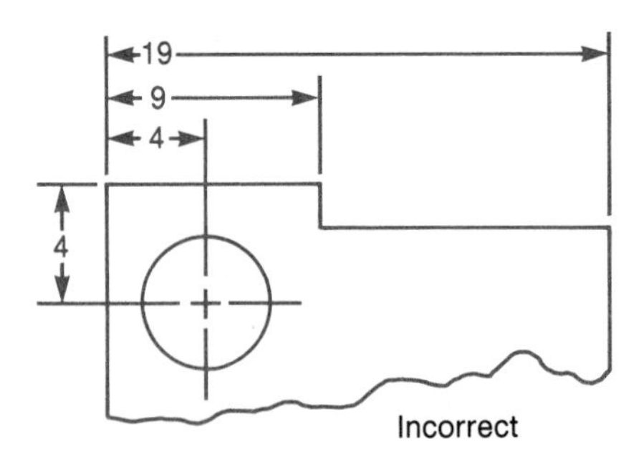

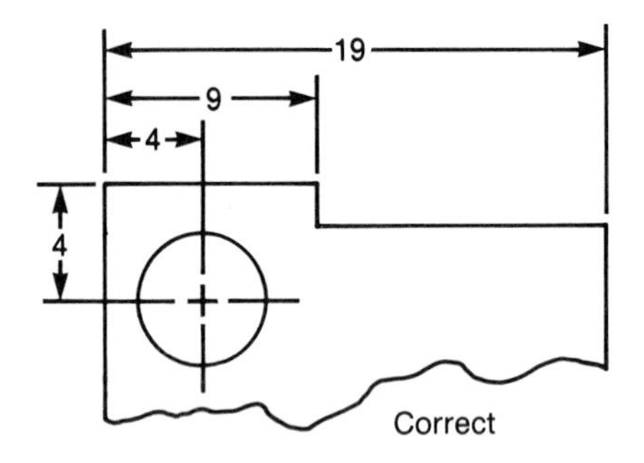

Figure 8.12 Parallel dimensions should be staggered, not aligned.

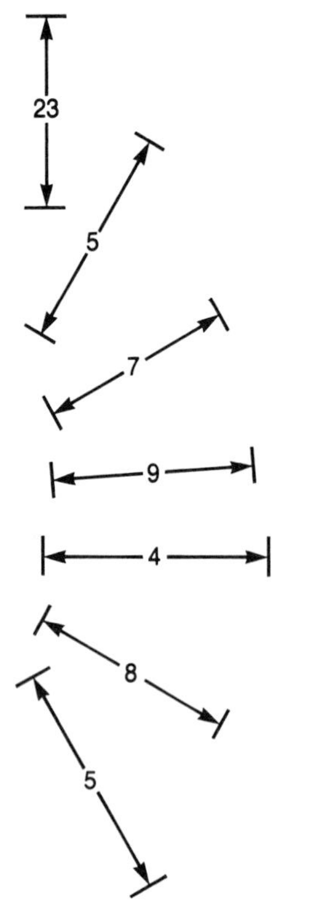

Figure 8.13 Numerical values should be unidirectional.

through or near an arrowhead. A dimension line may, however, cross an extension line.

- Rule 9. Dimension lines should not cross one another unless unavoidable.
- Rule 10. Extension lines may cross one another; furthermore, a dimension line may cross an extension line.

Notice in the figure that dimensioning information is located outside the given views in order to avoid crowding.

- Rule 11. Do not crowd dimensions in a set of orthographic views.

Notice also (in Figure 8.11) that any number of parallel dimension lines may be run from a given extension line.

Figure 8.12 presents a portion of a view in which three parallel (horizontal) width dimensions are shown with the numerical values *improperly aligned* (a) and properly *staggered* (b) in order to increase the ease of locating information by the reader of the drawing.

- Rule 12. Numerical values of parallel dimensions should be staggered.

8.6 Unidirectional and Aligned Values

Numerical values should be **unidirectional,** as shown in Figure 8.13, meaning that all dimensions should be read from the bottom of the page. It is permissible for inclined dimensions to be aligned with the corresponding dimension line, but such an alignment is not preferable. **Aligned** values are used in those cases where a unidirectional format would require an unreasonable amount of drawing space for the separation of parallel dimension lines. Also, dimensional values should never be placed so that they must be read from the left or from the top of a drawing. In summary, then:

- Rule 13. Dimensional values should be given in either a unidirectional or aligned format.

8.7 Angular Dimensions

Figure 8.14 presents front and right-side views of an object in which an inclined cut is to be formed in the top right cor-

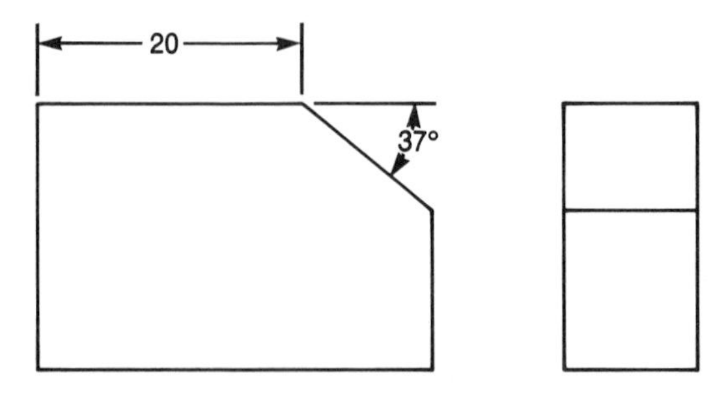

Figure 8.14 Angular dimensioning of an inclined cut.

ner. The angle of the cut is given, but the vertex of the angle must be located with respect to the known datum.

An alternative method for dimensioning this inclined cut is shown in Figure 8.15 in which only rectangular coordinate (linear) dimensions are used.

In summary, then:

- Rule 14. Locate the vertex of an angular dimension with respect to a datum.

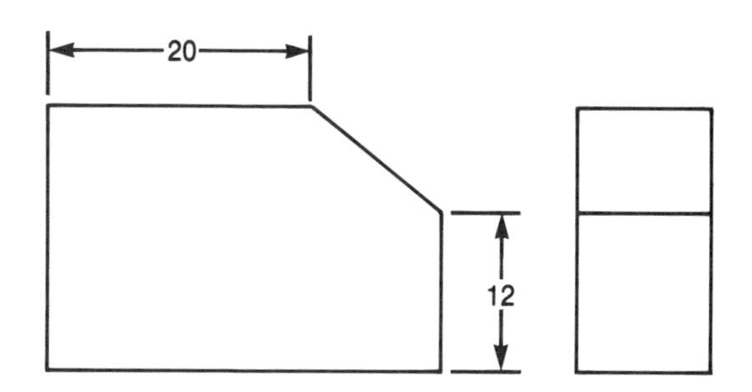

Figure 8.15 Rectangular dimensioning of an inclined cut.

8.8 Limited Drawing Space, Oblique Extension Lines, and Point Location

If there is a limited amount of space for dimensioning information, one should move the dimension line outside the extension lines with arrowheads directed toward the extension lines (Figure 8.16). If the space to be dimensioned is too small even to include the value between the extension lines, then the value itself is taken outside the extension lines.

Extension lines may be used in non-traditional formats in order to enhance the clarity of a drawing. For example, one may need to specify the location of a point at the intersection of two extension lines projecting from the appropriate surface as shown in Figure 8.17.

If there is limited drawing space, extension lines should be drawn at an oblique angle as shown in Figure 8.18. Notice that the dimension lines associated with the oblique extension lines are drawn parallel to the direction of the specified dimension (a height/vertical dimension in Figure 8.18) for greater clarity.

In summary, then:

- Rule 15. If there is limited space between extension lines, one should place the dimension line outside the extension lines with arrowheads directed toward the extension lines. For even smaller spaces, the numerical value is also placed outside the extension lines.
- Rule 16. Oblique extension lines should be used if drawing space is very limited (Figure 8.18). Extension lines may also be used to locate points of interest that lie at positions external to a part (Figure 8.17).

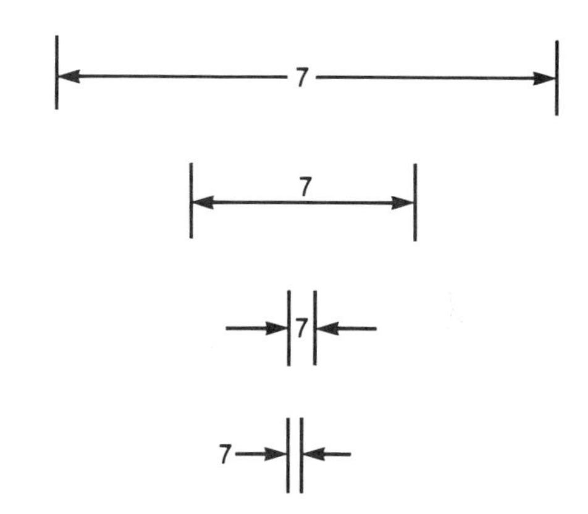

Figure 8.16 Proper changes in dimension linework are shown as drawing space becomes increasingly limited.

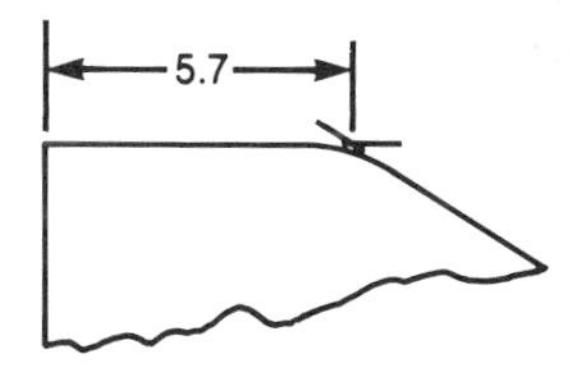

Figure 8.17 Locating a key point that is external to a part.

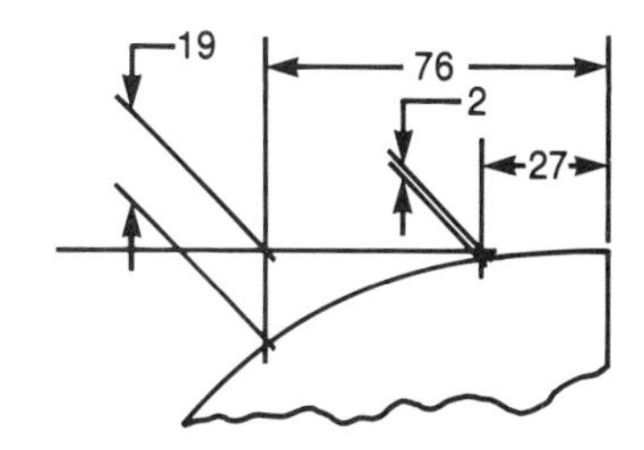

Figure 8.18 Oblique extension lines should be used if drawing space is very limited.

8.9 Leaders—Drilled, Counterbored, and Counterdrilled Holes

Figure 8.19 presents a section of an orthographic view that shows a leader directed toward a circle. (We know that the circle represents a hole because of the leader's presence.) The diameter of the hole is given in the figure, but the depth of the hole is not indicated.

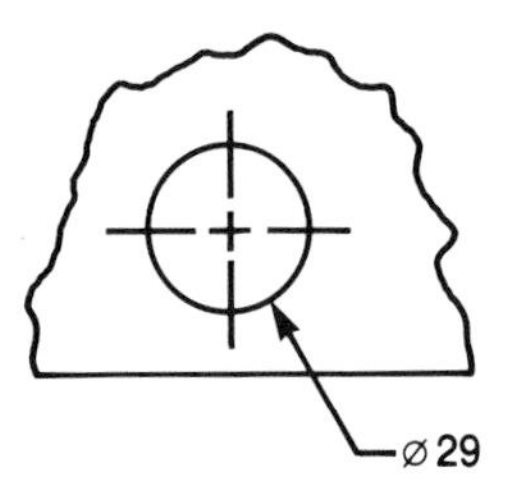

Figure 8.19 In this format for dimensioning a hole, the diameter but not the depth is specified.

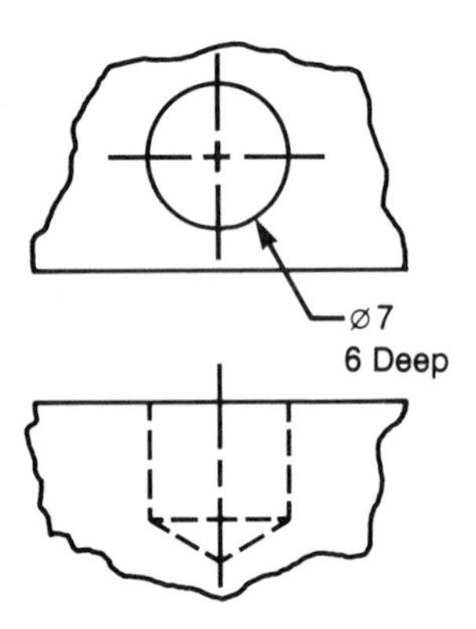

Figure 8.20 This is the preferred dimensioning format for a blind, unfinished hole.

Figures 8.20 and 8.21 present equivalent representations of a blind, unfinished hole in a part.

In these two figures both the diameter and the depth of the hole are specified. Although the format in Figure 8.21 is acceptable, the format shown in Figure 8.20 is *preferred* because both bits of information (diameter and depth) are given in a simple note. The machinist is expected to read the note in order to learn the specification for the hole diameter; inclusion of the depth specification in the same note ensures that the machinist will not need to search the drawing for this additional information. Furthermore, dimensioning linework is minimized with the format of Figure 8.20.

A drilled and bored hole is shown in Figure 8.22. Notice that the size specifications for this hole are identical to those given for the unbored hole in Figure 8.20 (see Section 5.6).

A note associated with a leader can be used to efficiently describe other specifications for a feature. For example, Figures 8.23 and 8.24 present counterbored and counterdrilled holes, respectively. A leader for a counterbored hole may include specifications for the hole diameter, the diameter of the counterbore, and the depth of the counterbore. (Notice that the radius of the fillet between the counterbore and the hole is also specified by a leader in Figure 8.23.) Similarly, a leader can be applied to a counterdrilled hole in order to specify the hole diameter, together with the depth and the diameter of the counterdrill. One may also wish to specify the included angle of the counterdrill, as shown in Figure 8.24.

A leader may also be used to indicate manufacturing specifications (Figure 8.25) or other types of information. Notice that the leader in Figure 8.25 does not include an arrowhead (since it is not being used as a dimension line) as it is directed toward a visible surface, not an edge view of the surface. A small but visible dot is used to clearly specify the endpoint of the leader and the surface to which the note applies.

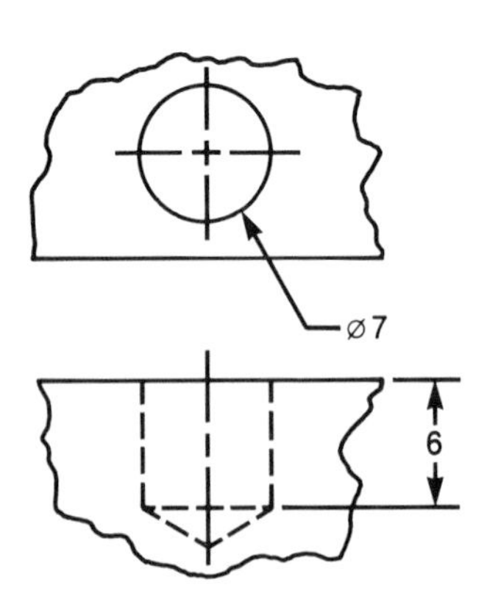

Figure 8.21 This is an acceptable, but not preferred, dimensioning format for a blind, unfinished hole.

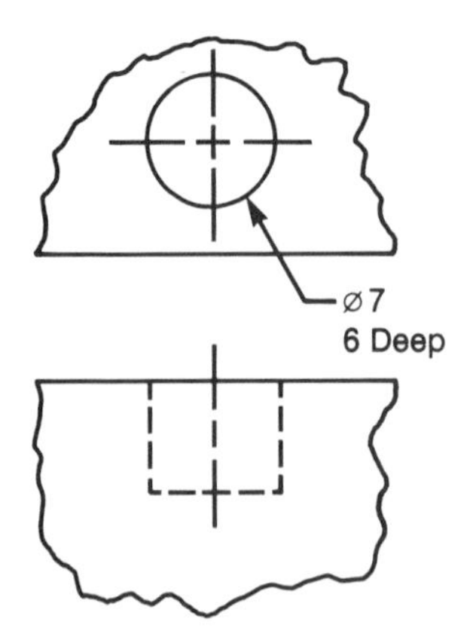

Figure 8.22 The size specifications for this drilled and bored hole are identical to those given for the unbored hole in Figure 8.20.

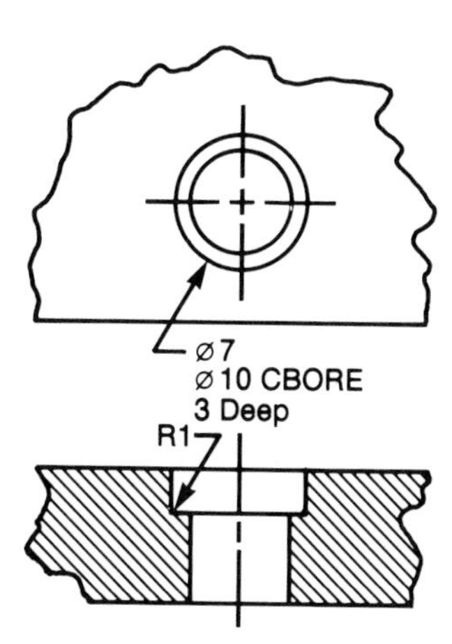

Figure 8.23 A leader with a note is used for this counterbored hole.

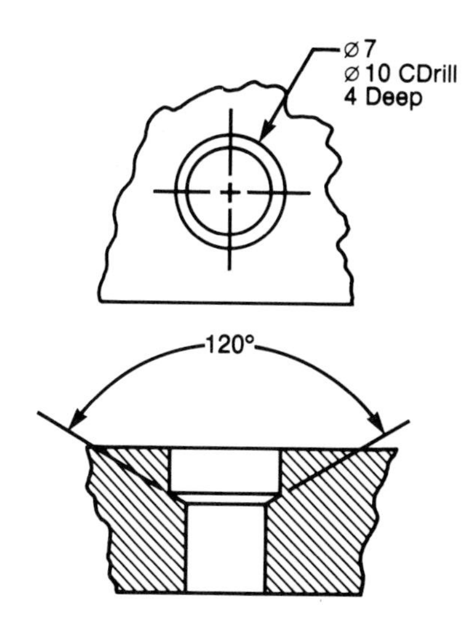

Figure 8.24 A leader with a note is also used for this counterdrilled hole.

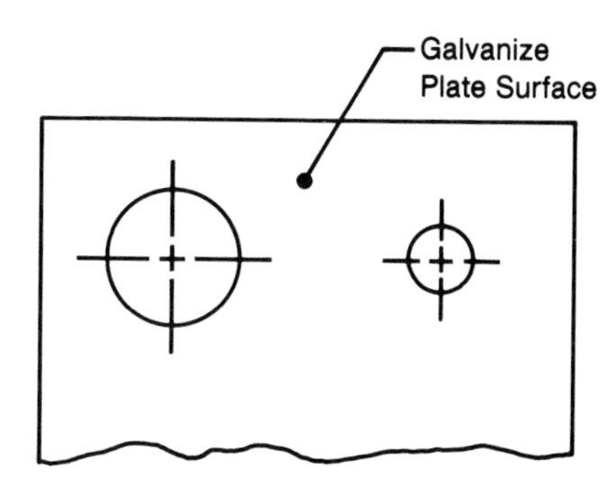

Figure 8.25 A leader may be used to indicate manufacturing/fabrication specifications.

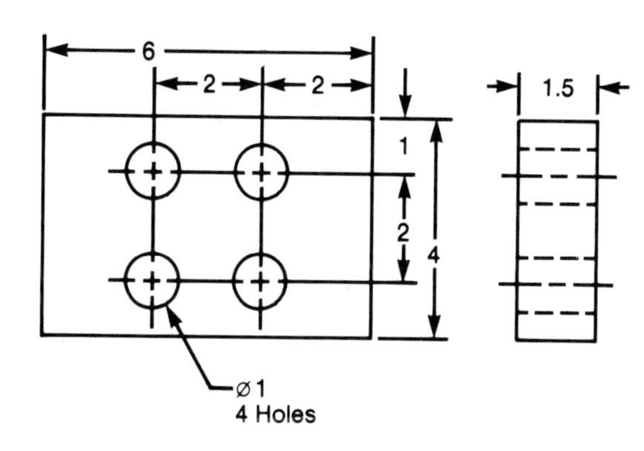

Figure 8.26 A repetitive feature in a part is indicated by a note.

In summary, then:

- Rule 17. Leaders should be used to indicate dimensioning specifications with clarity and minimal dimensioning linework on a drawing. Nondimensional leaders may be used to indicate other types of data such as manufacturing specifications.

8.10 Repetitive Features and Minimal Dimensioning Linework

A note may be used to indicate that a particular feature appears more than once. See Figure 8.26, in which a single leader is used to indicate the diameter of four identical through holes in a part.

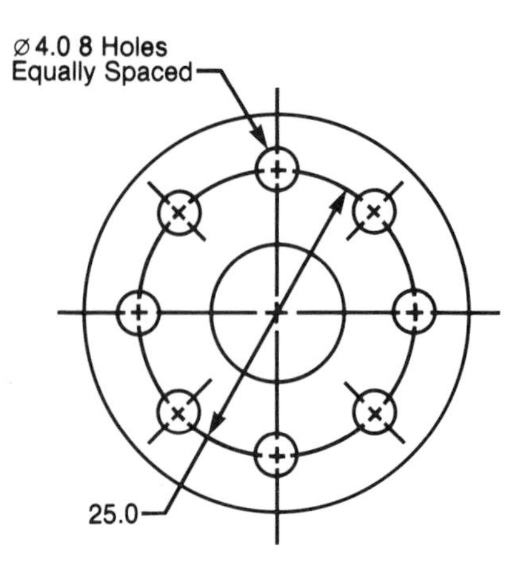

Figure 8.27 A bolt circle is used as a circular centerline for dimensioning this part with eight identical, equally spaced holes.

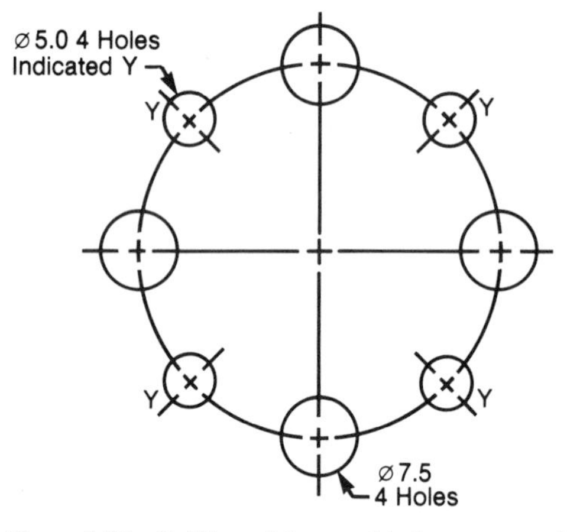

Figure 8.28 If different types of holes appear along a bolt circle, identifying letters are used.

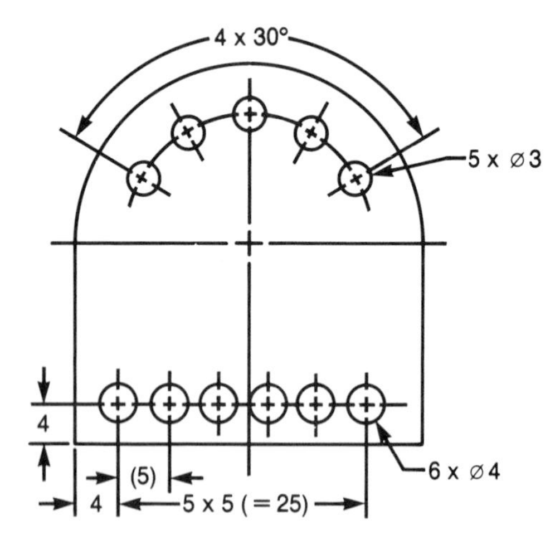

Figure 8.29 The symbol X can be used to represent repetitive features that appear at locations forming a consistent pattern.

Figure 8.27 presents a view of a part in which eight identical holes lie along a **bolt circle.** A bolt circle forms a circular path about a centerpoint or axis; it is used as a circular centerline for dimensioning purposes. In Figure 8.27, one leader is used to indicate the total number of holes that lie along the bolt circle, together with the relative spacing between these holes and the hole diameter. A bolt leader (with two arrowheads) is used to specify the diameter of the bolt circle. Notice that the symbol ∅ is not used for the bolt circle; this symbol should be used only in conjunction with actual features on a part, such as holes and shafts.

If different types of holes appear along a bolt circle, an identifying letter (such as Y) may be used to distinguish one set of holes from another set. A minimal number of leaders (one per set of holes) may then be used for dimensioning purposes (Figure 8.28).

Figure 8.29 presents an object in which repetitive features appear at locations forming a consistent pattern. In such cases dimensioning linework may be minimized by using the symbol X to indicate that a repetition of a linear or angular dimension is to occur at equal spacing. For example, in Figure 8.29 five holes of equal depth lie along an arcing centerline; each hole is separated by 30° from its neighbor, as indicated by the dimension for equal angular spacing:

4 X 30°

A similar note for equal linear spacing is used for other holes that appear in a pattern on this part:

5 X 5

A reference dimension, noted in parentheses, is used to clearly indicate the single dimension that is to be repeated in the linear pattern for these holes. For added clarity , the total linear dimension for this hole spacing is given as a reference dimension.

The symbol X may also be used to indicate the feature (and its size) that is to be repeated with equal spacing. For example, the leader notes in Figure 8.29 indicate the number of repetitions for a feature, followed by the symbol X and the dimension that specifies the required size for this feature.

In summary, then:

- Rule 18. A note should be used in conjunction with a leader to indicate that a feature is to be repeated in a part.
- Rule 19. The symbol X should be used to indicate equal linear or angular spacing between repetitive features.
- Rule 20. A bolt circle forms a circular path about a centerpoint or axis; it should be used as a centerline for dimen-

sioning purposes. A bolt leader is used to indicate the diameter of a bolt circle (without the symbol ⌀).

- Rule 21. Identifying letters (such as Y and Z) should be used to distinguish between different types of holes in a part.

8.11 Coordinate Dimensioning—Rectangular and Polar

Rectangular coordinates specify the location of a feature—either with respect to another feature or relative to a datum. Linear distances are specified from mutually perpendicular reference planes (Figure 8.30).

For cases in which dimensioning linework must be minimized in order to achieve maximum clarity in a drawing, rectangular coordinates may be used to specify the location of various features *without dimension lines* (or arrowheads). Figure 8.31 presents an example of a dimensional view in which distances are associated directly with the appropriate extension lines; these dimensions are given relative to **base lines** labeled as reference origins by the symbol 0 (or alternatively, by identifying letters such as Y or Z). In order to eliminate the need for numerous leaders, a table should be provided in which the depth dimension for each hole is specified. Each set of identical holes is labeled with an identifying **size symbol** in the table (Figure 8.31).

Figure 8.32 presents an example of dimensioning in tabular form wherein three mutually perpendicular reference planes (labeled X, Y, and Z) are used to specify the location of various holes. Such extended tabular dimensioning eliminates the need for confusing (extension and dimension) linework from the views. In addition, notice that hidden linework for the holes is not shown in the right-side view, since such linework would overlap in a somewhat ambiguous way.

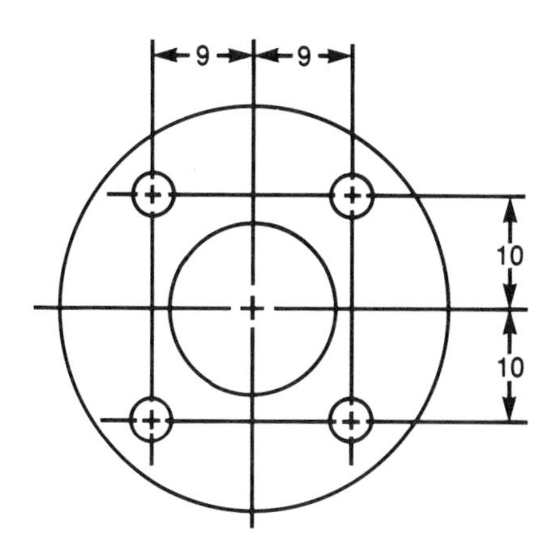

Figure 8.30 Rectangular coordinates specify the location of a feature with respect to mutually perpendicular reference planes.

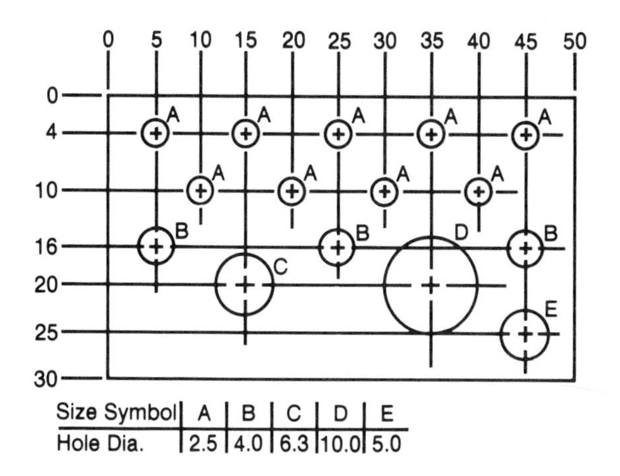

Size Symbol	A	B	C	D	E
Hole Dia.	2.5	4.0	6.3	10.0	5.0

Figure 8.31 In base line dimensioning, dimensions are given relative to baselines. A table should be used to give the depth dimensions for each hole, eliminating the need for numerous leaders.

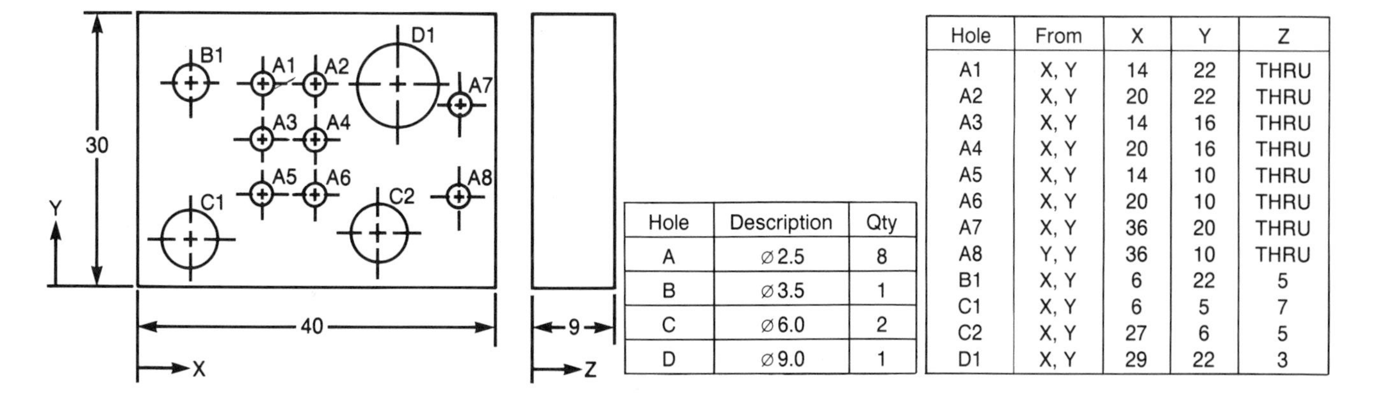

Hole	Description	Qty
A	⌀2.5	8
B	⌀3.5	1
C	⌀6.0	2
D	⌀9.0	1

Hole	From	X	Y	Z
A1	X, Y	14	22	THRU
A2	X, Y	20	22	THRU
A3	X, Y	14	16	THRU
A4	X, Y	20	16	THRU
A5	X, Y	14	10	THRU
A6	X, Y	20	10	THRU
A7	X, Y	36	20	THRU
A8	Y, Y	36	10	THRU
B1	X, Y	6	22	5
C1	X, Y	6	5	7
C2	X, Y	27	6	5
D1	X, Y	29	22	3

Figure 8.32 An example of base line dimensioning with an extended table.

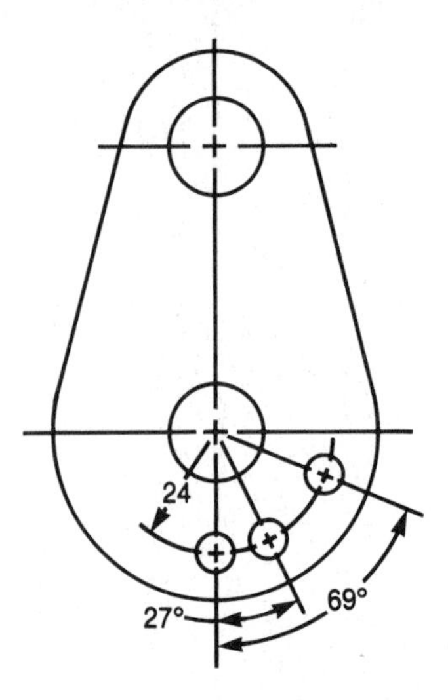

Figure 8.33 Polar coordinates specify the location of a feature that lies along a radial arc.

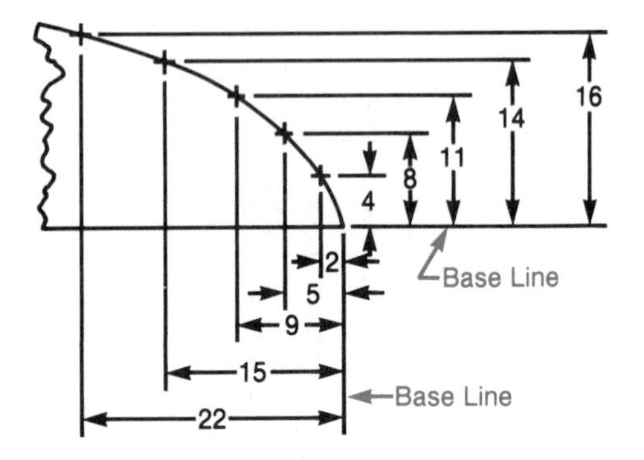

Figure 8.34 In the offset method, points are located along the outline of a surface with respect to appropriate base lines.

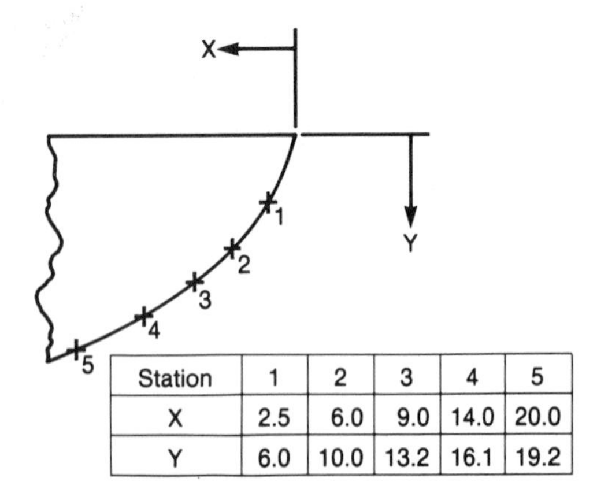

Station	1	2	3	4	5
X	2.5	6.0	9.0	14.0	20.0
Y	6.0	10.0	13.2	16.1	19.2

Figure 8.35 A tabular format may be used in the offset method to minimize linework.

Polar coordinates can be used to effectively specify the location of features that lie along a radial arc; locations may be specified in terms of a linear (radial) dimension and an angular dimension (Figure 8.33).

In summary, then:

1. *Rectangular coordinates* may be used to specify the location of a feature with respect to mutually perpendicular reference planes.
2. Rectangular coordinate dimensioning in *tabular form* allows one to eliminate confusing (extension and dimension) linework from a drawing through the use of appropriate *base lines.*
3. *Polar coordinates* may be used to specify the location of a feature that lies along a radial arc. Such a feature may then be located through the use of a linear (radial) dimension and an angular dimension.

8.12 Offset Method (Rectangular Coordinates)

Rectangular coordinates may be used to specify the outline of a surface by the **offset method.** The locations of points along the outline are given with respect to appropriate base lines (Figure 8.34). Dimensions may also be given in tabular form in order to minimize (extension and dimension) linework (see Figure 8.35).

In summary, then:

- Rule 22. The offset method should be used to specify the location of points that lie along the outline of a surface. Tabulated dimensions may be used to minimize dimensioning linework in a drawing.

8.13 Dimensioning of Special Features

Dimensioning formats for some special, though not uncommon, features are:

- *Parts with partially rounded ends* are dimensioned with overall numerical values, together with specifications for radii (Figure 8.36).

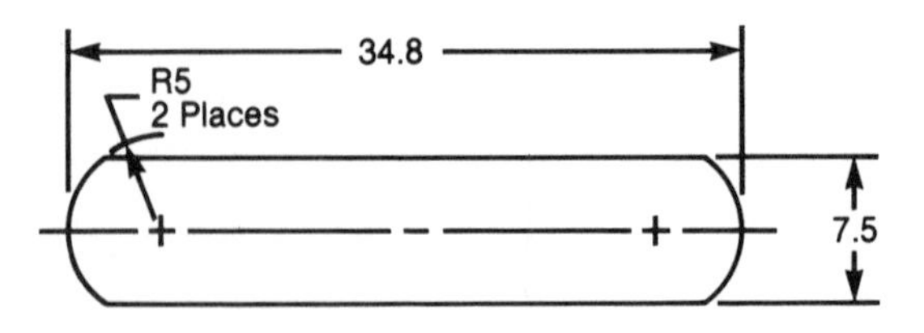

Figure 8.36 Overall numerical values are used to dimension parts with partially rounded ends. Radii are specified.

- *Parts with fully rounded ends* are dimensioned with overall numerical values, and the radii are noted with a leader but remain undimensioned (Figure 8.37).
- *Slotted holes* are dimensioned by any of the three equivalent formats presented in Figure 8.38. The radii of the ends are noted with a leader but remain undimensioned.
- *Fillets* are dimensioned with leaders (Figure 8.39) or by an appropriate note if the radii of many fillets are identical (Figure 8.40).
- *Chamfers* are dimensioned with an angle and a linear dimension (Figure 8.41). Two linear dimensions may be

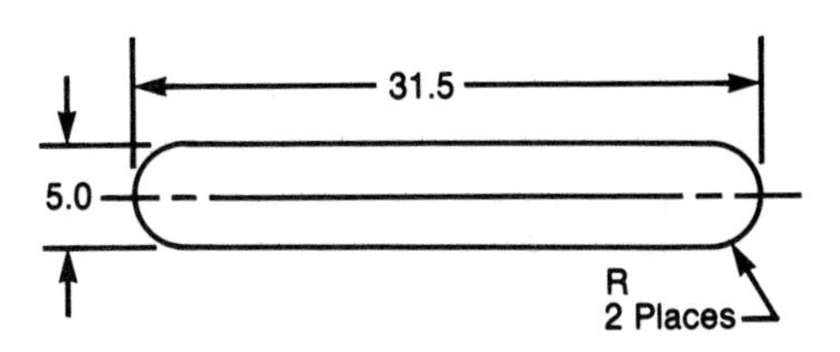

Figure 8.37 Overall numerical values are used to dimension parts with fully rounded ends, but radii remain undimensioned.

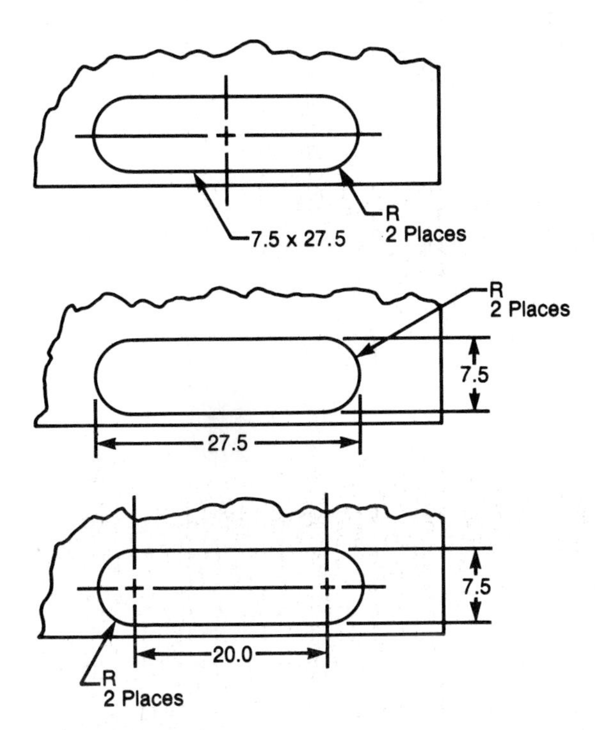

Figure 8.38 Three equivalent formats can be used for dimensioning slotted holes.

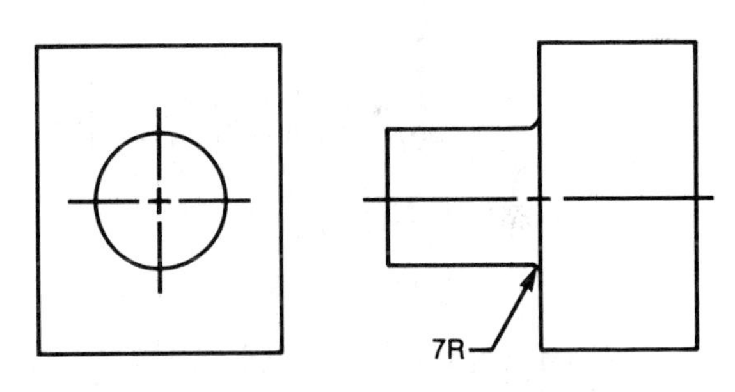

Figure 8.39 Fillets are dimensioned with leaders.

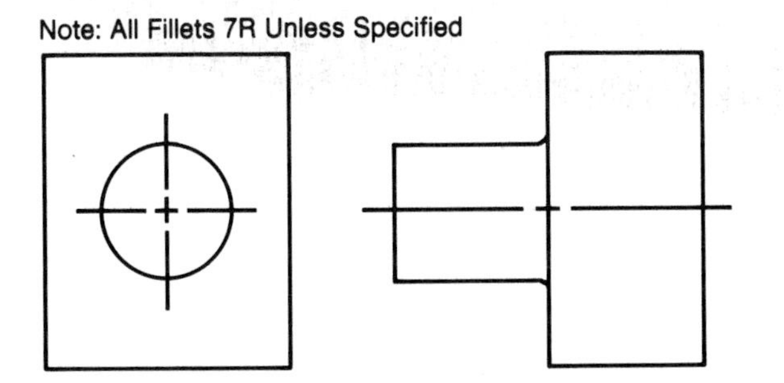

Figure 8.40 For multiple identical fillets, a note may be used for dimensioning purposes.

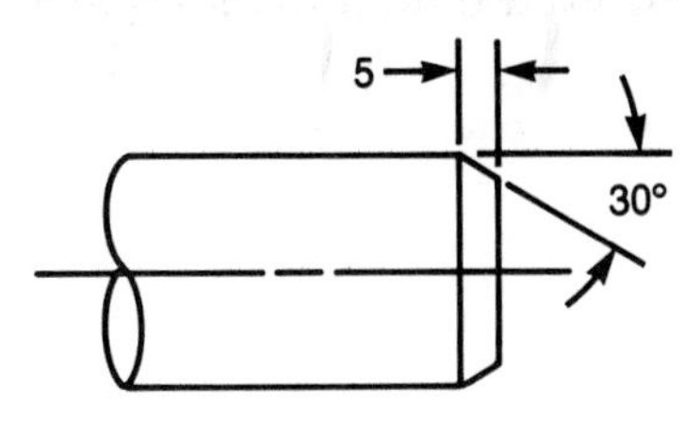

Figure 8.41 Chamfers are dimensioned with an angle and a linear dimension.

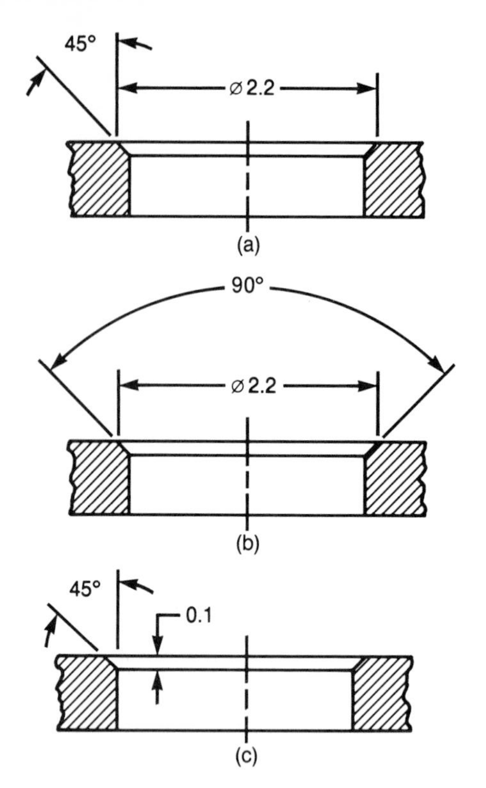

Figure 8.42 Three equivalent formats can be used for dimensioning internal chamfers.

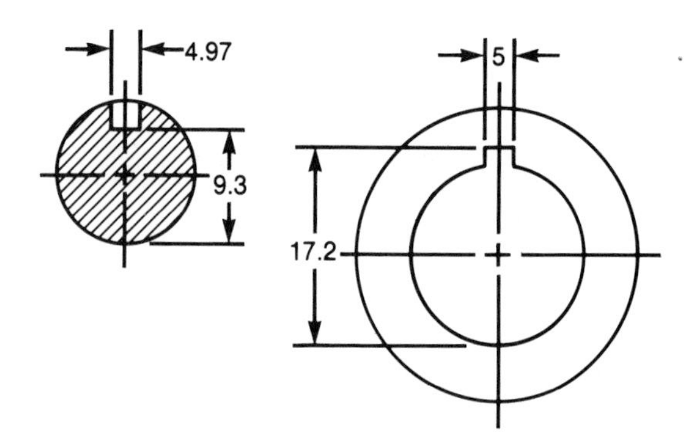

Figure 8.43 Depth, width, location, and length (if needed) are specified in keyseat dimensioning.

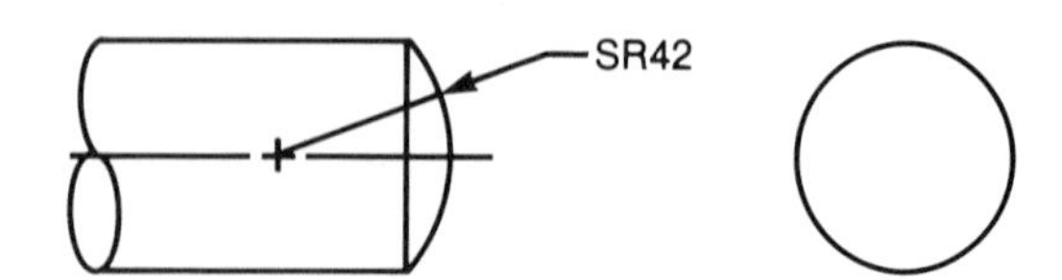

Figure 8.44 A numerical value, preceded by SR, is specified for the radius in dimensioning a spherical surface.

used if the clarity of the drawing will be enhanced. Internal chamfers may be dimensioned in any of the equivalent formats presented in Figure 8.42.

- *Keyseats* are dimensioned as indicated in Figure 8.43; the depth (relative to the opposite or far side of the shaft or hole), width, location, and length (if needed) should be specified.
- *Knurling* is dimensioned by providing both the knurling type and its pitch. In addition, if the knurling is included to provide a press fit between parts, the minimum acceptable diameter should be indicated both prior to knurling and after knurling.
- *Spherical surfaces* are dimensioned with a numerical value for the radius, preceded by the symbol SR (Figure 8.44).
- *Angles, arcs, and chords* are dimensioned as shown in Figure 8.45.
- *Cones and pyramids* are dimensioned as shown in Figures 8.46 and 8.47.
- *Conical tapers* are defined as the ratio of the difference in the diameters of two sections of the cone (where both sections are perpendicular to the conical axis) and the linear distance, L, between these sections.

$$\text{taper} \equiv \frac{(D - d)}{L}$$

(according to ANSI Y14.5M 1982).

They are dimensioned as shown in Figure 8.48.

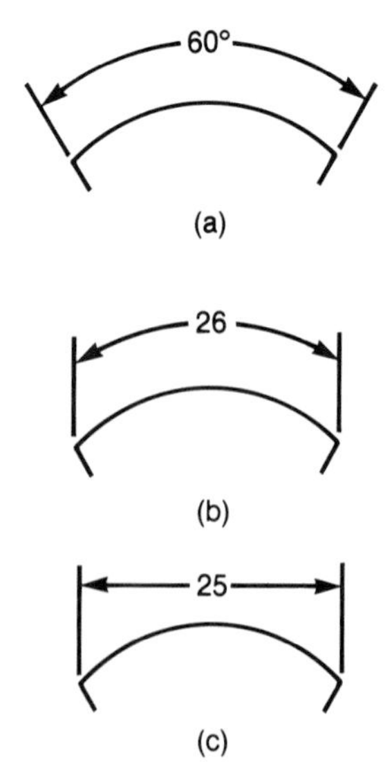

Figure 8.45 Dimensioning angles (a), arcs (b), and chords (c).

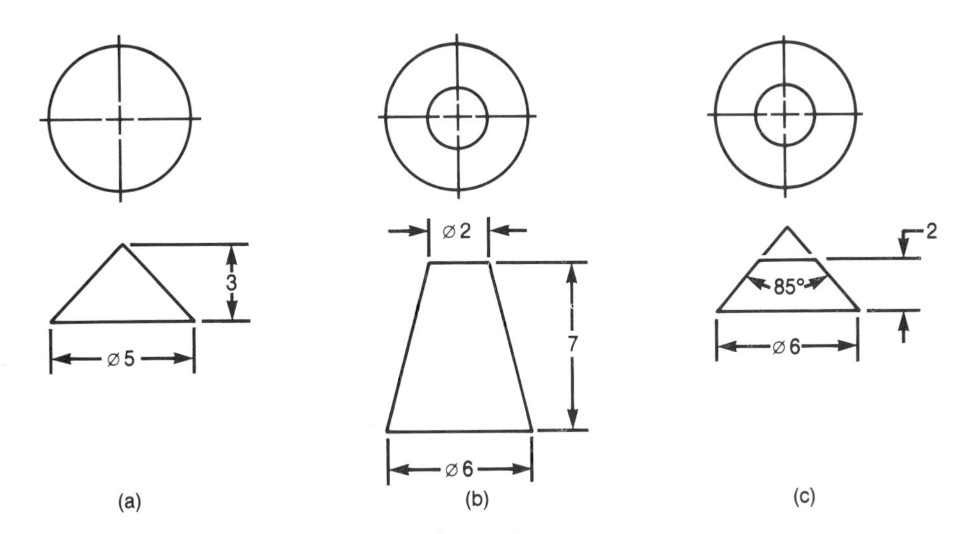

Figure 8.46 Dimensioning full cones (a) and truncated cones (b and c).

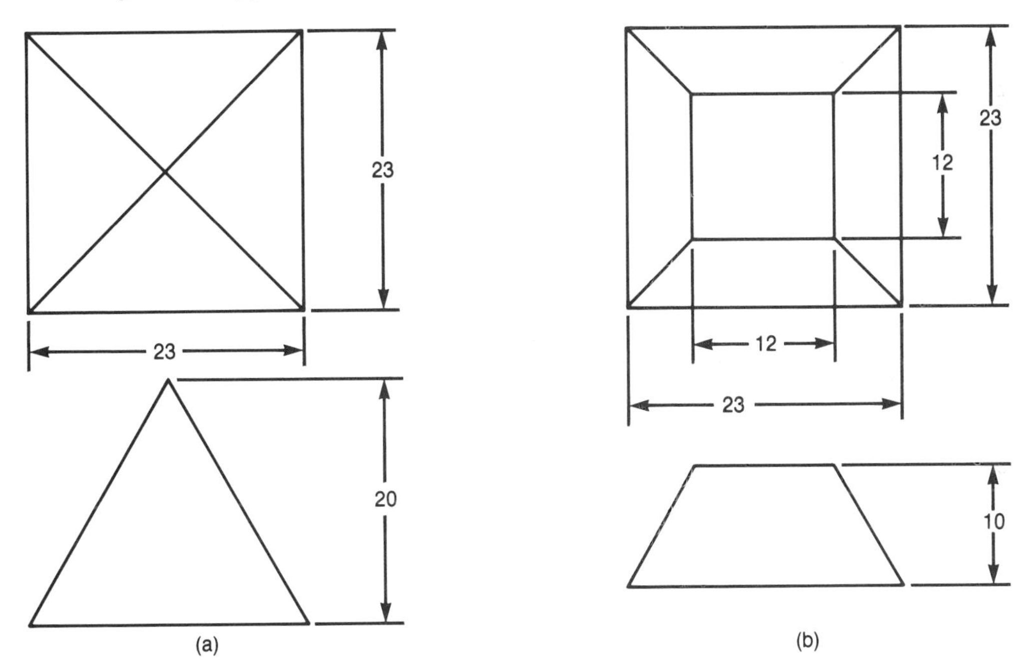

Figure 8.47 Dimensioning full pyramids (a) and truncated pyramids (b).

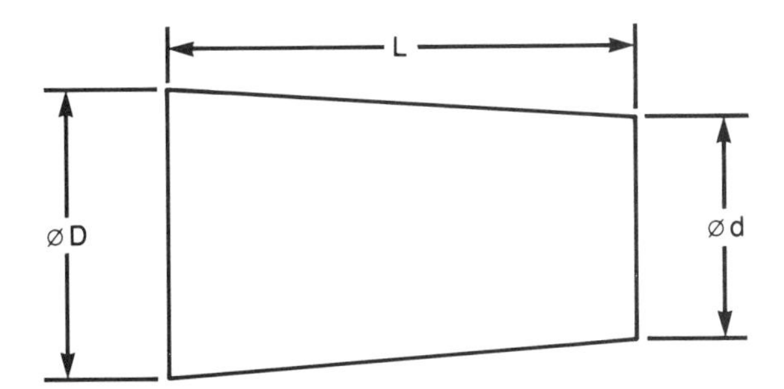

Figure 8.48 Dimensioning a conical taper.

8.14 Surface Texture

Finished, or machined, surfaces should be indicated by a **finishing symbol,** as shown in Figure 8.49. Note that a symbol often used (a) is not the preferred symbol (b). Note also that still another finishing symbol was once used (c).

Finishing symbols should be used on all edge views (both visible and hidden) of a surface to be machined. Surface textures (and fabrication methods) can be further specified by applying the symbology given in Figures 8.50 and 8.51.

8.15 A Suggested Procedure for Dimensioning Orthographic Views

Dimensioning a set of orthographic views may be smoothly performed following these steps:

- Step 1. *Draw all necessary extension and dimension lines* (with the appropriate spacing between parallel dimension lines, each view and its nearest dimension line, and so forth). *Check* to make sure that all dimensioning information has been included in the layout to allow proper fabrication of the part.
- Step 2. Insert all numerical values and dimensioning notes in the drawing. Inserting all values and notes simultaneously aids consistency in the lettering.
- Step 3. Finally, draw all arrowheads.

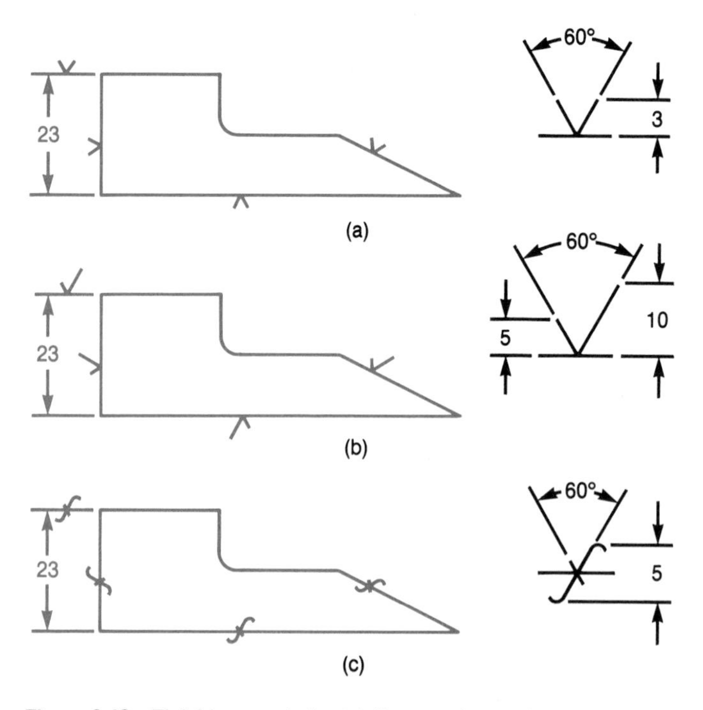

Figure 8.49 Finishing symbols. (a) Frequently used symbol. (b) Preferred symbol. (c) Earlier symbol.

Symbol	Meaning
(a) 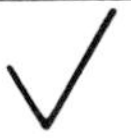	Basic Surface Texture Symbol. Surface may be produced by any method except when the bar or circle (Figure b or d) is specified.
(b) 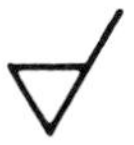	Material Removal By Machining Is Required. The horizontal bar indicates that material removal by machining is required to produce the surface and that material must be provided for that purpose.
(c) 3.5 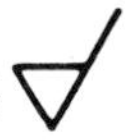	Material Removal Allowance. The number indicates the amount of stock to be removed by machining in millimeters (or inches). Tolerances may be added to the basic value shown or in a general note.
(d) 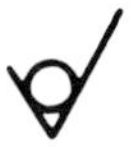	Material Removal Prohibited. The circle in the vee indicates that the surface must be produced by processes such as casting, forging, hot finishing, cold finishing, die casting, powder metallurgy or injection molding without subsequent removal of material.
(e) 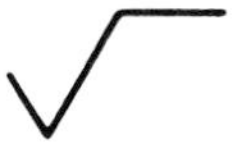	Surface Texture Symbol. To be used when any surface characteristics are specified above the horizontal line or to the right of the symbol. Surface may be produced by any method except when the bar or circle (Figure b and d) is specified.
(f) 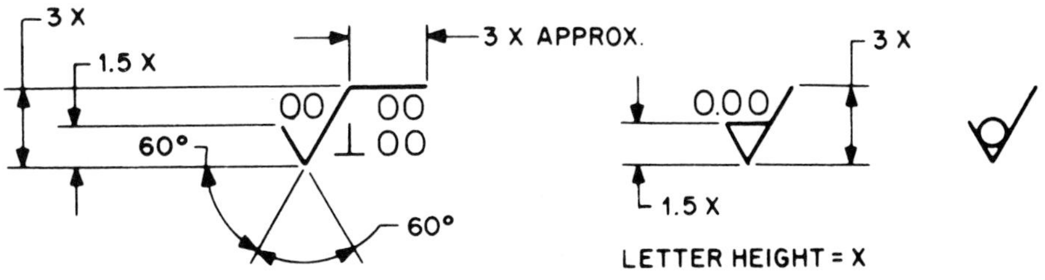 3 X; 1.5 X; 00; 00; 00; 60°; 60°; 3 X APPROX.; 0.00; 3 X; 1.5 X; LETTER HEIGHT = X	

Figure 8.50 Surface texture symbols and construction. (From ANSI Y14.36–1978. American National Standard Surface Texture Symbols.)

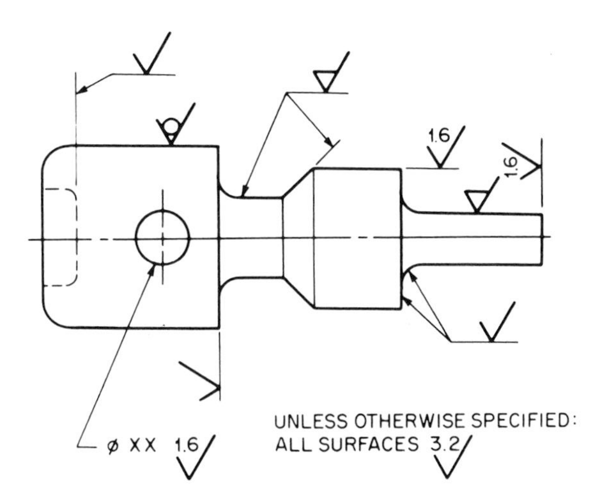

Figure 8.51 Application of surface texture symbols. (From ANSI Y14.36–1978. American National Standard Surface Texture Symbols.)

Without adherence to the preceding construction procedure, one often follows the natural tendency to insert each dimension in a drawing sequentially by first drawing the extension and dimension lines, followed by the numerical values, notes, and arrowheads—after which another dimension is added to the drawing in the same sequence of linework, lettering, and arrowhead construction. As a result, a necessary dimension may be inadvertently omitted because the designer fails to check the total amount of information contained in a drawing. Furthermore, the drawing may lack consistency in linework and lettering.

The procedure just outlined is preferred because it allows the designer to gain a perspective of the dimensioning information that is contained in a drawing and also aids consistency in both linework and lettering.

In summary, then:

1. A step-by-step procedure should be used to add dimensioning information to a drawing. All linework should be inserted and then checked for completeness, after which numerical values and notes should be added, and finally, all arrowheads drawn.

2. The suggested procedure helps the designer gain a perspective of the dimensioning information that is contained in a drawing and also achieve greater consistency in both linework and lettering.

Learning Check

What is indicated by a scale on a drawing?

Answer A scale on a drawing indicates the ratio of the length of a *line* to the length of the corresponding *feature* of the object.

In Retrospect

- Dimensions are used to specify both the size of a feature and its location.
- Basic dimensions specify the theoretically exact values for the size, orientation, and location of a feature, whereas reference dimensions provide information meant to enhance clarity and help the reader correlate dimensional data. Reference dimensions are either repetitions of basic dimensions or derived from basic dimensions. They should be enclosed within parentheses to distinguish them from basic dimensions.
- The tolerance associated with a dimension specifies the total range of acceptable variation in the actual size of the fabricated part (see Chapters 9 and 10).
- A datum is the exact reference point, axis, or plane from which one may specify the location of a feature.
- A scale should be included on a drawing in order to indicate the ratio between the length of a line on the drawing and the size of the corresponding feature of the part.

- Units of measurements must be specified on a drawing; they may be given according to the SI system or the English system.
- Extension lines specify the extent of a feature in a given linear or angular direction. Dimension lines run between extension lines (in most cases); arrowheads are drawn at the end of dimension lines with the numerical value for the dimension associated with the dimension line. Crowding of dimensioning should be avoided.
- Every dimension that is needed for the fabrication of a part must be shown in a drawing; a basic dimension should never be repeated (except as a reference dimension included to aid the reader in correlating information).
- Rectangular coordinates are used to specify the location of a feature with respect to mutually perpendicular reference planes. Polar coordinates are used to locate features that lie along a radial arc.
- Rectangular coordinates may be given in tabular form in order to eliminate (or at least minimize) confusing linework in a drawing. The tabular format is preferred, in fact, if numerically controlled (NC) machinery will be used to manufacture the part; see Section 18.1.
- Traditional formats exist for dimensioning certain special features, such as parts with rounded ends, slotted holes, fillets, chamfers, keyseats, knurling, spherical surfaces, angles, arcs, chords, cones, pyramids, and conical tapers.
- A finishing symbol should be used on all edge views of a surface to be machined (finished).
- In a recommended step-by-step procedure for dimensioning a set of orthographic views, all linework is inserted and checked for completeness before numerical values and notes are added, and all arrowheads are drawn last. This procedure facilitates a proper perspective of the dimensioning information that is included in a drawing and also aids consistency in linework and lettering.
- Dimensions describe the size of the object, not of the drawing.
- Important rules for properly dimensioning a set of orthographic drawings are:
  - Rule 1. Show every necessary dimension in a drawing.
  - Rule 2. Never repeat a basic dimension in a drawing.
  - Rule 3. Locate dimensions that are linked to one another (for example, dimensions for the location of a centerpoint) with respect to the same datum if possible.
  - Rule 4. Place larger dimensions farther from an orthographic view than smaller dimensions.
  - Rule 5. Place dimensioning information outside the orthographic views if possible.
  - Rule 6. Leaders should be used to dimension arcs (radii) and holes (diameters) but not cylindrical solids.
  - Rule 7. The direction of leaders used to dimension circles (holes) and arcs should be radial.
  - Rule 8. Specify the diameter, not the radius, of a hole, together with the symbol ∅, which should precede the diametral value. The symbol R follows the radial value of an arc.
  - Rule 9. Dimension lines should not cross one another unless unavoidable.
  - Rule 10. Extension lines may cross one another; furthermore, a dimension line may cross an extension line.
  - Rule 11. Do not crowd dimensions in a set of orthographic views.
  - Rule 12. Numerical values of parallel dimensions should be staggered.
  - Rule 13. Dimensional values should be given in either a unidirectional format or an aligned format.
  - Rule 14. Locate the vertex of an angular dimension with respect to a datum.
  - Rule 15. If there is limited space between extension lines, place the dimension line outside the extension lines with arrowheads directed toward the extension lines. For even smaller spaces, the numerical value is also placed outside the extension lines.
  - Rule 16. Oblique extension lines should be used if drawing space is very limited. Extension lines may also be used to locate points of interest that lie at positions external to a part.
  - Rule 17. Leaders should be used to indicate dimensioning specifications with clarity and minimal linework on a drawing. Non-dimensional leaders may be used to indicate other types of data such as manufacturing specifications.
  - Rule 18. A note should be used in conjunction with a leader to indicate that a feature is to be repeated in a part.

COMPUTER GRAPHICS IN ACTION

2D Computer Graphics: Automatic Dimensioning

In Section 3.19 we introduced the DIMENSION command (or its equivalent), which is included in most 2D computer graphics menu-driven software. This command can be used to insert the proper dimensioning data in a set of orthographic views. The user can (usually) specify the placement of dimension linework in the drawing while the computer graphics software *automatically determines the proper numerical values to be inserted* (the user may also specify the number of significant digits to be used for such values). As a result, dimensioning becomes a very simple task. In addition, the dimensioning information can be easily modified without redrawing the entire set of views.

In the example shown here of a dimensioned set of orthographic views, one dimension needed to completely specify the size of the object is not included in the final drawing. Can you identify this missing bit of needed information?

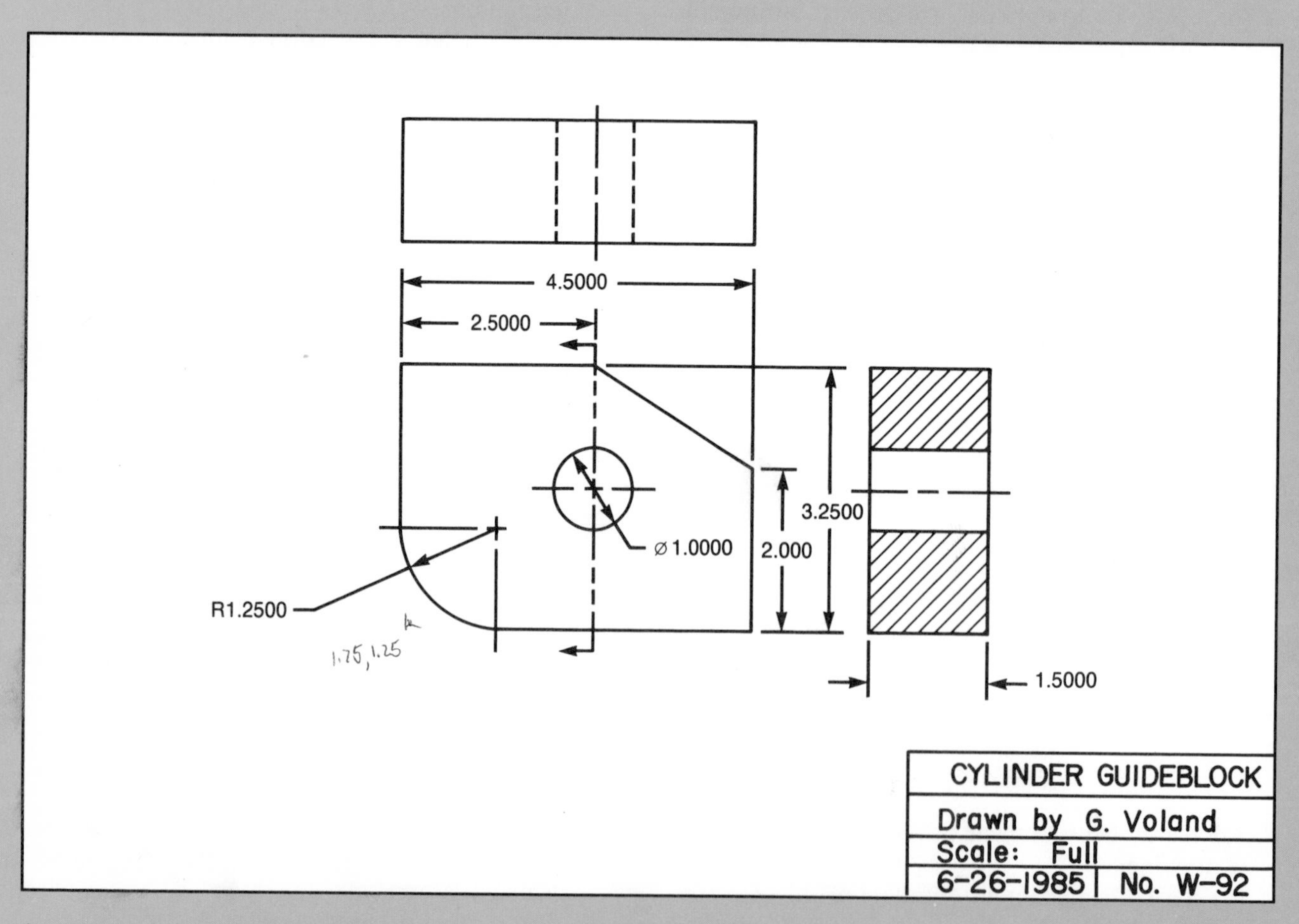

An example of an automatically dimensioned drawing. (One necessary dimension is missing.)

- Rule 19. The symbol X should be used to indicate equal linear or angular spacing between repetitive features.
- Rule 20. A bolt circle forms a circular path about a centerpoint or axis; it should be used as a centerline for dimensioning purposes. A bolt leader is used to indicate the diameter of a bolt circle (without the symbol ∅).
- Rule 21. Identifying letters (such as Y and Z) should be used to distinguish between different types of holes in a part.

Exercises

Your instructor will specify the appropriate scale for each exercise and also the *system of units (metric or English)* that should be used.

EX8.1 through EX8.12 Properly dimension a complete set of orthographic views (that is, the *best and necessary views* to describe the object) for the objects shown in these exercises. If an additional view (or views) not given in the exercise is needed to fully describe the object, then include such *additional views* in your finished and fully dimensioned drawing.

EX8.13 through 8.84 refer to specific exercises in preceding chapters. In each case draw an appropriate set of the *best and necessary views* of the object, including any *auxiliary views* and/or *section views* (in accordance with the directions of your instructor). *Properly dimension the resulting set of views* so that the object is fully described in terms of its shape and size. Again, your instructor will specify the scale of the drawing and the units of measurement that should be used.

EX8.13. See EX4.1.

EX8.14. See EX4.2.

EX8.15. See EX4.3.

EX8.16. See EX4.4.

EX8.17. See EX4.5.

EX8.18. See EX4.6.

EX8.19. See EX4.7.

EX8.20. See EX4.8.

EX8.21. See EX4.9.

EX8.22. See EX4.10.

EX8.23. See EX4.11.

EX8.24. See EX4.12.

EX8.25. See EX4.13.

EX8.26. See EX4.14.

EX8.27. See EX4.15.

EX8.28. See EX4.16.

EX8.29. See EX4.17.

EX8.30. See EX4.18.

EX8.31. See EX4.19.

EX8.32. See EX4.20.

EX8.33. See EX4.21.

EX8.34. See EX4.22.

EX8.35. See EX4.23.

EX8.36. See EX4.24.

EX8.37. See EX4.25.

EX8.38. See EX4.26.

EX8.39. See EX4.27.

EX8.40. See EX4.28.

EX8.41. See EX4.29.

EX8.42. See EX4.30.

EX8.43. See EX4.31.

EX8.44. See EX4.32.

EX8.45. See EX4.33.

EX8.46. See EX4.34.

EX8.47. See EX4.35.

EX8.48. See EX4.36.

EX8.49. See EX4.37.

EX8.50. See EX4.38.

EX8.51. See EX4.39.

EX8.52. See EX4.40.

EX8.53. See EX4.41.

EX8.54. See EX4.42.

EX8.55. See EX4.43.

EX8.56. See EX4.44.

EX8.57. See EX4.45.

EX8.58. See EX4.46.

EX8.59. See EX4.47.

EX8.60. See EX4.48.

EX8.61. See EX4.49.

EX8.62. See EX4.50.

EX8.63. See EX4.51.

EX8.64. See EX4.52.

EX8.65. See EX4.53.

EX8.66. See EX4.54.

EX8.67. See EX4.55.

EX8.68. See EX4.56.

EX8.69. See EX4.57.

EX8.70. See EX4.58.

EX8.71. See EX4.59.

EX8.72. See EX5.1.

EX8.73. See EX5.2.

EX8.74. See EX5.3.

EX8.75. See EX5.4.

EX8.76. See EX5.5.

EX8.77. See EX5.6.

EX8.78. See EX5.7.

EX8.79. See EX5.8.

EX8.80. See EX5.9.

EX8.81. See EX6.1.

EX8.82. See EX6.2.

EX8.83. See EX6.3.

EX8.84. See EX6.4.

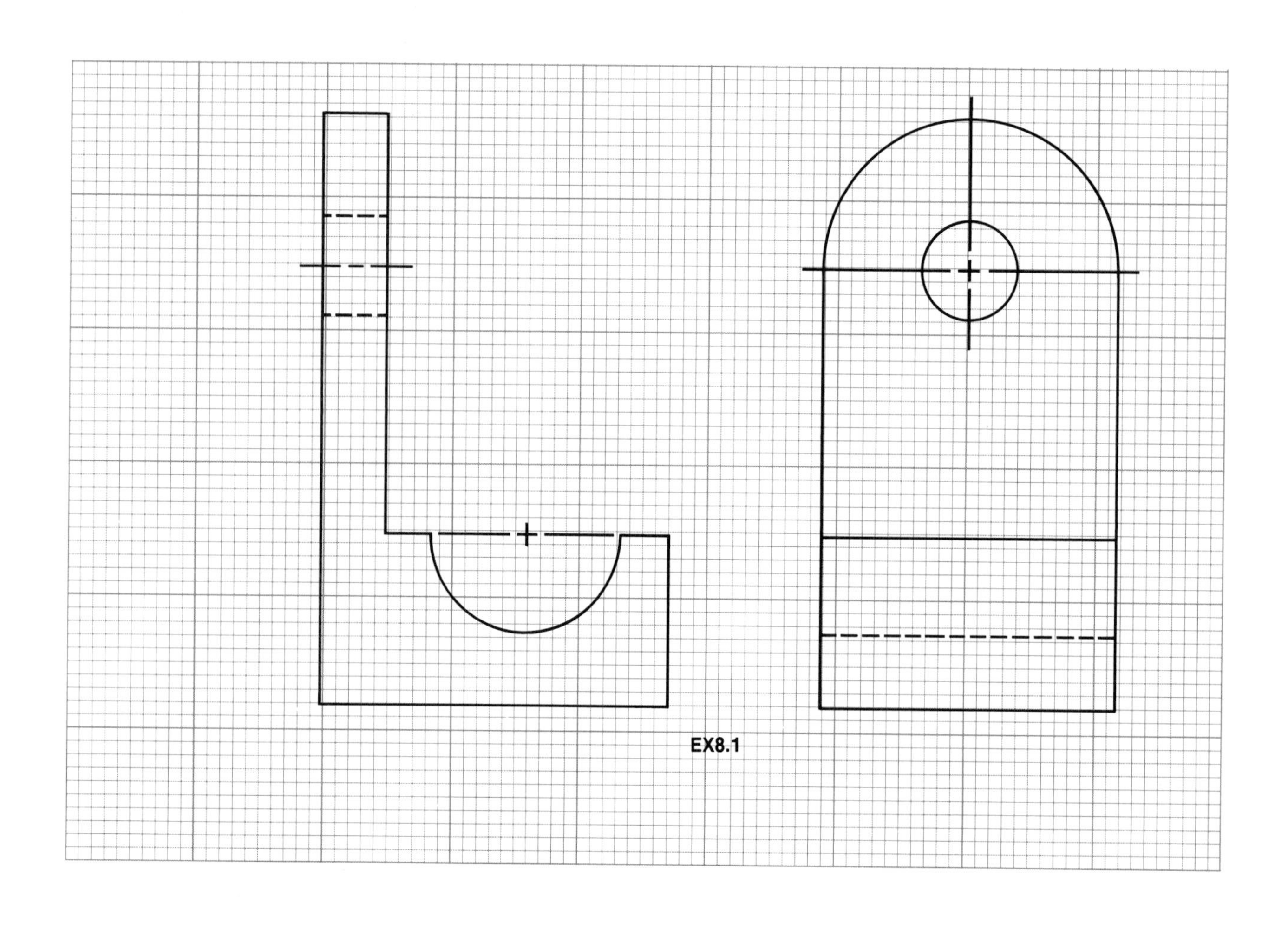

EX8.1

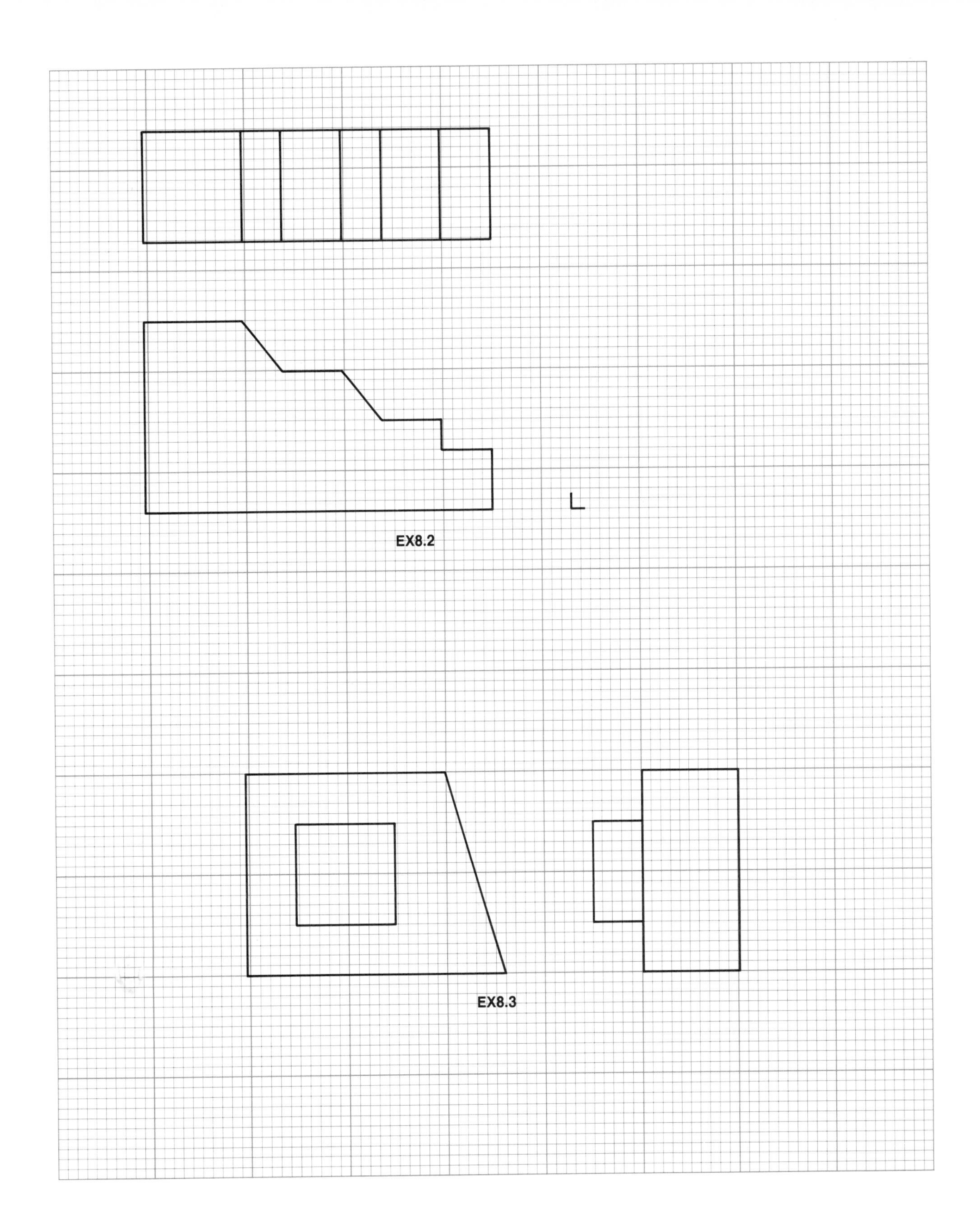
EX8.2
EX8.3

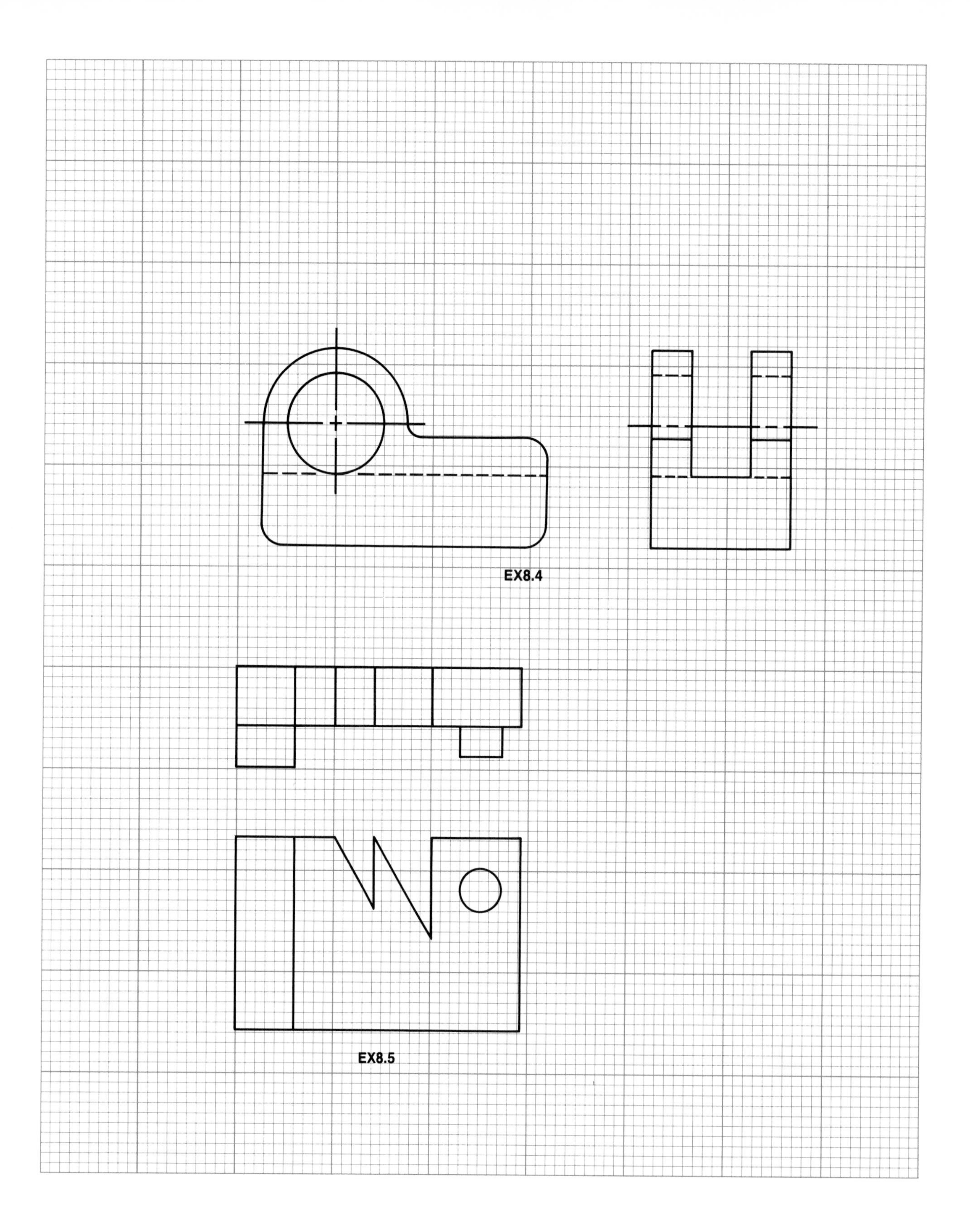
EX8.4
EX8.5

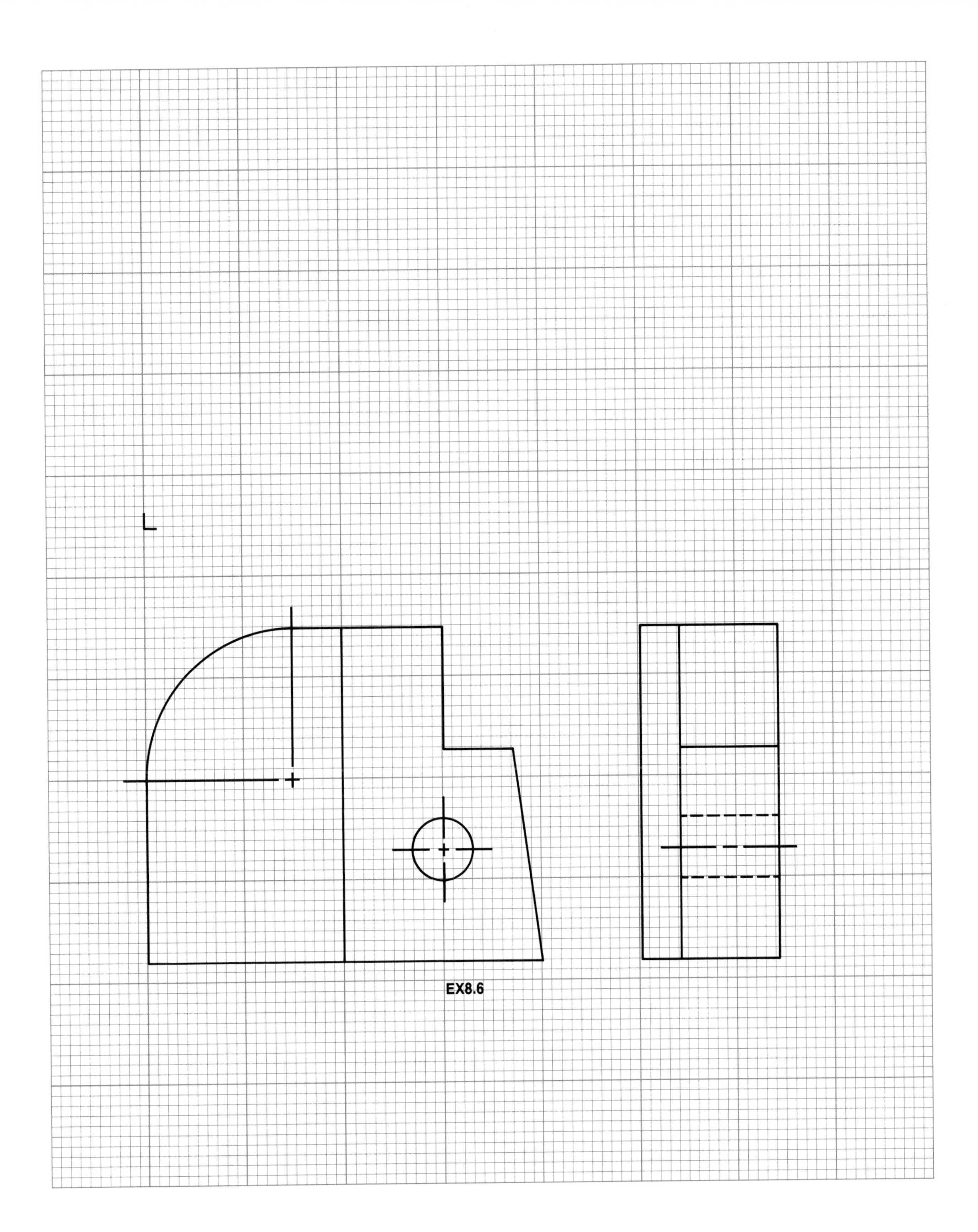
EX8.6

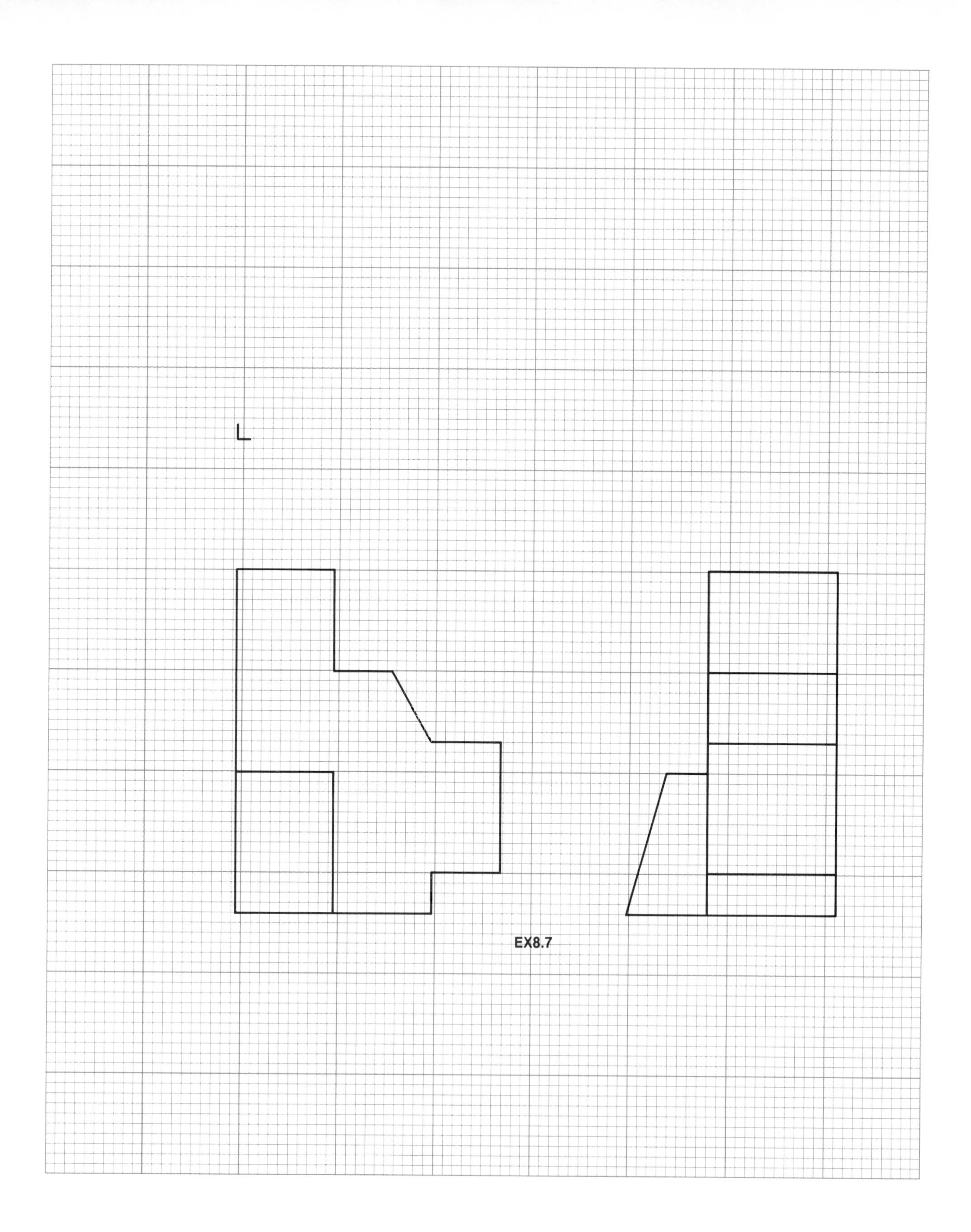
EX8.7

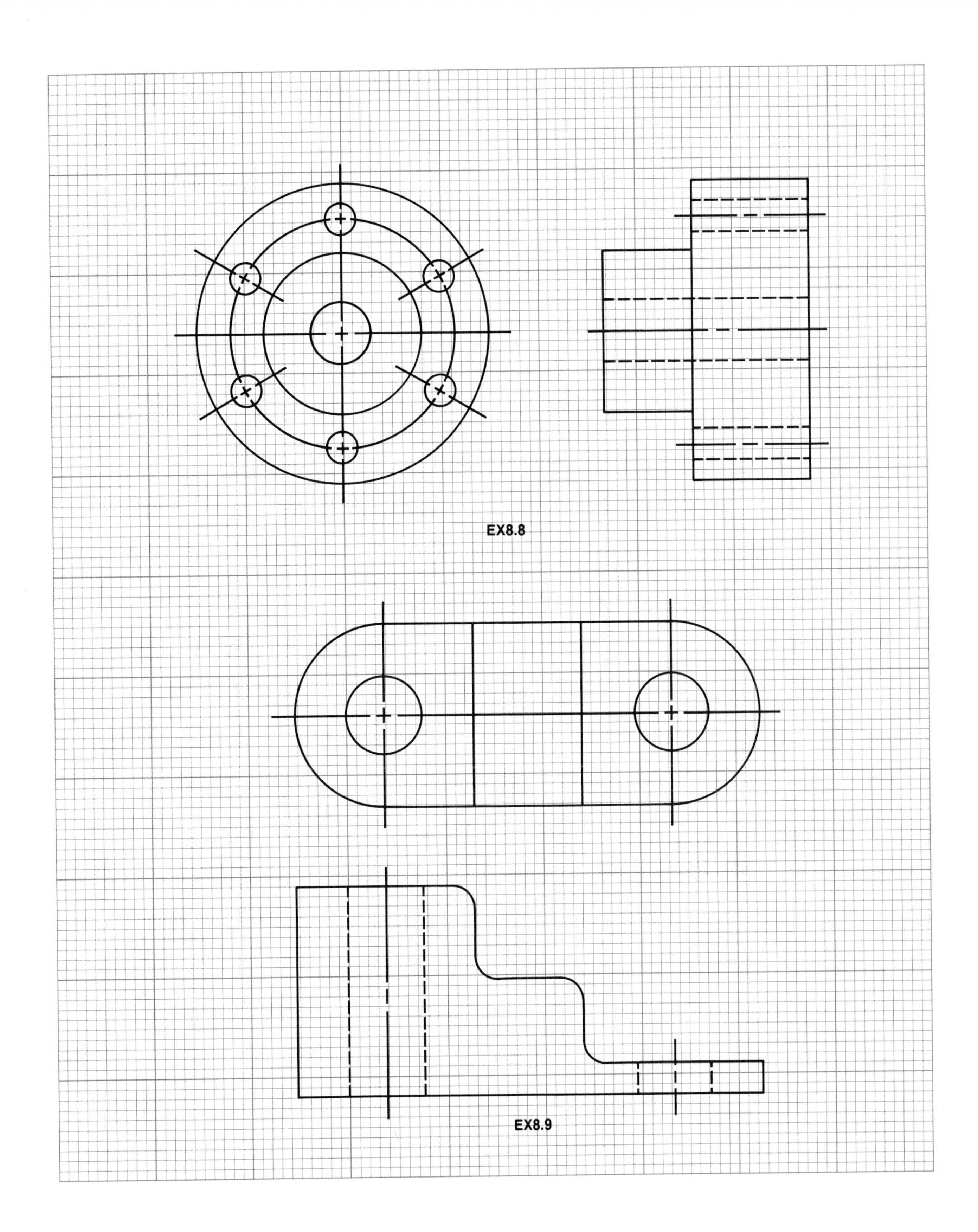
EX8.8
EX8.9

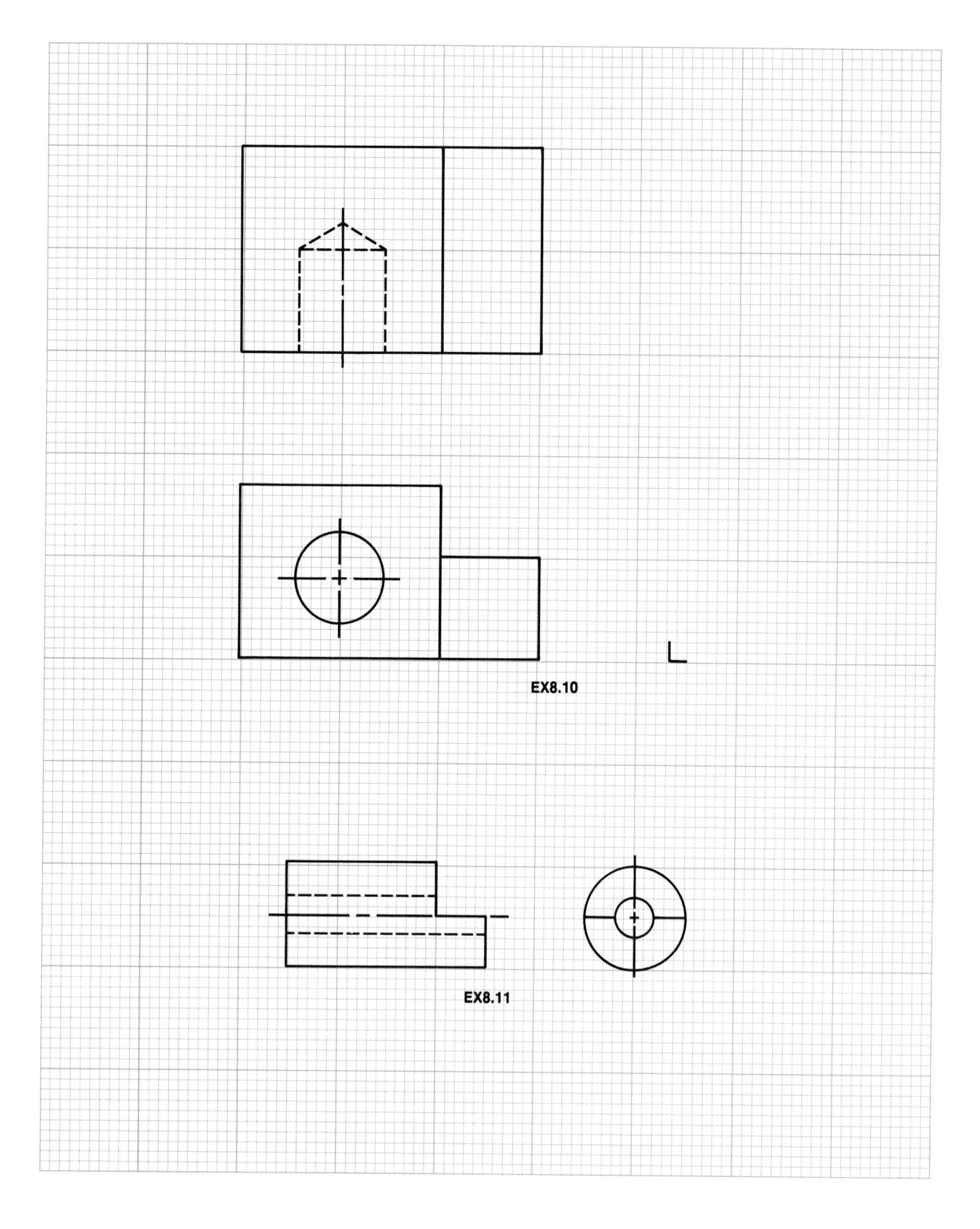
EX8.10
EX8.11

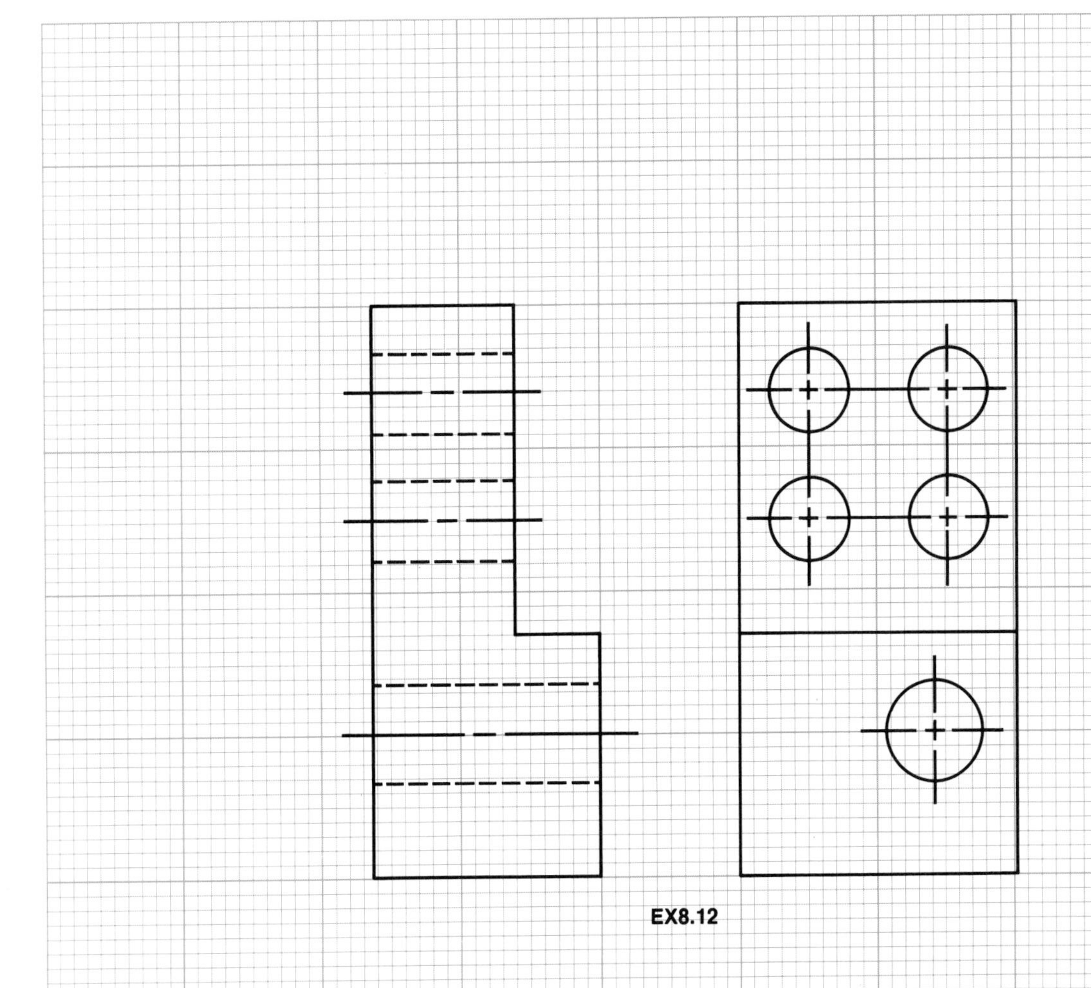

EX8.12

9 Size Tolerancing

Preview Engineering drawings specify the theoretically perfect specimen of a part; the actual fabricated part cannot be expected to correspond exactly to these theoretical basic dimensions due to the imprecise nature of fabrication processes. As a result, the designer must specify—in addition to the desired basic dimensions—the range of acceptable variation in the final dimensions. This range of acceptable variation is the **tolerance** associated with that dimension.

Tolerances may be specified in a variety of ways (for example, as *bilateral tolerancing* or *limit tolerancing*). The designer must be able to identify the **tolerance zone** that corresponds to a given tolerance specification for a dimension. In addition, the designer should seek to minimize **tolerance accumulation** in dimensioning specifications for a part.

Manufacturing is largely based upon the concept of interchangeable parts, which in turn is dependent upon the machinist's ability to fabricate such nearly identical objects. The designer uses tolerance specifications to identify the range of variation in the size and shape of interchangeable parts that is acceptable—that is, the range of variation that will allow mating between parts and interchangeability. Obviously, tolerance specification is critical to manufacturing and industrial processes and deserves our careful consideration.

Learning Objectives

Upon completion of this chapter, the reader should be able to:

- Recognize that a fabricated part cannot be expected to correspond *exactly* in its actual size to the theoretical basic dimensions specified in an orthographic description of the part.
- Define tolerance, allowance, and fit.
- Differentiate between bilateral tolerancing, unilateral tolerancing, and limit dimensioning.
- Use either the basic hole system or the basic shaft system in dimensioning mating parts in accordance with the expected functions of these parts.
- Specify the tolerance zone that corresponds to a given tolerance specification for a dimension.
- Minimize tolerance accumulation through the use of direct dimensioning where appropriate (if such accumulation is significant).
- Avoid chain line dimensioning since it produces maximum tolerance accumulation (if such accumulation is significant).
- Judiciously apply tolerance specifications to both internal and external features of a part.

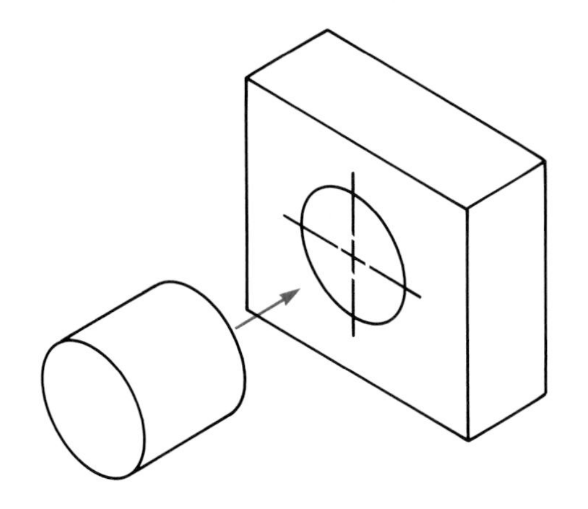

Figure 9.1 Mating parts to be the subject of successive figures in this chapter.

9.1 Tolerance, Allowance, and Fit

We recognize that any manufactured part should not be expected to correspond *exactly* in its actual size to the theoretical *basic dimensions* specified in an orthographic representation of the part. Inaccuracies will occur due to the uncertainties associated with any manufacturing process.

Tolerances are associated with dimensions in a drawing in order to specify the *range of acceptable variation* in the actual size of the manufactured part. Thus, a tolerance is the *difference between the maximum and minimum allowable sizes* for a part. (Tolerances may also be specified for the position and geometric form of a feature; such *positional and geometric form tolerances* are the focus of the next chapter.)

Clearances need to be specified for mating parts (parts that will contact each other in the finished product). Any difference between the sizes of such mating parts is referred to as their clearance.

Figure 9.1 presents two mating parts: a solid cylinder and a rectangular prism with a hole into which the cylinder is to be inserted. The range of tightness, or the **fit,** between these parts can be of several distinct types:

- **Clearance fit** (corresponding to a positive clearance), in which there is an intentional air space between the two parts, as shown in Figure 9.2. (Please note that Figure 9.2 is a side view of both mating parts; it is *not a set of orthographic views* for a single part.)
- **Interference fit** (corresponding to a negative clearance), in which there is an intentional overlap, or interference, between mating parts (Figure 9.3). This interference ensures that a tight (pressure) fit will be achieved if the parts are forced together. (The edges of the cylinder and the hole will need to be partially tapered or a chamfer will be needed in order to allow an initial joining of the parts; such modifications are not shown in Figure 9.3.)
- **Transitional fit,** in which either a clearance fit or an interference fit is allowed.
- **Line fit,** in which the upper limit of the shaft is equal to the lower limit of the hole.

In summary, then:

1. Tolerances are associated with dimensions in an engineering drawing in order to specify the range of acceptable variation in the actual size of the fabricated part.

2. The clearance is the difference between the sizes of mating features.

3. The fit is the range of tightness between mating features. Several types of fit can be specified: a clearance fit corresponding to a positive allowance between parts, an interference fit corresponding to a negative allowance between

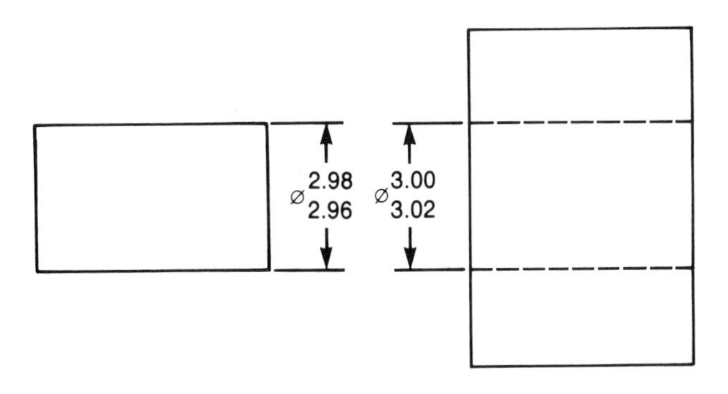

Figure 9.2 In a clearance fit between mating parts, there is an air space.

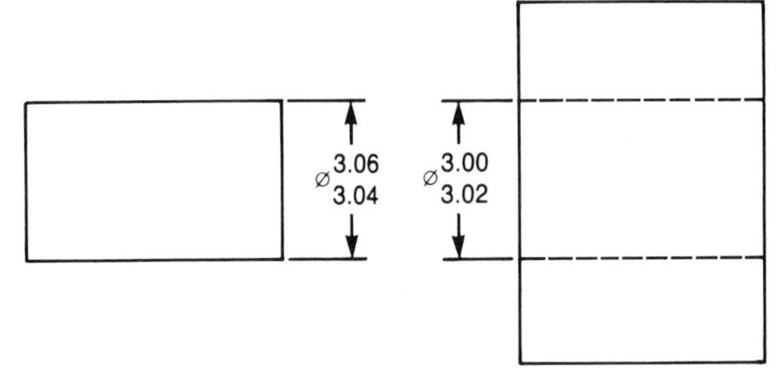

Figure 9.3 In an interference fit between mating parts, there is an overlap.

parts, a transitional fit in which either a clearance fit or an interference fit is allowed, and a line fit in which the upper limit of the shaft is equal to the lower limit of the hole.

Learning Check

What is a tolerance?

Answer The *total allowable variation* in a size is its tolerance.

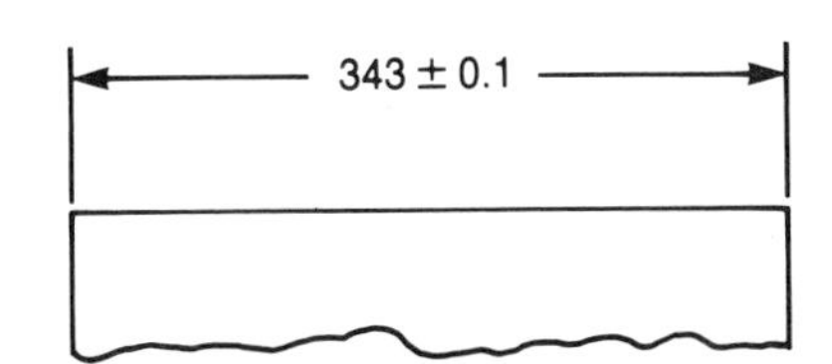

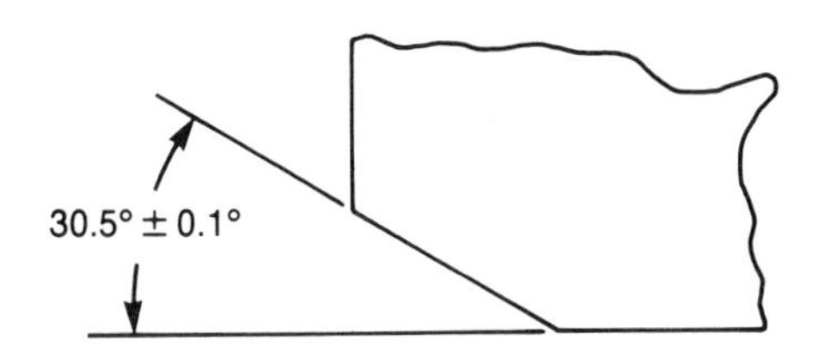

Figure 9.4 Bilateral tolerancing specifies a symmetric range of acceptable variation in terms of plus and minus values.

9.2 Bilateral, Unilateral and Limit Tolerancing

Bilateral tolerancing (Figure 9.4) specifies a symmetric range of acceptable variation on either side of the basic dimension in terms of plus and minus values (each of which is equal to one-half of the total range of variation). Bilateral tolerancing is sometimes called **plus-and-minus tolerancing.**

Unilateral tolerancing indicates the range of acceptable variation in a single direction of measurement (either entirely positive or entirely negative), as shown in Figure 9.5.

Limit dimensioning specifies the range of acceptable variation in terms of the minimum and maximum values of the dimension (Figure 9.6). The maximum value should be placed above the lower limit in the case of external features (linear and angular). If both limits are given in a single line, the minimum value should precede the upper limit, and the two should be separated with a dash.

Tolerances must be carefully indicated for external features and internal features. The basic guideline is that the least amount of material possible should be removed during

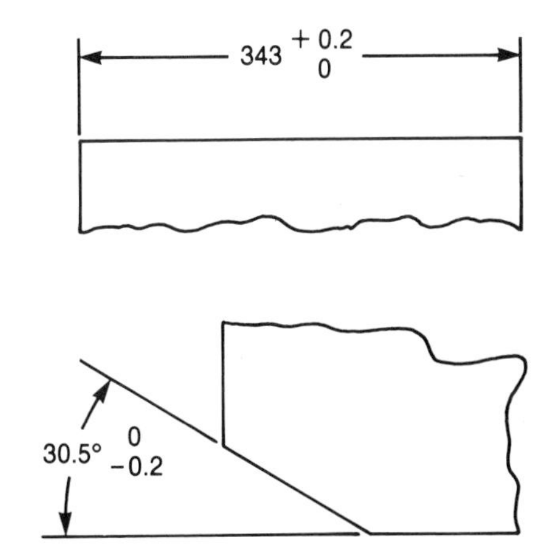

Figure 9.5 Unilateral tolerancing specifies a range of acceptable variation in a single direction of measurement.

Highlight

Maudslay, Whitney and Interchangeable Parts

Specification of tolerances for mating parts allows interchangeability of manufactured components of a product. Manufacturing is based upon this concept of interchangeability of fabricated components.

Eli Whitney is generally credited with introducing the concept of interchangeability to the world (although Honore Blanc had earlier applied the concept to the fabrication of guns) when, at the beginning of the nineteenth century, he was awarded a contract with the government of the United States to produce 10,000 muskets with interchangeable parts. Whitney never quite achieved truly interchangeable parts for his muskets, but the concept was established as a valuable asset in manufacturing and production. Interchangeability among manufactured parts depends upon the ability of the machinist to produce these parts within specified tolerance zones through the use of available manufacturing equipment. Henry Maudslay (early nineteenth century) influenced an entire generation of machine-tool designers through his efforts to improve the accuracy of machining equipment. Among those influenced by Maudslay were Clement (lathe and planar design work), Whitworth (standard screw thread, standard gauges), Nasmyth (special-purpose milling machine, shaper, and steam forge hammer) and Roberts (metal planar, lathe gearing, and drilling machines). Without the work of Maudslay and those who followed his example, accuracy in fabrication processes would not have been achieved.

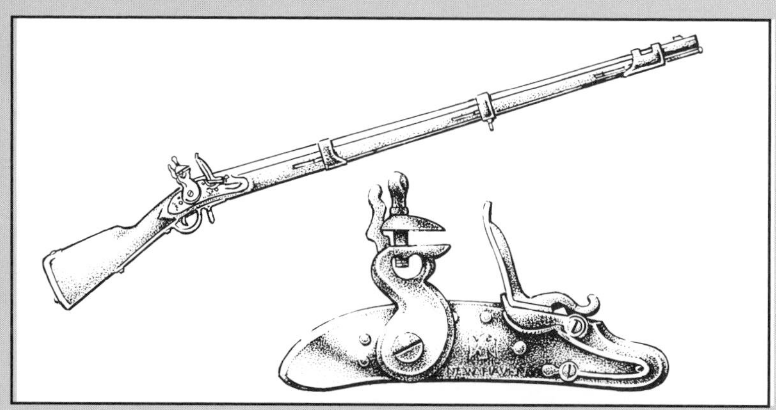
The Whitney musket

Source: Stephan Konz, Work Design: Industrial Ergonomics, *2d. ed. (Columbus, Ohio.: Grid Publishing, Inc., 1983).*

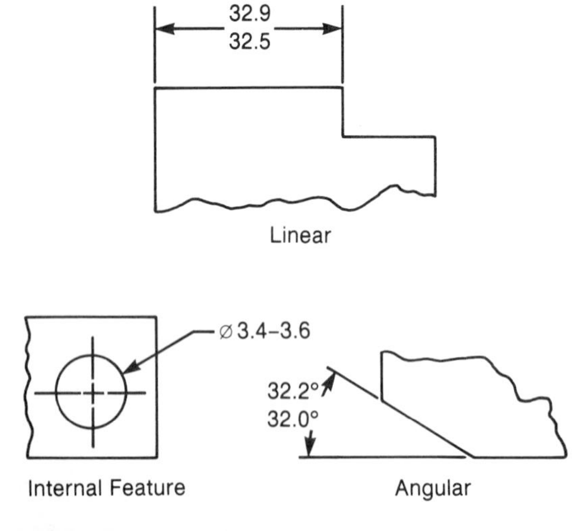

Figure 9.6 Limit dimensioning specifies a range of acceptable variation in terms of the minimum and maximum values of the dimension.

the fabrication of a part. Consequently, the limit initially achieved in the fabrication of a feature is given first in tolerancing specification on a drawing.

For an **internal** feature (such as the hole shown in Figure 9.7), the minimum limit is achieved first during fabrication (material removal), so it is listed above the maximum limit in a tolerancing note. For an **external** feature (Figure 9.8), the maximum limit is achieved first during fabrication (by cutting material from the raw supply), so it is listed above the minimum limit. The preceding guidelines are based on the **maximum material system.** Another practice sometimes used, known as the **maximum number system,** places the larger value above the smaller number in all dimensions.

9.3 Tolerance Zones

In order to appreciate the possible variation that may result from tolerance specifications, called the **tolerance zone,** consider the following three distinct cases.

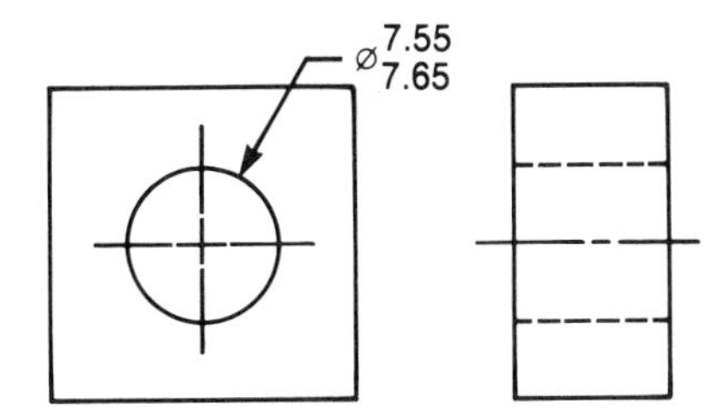

Figure 9.7 In tolerancing of an internal feature, the minimum limit is listed above the maximum limit (if the maximum material system is used).

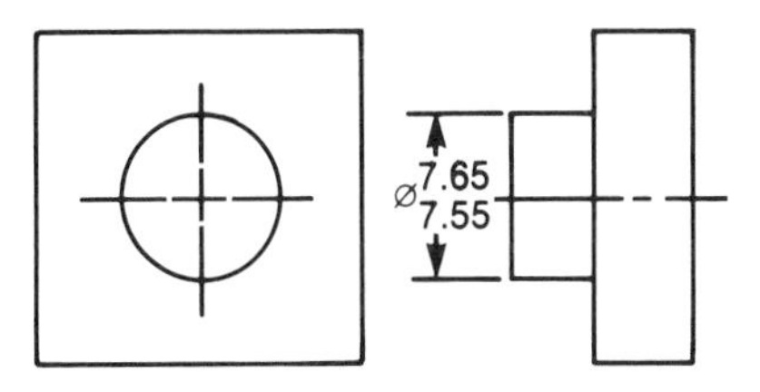

Figure 9.8 In tolerancing of an external feature, the maximum limit is listed above the minimum limit (if the maximum material system is used).

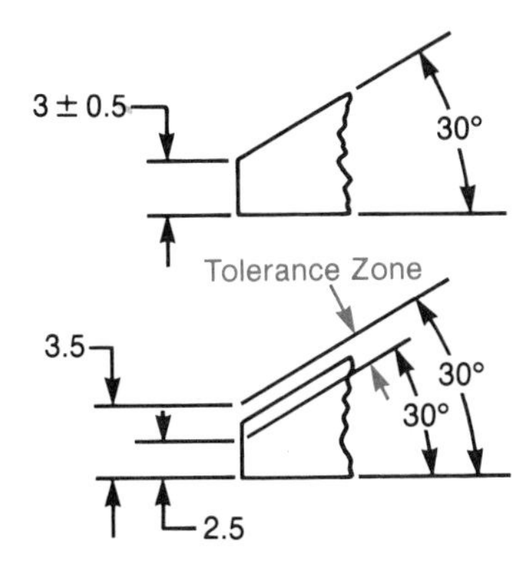

Figure 9.9 Tolerance zone for a part with a linear dimension that includes a tolerance specification and an angular dimension that does not.

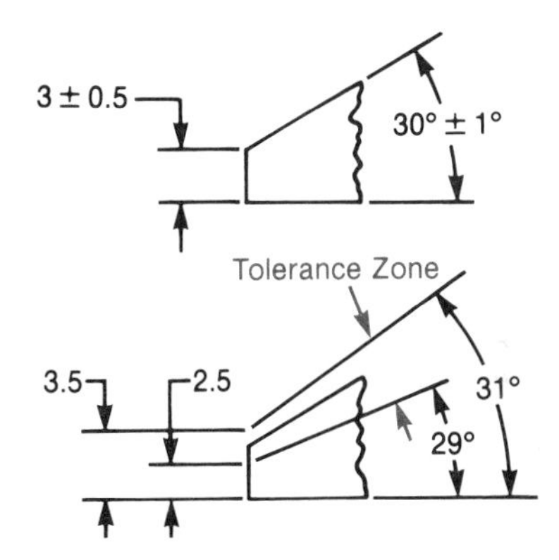

Figure 9.10 Tolerance zone for a part with tolerance specifications for both its linear and its angular dimensions.

1. Figure 9.9 presents the tolerance zone (of parallel boundaries) for possible variation in the size of a part that has two basic dimensions: a linear dimension with a tolerance specification and an angular dimension with no tolerance specification.

2. Figure 9.10 presents the tolerance for a part that has tolerance specifications for both its linear and angular dimensions.

3. Figure 9.11 presents a set of views in which a hole is described with a tolerance specification for its diameter (a). In the two possible corresponding tolerance zones for this hole (b and c), notice that the given tolerance specification simply means that the minimum and maximum distances—measured perpendicularly with respect to the axis of the hole—between points on the inner surface of the hole must lie between the tolerance limits. Therefore, either a conical form (b) or a noncylindrical form (c) is acceptable if the tolerance specifications are met.

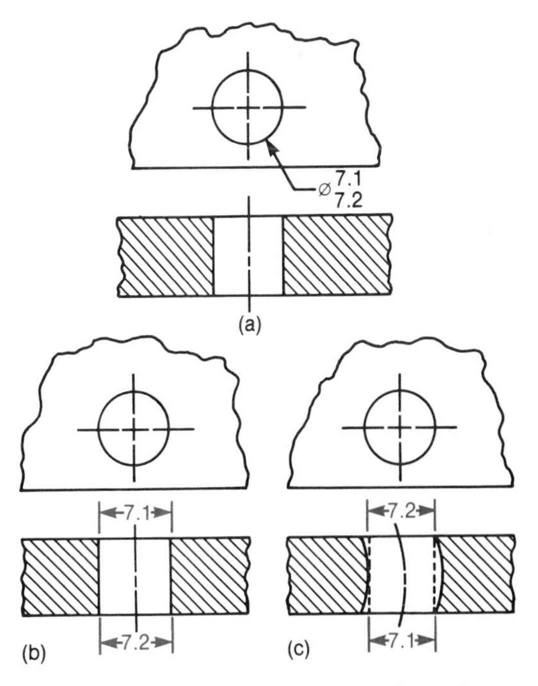

Figure 9.11 Tolerance zone for a hole with a tolerance specification for its diameter (a) could accommodate either a conical form (b) or a noncylindrical form (c).

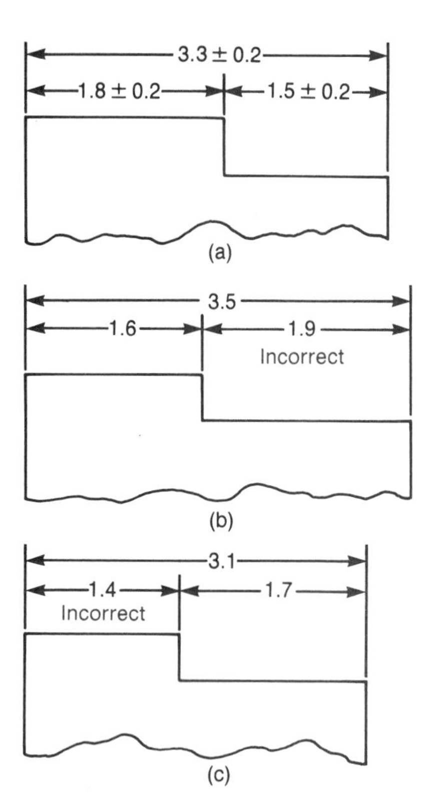

Figure 9.12 Repetition of basic dimensions and tolerance specifications (a) violates Rule 2 and can produce two possible incorrectly fabricated (out-of-tolerance) parts (b and c).

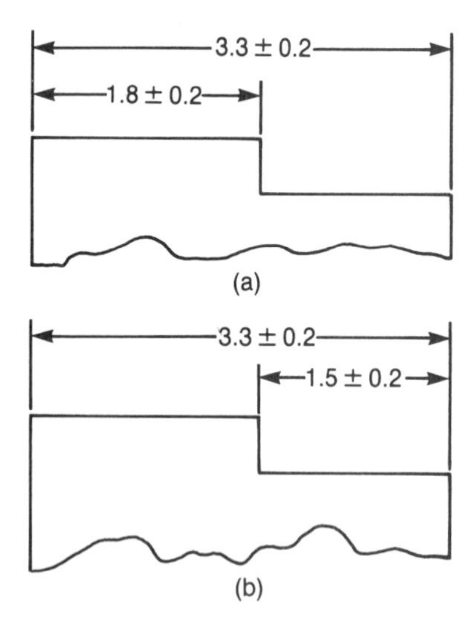

Figure 9.13 Properly dimensioned views of the incorrectly dimensioned layout in Figure 9.12. In one possible view (a), the left-side feature has a critical tolerance specification. In another possible view (b), the right-side feature has a critical tolerance specification.

Learning Check

Why are tolerance specifications added to a drawing?

Answer Tolerance specifications are added to a drawing in order to clearly identify the allowed ranges of variation in the size and shape of interchangeable and/or mating parts.

9.4 Redundant Dimensions and Out-of-Tolerance Parts

In order to emphasize the importance of Rule 2 for dimensioning in Chapter 8 (which states that a basic dimension should not be repeated in a set of orthographic views), consider the dimensioning layout shown in Figure 9.12(a), in which basic dimensioning information is repeated. Notice the two possible results, both incorrect (b and c). In each case the machinist has obeyed the instructions in the drawing by manufacturing the part within two of the three given tolerance specifications. However, the third tolerance specification is violated in these fabricated parts. If the third tolerance specification is a crucial value that must be achieved in the actual part, then either of the possible results shown here is unacceptable; these parts will not mate properly with other parts in the final product. The machinist is not responsible for this type of out-of-tolerance error; the designer is responsible due to the repetition of basic dimensioning information in the orthographic drawing. Figure 9.13 presents two possible ways of correctly dimensioning this part.

9.5 Tolerance Accumulation

A major consideration in the proper dimensioning and tolerance specification of a part is the possibility of **tolerance accumulation,** (Figure 9.14) which can occur as a result of **base line dimensioning** (a), a very common method for specifying the location of various features in a part. In this method, dimensions are specified with respect to a common base line, which enhances reading clarity in a drawing but also allows the possible variations in the basic dimensions to accumulate in a given direction (tolerance accumulation).

In **direct dimensioning** (b), less tolerance accumulation can occur. For example, in the figure, direct dimensioning specifies the maximum allowable tolerance for the distance between the reference positions X and Y to be equal to ± 0.2. In contrast, base line dimensioning will allow a possible tolerance accumulation of ± 0.4 between these positions.

Unless tolerance accumulation is not a significant concern (true in some cases), **chain line dimensioning** (c), sometimes used for economy in design layout, should be avoided since it results in the greatest tolerance accumulation (± 0.8 between positions X and Y in the figure).

9.6 Basic Size Systems

Two basic size systems may be used for mating holes and shafts: the **basic hole system** and the **basic shaft system.** Figure 9.15 shows a side view of two parts—a rectangular prism with a hole and a mating shaft. A clearance fit will be produced from the given tolerance specification. If the basic hole system is used, the smallest dimension for the hole (2.30 units in this case) is the basic (fundamental) limit dimension for the mating parts (a). The remaining values for the parts are then based upon this chosen dimension and specified in accordance with the desired limits of tolerance and allowance.

If the basic shaft system is used, the largest dimension for the shaft (equal to 2.32 units in this case) acts as the fundamental limit dimension for the mating parts (b).

The basic hole system is preferred because of the following reasons:

- Hole cutters (drills, reamers, and so forth) are available in standard diametral sizes.
- It is easier to machine external features to given specifications than it is to machine holes.
- Standard gauges may be used to inspect the accuracy of holes.

The basic shaft system should be used only if the shaft is to be used as a support for different parts.

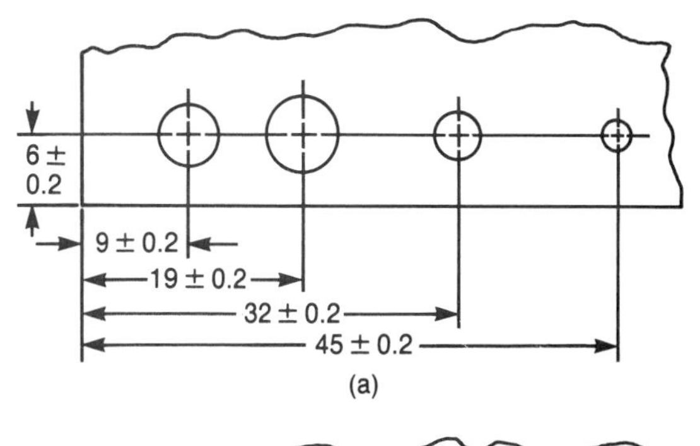

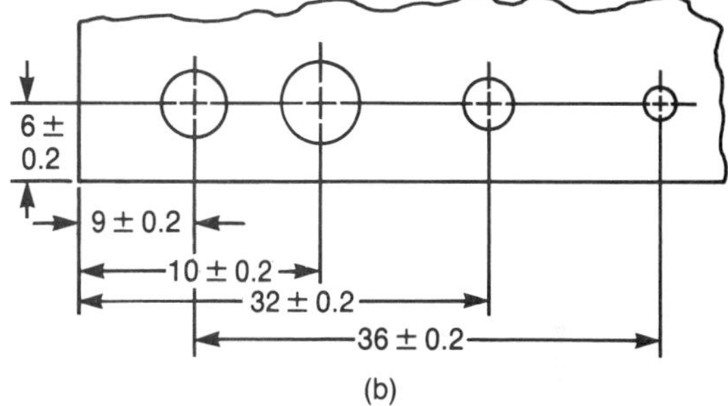

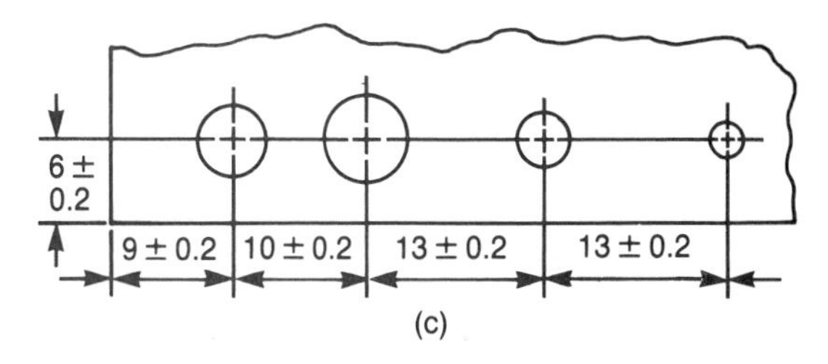

Figure 9.14 Tolerance accumulation (between the reference points X and Y) can amount to ± 0.4 in base line dimensioning (a), to ± 0.2 in direct dimensioning (b), and to ± 0.8 in chain line dimensioning in this example.

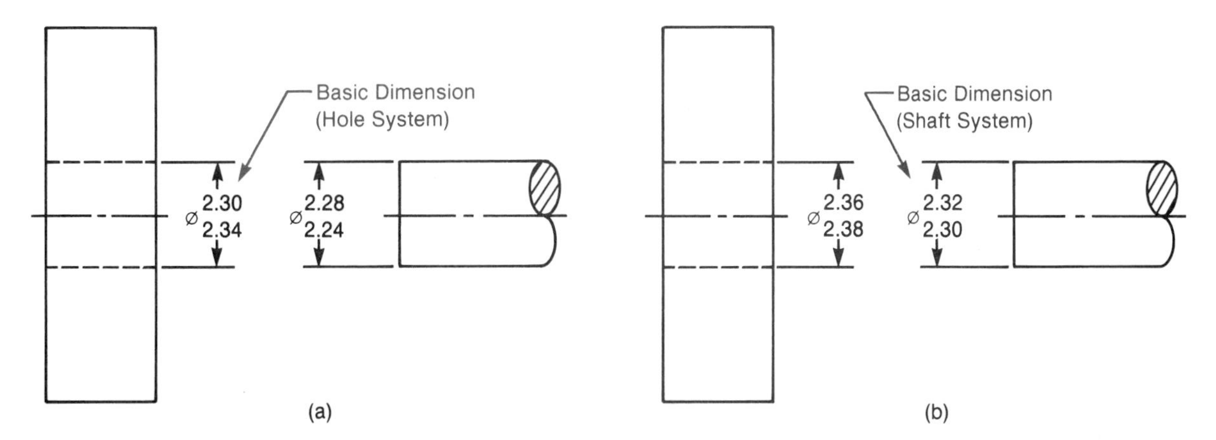

Figure 9.15 Two basic size systems may be used for mating holes and shafts, the basic hole system (a) and the basic shaft system (b).

In Retrospect

- A fabricated part should not be expected to correspond exactly in its actual size to the theoretical basic dimensions specified in an orthographic description of the part.
- Tolerances are associated with dimensions in a drawing in order to specify the range of acceptable variation in the actual size of the part.
- Tolerances need to be specified for mating parts. Any difference between the sizes of such mating parts is called the clearance for these parts.
- The range of tightness, or the fit, between mating parts can be of several types: a clearance fit, which corresponds to a positive clearance; an interference fit, which corresponds to a negative clearance, a transitional fit, in which either a clearance fit or an interference fit is allowed; or a line fit, in which the upper limit of the shaft is equal to the lower limit of the hole.
- Bilateral tolerancing specifies a symmetric range of acceptable variation on either side of the basic dimension in terms of plus and minus values.
- Unilateral tolerancing uses a single direction of measurement (entirely positive or entirely negative) to specify the range of acceptable variation.
- Limit dimensioning specifies the range of acceptable variation in the actual size of a feature in terms of the minimum and maximum values of the dimension.
- The smaller limit for a dimension of an internal feature should be given before the larger limit; the larger limit for the dimension of an external feature should be given before the smaller limit (if the maximum material system is used).
- The tolerance zone that corresponds to a given tolerance specification should be carefully considered during the design of a part.
- Two basic size systems may be used for mating parts: the basic hole system and the basic shaft system. The basic hole system is preferable in view of standardized hole cutters, measurement gauges, and machining processes.
- Tolerance accumulation must be considered during the dimensioning and tolerance specification of a part. Direct dimensioning (in which the required distance between two points is directly specified on a drawing) results in minimal tolerance accumulation. Base line dimensioning allows moderate tolerance accumulation to occur. Chain line dimensioning should be avoided since it produces maximum tolerance accumulation.

Exercises

Your instructor will specify the appropriate scale for each exercise and also whether bilateral tolerancing, unilateral tolerancing, or limit dimensioning should be used. *Unless otherwise specified by your instructor,* all dimensions given in millimeters should have tolerances equivalent to bilateral values of 0.2 (that is, ± 0.2 mm.); for dimensions given in inches, tolerances should be ± 0.01 in.

Each exercise refers to a specific exercise in a preceding chapter. In each case draw an appropriate set of the best and necessary views of the object, including any *auxiliary views* and/or *section views*—in accordance with the directions of your instructor. Properly dimension the resulting set of views using the appropriate system of units as specified by your instructor. Include the appropriate tolerance specifications in the finished drawing.

EX9.1. See EX4.7.

EX9.2. See EX4.8.

EX9.3. See EX4.9.

EX9.4. See EX4.10.

EX9.5. See EX4.11.

EX9.6. See EX4.12.

EX9.7. See EX4.14.

EX9.8. See EX4.15.

EX9.9. See EX4.19.

EX9.10. See EX4.20.

EX9.11. See EX4.21.

EX9.12. See EX4.22.

EX9.13. See EX4.23.

COMPUTER GRAPHICS IN ACTION

2D Computer Graphics: Automatic Tolerancing

Some menu-driven computer graphics software packages allow the user to specify the type of tolerancing to be used in a drawing. Bilateral tolerancing or limit dimensioning can be selected by the user, after which the appropriate values are inserted *automatically* in the drawing during the dimensioning phase. Once again, such software allows the engineer to construct or modify a drawing with both speed and accuracy.

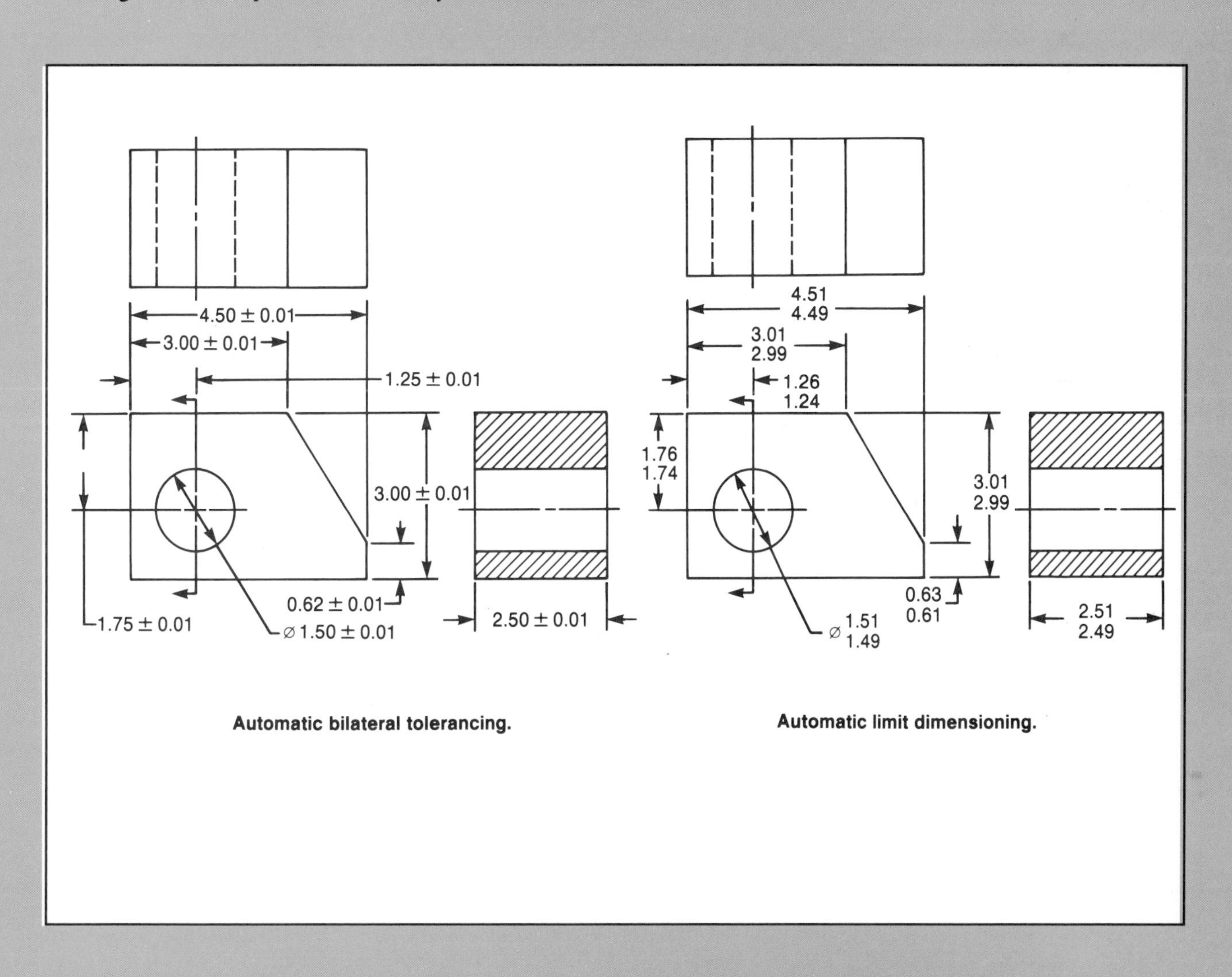

Automatic bilateral tolerancing.

Automatic limit dimensioning.

EX9.14. See EX4.26.

EX9.15. See EX4.28.

EX9.16. See EX4.31.

EX9.17. See EX4.32.

EX9.18. See EX4.33.

EX9.19. See EX4.35.

EX9.20. See EX4.36.

EX9.21. See EX4.37.

EX9.22. See EX4.38.

EX9.23. See EX4.39.

EX9.24. See EX4.40.

EX9.25. See EX4.41.

EX9.26. See EX4.43.

EX9.27. See EX4.44.

EX9.28. See EX4.45.

EX9.29. See EX4.47.

EX9.30. See EX4.48.

EX9.31. See EX4.49.

EX9.32. See EX4.50.

EX9.33. See EX4.51.

EX9.34. See EX4.52.

EX9.35. See EX4.53.

EX9.36. See EX4.54.

EX9.37. See EX4.55.

EX9.38. See EX4.56.

EX9.39. See EX4.57.

EX9.40. See EX4.58.

EX9.41. See EX4.59.

EX9.42. See EX5.1.

EX9.43. See EX5.2.

EX9.44. See EX5.3.

EX9.45. See EX5.4.

EX9.46. See EX5.5.

EX9.47. See EX5.6.

EX9.48. See EX5.7.

EX9.49. See EX5.8.

EX9.50. See EX5.9.

EX9.51. See EX6.1.

EX9.52. See EX6.2.

EX9.53. See EX6.3.

EX9.54. See EX6.4.

EX9.55. See EX8.1.

EX9.56. See EX8.2.

EX9.57. See EX8.3.

EX9.58. See EX8.4.

EX9.59. See EX8.5.

EX9.60. See EX8.6.

EX9.61. See EX8.7.

EX9.62. See EX8.8.

EX9.63. See EX8.9.

EX9.64. See EX8.10.

EX9.65. See EX8.11.

EX9.66. See EX8.12.

10 Positional and Geometric Form Tolerancing

Preview We focused upon simple size tolerancing in Chapter 9. As its name implies, size tolerancing allows us to clearly specify the acceptable range of variation in the *size* of a fabricated part and its features. Such tolerancing can be expected to result in actual (feature) sizes that range from the given minimum value to the given maximum value. Similarly, a fabricated part may also vary from its orthographic description in the location and geometric shape of a given feature. As a result, we must now consider positional tolerancing and geometric form tolerancing.

Positional tolerancing gives the acceptable range of variation in the location of a feature with respect to a **datum,** or reference location (point, axis, or plane). **Geometric form tolerancing** gives the acceptable range of variation in the shape of a feature. For example, it allows specification of the flatness of a surface, the perpendicularity between two surfaces, or the cylindricity of a shaft.

Positional tolerancing and geometric form tolerancing, together with standard dimensioning and size tolerancing practices, then provide us with the means to unambiguously describe the requirements for a part. Figure 10.1 presents the **symbology** that allows us to specify positional (locational) and geometric form requirements in standardized form (for complete information on these standards, see ANSI Y14.5M-1982 where ANSI is an abbreviation for the American National Standards Institute). Figure 10.2 demonstrates the use of such symbology in a set of orthographic views.

Learning Objectives

Upon completion of this chapter, the reader should be able to:

- Apply both positional tolerancing symbology and geometric form tolerancing symbology in engineering drawings in order to properly specify locational and shape requirements for a part.
- Recognize which geometric form tolerances require datum planes or axes.
- Explain why positional tolerance symbols require datum planes and axes.
- Specify datum planes with datum identification symbols.
- Justify the need for symbology in engineering drawings (to minimize the need for notes while optimizing both standardization and graphical communication in engineering).
- Interpret positional tolerances and geometric form tolerances in engineering drawings.

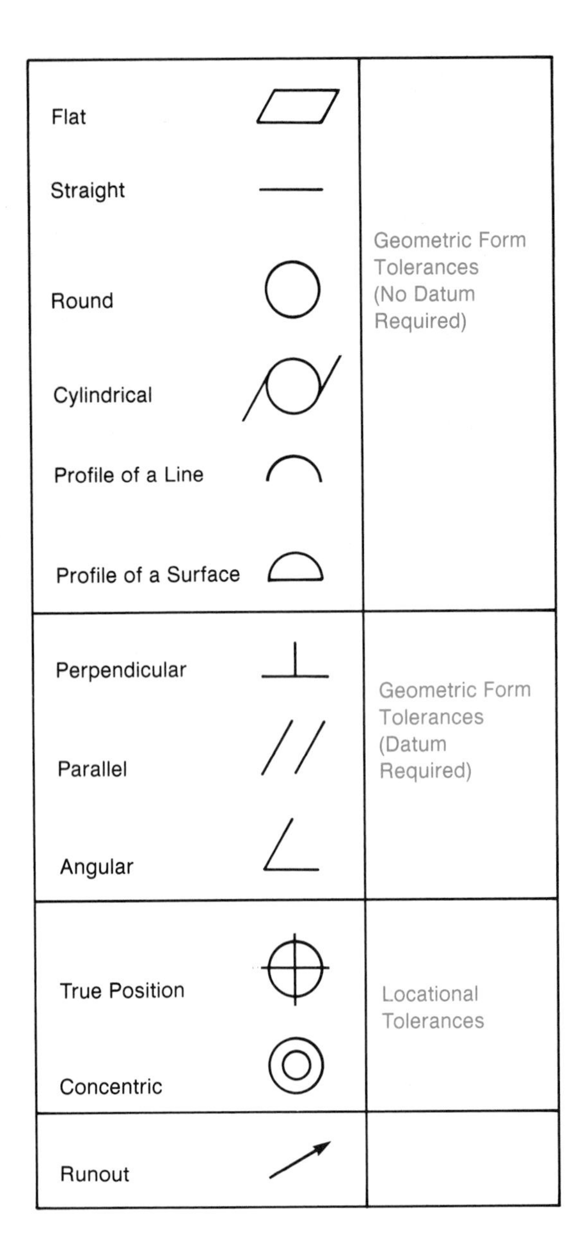

Figure 10.1 Symbology for geometric form tolerances and positional (locational) tolerances.

10.1 Datum Identifying Symbols

Positional, or **locational, tolerances,** together with some geometric form tolerances, require a theoretically exact reference point, line (axis), or plane from which specifications may be given (see Figure 10.1). Such a reference location is known as a **datum.** Only a surface that is finished (machined for accuracy), easily located, and measurable should be chosen as a datum; it is specified on an engineering drawing by a **datum identifying symbol** (Figure 10.3).

Any letter of the alphabet may be used as a datum symbol *except I, O, and Q,* which could be confused with other symbols and numerals. The letter should be enclosed in an identification block about 12 mm. long and about 6 mm. high. The block is attached to an extension line projecting from the reference surface (Figure 10.3).

10.2 Flatness

The tolerance specification symbol for surface **flatness** is shown in Figure 10.4, along with its **interpretation.** The symbol is attached to an extension line projecting from the surface under consideration. It defines a **tolerance zone** between given parallel planes—within which must lie every point on the surface. Notice that the tolerance zone is defined in terms of the distance between these parallel planes; this distance is the acceptable range of variation for the flatness of the surface. No datum is needed for the flatness symbol since it is attached directly to the extension line that is used to indicate the appropriate surface.

10.3 Straightness

The tolerance specification symbol for **straightness** is shown in Figure 10.5, together with its interpretation. Notice that the symbol is applied to a nonplanar surface, so the flatness symbol would not be applicable. The symbol defines a tolerance zone for the surface elements. A leader is used to directly identify the surface to which the symbol applies, and no datum is required.

10.4 Roundness

The tolerance specification symbol for **roundness** is shown in Figure 10.6, together with its interpretation. The symbol indicates that the boundary of a cross section of the feature (perpendicular to the axial direction) must remain between two concentric circles that define the given tolerance zone. A leader is used to directly identify the surface to which the symbol applies, and no datum is required.

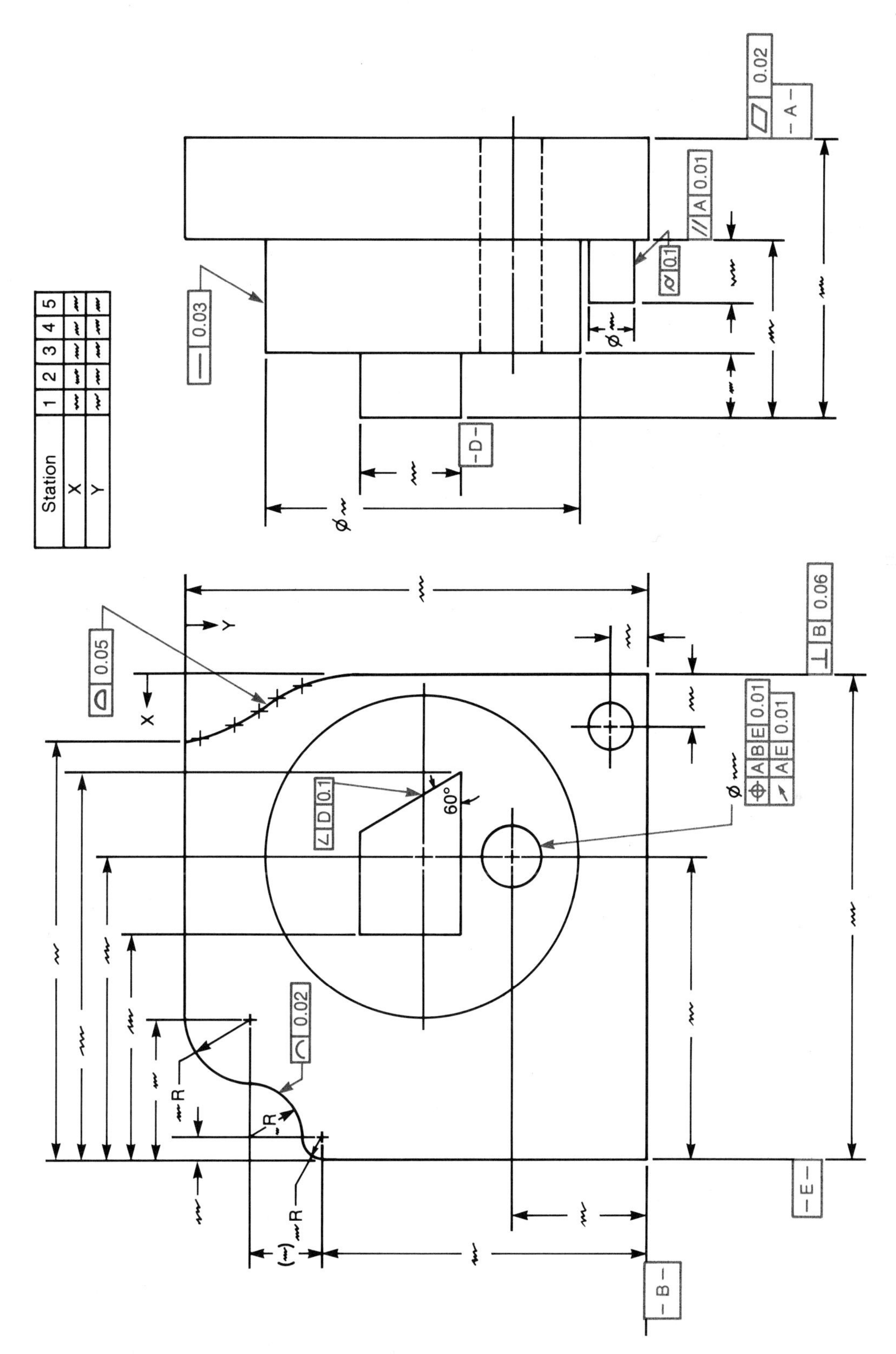

Figure 10.2 Tolerancing symbology in an engineering drawing.

Highlight

James Watt and the Steam Engine

James Watt designed and patented (in 1769) a revolutionary change in the steam engine. In the earlier Newcomen engine, the cylinder was cooled and then reheated during each stroke cycle. Watt's modification involved the use of a condenser that doubled the fuel efficiency of the engine by eliminating the need to cool and reheat the cylinder. Watt's design required accurate boring for the engine cylinders, which Wilkonson made possible by developing a boring mill with which concentric cylinders of the necessary accuracy could be produced.

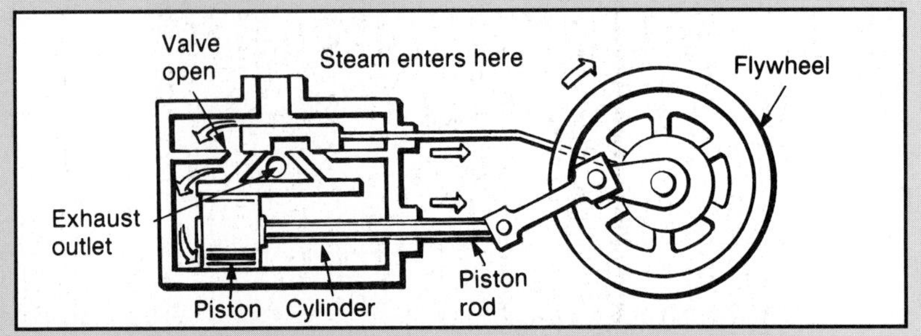

Beginning of the rightward power stroke of a double-acting steam engine.

Watt's design of the steam engine illustrates the importance of proper *geometric form tolerancing* in the design, manufacture, and application of machined parts within a system. Because of its portability and efficiency, Watt's engine provided impetus for the development of factories, transportation, and the industrialization of American society.

Sources: Stephan Konz, Work Design: Industrial Ergonomics, 2d ed., *(Columbus, Ohio, Grid Publishing, Inc., 1983) and Isaac Asimov,* Asimov's Biographical Encyclopedia of Science and Technology, 2d revised ed., *(Garden City, New York, Doubleday & Company, Inc., 1982).*

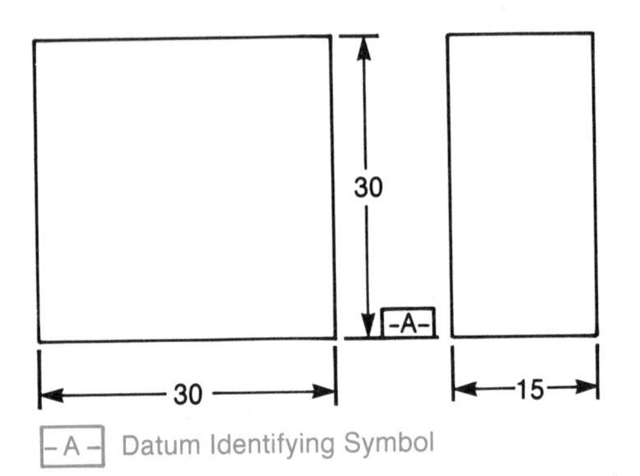

Figure 10.3 A datum is specified on an engineering drawing by a datum identifying symbol. Any letter may be used except I, O, and Q.

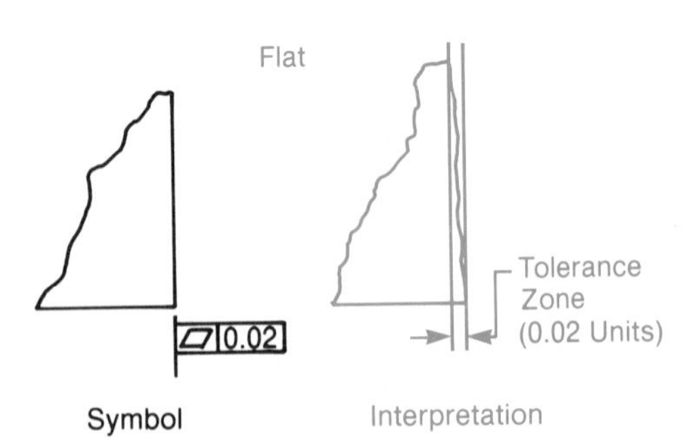

Figure 10.4 Geometric form tolerance for surface flatness.

10.5 Cylindricity

The tolerance specification symbol for **cylindricity** is shown in Figure 10.7, together with its interpretation. The symbol indicates that the feature has a surface that remains within a tolerance zone bounded by two concentric cylinders. As a result, the surface of a cylinder must meet a tolerance of

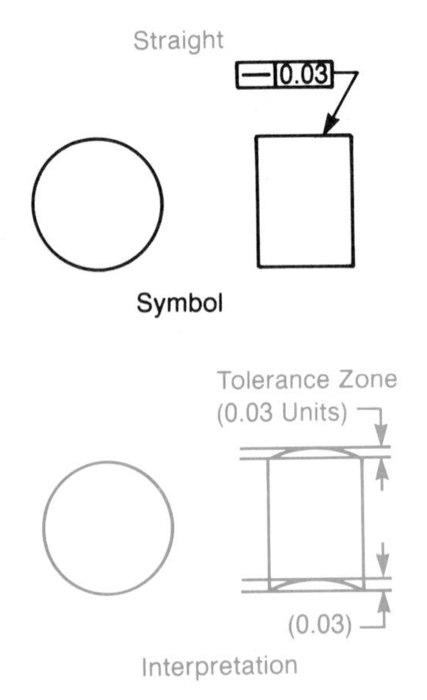

Figure 10.5 Geometric form tolerance for straightness.

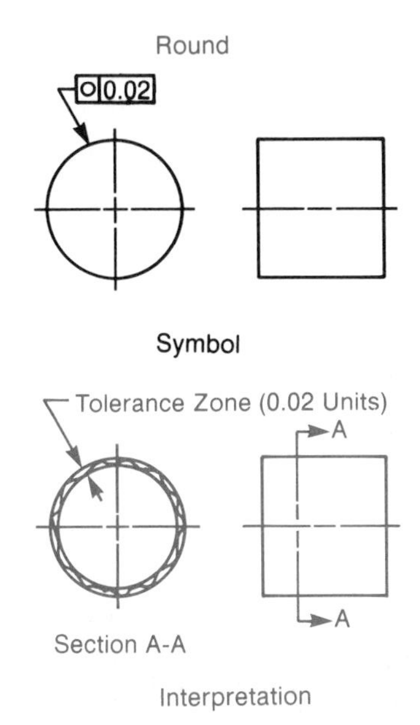

Figure 10.6 Geometric form tolerance for roundness.

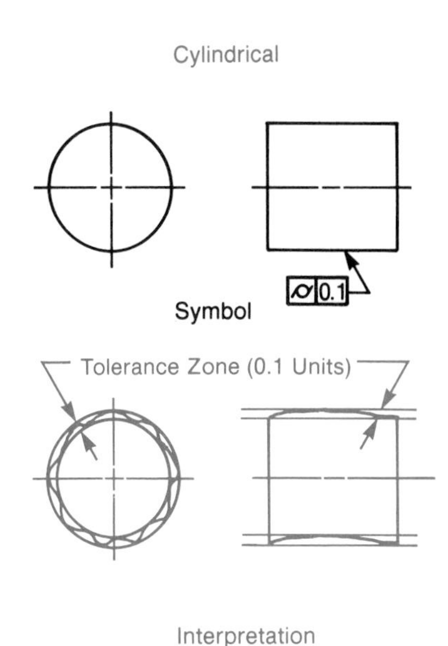

Figure 10.7 Geometric form tolerance for cylindricity.

form that is equivalent to simultaneous tolerances of straightness and roundness. No datum is required for this symbol.

10.6 Profiles (Lines and Surfaces)

The tolerance specification symbols for the **profile** of (1) a line and (2) a surface (representing the cross-sectional profile of a surface) are shown in Figures 10.8 and 10.9, respectively, together with their interpretations. The symbols indicate that the curving line or curving surface (defined by either a set of tangent arcs or by the coordinates of various points along an irregular curve) under specification must not vary from within the corresponding curving tolerance zone. No datum is required for these symbols.

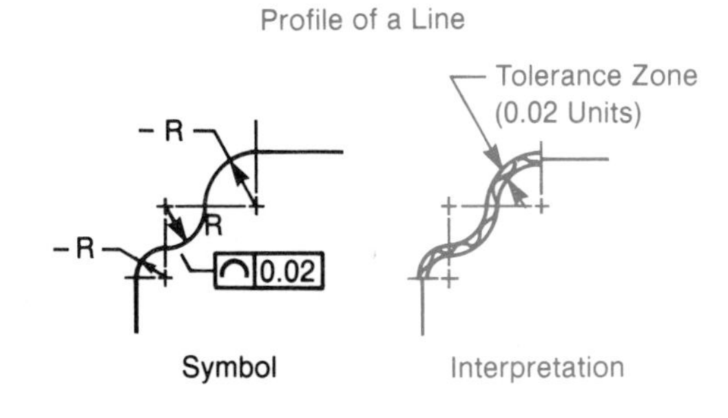

Figure 10.8 Geometric form tolerance for the profile of a line (outline of a surface cross section).

10.7 Perpendicularity

The tolerance specification symbol for **perpendicularity** between two surfaces is shown in Figure 10.10, together with its interpretation. Since the perpendicularity of a surface is defined with respect to another surface or reference plane, *a datum is required.* The symbol for perpendicularity is given first in the tolerance specification block, followed by the datum identification symbol and the appropriate tolerance value. Two planes—perpendicular to the datum surface and separated by a distance equal to the tolerance value—define the tolerance zone for the surface that is to be perpendicular to the datum.

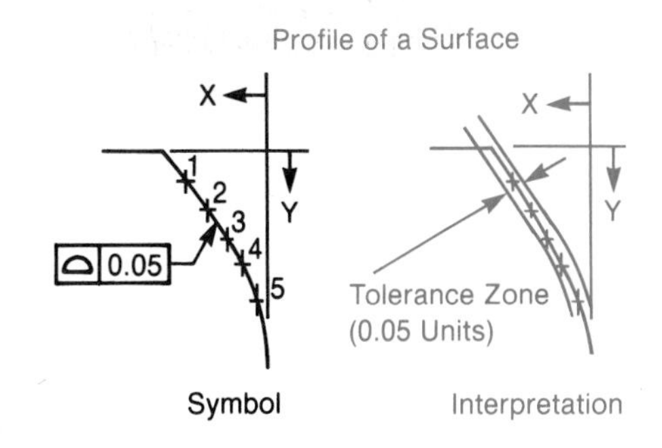

Figure 10.9 Geometric form tolerance for the profile of a surface.

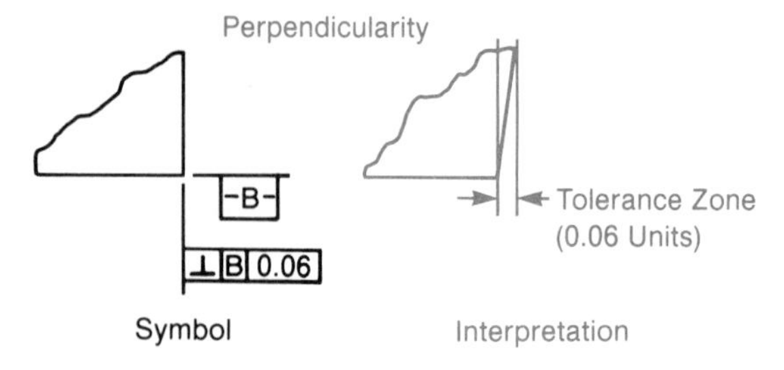

Figure 10.10 Geometric form tolerance for perpendicularity.

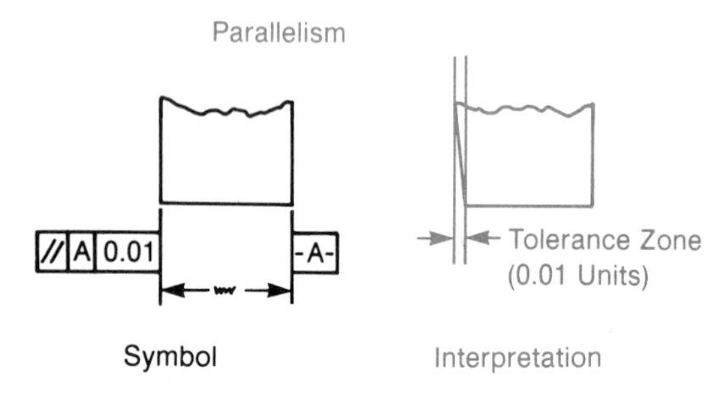

Figure 10.11 Geometric form tolerance for parallelism.

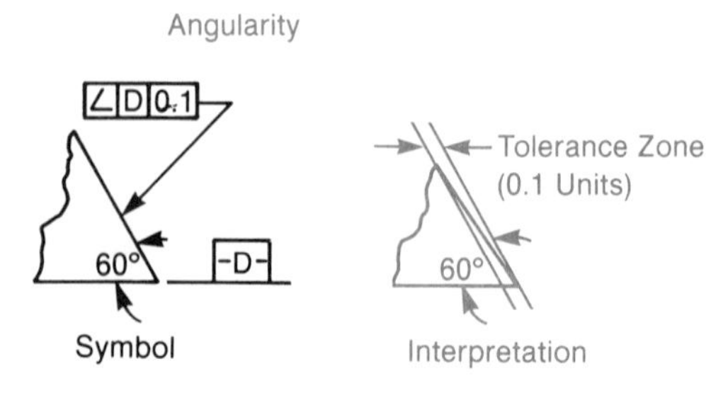

Figure 10.12 Geometric form tolerancing for angularity.

10.8 Parallelism and Angularity

The tolerance specification symbol for **parallelism** between two surfaces is shown in Figure 10.11, together with its interpretation. The symbol for **angularity** between two surfaces is shown in Figure 10.12, together with its interpretation. Both symbols require a reference location, so a datum is required. Two planes define the tolerance zone for the surface under consideration with respect to the given datum.

10.9 True Position

The locational tolerance symbol for the **true position** of a feature is shown in Figure 10.13, together with its interpretation. In this figure the true position of the feature is defined with respect to three datum planes (primary, secondary, and tertiary) that are mutually perpendicular. True position control symbols may use two or three datum planes in order to define the location of a feature. Three datum planes allow one to specify that greater accuracy is required in the fabrication of a part.

The symbol defines a circular tolerance zone for the location of the centerpoint for the feature; the *diameter* of this tolerance zone is equal to the tolerance value. (Allowed variation with respect to the third datum plane is not shown.)

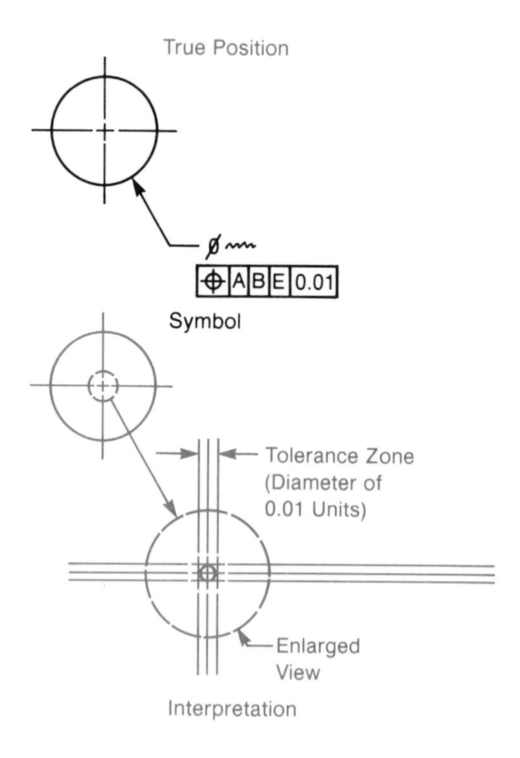

Figure 10.13 Positional tolerance for true position.

An enlarged view is shown in order to emphasize that the centerlines for the position of the feature must lie within the *circular* tolerance zone, not simply between the rectangular boundaries formed by sets of parallel lines tangent to the circular zone. The circular tolerance zone is then more accurate than rectangular coordinate tolerances of identical specification values, since the rectangular zone is larger in size than the circular zone (see the enlarged view). The basic dimensions for the location of a centerpoint may be contained within a *box* in the dimension line in order to emphasize that these values are indeed basic dimensions.

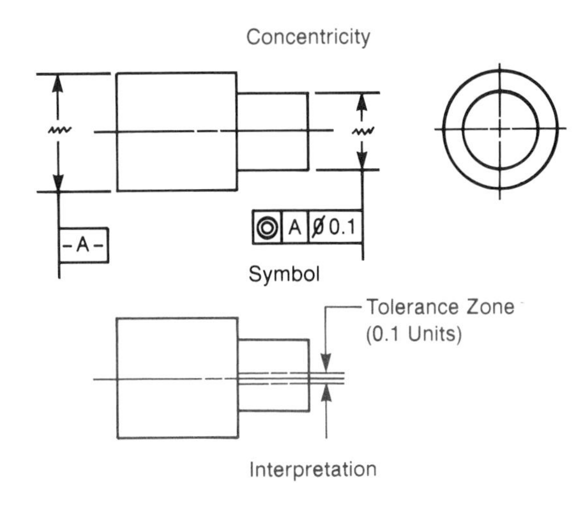

Figure 10.14 Positional tolerance for concentricity.

10.10 Concentricity

The locational tolerance symbol for the **concentricity** of two surfaces of revolution is shown in Figure 10.14, together with its interpretation. These surfaces must share a common axis in order to be concentric. A datum is therefore required in order to specify the concentricity of a surface with respect to another (datum) surface.

10.11 Runout

The tolerance specification symbol for **runout** is shown in Figure 10.15, together with its interpretation. It is used to control the relative variation between two or more features in terms of *perpendicularity, concentricity, parallelism,* and *alignment.* For example, in the interpretation, the datum plane, A, is a mounting surface for the feature with a datum axis, B, about which the feature is to be rotated. The variation in the surface associated with the runout symbol must remain within the resultant tolerance zone.

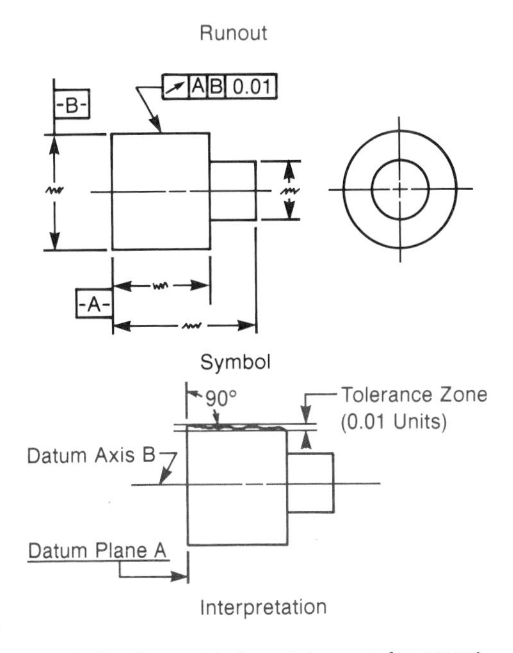

Figure 10.15 Geometric form tolerance for runout.

10.12 Working Drawings

Working drawings allow an engineering part or system to be fabricated (see Figures 1.1 and 1.2 in Chapter 1, and Appendix N). These drawings are constructed by using the various graphic techniques and principles reviewed in preceding chapters. They may be either assembly and subassembly drawings or detail drawings. Examples of working drawings are shown in Figures 10.16 and 10.17.

10.12.1 Assembly and Subassembly Drawings

In **assembly drawings,** the parts of an engineering system are shown in assembled form. Each part is identified, together with its relative location with respect to the other components of the system. **Subassembly drawings** are used for very complex systems. A master assembly drawing is broken into specific subsystems for greater clarity. Internal features may be shown by sectioning an assembly drawing.

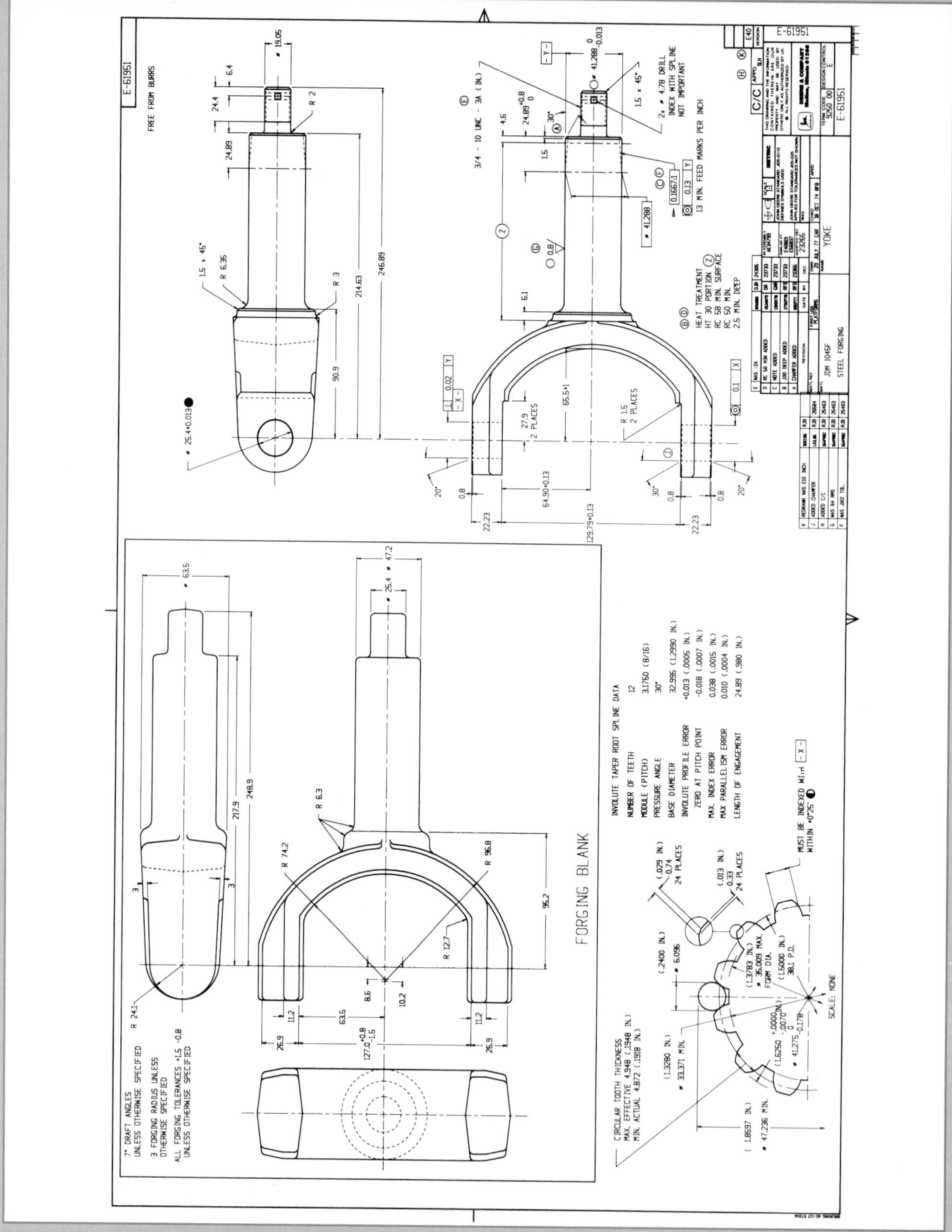

Figure 10.16 An example of a professional working drawing. Note the detailed information that is provided for the manufacture of this part, including dimensioning and tolerancing data. (Courtesy of Deere & Company.)

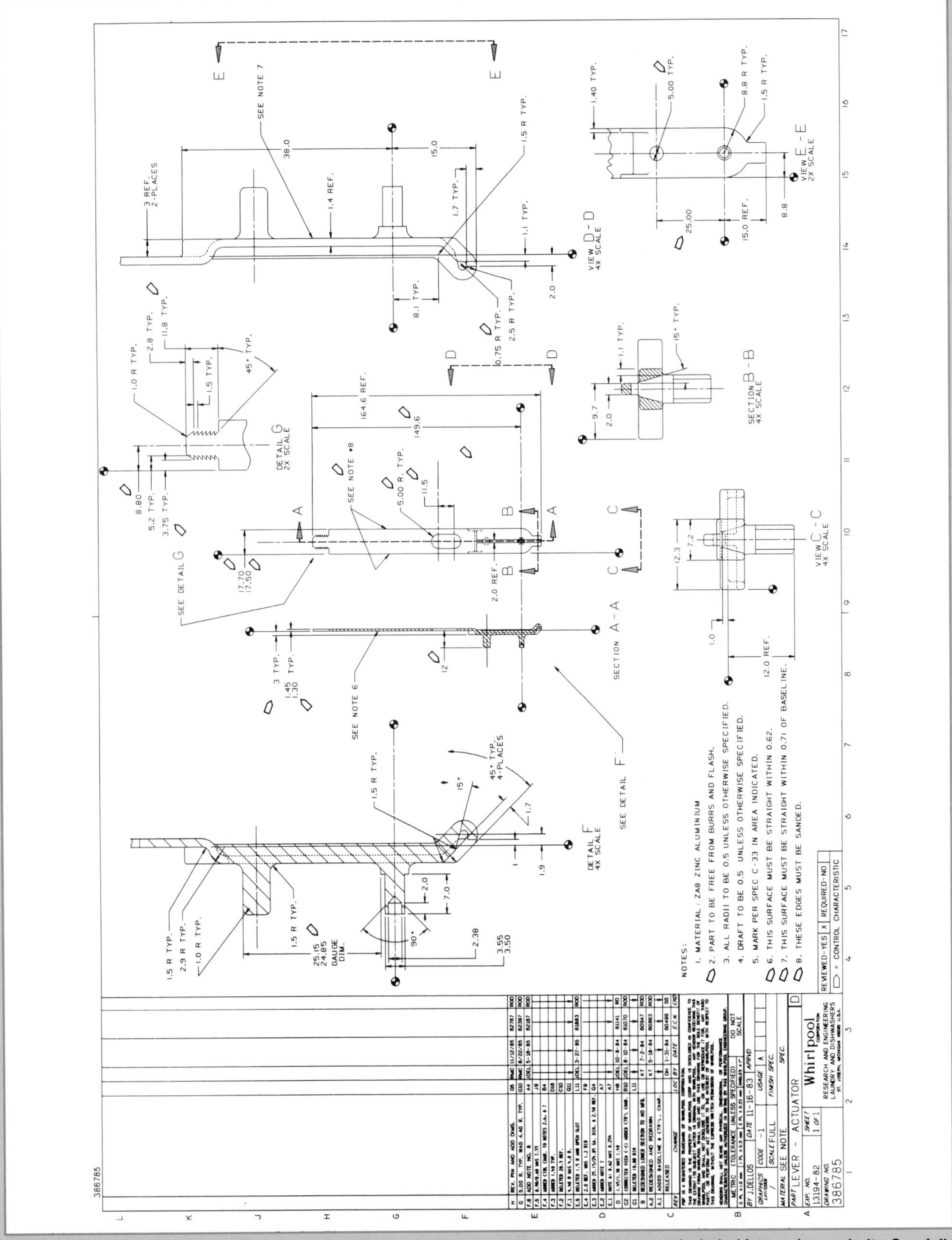

Figure 10.17 A professional working drawing in which sectional and enlarged detail views are included for maximum clarity. Carefully read and coordinate the various views that are shown in this drawing. (Drawing courtesy of Whirlpool Corporation, Benton Harbor, MI.)

Highlight

The Three-Datum Concept

Geometric form tolerancing involves the application of reference locations, or datums, which are used to specify distances. The three-plane datum system consists of three mutually perpendicular planes: the *primary* datum plane, the *secondary* datum plane, and the *tertiary* datum plane, as illustrated here. An object described with the use of these datum planes is assumed to be in contact with *at least three points* on the primary datum plane (a); otherwise, the object would not remain stable with respect to this plane (as also illustrated). Similarly, the position of the object with respect to the secondary datum plane is fixed if the object is in contact with at least two points on this plane (b). Finally, the object's location in space can be completely specified in terms of its relative position with respect to the three datum planes if it is assumed to also be in contact with *at least one point* on the tertiary datum plane (c). Lines (axes) and cylindrical planes can be used as datums (instead of a set of mutually perpendicular planes) in such tolerancing applications as concentricity and cylindricity.

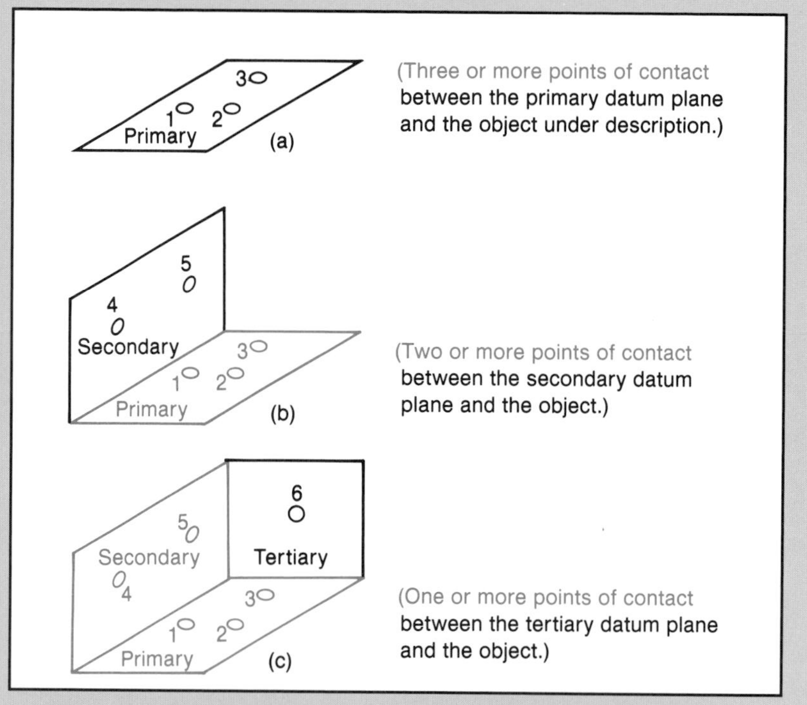

The three-plane datum system consists of three mutually perpendicular planes: the primary datum plane (a), secondary datum plane (b), and tertiary datum plane (c).

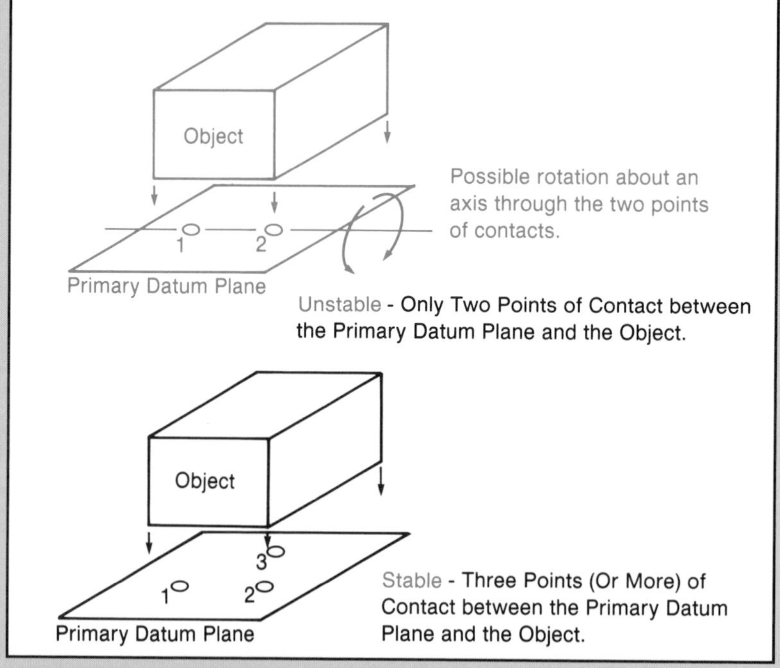

Three or more points of contact between the primary datum plane and the object are needed for stability.

Dimensions may be entirely omitted or only the principal (overall) dimensions may be retained in an assembly drawing. Hidden linework should be omitted unless it is necessary to properly identify specific components and their relative positions within the assembly. Fasteners are not sectioned in an assembly drawing, and clearances are not shown.

A *list of parts* is included in an assembly drawing in which each component within the assembly is identified by name, together with the number of such components required for a single assembly. (For an example, see Figure 1.1 in Chapter 1.)

Assembly drawings may be presented as exploded pictorials in order to clearly describe the relative positions of the components (see Figure 10.18). Figure 6.25 in Chapter 6 shows an exploded assembly pictorial, together with sectioned assembly views of a system.

10.12.2 Detail Drawings

Detail drawings describe each part of an assembly, one part per detail drawing. (Standard fasteners and other stock items are not shown in detail drawings; these standard components are shown only in the assembly or subassembly drawings.) Detail drawings include all of the information required to fabricate the parts under consideration, including dimensions, the materials to be used, the number of each part needed for a single assembly, the part name, and any identification number for the part (this identification number should also be shown on the assembly or subassembly drawings in which the component is shown). The assembly drawing with which the detail drawing corresponds should also be identified, together with the name(s) of the drafter(s) and any other persons who have been involved in the development of the drawing(s).

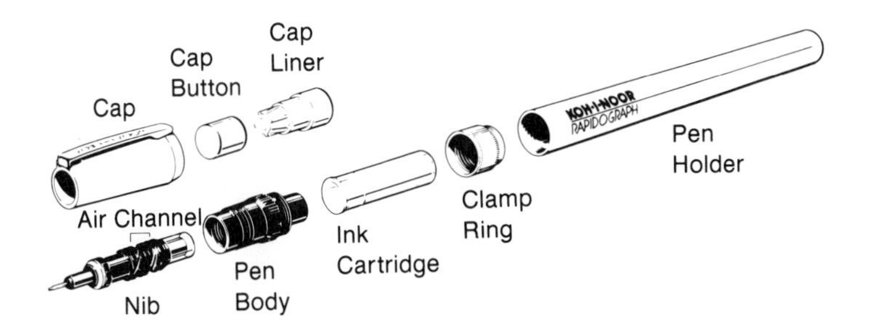

Figure 10.18 An exploded assembly view. (Illustration courtesy of Koh-I-Noor Rapidograph, Inc., Bloomsbury, NJ.)

ENGINEERING IN ACTION

Thinner Is Better for Aluminum-Can Design

Top: Elimination of the can opener through the development of "tear tops" has been a major convenience factor of the aluminum can. Top thickness can be reduced by profile changes. Newer cans will have more "necking" of the can body. This will save metal by reducing top diameter.

Labeling: Aluminum is naturally bright so transparent inks can be used. Because the aluminum can has no seam, it offers more decorating options.

Lining: The can's internal coating determines what it is to contain. There is a "conversion" coating to stop oxidation of the aluminum and an organic coating to keep the contents from reacting with the can walls. Heavier coatings are needed for soft drinks because of their composition.

Forming: Aluminum cans are formed by a "draw and iron" process. Water-based lubricant is of great importance during can fabrication.

Materials: Body is made from 0.014-in. 3004 H19 aluminum, which allows for a final wall thickness of approximately 0.0048 in. The top is of 5182 H19, approximately 0.013 in. thick.

Bottom: To keep costs down, can manufacturers use the thinnest practical gage aluminum. However, the can must survive pressures up to 95 psi without buckling. A concave shape prevents bottom deformation. More recent lightweight designs have expandable bottoms with dimples to maintain the can's stability.

Going to the Top

The aluminum can has reached its dominant position in the beer and soft-drink container market through a series of material, design, and manufacturing innovations. Among the major achievements:

- Aluminum rigid container sheet (RCS) that can be rolled economically to the needed thickness. A major innovation in the area was the development of lubricants that facilitate the high speed manufacture of thin sheet.
- Metal working techniques that form the can body as a single piece. This forming method is called "draw and iron." The metal is first drawn by stages into a cup, and then redrawn and ironed by a series of progressively smaller iron rings until the needed wall height is achieved.
- Designs that are as lightweight as possible. Significant weight reductions have been achieved by modifying the can-bottom configuration and by reducing the top diameter and by changing its profile.

These developments, along with customer appeal generated by the easy-open top, have caused aluminum-can production to increase dramatically.

YEAR	Billions of CANS	Beverage-can market %
1965	0.5	3.2
1970	4.6	13.7
1975	16.3	38.3
1980	41.6	76.3
1981	47.7	85.0

In the future, aluminum cans will grow in popularity with can manufacturers as 0.0100-in sheet becomes available.

Aluminum-can progress

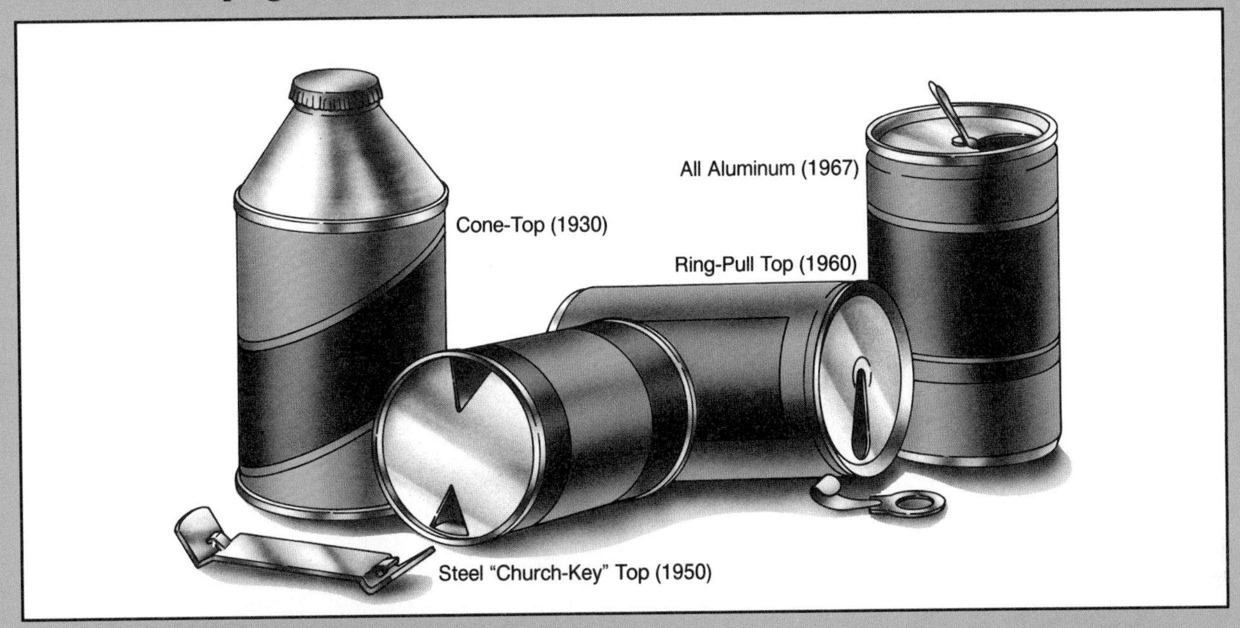

The cone-top steel can, which could be filled and capped on glass beer-bottle equipment, was introduced around 1930 and lasted until the mid-1950s. It was replaced by the flat-top, all-steel can, which was opened with a "church key." During 1954, 430 million flat-top cans were produced. The market grew to 820 million by 1960.

The first aluminum soft tops appeared on 3-piece steel cans in 1960. They also required an opener. In 1962 the easy-open "pull top" cans were introduced. By 1970, 90% of all soft drinks were in aluminum-top easy-open cans.

The first all-aluminum beverage can was introduced in 1967. By 1969 aluminum's share of the carbonated-drink market was 5%. Today 92% of the beer cans and 72% of the soft drink cans are aluminum.

Convenience With No Litter

The first easy-open tops were put on fruit-juice cans in 1960. A tab was riveted to a scored area of the can top. When the tab was pulled the metal "tore." To use this easy-open top on a beverage can presented some difficulties. Separate rivets caused sealing problems in the pressurized cans. Ultrasonic and impact-welded tabs also proved ineffective. The answer came from Ermal Fraze of Dayton Reliable Tool Co. He developed and patented the integral-rivet fastener. It was introduced in 1963 and is the only fastening technique used today. It involves a series of forming operations in which a rivet shank is created from the can-top metal. The tab is attached by heading the shank.

Early easy-open tops had throw-away rings and tabs which meant that some metal was lost even if the remainder of the can was recycled. Tabs also turned into litter. To help eliminate litter, the first captive-tab easy-open ends appeared a few years ago.

A number of patented aluminum can-top designs are now in use. There are only three basic designs, however: throwaway (ring and tab are discarded), capture pull top (top and ring stay with the can), and the rarely used pushbutton (two buttons on the can top are pushed in and stay within the can).

Source: Reprinted from MACHINE DESIGN, October 8, 1981. Copyright, 1981 by Penton/IPC., Inc., Cleveland, Ohio.

In Retrospect

- Positional tolerancing allows one to indicate the acceptable range of variation in the location of a feature with respect to a datum or reference location (point, axis, or plane).
- Geometric form tolerancing allows one to indicate the acceptable range of variation in the shape of a feature (flatness, perpendicularity, straightness, and so forth).
- Symbology has been developed in order to minimize the amount of information that must be included on a drawing in the form of notes. Symbology also allows tolerance specifications to be easily standardized.
- Geometric form tolerances that do not require a datum include:
  - flatness
  - straightness
  - roundness
  - cylindricity
  - profiles (lines and surfaces)

  Geometric form tolerances that do require datum planes include:
  - perpendicularity
  - parallelism
  - angularity
  - runout
- Locational tolerances—true position and concentricity—require datum planes.
- Datum identifying symbols are used to indicate reference locations on an engineering drawing. Any letter of the alphabet may be used as a datum symbol except I, O, and Q.

Exercises

Your instructor will specify the appropriate scale for each of the following exercises. In each case add suitable geometric form and positional tolerances to ensure that the part will be fabricated according to the described specifications. In addition, properly dimension the drawing.

EX10.1. Draw a set of orthographic views for the object shown in the figure. The inclined surfaces are to be flat within a tolerance of .004. The horizontal bottom surface should be denoted as datum surface A. Relative to this datum surface A, the horizontal top surface of the part is to be parallel within .002.

EX10.2. Draw a set of orthographic views for the object shown in the figure. Specify that the location of the hole is to have a true position within .004 with respect to the vertical left-side and horizontal bottom surfaces of the object (these two surfaces should be denoted as datum surfaces). In addition, specify that the curving top surface of this part is to be flat within .002. Finally, indicate that the hole is to be cylindrical in form within .004.

EX10.3. Draw a set of orthographic views for the object shown in the figure. Specify that the hole is to have a true position within .003 with respect to the horizontal bottom and vertical right-side surfaces of the part. In addition, specify that the hole is to be cylindrical in form within .002. Finally, indicate that the horizontal surfaces are to be parallel to one another within .002.

EX10.4. Draw a set of orthographic views for the object shown in the figure. Specify that the holes are to have true positions within .004 with respect to three mutually perpendicular datum surfaces of your choice. In addition, specify that these datum surfaces on the part are to be mutually perpendicular within .002.

EX10.5. Draw a set of orthographic views for the object shown in the figure. Specify that the outer nonplanar surfaces are to have profile tolerances within .003. In addition, specify that all holes are to have true positions within .002 with respect to datum surfaces of your choice. Each hole is also to be cylindrical in form within .002.

EX10.6. Draw a set of orthographic views for the object shown in the figure. Specify that all holes are to have true positions within .003 with respect to datum surfaces of your choice. In addition, specify that each hole is to be cylindrical in form within .002.

EX10.7. Draw a set of orthographic views for the object shown in the figure. Specify that all concentric cylindrical surfaces are to be concentric within .004.

EX10.8. Show the tolerance zones for the part described in EX10.1 with an appropriate set of orthographic views.

EX10.9. Show the tolerance zones for the part described in EX10.2 with an appropriate set of orthographic views.

EX10.10. Show the tolerance zones for the part described in EX10.3 with an appropriate set of orthographic views.

EX10.11. Show the tolerance zones for the part described in EX10.4 with an appropriate set of orthographic views.

EX10.12. Show the tolerance zones for the part described in EX10.5 with an appropriate set of orthographic views.

EX10.13. Show the tolerance zones for the part described in EX10.6 with an appropriate set of orthographic views.

EX10.14. Show the tolerance zones for the part described in EX10.7 with an appropriate set of orthographic views.

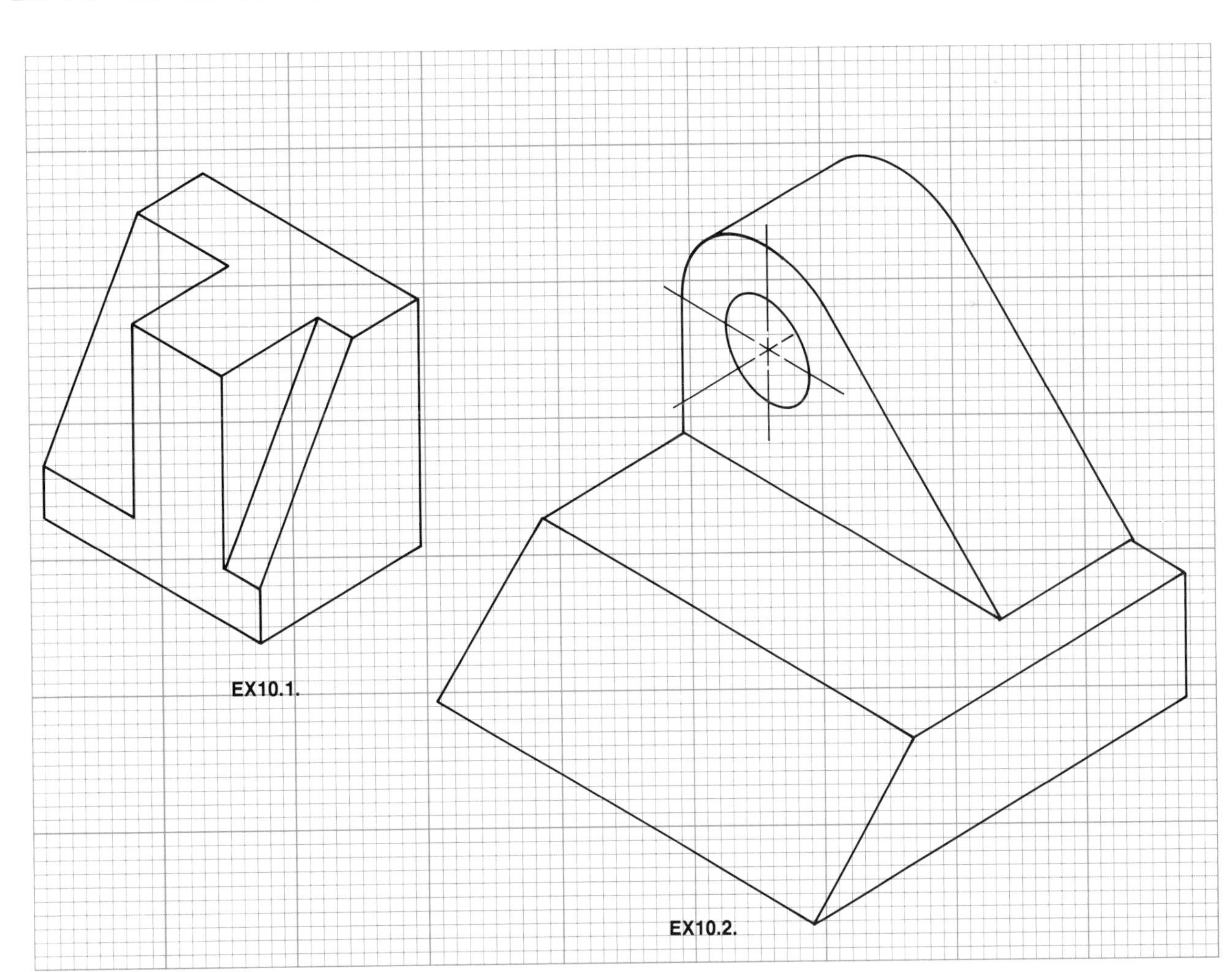
EX10.1.
EX10.2.

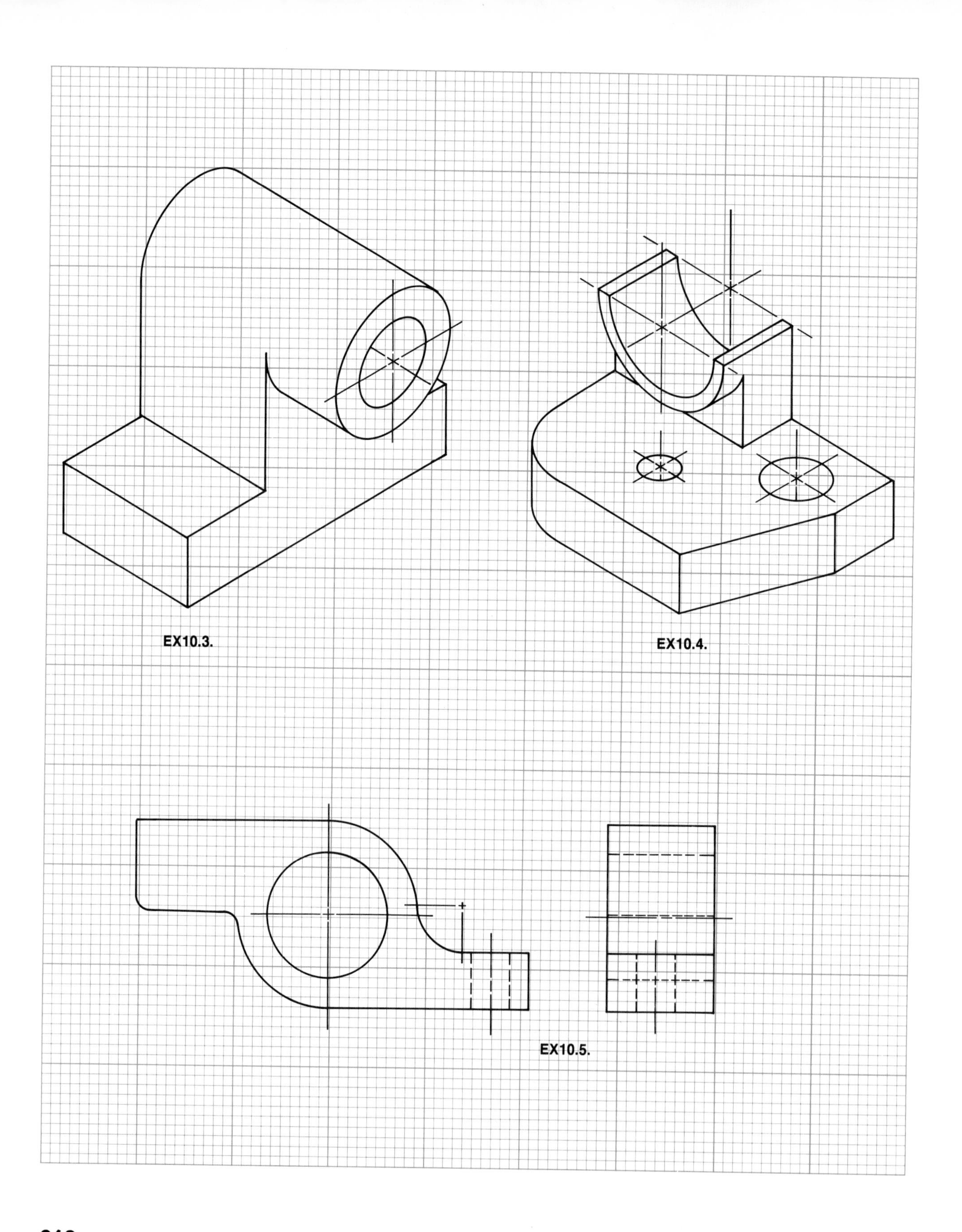
EX10.3.
EX10.4.
EX10.5.

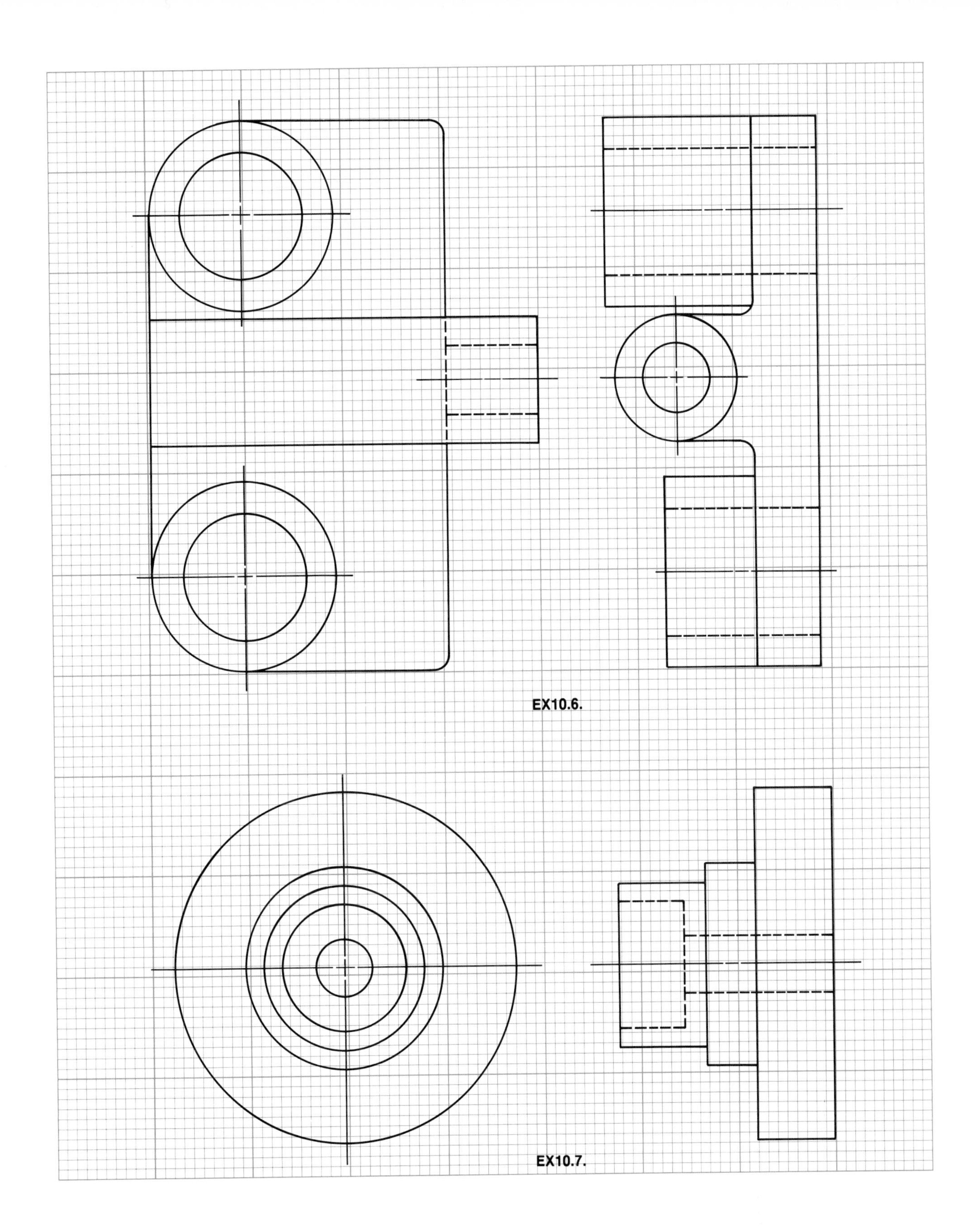

EX10.6.

EX10.7.

11 Descriptive Geometry, Part 1: True Lengths and True-Size Areas

Preview **Descriptive geometry** is the science of spatial relations and analysis of points, lines, and planes. It allows us to describe and design three-dimensional objects. In this introductory chapter about descriptive geometry, we focus upon the development of orthographic projections in which **true-length** images of lines and **true-size area** images of surfaces are shown. In Chapter 12 we will direct our attention to the challenges associated with the determination of the **relative positions** of points, lines and planes in space, together with the **intersections** of planes and solids. Our ability to graphically describe, analyze, and design three-dimensional objects will be significantly increased as a result of our studies in descriptive geometry.

Learning Objectives

Upon completion of this chapter, the reader should be able to:

- Define a line in space by specifying its orientation, together with its location, with respect to a reference system.
- Differentiate between normal lines, inclined lines and oblique lines.
- Construct true-length projection images of normal, inclined, and oblique lines in orthographic views.
- Define a plane in terms of lines and/or points in space.
- Differentiate between normal surfaces, inclined surfaces, and oblique surfaces.
- Construct true-size area projection images of normal, inclined, and oblique surfaces in orthographic views.

11.1 Defining a Point in Space

Descriptive geometry allows us to properly relate and analyze points, lines, and planes. In Section 4.1, we introduced Monge's planes of projection and the related concept of the glass box. These planes provided the basis for our develop-

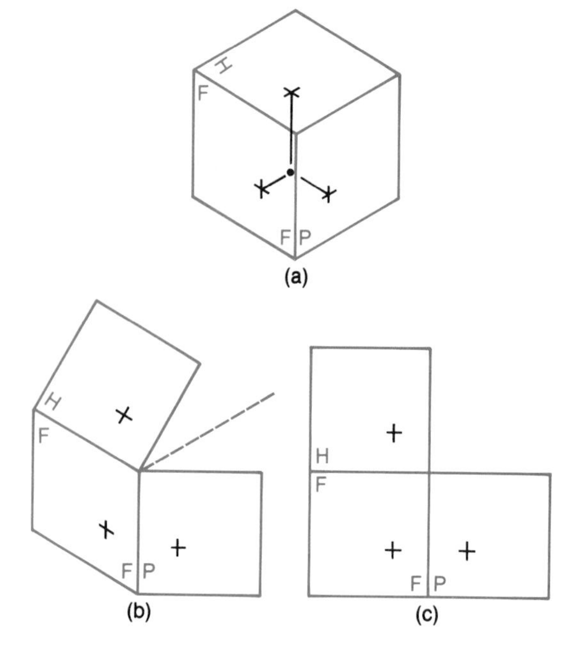

Figure 11.1 Projection and orthographic description of a point. The point is projected onto the walls of the glass box (a), and the glass box is unfolded (b), resulting in three orthographic views (c). The walls of the box are labeled H for horizontal, F for frontal, and P for profile projection plane.

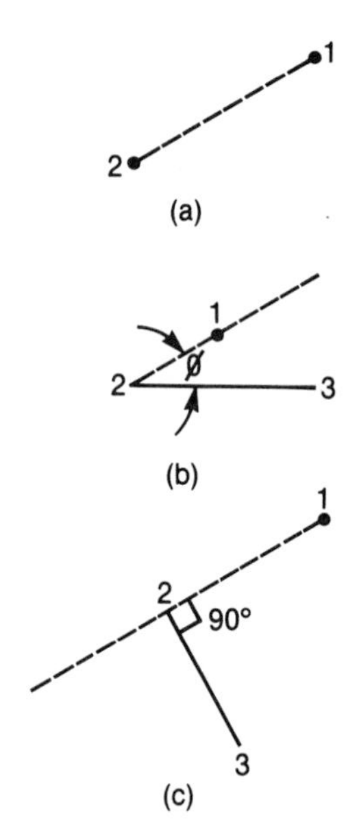

Figure 11.2 Defining a line is space by specification of two points on the line (a), one point on the line and the angle Φ between a given line ($\overline{23}$) and the line under construction (b), or, in the special case shown here, one point on the line and a line ($\overline{23}$) that is perpendicularly oriented to the line under construction (c).

ment of orthographic projection principles. We will now use these planes to develop the analytical procedures of descriptive geometry.

Recall that the spatial position of a point can be specified via projection onto the walls of a glass box. In Figure 11.1, these walls are labeled H, F, and P, corresponding to the horizontal, frontal, and profile projection planes, respectively, from which the box is formed. The walls of the box are then "unfolded" into a single plane, thereby forming a set of principal orthographic views (b and c). Any two of these views serve to specify the three dimensions, and hence the location, of the point with respect to an arbitrarily chosen origin (or reference point).

Monge's planes of projection and the resulting orthographic views may be similarly used to specify the location of points used to define lines and planes. Three-dimensional objects may then be described and analyzed in terms of their constituent surface boundary lines and surface planes.

11.2 Defining a Line in Space

A line is described by two distinct characteristics: its **orientation,** which is its direction with respect to a reference frame, and its **location** in space. These characteristics can be specified (Figure 11.2) by identifying the location of *two points* that lie on the line (a) or by identifying a *single point* that lies on the line, together with the *angle* Φ between this line and a given reference direction in space (b). A special case is also shown in Figure 11.2(c), where $\Phi = 90°$.

11.3 True Length of a Normal Line

A **normal line** is one that is perpendicular to a principal projection plane. Projection of a normal line onto the walls of the glass box (or, equivalently, onto the principal planes of projection) results in either (1) a *true-length* image of the line on those walls (or projection planes) that are parallel to the line or (2) a *point* image of the line onto the wall (projection plane) that is perpendicular to the line.

In Figure 11.3 a line that is normal to the frontal reference plane (represented by the front surface of the glass box) is projected onto the box, resulting in an equivalent set of orthographic views in which it is shown in true length (top and right-side views) and as a single point (front view).

11.4 True Length of an Inclined Line

An **inclined line** is parallel to one principal projection plane (either frontal, horizontal, or profile) but inclined to the other projection planes. Projection of an inclined line onto the principal planes will result in a true-length image on the

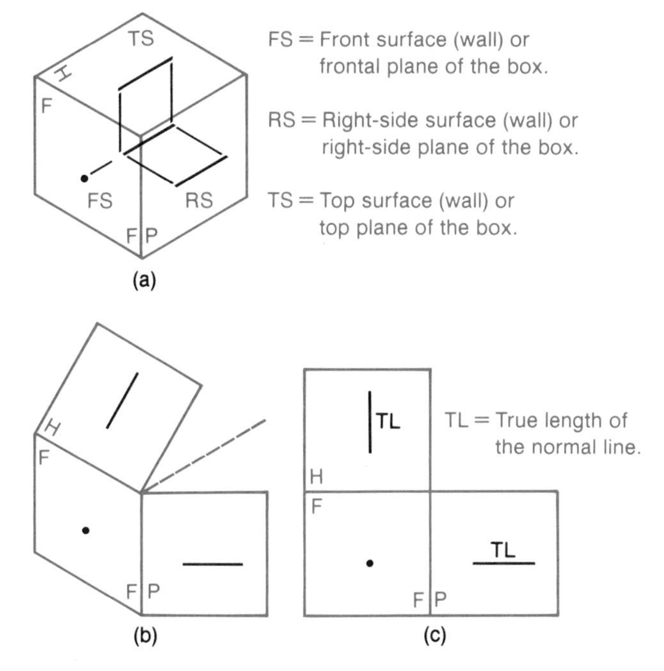

Figure 11.3 Projection of a normal line onto the walls of the glass box. (The line under consideration is *normal,* meaning it is perpendicular to the frontal plane of the box.) The line is projected (a) and the box unfolded (b), resulting in three orthographic views (c).

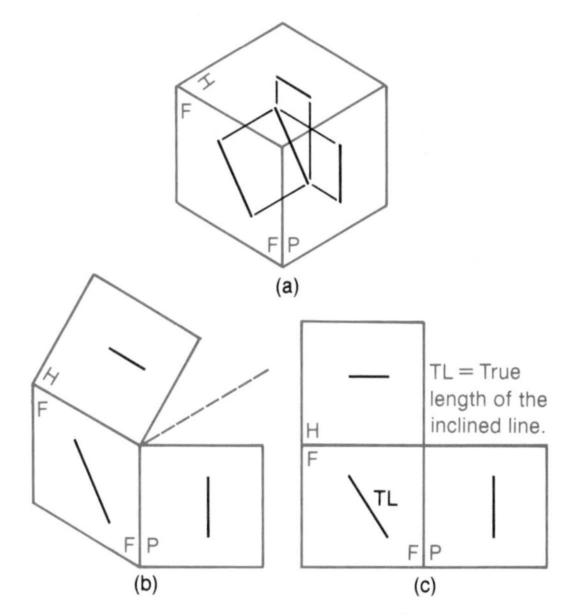

Figure 11.4 Projection of an inclined line onto the walls of the glass box. The line is projected (a) and the box unfolded (b), resulting in three orthographic views (c). This inclined line is contained within a plane that is parallel to the frontal reference plane, or the front wall of the box; as a result, this line is seen in true length in the given front view.

plane that is parallel to the line. Foreshortened (reduced) images of the line are projected onto the other principal planes.

The results of projection of an inclined line contained within a plane parallel to the frontal reference plane are shown in Figure 11.4. The line is shown in true length in the front orthographic view, whereas it appears foreshortened in the given right-side and top views.

11.5 True Length of an Oblique Line

An **oblique line** is not parallel to any principal projection plane. Projection of an oblique line onto a principal projection plane results in a foreshortened image (Figure 11.5). In order to generate a true-length projection image of an oblique line, we must specify an auxiliary projection plane in space that is parallel to the oblique line. We know that a normal line will project in true length onto those principal planes that are parallel to it. Similarly, an inclined line will project in true length onto the principal plane that is parallel to it. Since none of the principal planes of projection are parallel to an oblique line, an auxiliary view of such a line must be established in order to obtain a true-length image.

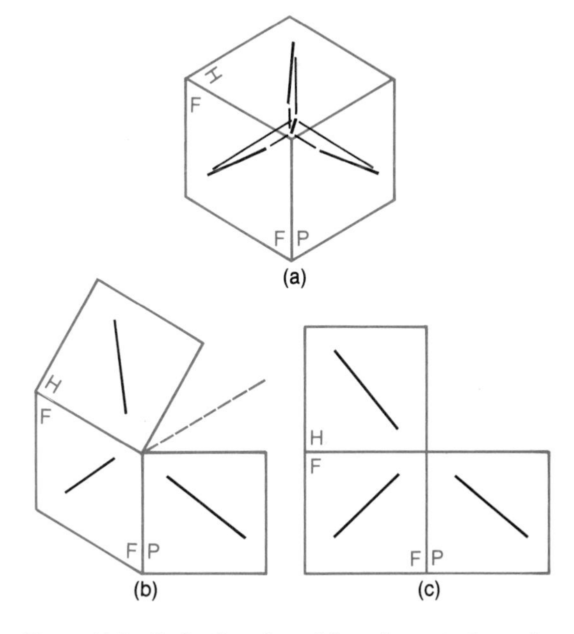

Figure 11.5 Projection of an oblique line onto the walls of the glass box. The line is projected (a) and the box unfolded (b), resulting in three orthographic views (c). The oblique line appears foreshortened (reduced) in all principal orthographic views.

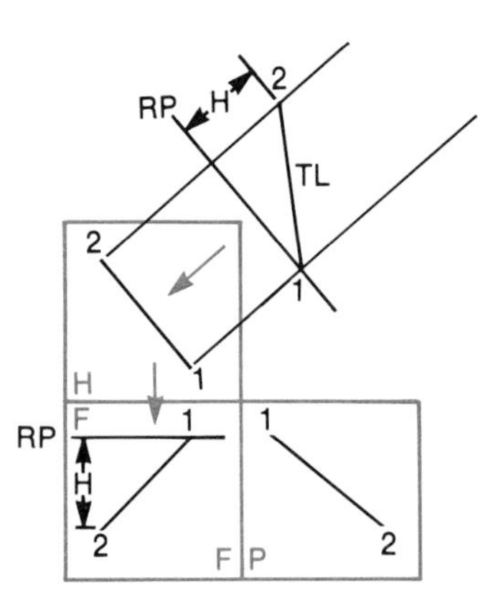

Figure 11.6 Development of an auxiliary view in which an oblique line, $\overline{12}$, appears in its true length. The auxiliary view is based upon (or projected from) the top view of the line.

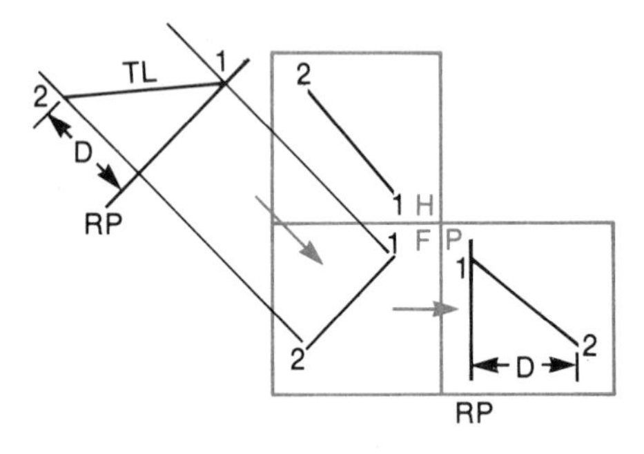

Figure 11.7 Development of an auxiliary view in which an oblique line, $\overline{12}$, appears in its true length. The auxiliary view is based upon (or projected from) the front view of the line.

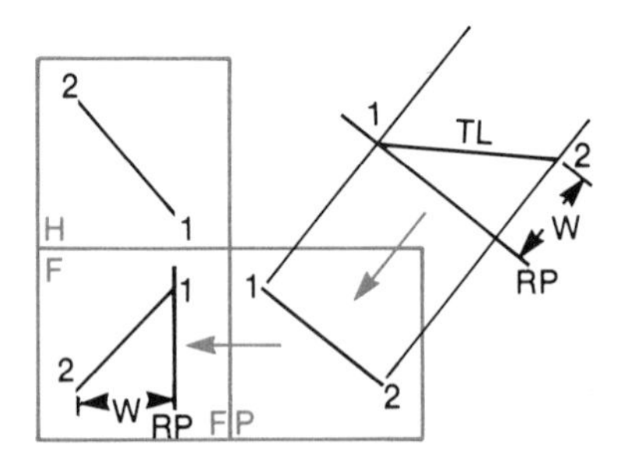

Figure 11.8 Development of an auxiliary view in which an oblique line, $\overline{12}$, appears in its true length. The auxiliary view is based upon (or projected from) the right-side view of the line.

An auxiliary (true-length) image of the oblique line $\overline{12}$ is developed in Figure 11.6. (The reader should review our discussion of auxiliary orthographic views in Chapter 7 if necessary.) In this case, the given top view acts as the basic view (toward which the auxiliary view sight-line is directed and upon which the auxiliary view will be based.)

Notice that a reference plane (denoted as *RP*) is used to transfer the height distance, *H*, between the line endpoints 1 and 2 from the given front view to the auxiliary-view construction region. This reference plane, seen on edge as a reference line in the construction work of Figure 11.6, is perpendicular to the auxiliary viewing direction and parallel to the principal plane onto which the basic view has been projected. The reference plane *RP* has been conveniently defined in Figure 11.6 so that it contains point 1.

The oblique line $\overline{12}$ may also be shown in true length in an auxiliary view that is based upon the front view (Figure 11.7). The depth distance, *D*, between the line endpoints 1 and 2 is transferred from the right-side view or the top view to the auxiliary view construction region.

As a final example of an oblique line, consider Figure 11.8, in which a true-length auxiliary view (based upon the given right-side view) of the line has been constructed. The width distance, *W*, between line endpoints 1 and 2 is transferred from the front view to the auxiliary-view construction region.

In summary, then, the true length of an oblique line can be shown in an auxiliary view that corresponds to projection of the line onto an auxiliary plane parallel to it. Such projection corresponds to a viewing direction that is *perpendicular* to the line in any view upon which the auxiliary view is based.

11.6 Point View of an Oblique Line

We have seen that a normal line will project as a single point onto the principal plane that is perpendicular to it (Figure 11.3). An inclined line will project as a single point onto an auxiliary plane that is perpendicular to it. Similarly, an oblique line will project as a single point onto an auxiliary plane that is perpendicular to it. In order to develop this point view of an oblique line described by a set of given principal orthographic views, a secondary auxiliary view of the line should be constructed.

As Figure 11.9 shows, a primary auxiliary view of the oblique line is first constructed in which the line is shown in true length. The secondary auxiliary viewing direction is then identified in such a way that it is parallel to the direction of the line in the true-length primary auxiliary view. The secondary auxiliary view is then formed to produce a point view of the oblique line. (Notice that two reference planes, denoted as RP_1 and RP_2, both seen on edge, as reference lines, in Figure 11.9, are used in the development of the secondary auxiliary view.)

The reader may be wondering if a point view of a line has any value or utility in graphics. In answer, consider the need to specify the true distance between a line and another line or plane; a point view of the line can be very useful in the determination of such *true relative distances*—as we will discover in the next chapter.

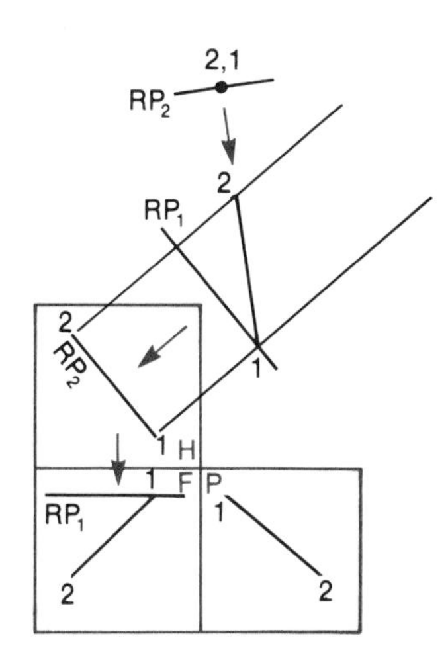

Figure 11.9 Development of a secondary auxiliary view in which the oblique line $\overline{12}$ appears as a point. A point view of the line is developed in two stages: First, a primary auxiliary view of the oblique line—in which the line appears in its true length—is constructed. Next, a secondary auxiliary view is constructed along a viewing direction parallel to the direction of the line in the primary view.

Learning Check

Are oblique lines seen in true length in any principal orthographic view?

Answer No. An auxiliary view is needed to show an oblique line in true length.

11.7 Defining a Plane in Space

We have seen that two points may be used to define a unique line in space. Similarly, *three noncollinear points* within a plane may be used to specify that particular plane in space, as shown in Figure 11.10(a). A plane may also be defined by the following:

- *Two intersecting, nonparallel lines* contained within that plane (b).
- A *noncollinear line and point* contained within that plane (c).
- *Two parallel, noncollinear* lines contained within that plane (d).

11.8 True-Size Projection of a Normal Surface

A normal (planar) surface of an object is, by definition, parallel to one of the principal projection planes (frontal, horizontal, or profile). Such a surface will project in its true size onto the projection plane parallel to it. That is, the projection will be identical in area and form to the surface itself.

A normal plane projects as a line onto the principal projection planes that are perpendicular to it. Figure 11.11 describes the projection of a normal surface to the frontal reference plane (FRP). A true-size (denoted as TS) image of the surface appears in the resultant front orthographic view.

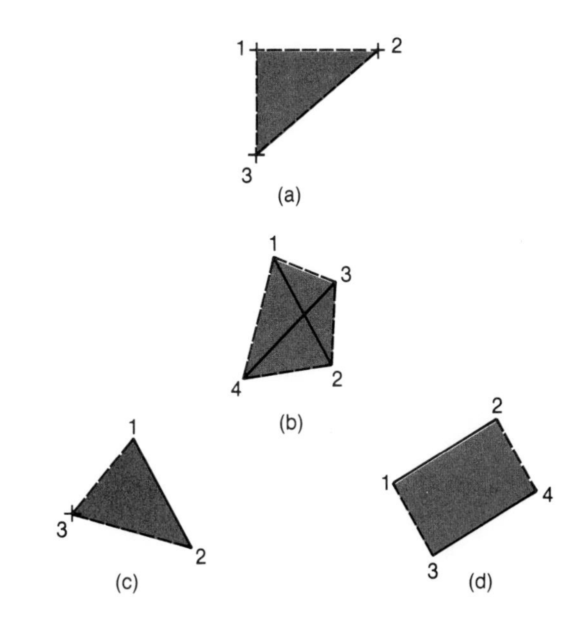

Figure 11.10 A particular plane in space may be specified in terms of three noncollinear points contained within that plane (a), two intersecting, nonparallel lines contained within that plane (b), a noncollinear line and point contained within that plane (c) or two parallel, noncollinear lines contained within that plane (d).

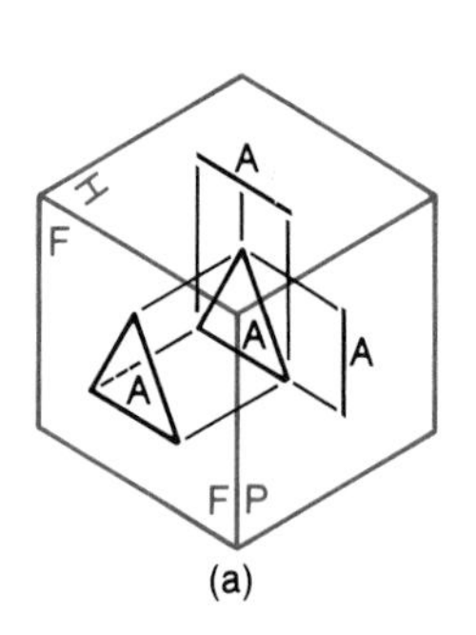

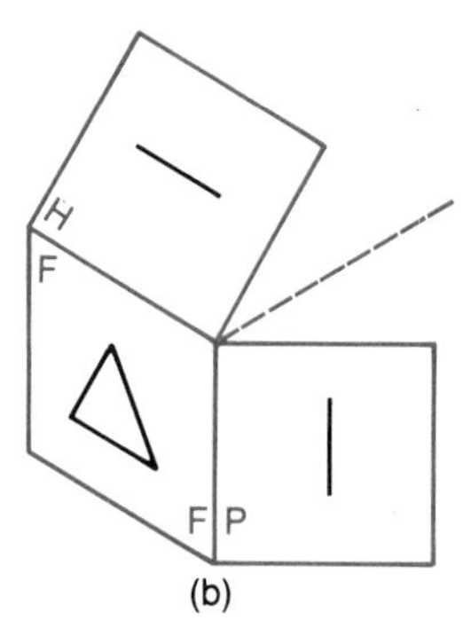

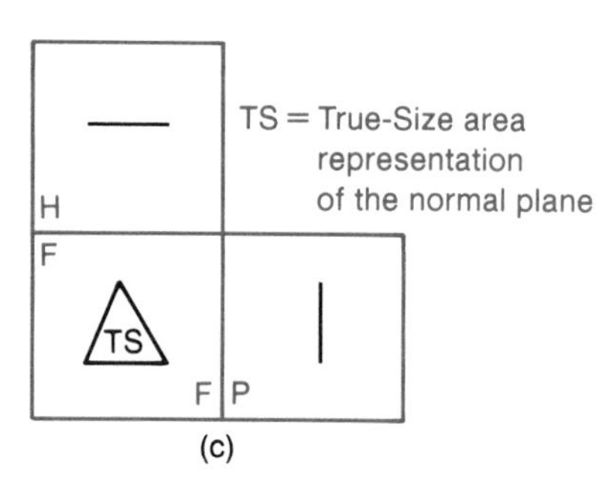

Figure 11.11 Projection of a normal plane onto the walls of the glass box. The plane is projected (a) and the box unfolded (b), resulting in three orthographic views (c). The normal plane appears as a true-size area in one (front) view and on edge, as a line, in the other two given principal views.

Figure 11.12 Projection of an inclined plane onto the walls of the glass box. The plane is projected (a) and the box unfolded (b), resulting in three orthographic views (c). The inclined plane in this example appears on edge in one (front) view and as a foreshortened area in the other two given principal views.

11.9 True-Size Projection of an Inclined Surface

An inclined (planar) surface of an object is perpendicular to one of the principal projection planes and inclined to the other principal planes. Such a surface will appear on edge, as a line, in that projection plane to which it is perpendicular. Furthermore, it will appear as a foreshortened, or reduced, area (denoted as FS) in those projection planes to which it is inclined.

Figure 11.12 presents the projection results of an inclined surface that is perpendicular to the frontal reference plane (FRP). The surface appears on edge, as a line, in the resultant front orthographic view and as foreshortened areas in the top and right-side views.

In order to generate a true-size area representation of any surface, projection of the surface must be performed onto a plane that is parallel to the surface. Projection onto this plane occurs along a direction that is perpendicular to the surface—that is, an observer must view the surface along a sight-line direction that is perpendicular to it in order to view the entire surface without distortion. As a result, an *auxiliary orthographic view* must be constructed in order to show an inclined surface as a true-size area (Figure 11.13). (Recall our discussion of auxiliary views in Chapter 7.)

11.10 True-Size Projection of an Oblique Surface

An oblique surface is inclined to all principal projection planes, so it will project as a foreshortened area onto each of these planes (Figure 11.14). In order to generate a projected image of an oblique surface in which the surface is represented by a true-size area, we will need to construct an auxiliary view corresponding to a sight-line that is perpendicular to the surface. This will require the construction of a primary auxiliary view in which the oblique surface appears on edge, as a line, together with a *secondary auxiliary view* in which the surface is represented by a true-size (TS) area (Figure 11.15). (Secondary auxiliary views were discussed in Chapter 7.)

Learning Check

Can an inclined surface be seen in true size in a primary auxiliary view?

Answer Yes.

Can an oblique surface be seen in true size in a primary auxiliary view?

Answer No.

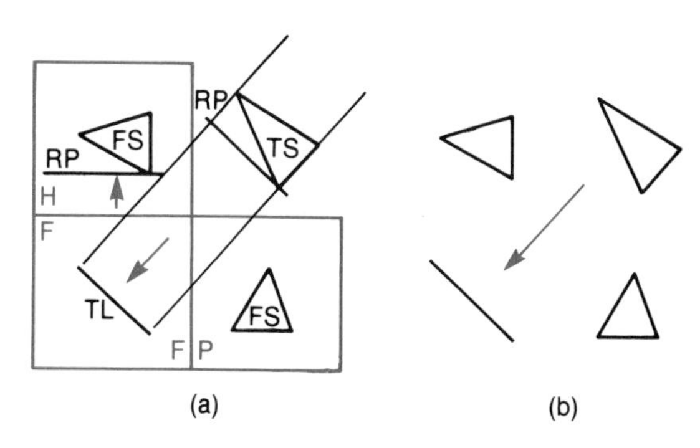

Figure 11.13 Construction of a primary auxiliary view in which the inclined surface A appears as a true-size area (a) and the final set of views (b).

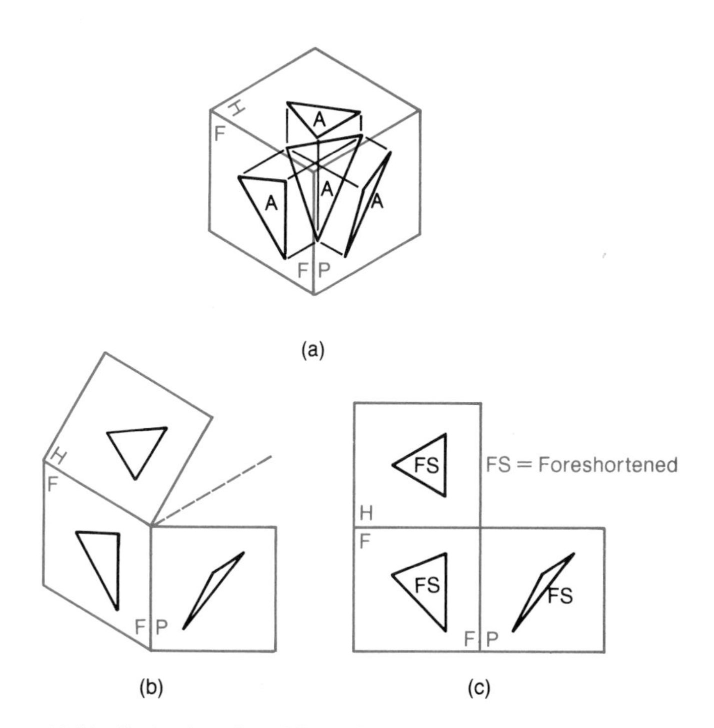

Figure 11.14 Projection of an oblique plane onto the walls of the glass box. The plane is projected (a) and the box unfolded (b), resulting in three orthographic views (c). The oblique plane in this example appears as a foreshortened area in all three principal views. A secondary auxiliary view will need to be constructed in order to present surface A in its true size.

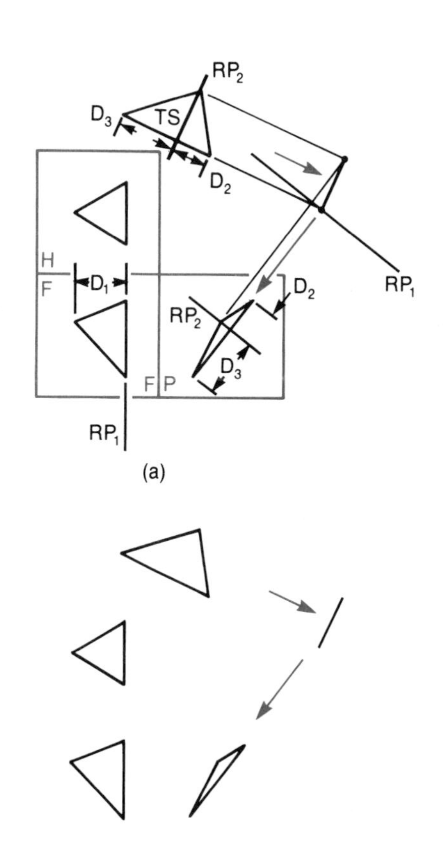

Figure 11.15 Construction of a secondary auxiliary view in which the oblique surface A appears as a true-size area (a) and the final set of views (b).

ENGINEERING IN ACTION

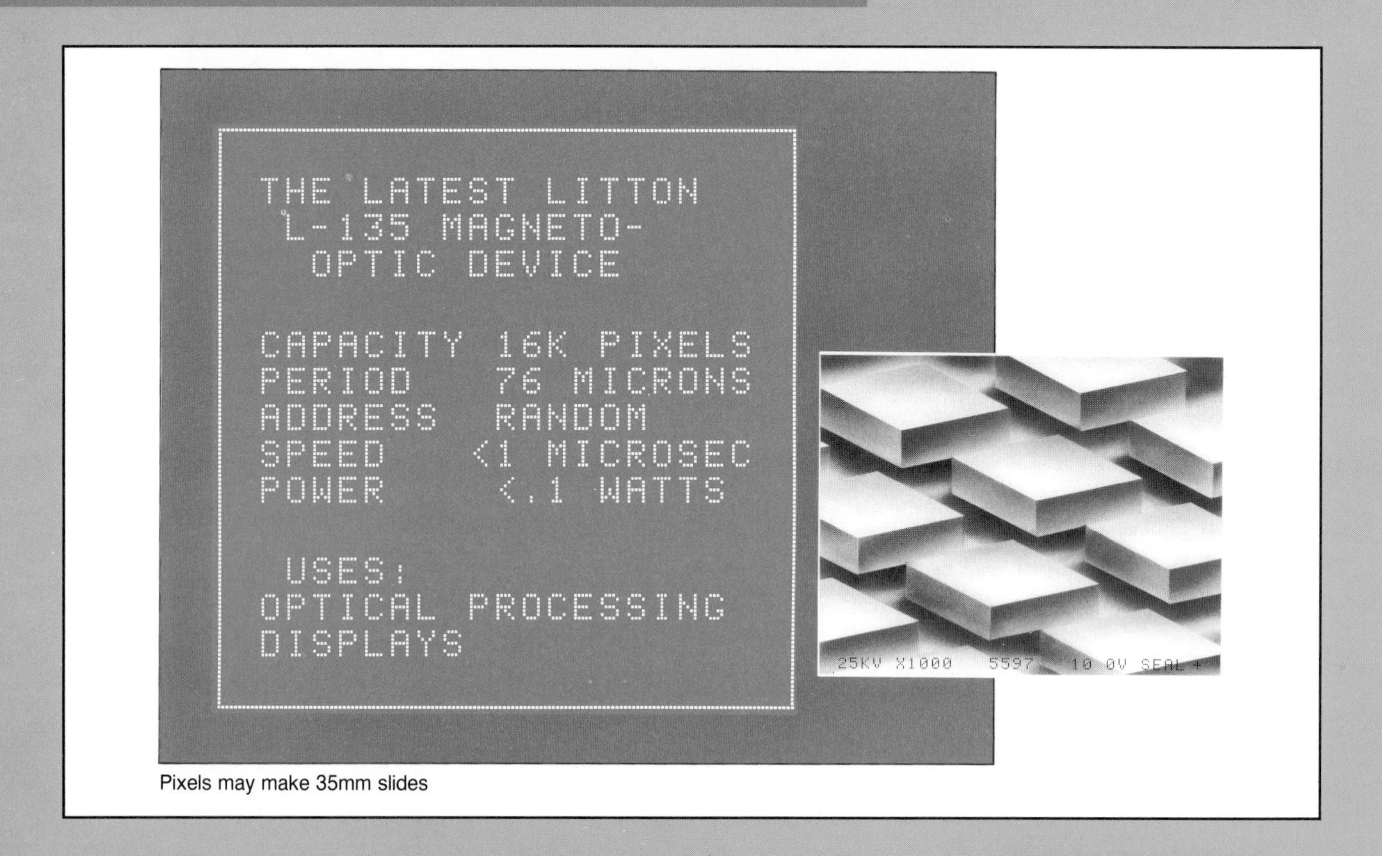

Pixels may make 35mm slides

Solid-State Magneto-optical Display Designed for Mass Production

Display and electronic imaging technology has been advancing rapidly in recent years, but there has never been a true solid state display which could be mass produced using semiconductor process technology and was fully compatible with VLSI. However, low-cost, non-volatile, solid-state displays may be made practical by a light modulating device called the Light-Mod. It is the invention of William E. Ross, Director of Light-Mod Electro Optics Engineering, Litton Data Systems, Van Nuys, CA. and is a joint development with Dr. A. Tanielian President of Semetex, Torrance, CA.

Key to the new design is an array of iron garnet pixels which act as light valves. They pass or block light by using the Faraday effect on plane polarized light. The orientation of the plane of polarization of transmitted light is determined by the direction of magnetization of the pixel. The pixel magnetic field orientation is changed by passing an electric current through conductors which intersect at the pixel. The magnetic field produced by the sum of the two electric currents switch the direction of magnetization of the individual pixel. The effect can be compared to manually rotating a polarizing filter.

An image can be electronically generated on the array in near real time by programmed control of the drive line currents setting the individual pixel orientation. This essentially provides the capability of producing a picture on a 35mm slide under computer or microprocessor control.

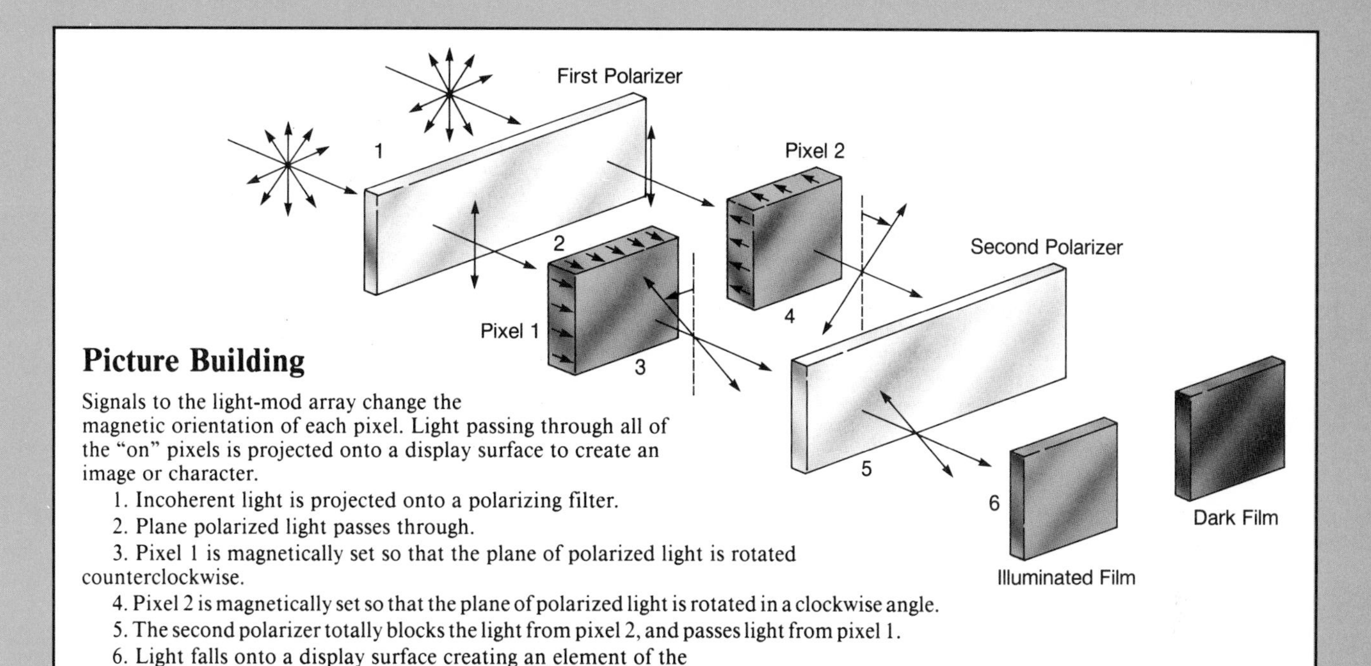

Picture Building

Signals to the light-mod array change the magnetic orientation of each pixel. Light passing through all of the "on" pixels is projected onto a display surface to create an image or character.

1. Incoherent light is projected onto a polarizing filter.
2. Plane polarized light passes through.
3. Pixel 1 is magnetically set so that the plane of polarized light is rotated counterclockwise.
4. Pixel 2 is magnetically set so that the plane of polarized light is rotated in a clockwise angle.
5. The second polarizer totally blocks the light from pixel 2, and passes light from pixel 1.
6. Light falls onto a display surface creating an element of the total picture, dark or light depending on pixel magnetization.

Triggering a Change

Half mil thick bismuth iron garnet film is structured on non magnetic substrates. Drive conductors are placed between the magnetically isolated pixels on X-Y grids so that they intersect at one corner of each pixel. When the conductors at a given pixel are both energized the pixel magnetic field orientation can be switched.

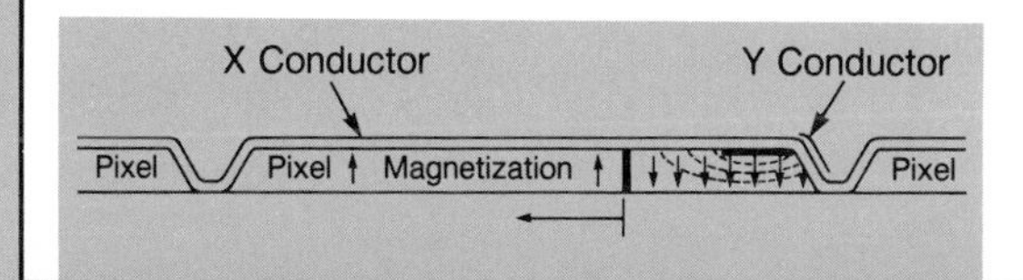

To date, solid state arrays as large as 128 X 128 have been successfully tested and 256 X 256 arrays are now under test. Both of these arrays use 3 mil (76 micron) square pixels. Evaluation devices with 48 X 48 5 mil pixels and 64 X 64 0.4mm pixels have also been produced. Present experimental models may be any single color of the spectrum from green through red but multicolor displays may be made practical by adding color filters to the optical system.

Source: Reprinted from MACHINE DESIGN, January 26, 1984. Copyright, 1984, by Penton/IPC., Inc., Cleveland, Ohio.

Big and Small Screens

The Light-Mod can best be used as a real-time, high-resolution physically small display with image magnification for alphanumeric or graphics. Typical applications of this type would be helmet or headset displays, hand held displays, FLIR viewers, and instrument readouts. Large screen projection of cockpit head up displays may be implemented with the Light-Mod used in place of conventional film slides.

In Retrospect

- Descriptive geometry is the science of spatial relations and analysis. It allows one to properly relate and analyze points, lines, and planes used to describe three-dimensional objects.
- A line is described by two distinct characteristics: its orientation and its location in space. These characteristics can be specified by identifying (1) the location of two points that lie on the line or (2) a single point that lies on the line and the angle between this line and a given reference direction in space.
- A normal line is one that is perpendicular to a principal projection plane. The line projects as a point onto the plane to which it is perpendicular; it projects in its true length onto the other principal projection planes.
- An inclined line is parallel to one principal projection plane and inclined to the other principal planes. A true-length image of an inclined line will be projected onto the projection plane that is parallel to it. Foreshortened (reduced) images of the line are projected onto the other planes of projection.
- An oblique line is not parallel to any principal plane of projection. A true-length image can only be shown in an auxiliary view corresponding to a sight-line direction that is perpendicular to a plane containing the line. The point view of an oblique line may be constructed as a secondary auxiliary view based upon the primary auxiliary view in which the line appears in true length.
- A plane may be defined in terms of the following:
  - Two intersecting, nonparallel lines contained in the plane.
  - A line and a (noncollinear) point contained in the plane.
  - Two parallel, noncollinear lines contained in the plane.
- A normal surface is parallel to a principal plane of projection onto which it will project as a true-size area. This surface will appear on edge, as a line, in the other principal projection planes (to which it is perpendicular).
- An inclined surface is perpendicular to one of the principal projection planes, onto which it will project on edge, as a line. Such a surface will project as a foreshortened area onto the other principal projection planes to which it is inclined. A true-size area projection of an inclined surface can be presented in an auxiliary view.
- An oblique surface is inclined to all principal planes of projection onto which it will project as foreshortened areas. A true-size area projection can be constructed in a secondary auxiliary view. The secondary view is based upon a primary auxiliary view in which the oblique surface is seen on edge, as a line.

Table 11.1 summarizes some information regarding the projected images of lines and surfaces.

Table 11.1 Projection Images of Lines and Surfaces

	Principal Plane[a] 1	Principal Plane[a] 2	Principal Plane[a] 3	Primary Auxiliary Plane	Secondary Auxiliary Plane
Normal Line	Point	[TL]	[TL]	—	—
Inclined Line	[TL]	FL	FL	(Point)[b]	—
Oblique Line	FL	FL	FL	[TL]	(Point)[b]
Normal Planar Surface	[TS]	Edge	Edge	—	—
Inclined Planar Surface	Edge	FS	FS	[TS]	—
Oblique Planar Surface	FS	FS	FS	Edge	[TS]

[a] Principal plane may be frontal, horizontal, or profile.
[b] If this auxiliary view is based upon the view in which the line appears in true length, with the auxiliary view sight-line directed parallel to the line in this true-length basic view.

TL ≡ True-length view of line
FL ≡ Foreshortened-length view of line
TS ≡ True-size area representation of surface
FS ≡ Foreshortened-size area representation of surface
Edge ≡ Surface appears on edge, as a line.
Point ≡ Line appears as a single point.

Exercises

Your instructor will specify the appropriate scale for each exercise.

EX11.1. A line, $\overline{12}$, is shown in two given views. Complete the missing front view of this line.

EX11.2. Complete the missing top view of line $\overline{12}$.

EX11.3. Complete the missing right-side view of line $\overline{12}$.

EX11.4. Complete the missing right-side view of line $\overline{12}$.

EX11.5. Complete the missing top view of line $\overline{12}$.

EX11.6. Complete the missing front view of line $\overline{12}$.

EX11.7. Complete a view of line $\overline{12}$ in which it is shown in true length.

EX11.8. Complete the view of line $\overline{12}$ in which it is shown in true length.

EX11.9. Complete the missing front view of the plane A.

EX11.10. Complete the missing top view of plane A.

EX11.11. Complete a view of plane A in which it is shown in true size.

EX11.12. Complete a view of plane A in which it is shown in true size.

EX11.13. Complete a view of line $\overline{12}$ in EX11.4 in which it is shown in true length.

EX11.14. Complete a view of line $\overline{12}$ in EX11.5 in which it is shown in true length.

EX11.15. Complete a view of line $\overline{12}$ in EX11.6 in which it is shown in true length.

EX11.16. Complete a view of plane A in EX11.9 in which it is shown in true size.

EX11.17. Complete a view of plane A in EX11.10 in which it is shown in true size.

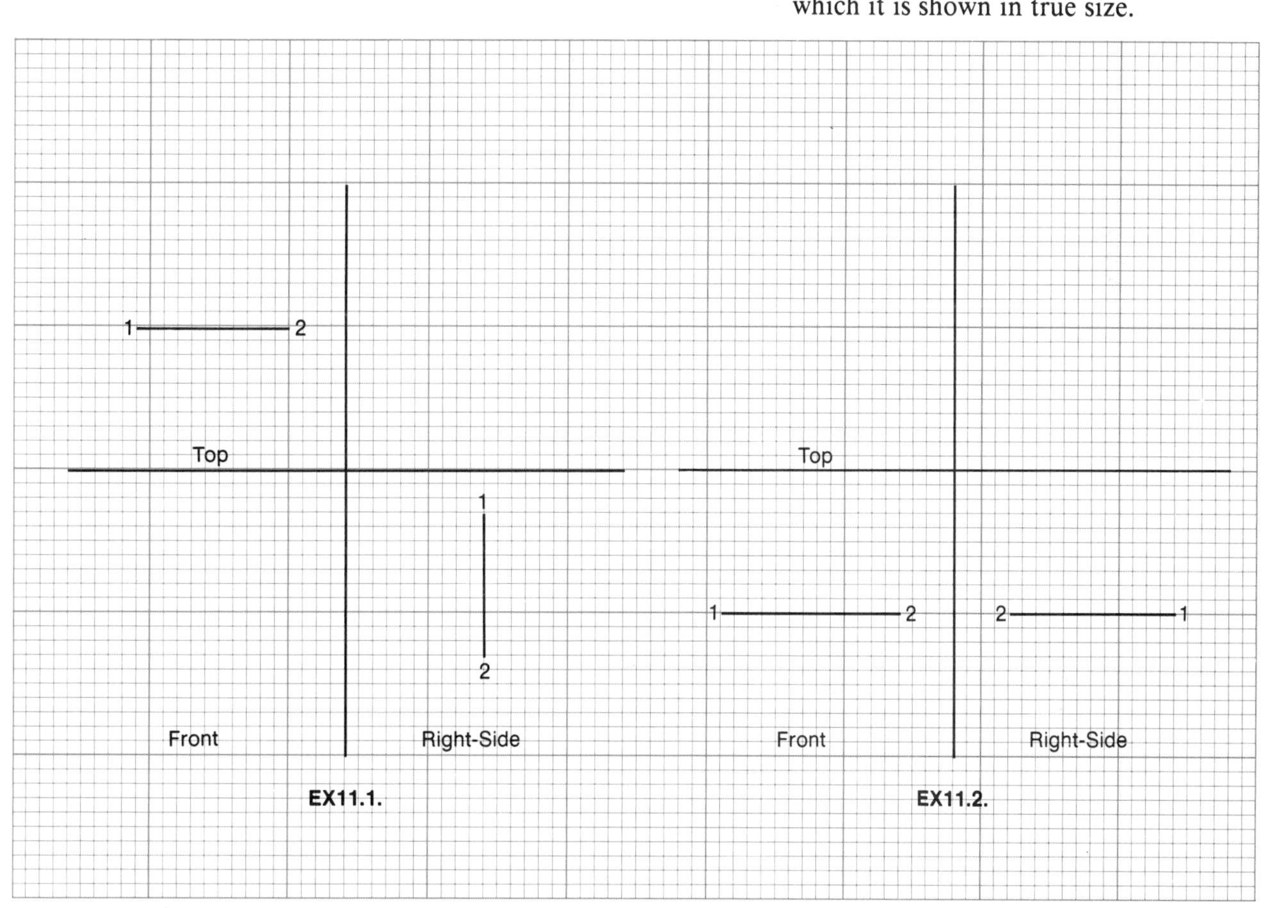

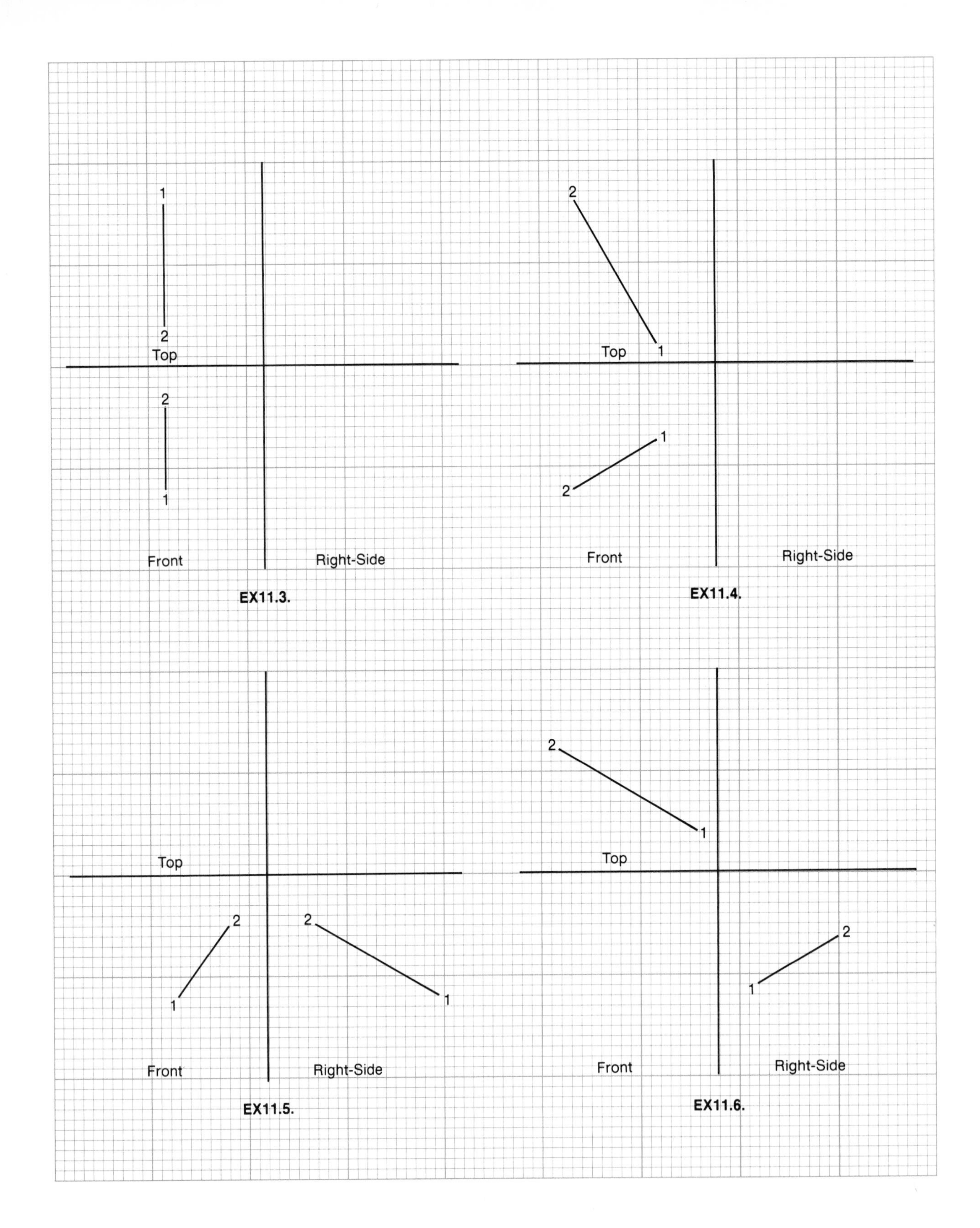
1
2
Top
2
1
Front
Right-Side
EX11.3.
2
Top
1
1
2
Front
Right-Side
EX11.4.
Top
2
1
2
1
Front
Right-Side
EX11.5.
2
1
Top
2
1
Front
Right-Side
EX11.6.

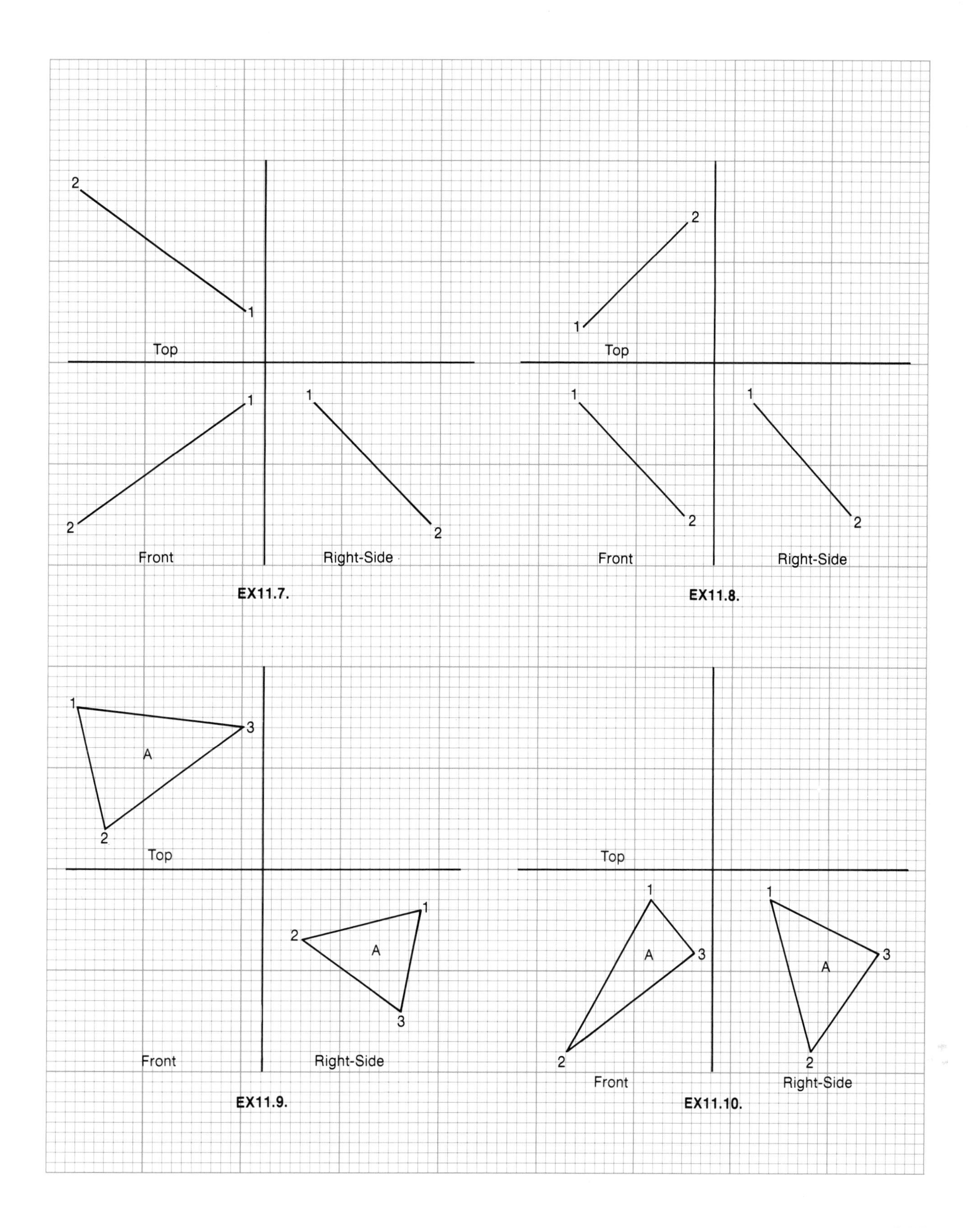
2
1
Top
1
1
2
2
Front
Right-Side
EX11.7.
2
1
Top
1
1
2
2
Front
Right-Side
EX11.8.
1
3
A
2
Top
2
1
A
3
Front
Right-Side
EX11.9.
Top
1
1
A
3
3
A
2
2
Front
Right-Side
EX11.10.

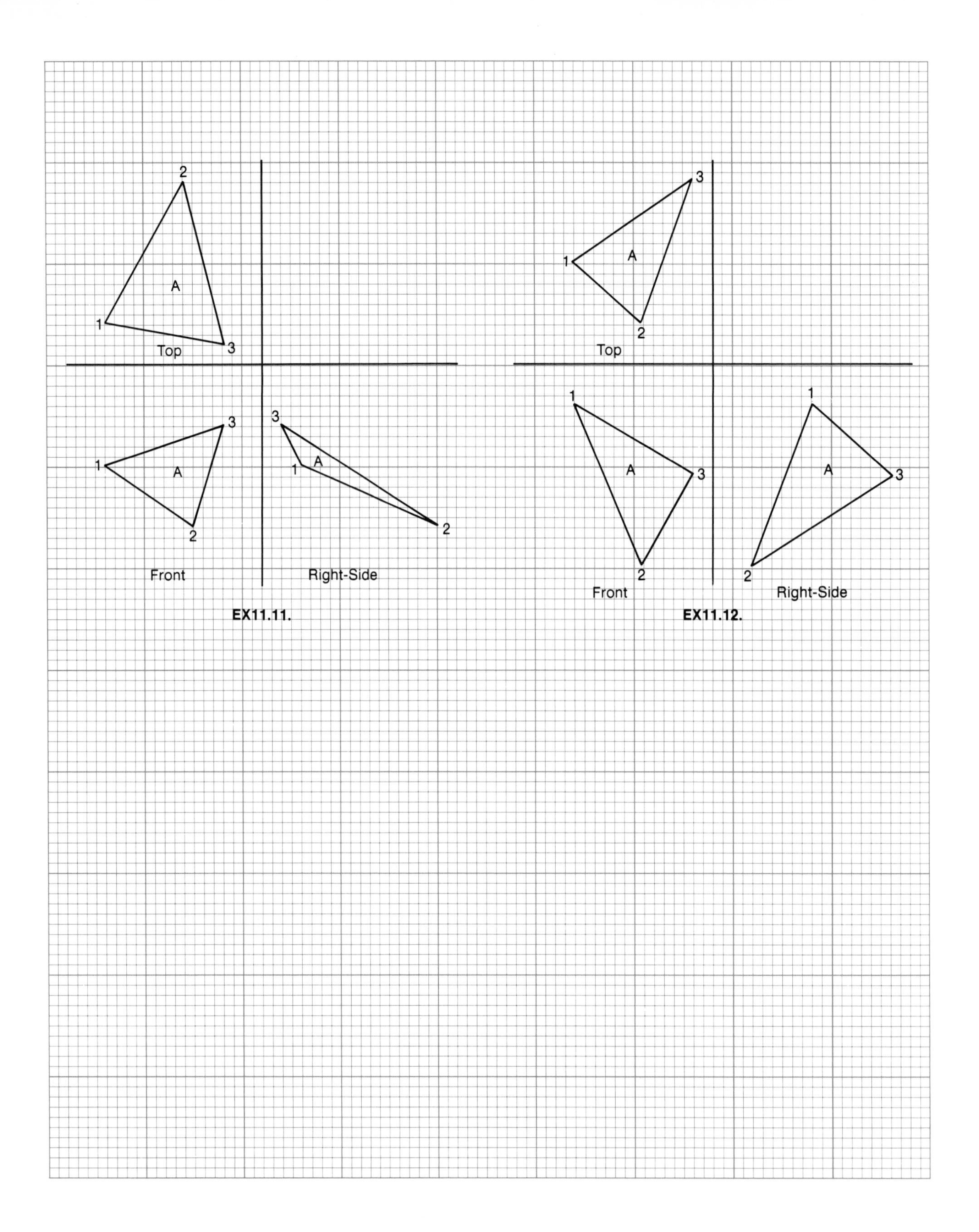
2
A
1
3
Top
3
1
A
2
Front
3
1
A
2
Right-Side
EX11.11.
3
1
A
2
Top
1
A
3
2
Front
1
A
3
2
Right-Side
EX11.12.

12 Descriptive Geometry, Part 2: Relative Distances and Orientations

Preview The previous chapter focused upon the development of orthographic projections in which true-length images of lines and true-size area images of surfaces are shown. This chapter explains how to determine the *relative positions* of points, lines, and planes in space.

The intersections between points and lines in orthographic views allow us to specify the relative locations of graphic elements. In addition, we may determine the *relative orientation* of lines and planes via the application of descriptive geometry principles as outlined in this chapter. The ability to analyze the relative locations and orientations of graphic elements is necessary if we are to represent engineering devices and designs with orthographic drawings; in particular, we must be able to determine the *relative visibility* of lines and surfaces in orthographic views. In addition, descriptive geometry allows us to develop flat-surface patterns, called developments, from which hollow objects may be fabricated by simply folding or rolling the pattern into the desired form.

Learning Objectives

Upon completion of this chapter, the reader should be able to:

- Differentiate between horizontal, profile, and frontal lines.
- Recognize that the intersections of points and lines in orthographic views allow us to specify the relative locations of lines, points, and planes in space.
- Construct such graphic elements as:
  - A line perpendicular to a given line.
  - A line perpendicular to a given plane.
  - A plane perpendicular to a given line.
  - A plane perpendicular to a given plane.
- Determine when parallelism and perpendicularity exist between given graphic elements.
- Identify the intersection of lines in space.
- Determine the shortest distance between a point and a line.
- Determine the shortest distance between two lines.
- Develop an orthographic view showing the dihedral angle between two planes without distortion.
- Determine the relative visibility of lines and planes in orthographic views.

- Construct the intersection between two or more surfaces by applying the point-by-point, surface-by-surface procedure.
- Define development and explain the application of a development in the fabrication of a design.
- Construct a development, or flat-surface pattern, for such geometric forms as right prisms, cylinders, pyramids, and cones.
- Use revolution techniques to obtain true lengths of lines and true-size area representations of inclined and oblique surfaces.

12.1 Horizontal, Frontal, and Profile Lines

A **horizontal line,** as its name implies, is parallel to the horizontal reference plane (HRP) onto which the top view of the line may be projected. As a result, the top view (and the bottom view) of a horizontal line will show the line in its *true length* (and with its *true orientation,* or direction in space, relative to vertical reference planes).

Consider Figure 12.1 in which a set of two views (top and front) of a triangular surface, A, is shown (a). The three corner points, 1, 2, and 3, define a plane in space in which surface A is contained. Two horizontal lines (denoted as $\overline{45}$ and $\overline{67}$) are then defined to lie within this triangular surface (b). We know that these lines are horizontal because they appear as horizontal lines in the given front view. The top view is incomplete in Figure 12.1(b) since it does not include lines $\overline{45}$ and $\overline{67}$. Upon completion, the top view will show these horizontal lines in *true length*—that is, the linear distance between points 4 and 5 will be accurately shown in this top view, as will the distance between points 6 and 7.

The endpoints 4, 5, 6, and 7 of the horizontal lines $\overline{45}$ and $\overline{67}$ have been conveniently chosen to lie within the boundaries of the triangular surface A; as a result, each of these points may be projected from the front view to the top view (c). For example, point 4 is shown to lie within the boundary line $\overline{12}$ of surface A; point 4 must therefore be located at the same relative width position on line $\overline{12}$ in the top view. The width position of point 4 is simply projected from the front view to the top view of line $\overline{12}$, thereby locating point 4 in the top view. Points 5, 6, and 7 may be similarly located in the top view. Lines $\overline{45}$ and $\overline{67}$ are then drawn in the top view (d). The final set of views (e) includes true-length images of the horizontal lines $\overline{45}$ and $\overline{67}$ in the completed top view.

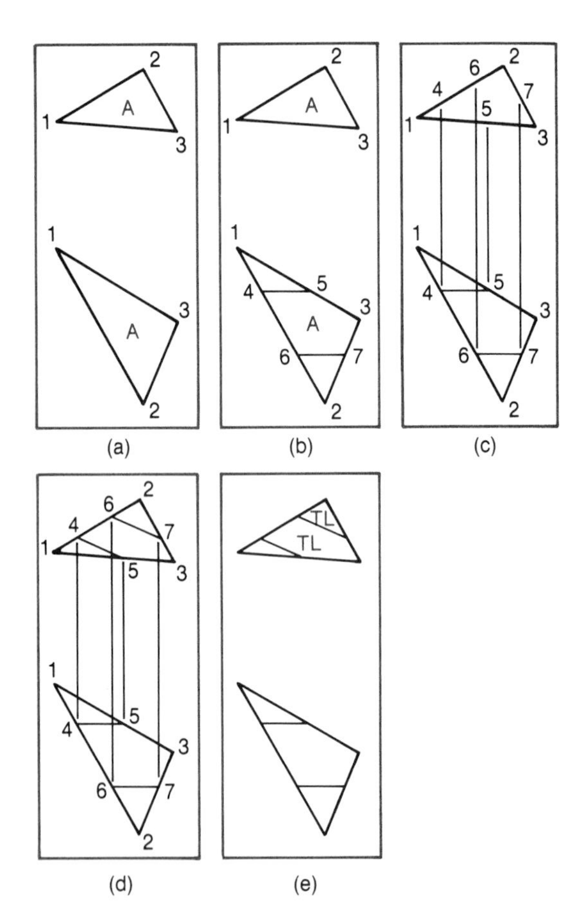

Figure 12.1 In two given views of a triangular surface (top and front views) (a), two horizontal lines are introduced (b), then points are projected (c), and lines $\overline{45}$ and $\overline{67}$ are constructed in the top view (d). In the final set of views, the horizontal lines appear in their true length (TL) in the top view (e).

A **frontal line** is parallel to the frontal reference plane (FRP) and thus projects onto this plane in its true length. Figure 12.2 presents a set of two views (top and front) of a triangular surface, A, defined by its three corner points, 1, 2, and 3. Two frontal lines $\overline{45}$ and $\overline{67}$ are identified in the top view of surface A (b). The front view must now be completed in order to show lines $\overline{45}$ and $\overline{67}$ in their true lengths and with their true orientations, or directions in space. After projection of the endpoints 4, 5, 6, and 7 from the top view to the

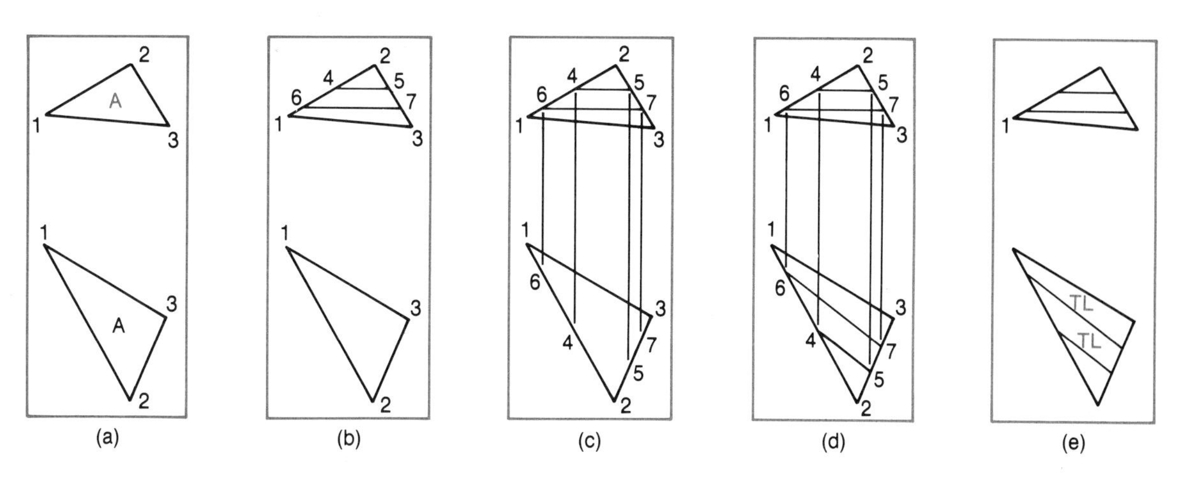

Figure 12.2 In two given views of a triangular surface (top and front views) (a), two frontal lines are introduced (b), then points are projected (c), and lines $\overline{45}$ and $\overline{67}$ are constructed in the front view (d). In the final set of views, the frontal lines appear in their true length (TL) in the front view.

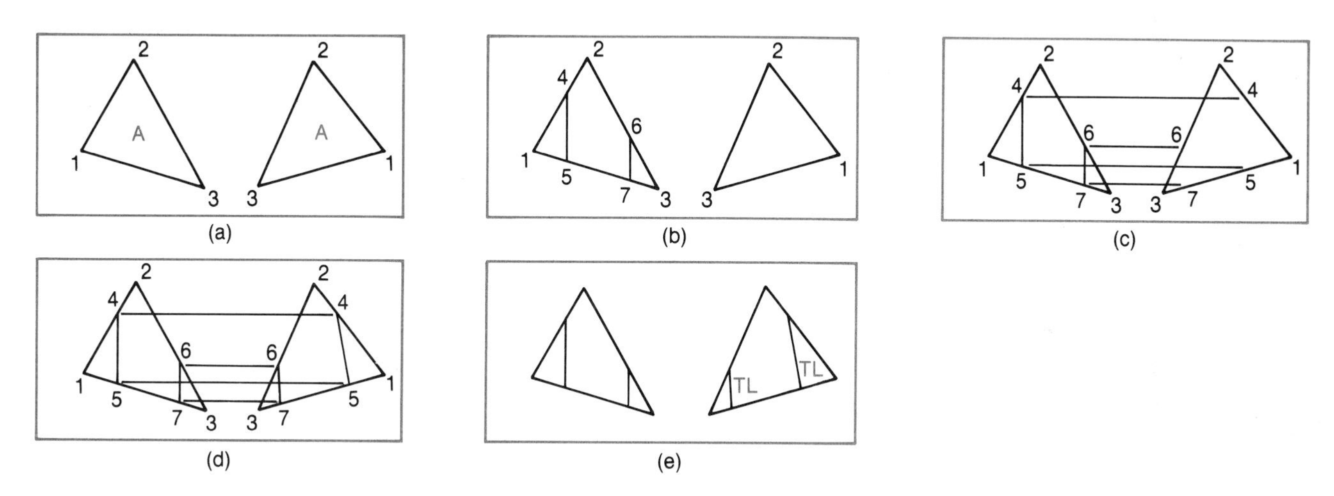

Figure 12.3 In two given views of a triangular surface (front and right-side views) (a), two profile lines are introduced (b), then points are projected (c), and lines $\overline{45}$ and $\overline{67}$ are constructed in the right-side view (d). In the final set of views, the profile lines appear in their true length (TL) in the right-side view.

front view (c), the lines $\overline{45}$ and $\overline{67}$ are drawn in true length in the final front view (d and e).

A **profile line** is parallel to the profile reference plane (PRP) and thus projects onto this plane in its true length. Figure 12.3 treats profile lines the same way horizontal and frontal lines were treated in Figures 12.1 and 12.2 respectively.

12.2 Specifying a Line within a Plane

Figure 12.4 shows how to specify a line within a plane. The line $\overline{45}$ is contained within the plane A, as shown in the completed front view of this figure (a). (The plane A is defined

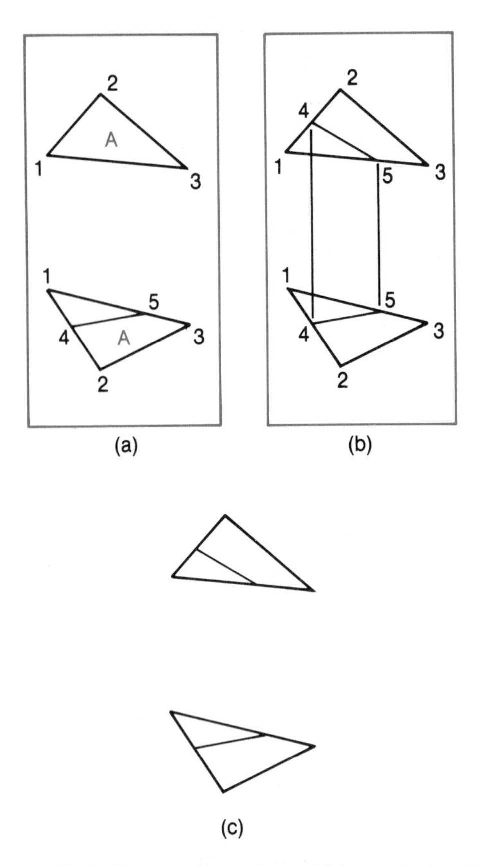

Figure 12.4 To specify a line within a plane, given plane A and the line $\overline{45}$ within the front view (a), points 4 and 5 are projected from the given view to the construction region of the top view (b), resulting in completed front and top views (c).

in terms of its three corner points, 1, 2, and 3.) To locate the line $\overline{45}$ in the incomplete top view, points 4 and 5 are projected from the front view to their appropriate locations in the top view (b). Since point 4 lies on the line $\overline{12}$ in the given front view, it must also lie on $\overline{12}$ in the top view, thereby restricting its location in this view to that *unique* position on line $\overline{12}$ that is at the same width position as point 4 in the front view. Similarly, point 5 can be uniquely specified on line $\overline{13}$ in the top view via projection from the front view. In the completed front and top views (c), line $\overline{45}$ has been properly located within the given plane.

12.3 Specifying a Point within a Plane

Figure 12.5 shows how to specify a point within a plane. The point 4 is contained within the plane A, as shown in the front view (a). (Plane A is defined in terms of its three corner points, 1, 2, and 3.) In order to completely specify the location of point 4 within plane A, we must identify its depth location—that is, we must locate point 4 in the given top view.

We need to construct a line, $\overline{56}$, that contains point 4 (b). Projection of this line from the front view to the top view, in accordance with our discussion in Section 12.2, then allows us to locate point 4 on line $\overline{56}$ in the top view. Removal of the contruction line $\overline{56}$ results in a final set of views (c) in which point 4 is completely specified within the given plane A.

12.4 Parallel Lines

Parallel lines, such as the inclined lines $\overline{12}$ and $\overline{34}$ in Figure 12.6, always appear as parallel lines in orthographic views.

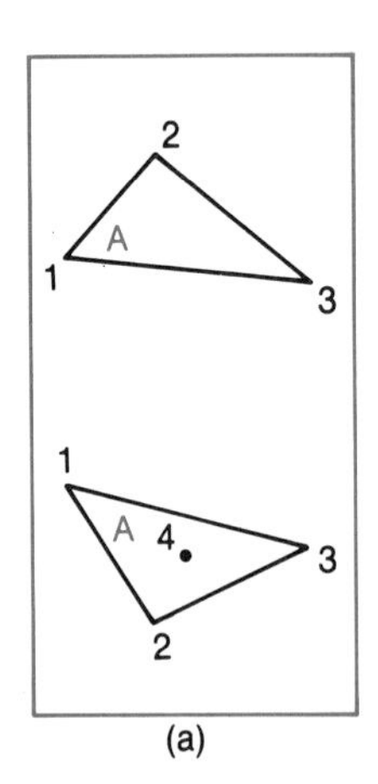

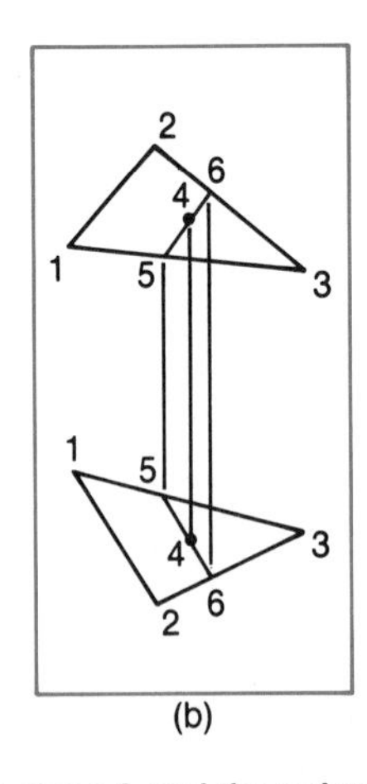

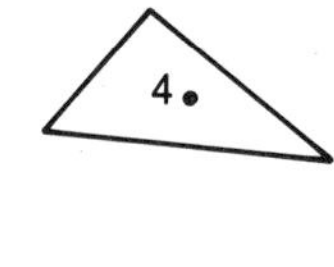

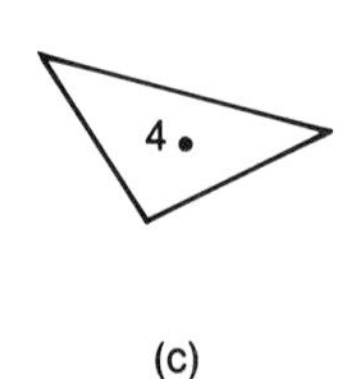

Figure 12.5 To specify a point within a plane, given plane A and the point 4 within the front view (a), points 5 and 6 (which define the construction line $\overline{56}$) are projected from the given view to the construction region of the top view (b), followed by projection of point 4, resulting in completed front and top views (c).

Notice that line $\overline{34}$ is visible in the right-side view of the figure, whereas $\overline{12}$ is hidden by $\overline{34}$ in this view; nevertheless, both lines are indeed parallel to one another in this view.

12.5 Perpendicular Lines

Perpendicular lines appear as perpendicular (at 90° to one another) in any view in which at least one of the lines is shown in its true length. Lines $\overline{13}$ and $\overline{24}$ are shown in true length (TL) in both the top and front views of Figure 12.7 (b) (both lines are normal to the profile reference plane). Thus, we may conclude that lines $\overline{12}$ and $\overline{34}$ are indeed perpendicular to $\overline{13}$ and $\overline{24}$, as shown in these views. This is an extremely useful characteristic of perpendicular lines, as we will discover in the following sections.

12.6 A Line Parallel to a Plane

A line and a plane are parallel if the line is parallel to any other line contained within the plane. Consider line $\overline{12}$ and plane A shown in the top and front views of Figure 12.8 (where plane A has been defined in terms of its three corner points, 3, 4, and 5). Line $\overline{12}$ is parallel to the line $\overline{35}$ of plane A in both of the given views; therefore, these two lines must be truly parallel in space. Accordingly, line $\overline{12}$ and plane A must also be parallel.

12.7 Parallel Planes

Recall that a plane can be defined in terms of two intersecting, nonparallel lines that are contained within the plane. Parallelism between two planes can be specified by two intersecting lines in one plane that are parallel to two intersect-

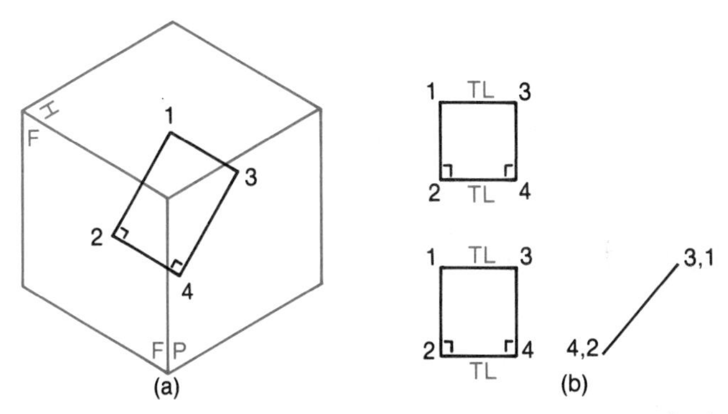

Figure 12.7 Perpendicular lines always appear as perpendicular lines in orthographic views in which at least one of the lines is shown in true length. Lines $\overline{12}$, $\overline{34}$, $\overline{13}$, and $\overline{24}$ are shown within the glass box (a). In three principal views of these lines, lines $\overline{13}$ and $\overline{24}$ appear in true length (TL) in the top and front views (b). As a result, we know that lines $\overline{13}$ and $\overline{24}$ must be truly perpendicular to lines $\overline{12}$ and $\overline{34}$, respectively, as shown in these two views.

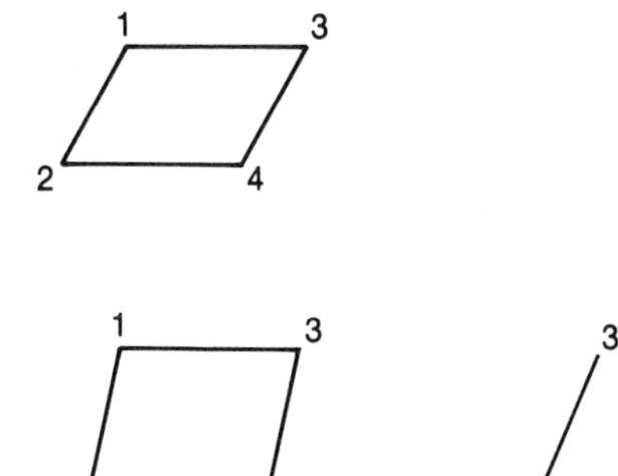

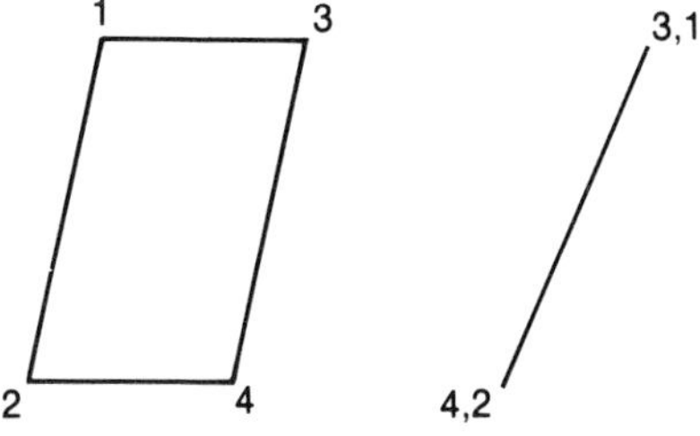

Figure 12.6 Parallel lines always appear as parallel lines in orthographic views. Two sets of parallel lines ($\overline{12}$ and $\overline{34}$, $\overline{13}$ and $\overline{24}$) are shown in this figure.

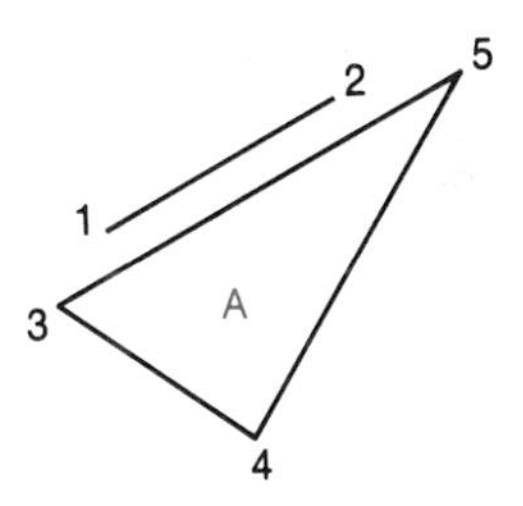

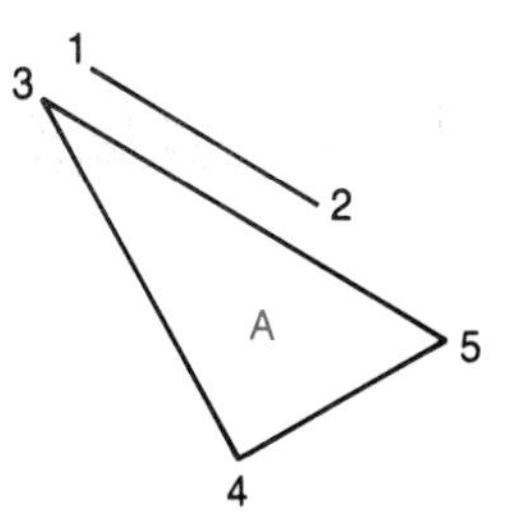

Figure 12.8 A line parallel to a given plane is also parallel to a line within the plane. Two (top and front) views of a line $\overline{12}$ and a plane A (defined in terms of three points, 3, 4, and 5, within the plane) are shown.

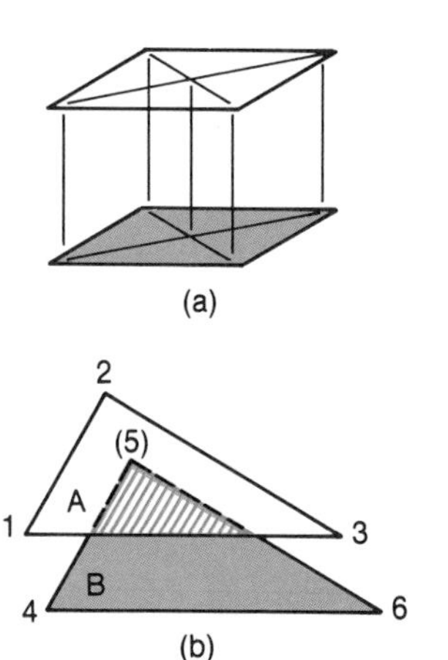

Figure 12.9 Parallel planes may be specified in terms of sets of intersecting lines that are parallel to one another. For two planes defined in terms of intersecting lines, projection demonstrates that one set of lines is parallel to the other set, so the planes must be parallel to one another (a). The intersecting lines $\overline{12}$, $\overline{23}$, and $\overline{13}$ in plane A are parallel to the intersecting lines $\overline{45}$, $\overline{56}$, and $\overline{46}$ in plane B, respectively; planes A and B must therefore be parallel (b).

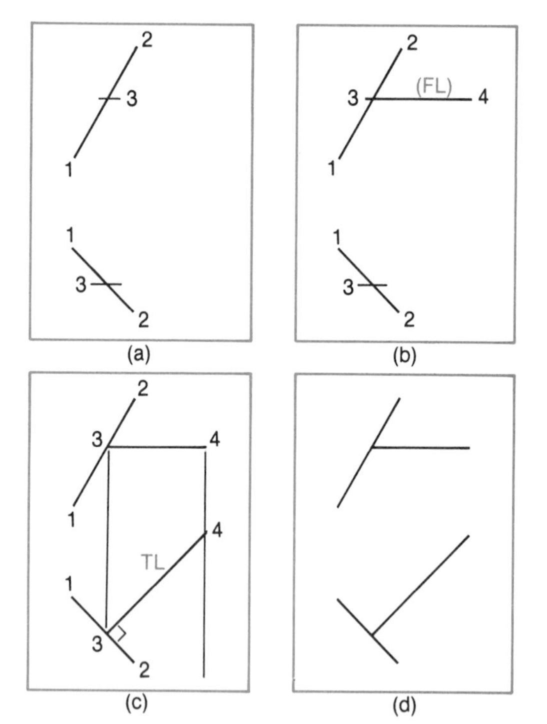

Figure 12.10 To construct a line perpendicular to the given line shown in two (top and front) views and passing through point 3 (a), construct a line that will appear in true length (TL) in one of the given views—for example, the frontal line (FL) $\overline{34}$ shown here (b) appears in true length in the given front view (c). In the final views, the lines $\overline{12}$ and $\overline{34}$ are mutually perpendicular (d).

ing lines in the other plane. In the two parallel planes shown in Figure 12.9, notice that the intersecting lines $\overline{12}$, $\overline{23}$ and $\overline{13}$ in plane A are parallel to the intersecting lines $\overline{45}$, $\overline{56}$ and $\overline{46}$, respectively, in plane B.

12.8 Constructing a Line Perpendicular to a Given Line

In descriptive geometry efforts, it is frequently necessary to construct a line that is perpendicular to another given line and that passes through a given point. (Perpendicularity with respect to a given line defines the *orientation* of the line under construction; inclusion of the given point within the line under construction specifies the *location* of the line in space.)

In Figure 12.10, a line $\overline{12}$ is specified in two (top and front) views (a). We wish to construct a line that passes through point 3 (shown in the figure) and that is perpendicular to the line $\overline{12}$. In order to accomplish this goal, recall that two perpendicular lines will appear perpendicular (at 90° to one another) in any view where at least one of the lines is shown in true length (see Section 12.5). As a result, we need only construct a true-length line through point 3 that is drawn at 90° to line $\overline{12}$ in one of the given views of Figure 12.10(a). For example, we could draw a construction *frontal* line, $\overline{34}$, through point 3 (b), that will then appear in true length in the front view. Adding the frontal line $\overline{34}$ to the top view (b) does not restrict point 4 to a particular height position in the front view; as a result, we may simply draw $\overline{34}$ at 90° to $\overline{12}$ in the front view (c). Line $\overline{34}$ is then completely specified: it is perpendicular to line $\overline{12}$ and it passes through point 3 as required (d).

Learning Check

Identify the reference planes (horizontal, frontal, and profile) in which each of the following types of lines is shown in true length: horizontal line (HL), frontal line (FL), and profile line (PL).

Answer Horizontal, frontal, and profile lines appear in true length in horizontal, frontal, and profile reference planes, respectively.

12.9 Constructing a Line Perpendicular to a Given Plane

Figure 12.11 shows how to construct a line that is perpendicular to a given plane, surface A, shown in two (top and front) views and that passes through a given point, 4 (a). (The plane is defined in terms of its three corner points, 1, 2, and 3.) Again, we will use the fact that two perpendicular lines appear perpendicular in any view where at least one line is shown in true length. A *horizontal* line, $\overline{56}$, is constructed through point 4 in the front view (b). This line, $\overline{56}$, is then projected to the top view in which it will appear in true length (c). The line $\overline{47}$ is then drawn in the top view so that it is perpendicular to $\overline{56}$ (d). Finally, line $\overline{47}$ is determined in the front view by first defining a *frontal* line, $\overline{89}$, in the top view so that $\overline{89}$ is contained in plane A and intersects $\overline{47}$ at point 4. Projection of line $\overline{89}$ to the front view then uniquely specifies this line in space. If we further specify that $\overline{47}$ is perpendicular to $\overline{89}$, $\overline{47}$ must appear at 90° to the true-length image of $\overline{89}$ in the front view (e). Line $\overline{47}$ is then drawn perpendicular to line $\overline{89}$ in the front view. Since (1) line $\overline{89}$ is seen in true length in the front view, (2) line $\overline{56}$ is seen in true length in the top view, and (3) line $\overline{47}$ is perpendicular to these two lines in their true-length views, we know that line $\overline{47}$ is indeed perpendicular to both lines $\overline{56}$ and $\overline{89}$. Since lines $\overline{56}$ and $\overline{89}$ are both contained within plane A, we then know that line $\overline{47}$ is perpendicular to plane A (f).

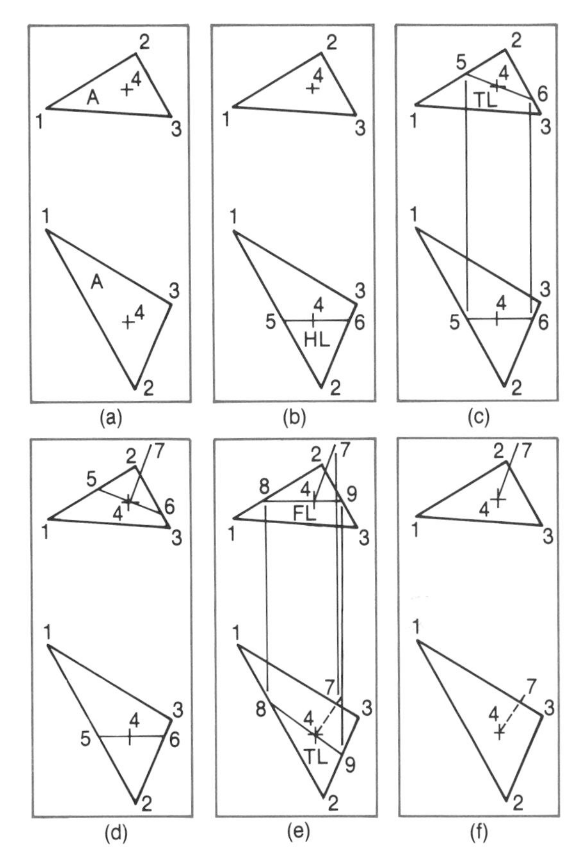

Figure 12.11 To construct a line perpendicular to the given plane A shown in two (top and front) views, construct a line that will appear in true length (TL) in one of the given views—for example, the horizontal line (HL) shown here (b), through point 4, appears in true length in the given top view (c). Another line, $\overline{47}$, is drawn perpendicular to the horizontal line, $\overline{56}$ (d). Finally, line $\overline{47}$ is constructed in the other view by first defining a frontal line, $\overline{89}$ in the top view, intersecting line $\overline{47}$ at point 4 and then projecting line $\overline{89}$ to the other view (e). In the final front view, line $\overline{47}$ is hidden by plane A (f).

12.10 Constructing a Plane Perpendicular to a Given Line

Figure 12.12 shows how to construct a plane that contains a particular point (thereby specifying the plane's location in space) and that is perpendicular to a given line (thereby constraining the orientation of the plane to a single possibility). Once again, we will use the fact that two perpendicular lines will appear perpendicular in any view where at least one line is shown in true length.

Figure 12.12 presents two views (top and front) of a line, $\overline{12}$, and a point, 3 (a). Construction of a plane perpendicular to line $\overline{12}$ and containing point 3 begins with drawing a true-

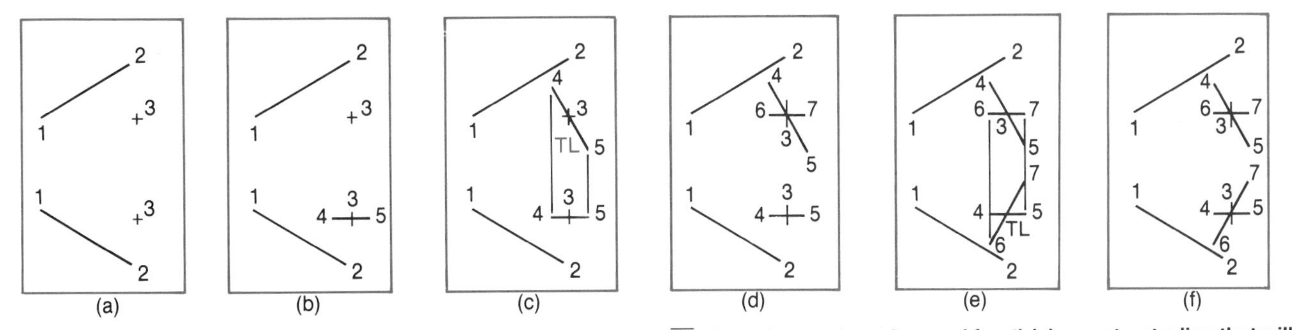

Figure 12.12 To construct a plane perpendicular to the given line $\overline{12}$ shown in two views (top and front) (a), construct a line that will appear in true length (TL) in one of the given views—for example, the horizontal line (HL) shown here (b) appears in true length in the given top view (c). Another line, $\overline{67}$, is introduced in the top view (d) and projected to the front view (e), where it is shown in true length (TL). The intersecting lines $\overline{45}$ and $\overline{67}$ therefore define a plane that is perpendicular to line $\overline{12}$ (f).

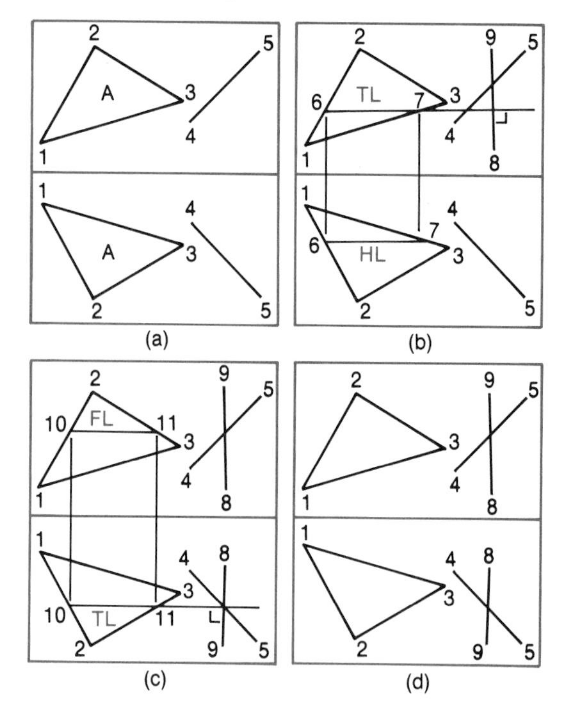

Figure 12.13 To construct a plane perpendicular to the given plane A shown in two views (top and front) (a), construct a line that will appear in true length (TL) in one of the given views—for example, the horizontal line (HL) shown here appears in true length in the given top view. Another line, $\overline{89}$, is introduced also in the top view, and then a frontal line, $\overline{10\text{-}11}$, as well (c), which is projected to the front view, where it is shown in true length (TL). The intersecting lines $\overline{45}$ and $\overline{89}$ therefore define a plane that is perpendicular to plane A (d).

length horizontal line, $\overline{45}$, through point 3 in the front view (b). If we define $\overline{45}$ to be perpendicular to $\overline{12}$, then these lines must appear at 90° to one another in the top view, since the horizontal line $\overline{45}$ appears in true length in this view (c).

Similarly, we may introduce a frontal line, $\overline{67}$, in the top view and define this line to be perpendicular to $\overline{12}$ (d). As a result, $\overline{67}$ and $\overline{12}$ will appear at 90° to one another in the front view where the frontal line $\overline{67}$ is shown in true length (e). The intersecting lines $\overline{45}$ and $\overline{67}$ then define a plane that is perpendicular to the given line $\overline{12}$ (f).

12.11 Constructing a Plane Perpendicular to a Given Plane

Figure 12.13 shows how to construct a plane, B, perpendicular to the given plane A (which specifies the orientation of plane B in space) and containing line $\overline{45}$. (Plane A is defined by its three corner points, 1, 2, and 3.) Again, we use the fact that two perpendicular lines will appear perpendicular in any view where at least one of these lines is shown in true length.

A horizontal construction line, $\overline{67}$, is drawn in the front view of the figure (b) and then projected to the top view, where it appears in true length. Another construction line, $\overline{89}$, is drawn at 90° to $\overline{67}$ in the top view; lines $\overline{89}$ and $\overline{67}$ are then truly perpendicular to one another in space. A frontal construction line, $\overline{10\text{-}11}$, is then drawn in the top view (c) and projected to the front view, where it appears in true length. The line $\overline{89}$ is projected from the top view to the front view and drawn at 90° to $\overline{10\text{-}11}$ in the front view. Lines $\overline{89}$ and $\overline{10\text{-}11}$ are then truly perpendicular to one another in space. Line $\overline{89}$ is thus perpendicular to the plane A since it is perpendicular to two lines ($\overline{67}$ and $\overline{10\text{-}11}$) contained within A. Together with line $\overline{45}$, line $\overline{89}$ defines plane B, which is perpendicular to A and which contains line $\overline{45}$ as required.

12.12 Intersecting Lines

Two **intersecting lines** will always intersect at the same point in all views. Therefore, to determine whether two lines that *appear* to cross each other in a set of views *truly* intersect, we need only verify that the point at which the lines cross in each view is a single point in space.

Figure 12.14 presents two lines, $\overline{12}$ and $\overline{34}$, enclosed within the glass box and intersecting at point 5 (a). When these lines and the point are projected onto the walls of the glass box and the walls are unfolded, notice that the location of the intersection point 5 can be projected with common dimensions between adjacent views, verifying that point 5 is indeed a single point in space (b). In other words, the point

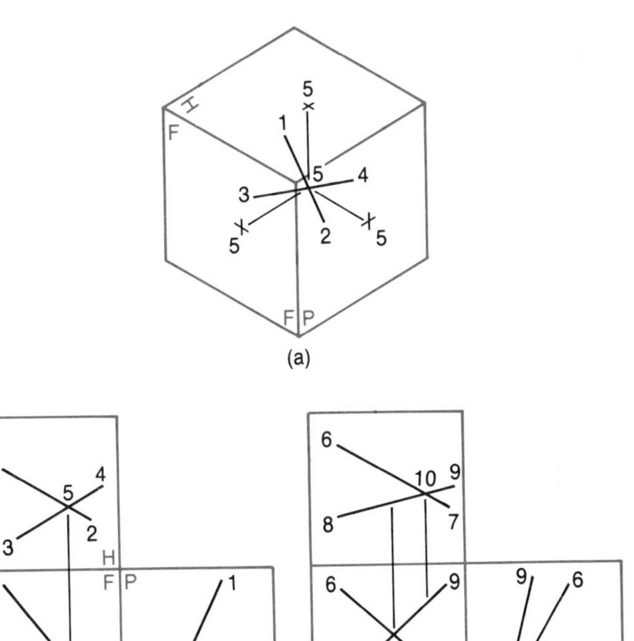

Figure 12.14 Intersecting and nonintersecting lines. Intersecting lines $\overline{12}$ and $\overline{34}$ within the glass box (a) are proven to intersect at point 5 (b). Lines $\overline{67}$ and $\overline{89}$ do not intersect (c).

5 at which lines $\overline{12}$ and $\overline{34}$ cross in each view is in fact the point of intersection between these lines.

Next consider the lines $\overline{67}$ and $\overline{89}$ shown in a set of orthographic views (c). These lines cross at point 10 in the top view and point 11 in the front view. We know that points 10 and 11 are distinct, since they do not share a common dimension of width in the top and front views (that is, points 10 and 11 cannot be projected between these two views along a single vertical construction line). As a result, we also know that although lines $\overline{67}$ and $\overline{89}$ *appear* to intersect in the top and front views, they do not truly intersect in space. And inspection of the right-side view of these lines verifies that $\overline{67}$ and $\overline{89}$ do not intersect.

12.13 Shortest Distance between a Point and a Line

In order to determine the shortest distance between a point and a line, we need to develop a point view of the line in which the point also appears. Consider the three (top, front, and right-side) orthographic views of the line $\overline{12}$ and the (noncollinear) point 3 in Figure 12.15. As discussed in the previous chapter, the point view of an oblique line such as $\overline{12}$

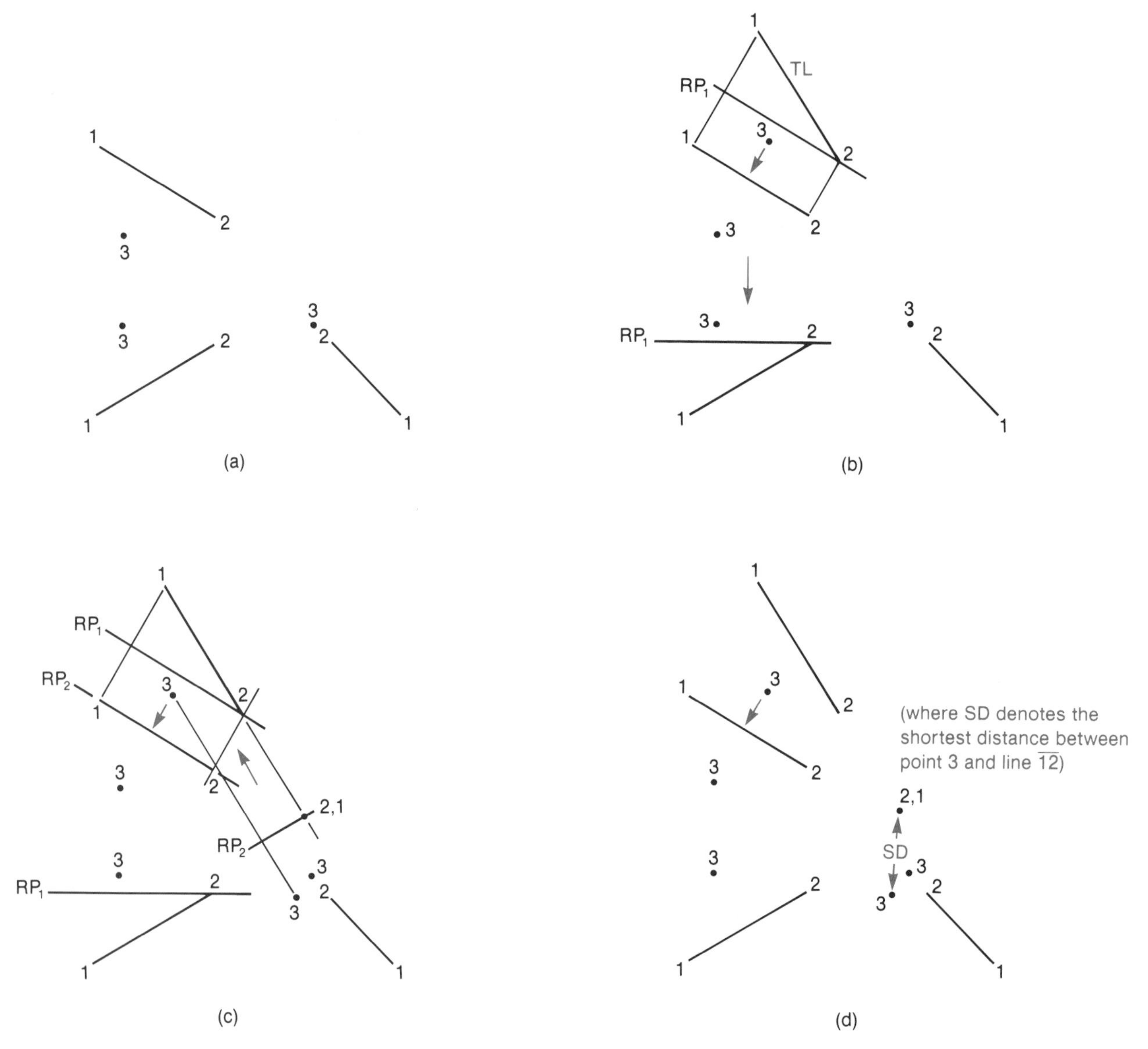

Figure 12.15 To determine the shortest distance between point 3 and line $\overline{12}$ (a), first construct a primary auxiliary view of the point 3 and a true-length image of line $\overline{12}$ (b). Then construct a secondary auxiliary (point) view of line $\overline{12}$ and point 3 (c). The final set of views shows the shortest distance (SD).

corresponds to a particular secondary auxiliary view of the line. So first a primary auxiliary view in which line $\overline{12}$ appears in true length is obtained (b). (The location of point 3 is also identified in this primary auxiliary view.) Then the secondary auxiliary view—corresponding to a sight-line that is parallel to line $\overline{12}$—is generated (c). Line $\overline{12}$ appears as a point in this view. The distance between the point image of $\overline{12}$ and point 3 in this secondary auxiliary view is the shortest distance (SD) between line $\overline{12}$ and point 3 (d).

12.14 Shortest Distance between Two Lines

In order to determine the shortest distance between two lines, we need to construct a secondary auxiliary view of these lines in which one of them appears as a point. Consider the three orthographic views of the oblique lines $\overline{12}$ and $\overline{34}$ in Figure 12.16. First a primary auxiliary view of lines $\overline{12}$ and $\overline{34}$ is developed in which line $\overline{34}$ is shown in true length (b). A secondary auxiliary view is then constructed in which line $\overline{34}$ appears as a point (c). The shortest distance between lines $\overline{12}$ and $\overline{34}$ (seen as a point) may then be obtained from this secondary auxiliary view.

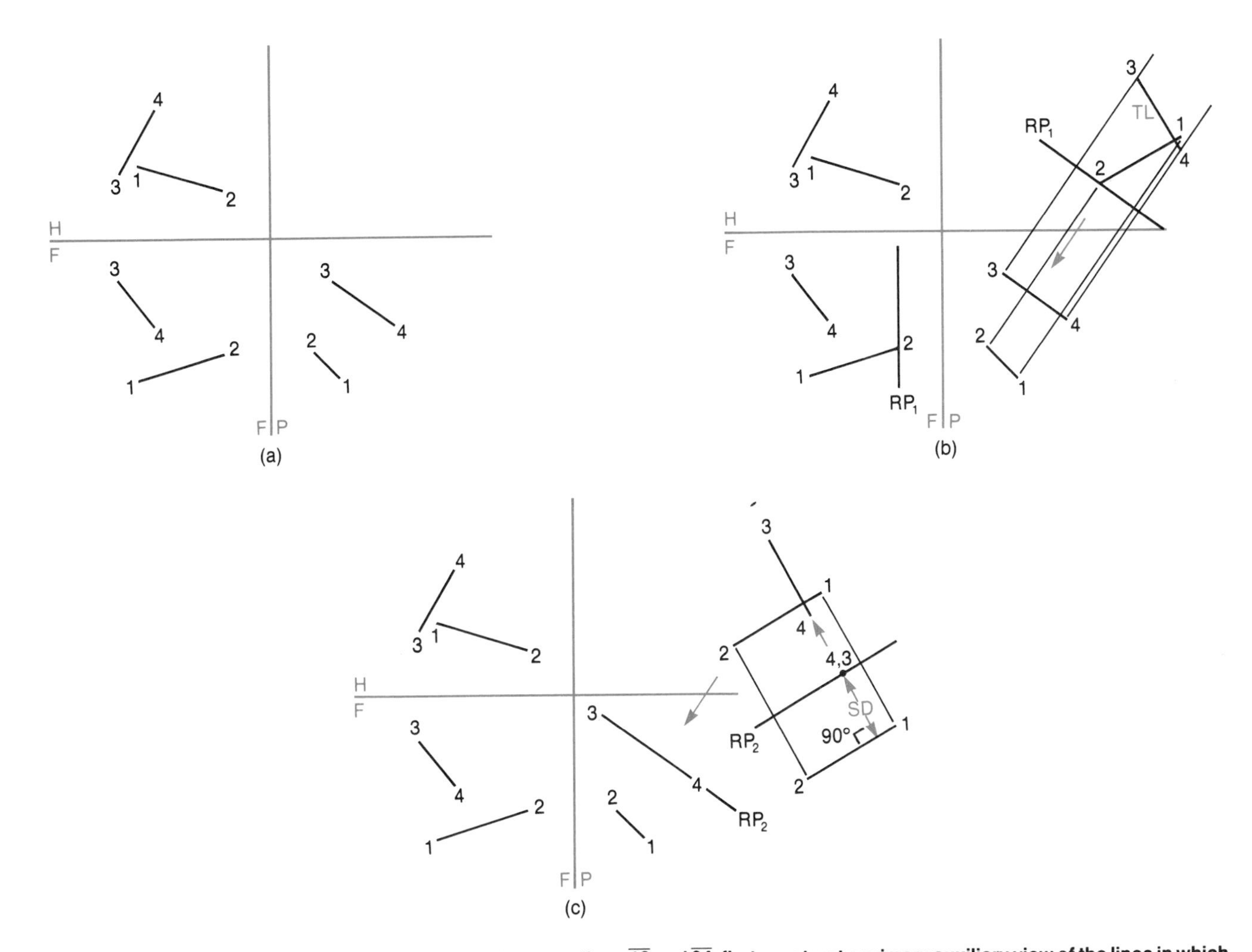

Figure 12.16 To determine the shortest distance between lines $\overline{12}$ and $\overline{34}$, first construct a primary auxiliary view of the lines in which line $\overline{34}$ is shown in true length (b). Then construct a secondary auxiliary view in which line $\overline{34}$ appears as a point (c). The shortest distance (SD) between lines $\overline{12}$ and $\overline{34}$ may be obtained from this view.

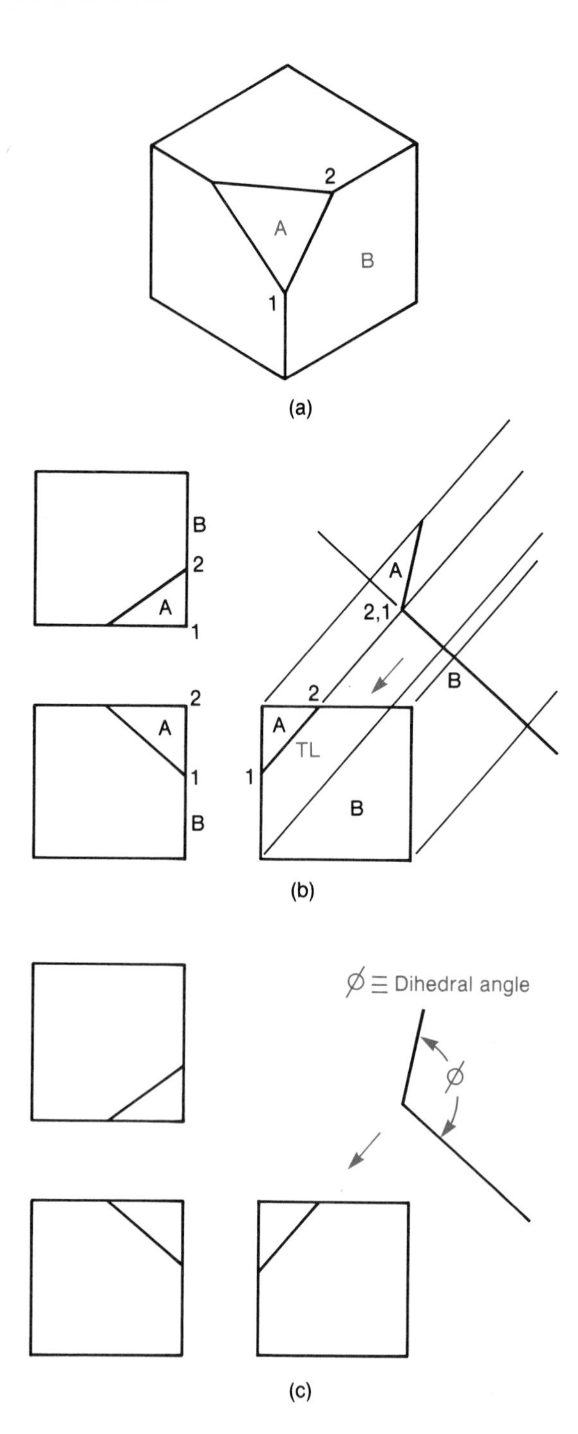

Figure 12.17 To determine the dihedral angle between planes A and B, which intersect along line $\overline{12}$ (a), first construct a primary auxiliary view in which line $\overline{12}$ appears as a point and planes A and B appear on edge, as a line (b). Note that the profile line $\overline{12}$ appears in true length (TL) in the given right-side view. The final set of views includes a primary auxiliary view in which the dihedral angle Φ between planes A and B is accurately shown (c).

12.15 Dihedral Angle between Two Planes

The angle between two planes is known as the **dihedral angle.** It will be shown accurately in the orthographic view where the line of intersection between the planes appears as a point. Figure 12.17 presents an object in which an oblique surface, A, and a vertical right-side surface, B, have been identified (a), along with a set of principal (top, front, and right-side) orthographic views of this object (b). To determine the dihedral angle between planes A and B, first construct a primary auxiliary view in which the intersection line, $\overline{12}$, between planes A and B appears as a point. (Note that the auxiliary viewing direction is parallel to line $\overline{12}$ in the basic view so that line $\overline{12}$ will be seen as a point in the auxiliary view.) The final set of views identifies the dihedral angle between planes A and B (c).

Figure 12.18 presents another object in which an oblique surface, A, and an inclined surface, B, have been identified (a), along with a set of principal orthographic views of this object (b). To develop a view in which the dihedral angle between surfaces A and B is correctly shown, we need to first construct a primary auxiliary true-length view of the intersection line $\overline{12}$ between these surfaces (c). A secondary auxiliary view in which the dihedral angle is correctly shown may then be constructed (d). The secondary auxiliary view sightline must be parallel to line $\overline{12}$ in the primary auxiliary view so that $\overline{12}$ will appear as a point in the secondary view. The final views for this object identifies the dihedral angle between planes A and B (e).

12.16 Visibility of Lines

In addition to determining true relative positions and orientations of graphic elements (points, lines, and planes), descriptive geometry allows us to determine the relative **visibility** of such elements that cross one another in an orthographic view. Consider the top and front views of two crossing lines ($\overline{12}$ and $\overline{34}$) in Figure 12.19. We recognize that these lines do not intersect, since the two points (5 and 6) at which they cross in the given views are distinct—as indicated by their distinct width positions (point 5 in the top view does not lie directly above point 6 in the front view).

To determine the relative visibility of these lines, it is necessary to determine which line is nearer to the observer in each of these views. We begin our analysis by projecting the crossing point 5 between lines from the top view to the front view (b). The projection line intersects with line $\overline{12}$ in the front view before it can intersect line $\overline{34}$ in this view. We may then conclude that where the lines cross at point 5 in the top

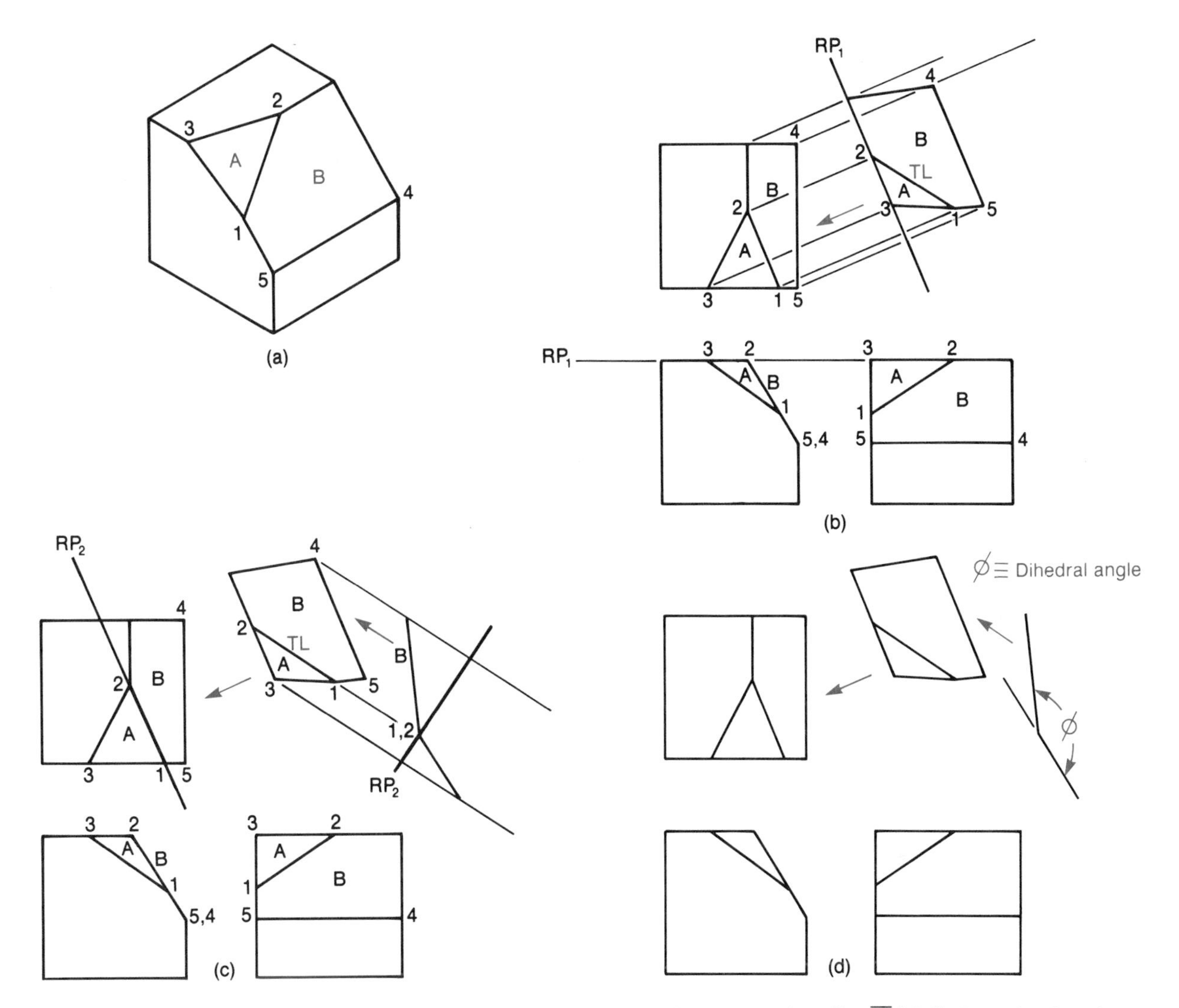

Figure 12.18 To determine the dihedral angle between planes A and B, which intersect along line $\overline{12}$ (a), first construct a primary auxiliary view in which the intersection line $\overline{12}$ appears in true length (TL) (b). Then construct a secondary auxiliary view in which the intersection line $\overline{12}$ appears as a point and planes A and B appear on edge, as lines (c). The final set of views includes a secondary auxiliary view in which the dihedral angle Φ between planes A and B is accurately shown (d).

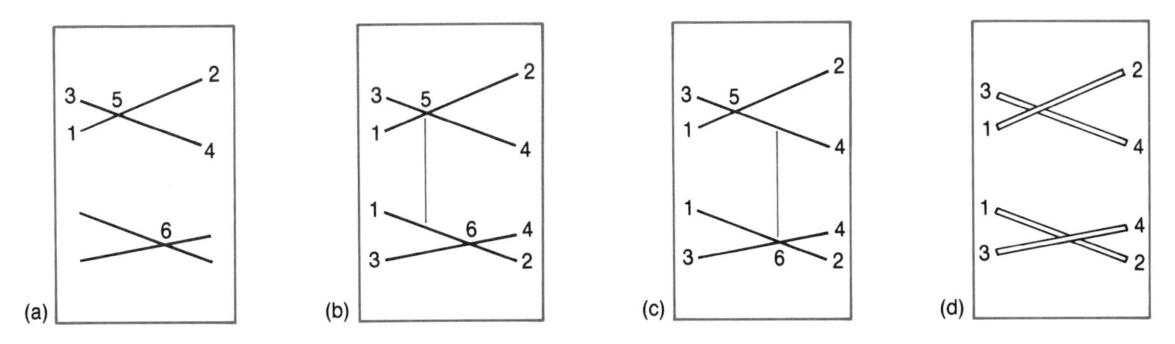

Figure 12.19 To determine the relative visibility of lines $\overline{12}$ and $\overline{34}$ (a), project the crossing point 5 from the top view to the front view (b). This indicates that line $\overline{12}$ lies above line $\overline{34}$, so line $\overline{12}$ is entirely visible in the top view. Projection of crossing point 6 from the front view to the top view indicates that line $\overline{34}$ lies in front of line $\overline{12}$ (c), so line $\overline{34}$ is entirely visible in the front view. When the lines are shown as two crossing rods, rod $\overline{34}$ is visible in the front view, and rod $\overline{12}$ is visible in the top view.

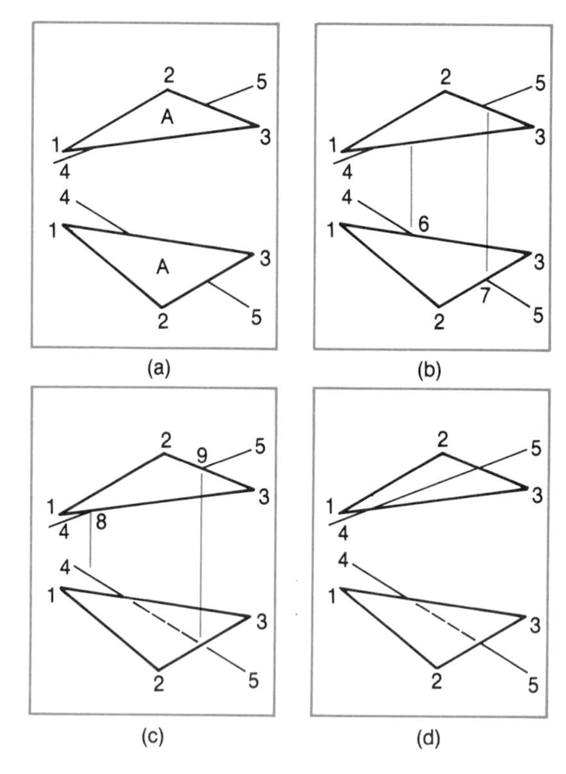

Figure 12.20 **To determine the relative visibility of line $\overline{45}$ and plane A (defined in terms of its three corner points, 1, 2, and 3) (a), project crossing points 6 and 7 from the top view to the front view (b). This indicates that surface A lies in front of line $\overline{45}$. Projection of points 8 and 9 indicates that line $\overline{45}$ lies above surface A (c). The final set of views accurately portrays the relative visibility (d).**

view, line $\overline{12}$ lies above line $\overline{34}$. Consequently, line $\overline{12}$ is entirely visible in the top view.

We continue our analysis by projecting the crossing point 6 between lines from the front view to the top view (c). The projection line is seen to intersect line $\overline{34}$ in the top view before it can intersect line $\overline{12}$ in this view. Thus, line $\overline{34}$ lies in front of line $\overline{12}$ at the location of point 6. Consequently, line $\overline{34}$ is entirely visible in the front view.

To emphasize the importance of visibility considerations in orthographic views of three-dimensional objects, lines $\overline{12}$ and $\overline{34}$ are portrayed as slender rods in part d of the figure. Obviously, such rods, unlike lines, could not be portrayed accurately unless it is known which rod is entirely visible in each view.

12.17 Visibility of a Line and a Plane

Figure 12.20 shows how to determine the relative visibility of a line, $\overline{45}$, and a plane, triangular surface A, in top and front views.

Two points (denoted as 6 and 7) at which line $\overline{45}$ contacts surface A in the front view are projected to the top view (b) in order to determine if these points actually lie on the surface A or on the line $\overline{45}$.

Point 6 appears to be on the surface boundary edge $\overline{13}$ in the front view. The projection line for this point intersects line $\overline{13}$ in the top view before it can intersect line $\overline{45}$. Thus, $\overline{13}$ must lie front of $\overline{45}$ at the width position of point 6. Consequently, point 6—which, by definition, is visible in the front view—must lie on line $\overline{13}$.

Point 7 appears to be on the surface boundary edge $\overline{23}$ in the front view (b). The projection line for point 7 intersects line $\overline{23}$ in the top view before it can intersect line $\overline{45}$. Thus, $\overline{23}$ must lie in front of $\overline{45}$ at the width position of point 7. Consequently, point 7—which, by definition, is visible in the front view—must lie on line $\overline{23}$.

Points 6 and 7 both lie on surface A (on the two edges $\overline{13}$ and $\overline{23}$, respectively), so both points are visible in the front view. Surface A is therefore visible in the front view at points 6 and 7 with line $\overline{45}$ lying behind it, as indicated by hidden linework (c).

A similar analysis may be performed for the top view (c). Points 8 and 9 are identified in this view. These points appear to lie on surface A—on edges $\overline{13}$ and $\overline{23}$, respectively. However, projection of point 8 from the top view to the front view results in the intersection of the projection line with line $\overline{45}$ in the front view. Therefore, line $\overline{45}$ must lie above surface A at the position of point 8 (and point 8 must, in fact, lie on $\overline{45}$—not $\overline{13}$ on surface A).

Projection of point 9 from the top view to the front view results in the intersection of the projection line with the hidden line $\overline{45}$—before this projection line can intersect edge $\overline{23}$. Point 9 must then lie on line $\overline{45}$.

As a result of our analysis of points 8 and 9—which are visible in the top view and which lie on line $\overline{45}$—we conclude that line $\overline{45}$ is located above surface A. Consequently, line $\overline{45}$ must be entirely visible in the top view (d).

We have determined the relative visibility of line $\overline{45}$ and surface A in the given views of Figure 12.20. The relative visibility of two planes in a set of orthographic views can be analyzed in an analogous manner by focusing upon the visibility of the boundary edges of one surface with respect to the other.

12.18 Intersections between Surfaces

The intersection between two (or more) surfaces in a set of orthographic views can be constructed by using the point-by-point, surface-by-surface procedure presented in Chapter 4. Figure 12.21 shows the point-by-point construction of the intersection between two right circular cylinders. As shown in Figure 12.22, a similar point-by-point approach can be applied to quickly construct the intersection between a right circular cone and a horizontal plane; such an intersection is a circle.

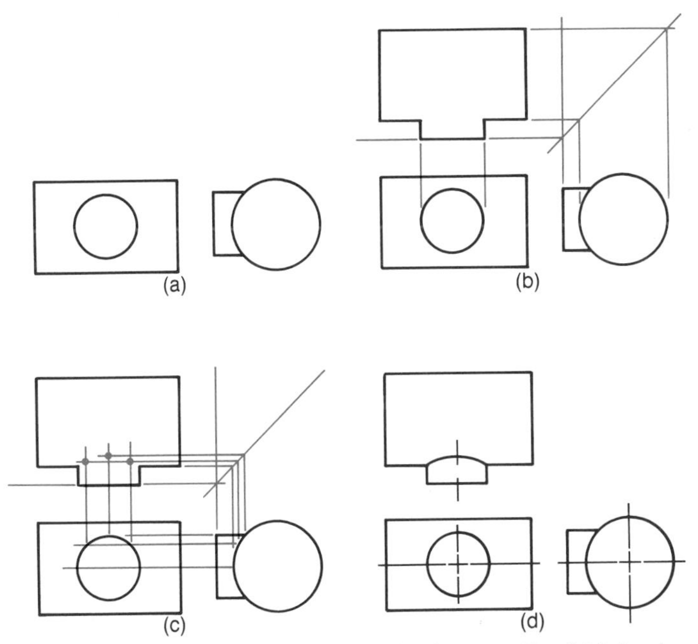

Figure 12.21 To construct an intersection between the right circular cylinders in the given front and right-side views (a), construct outlines of the outer boundary of the missing top view by projection (b). Then develop the intersection between the cylinders via projection of key points from the given views (c). The final set of views shows the intersection (d).

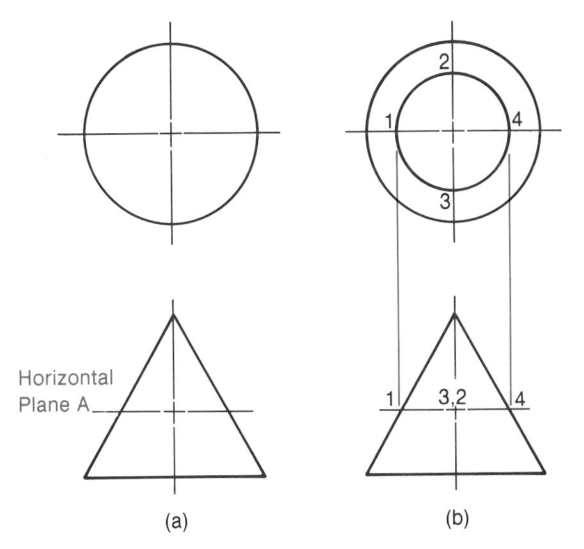

Figure 12.22 To construct an intersection between the right circular cone and horizontal plane in the given top and front views (a), project key points from the intersection of the cone and the horizontal plane in the front view to the top view (b). A circle, representing the intersection in the top view, is then drawn.

The intersection between a right circular cone and an inclined plane, as shown in Figure 12.23(a), is constructed with the aid of several imaginary horizontal construction planes (b). We know that the intersection of each horizontal construction plane with the cone is a circle. Thus, we simply select a set of arbitrary points that lie on the intersection between the cone and the inclined plane, as seen in the given front view of the figure (b). These points are then defined to lie on the horizontal construction planes. It is therefore possible to project these points from the given front view to the top view, since each point must lie on the appropriate circle in the top view that represents the intersection of the cone with the horizontal construction plane containing this point. After a suitable number of points have been determined in the top view, the intersection between the cone and the inclined plane can be drawn (c). Figure 12.24 shows a similar example in which the intersection between an inclined plane and a right circular cone is constructed.

A point-by-point procedure is used in Figure 12.25 to construct the intersection between an oblique cone and an inclined plane in top and front views (a). An arbitrarily selected line is drawn from point 1 (at the tip of the cone) to a point 2 on the boundary of its base (b). The cone and the inclined plane intersect at point 3, which lies on line $\overline{12}$. Points 1 and 2 are then projected to the top view, and the construction line $\overline{12}$ is drawn in this view. Point 3 is then projected to line $\overline{12}$ in the top view. This process is repeated to locate a sufficient number of points in the top view that lie on the intersection between the cone and the inclined plane so that this intersection line can be drawn (c) and shown in the final set of views (d).

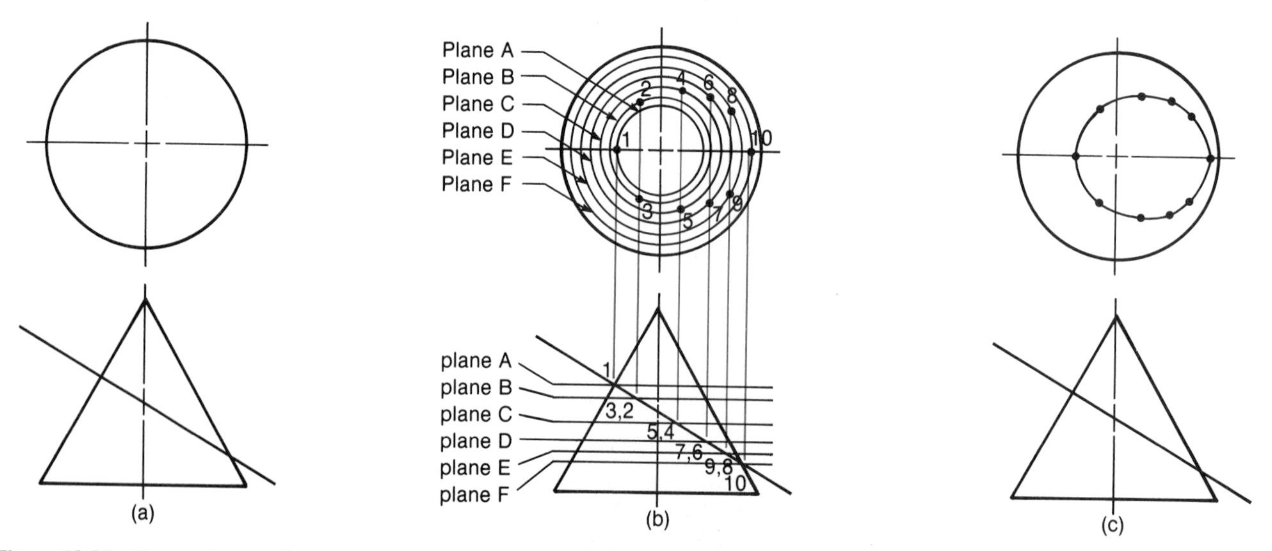

Figure 12.23 **To construct an intersection between the plane and the right circular cone in the given top and front views (a), imaginary horizontal planes are used and their intersection points identified (b). Points projected from the front view to the top view allow construction of the intersection (c).**

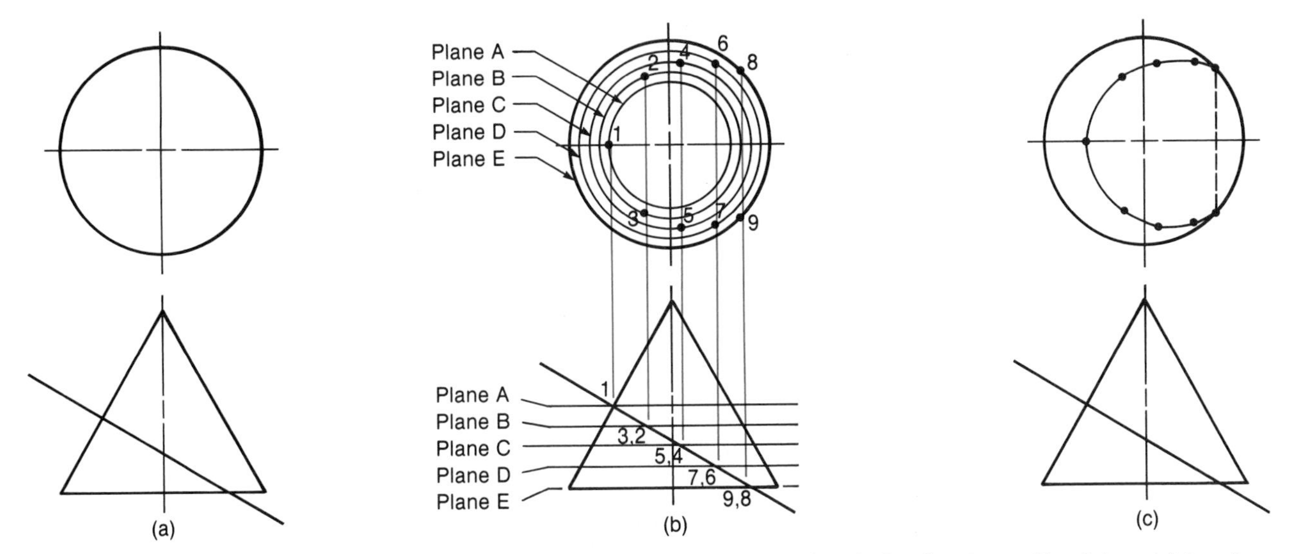

Figure 12.24 To construct an intersection between the right circular cone and the plane in the given top and front views (a), imaginary horizontal planes are used and their intersection points identified (b). Points projected from the front view to the top view allow construction of the intersection, which is in the form of a partial ellipse (c).

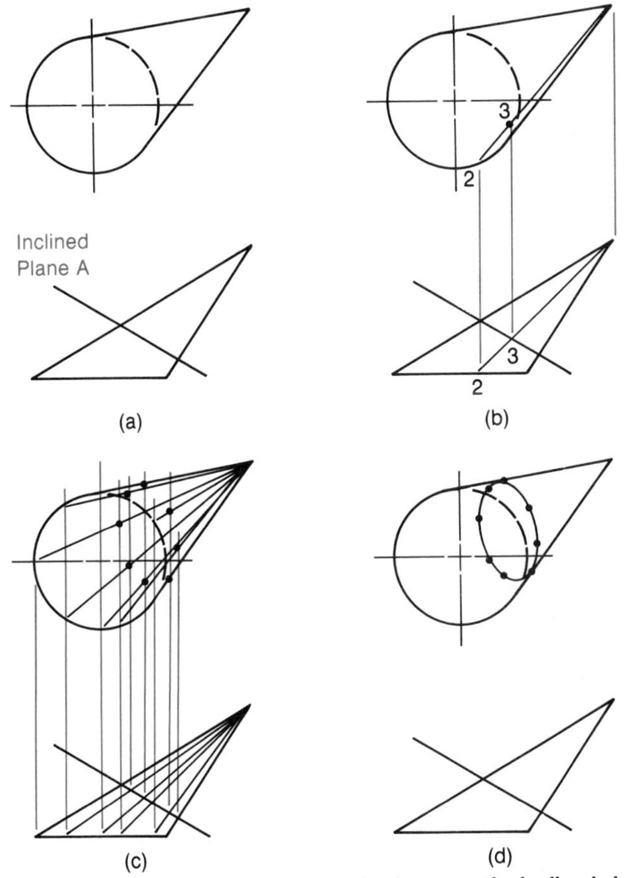

Figure 12.25 To construct an intersection between the inclined plane and the oblique cone in the given top and front views (a), line $\overline{12}$ is drawn in the front view and projected to the top view (b). Point 3 is then projected to the top view, and the process is repeated until the intersection can be constructed (c) and shown in the final set of views (d).

12.19 Revolution

Revolution techniques are used by drafters to quickly obtain true lengths of lines and true-size area representations of surfaces. In Figure 12.26, line $\overline{12}$ is foreshortened in both given (top and front) views (a). This line may be rotated about point 1 in the top view so that it becomes equivalent to a frontal line, seen in true length in the resulting front view (b). Alternatively, line $\overline{12}$ may be rotated about point 1 in the given front view so that it becomes equivalent to a horizontal line—seen in true length in the resulting top view (c).

Auxiliary views were used in Chapter 7 to obtain true-size area representations of inclined and oblique surfaces. Revolution drawings can also be used to obtain true-size views of such surfaces. For example, an object may be rotated about any given axis to obtain a new orientation (Figure 12.27).

An inclined surface can be seen in true size in a principal view if the object is rotated so that this surface becomes parallel to a principal plane of projection. In Figure 12.28 a revolution is performed so that the inclined surface A (a) is seen in true size in the new right-side view (b), thereby eliminating the need for an auxiliary view (a). Similarly, an object is

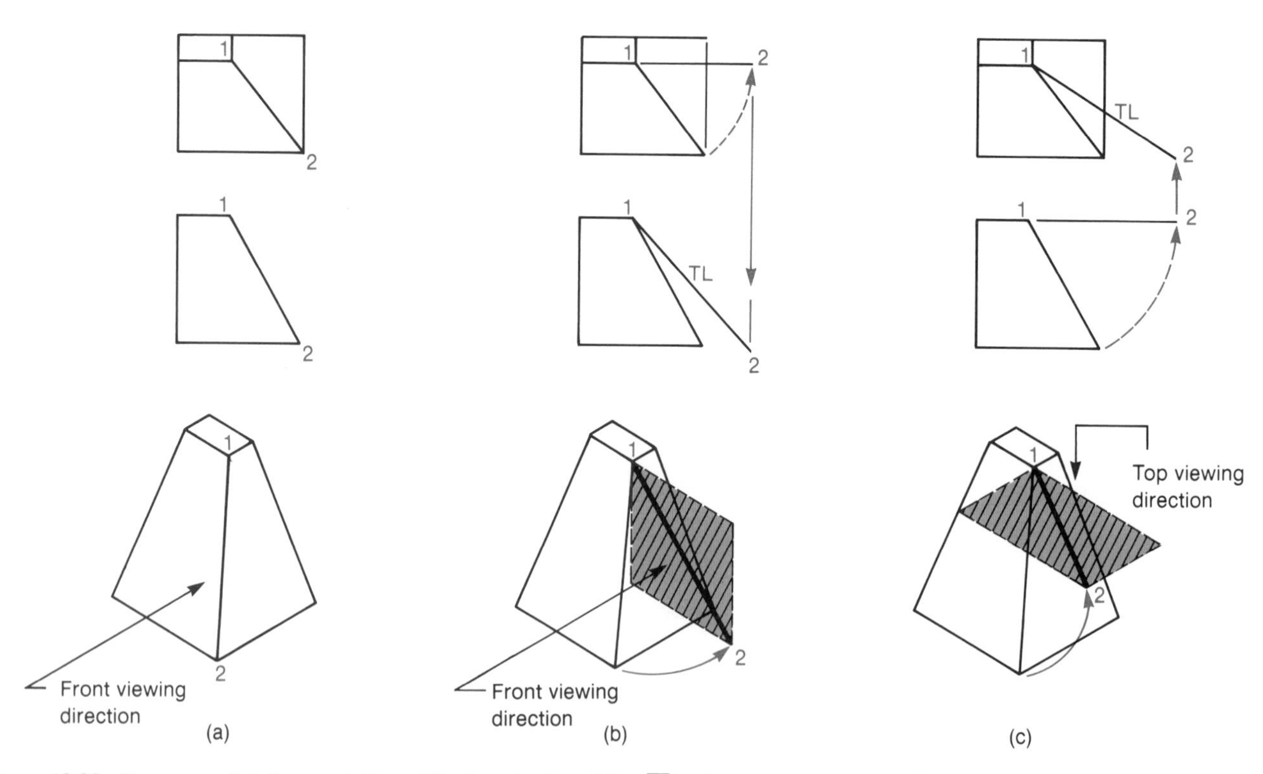

Figure 12.26 To accomplish the revolution of the foreshortened line $\overline{12}$ in the given top and front views (a), in order to obtain a true-length representation in a principal view, the line may be rotated to become equivalent to a frontal line, seen in true length in the front view (b), or a horizontal line, seen in true length in the top view (c).

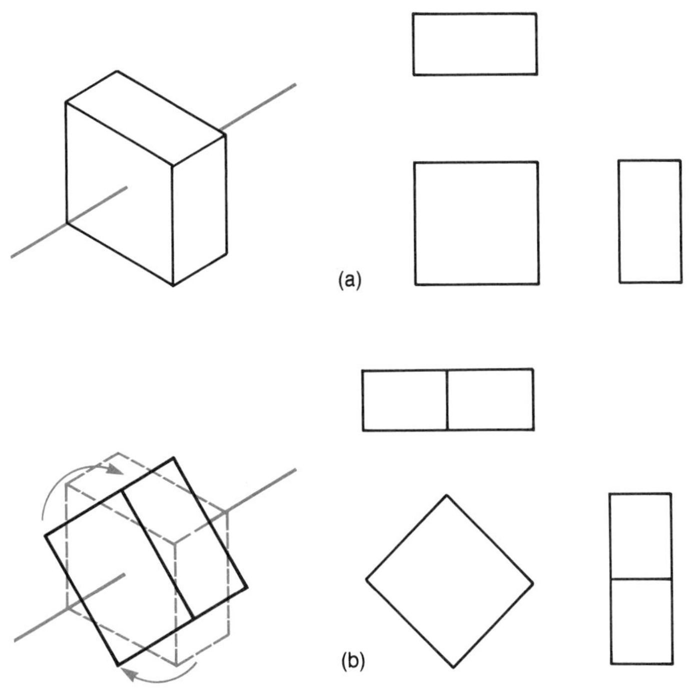

Figure 12.27 An entire object can be revolved about an axis that is perpendicular to a principal reference plane. Compare the given views of this rectangular prism (a) with the principal views corresponding to rotation of 45° about an axis perpendicular to the frontal reference plane (FRP) (b).

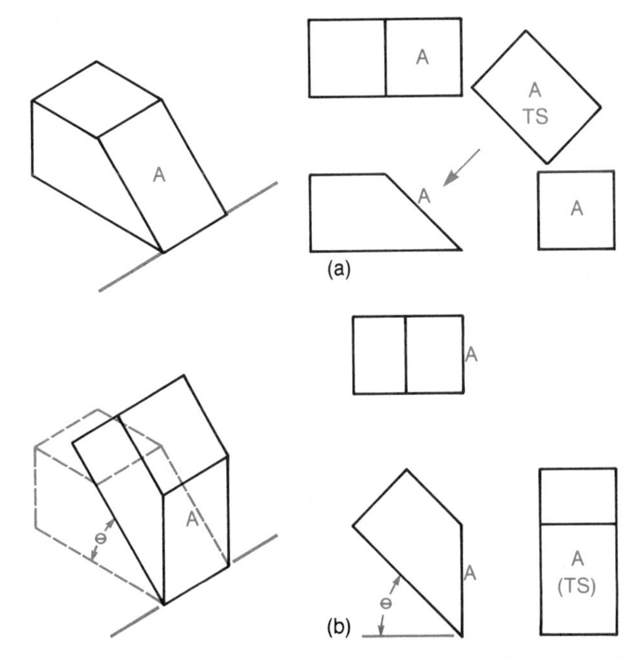

Figure 12.28 A true-size view of the inclined surface A can be obtained via construction of a primary auxiliary view (a) or rotation of the object so that the (originally inclined) surface A becomes parallel to one to the principal reference planes—in this case, the PRP (b).

revolved in Figure 12.29 so that an initially inclined surface (a) becomes parallel to the frontal reference plane, resulting in a new front view where this surface is seen in true size (b). Figure 12.30 shows another example of a revolution of an initially inclined surface (a) to obtain a true-size area representation (b) without the construction of an auxiliary view.

Figure 12.31 presents a set of orthographic views of an object that includes an oblique surface, A. A secondary auxiliary view is developed in order to show this surface A in true size. Alternatively, revolution techniques can be used to construct a true-size view of this oblique surface, as shown in Figure 12.32. An initial revolution is used to reposition the object so that one edge of the oblique surface A is parallel to a principal reference plane—in this case, the profile reference plane (PRP). As a result, the oblique surface A appears on edge, as a line, in one of the resulting principal views—in this case, the top view (a). A second rotation is then applied to the object so that the edge view of the surface A becomes parallel to a principal reference plane; as a result, the surface will be seen in true size in that principal view corresponding

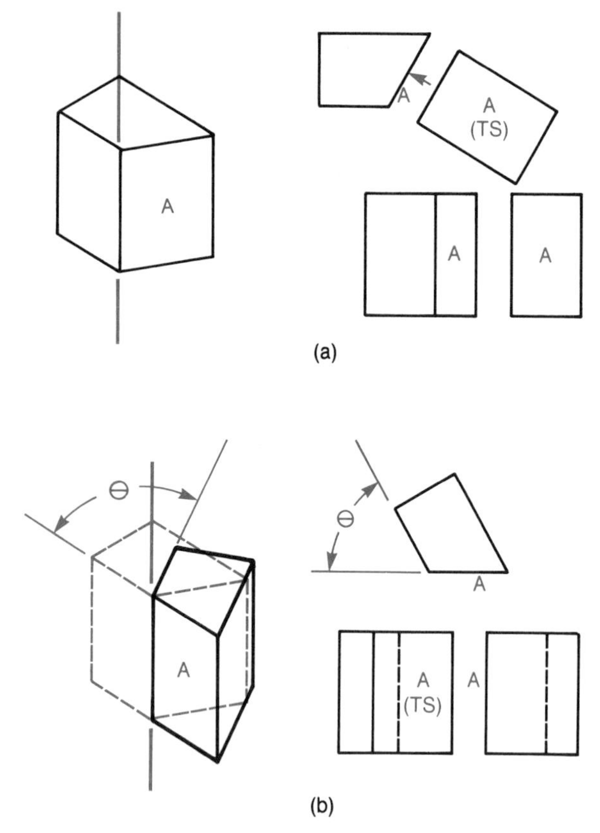

Figure 12.29 A true-size view of the inclined surface A is obtained via construction of a primary auxiliary view (a) or rotation of the object so that the surface A becomes parallel to a principal reference plane—in this case the FRP (b).

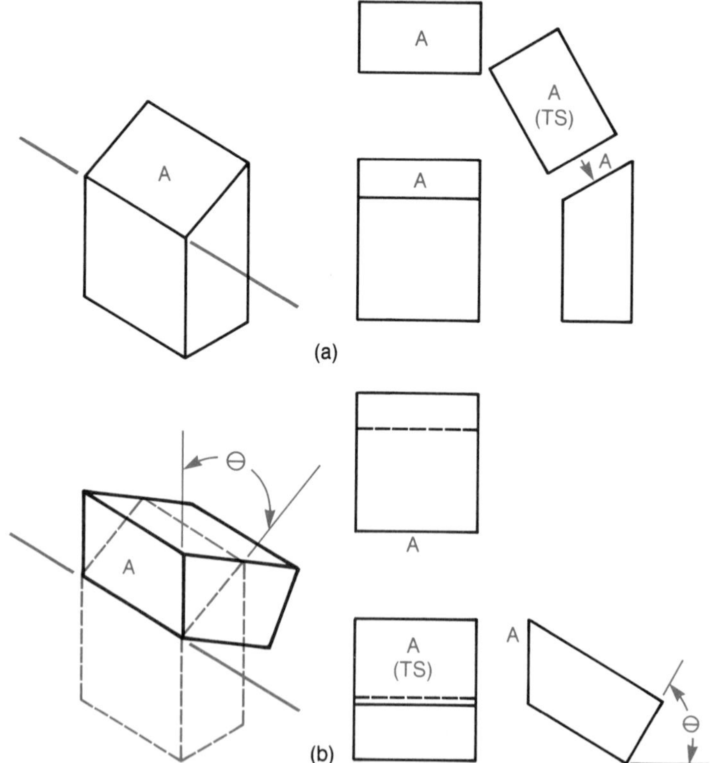

Figure 12.30 A true-size view of the inclined surface A is obtained via construction of a primary auxiliary view (a) or rotation of the object so that the surface A becomes parallel to a principal reference plane—in this case, the FRP (b).

to projection onto this plane (b). Surface A appears in true size in the resulting front view.

Revolution techniques are very useful if one needs to quickly determine the true length of a line. These techniques are less useful for true-size area representations of inclined and oblique surfaces, since the resulting principal views may or may not present the form of the object with maximum clarity. Recall that a set of principal views is initially chosen so that the shape of the object is shown clearly; an auxiliary view or views can then be used for true-size descriptions of inclined and oblique surfaces. Revolutions should be applied judiciously so that needed data regarding true lengths and true sizes are obtained with ease and speed—but without making interpretation of an orthographic description of an object more difficult.

12.20 Developments

A **development** is a flat pattern of an object that can be folded or rolled to form the part (Figure 12.33). Such patterns are also called *templates* or *stretchouts.* Developments are used in the packaging industry, the sheet metal industry, and the clothing industry, among *many* others.

Planar and single-curved surfaces can be accurately fabricated from flat-surface patterns that are folded or rolled to form the appropriate form. However, warped surfaces (spheres or paraboloids) can only be approximated by such flat patterns; one must stretch or press the material of the pattern into the required shape.

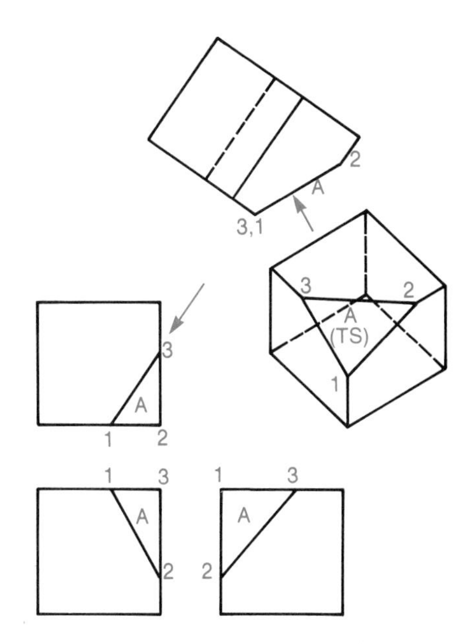

Figure 12.31 A true-size view of the oblique surface A can be obtained by constructing a secondary auxiliary view.

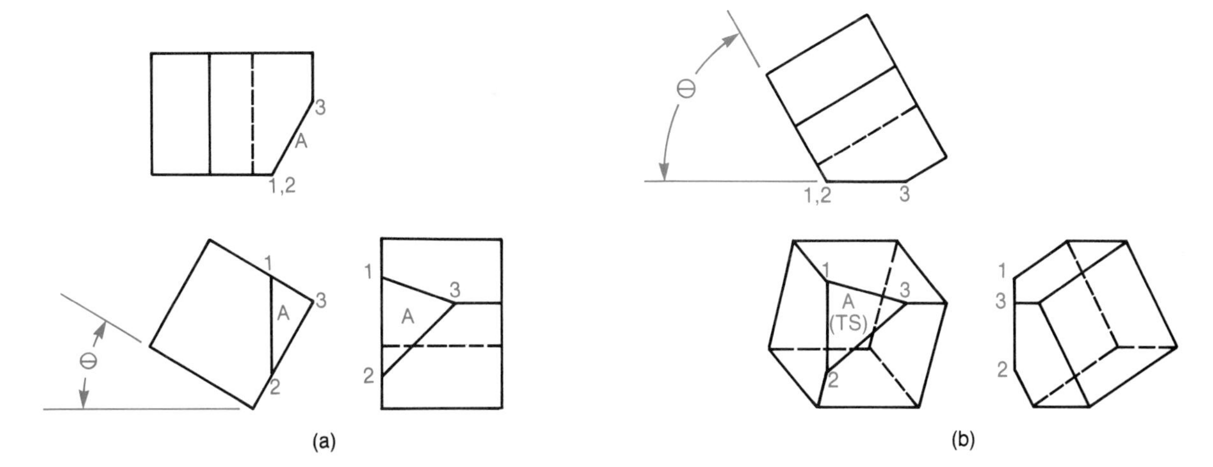

Figure 12.32 A true-size view of the oblique surface A can be obtained via revolution drawings (a). The first revolution rotates the object so that one edge of the oblique surface A is parallel to a principal reference plane—in this case, the PRP. As a result, the oblique surface appears on edge, as a line, in one of the principal views—in this case, the top view. The edge view of the oblique surface A is next rotated so that this surface is parallel to a principal reference plane—in this case, the FRP (b). A true-size view of the surface will then be obtained by projection onto this reference plane. Thus, surface A appears in true size in the resulting front view.

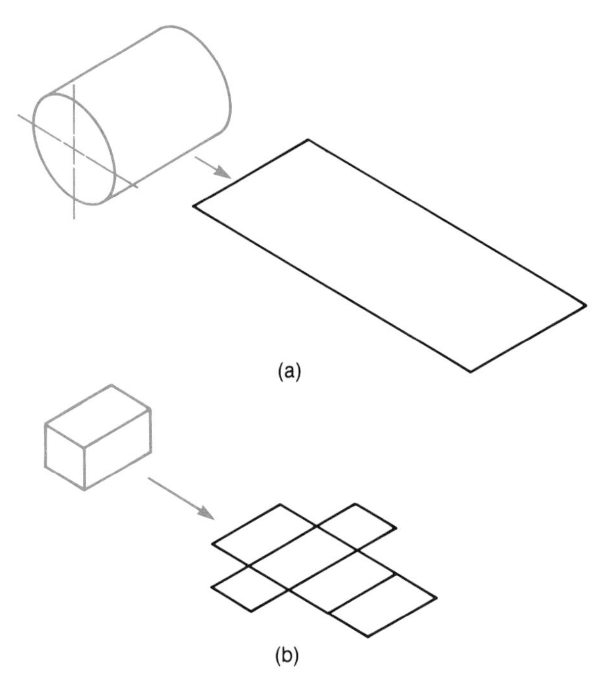

Figure 12.33 Developments are surfaces that can be unrolled or unfolded into a single plane to form a pattern from which the surface can be fabricated. Examples of a cylinder (a) and a right rectangular prism (b) are shown.

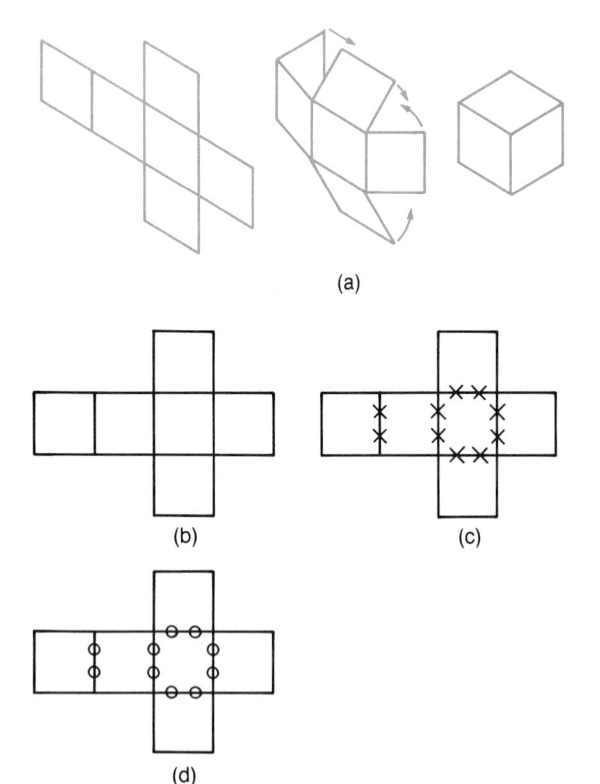

Figure 12.34 Fold lines may be specified in several ways. The pattern may be folded into the desired form (a) or the fold lines may be represented on the pattern by thin lines (b), by X marks (c), or by O marks (d).

Fold lines in a pattern can be denoted by thin lines, X marks on thin lines, or 0 marks on thin lines (see Figure 12.34). **Seams** between sections of a pattern can be joined by riveting, welding, sewing, gluing, soldering, and other methods of fastening. Figure 12.35 presents some of the more popular types of seams and edging hems (used to strengthen outer edges and eliminate sharp surface boundaries).

Many different forms can be fabricated as combinations of prisms, cylinders, pyramids, and cones. Three different types of development can be used to create patterns for these basic geometric shapes: parallel line, radial line, and triangulation.

12.20.1 Parallel Line Development

Parallel line development is used to construct patterns for right prisms and cylinders (Figures 12.36 and 12.37). A straight line (known as a **stretchout line**) is used as the base along which true-length distances of adjacent surfaces are aligned. Parallel lines and surfaces are then transferred from a given set of orthographic views to the development. Since all surfaces and lines appear in true size and true length, respectively, in (at least) one of these given orthographic views, the pattern can be constructed with relative ease.

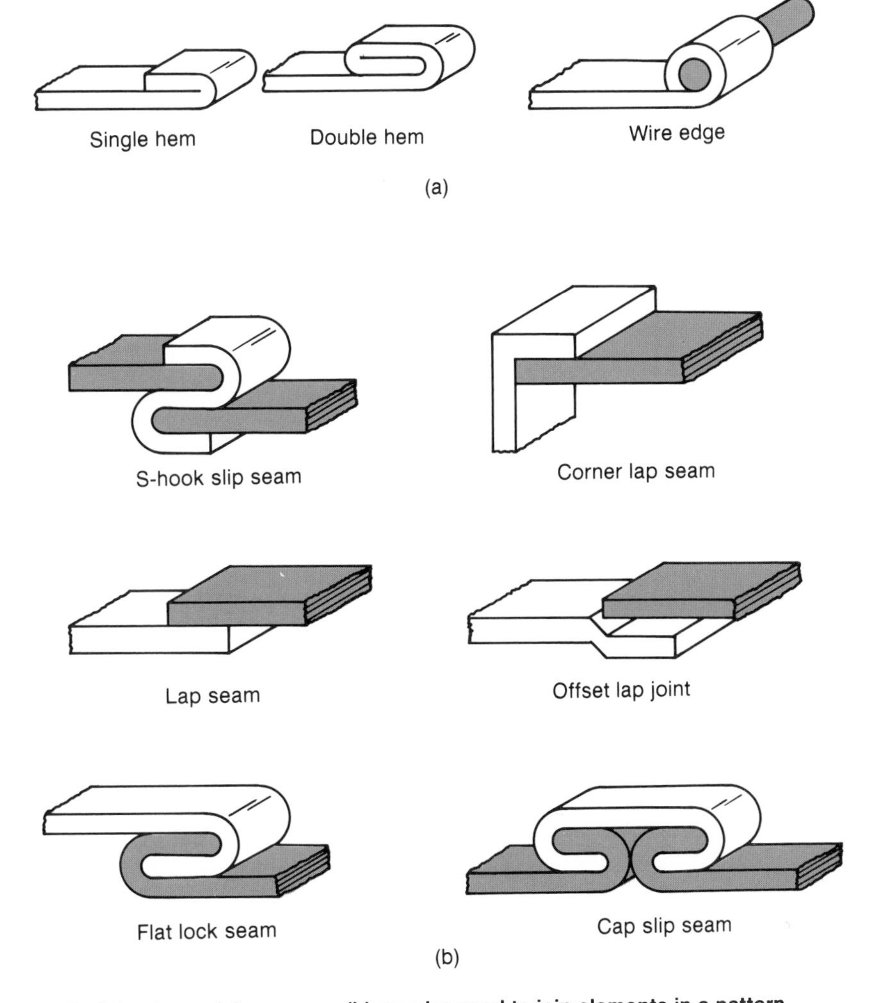

Figure 12.35 **A variety of edging hems (a) or seams (b) may be used to join elements in a pattern.**

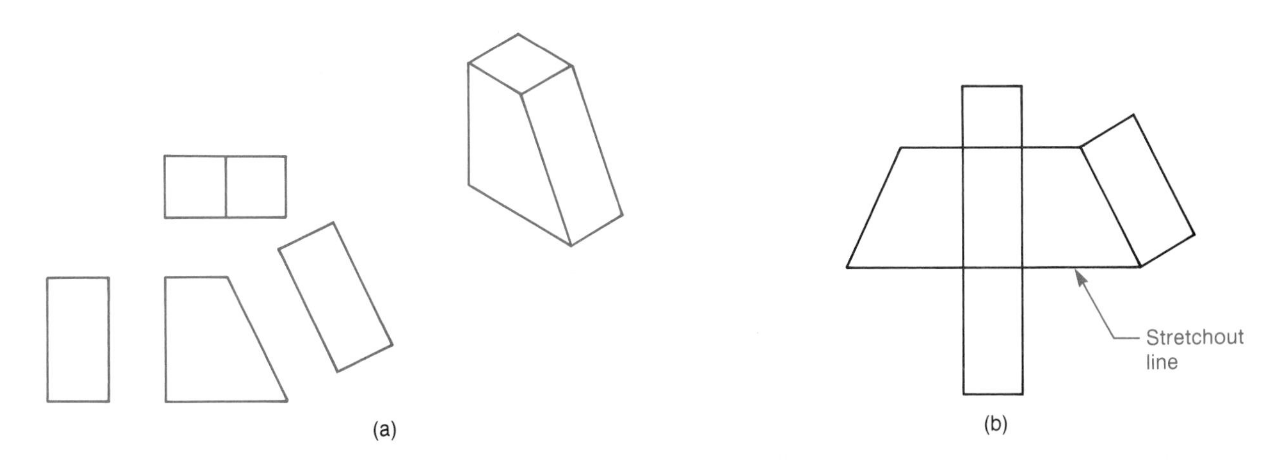

Figure 12.36 **Parallel line development may be used to construct a pattern for a right prism. Compare the given orthographic views (a) with the pattern (b).**

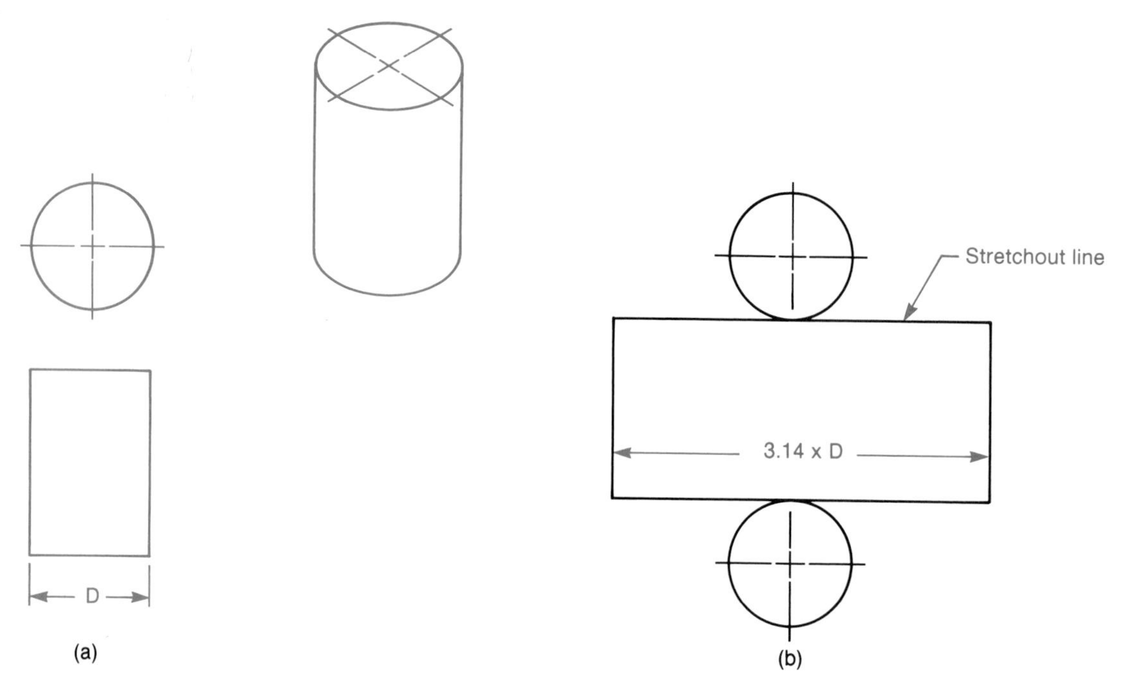

Figure 12.37 Parallel line development may be used to construct a pattern for a cylinder. Compare the given orthographic views (a) with the pattern (b).

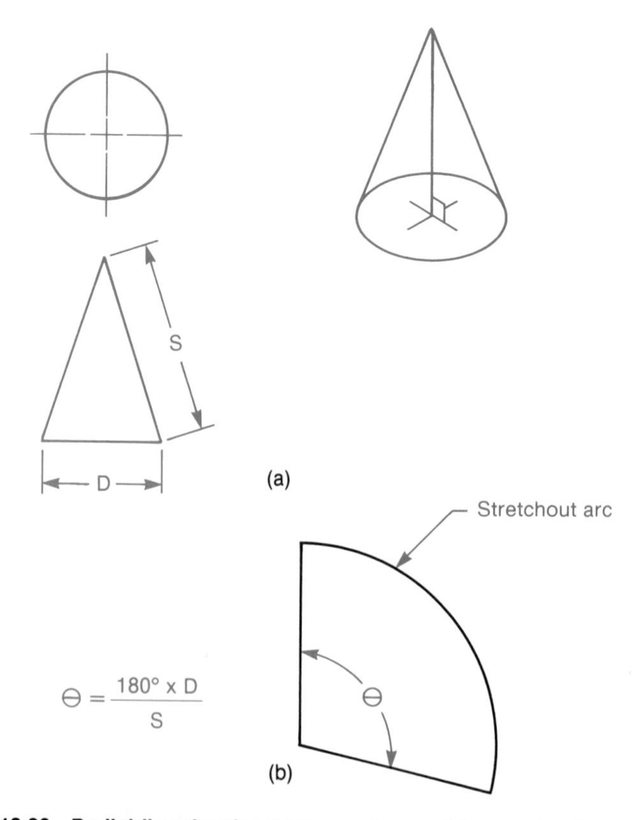

Figure 12.38 Radial line development may be used to construct a pattern for a cone. Compare the given orthographic views (a) with the pattern (b).

12.20.2 Radial Line Development

Radial line development is used to construct patterns for cones and pyramids (Figures 12.38 and 12.39). The inclined or tapered surfaces of such objects meet at a single point, or vertex, which then serves as a centerpoint for a construction arc (known as a **stretchout arc**) in the development. This stretchout arc has a radial distance equal to the true length, S, of the side of the cone (Figure 12.38) or the true length of the pyramid's tapered surface edge (Figure 12.39). In the case of a cone, the angle, θ, of this arc is equal to:

$$\theta = \frac{(180)(D)}{S}$$

where D denotes the diameter of the cone's base. In the case of the pyramid, true-length values for surface boundaries are drawn as chords along the stretchout arc, as seen in Figure 12.39.

12.20.3 Triangulation Development

Triangulation development is used to construct patterns for oblique or complex forms (Figure 12.40). Such forms are described in terms of appropriate triangular surfaces into which the forms may be divided. True-length measurements are obtained for the edges of each triangle. The triangles are then used to construct the composite pattern for the object.

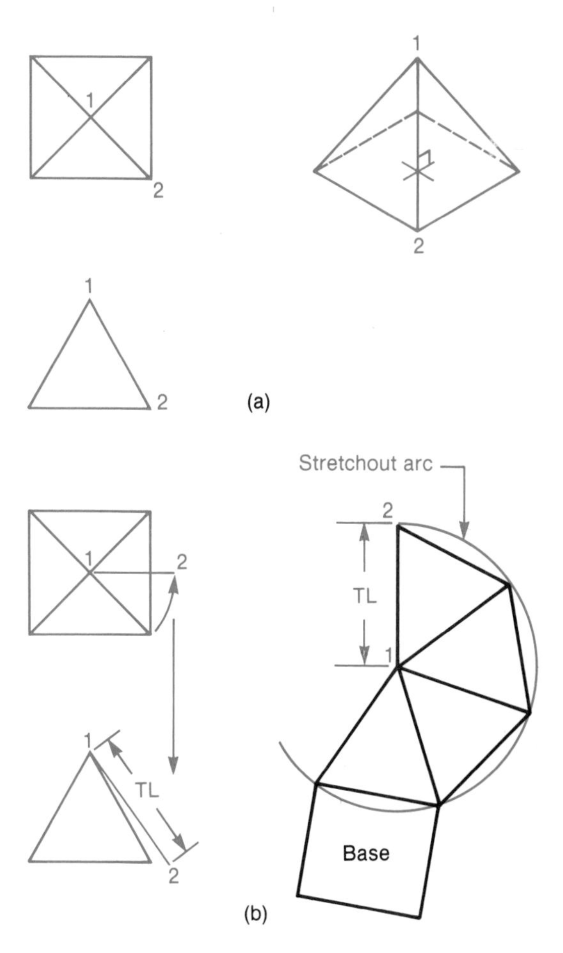

Figure 12.39 Radial line development may be used to construct a pattern for a pyramid. Compare the given orthographic views (a) with the pattern (b).

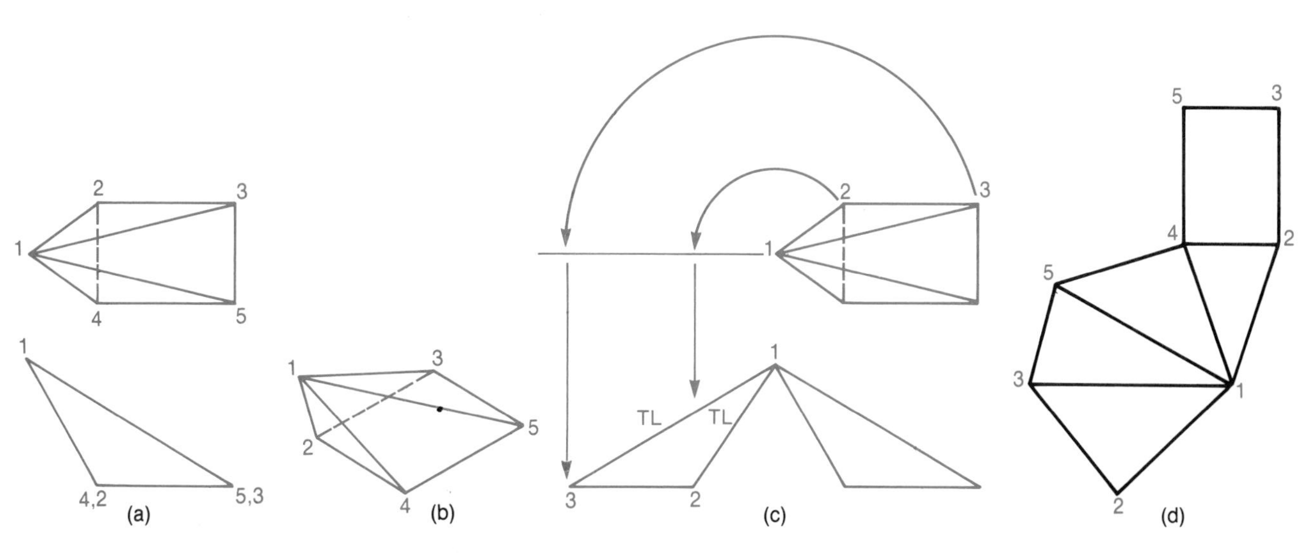

Figure 12.40 Triangulation development may be used to construct a pattern for oblique or complex forms. Compare the given orthographic views of an oblique pyramid (a) and the pictorial (b) with the pattern (d). True-length distances are obtained via revolution (c). One can obtain an inside pattern (a pattern viewed from the interior of the form) by simply reversing the outside or exterior pattern shown in (d).

In Retrospect

- Horizontal, frontal, and profile lines are, by definition, parallel to the horizontal, frontal, and profile reference planes, respectively; each type of line projects in true length onto that reference plane to which it is parallel.
- The intersection of lines (and points) in orthographic views allows us to specify the relative locations of lines, points, and planes in space.
- Parallel lines always appear as parallel lines in orthographic views.
- Lines that are truly mutually perpendicular will appear perpendicular (at 90° to one another) in any view in which at least one of the lines is shown in true length. This fact allows one to construct such graphic elements as:
  - A line perpendicular to a given line.
  - A line perpendicular to a given plane.
  - A plane perpendicular to a given line.
  - A plane perpendicular to a given plane.
- A line can be shown to be parallel to a plane if it is parallel to a line within the plane.
- One plane (A) is parallel to another plane (B) if plane A contains two intersecting lines that are parallel to two intersecting lines within B.
- Truly intersecting lines will always intersect at the same point in all orthographic views.
- One may determine the shortest distance between a particular point in space and a given line by developing (1) a primary auxiliary view in which the line is shown in true length and (2) a secondary auxiliary view in which the line appears as a point.
- The shortest distance between two lines can be determined by developing a secondary auxiliary view in which one of the lines appears as a point. The shortest distance may or may not be seen in other views; in addition, it will definitely be seen if the lines appear parallel to each other.
- The dihedral angle between two planes will be shown accurately (without distortion) in the view where the line of intersection between the planes appears as a point.
- The relative visibility of lines and planes can be determined through the application of descriptive geometry and the principles of orthographic projection.
- Intersections between surfaces can be constructed by applying the point-by-point, surface-by-surface procedure.
- Developments are surfaces that are unrolled or unfolded into a single plane to form a pattern from which the surface can be fabricated.
- Revolution techniques allow one to show true lengths of lines and true-size area representations of inclined and oblique surfaces without constructing auxiliary views.

ENGINEERING IN ACTION

Rolling Ball Generates Cartesian-like Coordinates

High-Resolution Digital Device Uses Precision Motion-Transfer System to Steer the Cursor on a Computer Screen

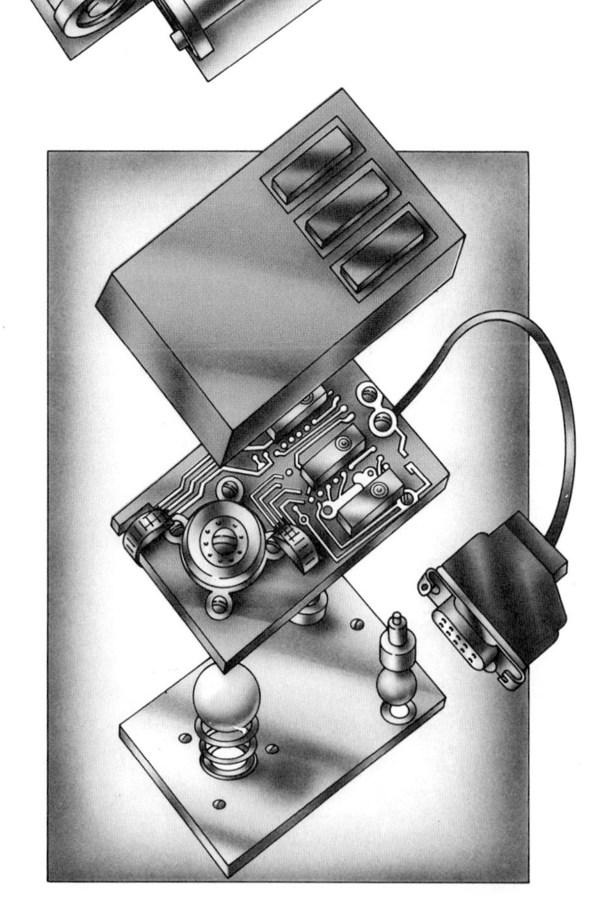

A 0.75-in.-diam. ball rolling across a smooth surface provides the input motion to move a cursor in any direction on a computer display screen. Projecting through the base plate of a computer mouse, or digital input device, the ball rotates against two perpendicular shafts. Each shaft turns a 0.5-in.-diam., 25-bar commutator wheel that is energized through an inboard wire finger, then decoded as four wire fingers of varying lengths wipe across the passing bars. This assembly generates electrical pulses corresponding to the extent and direction of ball travel, thereby producing digital signals sent to a computer for moving the cursor. The mouse has three buttons for actuating microswitches to select menus, edit text, and move symbols. The Mouse House Div. of Hawley Labs., Berkeley, CA.

Source: Reprinted from MACHINE DESIGN, January 12, 1984. Copyright, 1984, by Penton/IPC., Inc., Cleveland, Ohio, p. 36.

Exercises

Your instructor will specify the appropriate scale for each exercise.

EX12.1. Construct a true-length view of line $\overline{45}$ in the top view.

EX12.2. Construct true-length views of lines $\overline{45}$ and $\overline{67}$ in the front and top views, respectively.

EX12.3. Construct true-length views of lines $\overline{45}$ and $\overline{67}$ in the front and top views, respectively.

EX12.4. Construct true-length views of lines $\overline{45}$ and $\overline{67}$ in the front and top views, respectively.

EX12.5. Construct line $\overline{45}$ in the front view.

EX12.6. Construct line $\overline{45}$ in the front view.

EX12.7. Construct line $\overline{45}$ in the right-side view.

EX12.8. Construct line $\overline{45}$ in the right-side view.

EX12.9. Construct line $\overline{45}$ in the front view.

EX12.10. Construct line $\overline{45}$ in the top view.

EX12.11. Given that point 4 lies in plane A, locate point 4 in the given top view.

EX12.12. Construct a line that is perpendicular to plane A and that contains point 4.

EX12.13. Construct a line that is perpendicular to plane A and that contains point 4—given that point 4 lies within plane A.

EX12.14. Construct a line that is perpendicular to plane A and that contains point 4.

EX12.15. Construct a line that is perpendicular to line $\overline{12}$ and that contains point 3.

EX12.16. Construct a line that is perpendicular to line $\overline{12}$ and that contains point 3.

EX12.17. Construct a plane that is perpendicular to line $\overline{12}$ and that contains point 3.

EX12.18. Construct a plane that is perpendicular to line $\overline{12}$ and that contains point 3.

EX12.19. Construct a plane that is perpendicular to plane A and that contains line $\overline{45}$.

EX12.20. Determine the relative visibility of line $\overline{45}$ and plane A in the given views.

EX12.21. Construct a plane that is perpendicular to plane A and that contains line $\overline{45}$.

EX12.22. Determine the relative visibility of line $\overline{45}$ and plane A in the given views.

EX12.23. Determine the shortest distance between line $\overline{12}$ and point 3.

EX12.24. Determine the shortest distance between line $\overline{12}$ and point 3.

EX12.25. Determine the shortest distance between lines $\overline{12}$ and $\overline{34}$.

EX12.26. Determine the shortest distance between lines $\overline{12}$ and $\overline{34}$.

EX12.27. Determine the dihedral angle between surfaces A and B.

EX12.28. Determine the dihedral angle between surfaces A and B.

EX12.29. Determine the intersection between plane A (seen on edge in the given front view), and the right circular cone. Draw the intersection in the given top view.

EX12.30. Determine the intersection between plane A (seen on edge in the given front view), and the right circular cone. Draw this intersection in the given top view.

EX12.31. Determine the intersection between plane A (seen on edge in the given front view), and the oblique circular cone. Draw this intersection in the given top view.

EX12.32. Determine the intersection between plane A (seen on edge in the given front view), and the oblique circular cone. Draw this intersection in the given top view.

EX12.33. Obtain a true-size area representation of surface A in EX12.33 via revolution.

EX12.34. Obtain a true-size area representation of surface A in EX12.34 via revolution.

EX12.35. Obtain a true-size area representation of surface A in EX12.35 via revolution.

EX12.36. Construct a development for the object shown in EX12.36.

EX12.37. Construct a development for the object shown in EX12.37.

EX12.38. Construct a development for the object shown in EX12.33.

EX12.39. Construct a development for the object shown in EX12.34.

EX12.40. Construct a development for the object shown in EX12.34.

EX12.41. Construct a development for the object shown in EX12.35.

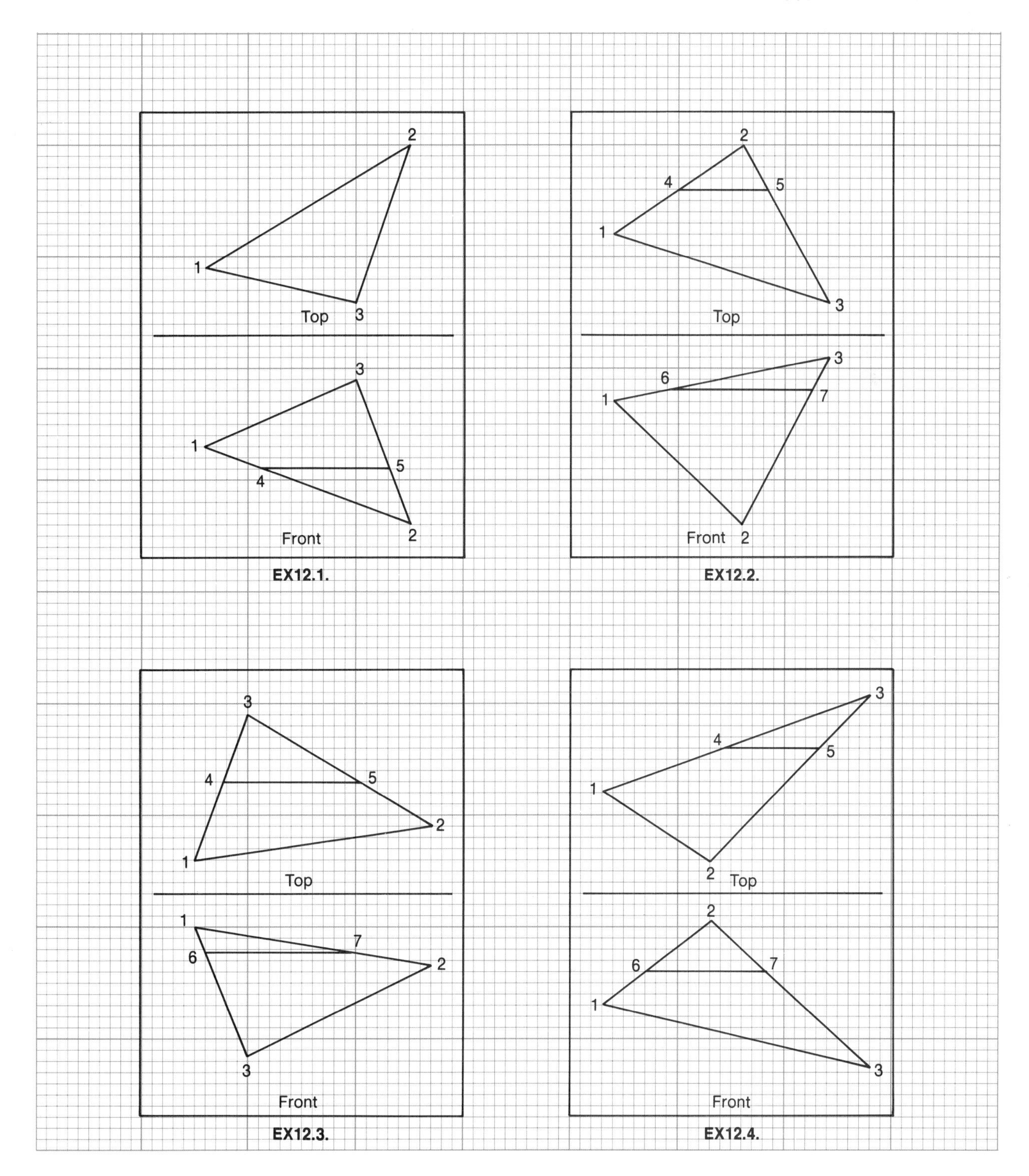

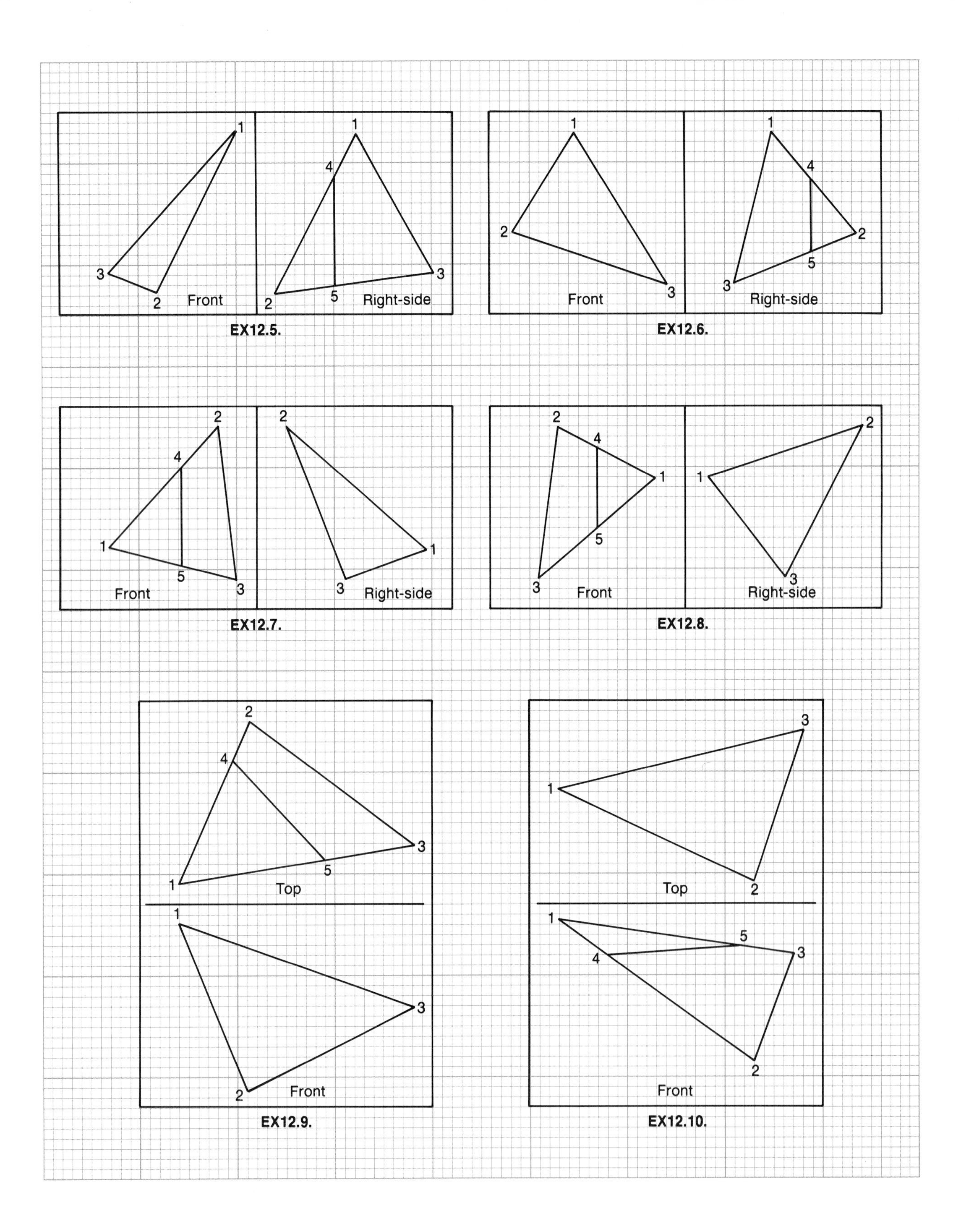
1
3
2
Front
1
4
3
2
5
Right-side
EX12.5.
1
2
3
Front
1
4
2
5
3
Right-side
EX12.6.
2
4
1
5
3
Front
2
1
3
Right-side
EX12.7.
2
4
1
5
3
Front
2
1
3
Right-side
EX12.8.
2
4
3
1
5
Top
1
3
2
Front
EX12.9.
3
1
Top
2
1
5
4
3
2
Front
EX12.10.

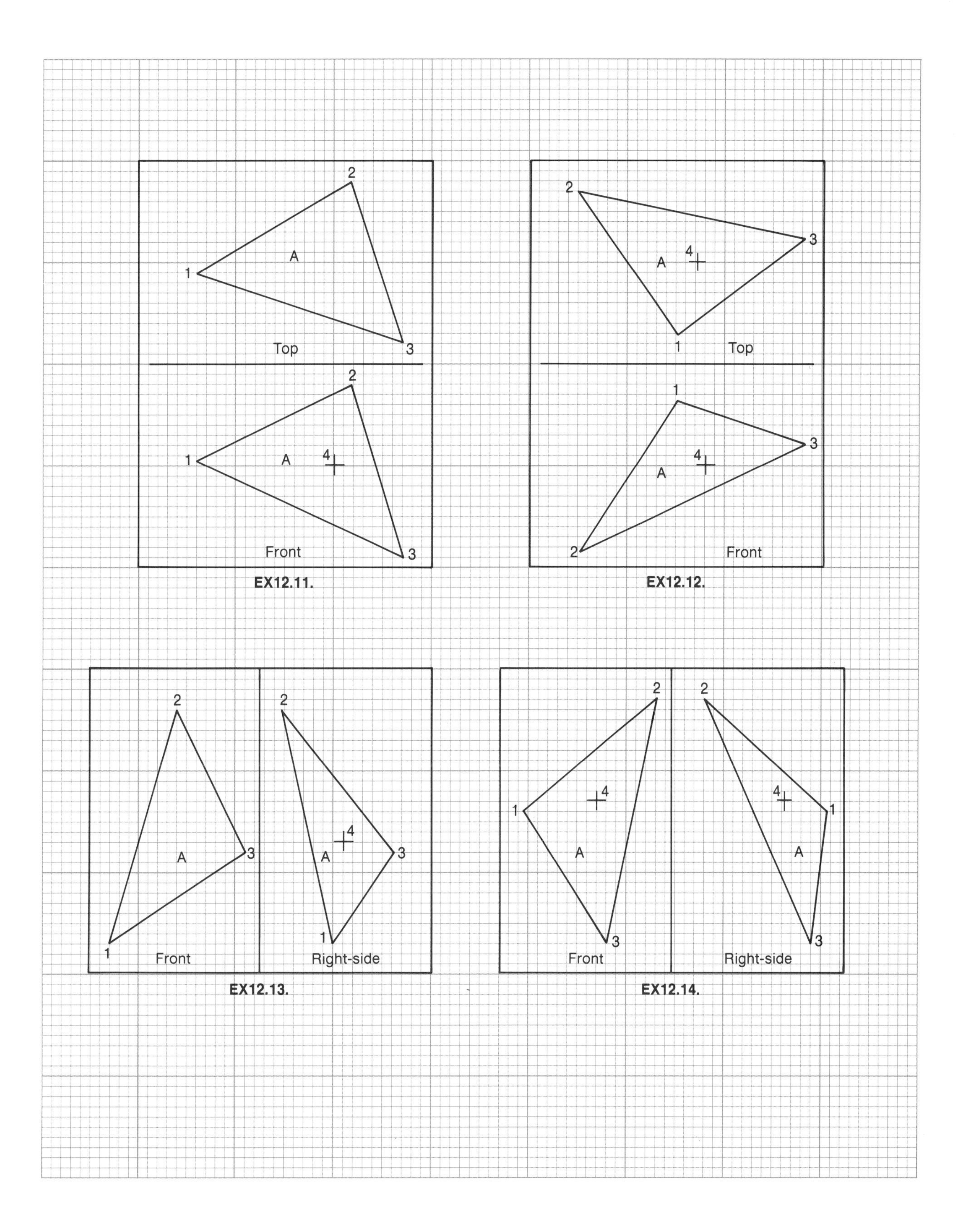
2
A
1
Top
3
2
1
A
4
Front
3
EX12.11.
2
3
4
A
1
Top
1
3
4
A
2
Front
EX12.12.
2
A
3
1
Front
2
4
A
3
1
Right-side
EX12.13.
2
4
1
A
3
Front
2
4
1
A
3
Right-side
EX12.14.

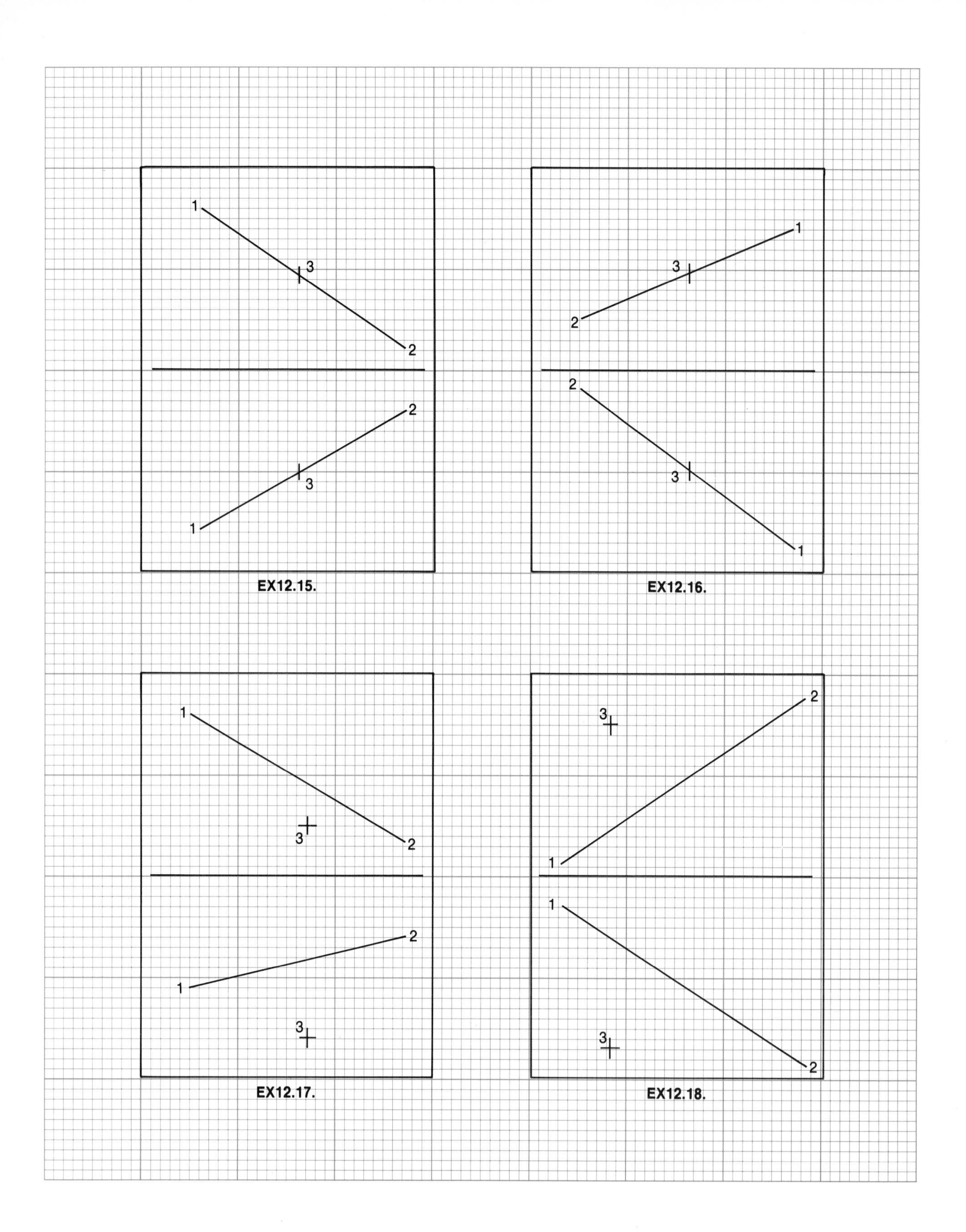
1
3
2
2
3
1
EX12.15.
1
3
2
2
3
1
EX12.16.
1
3
2
2
1
3
EX12.17.
2
3
1
1
3
2
EX12.18.

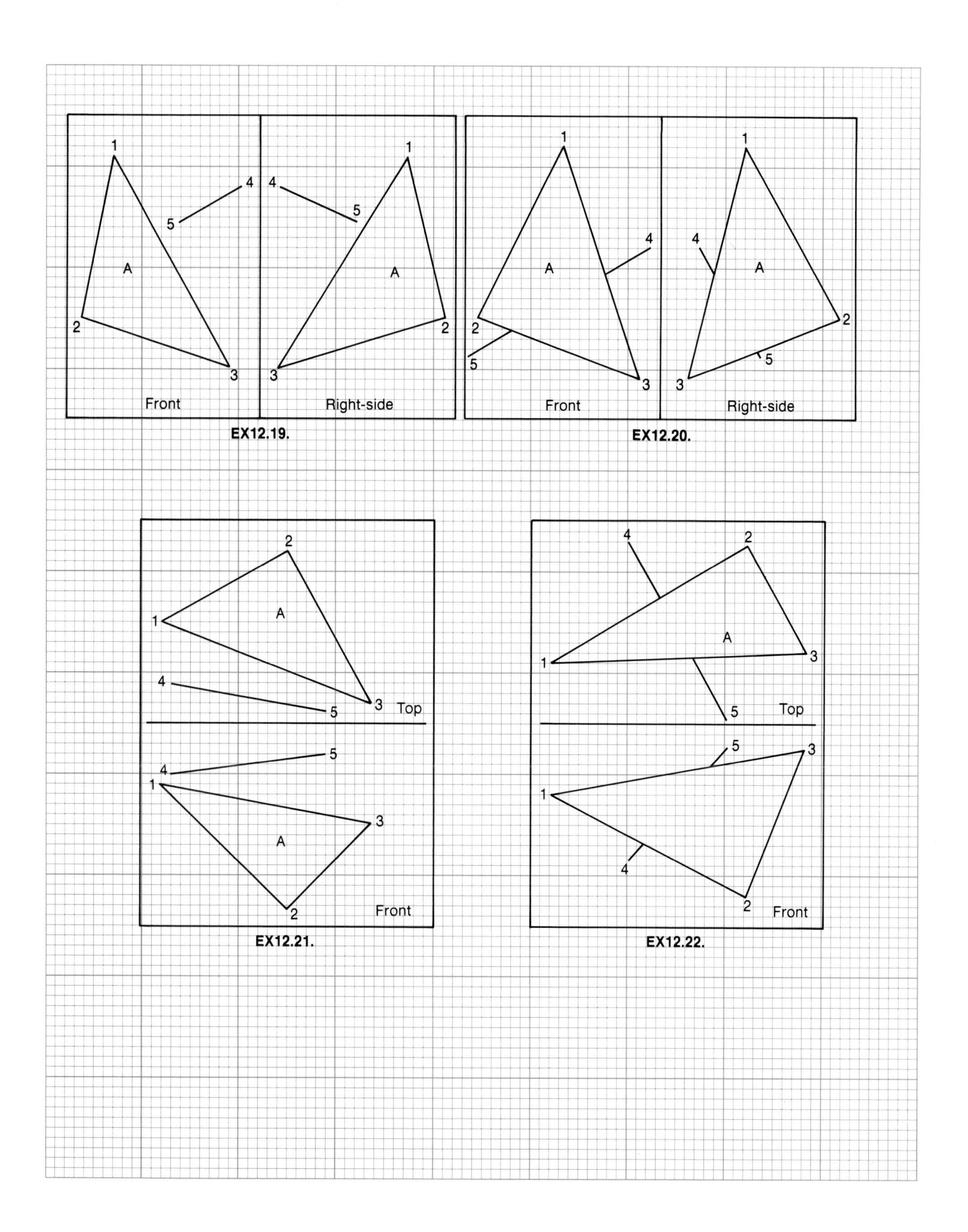
1
4
4
5
5
A
A
2
2
3
3
Front
Right-side
EX12.19.
1
1
4
4
A
A
2
2
5
5
3
3
Front
Right-side
EX12.20.
2
A
1
4
5
3
Top
5
4
1
3
A
2
Front
EX12.21.
4
2
A
1
3
5
Top
5
3
1
4
2
Front
EX12.22.

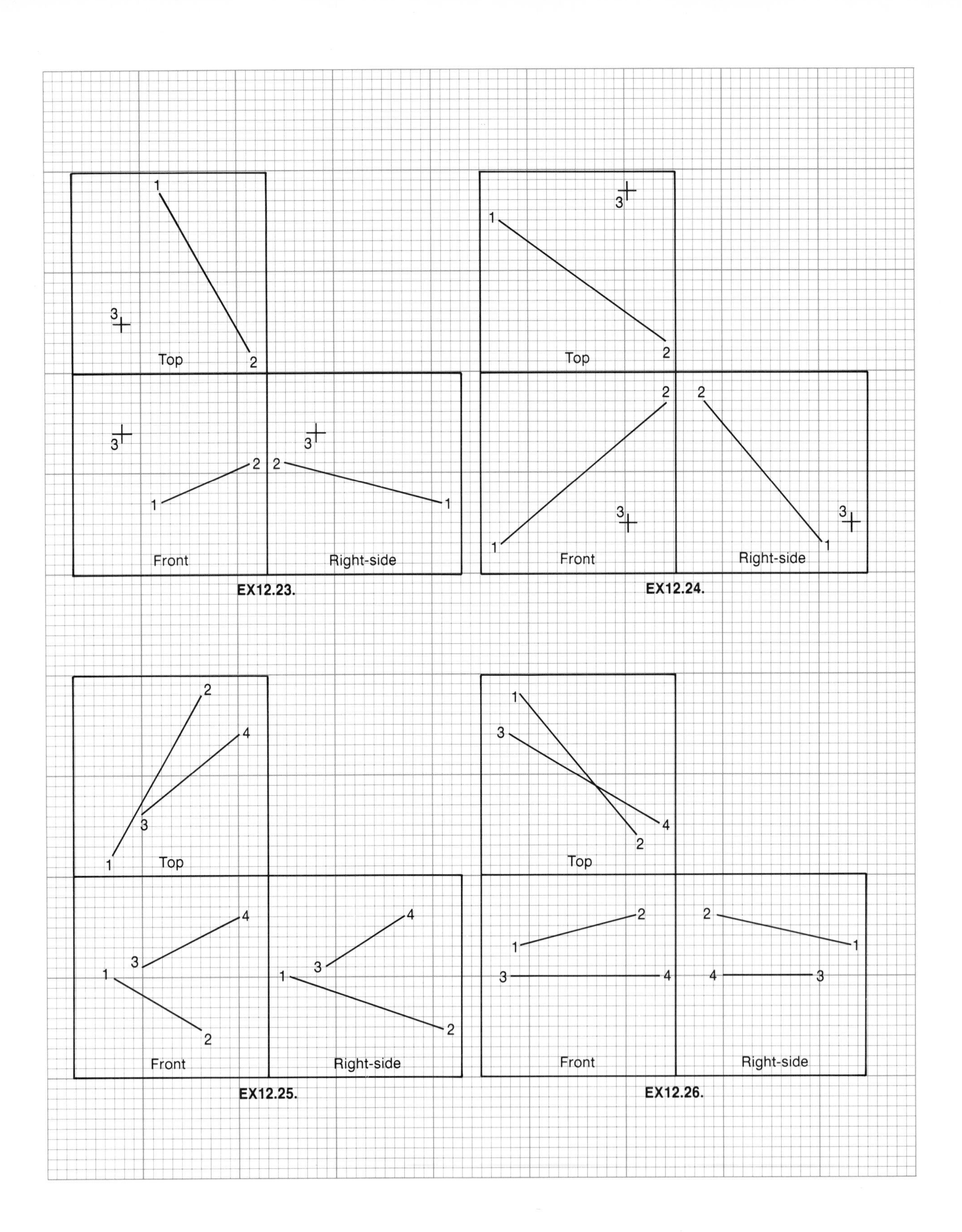
1
3
Top
2
3
2
1
Front
3
2
1
Right-side
EX12.23.
3
1
Top
2
2
1
3
Front
2
3
1
Right-side
EX12.24.
2
4
3
1
Top
4
3
1
2
Front
4
3
1
2
Right-side
EX12.25.
1
3
4
2
Top
2
1
3
4
Front
2
1
4
3
Right-side
EX12.26.

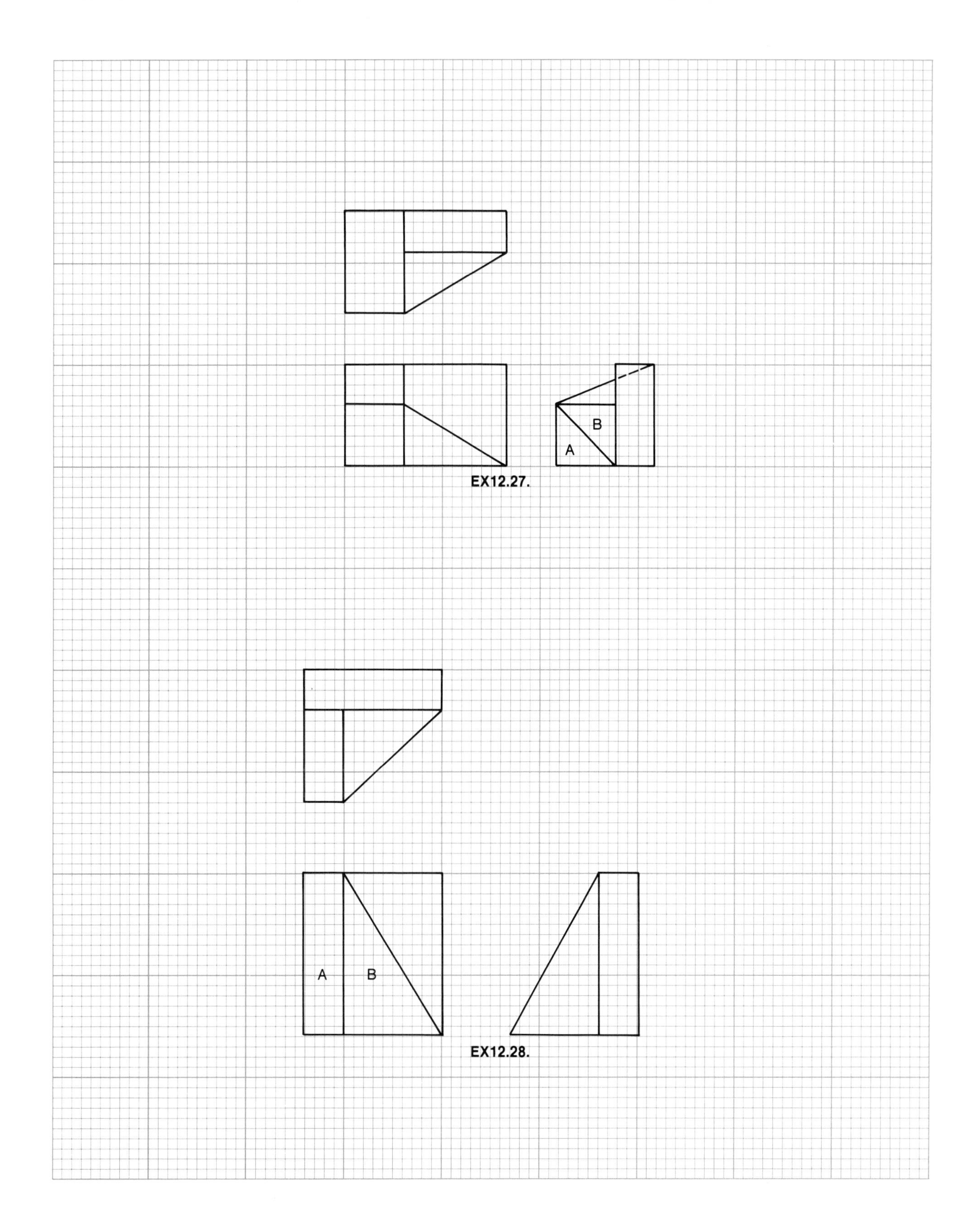
B
A
EX12.27.
A
B
EX12.28.

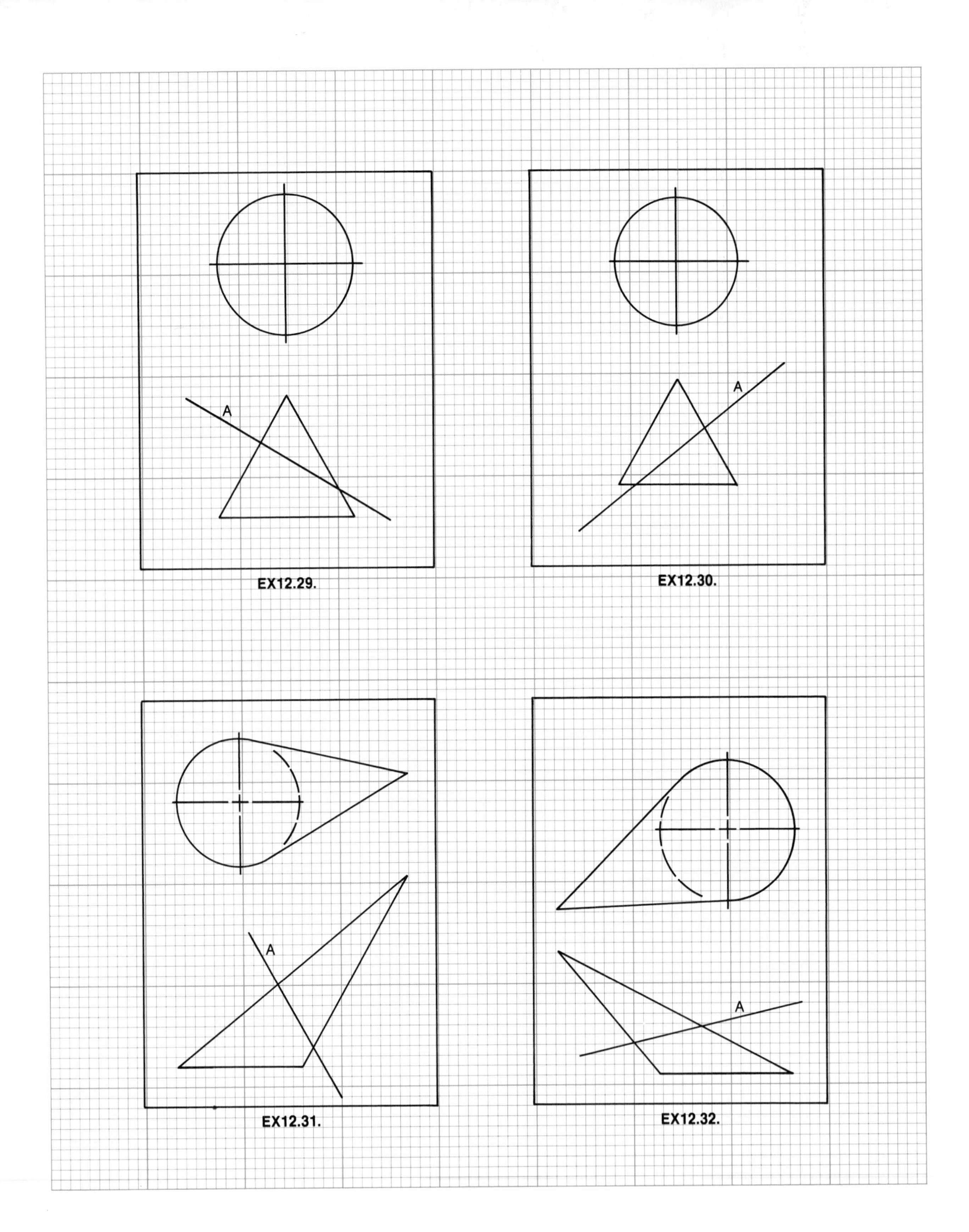

EX12.29.
EX12.30.
EX12.31.
EX12.32.

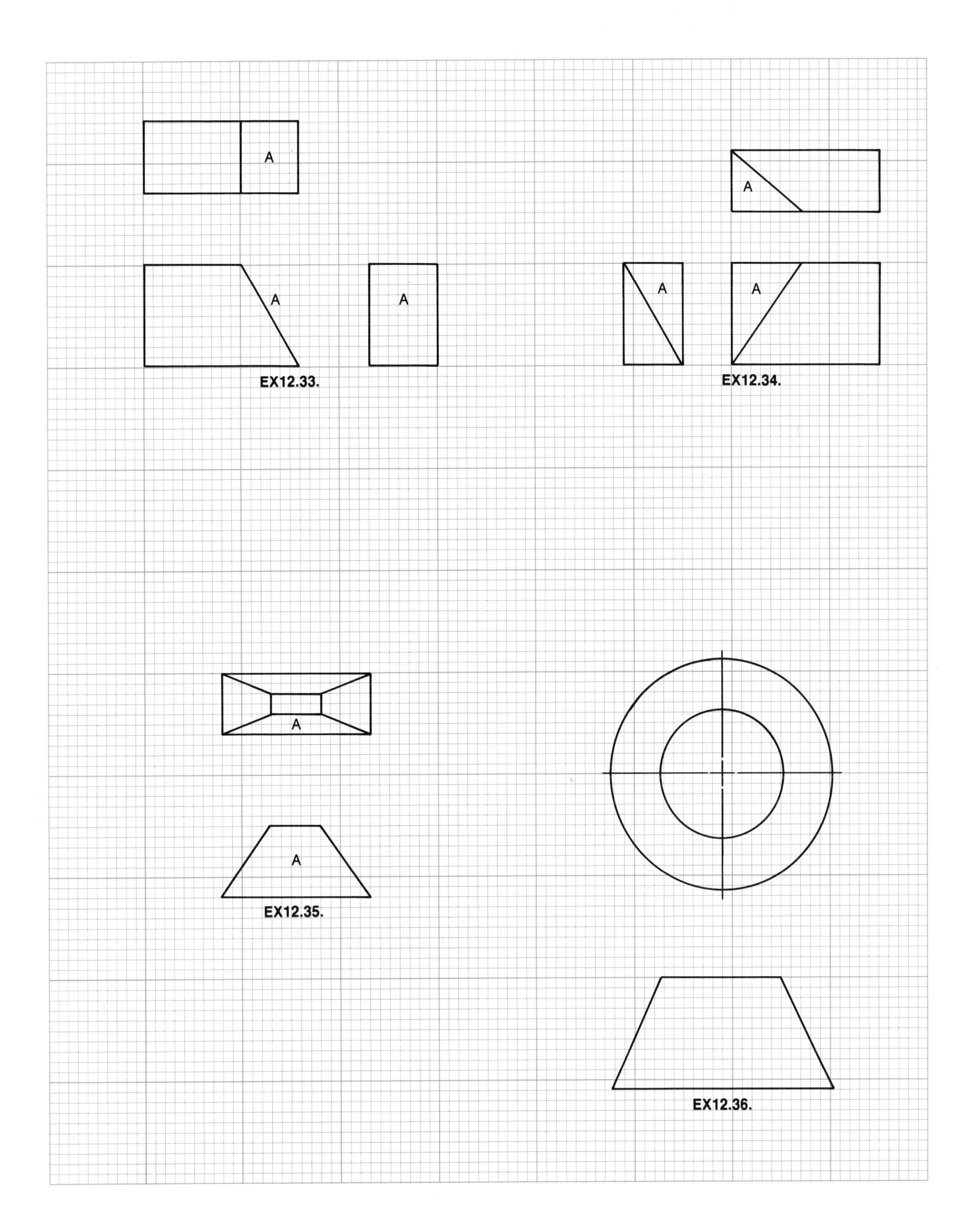
A
A
A
EX12.33.
A
A
A
EX12.34.
A
A
EX12.35.
EX12.36.

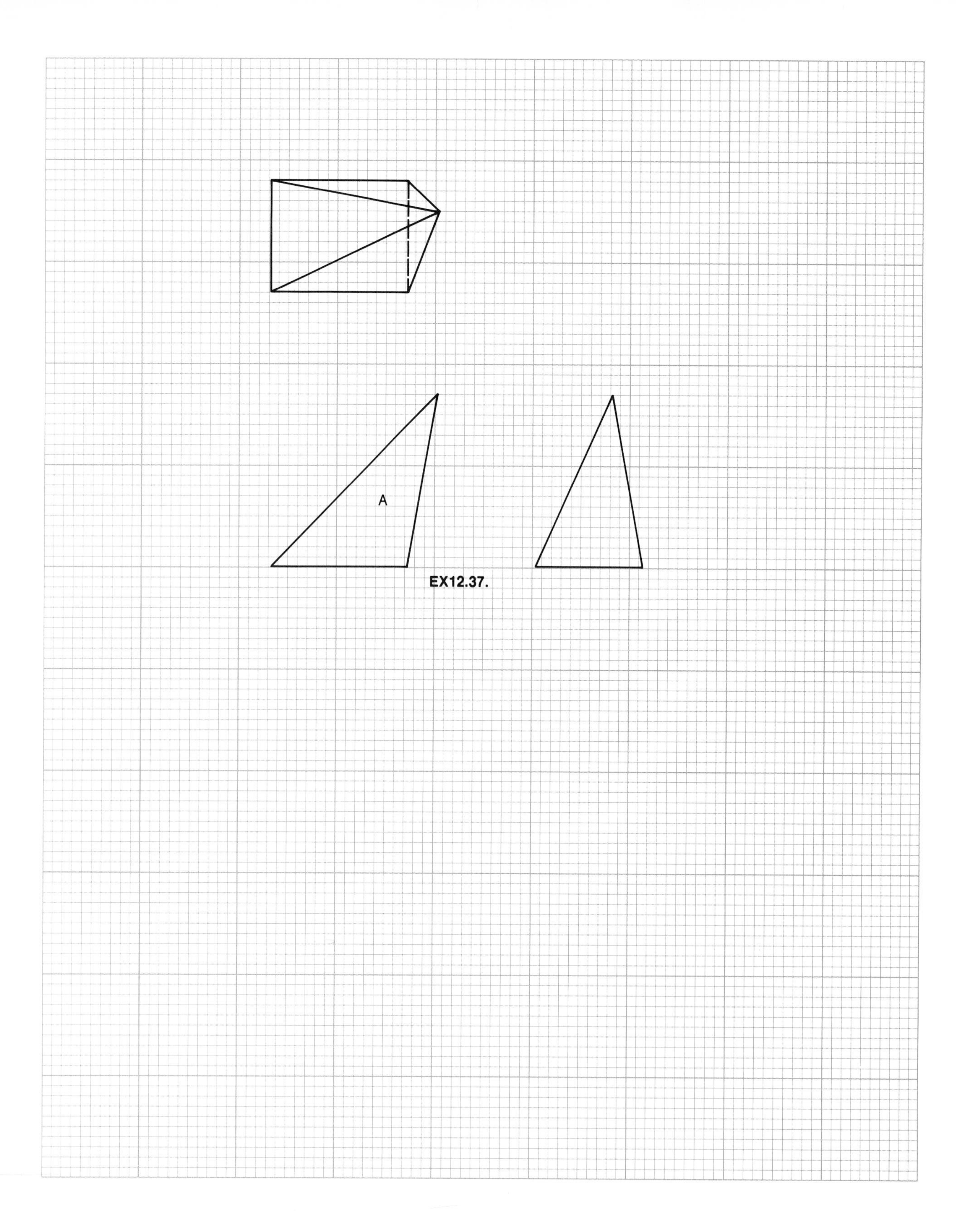

EX12.37.

13 Fasteners

Preview Fasteners are used to join components within an engineering system. Many different types of fasteners are available from manufacturers: screws, bolts, pins, rivets, retaining or snap rings, and so forth. Depending on the type of fasteners used, components may be positioned in fixed relative locations and orientations in either a permanent or semipermanent fashion.

This chapter briefly reviews common types of threaded and nonthreaded fasteners, including springs, and their significant characteristics.

Learning Objectives

Upon completion of this chapter, the reader should be able to:

- Differentiate between screws and bolts and identify each.
- Identify nonthreaded fasteners such as clevis pins, cotter pins, rivets, retaining or snap rings, and washers.
- Specify such fastener features as the head, neck, body, collar, bearing surface, shank, shoulder, thread, and point.
- Recognize the principal dimensions of fasteners, including body diameter, length, angularity, and bow, and such thread parameters as pitch, thread angle, series, class, and (major, minor, and pitch) diameters.
- Identify nut parameters, such as thickness, width, and thread specifications.
- Explain the application of springs to fasten components within a system, and identify various types of springs that are commonly used.
- Differentiate between common thread forms such as Sharp V, Acme, Worm, Unified, American National, and Knuckle.
- Insert thread specifications in both the metric and English unit systems (as required) in engineering drawings.

13.1 Screws and Bolts

Fasteners are used to hold two or more components together in a specific configuration or orientation. Screws and bolts are **externally threaded** fasteners (Figure 13.1 and 13.2). They are manufactured in a variety of shapes and sizes.

Screws are designed to (1) create, or cut, internal threads in the material to which they are attached (known as **self-tapping** screws) or (2) mate with preformed internal threads in such material. **Bolts** are designed to mate with a **nut** (which is a sleeve with an internal thread; nuts will be discussed later in this chapter). Some bolts (such as plow bolts) are designed so that rotation cannot occur; such bolts cannot be used as screws. Similarly, self-tapping screws cannot be used as bolts. Otherwise, bolts and screws are very similar in

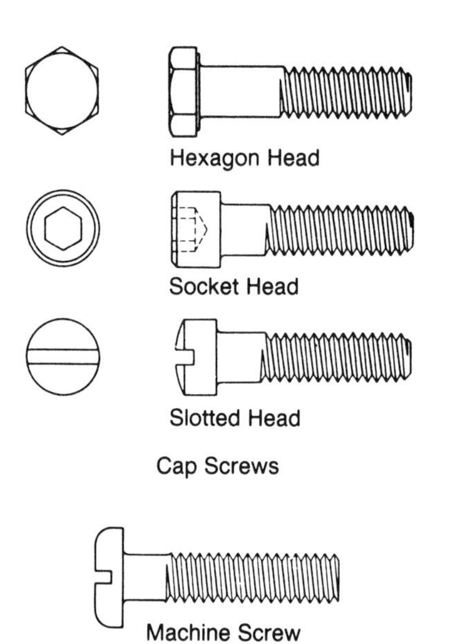

Figure 13.1 Examples of screws. (ASA B18.12-1962 (R1981), *Glossary of Terms for Mechanical Fasteners.*)

form and are somewhat interchangeable in use. The significant difference between screws and bolts is that bolts are designed to mate with a nut.

Figure 13.1 presents a few examples of two common screw styles: **cap screws** (in which all surfaces are machined, or finished, and in which the body diameter is well-defined, or toleranced) and **machine screws.** Both styles are manufactured with a variety of heads: recessed, slotted, wrenching, and so forth (see Appendices H and I). Different head styles are chosen in accordance with such factors as:

- The desired appearance of a product.
- The type of loading (application of forces) to be applied to the components fastened by a screw or bolt.
- The particular type of driving equipment (with which the fastener and the components are assembled) to be used in the fastening process (for example, crescent wrench, socket wrench, or screwdrivers).

Examples of bolt styles are shown in Figure 13.2. A **bent bolt** consists of a cylindrical rod with one or both of its ends threaded for use as a fastener. Examples of bent bolts include:

- **Eye bolts,** in which the unthreaded end of the rod is curved into the form of a ring.
- **Hook bolts,** in which the unthreaded end of the rod is "hooked," or bent, at either an acute or right angle.
- **J-bolts,** in which the unthreaded end is formed into a semicircular shape.
- **U-bolts,** which are threaded at both ends; U-bolts may be round (with an arcing bend) or square (in which the bend approximates two right angles).

Another common type of bolt shown in Figure 13.2 is the **battery bolt,** which has a square head and is used for clamping or fastening storage batteries (of course, this type of bolt may be used for fastening other components in position).

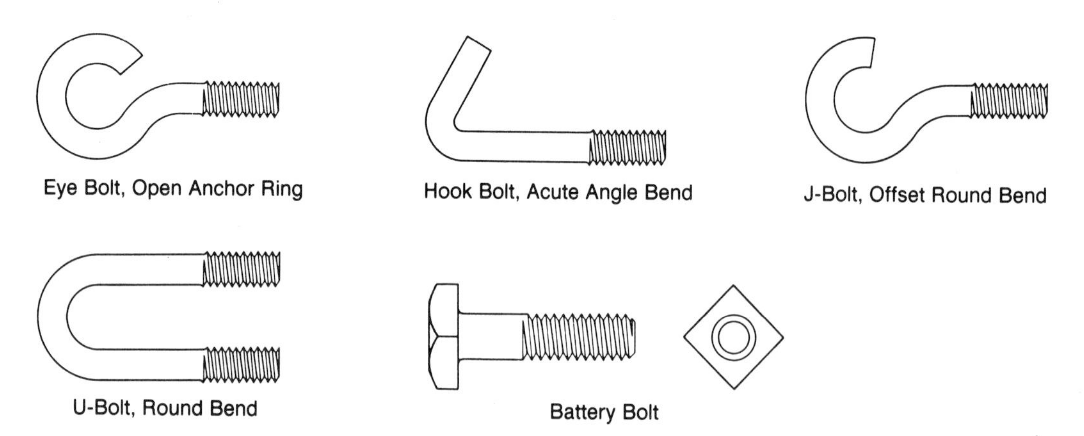

Figure 13.2 Examples of bolts. All types are designed to mate with a nut. (ASA B18.12-1962 (R1981), *Glossary of Terms for Mechanical Fasteners.*)

13.2 Pins and Other Nonthreaded Fasteners

Pins are tapered or cylindrical fasteners used to lock two or more components in position to one another. Pins are generally less expensive than threaded fasteners; however, the fastening achieved with them may not be as strong or as load-resistant. Figure 13.3 presents examples of **clevis** and **cotter** pins. Clevis pins have cylindrical heads, chamfered points, and a small hole at one end in which a cotter pin may be inserted to achieve fastening between two parts. Cotter pins consist of semicircular wire that has been bent to form a loop; this loop acts as the head of the cotter pin and prevents the pin from passing completely through the clevis pin hole in which it is to be inserted. Cotter pins are manufactured in a wide variety of point styles, as shown in the figure. After insertion in the clevis pin's hole, the end of the cotter pin may be bent to complete the fastening effort.

Other types of nonthreaded fasteners include **rivets** (which are used to achieve inexpensive and permanent fastening in automatic or semiautomatic assembly processes), **retaining or snap rings** (used to position a component on a shaft), and **washers.** Washers are used in conjunction with bolts and nuts in fastening operations in order to:

- Reduce the clearance hole through which the bolt is to be inserted.
- Allow the applied load to be distributed across a larger surface area.
- Protect the surface of the components to be fastened.
- Provide insulation.
- Provide greater electrical conduction.
- Increase the gripping action between the bolt and the nut via greater surface friction.

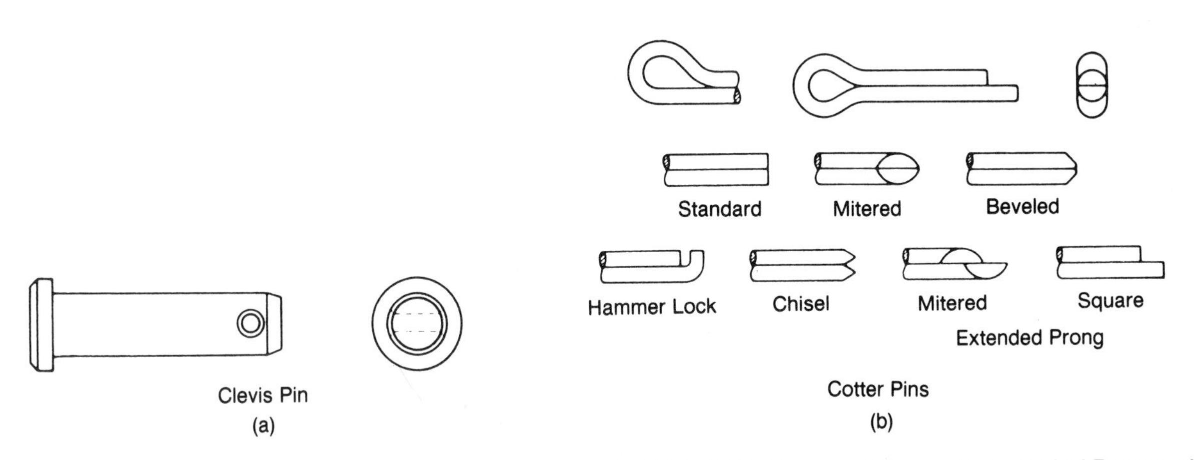

Figure 13.3 Examples of clevis (a) and cotter (b) pins. (ASA B18.12-1962 (R1981), *Glossary of Terms for Mechanical Fasteners.*)

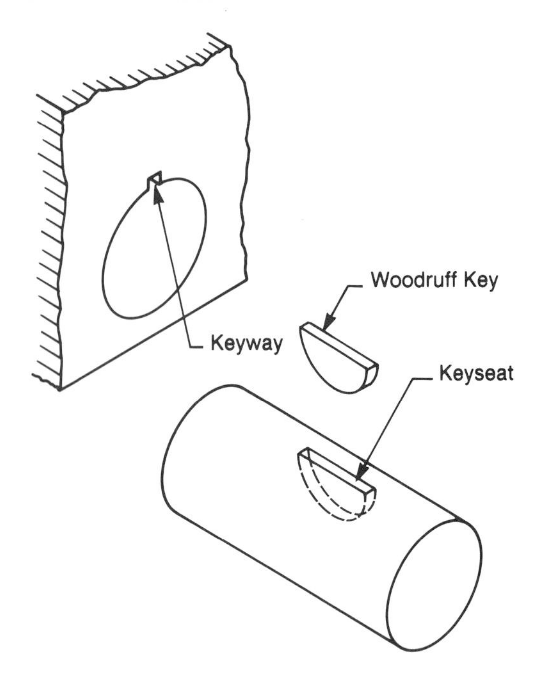

Figure 13.4 Example of a key system: a Woodruff key, keyseat, and keyway.

In addition to such mass-produced fasteners, there are many special-purpose fasteners, including self-sealing fasteners, plastic fasteners, stamped spring-steel fasteners, quick-operating fasteners, and formed metal fasteners. Each type of special-purpose design allows a specific set of functions in a fastening operation. These special-purpose fasteners are beyond the scope of this text. However, the interested reader is referred to the professional engineering magazine *Machine Design* (which has devoted special issues to the topic of fasteners) and such publications as ASA B18.12-1962 (Reaffirmed 1981), *Glossary of Terms for Mechanical Fasteners* and the American National Standards Institute (ANSI) catalogs for screw threads, nuts, rivets, and other fasteners.

Other fasteners used to lock two components in a fixed relative position are splines and keys. **Splines** consist of cylindrical shafts in which longitudinal grooves are used for locking. Splines may be external or internal in form. **Keys** are used in conjunction with keyways and keyseats. Figure 13.4 shows a Woodruff key system (also see Figure 8.43 and Appendix K).

Rivets are used to lock components in a fixed, permanent configuration. A rivet is a cylinder with a single head. It is inserted through holes in each component, after which a second head is created by hammering the rivet. The components are thereby permanently fastened. Figure 13.5 shows examples of different types of rivets.

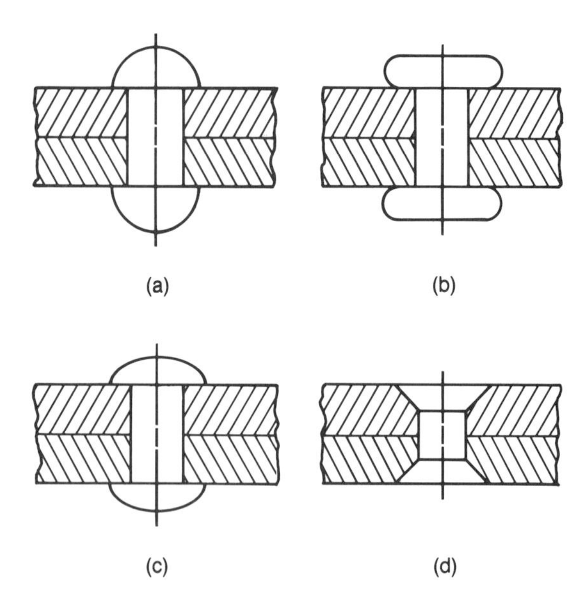

Figure 13.5 Examples of rivets. They may have a button head (a), flat head (b) pan head (c), or flat-top countersunk head (d).

13.3 Characteristics of Fasteners

A detailed list and corresponding definitions for the various features, or characteristics, of fasteners can be found in ASA B18.12-1962(R1981), *Glossary of Terms for Mechanical Fasteners.* In this section we will review only those fastener characteristics that are very common and that the reader should be familiar with in order to (1) specify particular types of fasteners for a given purpose or (2) interpret catalog information about the various types of fasteners available from manufacturers.

Figure 13.6 presents these common characteristics of mechanical fasteners:

- **Head,** which provides a bearing surface against which a load may be applied; it is an enlarged portion located at one end of the fastener. As noted earlier, a variety of head styles are available to meet different needs in appearance, driving equipment, and expected loading.
- **Neck,** which is (1) a reduced portion of the fastener shank or (2) a portion of the fastener located near the head and designed for a specific function.

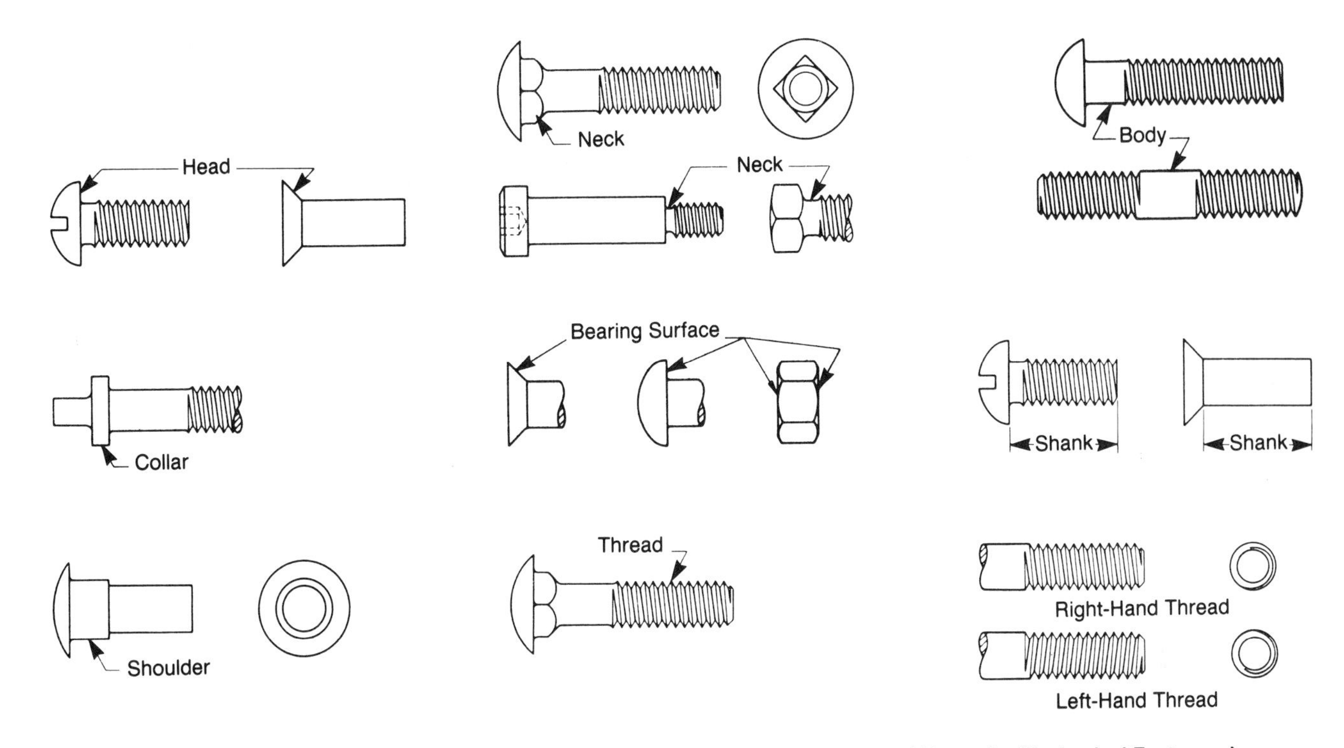

Figure 13.6 Common characteristics of fasteners. (ASA B18.12-1962 (R1981), *Glossary of Terms for Mechanical Fasteners.*)

- **Body,** which is the unthreaded portion of a fastener.
- **Collar,** which is a flange or enlarged ring near the head or on the shank.
- **Bearing surface,** which is the primary surface of contact between the fastener and the component to which it is connected.
- **Shank,** which is the portion of a headed fastener that lies between its head and its point.
- **Shoulder,** which is an enlarged portion of the shank or the body.
- **Thread,** which consists of a **helix-shaped ridge** located on the external or internal surface of the cylindrical portion of the device, or a **conical-shaped ridge** on some fasteners that are conical in form (such as self-tapping screws). A thread may be **right-handed** or **left-handed** in form. With respect to an axial viewing direction (*from* head *to* point of the fastener), a right-handed thread moves linearly away from the observer as it is turned in a clockwise direction (that is, as it is screwed into position). A left-handed thread must be rotated in a counterclockwise direction for receding linear motion. Standard threads are right-handed in form *unless otherwise specified.*

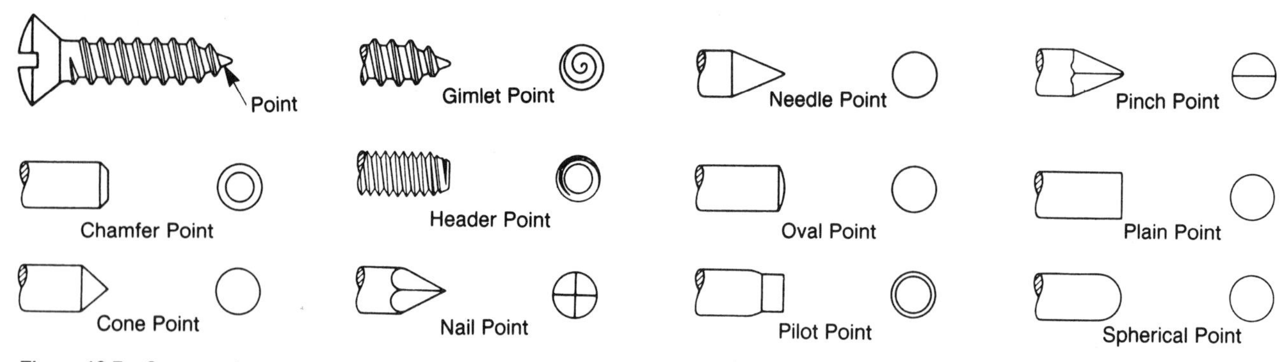

Figure 13.7 Common fastener point styles. A wide range of styles exists to meet various specific purposes. (ASA B18.12-1962 (R1981), *Glossary of Terms for Mechanical Fasteners.*)

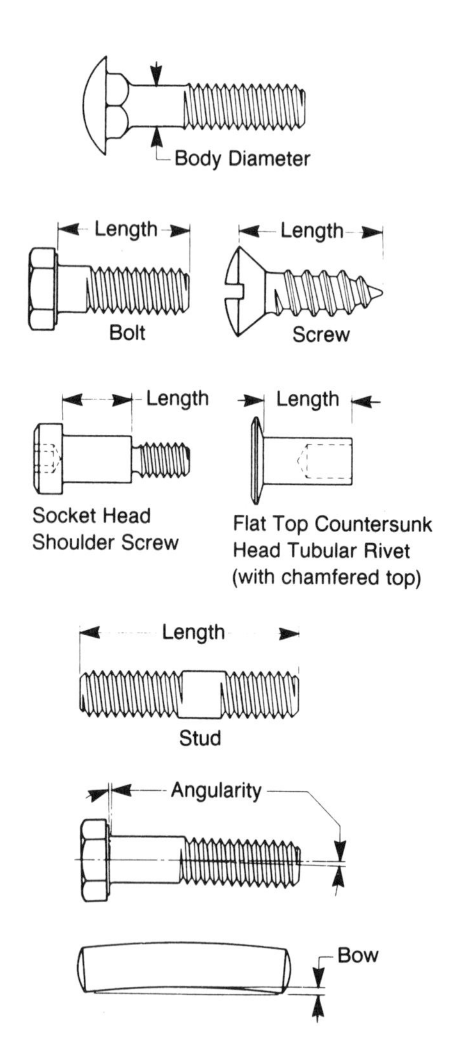

Figure 13.8 Dimensions and size characteristics of fasteners. (ASA B18.12-1962 (R1981), *Glossary of Terms for Mechanical Fasteners.*)

- **Point,** which is the tip or geometric form located at one end (or at both ends in the case of a headless fastener). See Figure 13.7. Points are manufactured in a wide range of styles for specific purposes, including (to name only four examples):
  - **Chamfer** points to ease assembly.
  - **Cone** points for perforation operations.
  - **Gimlet** points for self-tapping fasteners.
  - **Nail** points for piercing applications.

13.4 Dimensions and Size Specifications of Fasteners

Threaded and nonthreaded mechanical fasteners, of course, must be dimensioned in a drawing. The principal dimensions are shown in Figure 13.8 and include:

- **Body diameter,** which is the diameter of the cylindrical portion of a threaded fastener.
- **Length,** which refers to the following measurements:
  - *For headed fasteners,* the distance from the point of the intersection between the fastener's bearing surface and the largest-diameter portion of the head to the extreme tip of the fastener's point.
  - *For headless fasteners,* the distance from one end (point) to the other (point).
  - *For shoulder screws,* the length of the shoulder.
- **Angularity,** which refers to the angle between the axes of two surfaces on the fastener.
- **Bow,** which refers to the amount of deviation exhibited by a surface from linearity (straightness); it is also known as the **camber.**

The thread dimensions for a threaded fastener must also be specified in a drawing. Three generally acceptable formats for representing a threaded fastener within a drawing are (Figure 13.9):

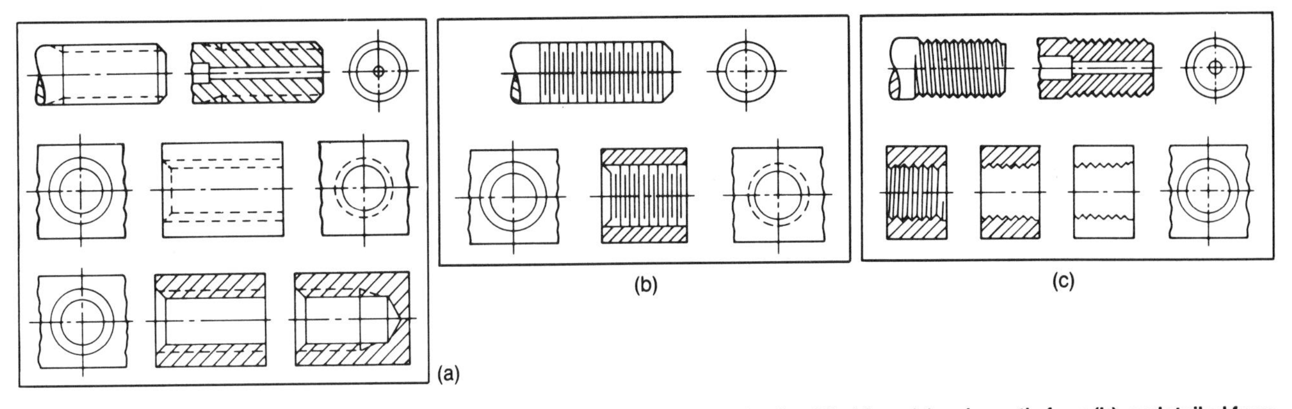

Figure 13.9 **Representation of threads in engineering drawings may be made in simplified form (a), schematic form (b), or detailed form (c). (ANSI Y14.6-1978, American National Standard Screw Thread Representation.)**

- Detailed.
- Schematic.
- Simplified.

The **detailed** representation most accurately represents the appearance of a threaded fastener in a sectional view. Helices are shown as straight slanting lines (which is a simplification of the actual appearance), and the crests (external edges of the thread) and roots (the internal ridges) are shown as sharp Vs. This format requires more time and effort than the simpler **schematic** and **simplified** representations; however, it may be required for clarity within a drawing.

Schematic representations may be used for greater ease in construction. Threads are shown as vertical or inclined lines in this format. Finally, the simplified format is easily applied in a drawing and is recommended for many common threaded fasteners (such as straight and tapered V-Form, Acme, Stub Acme, Buttress, and Helical Coil Insert) by the American National Standards Institute (ANSI Y14.6-1978).

Figure 13.10 presents the following common thread types:

- **Sharp V,** which has pointed crests and roots, very simple to draw but difficult to manufacture or use without damage to the thread. As a result, it is infrequently used in engineering applications.
- **Unified External** and **American National,** which are more common threads than the Sharp V because of their more rounded roots and crests.
- **Worm, Square, and Acme,** which are very useful for transmitting power between components within a system. The Worm and Acme thread forms are more popular (both in use and in manufacture) than the square thread because of their less discontinuous (less sharp-edged) boundaries.

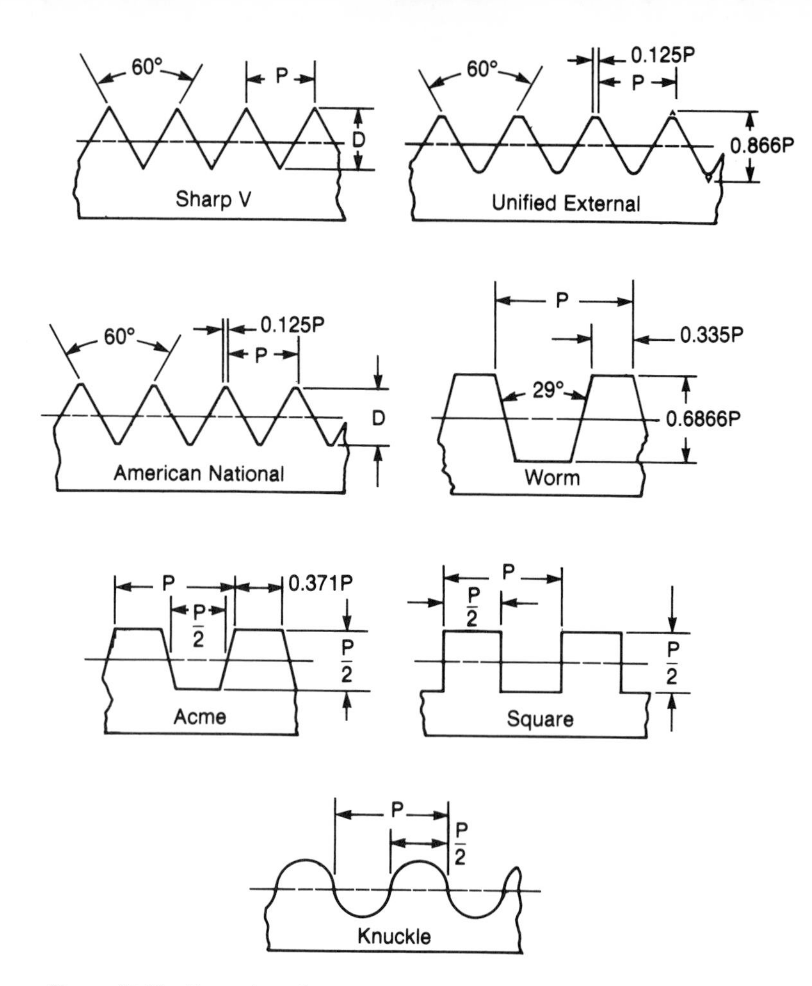

Figure 13.10 Examples of screw thread types.

- **Knuckle,** which is formed by *molding* plastic, sheet metal, or glass (as in the Knuckle threads on some glass jars).

In general, then, the Sharp V thread is more popular in drawings than in applications, the Worm and Acme threads are used in most general applications requiring significant power transmission, and the Knuckle thread is used for molded threaded fasteners. The Unified and American National threads are used for general-purpose applications.

Figure 13.11 identifies the following basic thread characteristics that are used in specifications:

- **Pitch,** which is the distance between adjacent crests, measured along the axial direction of the thread.
- **Thread angle,** which is the angle between threads.
- **Major diameter,** which is the largest diameter on a thread.
- **Minor diameter,** which is the smallest diameter on a thread.
- **Pitch diameter,** which is equal to twice the distance from the axis of the thread to a point at which the thread width

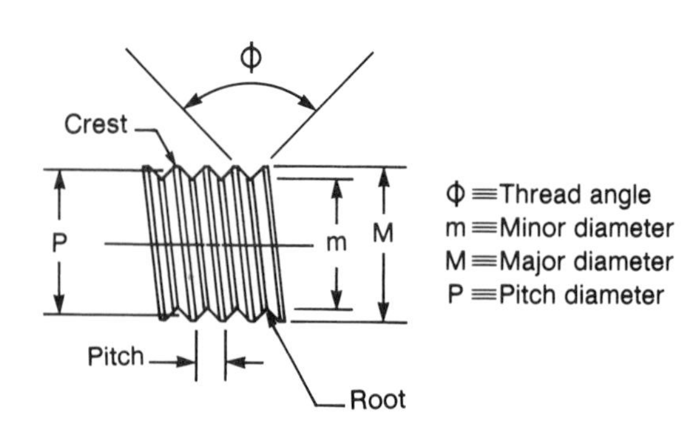

Figure 13.11 Screw thread characteristics used in specifications.

is identical to the distance between the threads at this location.

- **Thread series,** which specifies the number of threads per unit length.
- **Thread class,** which specifies the closeness of fit between mating threads.

Threads may be specified according to either the metric (SI) system or the English unit system. In the **metric** system, there is both a coarse and a fine series, each extending from a *combination of diameter and pitch* equal to 1.6 mm. (diameter) and 0.35 mm. (pitch) to 100 mm. (diameter) and 6.0 mm. (pitch); see Appendix I.

Several thread series are identified within the **English** unit system. Of these, the **Unified** *screw thread series* is the one most frequently applied in specifications. In this series, the number of threads per unit length is specified instead of the pitch (as in the metric system). The inverse of the pitch is equal to the number of threads per unit length.

Three distinct categories are identified within the **Unified** system for series with **graded pitchs** (see Appendix D): Coarse (denoted by UNC for Unified Coarse or by NC for National Coarse), Fine (UNF or NF), and (Extra Fine (UNEF or NEF). The number of threads per unit length should be specified in accordance with design needs as follows:

A **Coarse** thread specification corresponds to relatively few threads per unit length; such a specification is appropriate for (1) quick assembly of parts in a design and (2) long engagement; that is, the length of the threaded shaft should be relatively large so that a coarse specification will nevertheless correspond to a large *total* number of threads, thereby ensuring secure fastening between components.

A **Fine** thread specification is appropriate for (1) short engagement and (2) cases in which vibrations may be substantial; such vibrations may result in loosening and/or significant relative motion between components in a design if the total number of threads is not large.

An **Extra-Fine** thread is appropriate for (1) extremely short engagement and (2) cases in which vibrations may be very substantial.

An alternative specification in the Unified system is used for series with **constant pitches** (Appendix D): the designations extend from 4 UN to 32 UN, where the value (4, 6, 8, . . ., 32) denotes the number of threads per inch.

The thread class, as noted earlier, specifies the closeness of fit between mating threads. In the **metric** system, there are two distinct classes: **close fit** and the **general-purpose fit.** The general-purpose fit is always assumed unless the close fit is clearly identified in the specification. In the **Unified** series, three classes have been identified: **1** (1A or 1B), **2** (2A or 2B) and **3** (3A or 3B), in which the corresponding letter designation identifies the thread as external **(A)** or internal **(B). Class 1** corresponds to very generous tolerances (or a loose fit) for easy assembly. **Class 3** corresponds to precision tolerances (tight fits) for fine adjustments in an assembly. **Class 2** is used in most applications.

Threads are specified on a drawing in terms of the following data if the **metric** system is used:

NOMINAL DIAMETER (in millimeters)—
PITCH—TOLERANCE CLASS

for example,

M18 x 2.5 — 6e8e

in which the metric system is identified by the letter designation M, the nominal diameter is 18 mm., the pitch is 2.5 mm., and the tolerance class is given as *6e8e*. The numbers in the tolerance class refer to the **tolerance grade** (4 denotes a close tolerance, and 8 identifies a large tolerance); the **tolerance position,** or the material thread allowance, is specified by a corresponding letter (e, g, and h denote large, small, and zero allowances, respectively; uppercase letters (G, H) are used for internal threads). In the example, 6e8e indicates that the **pitch diameter** has a tolerance of **6e** and that the **major diameter** of the thread has a tolerance specification of **8e.**

Threads are specified on a drawing in terms of the following data if the **Unified** series is used:

NOMINAL DIAMETER (in inches)—
THREADS PER INCH—THREAD FORM and
THREAD CLASS—THREAD CLASS SYMBOL

for example

½-20UNF-2A

in which the corresponding specification is for a diameter equal to 0.5 in., 20 threads per in., (Unified) fine, and a Class 2 (general-purpose) fit for an external (A) thread. A tolerance note may be added, if necessary, for the major diameter and pitch diameter limits. All of the preceding information is supplied in a thread note on the drawing, together with a leader directed to the thread representation.

Thread specifications for **screws** should include the name of the screw, the type of head desired, the thread designation, length, and any other specifications necessary in the form of an additional note. For example, a **metric** specification would read:

HEX HEAD MACHINE SCREW
M16 x 2 x 30, STAINLESS STEEL

(corresponding to a machine screw with a hexagonal head, a nominal diameter of 16 mm., a pitch of 2 mm. and a length of 30 mm.)

A **Unified** series specification would read:

¼-20 UNC-2A x 2
FILLISTER HEAD MACHINE SCREW, STEEL

(nominal diameter of 0.25 in., 20 threads per in., Unified coarse, Class 2 fit, external (A) thread with a length of 2 in.).

For a **metric** specification for **bolts,** the bolt type, thread designation, length, and other necessary notes should be included as follows:

HEX BOLT, M10 X 1.5 X 60
STAINLESS STEEL

For a **Unified** specification for bolts, the nominal diameter, thread designation, length, name, and necessary notes should be included as follows:

½-13 UNC x 3½ SQUARE BOLT, STEEL

Learning Check

Interpret the meaning of each of the following specifications:

M30 x 3.5
M14 x 1.5
M14 x 1.5 x 80

Answer M30 x 3.5 is a metric specification corresponding to:

diameter = 30 mm.
pitch = 3.5 mm. (coarse)

(See Table D.2 in Appendix D.) M14 x 1.5 is a metric specification corresponding to:

diameter = 14 mm.
pitch = 1.5 mm. (fine)

(See Table D.2 in Appendix D.) M14 x 1.5 x 80 is a metric specification corresponding to:

diameter = 14 mm.
pitch = 1.5 mm. (fine)
length = 80 mm.

In the Unified system, identify the possible (graded pitch) specifications for a thread with a nominal diameter equal to ⅝ inch (see Table D.1 in Appendix D).

Answer The three possible (graded pitch) specifications are:

⅝-11 UNC

corresponding to a coarse specification (11 threads per inch).

⅝-18 UNF

corresponding to a fine specification (18 threads per inch).

5/8-24 UNEF

corresponding to an extra-fine specification (24 threads per inch).

Interpret the meaning of the following specification:

1½-12 UNF x 4½, HEX BOLT, STEEL

Answer The specification corresponds to:

diameter = 1½ inches
thread/in = 12 (Fine)
length = 4½ inches

13.5 Nuts

A **nut** is a sleeve with an internal thread for fastening to a mating bolt. (Bolts are discussed in Section 13.1.) The principal dimensions of a nut are shown in Figure 13.12. The *thickness* of a nut is the distance between the surfaces that are perpendicular to the axial direction, measured along that direction. The *width* of a nut is the distance between opposite sides (or flats) that are parallel to the nut axis. Rectangular nuts are dimensioned in terms of widths and lengths (since all four sides are not identical in length). Finally, one needs to specify the internal thread dimensions for the nut.

Thread specification in the **metric** system should include the name of the nut, the thread designation, and any other features that should be noted, as follows:

HEX FLANGE NUT, M8 X 1.25, STAINLESS STEEL

Thread specification in the **Unified** system should include the nominal diameter (or the basic major diameter of the thread), the thread designation, the type of nut, and other necessary notes, as follows:

7/8-9UNC, SQUARE NUT, STAINLESS STEEL

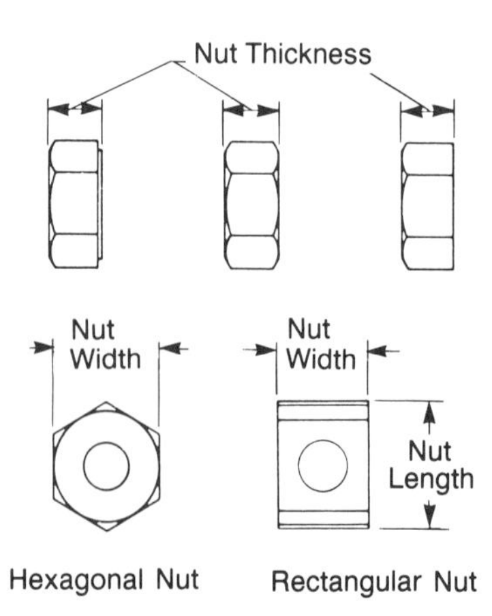

Figure 13.12 Principal dimensions of nuts. (ASA B18.12-1962 (R1981), *Glossary of Terms for Mechanical Fasteners*.)

13.6 Springs

Springs are used to fasten components together in a flexible fashion. Energy is absorbed by the spring and then released to the system as needed. Parts may move linearly or rotationally with respect to one another (through small distances) if fastened together with springs.

Figure 13.13 presents a series of helical extension springs that may be used in a system. **Compression, torsion, constant-force,** and **extension** springs are commonly used in industrial applications. Springs are shown in working drawings with **double-line** representations (as shown in Figure 13.13) or in schematic form by a set of inclined lines.

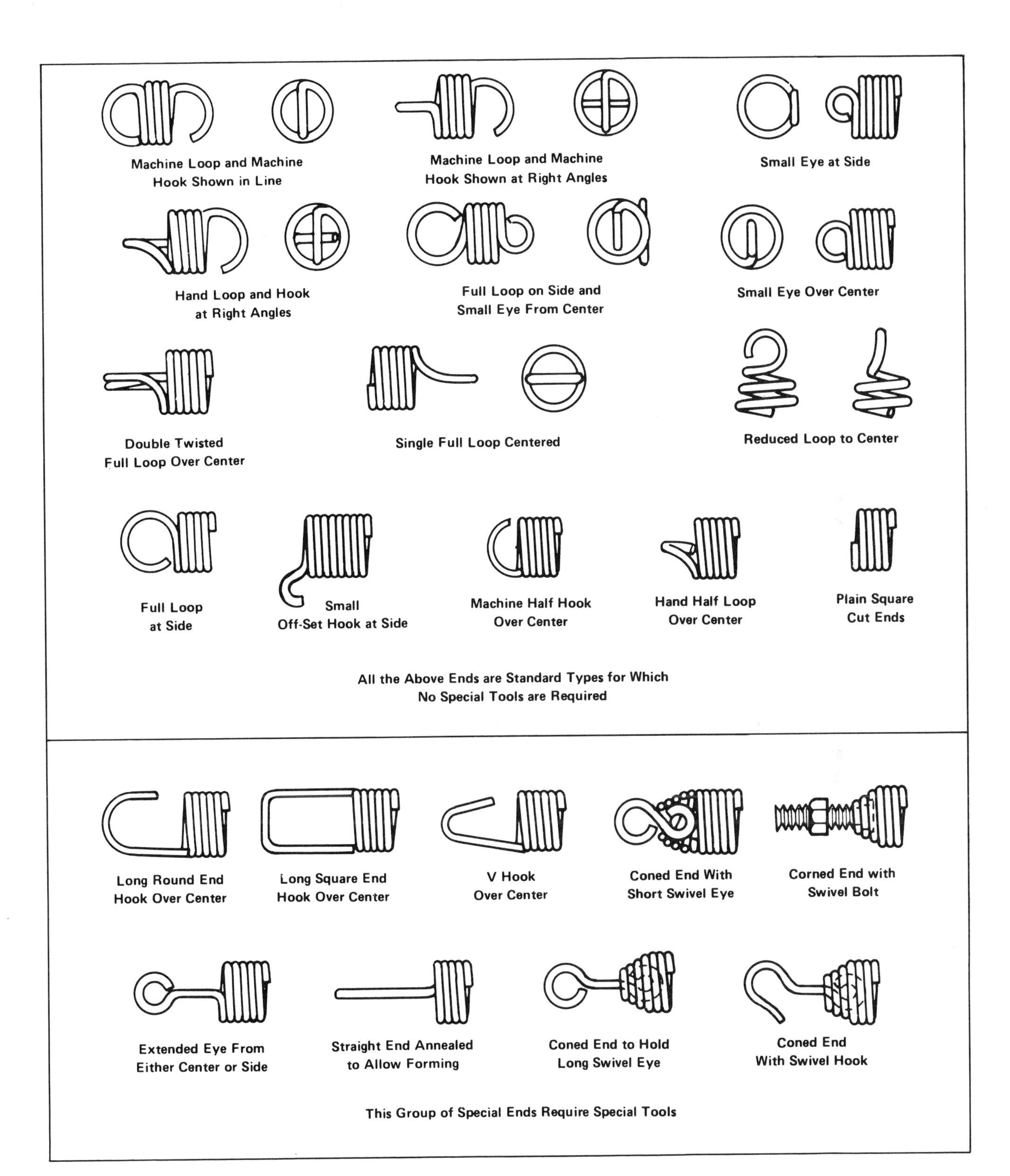

Figure 13.13 Types of helical extension spring ends. (ANSI Y14.13M-1981, American National Standard Mechanical Spring Representation.)

Reflector-Tipped Arm Reveals Nut Tightness

A reflector-tipped flag arm pops into view when a truck wheel's lug nut loosens enough to require tightening. The warning device eliminates the routine maintenance task of tightening all wheel nuts on multiaxle rigs, reducing the hazard of wheel loss. Tightening a wheel nut forces a disc spring to rotate and deflect, lowering the flag arm and clamping it between the spring and a washer. This operation conceals the reflector-tipped element. As the nut loosens and releases pressure on the spring, centrifugal force allows the flag arm to swing out, revealing the reflector and indicating that the nut should be tightened. M.D. Chataway Ltd., Great Britain.

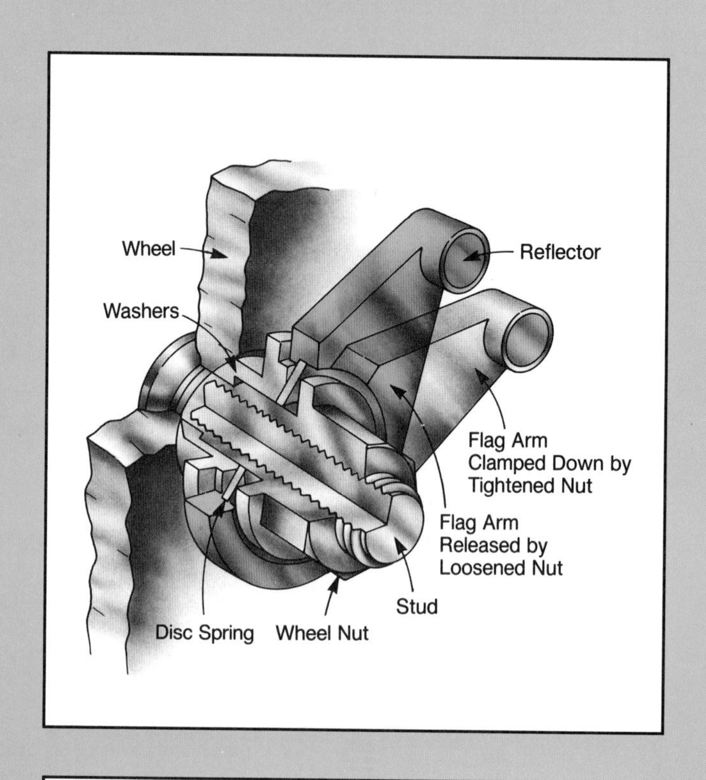

Source: Reprinted from MACHINE DESIGN, January 26, 1984. Copyright, 1984, by Penton/IPC., Inc., Cleveland, Ohio, p. 32.

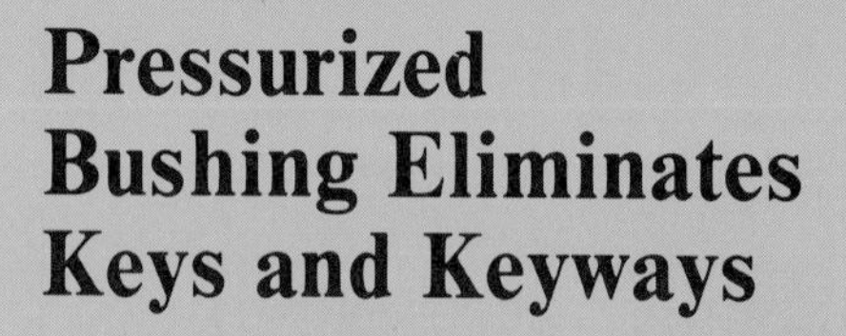

Pressurized Bushing Eliminates Keys and Keyways

Friction-grip bushing requires no keys, keyways, or tapped holes. As a result the component can be positioned freely on the shaft in both radial and axial directions. The bushing has a double wall that is closed at one end. The resulting cavity is filled with a semi-solid pressure medium, retained by a sealing ring and a thrust ring that acts as a piston. When the clamping screws are tightened, the walls flex and grip the shaft and hub. Pressure can be retained indefinitely or released by loosening the clamping screws. Eagle-Picher Bearings, South Bend, Ind., makes the ETP bushings in both metric and inch sizes.

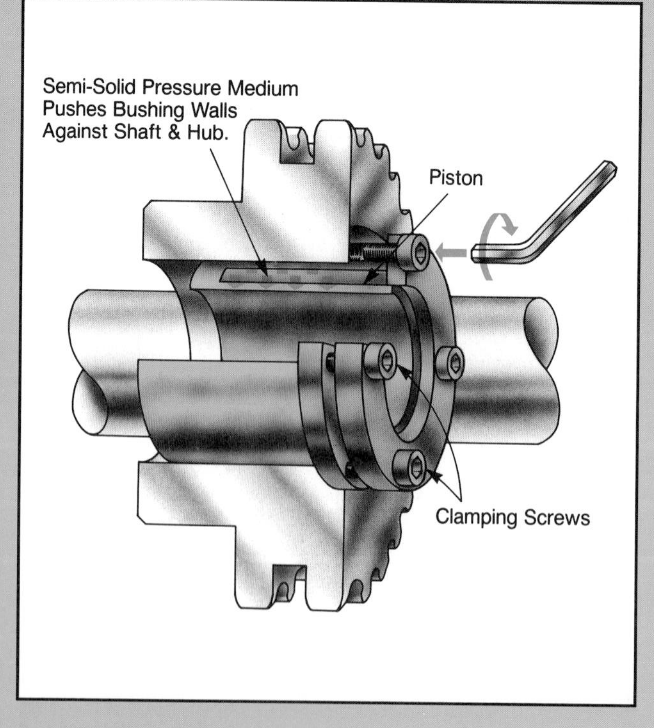

Source: Reprinted from MACHINE DESIGN, September 9, 1982. Copyright, 1982, by Penton/IPC., Cleveland, Ohio, p. 34.

A **specification table** must be included within the drawing to provide detailed information (such as extension under loading, initial tension, direction of the helix, and maximum deflection) for the spring(s). In addition, one should specify the material, the wire size, the number of coils, the outside diameter, the length, and the style of the ends of the spring.

In Retrospect

- Screws and bolts are externally threaded fasteners. Screws are designed to mate with preformed internal threads in the material of components to be fastened, or to cut such threads in the material (self-tapping screws). Bolts are designed to mate with a nut, which is a sleeve with an internal thread.
- Examples of screws are cap and machine screws. Examples of bent bolts are eye bolts, hook bolts, J-bolts, and U-bolts.
- Nonthreaded fasteners include pins (clevis and cotter), rivets, retaining or snap rings, and washers.
- Characteristics, or features, of fasteners include the head, neck, body, collar, bearing surface, shank, shoulder, thread, and point.
- Principal dimensions of fasteners include:
  - Body diameter.
  - Length.
  - Angularity.
  - Bow.
  - Thread parameters.

  Thread parameters include the pitch, thread angle, major diameter, minor diameter, thread series, and class.
- Nut dimensions include the thickness, width, length (for rectangular nuts), and thread specifications.
- Springs are used to fasten components together in a somewhat flexible, or elastic, fashion; linear and/or rotational displacement is allowed over small distances.

Exercises

Your instructor will specify the appropriate scale for each exercise. In each case use the suggested layout for the drawing.

EX13.1. Draw each of the following bolts to be used to fasten two parts together. Use the layout suggested in the figure for this exercise (a section view and an end view). Refer to Appendices F and G for specific data. Include all thread specifications in a thread note; also include a drawing of the nut (with all necessary specifications) in the section view.

a. Square bolt and square nut.
b. Hex bolt and hex nut.
c. Round head bolt and square nut.
d. Round head bolt and hex nut.

Use a detailed representation for each drawing.

EX13.2. Same as exercise EX13.1, but use a schematic representation.

EX13.3. Same as exercise 13.1, but use a simplified representation.

EX13.4. Same as exercise 13.1, but draw the following cap screws instead of bolts. Refer to Appendix H for specific data. Use a detailed representation.

a. Slotted round head cap screw and square nut.
b. Slotted round head cap screw and hex nut.
c. Slotted flat countersunk head cap screw and square nut.
d. Slotted fillister head cap screw and hex nut.

EX13.5. Same as exercise EX13.4, but use a schematic representation.

EX13.6. Same as exercise EX13.4, but use a simplified representation.

EX13.7. Draw each of the following bolts in the layout shown in the figure for EX13.1. Refer to appendices F and G for specific data. Include all thread specifications in a thread note; also include a drawing of an appropriate mating nut (with specs) in the section view. Use a detailed representation.

a. ⅜-16 UNC x 3 SQUARE BOLT, STEEL
b. ¾-16 UNF x 3½ SQUARE BOLT, STEEL
c. 7/16-14 UNC x 3 HEX BOLT, STEEL
d. 1-12 UNF x 4 HEX BOLT, STEEL
e. 5/16-18 UNC x 4 ROUND BOLT, STEEL
f. HEX BOLT, M12 x 1.75 x 50, STEEL
g. SQUARE BOLT, M20 x 2.5 x 80, STEEL
h. ROUND BOLT, M16 x 1.5 x 50, STEEL

EX13.8. Same as exercise EX13.7, but use a schematic representation.

EX13.9. Same as exercise EX13.7, but use a simplified representation.

EX13.10. Draw each of the following cap screws in the layout shown in the figure for EX13.1. Refer to appendix H for specific data. Include all thread specifications in a thread note; also include a drawing of an appropriate mating nut (with specs) in the section view. Use a detailed representation.

a. 5/16-24 UNF x 3 SLOTTED FLAT COUNTERSUNK CAP SCREW, STEEL
b. ½-13 UNC x 3½ SLOTTED ROUND CAP SCREW, STEEL
c. FILLISTER CAP SCREW, M8 x 1.25 x 40, STEEL

EX13.11. Same as exercise EX13.10, but use a schematic representation.

EX13.12. Same as exercise EX13.10, but use a simplified representation.

EX13.13. Draw the following threaded fasteners. Include a thread note for all specifications. Use the layout suggested in the figure for this exercise. Refer to the appropriate appendices for additional information if necessary. Use a detailed representation.

a. HEX HEAD MACHINE SCREW, M20 x 2.5 x 40, STEEL
b. SLOTTED ROUND HEAD CAP SCREW, M12 x 1.75 x 30, STEEL
c. SLOTTED FILLISTER HEAD MACHINE SCREW, M24 x 3 x 60, STEEL
d. PAN HEAD MACHINE HEAD, M16 x 2 x 40, STEEL
e. OVAL COUNTERSUNK HEAD MACHINE SCREW, M12 x 1.75 x 30, STEEL
f. ⅜-24 UNF-2A x 2, ROUND HEAD CAP SCREW, STEEL
g. ¾-10 UNC-2A x 3½, TRUSS HEAD MACHINE SCREW, STEEL
h. ¼-28 UNF-2A x 1½, FLAT COUNTERSUNK HEAD MACHINE SCREW, STEEL
i. 5/16-18 UNC-2A x 1¼, ROUND HEAD BOLT, STEEL

EX13.14. Same as exercise EX13.13, but use a schematic representation.

EX13.15. Same as exercise EX13.13, but use a simplified representation.

EX13.16. Draw two orthographic views of each of the following nuts specified. Include all necessary dimensioning information and thread specifications.

a. ⅞-14 UNF, SQUARE NUT, STAINLESS STEEL
b. ½-13 UNC, SQUARE NUT, STAINLESS STEEL
c. 1¼-7 UNC, HEX NUT, STEEL
d. ¼-28 UNF, HEX MACHINE SCREW NUT, STEEL
e. SQUARE NUT, M10 x 1.5, STAINLESS STEEL
f. HEX NUT, M16 X 2, STAINLESS STEEL

EX13.17. Draw the following cap screws in a layout similar to that shown in the figure for this exercise. Include all thread specifications and necessary dimensions. Refer to appendix H for additional information if necessary. Use a detailed representation.

a. ROUND HEAD CAP SCREW, M16 x 2 x 30, STEEL
b. FLAT COUNTERSUNK HEAD CAP SCREW, M24 x 3 x 80, STEEL
c. 9/16-12 UNC-2A x 1¾, FLAT COUNTERSUNK HEAD CAP SCREW, STEEL
d. ½-20 UNF-2A x 2½, SLOTTED ROUND HEAD CAP SCREW, STEEL
e. ⅜-16 UNC-2A x 2, SLOTTED FILLISTER HEAD CAP SCREW, STEEL

EX13.18. Same as exercise EX13.17, but use a schematic representation.

EX13.19. Same as exercise EX13.17, but use a simplified representation.

EX13.20. Draw internal and external detailed representations of the following threads. Specify the diameter value of your choice. Use a layout similar to that shown in the figure for this exercise.

a. Sharp V.
b. American National.
c. Acme.
d. Square.

EX13.21. Same as exercise EX13.20, but use schematic representations.

EX13.22. Same as exercise EX13.20, but use simplified representations.

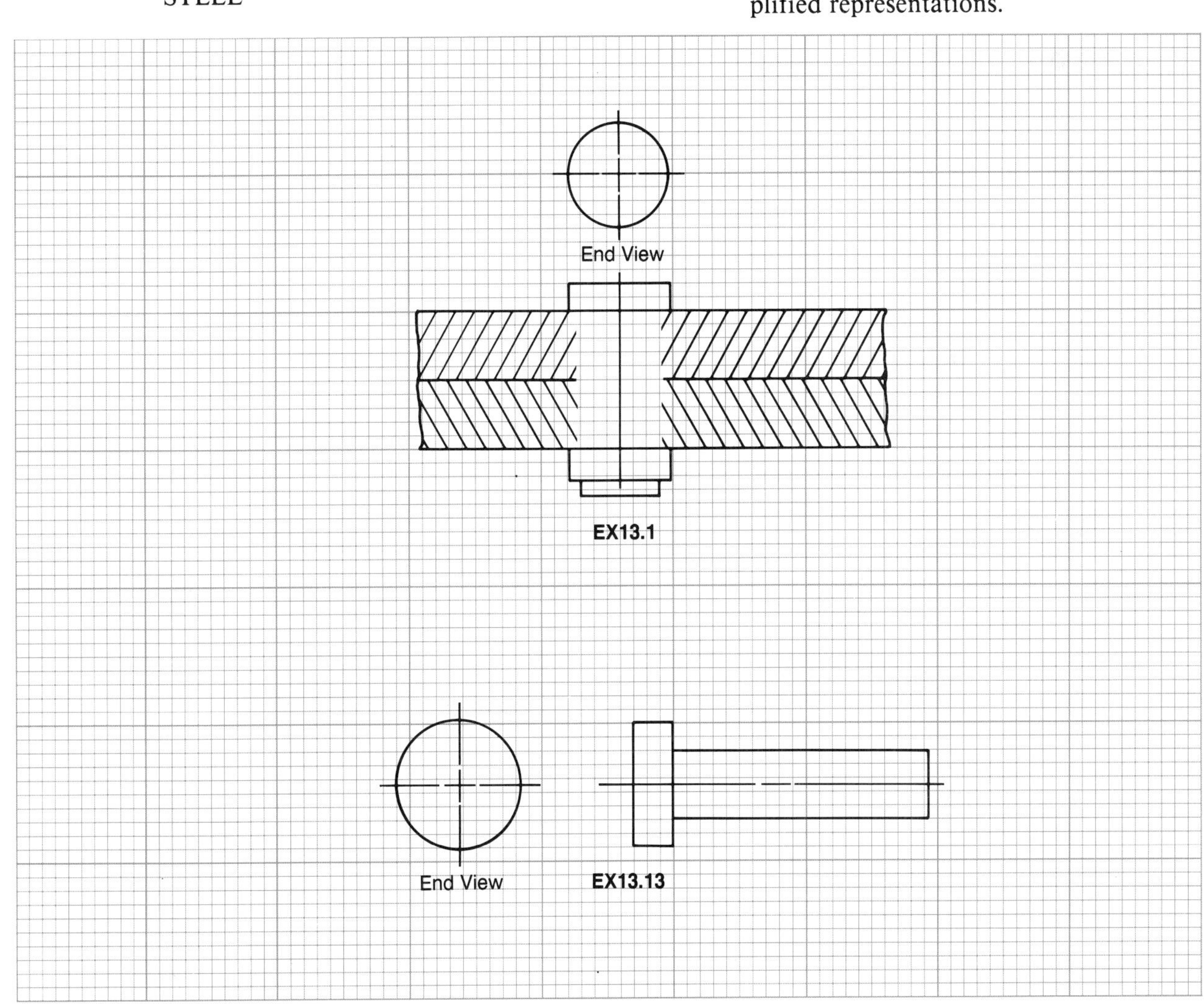

EX13.1

EX13.13

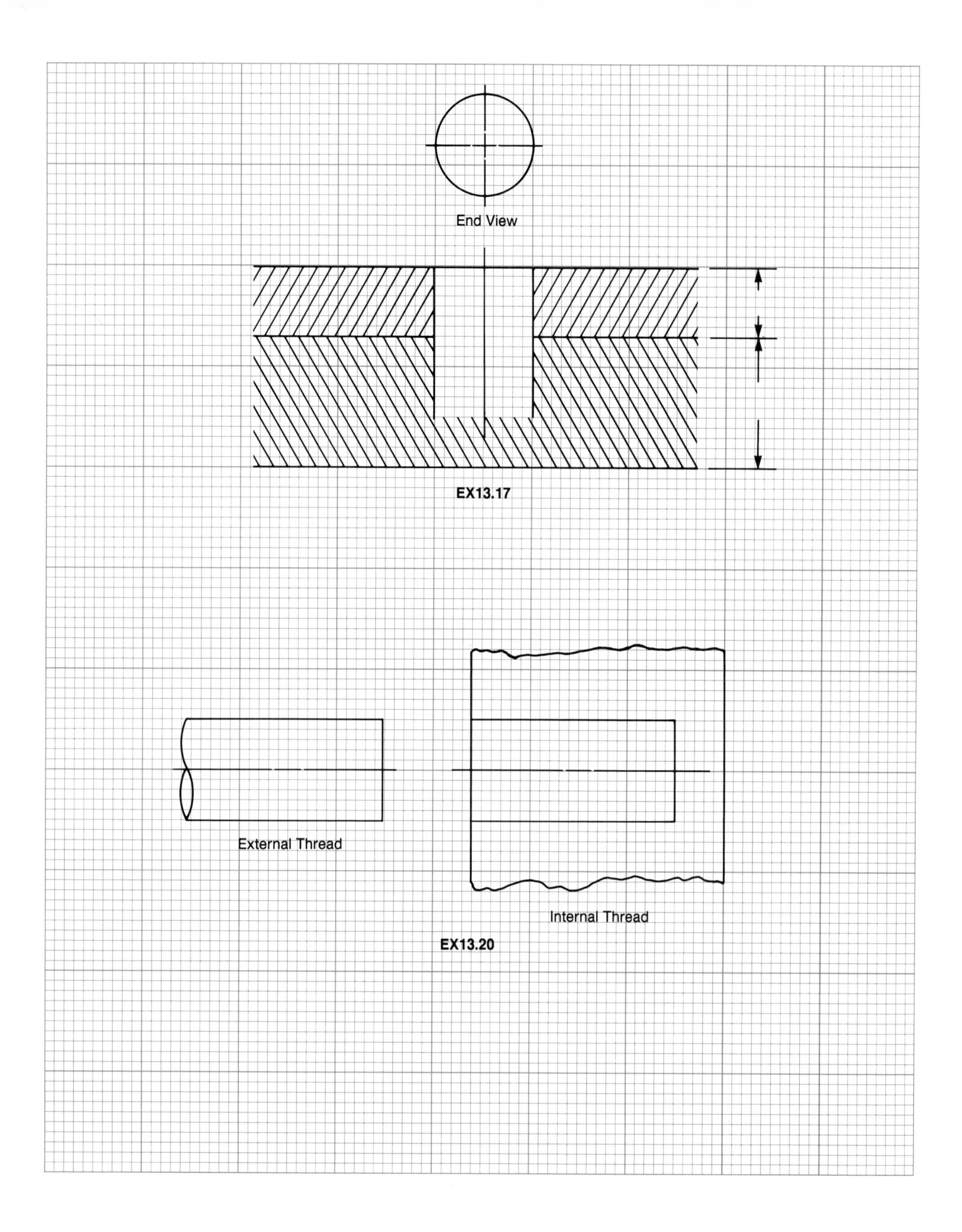

EX13.17

EX13.20

14 Pictorials, Part 1: Pictorial Projection Theory

Preview A pictorial is an approximate representation of an object as it actually appears to an observer (Figure 14.1). Put another way, it is a graphic description of the three-dimensional appearance of an object.

Perspective (in which depth-directed or receding lines appear to converge at a vanishing point on the horizon) can be accurately shown in a pictorial, as we will demonstrate in this chapter, or an approximate **('pseudo-perspective')** description of an object (in which receding linework is *not* shown converging toward a single point on the horizon) may be generated as a pictorial. Both types of graphic descriptions can be constructed from a given set of orthographic views with relative ease and accuracy. Pseudo-perspective pictorials are frequently used in engineering applications and will be the focus of the next chapter, where direct construction techniques are discussed.

In this chapter we will develop the theoretical foundation for both perspective and **axonometric** pictorials. (A similar development for **oblique** pictorials, which are discussed in the next chapter, can be derived from this explanation.) In addition, we will apply construction techniques for generating these pictorials from principal orthographic views of an object and then continue that discussion, focusing on pseudo-perspective pictorials, in Chapter 15.

Learning Objectives

Upon completion of this chapter, the reader should be able to:

- Define an axonometric pictorial and name the three distinct categories of such pictorials.
- Determine the foreshortening ratio of a line in an axonometric projection.
- Differentiate between an axonometric projection and an axonometric drawing.
- Construct axonometric projections from a set of given orthographic views of an object.
- Develop perspective projections of an object from a set of given orthographic views by specifying the elevation angle and the lateral angle of the observer's viewing point with respect to the object and its picture plane.

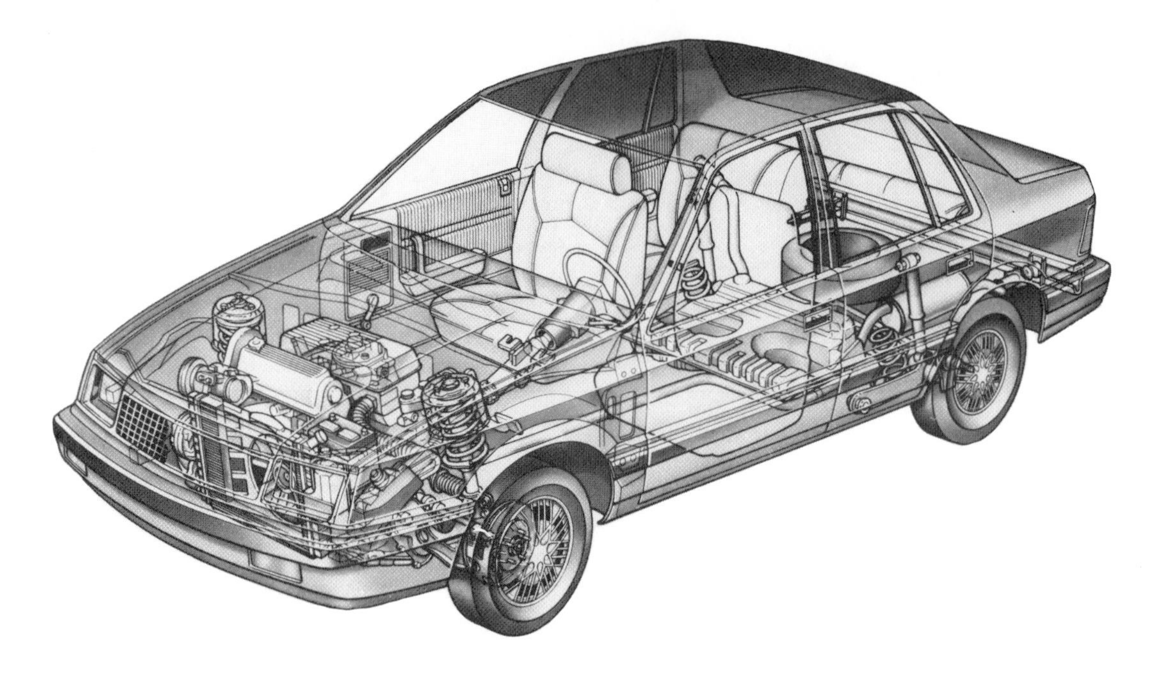

Figure 14.1 Example of a pictorial—in this case, a type of pictorial known as a *phantom illustration,* which shows the front-wheel drive powertrain layout of the four-door 1987 Plymouth Sundance. (Courtesy of Chrysler Motors.)

14.1 Axonometric Projections

An **axonometric pictorial** is a projected image of an object onto a projection plane wherein three principal faces (surfaces) of the object are shown in the pictorial. In other words, the object is rotated into a position (with respect to the projection plane) in such a way that three of its principal surfaces (for example, perhaps the front, side, and top surfaces) may be simultaneously projected onto the plane.

Figures 14.2 through 14.11 demonstrate the basis of such a projected image. In Figure 14.2, an example object is shown together with a projection plane. One vertical surface of the object is parallel to this projection plane and is identified as the front face of the object. In Figure 14.3 projectors are used to produce a projected image of the object in the projection plane. Figure 14.4 presents the resulting projected view (a view that is simply the front orthographic view of this object). This orthographically projected view of the object is a true representation (shown full scale) of its front surface.

The object now undergoes rotation about a vertical axis in Figure 14.5. Projection of the object (Figure 14.6) then results in a graphic representation in which two surfaces (front and right-side) are shown. Each surface has been distorted, or foreshortened, in this view; that is, some boundary lines

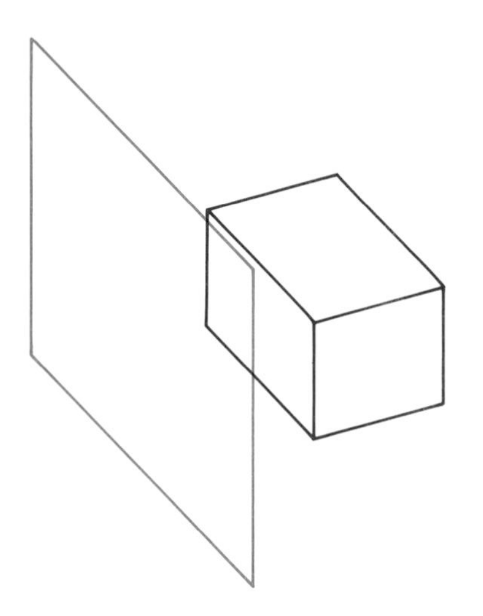

Figure 14.2 An example object and a projection plane to be the subject of successive figures in this chapter.

of each surface do not appear in true-length size because they do not lie in spatial planes that are parallel to the projection plane. We may then define a quantity known as the **foreshortening ratio,** according to the following relation:

$$\text{Foreshortening Ratio of a line} = \frac{\text{Length of line in projection}}{\text{True length of line segment}}$$

Figure 14.7 shows the resulting projected view of the object.

Finally, the object is rotated about a horizontal axis (which lies in a plane parallel to the projection plane) so that the top surface may now be projected onto the projection plane together with the front and right-side surfaces (Figures 14.8 through 14.11). The resulting projection (Figure 14.11) includes foreshortened representations of the front, top, and right-side surfaces of the object; such a drawing is an axonometric projection.

An infinite number of rotations can be applied to an object with respect to a given projection plane. Consequently, an infinite number of axonometric projections of an object can be generated through such rotations. However, there are three common *categories* of axonometric projections:

- **Isometric Projections,** in which distances along all three mutually perpendicular axes are equally foreshortened.
- **Dimetric Projections,** in which distances along two mutually perpendicular axes are equally foreshortened.
- **Trimetric Projections,** in which distances along all three mutually perpendicular axes are unequally foreshortened.

Isometric projections are the least difficult to construct; however, they are also the most distorted (or unrealistic)

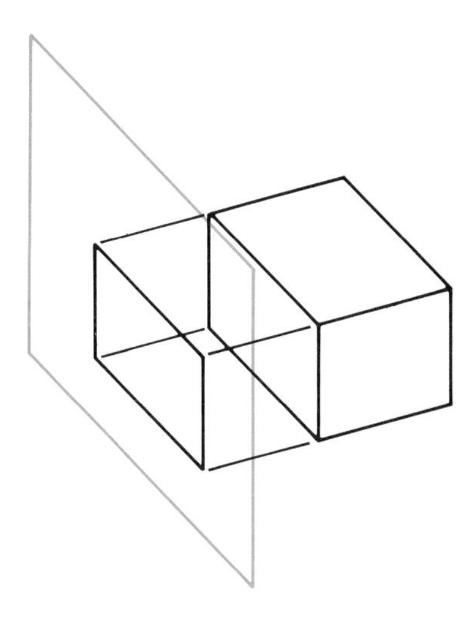

Figure 14.3 Projection of the object in Figure 14.2 onto the given plane.

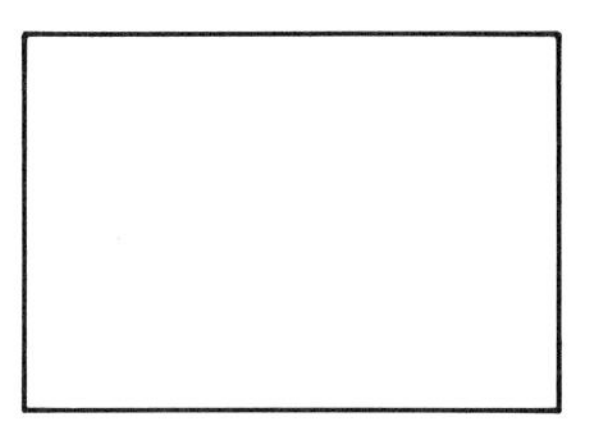

Figure 14.4 Resulting projection (front view) of the example object in Figures 14.2 and 14.3.

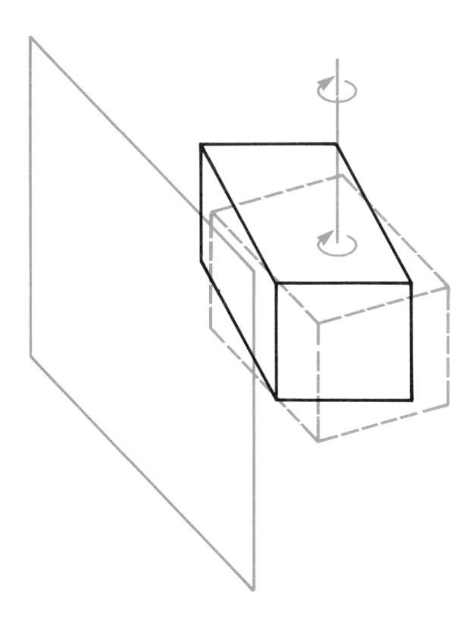

Figure 14.5 Rotation of the object in Figure 14.2 about a vertical axis.

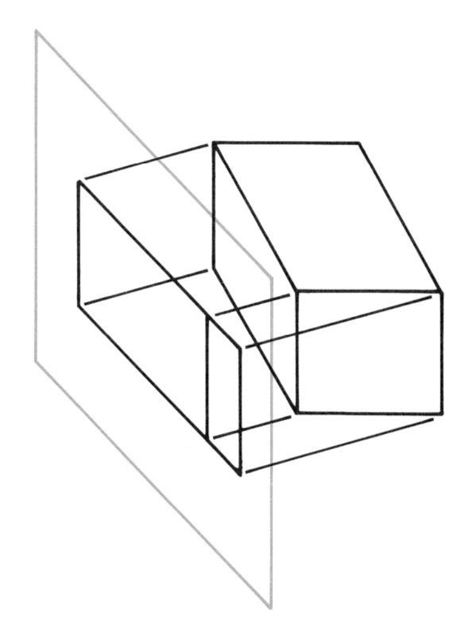

Figure 14.6 Projection of the vertically rotated object in Figure 14.5 onto the given plane.

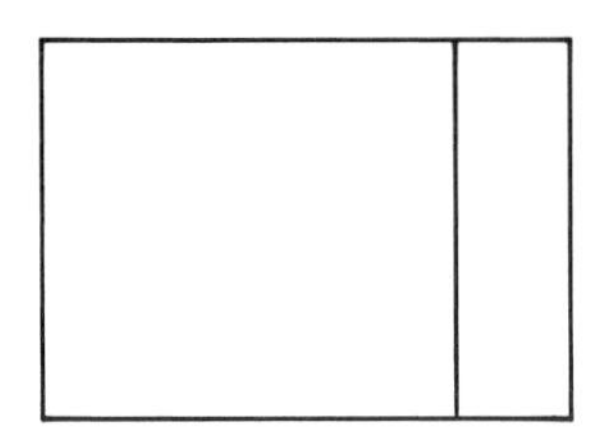

Figure 14.7 Resulting projection of the vertically rotated object in Figures 14.5 and 14.6. Foreshortened representations of the front and right-side surfaces appear in the projection.

graphic pictorials among these three categories. Trimetric projections are the most difficult, but least distorted, graphic representations. Ease of construction usually outweighs the relative distortion associated with isometric projections, so most axonometric pictorials are isometric.

In summary, then:

1. An axonometric pictorial is a projected image of an object onto a projection plane wherein three principal faces (surfaces) of the object under consideration are shown.

2. The foreshortening ratio is defined by the expression:

$$\text{Foreshortening Ratio of a line} \equiv \frac{\text{Length of line in projection}}{\text{True length of line segment}}$$

3. Three distinct categories of axonometric projections are defined in terms of the distances that are foreshortened in each: isometric projections, dimetric projections, and trimetric projections.

4. Isometric projections (in which distances along all three mutually perpendicular axes are equally foreshortened) are the least difficult axonometric pictorials to construct but are somewhat unrealistic in appearance. Nevertheless, isometric projections are the most popular of the three types of axonometric projections because of their ease of construction.

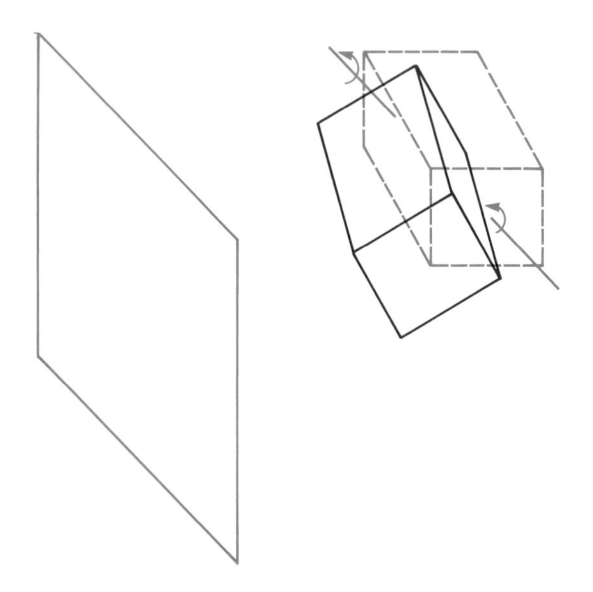

Figure 14.8 Second rotation of the object in Figure 14.6 about a horizontal axis that is parallel to the projection plane.

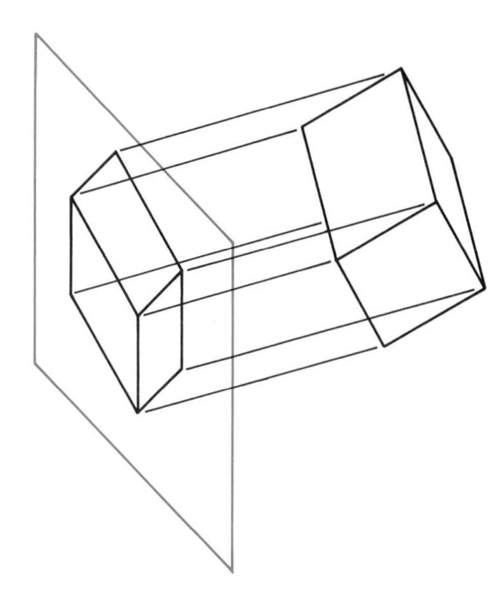

Figure 14.9 Projection of the image of the object in Figure 14.8 onto the projection plane.

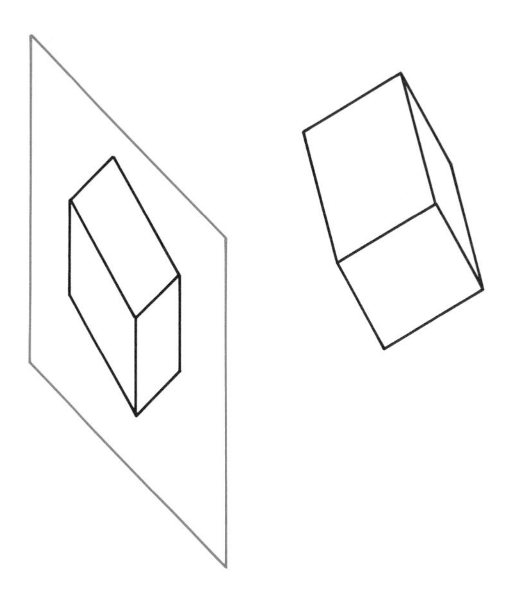

Figure 14.10 Projected image of the object in Figure 14.9 onto the given plane, shown without projector linework.

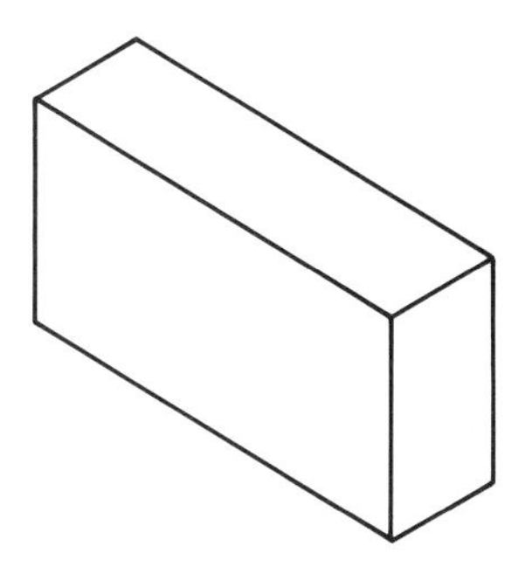

Figure 14.11 Axonometric projection of the example object in Figure 14.10. Three principal faces (surfaces) are shown in the projection. All three surfaces are foreshortened.

14.2 Construction of Axonometric Projections from Orthographic Views

Axonometric projections may be formed from a set of principal orthographic views. Consider the object shown enclosed in a glass box in Figure 14.12. One corner (top, front, right-side) of the glass box has been removed. The resulting oblique surface of the box represents the **axonometric projection plane** onto which a pictorial image of the object is to be formed. This pictorial image will represent the view of the object that would be seen by an observer who is viewing the object along a sight-line direction that is *perpendicular* to the axonometric projection plane. That is, the sight-line of the observer is perpendicular to the inclined (top, front, right corner) surface of the box.

In order to construct the appropriate axonometric projection of the object, we will need two principal orthographic views of the object. Figure 14.13 presents the object enclosed within the glass box with two views (front and top) of the object projected onto the surfaces of the box. These two surfaces of the box, together with the inclined surface that represents a portion of the axonometric projection plane, are then "unfolded" so that they lie in a single spatial plane (Figure 14.14). Note that the relative orientation of the two principal surfaces (top and front) of the box in Figure 14.14 is dependent upon the direction of the intersection lines ($\overline{AB}$ and $\overline{AC}$) between each of these surfaces and the inclined (projection) surface.

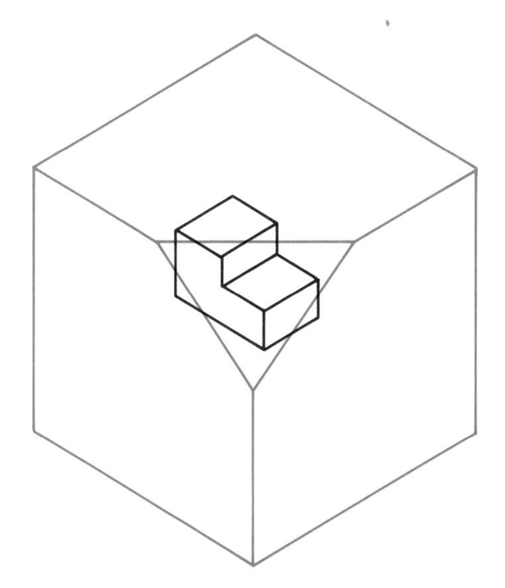

Figure 14.12 Example object, to be the subject of successive figures in this chapter. Here it is enclosed in a glass box that has had one corner removed. The oblique surface of the box is contained within the projection plane onto which a pictorial is to be constructed.

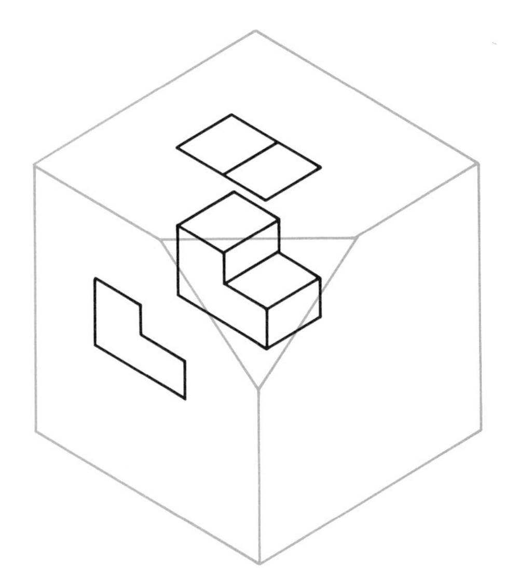

Figure 14.13 The front and top views of the object in Figure 14.12 have been projected onto the appropriate walls of the glass box.

Projectors from point 1 in the top and front views of the object are then drawn to the axonometric projection plane (Figure 14.15). Each of the projectors is perpendicularly oriented with respect to boundary line $\overline{AB}$ or boundary line $\overline{AC}$ of the axonometric projection plane. The intersection of these projectors represents the projection of point 1 (along the sight-line direction that is perpendicular to the axonometric projection plane) to the projection plane. (Each projector represents the view of this projection as seen from the front or top viewing direction.) In other words, the intersection of these projectors is the location of point 1 in the axonometric pictorial under construction.

Additional points of the object are projected to the axonometric plane in Figure 14.16, thereby locating these points in the pictorial construction area. Figure 14.17 shows the pictorial that is formed by connecting these points with appropriate linework. The completed pictorial (without projectors) is shown in Figure 14.18, together with the surfaces of the box that were used in the construction of this pictorial.

Figure 14.19 presents the axonometric pictorial, together with a partial view of the projection planes used in its construction. If we enlarge the glass box, the pictorial then appears as a projection of the object onto the inclined surface of the box (as shown in Figure 14.20).

Figures 14.12 through 14.20 provide the theoretical basis for the construction of axonometric projections from a set of

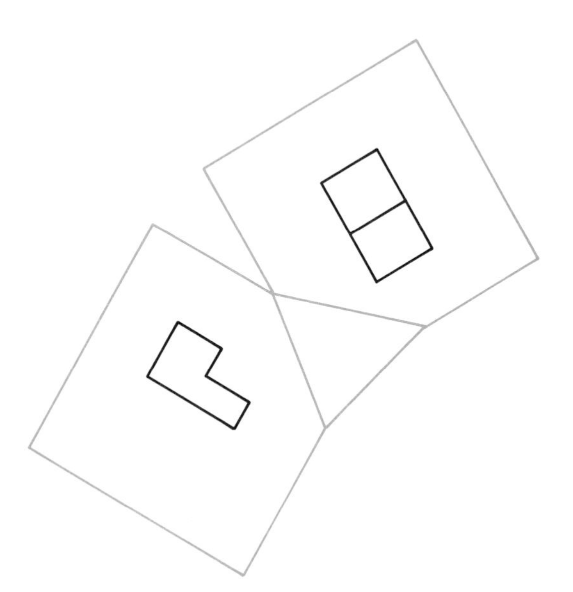

Figure 14.14 Selected surfaces of the glass box, unfolded so that they lie within a single spatial plane.

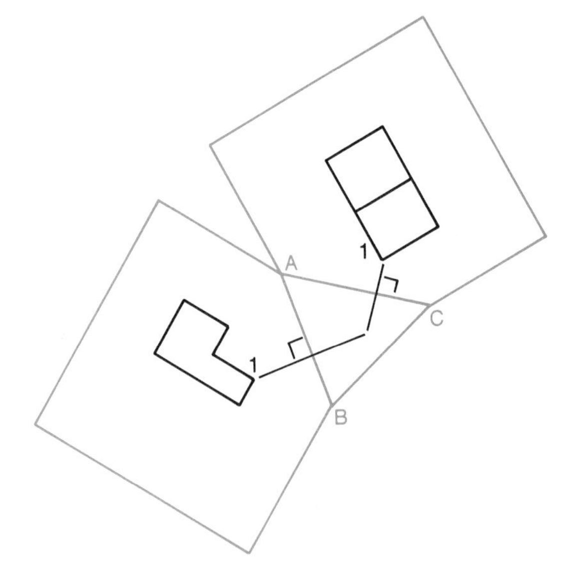

Figure 14.15 Point 1 is projected into the axonometric pictorial construction area.

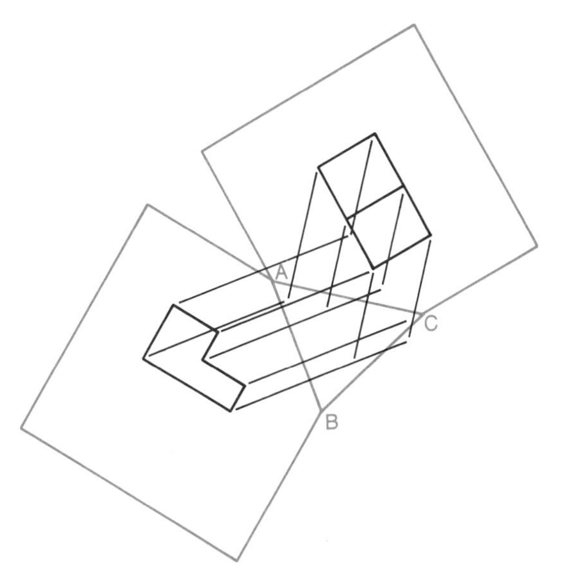

Figure 14.16 Additional points are projected into the axonometric pictorial construction area.

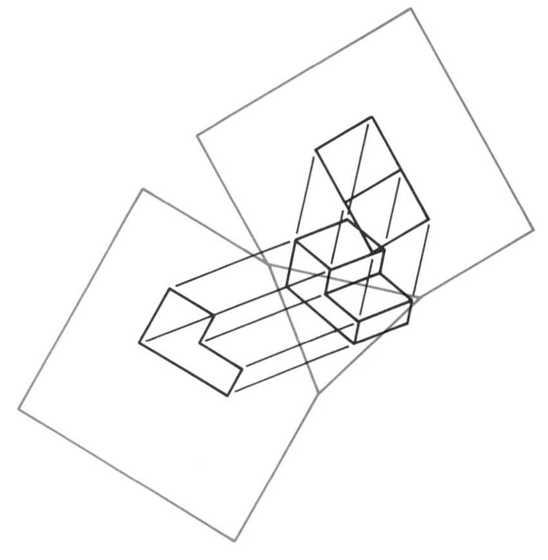

Figure 14.17 Axonometric projection is developed from the intersection of appropriate projectors.

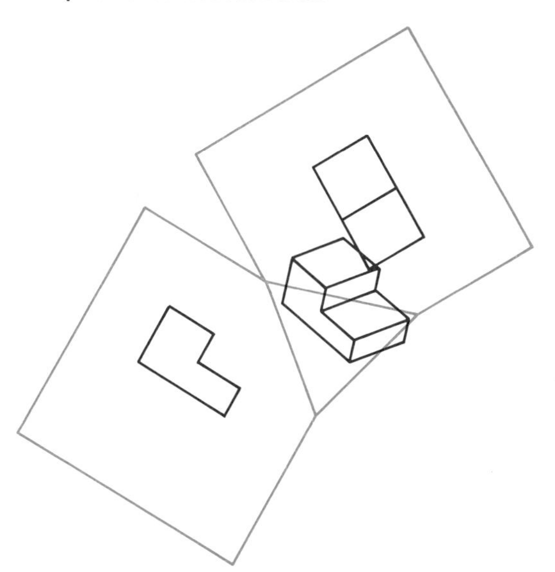

Figure 14.18 Axonometric pictorial of the object in Figure 14.17 and the given orthographic views, without construction projectors.

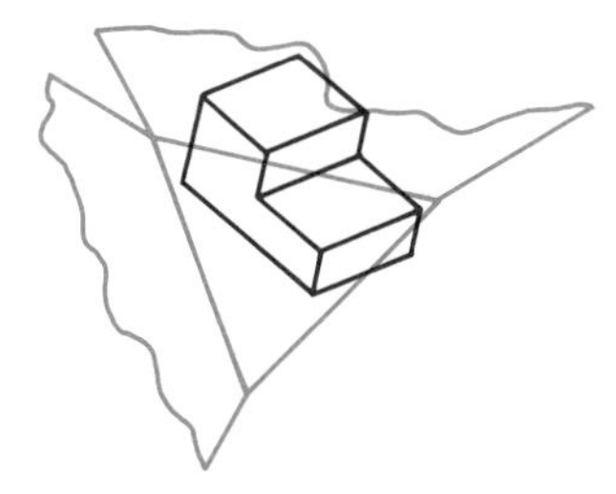

Figure 14.19 Axonometric pictorial of the object in Figure 14.17, together with a partial view of the construction planes of the glass box.

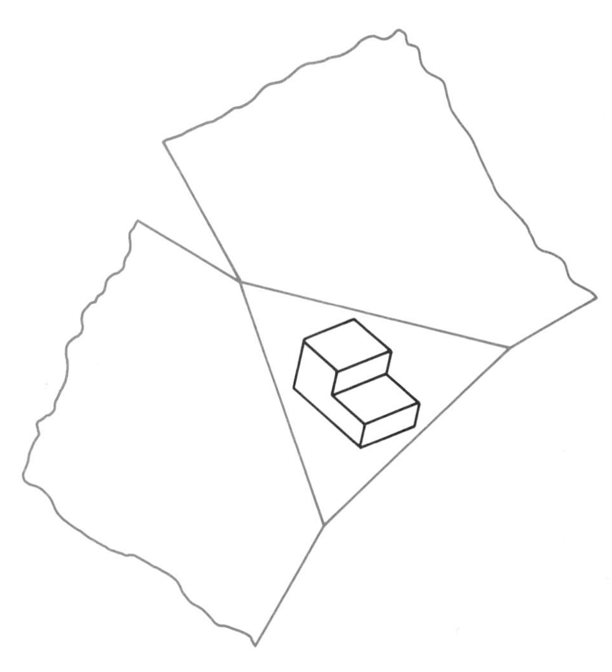

Figure 14.20 Axonometric pictorial of the object in Figure 14.19, together with *enlarged* partial views of the construction planes of the glass box.

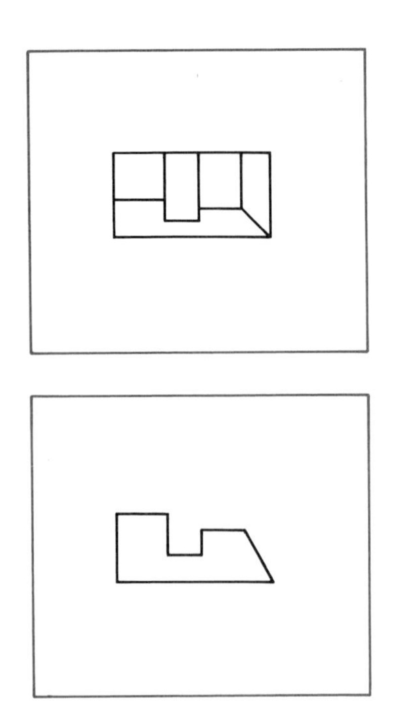

Figure 14.21 An example object, described by a set of front and top orthographic views, to be the subject of successive figures in this chapter.

given orthographic views. As another example of such a construction, consider the two orthographic views given in Figure 14.21 (top and front views). These two views are oriented with respect to each other—in terms of three orientation angles α, β, and Φ—in preparation for the construction of an axonometric pictorial (Figure 14.22). (Compare Figures 14.14 and 14.22.) The angle α represents the orientation of the given front view with respect to the horizontal direction of the construction sheet. The angles β and Φ define the orientation of the horizontal reference plane (HRP) and the frontal reference plane (FRP), respectively, with respect to the axonometric projection plane. The HRP is located so that it contains the bottom surface of the object (in this example); β defines the angle between the HRP and the projectors of points from the given front view to the axonometric projection plane. Similarly, Φ defines the angle between the FRP and the projectors from the given top view to the axonometric projection plane. Together, the angles α, β, and Φ actually define the spatial orientation of the axonometric projection plane or, equivalently, these three angles define the observer sight-line that corresponds to the resulting pictorial representation of the object. (An isometric projection will result if $\alpha = 15°$, $\beta = 45°$, and $\Phi = 45°$.)

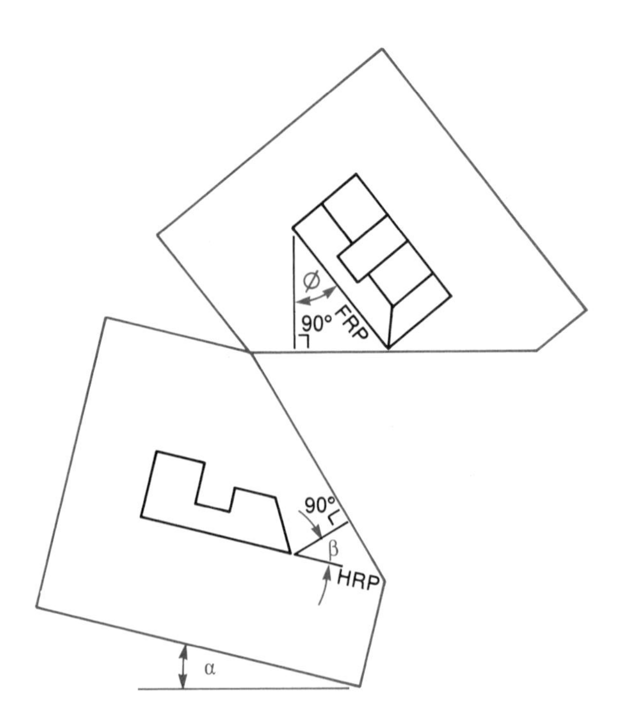

Figure 14.22 Defining the axonometric projection plane in terms of three specific angles and the given orthographic views of the object in Figure 14.21

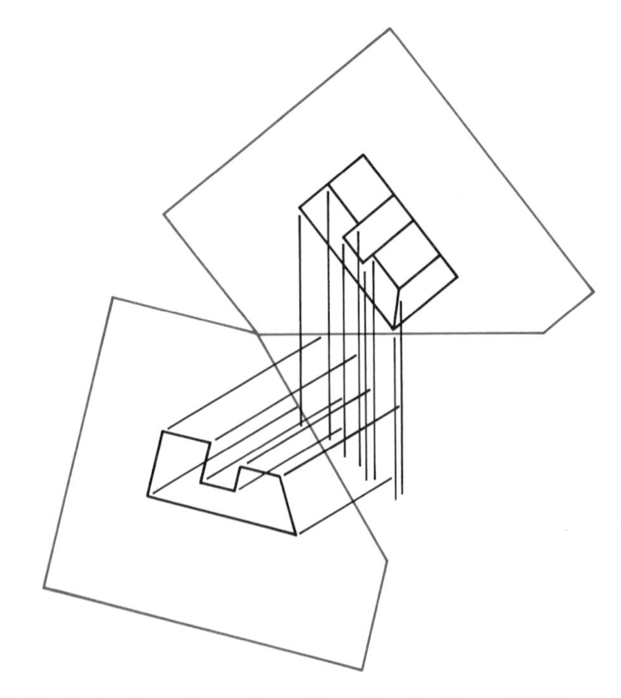

Figure 14.23 Projection of key points of the object in Figure 14.22 into the axonometric construction region.

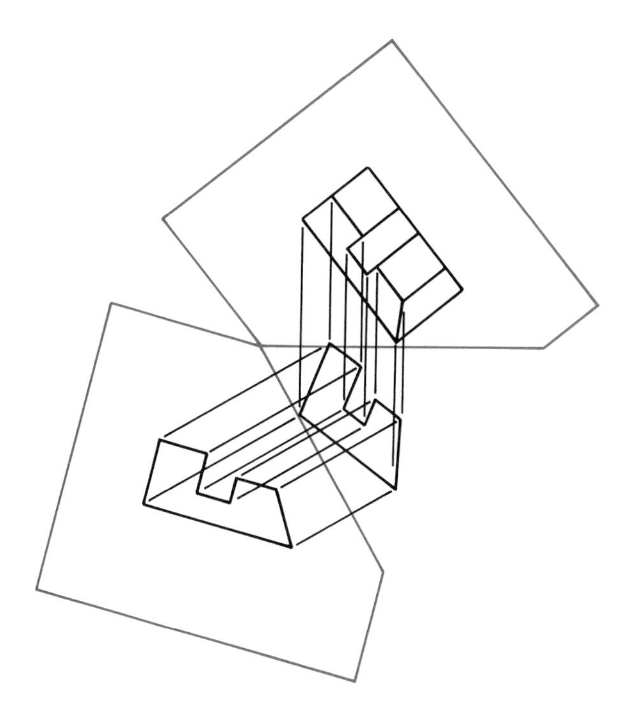

Figure 14.24 Construction of the axonometric projection of the front surface of the object in Figure 14.23.

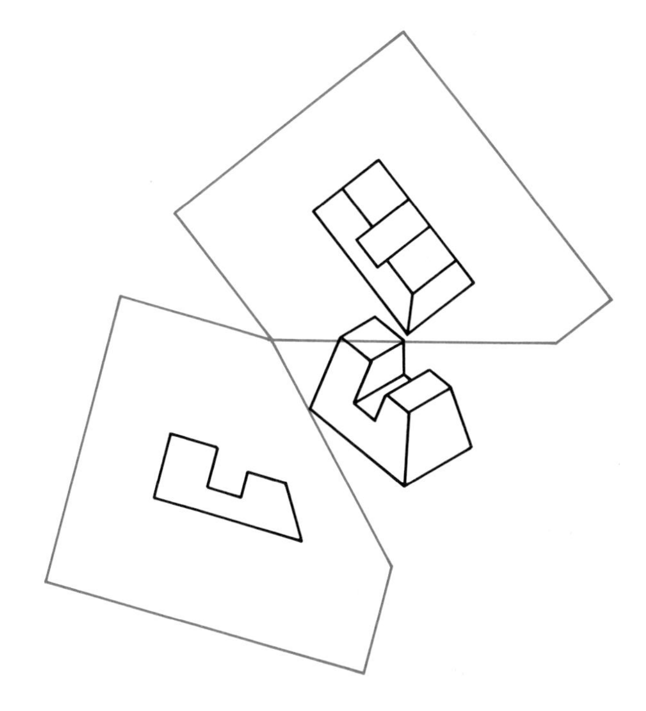

Figure 14.25 Completed axonometric projection of the object in Figure 14.24, together with the given orthographic views upon which it is based.

Figure 14.23 shows the projection of key points (which define the front surface of the object) to the axonometric construction area. Figure 14.24 illustrates the connection of these projected points by appropriate linework to form the pictorial view of the front surface of the object. Finally, the completed axonometric projection of the object is shown in Figure 14.25.

Figures 14.26 through 14.30 present the construction of an isometric pictorial ($\alpha = 15°$, $\beta = 45°$, $\Phi = 45°$) of another example object from a set of orthographic views. The completed isometric pictorial, together with the initial set of orthographic views, is shown in Figure 14.31.

In summary, then:

1. Axonometric projections may be constructed from a set of given orthographic views of an object. One need only define three angles (α, β, and Φ) of orientation between the orthographic views and the axonometric projection plane. (These three angles define the observer's sight-line that corresponds to the resulting pictorial description of the object.)
2. An isometric projection will result if $\alpha = 15°$, $\beta = 45°$, and $\Phi = 45°$.

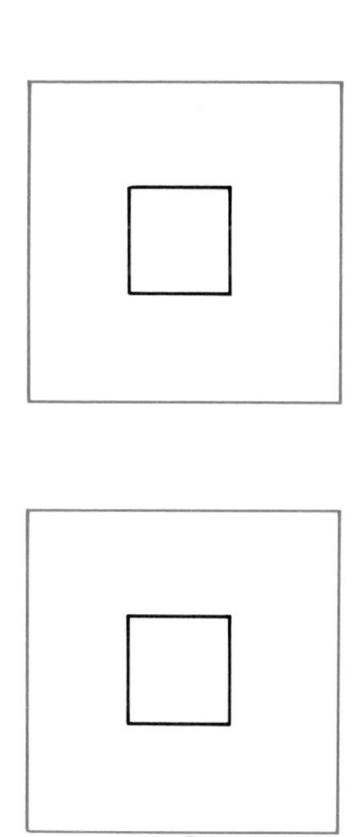

Figure 14.26 Two orthographic views of an object to be the subject of successive figures in this chapter.

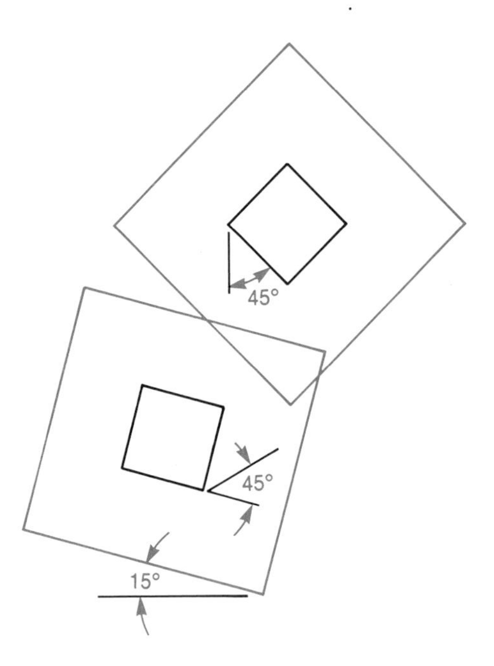

Figure 14.27 Orientation of given orthographic views of the object in Figure 14.26 in order to produce an isometric projection.

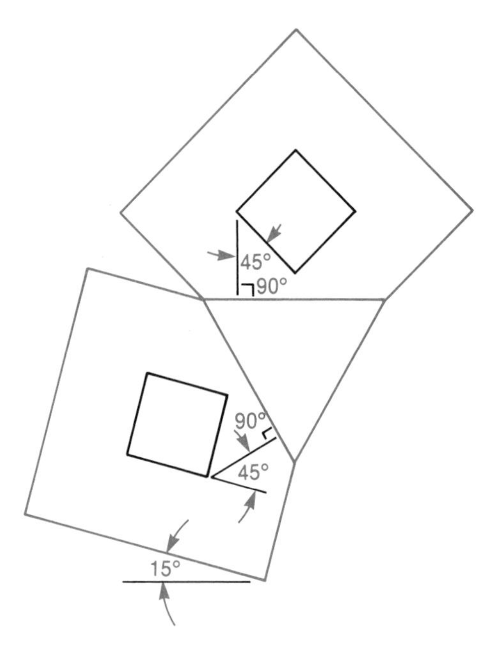

Figure 14.28 Formation of the isometric projection plane of the object in Figure 14.27.

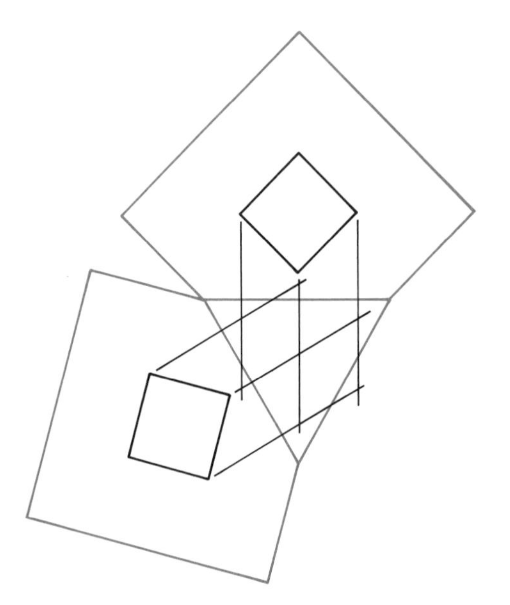

Figure 14.29 Projection of key points onto the isometric construction area for the object in Figure 14.28.

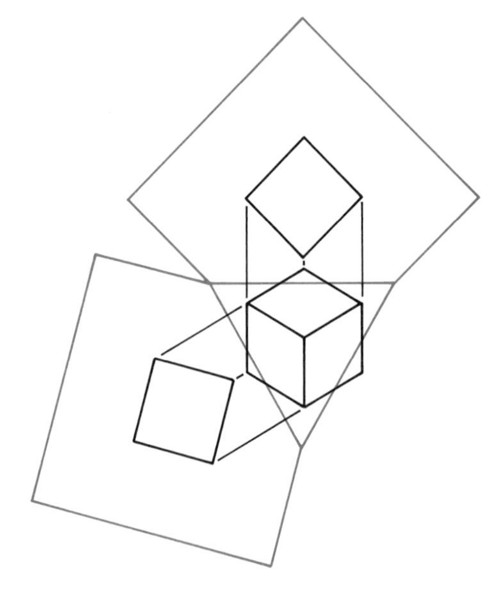

Figure 14.30 Isometric projection formed from the given orthographic views in Figure 14.26.

14.3 Foreshortened Projections and Full-Scale Drawings

The object shown in the isometric projection of Figure 14.31 has been foreshortened in its measurements. Figure 14.32 presents the **foreshortened isometric projection** (in which each of the three principle dimensions of the object has been foreshortened by a factor of 0.82), together with a **full-scale isometric drawing** of this object.

The foreshortened isometric projection was developed from the original orthographic views of this object. The full-scale drawing presents the object with principal dimensions (height, width, and depth) shown in true size—that is, with distances equal to those of the object as we would like it to be fabricated. An advantage of full-scale pictorials of objects to be fabricated is that the machinist may obtain true measurements from the pictorial itself if necessary. (The construction of full-scale pictorials will be discussed in the next chapter.)

In summary, then:

1. An isometric pictorial in which distances are foreshortened is known as a projection; full-scale pictorials are properly referred to as drawings.

2. True lengths along axial directions are shown in drawings (full-scale pictorials).

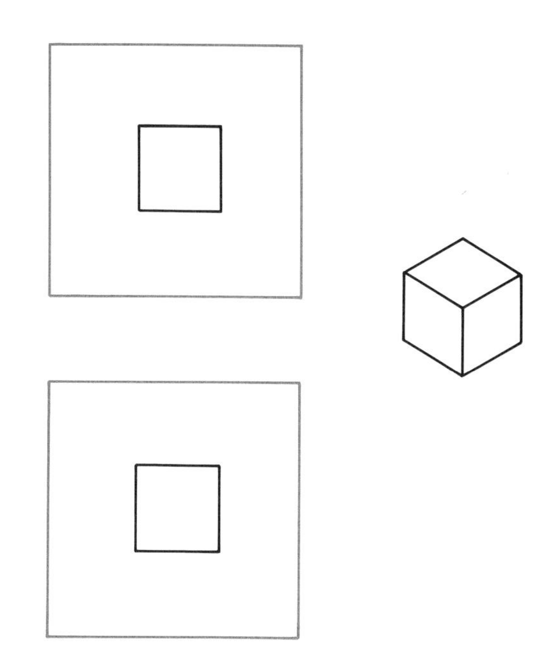

Figure 14.31 Completed isometric pictorial, together with the given orthographic views of the object.

14.4 Construction of Perspective Pictorials from Orthographic Views

Axonometric (and oblique[1]) pictorials are not true-perspective representations of the objects under consideration (unless the observer is assumed to be at such a large distance from the object, relative to its size, that distortion due to perspective becomes insignificant). Such pictorials can be called **pseudo-perspective** descriptions, because they represent approximations of true-perspective views of the objects. In these pictorials, depth-directed (receding) lines fail to converge toward some vanishing point on the horizon as required for true perspective.

Perspective pictorials can be generated via a number of different graphic techniques. We will focus upon the **ortho-**

1. An *oblique* projection is formed with one principal surface of the object remaining parallel to the projection plane. The observer then views the object at an angle that is oblique (not perpendicular) to this projection plane. Projection along this sight-line results in a pictorial of the object in which the surface parallel to the projection plane is shown without any distortion in shape. In oblique projection, the observer's position is rotated with respect to the projection plane, whereas in axonometric projection, the object is rotated. Oblique pictorials will be discussed further in Chapter 15.

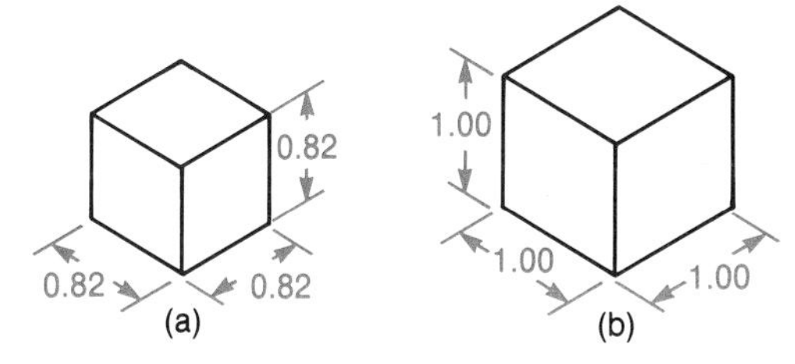

Figure 14.32 Foreshortened isometric projection of a cube (a) and full-scale isometric drawing of a cube (b).

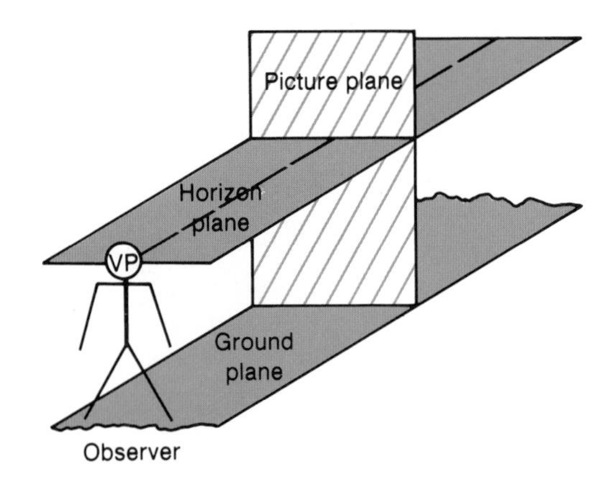

Figure 14.33 Elements for the construction of a perspective projection.

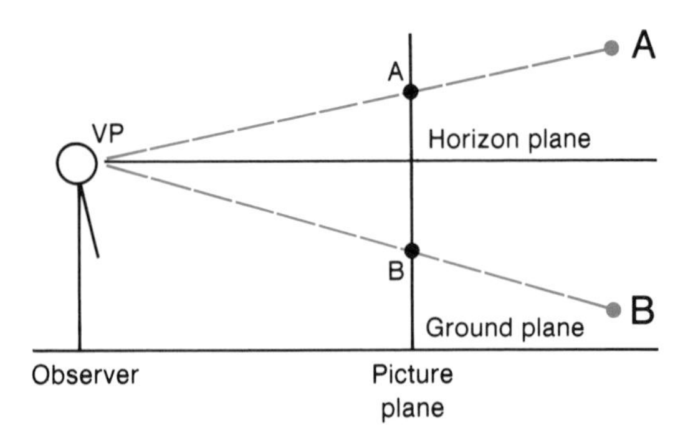

Figure 14.34 Side view of observer and relevant planes in a perspective projection.

graphic method in which the pictorials are constructed through the use of orthographic views. This technique is consistent with the one presented in Section 14.2 for constructing axonometric pictorials from a set of given orthographic views.

Figure 14.33 presents the relationship between several basic elements in perspective construction. An observer is shown with respect to three spatial planes: a **horizon plane** (located at the observer's eye level), a **ground plane,** and a **picture plane.** The perspective pictorial is projected onto the picture plane, which is perpendicularly oriented with respect to the horizon plane. Note the intersection of the horizon and picture planes and the axis of vision for the observer in the figure. The **viewing point** (denoted as VP) is the position of the observer's eye with respect to the horizon and picture planes. The viewing point is also called the **station point.**

Figure 14.34 presents a side view of the observer, together with the horizon, picture, and ground planes. Two points (A and B) are seen by the observer along sight-lines that pass through the picture plane. The projection of these two points along those sight-lines to the picture plane results in a projected vertical distance between these points that is less than the true vertical distance between points A and B; that is, the vertical distance between these points is foreshortened in the projected view in the picture plane.

Figure 14.35 presents a side view of the observer viewing four distinct points (A, B, C, and D) in space (a). The projection of these points along the observer's sight-lines to the picture plane again results in a foreshortening of distances (b). In addition, the distance between points C and D in the projection is greater than the distance between points A and B in this projection; that is, line $\overline{CD}$ appears to be larger than line $\overline{AB}$ in the projected view. Figure 14.35(a) indicates that points C and D lie *closer* to the picture plane (and the observer) than do points A and B; as a result, **perspective distortion** results in a projected view of these points: line $\overline{CD}$ appears to be greater than line $\overline{AB}$. ($\overline{AB}$ and $\overline{CD}$ are shown with finite widths in Figure 14.35 for greater clarity.)

To draw a simple rectangular block in true perspective, the location of the viewing point (VP) must be defined with respect to the object. Figure 14.36 shows the location of the VP with respect to the object in terms of an **elevation angle** (Φ). The elevation angle defines the height of the horizon plane with respect to the object under consideration. (Notice that VP must be located above the top of the object if the observer is to see the top surface of the object or, equivalently, if the perspective pictorial is to include a view of the top surface.)

Figure 14.37 defines the location of the VP with respect to the object in terms of the **lateral angle** (θ). The lateral and

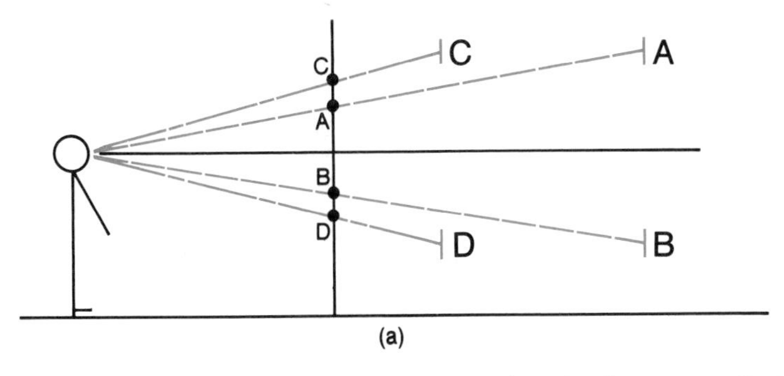

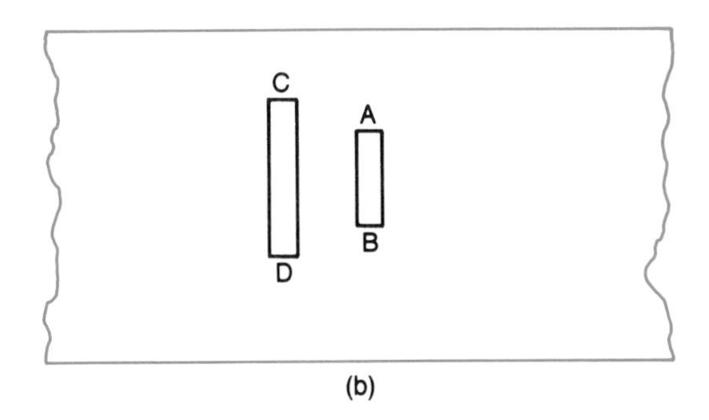

Figure 14.35 Perspective projection of four points that form two vertical lines in space. Side view of the projection of points onto the picture plane (a). Relative perspective projection of lines $\overline{AB}$ and $\overline{CD}$ onto the picture plane (b).

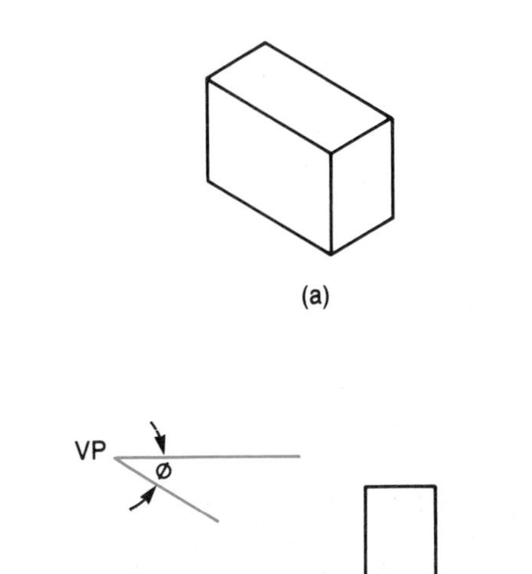

Figure 14.36 Rectangular prism (a) and the location of the viewing point (VP) in terms of the elevation angle Φ for the prism (b).

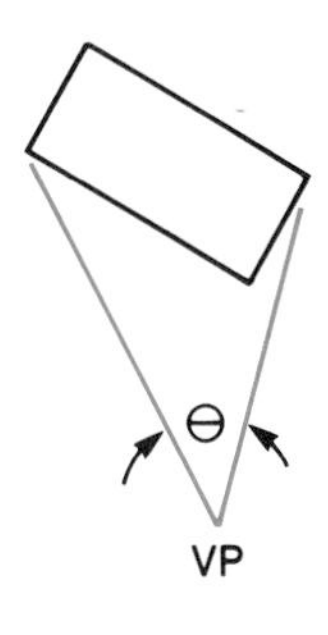

Figure 14.37 The viewing point (VP) and the lateral angle (θ) with respect to the top view of the rectangular prism in Figure 14.36.

elevation angles together completely specify the location of the VP with respect to the object; the corresponding perspective projection of the object is similarly completely specified by the choice of Φ and θ. (θ is shown relative to a top view.)

A particular choice of Φ and θ values is shown in Figure 14.38. Notice that the lateral orientation of the VP with respect to the top view of the object indicates that the observer will be able to see both the front and right-side surfaces of the object (the two orthographic views are defined here as the top and right-side views). Furthermore, the elevation angle, Φ, is shown with respect to the given right-side view of the object, indicating that the observer will be able to view the

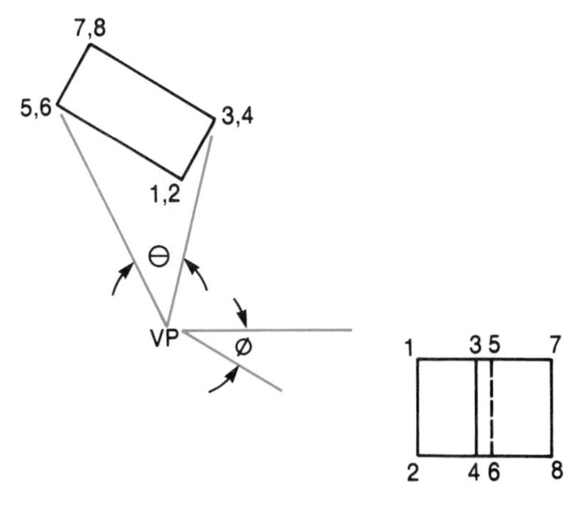

Figure 14.38 Top and right-side views of the rectangular prism in Figure 14.37, together with the location of the viewing point (VP) in terms of the elevation and lateral angles.

top surface of the rectangular block. The corresponding perspective pictorial will then include distorted views of the front, right-side, and top surfaces of this object.

In addition, notice that the given orthographic views in Figure 14.38 are mutually consistent. The choice of the lateral location of the VP with respect to the top view of the object necessitated the given right-side view in Figure 14.38.

Eight points on the object are identified in Figure 14.38. We need to locate each of these points in the perspective projection that corresponds to our choice of Φ and θ. Figure 14.39 defines the picture plane with respect to the given orthographic views of the object and the VP. In addition, several projectors are shown from the eight points on the object that were defined in Figure 14.38 to the VP. These projectors allow us to locate these eight points in the perspective projection under construction as follows:

Points 1 and 2 on the object are projected along the VP sight-lines onto the picture plane, as shown in Figure 14.40. The location of each point on the picture plane is then projected along a direction that is perpendicular to the picture plane, as shown. The intersection of these perpendicular projectors in the pictorial construction area defines the location of the point in the corresponding perspective projection of the object. In other words, we use a two-step projection process in which (1) each point used to define the object is projected along a sight-line from the object toward the VP, thereby locating the projection of the point in the picture plane and (2) the projected point in the picture plane is then projected along a direction that is perpendicular to the picture plane to the pictorial construction area.

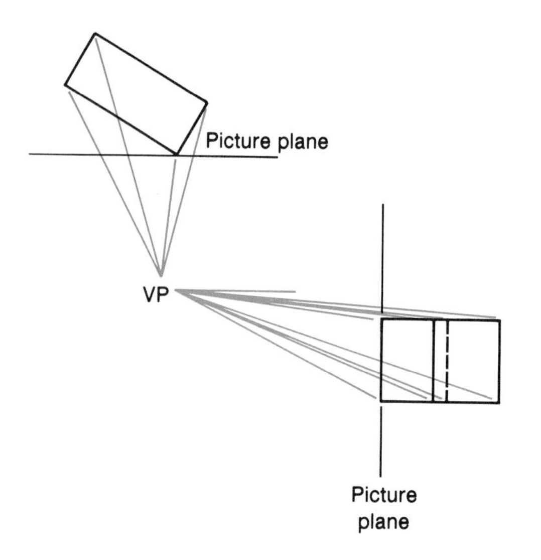

Figure 14.39 Projectors of key points from the given orthographic views in Figure 14.38 to the viewing point (VP). The picture plane is also shown with respect to the given views of the object.

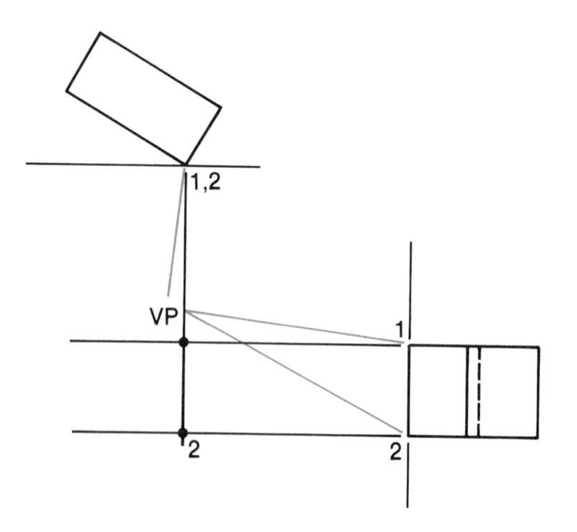

Figure 14.40 Two-step projection of points 1 and 2 into the perspective construction region (continuing from Figure 14.39).

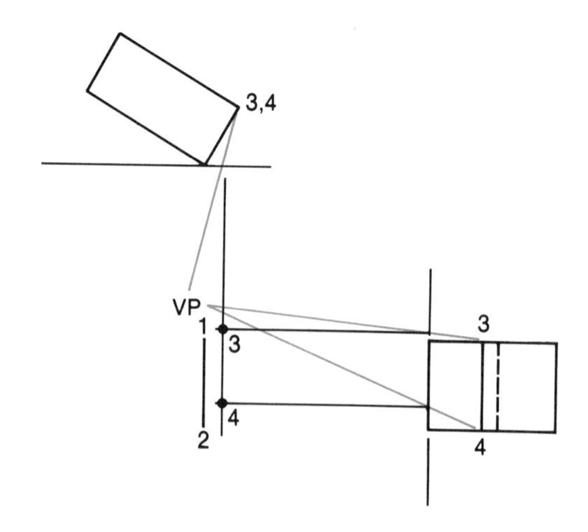

Figure 14.41 Two-step projection of points 3 and 4 into the perspective construction region (continuing from Figure 14.40).

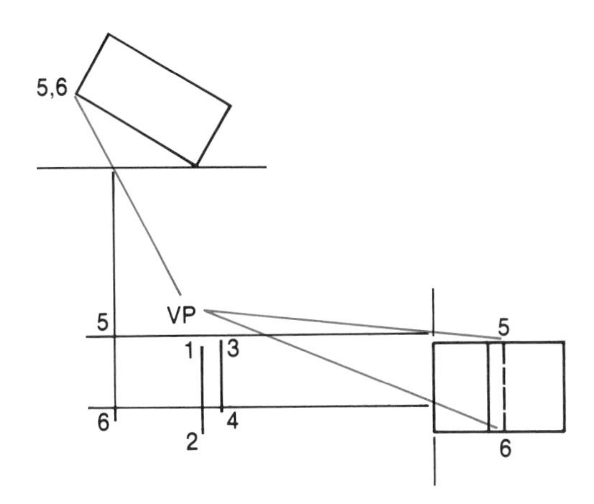

Figure 14.42 Two-step projection of points 5 and 6 into the perspective construction region (continuing from Figure 14.41).

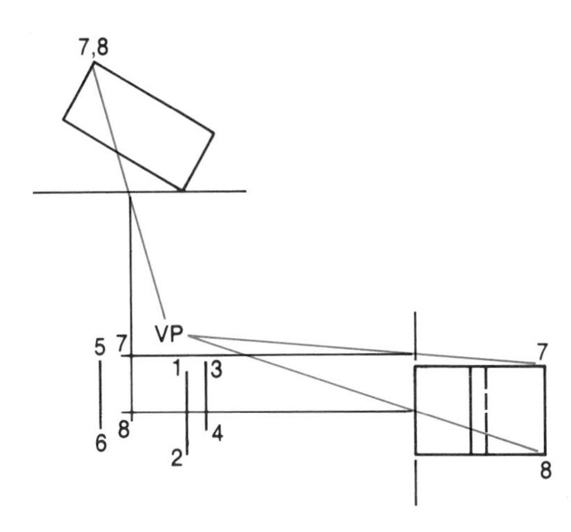

Figure 14.43 Two-step projection of points 7 and 8 into the perspective construction region (continuing from Figure 14.42).

This two-step projection process is used in Figures 14.41 through 14.43 to locate points 3 through 8. Finally, the properly projected points are connected with appropriate linework in order to form the completed perspective pictorial in Figure 14.44.

In order to emphasize that the location of the VP determines the resultant perspective projection, Figure 14.45 presents the object of Figure 14.38 with a slightly different specification of the VP than that chosen in Figure 14.38. The new location of the VP has a greater elevation than the earlier choice; in addition, it now lies more to the right of the object. Figures 14.46 through 14.49 illustrate the two-step projection of points 1 through 8 onto the pictorial construction area; the resulting perspective projection is shown in Figure 14.50. Figure 14.51 presents a comparison of the two perspective pictorials obtained from our two different choices of the VP location. With the foregoing technique, one may easily obtain a perspective pictorial of an object from a given set of orthographic views.

In summary, then:

1. Perspective pictorials may be generated from a set of given orthographic views of an object. The viewing point (VP) must be defined with respect to the object, the horizon plane, and the picture plane. The elevation angle, Φ, and the lateral angle, θ, may be used to specify the relative location of the viewing point.

2. A two-step projection process may be used to develop the desired perspective projection of the object.

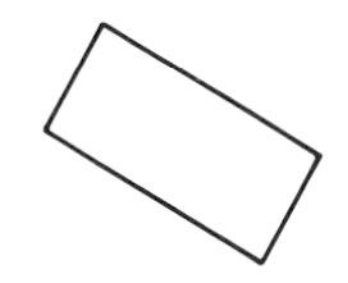

Figure 14.44 Completed perspective projection of the rectangular prism in Figure 14.36, together with the given orthographic views upon which it is based.

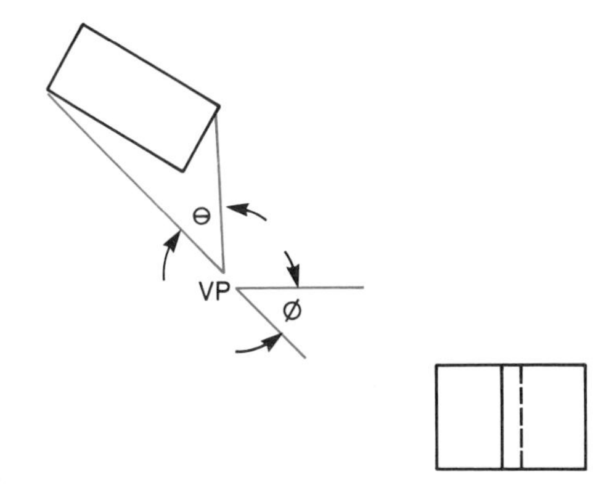

Figure 14.45 The rectangular prism with a different viewing point (VP) than the one in Figure 14.36.

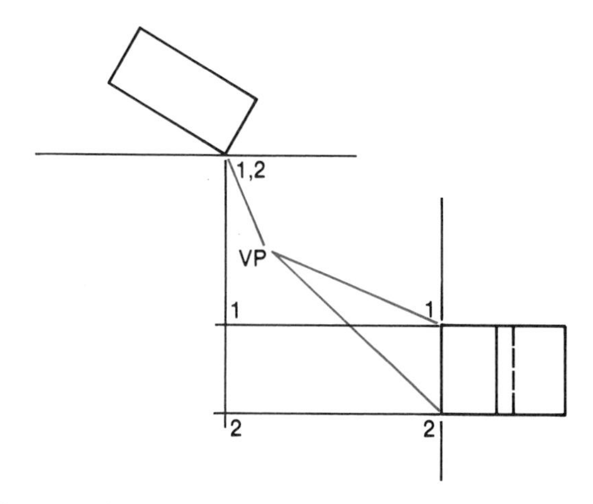

Figure 14.46 Two-step projection of points 1 and 2 into the perspective construction region (continuing from Figure 14.45).

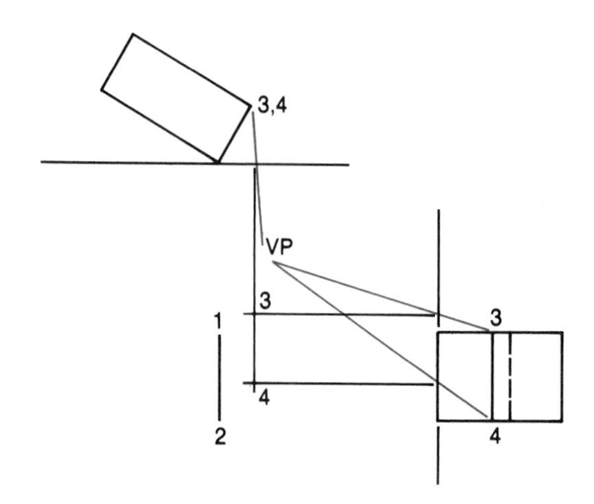

Figure 14.47 Two-step projection of points 3 and 4 into the perspective construction region (continuing from Figure 14.46).

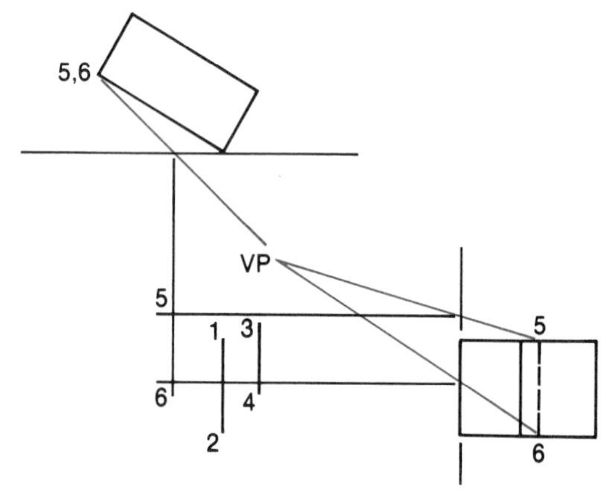

Figure 14.48 Two-step projection of points 5 and 6 into the perspective construction region (continuing from Figure 14.47).

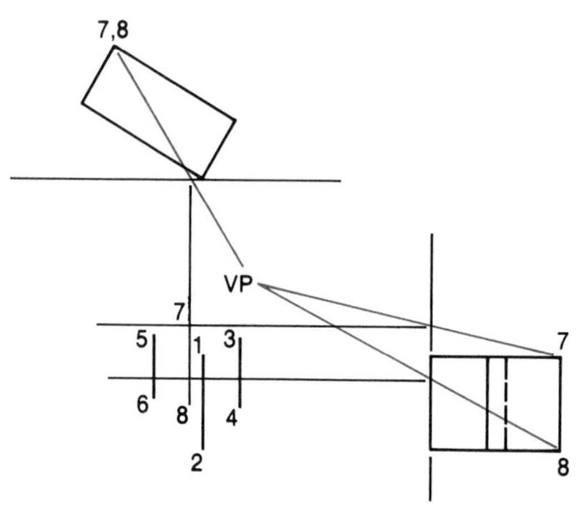

Figure 14.49 Two-step projection of points 7 and 8 into the perspective construction region (continuing from Figure 14.48).

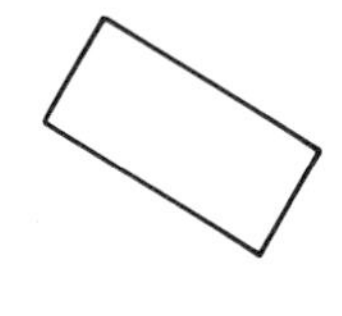

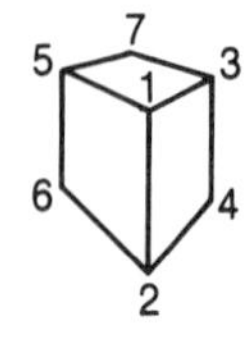

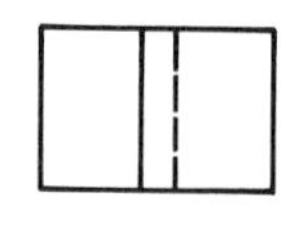

Figure 14.50 Completed perspective projection of the rectangular prism in Figure 14.45, together with the given orthographic views upon which it is based.

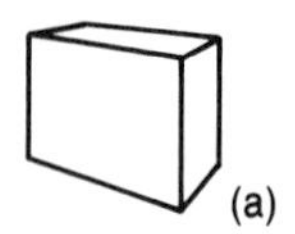

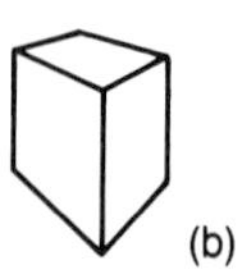

Figure 14.51 Two perspective projections of the same rectangular prism. The first projection is based upon the viewing point (VP) given in Figure 14.38 (a). The second projection is based upon the VP given in Figure 14.45 (b).

In Retrospect

- An axonometric pictorial is a projected image of an object onto a projection plane wherein three principal faces (surfaces) of the object are shown.
- Distances (and corresponding surface areas) are foreshortened in axonometric projections. The foreshortening ratio is defined in accordance with the following relationship:

$$\text{Foreshortening ratio of a line} = \frac{\text{Length of line in projection}}{\text{True length of line segment}}$$

- Three distinct categories of axonometric projections are defined in terms of the distances that are foreshortened in each: isometric projections, dimetric projections, and trimetric projections. Isometric projections (in which distances along all three mutually perpendicular axes are equally foreshortened) are the least difficult axonometric pictorials to construct but the most distorted, or unrealistic, in appearance. They are the most popular of the three types of axonometric pictorials because of their ease of construction.
- An axonometric pictorial in which distances are foreshortened is known as a projection. Full-scale pictorials are properly referred to as drawings.
- Axonometric projections may be constructed from a set of given orthographic views of an object. One need only define three angles (α, β, and Φ) of orientation between the orthographic views and the axonometric projection plane. An isometric projection will result if $\alpha = 15°$, $\beta = 45°$, and $\Phi = 45°$. (These three angles define the observer's sight-line corresponding to the resulting pictorial representation of the object.)
- Like axonometric projections, perspective pictorials may also be constructed from a set of given orthographic views of an object. The viewing point (VP) of the observer must be defined with respect to the object, the horizon plane, and the picture plane. The elevation angle, Φ, and the lateral angle, θ, may be used to specify the relative location of the viewing point. A two-step projection process is then used to develop the desired perspective projection of the object.

Exercises

Your instructor will specify the appropriate scale for each exercise. EX14.1 through EX14.20 refer to specific exercises in preceding chapters. In each case use the given set of orthographic views to develop a pictorial representation of the object that clearly shows its important features. You may wish to construct additional orthographic views to aid your development of the appropriate pictorial.

EX14.1. See EX4.8.

EX14.2. See EX4.9.

EX14.3. See EX4.12.

EX14.4. See EX4.13.

EX14.5. See EX4.14.

EX14.6. See EX4.20.

EX14.7. See EX4.22.

EX14.8. See EX4.28.

EX14.9. See EX4.31.

EX14.10. See EX4.33.

EX14.11. See EX4.35.

EX14.12. See EX5.7.

EX14.13. See EX5.9.

EX14.14. See EX6.3 (do not section).

EX14.15. See EX6.4 (do not section).

EX14.16. See EX8.4.

EX14.17. See EX8.5.

EX14.18. See EX8.6.

EX14.19. See EX8.11.

EX14.20. See EX8.12.

ENGINEERING IN ACTION

Mini-tractor Takes on Full-Size Jobs

For small-plot farmers and wood-lot operators, a unique tractor that looks like a motorized table may be ideal. The four-wheel-drive, four-wheel-steer vehicle, called a Quadractor, is a spinoff of an idea that didn't sell. Originally the designer, William Spence, built a powered nose wheel and offered it to aircraft manufacturers. The powered wheel would save much valuable fuel now wasted by taxiing aircraft. When the idea literally didn't fly, Spence used the powered wheel as the key element in a low-cost (about $3,600) light weight (900 lb) tractor.

Although the engine is slightly more powerful than a lawnmower, the Quadractor can move impressive loads over rough terrain and handle most farming chores. The design has been developing slowly since 1979 and is now being offered through dealers in the U.S. and Canada. Because the design stresses simple components and easy maintenance, Spence hopes the Quadractor will have major appeal in underdeveloped nations.

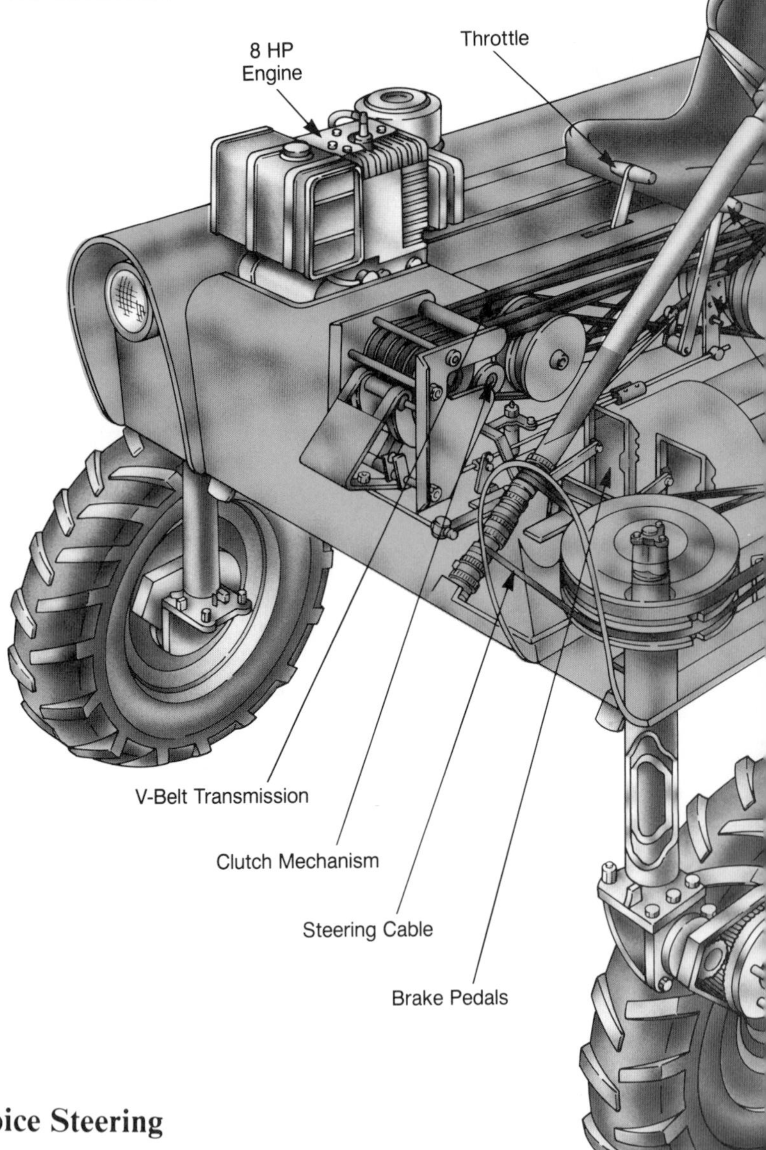

Two-Choice Steering

The Quadractor has two steering systems. For tight maneuvering the Quadractor's four-wheel steering is used. Control cables link the wheel-steering shafts to the steering wheel. In the second system, skid steering, either the left or the right-side wheels are braked.

With skid steering the turning radius can be zero. Usually the two steering systems are used in combination.

Climb on It

The drive-gear arrangement in each leg acts as both a shock absorber and traction control.

A gear in each wheel meshes with a drive pinion in a 4:1 ratio. When the vehicle starts, the pinion "climbs up" the gear until the wheels begin to turn. Normally, the pinion rides about half way up a 30° arc. When a wheel hits an obstacle, the gears shift relative to each other and absorb some of the shock.

At the same time, the added resistance causes the pinion to climb further up the gear. Traction is assisted by the increased downward pressure on the wheel.

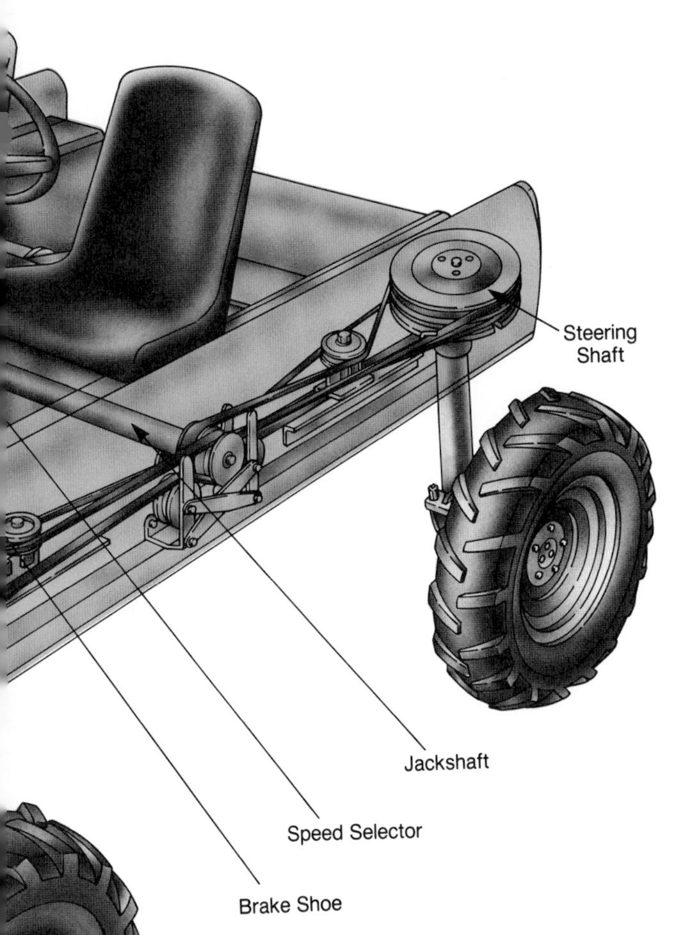

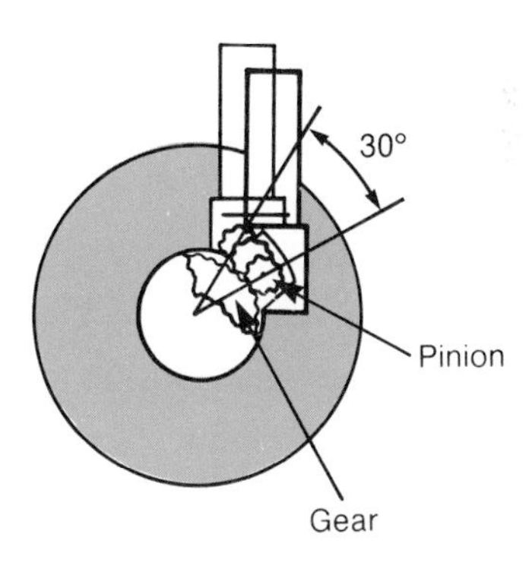

Multiple V-Belt Drive

Power goes from the single-cylinder, 8-hp gasoline engine through a V-belt transmission to two jackshafts. The jackshafts drive another set of belts that connect to the drive shaft on each wheel.

Although the transmission is complex looking with six V-belts (two each for slow forward, fast forward, and reverse), it utilizes simple, established principles. And, it meets the design criterion of minimizing the number of machine parts.

The Quadractor has a top speed of 8 mph and can climb a 42° slope or traverse a 32° incline. Measuring only 7 ft 6 in. by 5 ft 4 in., the vehicle is said to have a 2,000-lb drawbar pull and be able to drag 4,000-lb loads.

A 31-in. ground clearance helps with rough terrain mobility. Loads may be slung beneath the tractor, and farm implements can be mounted on the hitch.

Unlike the conventional 3-point hitch, which chiefly loads a tractor's rear wheels, the Quadractor's hitch attaches behind the front wheels. Load is therefore taken equally by all four wheels.

The Quadractor does not have a conventional suspension system: it relies on the drive-gear system and flexible chassis to absorb shock. Vehicle corners can flex as much as 12 in.

Source: Reprinted from MACHINE DESIGN, April 22, 1982. Copyright 1982, by Penton/IPC., Inc., Cleveland, Ohio, pp. 78, 79.

In EX14.21 through EX14.25, a specific viewing point (VP) is indicated. Construct the corresponding pictorial of the given object.

EX14.21. See the figure for this exercise.

EX14.22. See the figure for this exercise.

EX14.23. See the figure for this exercise.

EX14.24. See the figure for this exercise.

EX14.25. See the figure for this exercise. Two viewing points are indicated in the figure for this exercise. Draw each of the corresponding pictorials.

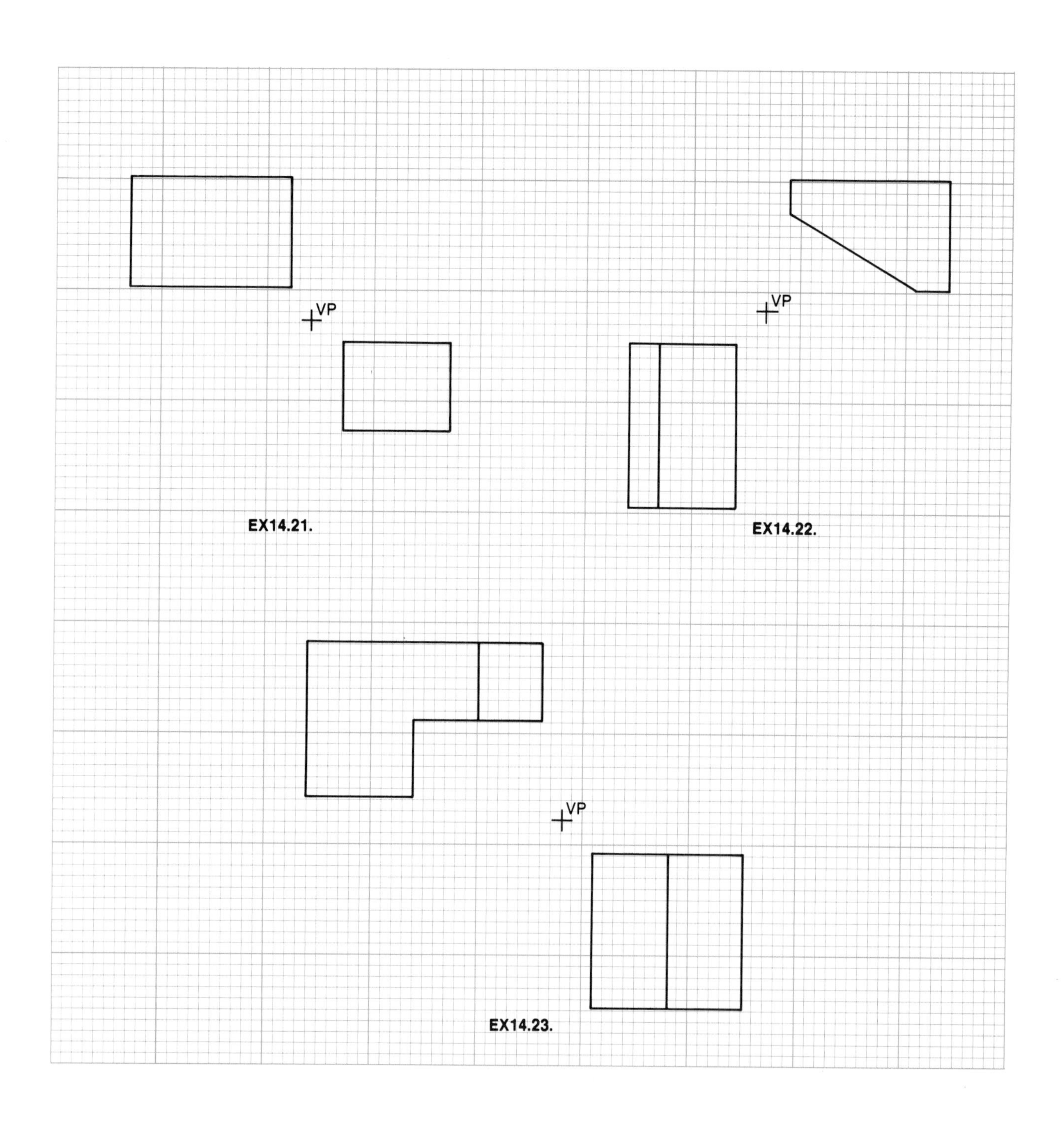

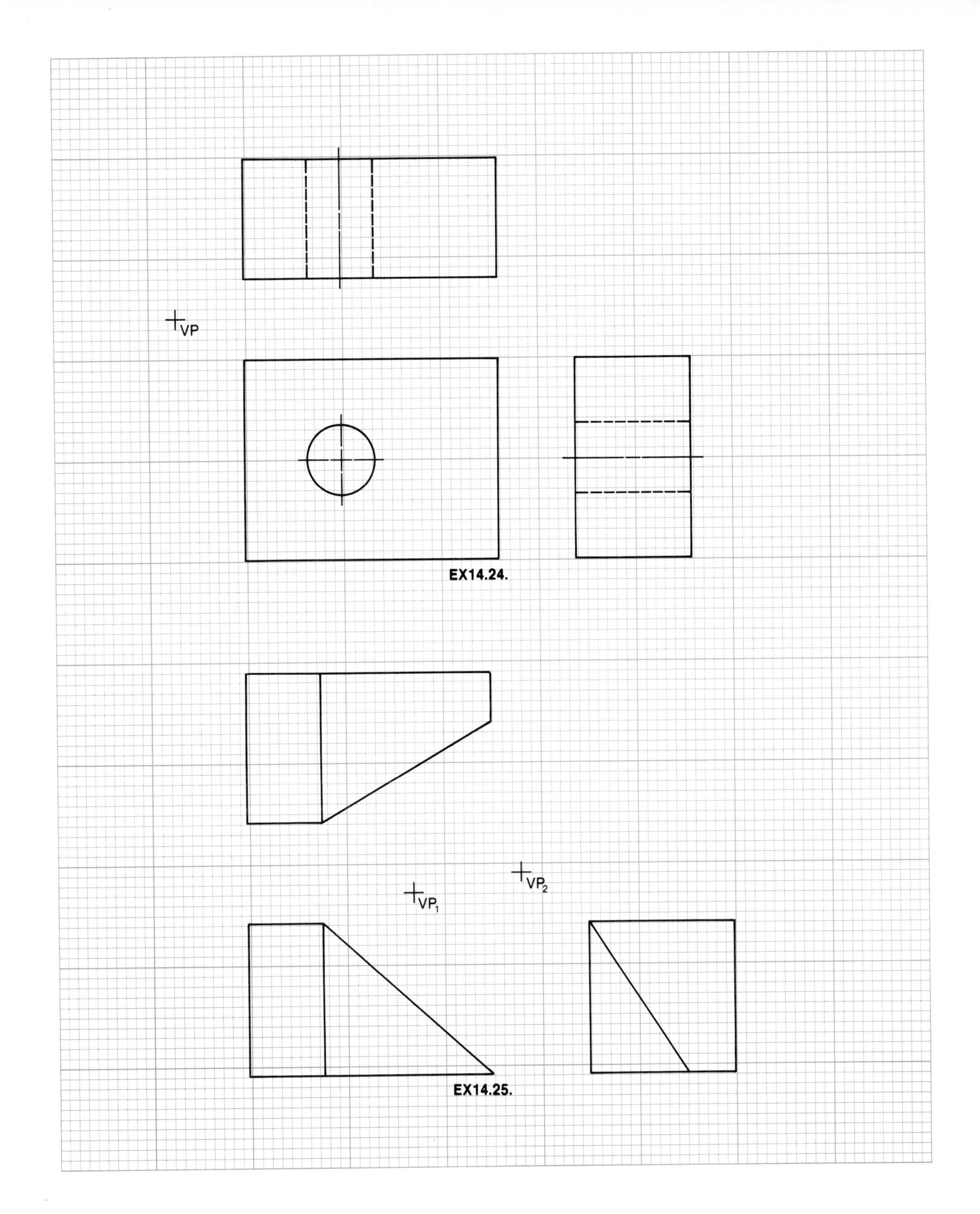
VP
EX14.24.
VP1
VP2
EX14.25.

15 Pictorials, Part 2: Constructing Pictorials in Practice

Preview In Chapter 14, pictorial descriptions of objects were produced through projection techniques from given orthographic views. In this chapter we will develop direct construction techniques that allow us to develop pictorials from orthographic views without the difficulties associated with projection.

The pictorials in this chapter (with the exception of cabinet oblique drawings) share one characteristic: true-length measurements along axial directions. As we noted in Chapter 14, pictorials with such true-length measurements along axial directions are properly referred to as *drawings,* not as *projections,* since foreshortening of axial distances occurs in projections.

The construction methods developed in this chapter may be applied with ease to generate the two most popular types of pictorials: isometric drawings and oblique drawings. Sectioning of pictorials to clarify the internal features of an object under consideration will also be demonstrated.

Learning Objectives

Upon completion of this chapter, the reader should be able to:

- Identify true-length measurements in axonometric drawings.
- Define isometric and oblique pictorials in terms of the relative orientation of the principal axes (height, width, and depth).
- Construct isometric and oblique drawings via the point-by-point, surface-by-surface technique.
- Judiciously choose the location of the origin for the axes in a pictorial so that the drawing will be properly centered within the available construction space.
- Recognize the need for consistency in the orientation of an object described in a pictorial and in any accompanying orthographic views.
- Apply the four-center method to quickly construct an arc or circle in an isometric drawing with an approximately realistic distortion and with true-length dimensions along axial directions.
- Explain the advantage and disadvantage of oblique pictorials compared to axonometric drawings.
- Differentiate between cabinet and cavalier oblique drawings, and identify the advantage associated with cabinet drawings.
- Section pictorials in order to clarify the internal features of the object under consideration.

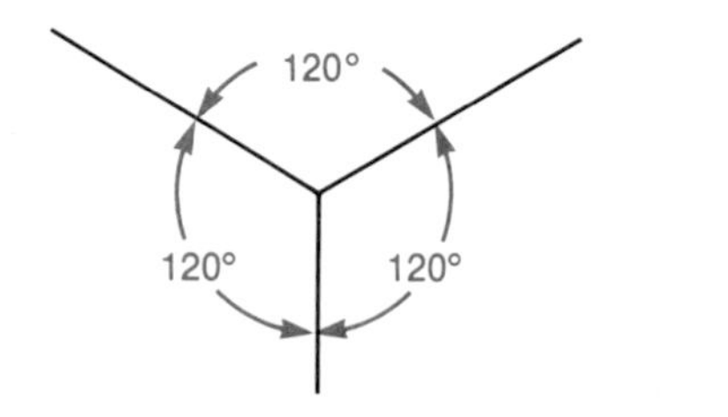

Figure 15.1 Orientation of axes in isometric pictorials.

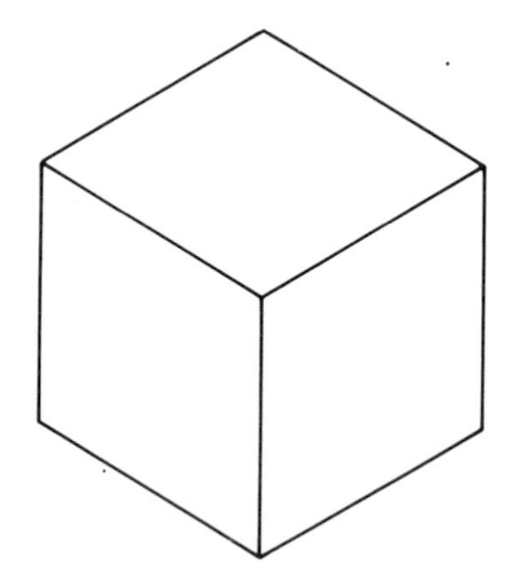

Figure 15.2 An isometric pictorial of a cube.

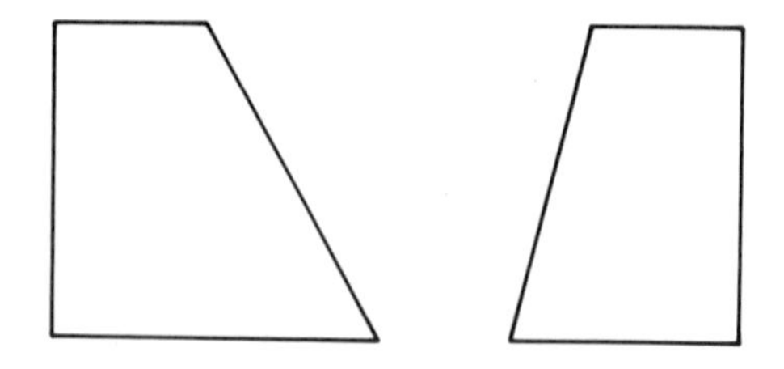

Figure 15.3 Two orthographic views (front and right side) of an example object to be the subject of successive figures in this chapter.

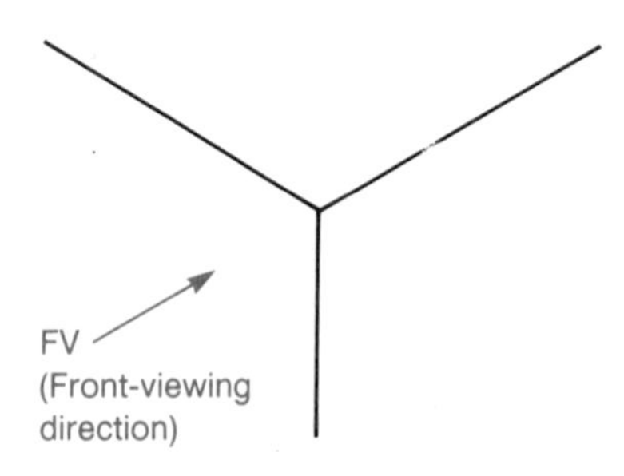

Figure 15.4 Choosing the front-viewing direction is critical in the orientation of an object in an isometric pictorial. This is the front-viewing direction for the object that is the subject of Figure 15.3.

15.1 Isometric Drawings

Isometric pictorials are the most common type of axonometric drawing. As explained earlier, in Section 14.3, axonometric drawings are pictorial representations of objects in which *true-length measurements* along the axial directions of the drawing are used. The three axes in an isometric pictorial are drawn with equal angles of 120° between each two of them (Figure 15.1).

The isometric drawing of a cube is shown in Figure 15.2. Recall that an axonometric *projection* results in the foreshortening of distances along the axial directions, whereas there is no such foreshortening in an axonometric *drawing.* Foreshortening and/or elongation of distances along *nonaxial* directions, however, can occur in axonometric drawings.

In summary, then:

1. Isometric drawings have true-length measurements along axial directions. Distances along nonaxial directions are foreshortened or elongated.

2. Isometric drawings are constructed with the three axes (height, width, and depth) oriented with equal (120°) angular spacing.

15.2 Planar Surfaces in Isometric Drawings

We will begin our discussion of direct construction techniques for isometric drawings with a consideration of planar surfaces on an object. For example, in order to develop an isometric drawing of the object for which two orthographic views (front and right side) are presented in Figure 15.3, one must first choose an appropriate origin for the axes and then draw the three axes with 120° between them (Figure 15.4). The front-viewing direction is then chosen to be consistent with the given orthographic views of the object; that is, if the orthographic views present the front and right-side surfaces of the object, then the isometric drawing should also be constructed so that the front and right-side surfaces of the object are visible. Such consistency in the orientation of the object in both its orthographic and pictorial representations will aid the reader in the interpretation of these drawings. Inconsistency here will make interpretation difficult.

After the front-viewing direction is chosen, the true-length measurements of total width, total height, and total depth for the object are transferred from the orthographic views to the pictorial construction area along the axial directions. Figure 15.5 presents the given orthographic views of the object with complete dimensioning information (units are not specified). The three principal (total) dimensions of

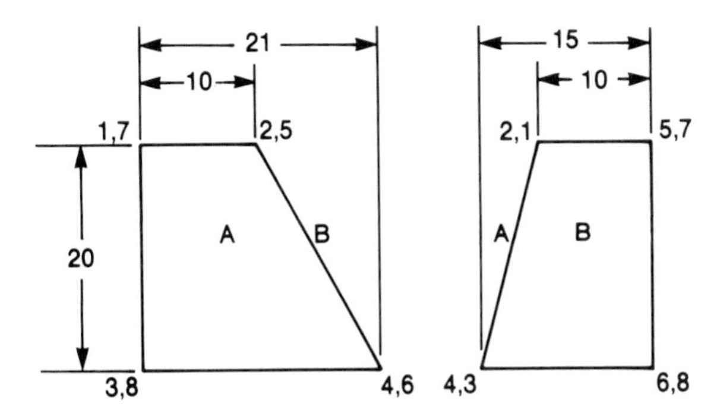

Figure 15.5 Dimensions and key points for the orthographic views in Figure 15.3.

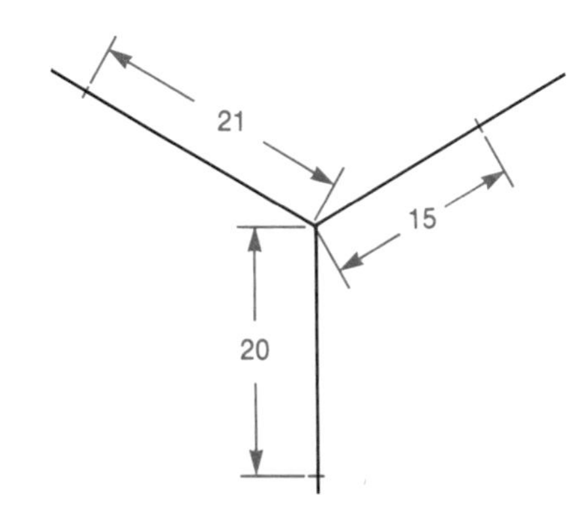

Figure 15.6 Principal (total height, total width, total depth) dimensions are transferred from the orthographic views in Figure 15.5 to the isometric pictorial construction axes.

the object are then transferred to the pictorial axes, as shown in Figure 15.6. One may then draw construction lines parallel to the axial directions to form a **construction box,** within which the isometric pictorial will be formed (Figure 15.7).

Each surface of the object is then developed by a **surface-by-surface, point-by-point** technique similar to the one used for construction of missing views in an orthographic description of an object. Surface A of the object shown in Figure 15.5 and defined by boundary points 1 through 4 can be developed by carefully locating each point at its proper position in the pictorial construction box. *The height, width, and depth of each point is measured along the axial directions,* as shown in Figures 15.8 and 15.9. The inclined surface, A, may then be drawn in the isometric pictorial (Figure 15.9).

Surface B (Figure 15.5) is developed in a similar way (Figure 15.10). Finally, point 7 (see Figure 15.5) is located in the

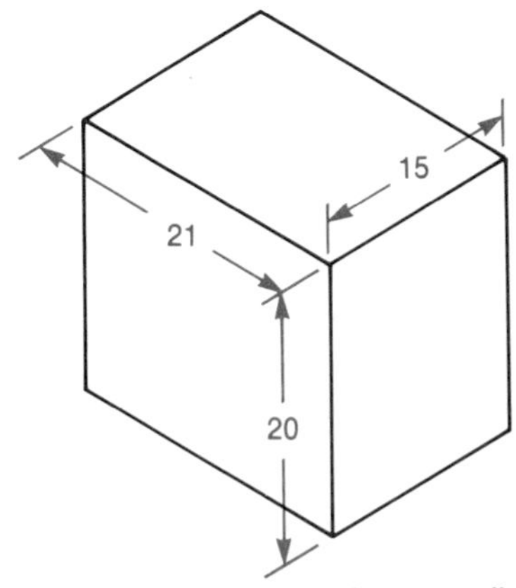

Figure 15.7 Construction lines drawn parallel to the axial directions in Figure 15.6 form a construction box for an isometric pictorial.

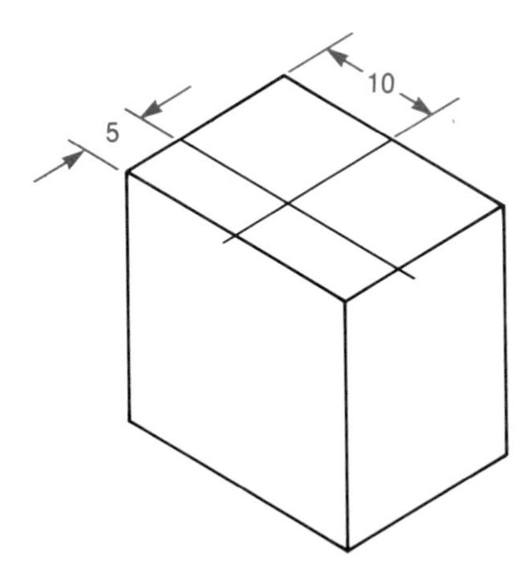

Figure 15.8 Additional dimensions are transferred to the isometric pictorial construction area (continuing from Figure 15.7); true-length measurements are always made along axial directions.

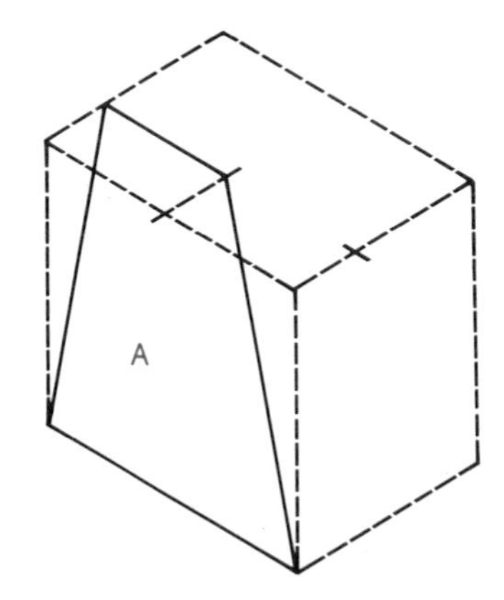

Figure 15.9 Surface A is drawn in the isometric pictorial (continuing from Figure 15.8).

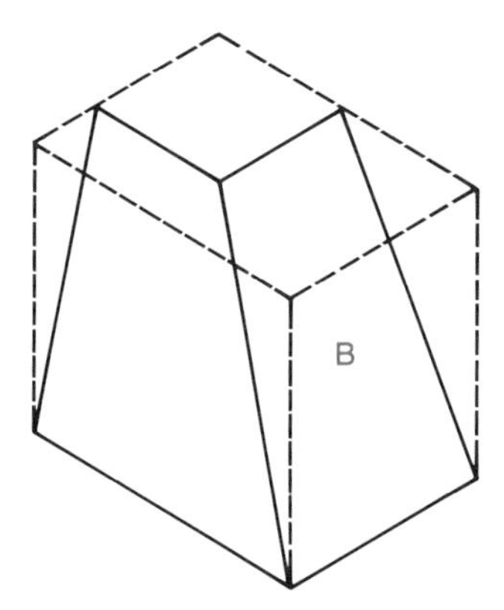

Figure 15.10 Surface B is drawn in the isometric pictorial (continuing from Figure 15.9) after dimensions are transferred in a manner similar to that used for surface A.

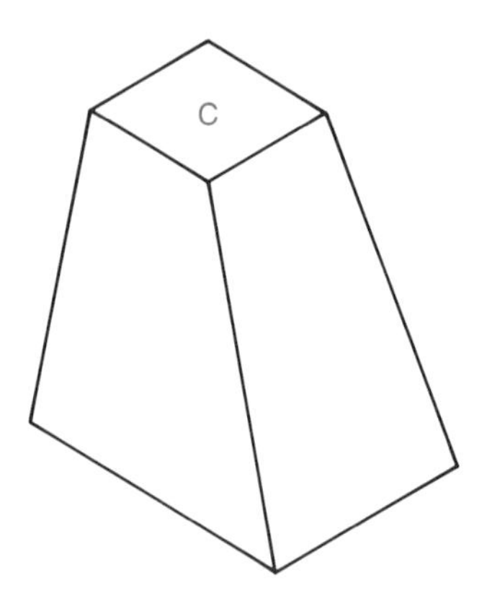

Figure 15.11 Top surface C is drawn (continuing from Figure 15.10), thereby completing the isometric pictorial for the object described by the orthographic views in Figure 15.3.

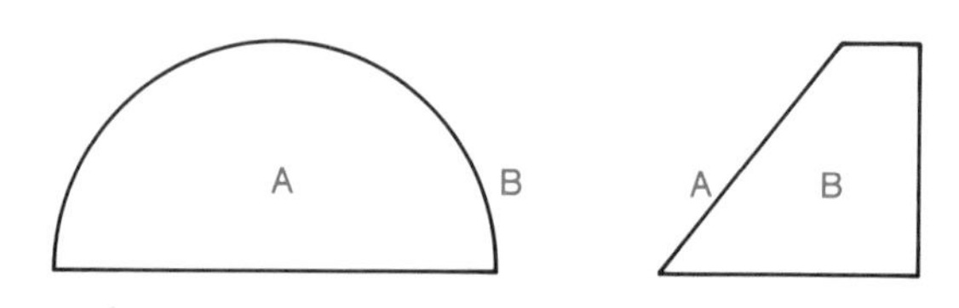

Figure 15.12 Two orthographic views (front and right side) of an example object with a curved-edge planar surface, A, and a nonplanar surface, B.

pictorial construction region, and the top surface, C, is drawn (Figure 15.11), thereby completing the isometric pictorial. (The construction work has been erased in Figure 15.11.)

In summary, then:

1. An isometric drawing of an object should be constructed in a methodical manner, point-by-point, surface-by-surface.
2. In constructing an isometric pictorial, one must carefully choose the origin for the axes so that the drawing will be properly centered within the available drawing space.
3. One must also carefully identify the front-viewing direction with respect to the isometric pictorial under construction. The choice of this viewing direction (which is parallel to the depth dimension) will determine which surfaces of the object will be shown in the pictorial. The orientation of the pictorial description of the object should be identical to the orientation of the object in any accompanying orthographic representation.
4. In pictorial *drawings,* true-length measurements are given along axial directions, in contrast to pictorial *projections* in which measurements along axial directions are foreshortened. Distances along *nonaxial* directions in pictorial drawings are either foreshortened or elongated.

15.3 Nonplanar Surfaces in Isometric Drawings

The point-by-point, surface-by-surface technique may also be used to construct nonplanar surfaces in pictorial drawings. A curved-edge planar surface, A, is shown in the orthographic views of Figure 15.12. The curved boundary in the front view of this figure represents an edge view of the nonplanar surface, B, of the object. The isometric pictorial of this object is developed in identical fashion to the way a set of purely planar surfaces is developed, except that the height, width, and depth locations of many more points along the curved boundary between surfaces A and B must be found in the pictorial construction box for a nonplanar surface.

In Figure 15.13 numerous points (1 through 26) have been identified along the surface boundaries of the given orthographic views. The construction box for the isometric drawing is shown in Figure 15.14. Points 1 and 2 have been located within the construction box in Figure 15.15 through the use of *true-length axial measurements* for each point from a reference origin (which in this case is the location of point 1, that is, the front, right, bottom corner of the construction box.)

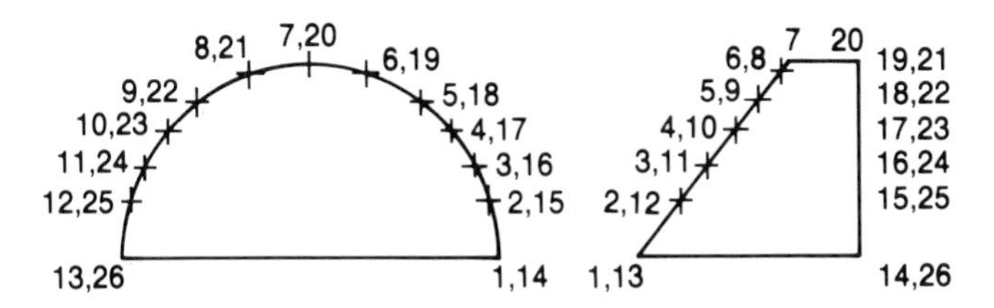

Figure 15.13 Arbitrarily chosen points along surface boundaries are identified in the orthographic views of Figure 15.12.

Figure 15.16 shows the location of points 1 through 13 within the construction box; once again, true-length measurements have been made along the axial directions in order to determine the locations of these points. An irregular curve is then used to produce a best-fit curve through these thirteen points; the resultant curve (shown in Figure 15.17) represents the curving boundary between surfaces A and B of the object. Additional points (14 through 25) have also been located in Figure 15.17. Finally, a smooth curve is drawn through these points in order to produce the rear curving boundary of the nonplanar surface, B. Figure 15.18 presents the completed isometric drawing of this object. (Notice that a horizon line, tangent to both curving boundaries of surface B and parallel to the depth axis, must be included in the finished pictorial.)

In summary, then:

1. The point-by-point, surface-by-surface technique may be used to construct both planar and nonplanar surfaces in an isometric pictorial drawing.
2. True-length axial measurements are used to determine the location of significant points on a surface in the isometric pictorial construction region.

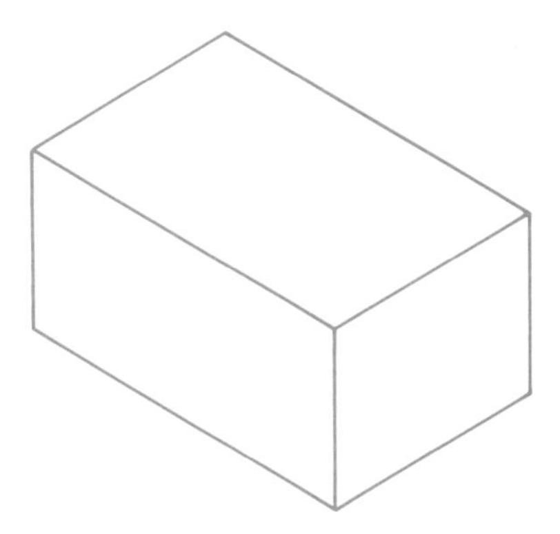

Figure 15.14 A construction box is formed for isometric pictorial development of the object described by the orthographic views in Figure 15.12.

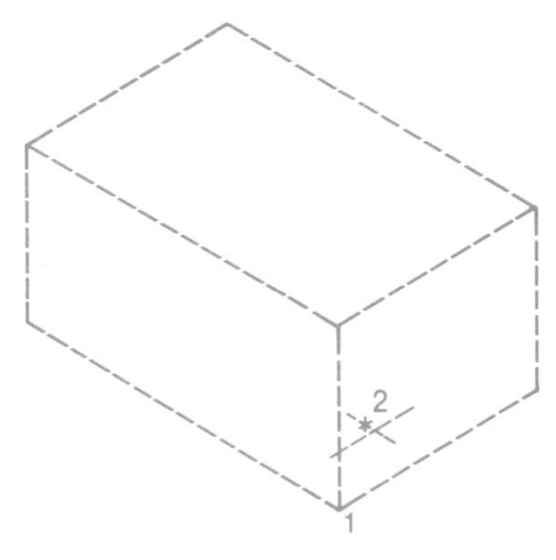

Figure 15.15 Points 1 and 2 are located via true-length measurements along axial directions (continuing from Figure 15.14).

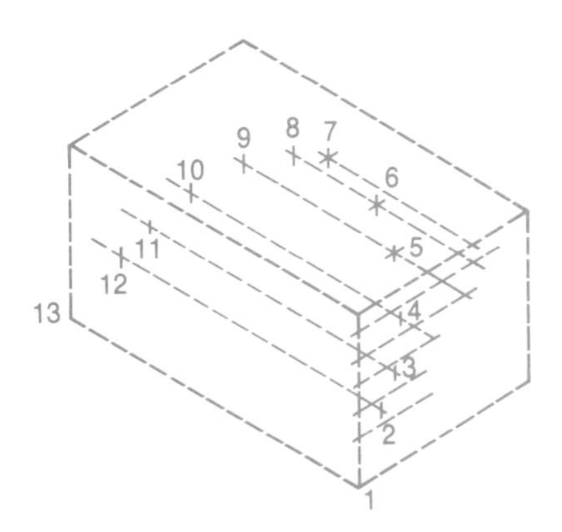

Figure 15.16 Points 1 through 13 are located via true-length measurements along axial directions (continuing from Figure 15.15).

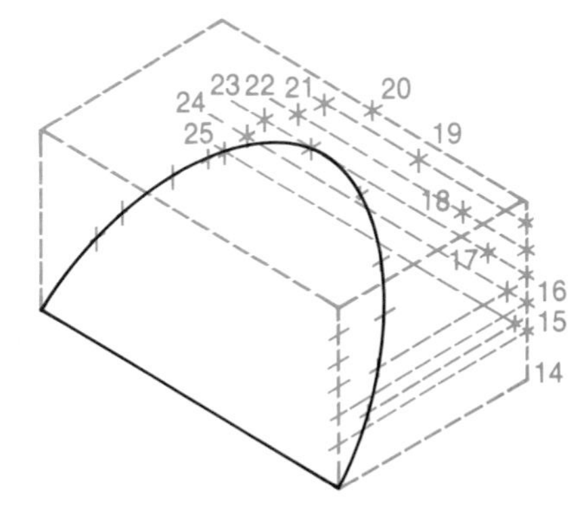

Figure 15.17 A smooth curve is drawn through points 1 through 13 (continuing from Figure 15.16). In addition, points 1 through 25 are located via true-length measurements along the axial directions (dashed linework indicates true-length construction measurements).

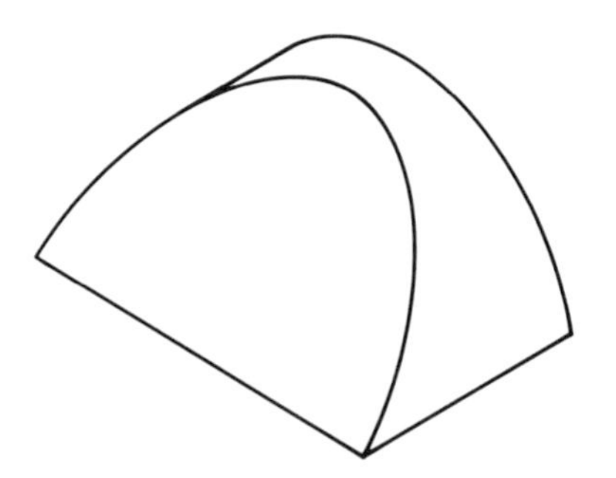

Figure 15.18 The completed isometric pictorial (continuing from Figure 5.17) of the object described by the orthographic views in Figure 15.12.

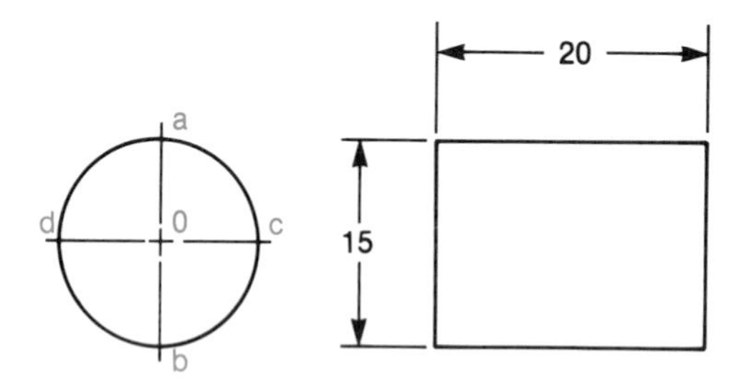

Figure 15.19 Two orthographic views (front and right side) of a cylinder, with several key points identified by letters.

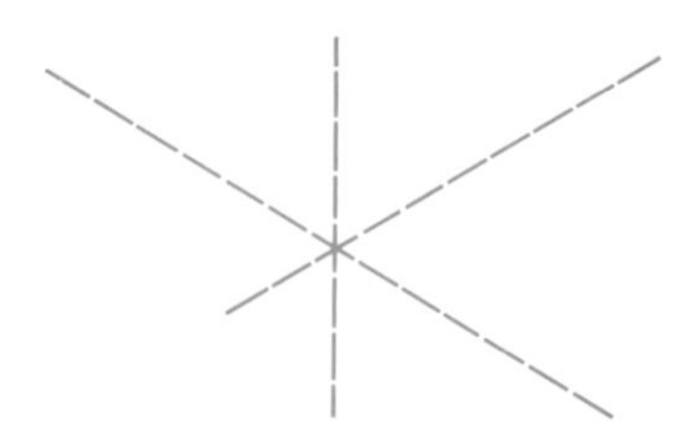

Figure 15.20 Construction axes are drawn in preparation for the isometric pictorial of the cylinder described by the orthographic views in Figure 15.19.

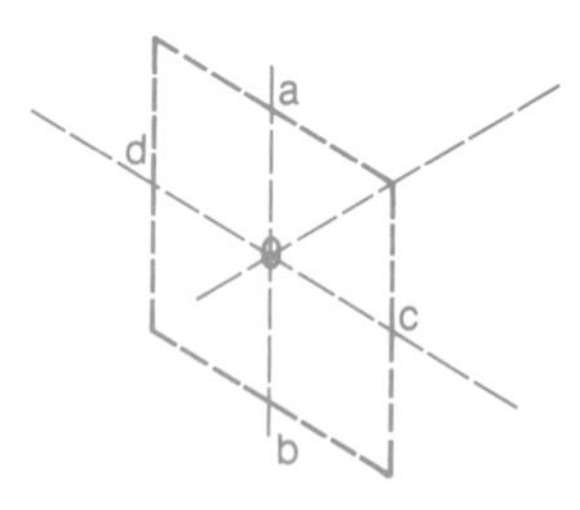

Figure 15.21 A parallelogram, is constructed by drawing construction lines through the four points (a, b, c, and d) that are transferred from the given front view of the object in Figure 15.19.

15.4 Four-Center Method

A circle or an arc (that is, a curve composed of points equally distant from a single center point) can be drawn in a pictorial either by using an irregular curve to obtain a best-fit of many points along the curve or (less accurately) by using the **four-center method.**

Figure 15.19 presents the front and right-side orthographic views of a cylinder. The corresponding isometric drawing of this cylinder can be developed via the four-center method as follows:

1. A set of construction axes are drawn (Figure 15.20) in preparation of the pictorial development. The intersection (origin) of these axes will also represent the centerpoint of the circular front surface of the cylinder.

2. A construction parallelogram is drawn (with sides parallel to the axial directions that define the spatial plane of the circular surface to be drawn; see Figure 15.21). The true relative distances (height, width, and depth) of the four points a, b, c, and d with respect to the axial origin are transferred from the given front view of the object to the pictorial construction area, after which four construction lines are drawn through these four points in order to form the sides of the parallelogram. As a result, the distances $\overline{Oa}$, $\overline{Ob}$, $\overline{Oc}$, and $\overline{Od}$ are identical and equal to the radius of the circle under construction.

3. Four more construction lines are drawn *within* the construction parallelogram. These additional lines are drawn through the points a, b, c, and d so that each line is perpendicularly oriented to the side of the parallelogram containing the point (a, b, c, or d) through which it passes (Figure 15.22). That is, each of these four lines within the parallelogram is perpendicular to an axial direction. The new points (1, 2, 3, and 4) at which these four interior construction lines intersect are the four centerpoints of four arcs

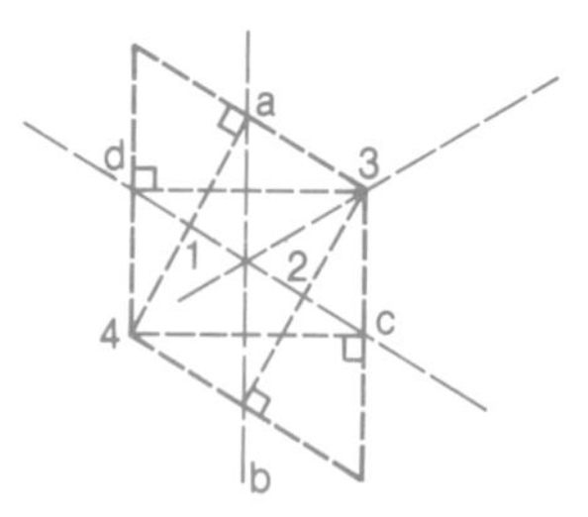

Figure 15.22 Four centerpoints are located for use in the four-center construction method by drawing four more construction lines through points a, b, c, and d so that each line is perpendicularly oriented to the side of the parallelogram containing the point through which the line passes.

that will be used to form an isometric representation with proper distortion of the circular front surface of the cylinder.

4. The first arc is centered at point 1 and drawn from point a to point d, with a radius equal to $\overline{1a}$. The second arc is centered at point 2 and drawn from point b to point c (Figure 15.23). (Theoretically, the distances $\overline{1a}$, $\overline{1d}$, $\overline{2b}$, and $\overline{2c}$ should be equal to one another; however, if one performs this construction with drawing instruments, some human error can be expected to be introduced into the results. One should then correct the settings of the compass so that the arcs centered at points 1 and 2 do indeed have endpoints at points a, d and b, c, respectively.)

5. Two more arcs, centered at points 3 and 4, are then drawn between points b, d and a, c, respectively, thereby completing the isometric drawing of the circular front surface of the cylinder (Figure 15.24).

With the four-center method, a properly distorted isometric representation of an arc may be quickly generated. In addition, the resulting isometric description of the arc will include *true-length* measurements along axial directions; for example, the distances $\overline{ab}$ and $\overline{cd}$ in Figure 15.24 are equal to the diameter of the circular front surface of the object under construction.

One may then continue with the construction of the isometric drawing of the cylinder by locating the total depth of the object (Figure 15.25), drawing the rear circular surface and appropriate horizon lines (Figure 15.26), and finally erasing all construction linework (Figure 15.27).

Figure 15.28 presents a set of orthographic views that describe an object with several circular surfaces. Figure 15.29 shows the identification of key depth positions of various surfaces on the object. These depth positions are then transferred to the isometric construction region (Figure 15.30). Figures 15.31 through 15.36 present the construction of the isometric drawing of this object.

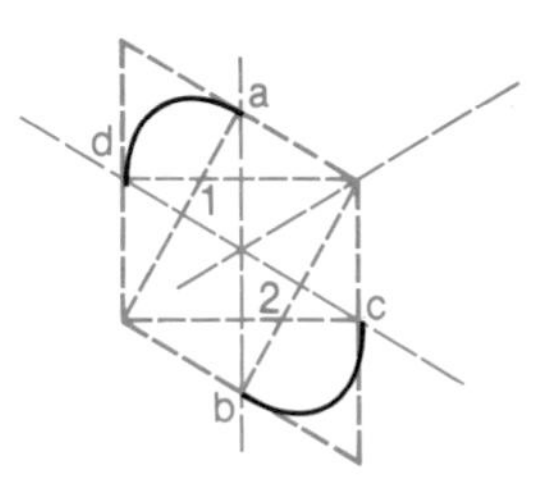

Figure 15.23 Two arcs, centered at points 1 and 2, are drawn (continuing from Figure 15.22).

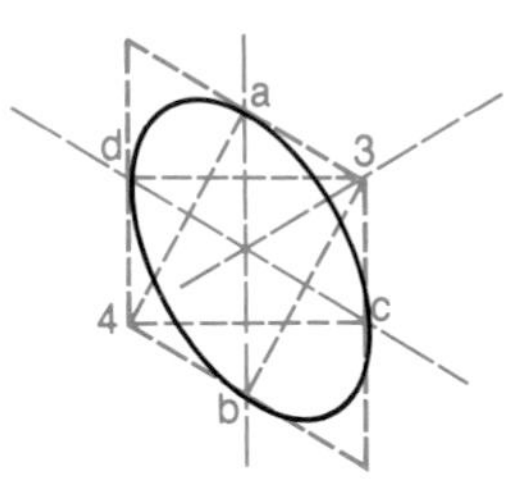

Figure 15.24 Two additional arcs are drawn, centered at points 3 and 4 (continuing from Figure 15.23).

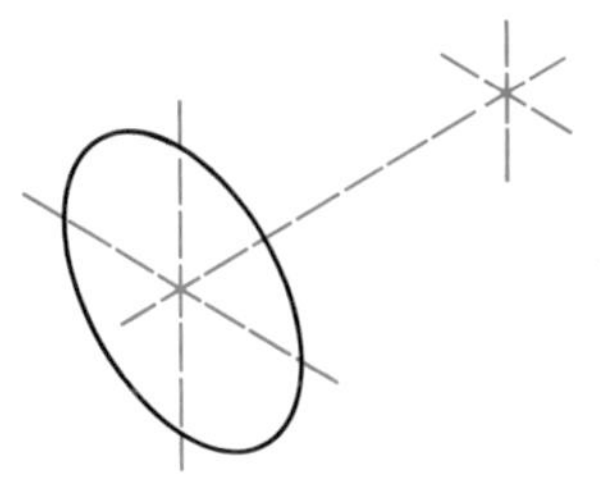

Figure 15.25 The total depth of the object is located (continuing from Figure 15.24).

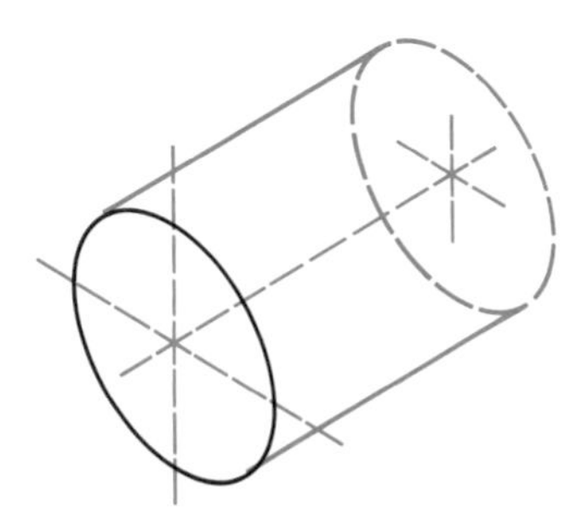

Figure 15.26 The rear circular surface and the horizon lines are drawn (continuing from Figure 15.25).

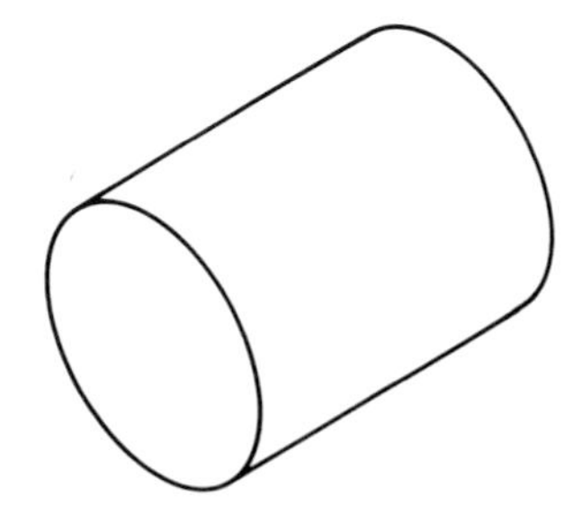

Figure 15.27 The completed isometric drawing (continuing from Figure 15.27) of the cylinder described by the orthographic views in Figure 15.19.

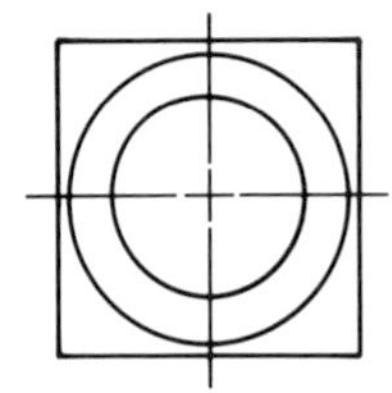
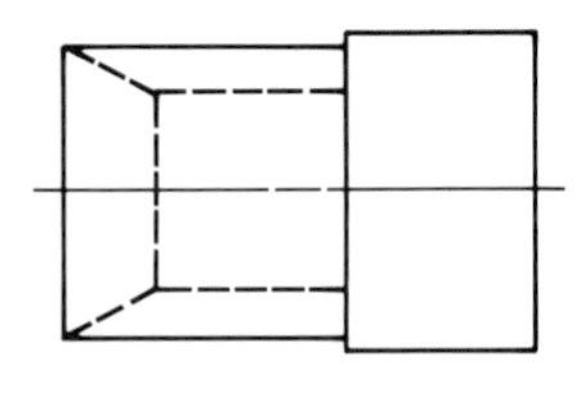

Figure 15.28 Two orthographic views of an object with several circular surfaces.

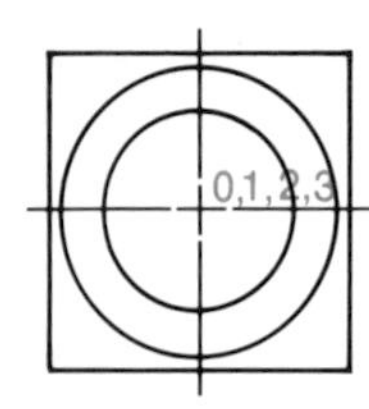

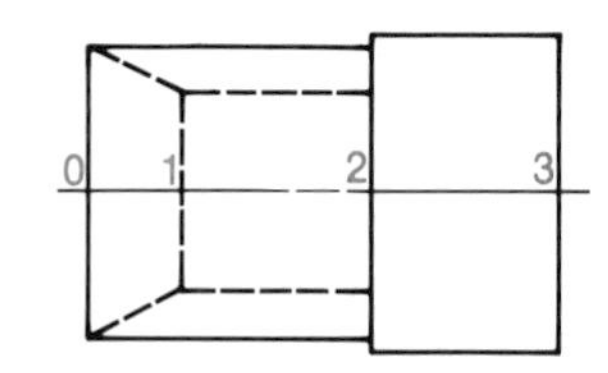

Figure 15.29 Key positions of depth are identified on the orthographic views from Figure 15.28.

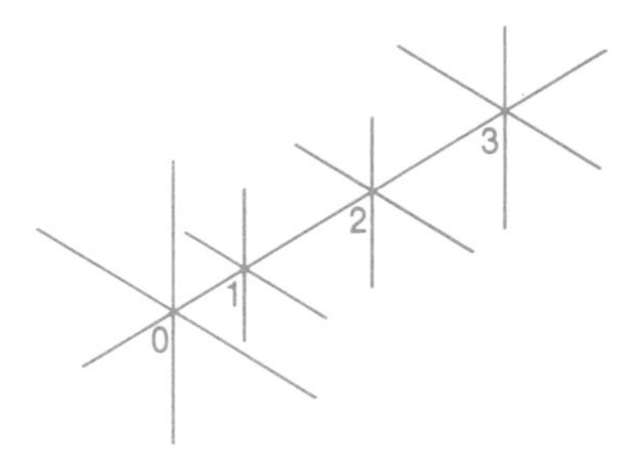

Figure 15.30 Key depth positions (continuing from Figure 15.29) are located in the isometric pictorial construction region.

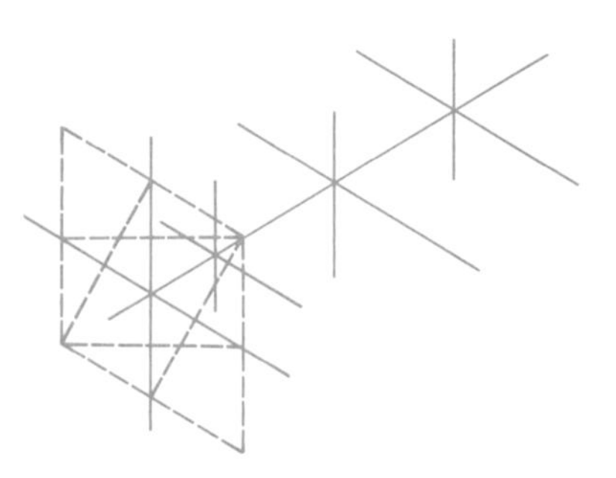

Figure 15.31 The circular boundary of the object in the frontal reference plane is constructed (continuing from Figure 15.30).

In summary, then, a circle or arc can be drawn in an isometric drawing through the use of the four-center method. This method produces an *approximately realistic distortion* of the circle or arc in the drawing, together with *true-length measurements along axial directions.*

15.5 Oblique Drawings

Oblique pictorials (see Section 14.4) are formed with two axes (height and width) drawn at a 90° orientation to one another and a third axis drawn at an angle, α, relative to the horizontal direction (Figure 15.37). The angle α is usually chosen so that $30° \leq \alpha \leq 45°$, depending upon the *emphasis* one wishes to place upon a particular surface. In Figure 15.38 several oblique pictorials are shown (a–c), together with an isometric pictorial (d).

The advantage offered by oblique pictorials is that circles and arcs appear *without distortion* in all surfaces that are parallel to the plane defined by the height and width axes (axes at a 90° orientation to each other). The cost paid for greater ease in drawing circles and arcs in oblique pictorials is a slight loss in the realistic, three-dimensional appearance of the object.

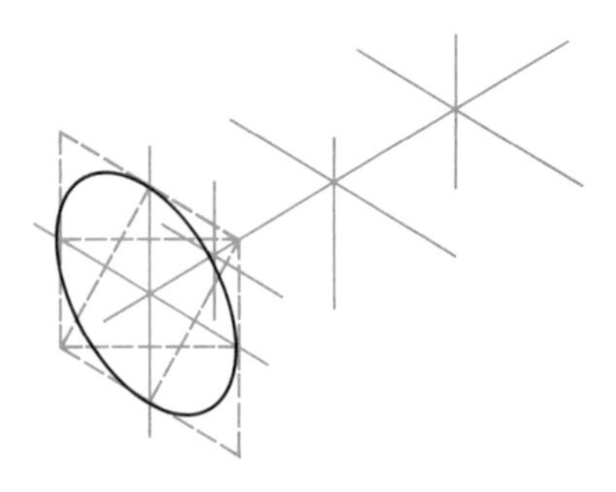

Figure 15.32 The circular boundary at the front of the object is constructed (continuing from Figure 15.31).

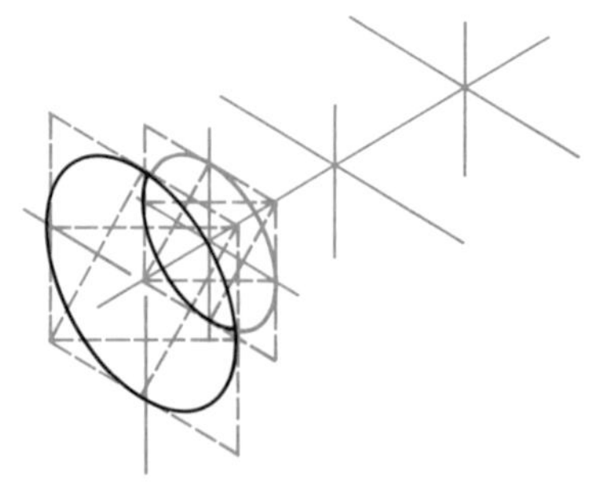

Figure 15.33 The circular feature at depth position 1 is constructed (continuing from Figure 15.32).

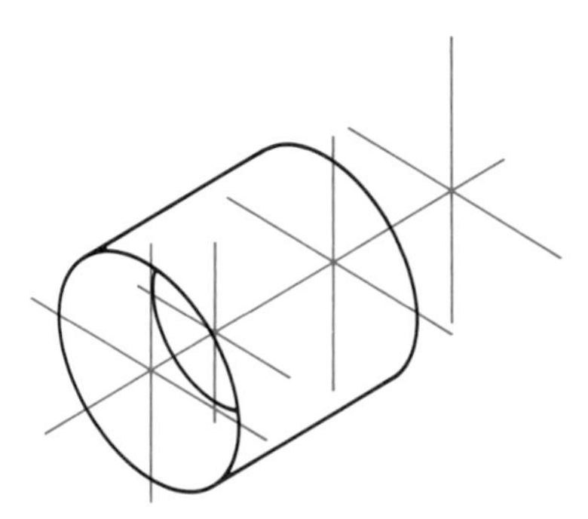

Figure 15.34 The cylindrical extension of object is developed (continuing from Figure 15.33).

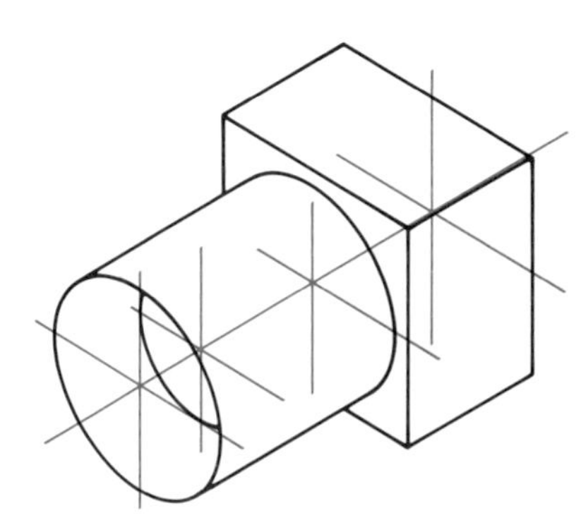

Figure 15.35 The rectangular base of object is developed (continuing from Figure 15.34).

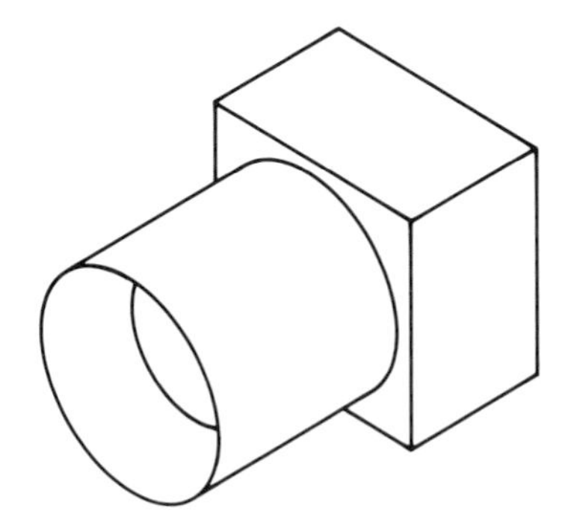

Figure 15.36 Construction linework is erased, thereby completing the isometric drawing (continuing from Figure 15.35) of the object described by the orthographic views in Figure 15.28.

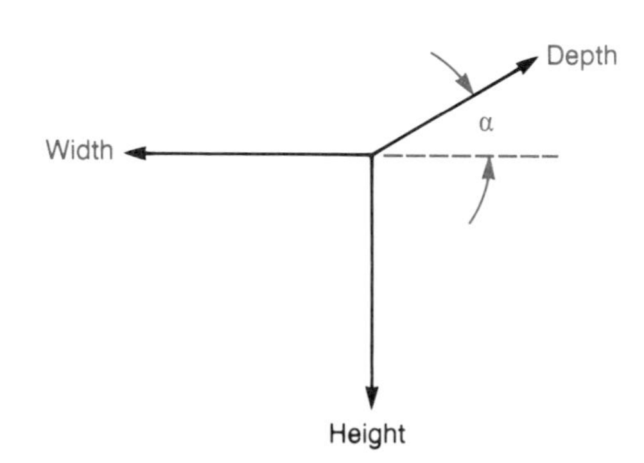

Figure 15.37 Orientation of principal axes for an oblique drawing. Height and width axes are perpendicular to each other. The depth axis is drawn at an angle, α, that varies according to the emphasis desired for each surface.

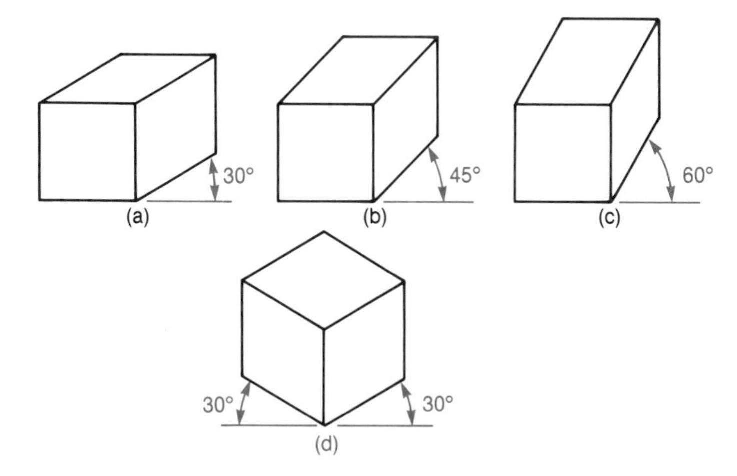

Figure 15.38 Oblique drawings with different choices for angle α, (a–c) and isometric drawing (d).

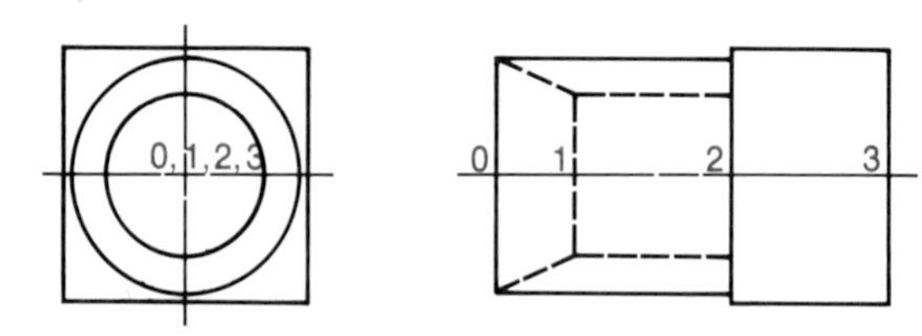

Figure 15.39 Two orthographic views of an object in which key positions of depth have been identified by numbers.

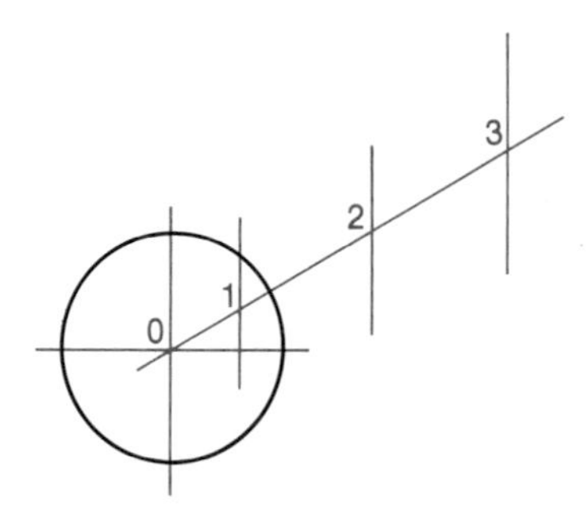

Figure 15.40 Development of the oblique drawing of the object described by the orthographic views in Figure 15.39.

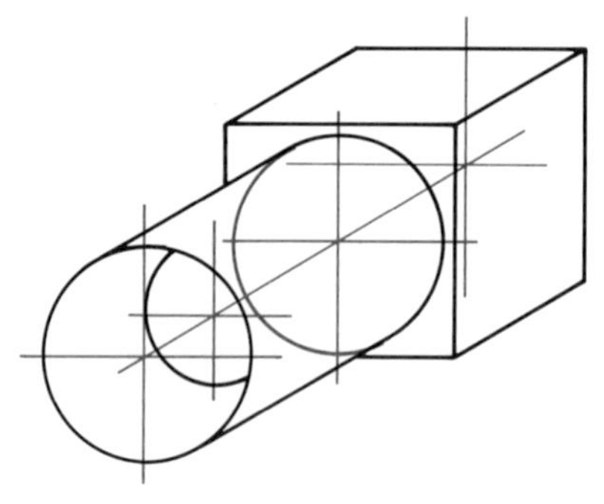
Figure 15.41 Additional construction of the oblique drawing (continuing from Figure 15.40).

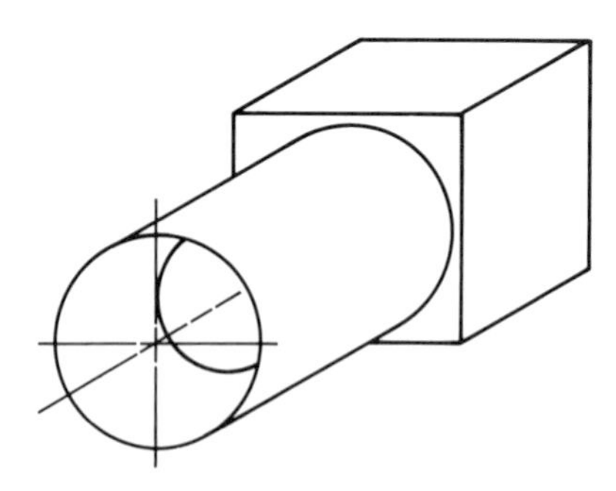
Figure 15.42 Construction linework is erased, thereby completing the oblique drawing (continued from Figure 15.41) of the object described by the orthographic views in Figure 15.39.

Figure 15.39 presents two orthographic views of an object in which key positions of depth have been identified (points *O* through 3). An oblique pictorial construction of this object is initiated, as shown in Figure 15.40, through a specification of the axes and the depth positions (points *O* through 3). The circular front surface of the object under construction has also been drawn in the oblique construction area, centered about the axial origin point, *O*. The circular boundary of this front surface is drawn without distortion. Figures 15.41 and 15.42 present additional construction of the drawing and the completed oblique pictorial, respectively.

Oblique drawings may be of the cavalier or cabinet type (Figure 15.43). In **cavalier** drawings, true-length measurements along all axial directions are used. (The drawing in Figure 15.42 is a cavalier type.) An optical illusion results from this type of oblique drawing because it is so unlike true-perspective illustrations; parallel lines *appear* to diverge, making the rear surface seem wider and higher than the front (nearest to the observer) surface. In order to minimize such an optical illusion, one can draw depth-directed distances with *one-half* of their true-length values, resulting in a **cabinet** drawing (Figure 15.44).

In summary, then:

1. Oblique pictorials are formed with two axes at a 90° orientation to each other; the third (depth) axis is drawn at an angle (α) with respect to the horizontal direction that is usually chosen so that $30° \leq \alpha \leq 45°$, depending upon the emphasis desired for a particular surface of the object under consideration.
2. The advantage of oblique pictorials is that circles and arcs appear without distortion in all surfaces that are parallel to the plane defined by the height and width axes.
3. Cabinet drawings are oblique pictorials in which depth-directed distances are drawn at half-scale in order to minimize the optical distortions associated with full-scale (cavalier) drawings.

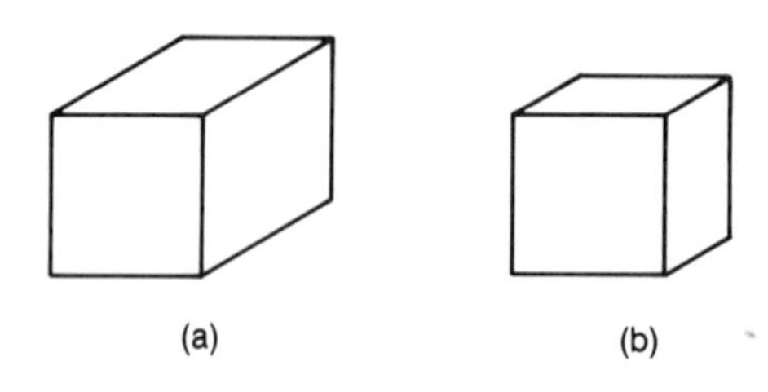

Figure 15.43 Oblique drawings may be of the cavalier (a) or cabinet (b) type.

Learning Check

Are oblique pictorials also axonometric pictorials?

Answer No. Recall (Section 14.1) that an axonometric pictorial is a projected image of an object onto a projection plane wherein three principal faces (surfaces) of the object are shown in the pictorial—with the object rotated into a position (with respect to the projection plane) in such a way that three of its principal surfaces may be simultaneously projected onto the plane. In an oblique pictorial (see Section 14.4), the observer views the object along a sight-line direction that is oblique (neither perpendicular nor parallel) to the plane of projection onto which the pictorial is drawn. One principal surface of the object remains parallel to the projection plane in an oblique pictorial, so it is shown without any shape distortion.

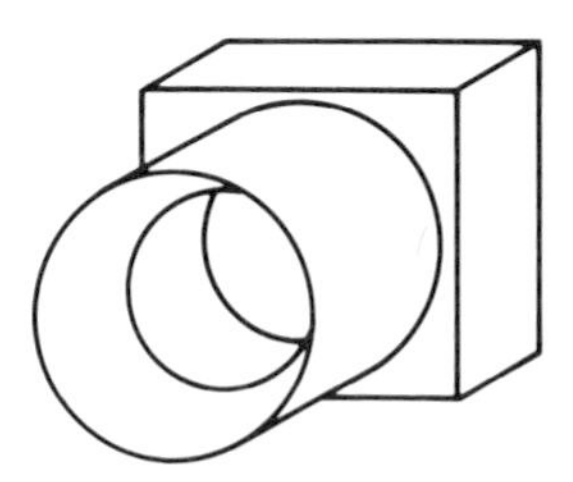

Figure 15.44 Cabinet drawing of the object described by the orthographic views in Figure 15.39.

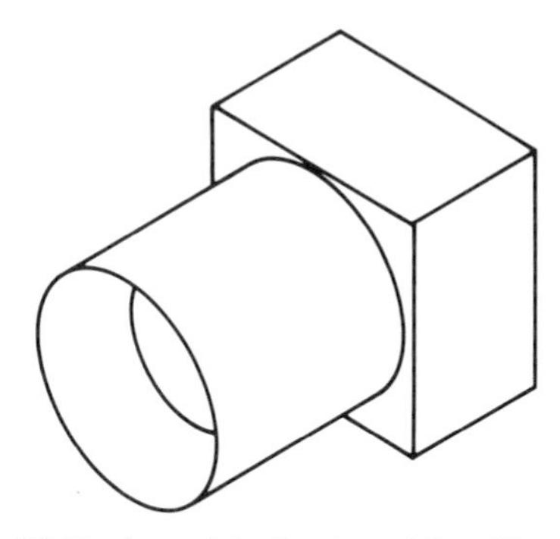

Figure 15.45 Isometric drawing of the object described by the orthographic views in Figure 15.39.

15.6 Sectioning in Pictorials

The isometric drawing shown in Figure 15.45 fails to clearly indicate that the hole (1) does not extend through the entire object and (2) is tapered at the front of the object. In order to show these details unambiguously, a **cutaway,** or **sectional,** view of the object is needed.

Chapter 6 focused upon sectioning in orthographic drawings. Full sections are formed by passing an imaginary cutting plane through the entire object and then removing the portion of the object that lies between the observer and the interior portion that is under consideration. A half section is formed by removing one-quarter of the object that lies between the observer and the two cutting planes that have intersected the object.

To develop a sectional pictorial of the object shown in Figure 15.45, first draw the principal axes (height, width, and depth) of the pictorial (Figure 15.46). Then locate significant depth locations in the figure—that is, depth positions at which surfaces or the intersection of surfaces appear.

In Figure 15.47 a cutting plane has been passed through one-half of the object in preparation for the construction of a half-sectional pictorial. Construction lines have been added to the original pictorial in order to indicate the intersection of the cutting plane with the surfaces of the object. (Notice that most of these intersection construction lines are directed either vertically or along the depth direction since the cutting plane is vertical in this case).

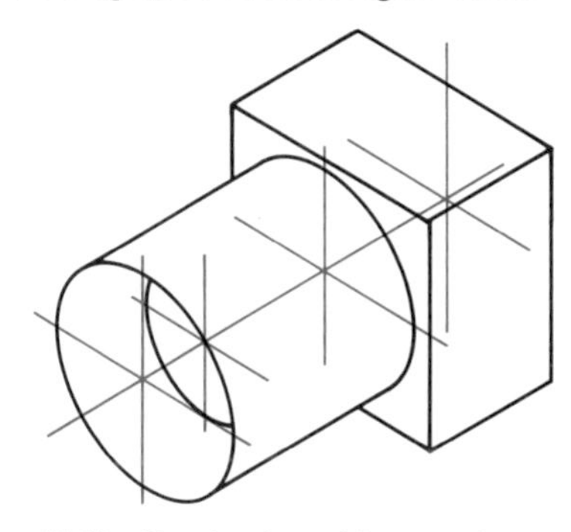

Figure 15.46 Key depth positions are located in the isometric drawing (continuing from Figure 15.45).

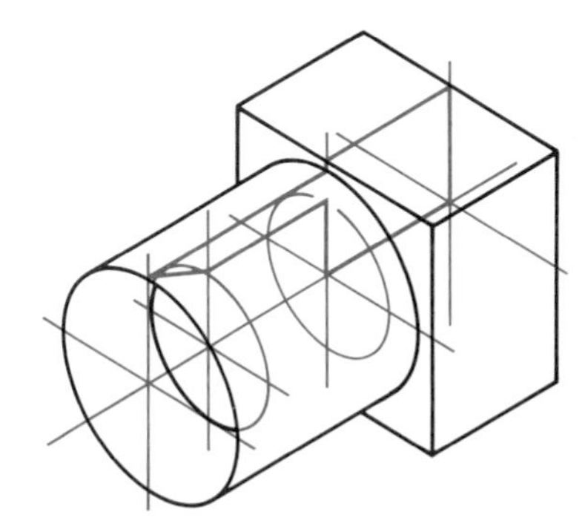

Figure 15.47 A vertical cutting plane intersects the object (continuing from Figure 15.46). Construction linework is used to represent the intersection of this cutting plane with the surfaces of the object.

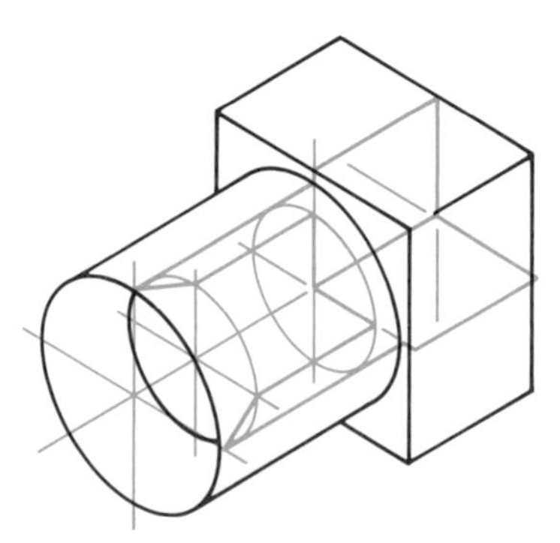

Figure 15.48 Construction linework is added to the drawing to indicate the intersection between a horizontal cutting plane and the surfaces of the object.

A second cutting plane (horizontal in this case) is then applied to the object (Figure 15.48). Once again, construction linework is added to the pictorial in order to indicate the intersection of this horizontal cutting plane with various surfaces of the object. (The intersection construction lines are primarily directed along the horizontal and depth directions since a horizontal cutting plane was used.)

The one-quarter of the object that lies between the observer and the intersections of the cutting planes with the object is now removed (Figure 15.49). Cross-hatching or sectional linework is added to indicate to the reader of the sectioned pictorial that it is indeed a sectional description of the object.

A similar procedure may be used to construct sectioned oblique pictorials, as shown in Figures 15.50 through 15.53. Figures 15.52 and 15.53 present sectioned cavalier and cabinet drawings, respectively. (Cross-hatching and other aspects of sectioning are discussed in detail in Chapter 6.)

In summary, then:

1. Pictorial drawings may be sectioned in order to clarify the internal features of an object.

2. Sectioning of pictorials may be methodically accomplished by passing imaginary cutting planes through the pictorial representation of the object, drawing construction lines to denote the intersection between a cutting plane and various surfaces of the object, and then removing (erasing) the portion of the object that lies between the observer and the cutting planes. Cross-hatching should then be used to emphasize that the resulting pictorial is a sectional representation of the object.

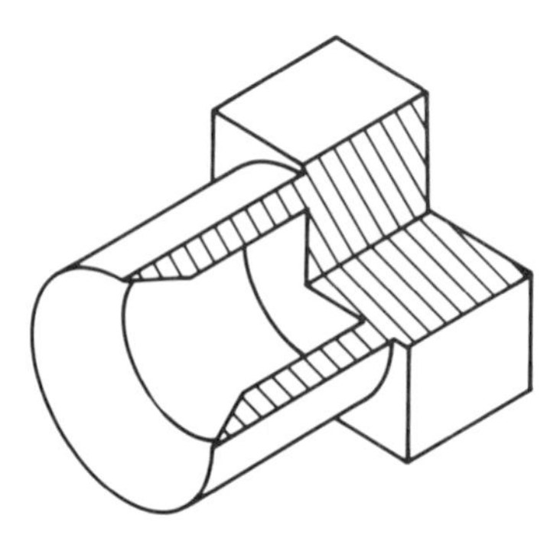

Figure 15.49 The portion of the object that lies between the observer and the cutting planes is removed. Cross-hatching is then added to indicate the internal surfaces of the object. A half-sectional isometric drawing is thus produced.

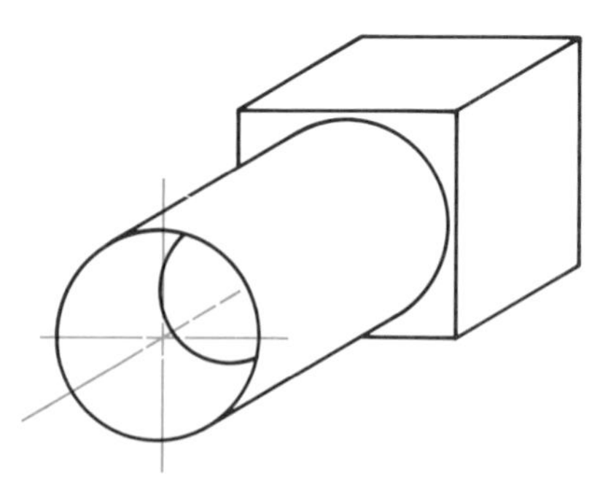

Figure 15.50 An oblique drawing of the object described by the orthographic views in Figure 15.39.

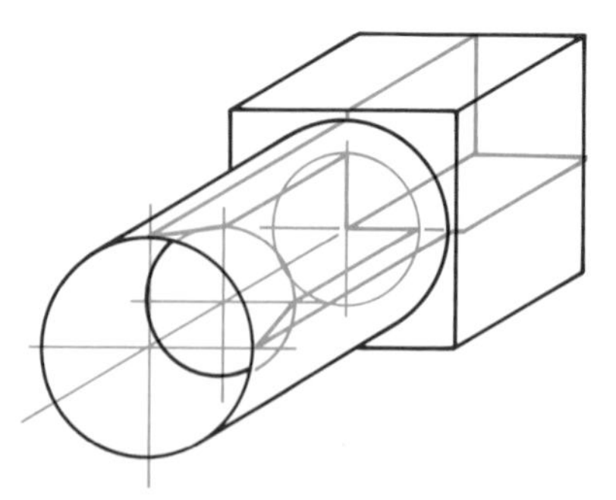

Figure 15.51 Construction linework is used to indicate the intersection between (vertical and horizontal) cutting planes and the surfaces of the object (continuing from Figure 15.50).

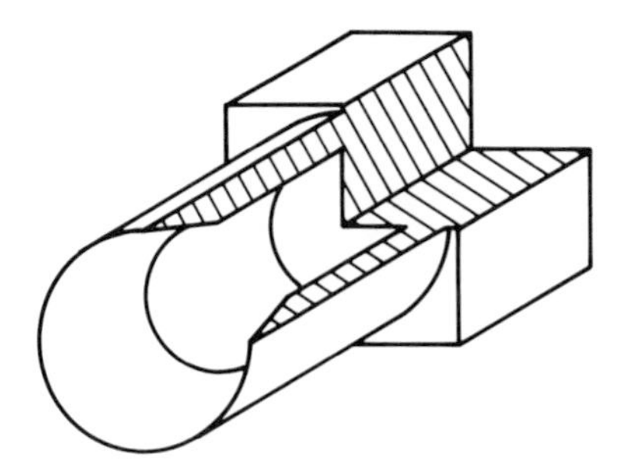

Figure 15.52 The completed half-sectional cavalier oblique drawing (continuing from Figure 15.51).

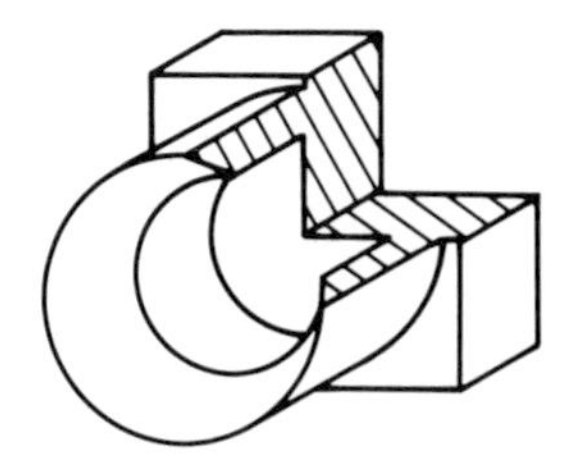

Figure 15.53 The completed half-sectional cabinet oblique drawing (continuing from Figure 15.52).

In Retrospect

- Isometric drawings have true-length measurements along axial directions. Distances along nonaxial directions are distorted (foreshortened or lengthened, depending upon the orientation of a particular line).
- Isometric drawings are constructed with the three principal axes (height, width, and depth) oriented with equal (120°) angular spacing.
- An isometric drawing should be constructed via the point-by-point, surface-by-surface approach.
- In the construction of a pictorial, the location of the origin for the axes must be judiciously chosen so that the drawing will be properly centered within the available drawing space. The front-viewing direction of the pictorial must be chosen so that the orientation of the pictorial description of an object and any accompanying orthographic representations will be mutually consistent.
- The four-center method may be used to quickly construct an arc or a circle in an isometric drawing with an approximately realistic distortion together with true-length dimensions along axial directions.
- Oblique pictorials are formed with two axes (height and width) at a 90° orientation to each other; the third axis (depth) is drawn at an angle (α) with respect to the horizontal direction. The angle α is usually chosen so that $30° \leq \alpha \leq 45°$, depending upon the emphasis one wishes to place upon a particular surface of the object under consideration.
- The advantage of oblique pictorials is that circles and arcs appear without distortion in all surfaces that are parallel to the plane defined by the height and width axes.
- Cabinet drawings are oblique pictorials in which depth-directed distances are drawn at half-scale in order to minimize the optical illusions associated with full-scale (cavalier) oblique drawings.
- Pictorial drawings may be sectioned to clarify the internal features of an object.

COMPUTER GRAPHICS IN ACTION

2D Computer Graphics: Developing an Oblique Pictorial

Menu-driven computer graphics software can be used to quickly construct an oblique pictorial of an object from a previously created set of orthographic views such as the front and right-side views shown here (a). Linework in the given front view can be translated about the drawing area with the aid of the MOVE and COPY commands (or their equivalents) (b). The results are then further modified by selectively applying the ERASE, LINE, and ARC (if necessary) commands until the desired oblique pictorial has been formed (c). As suggested by this example, 2D computer graphics software can be used to quickly construct different types of engineering drawings from a given set of stored data (in this case the stored data consists of the original orthographic views). The engineer need only be creative in applying such software.

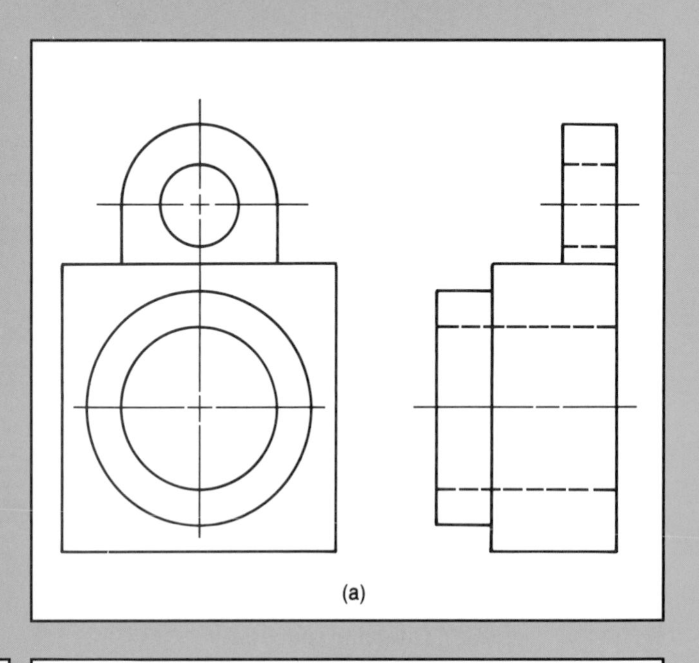
(a)

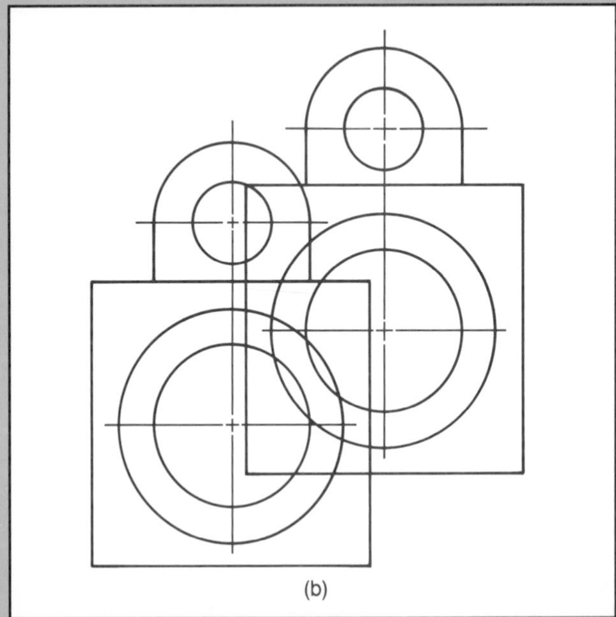
(b)

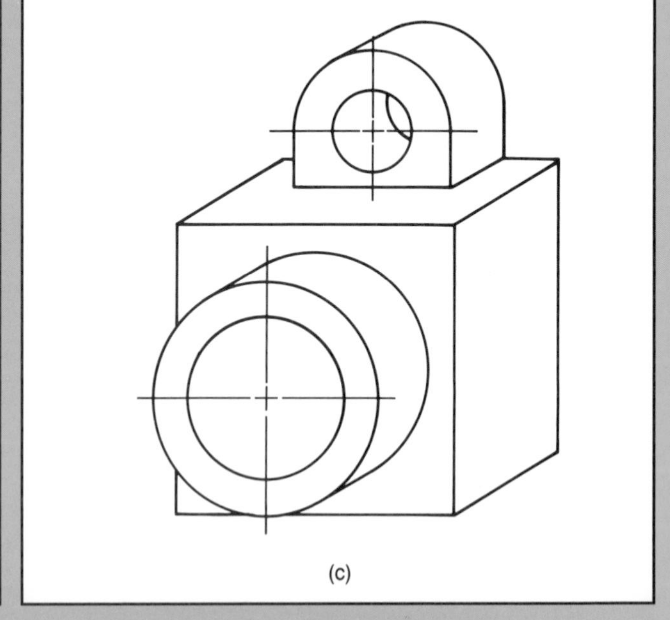
(c)

An oblique pictorial is constructed from a given set of orthographic views (a) by translating given data with the MOVE and COPY commands (b) and modifying the results with the LINE, ERASE, and ARC commands (c).

Exercises

Your instructor will specify the appropriate scale for each exercise and the particular type of pictorial to be constructed (for example, isometric, oblique, isometric fully sectioned). EX1 through EX50 refer to specific exercises in preceding chapters. In each case use the given set of orthographic views to develop a pictorial representation of the object in accordance with the directions of your instructor.

EX15.1. See EX4.3.

EX15.2. See EX4.7.

EX15.3. See EX4.8.

EX15.4. See EX4.9.

EX15.5. See EX4.10.

EX15.6. See EX4.12.

EX15.7. See EX4.13.

EX15.8. See EX4.14.

EX15.9. See EX4.15.

EX15.10. See EX4.16.

EX15.11. See EX4.17.

EX15.12. See EX4.18.

EX15.13. See EX4.20.

EX15.14. See EX4.22.

EX15.15. See EX4.23.

EX15.16. See EX4.24.

EX15.17. See EX4.25.

EX15.18. See EX4.26.

EX15.19. See EX4.28.

EX15.20. See EX4.29.

EX15.21. See EX4.31.

EX15.22. See EX4.32.

EX15.23. See EX4.33.

EX15.24. See EX4.35.

EX15.25. See EX4.55.

EX15.26. See EX4.56.

EX15.27. See EX4.57.

EX15.28. See EX4.58.

EX15.29. See EX4.59.

EX15.30. See EX5.5.

EX15.31. See EX5.6.

EX15.32. See EX5.7.

EX15.33. See EX5.8.

EX15.34. See EX5.9.

EX15.35. See EX6.1.

EX15.36. See EX6.2.

EX15.37. See EX6.3.

EX15.38. See EX6.4.

EX15.39. See EX8.1.

EX15.40. See EX8.2.

EX15.41. See EX8.3.

EX15.42. See EX8.4.

EX15.43. See EX8.5.

EX15.44. See EX8.6.

EX15.45. See EX8.7.

EX15.46. See EX8.8.

EX15.47. See EX8.9.

EX15.48. See EX8.10.

EX15.49. See EX8.11.

EX15.50. See EX8.12.

Draw the appropriate pictorial of the objects described in each exercise in accordance with the directions of your instructor.

EX15.51.

EX15.52.

EX15.53.

EX15.54.

EX15.55.

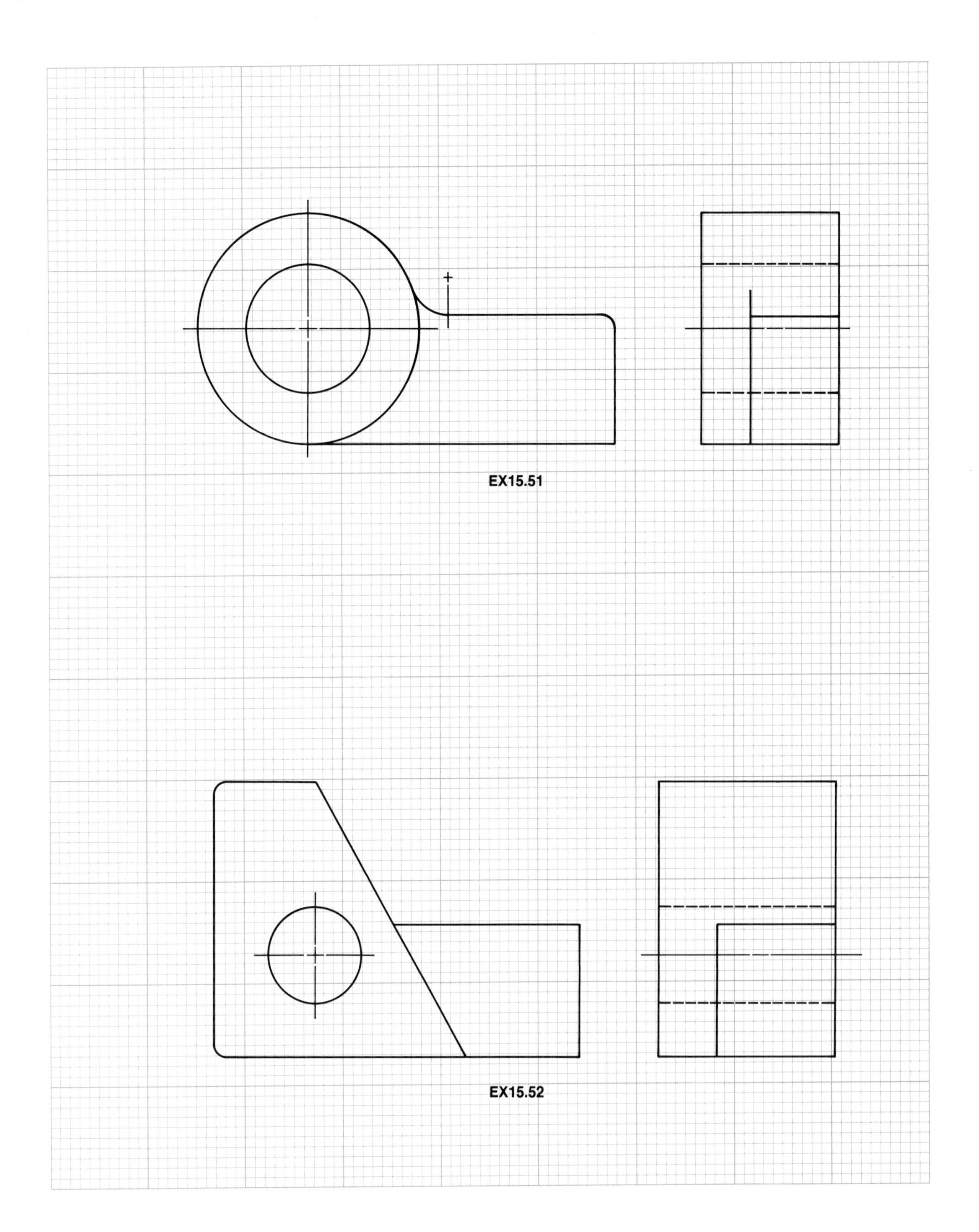
EX15.51
EX15.52

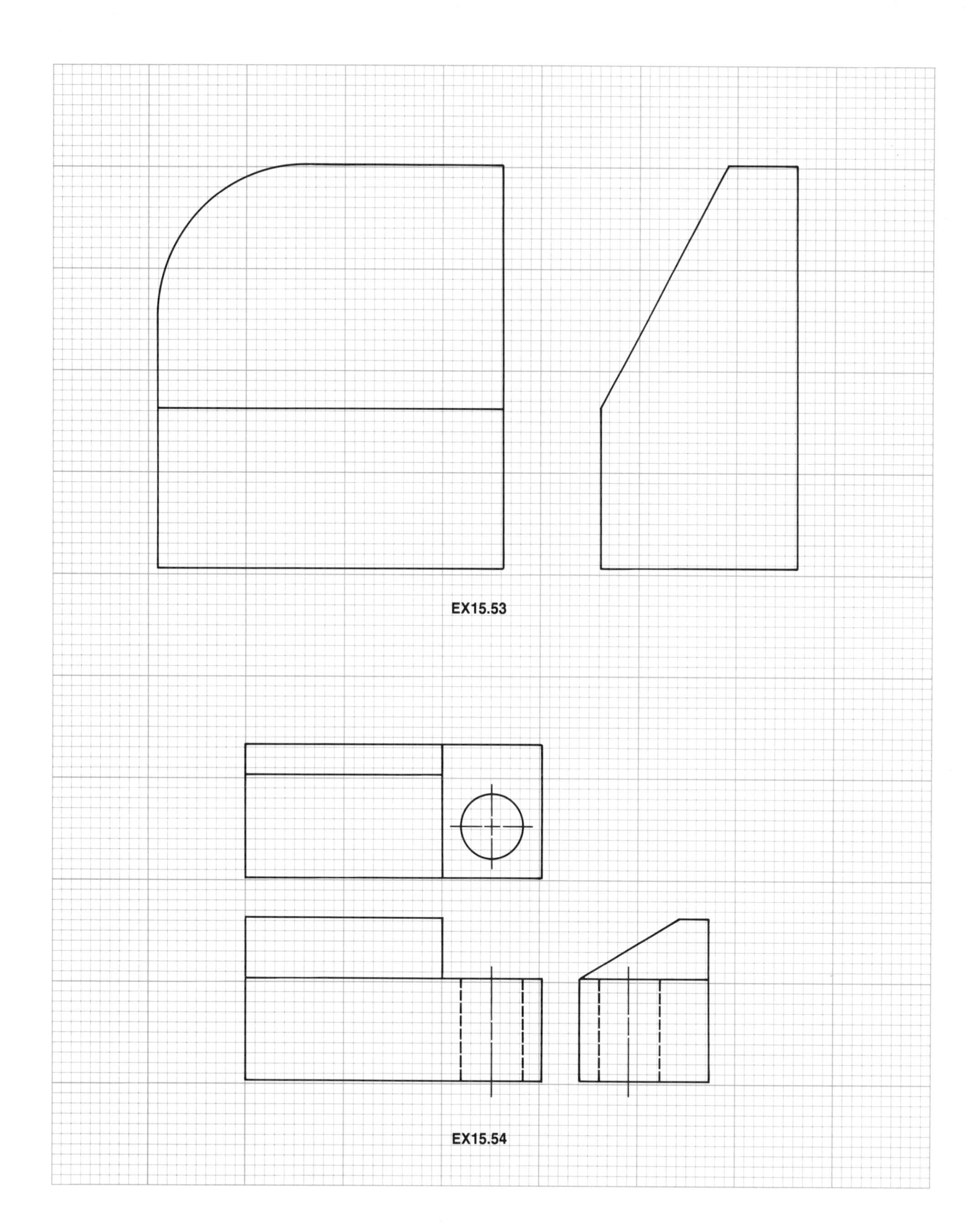
EX15.53
EX15.54

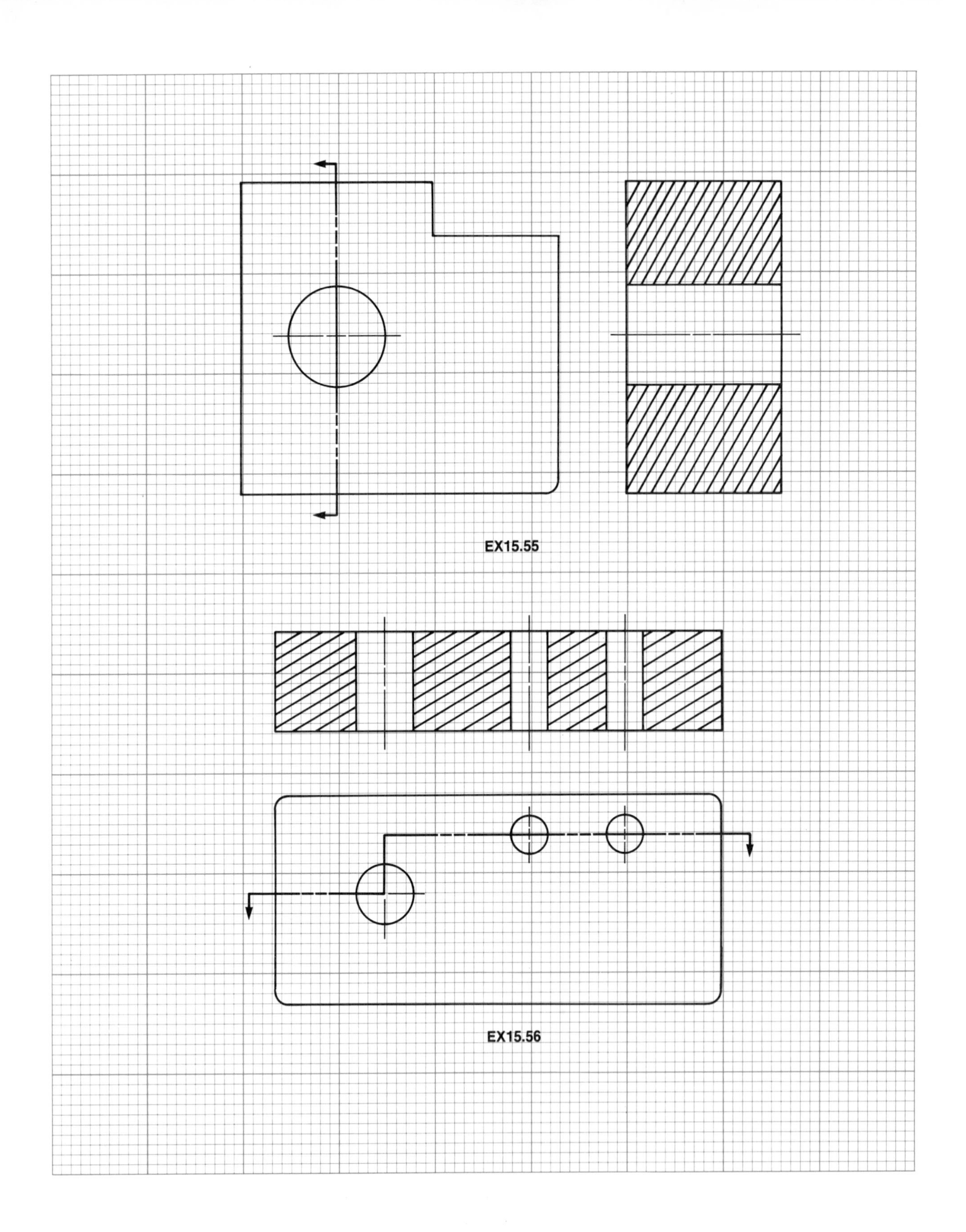
EX15.55
EX15.56

16 Sketching

Preview The preceding chapters were intended to help the reader develop skill in the use of manual and/or computer graphics instruments to produce accurate engineering drawings. However, the engineer must also be able to graphically record and communicate ideas with *speed—in the absence of drawing instruments!* Thus, the ability to **sketch** both orthographically and pictorially is a required engineering skill.

Sketching refers to **freehand drawing**—that is, drawing without the use of instruments. This skill is necessary when the engineer must graphically describe ideas and objects during conferences in offices and other environments where neither drawing equipment nor the time to use such equipment exists. A pencil or pen and a sheet of paper should be the only tools required for sketching.

In this chapter we will focus upon the fundamental techniques for creating freehand sketches. Skill in sketching can be acquired through diligent practice. The most challenging aspect of developing good engineering sketches is the ability to create two-dimensional graphic descriptions of three-dimensional objects (see Figure 16.1). Since preceding chapters on creating orthographic and pictorial drawings of objects should have developed the reader's ability to visualize three-dimensional objects in two dimensions, the reader should not find the sketching of orthographic or pictorial drawings difficult if the material in preceding chapters has been mastered.

Learning Objectives

Upon completion of this chapter, the reader should be able to:

- Define sketching.
- Recognize the need for sketching abilities in engineering.
- Create both orthographic and pictorial engineering sketches.
- Identify the corrections needed in sketches that have not been carefully or properly created.
- Draw straight lines and arcs with speed and accuracy.
- Identify and produce visible, hidden, center, and construction linework in sketches.
- Draw lines at specified angles to one another with sufficient accuracy.
- Draw circles in isometric sketches as appropriately formed ellipses.
- Know how to use shading in a sketch to enhance clarity and specify materials and surface textures.
- Apply a step-by-step procedure to develop both orthographic and pictorial sketches of objects.

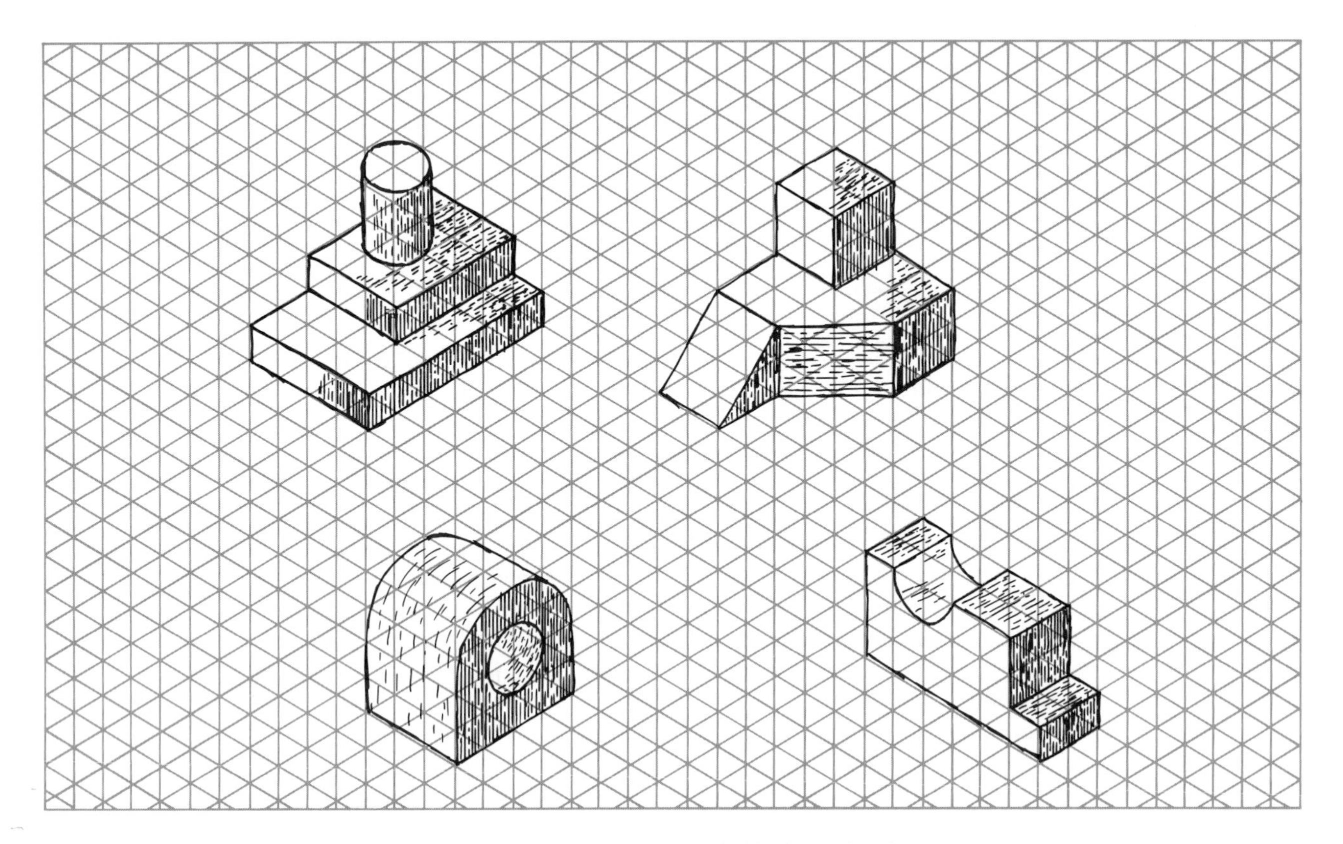

Figure 16.1 Examples of pictorial sketches; shading has been included in these drawings.

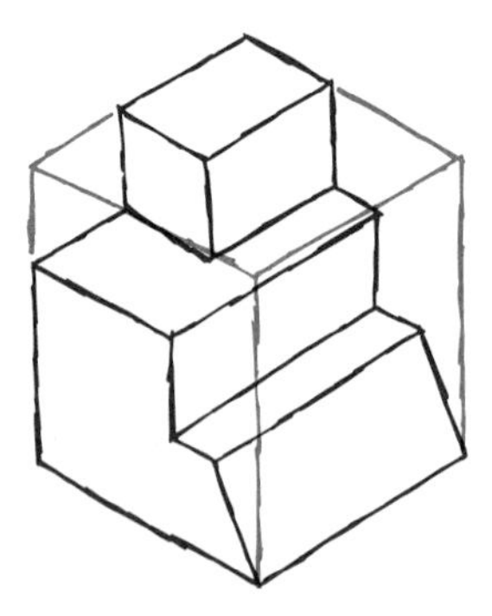

Figure 16.2 Construction linework usually does not need to be erased in a sketch. Such linework can even help in interpretation of the sketch.

16.1 Equipment

Sketching, as stated earlier, refers to **freehand drawing**—drawing without the use of manual or computer graphics equipment. A pencil or pen and a sheet of paper should be the only material required by the engineer in the production of a sketch.

An eraser is not necessary, since construction linework usually should remain in the sketch in order to add to the three-dimensional appearance of a pictorial (Figure 16.2). Such linework often aids interpretation of a sketch.

A pencil of medium weight (F, H, or HB grades) is appropriate for sketching. A pencil lead that is too soft can smudge the drawing, whereas a pencil that is too hard may not allow the user to draw lines of sufficiently variable widths and intensity.

A printed grid (which may be in a square pattern, isometric, and so forth—see Figure 16.3) aids the creation of parallel lines—particularly in the case of isometric pictorial sketches. In addition, the uniform spacing of such printed grids allows the user to control distances between lines, an aid to properly *proportioned* sketches.

16.2 Linework

Speed in sketching must be accompanied by the ability to generate reasonably accurate linework without the aid of drawing instruments. Lines in sketches do not need to be as straight and accurate as those shown in instrument drawings, of course, but a sketch must be carefully drawn (see Figure 16.4). Intersecting lines should be sure to intersect, not cross or fail to meet. Finished lines should be drawn with a consistent intensity, or darkness and width. A line intended to be straight should appear straight to the reader's eye. In Figure 16.5, two imperfectly drawn lines are shown, both intended to be straight. Line A is less irregular in its appearance than line B but it has not been drawn along a direction indicated by the (instrument-drawn) straight construction line. Line B is centered about the straight construction line; though more irregular in its detail, line B is preferable to line A because it is drawn more accurately in a linear (straight) direction between two specific endpoints on the sketch. Line A fails to follow the proper path between points. This example then leads us to two very important conclusions about the sketching of straight lines:

1. It is acceptable for a sketched line to include *slight* irregularities since it has not been drawn with the aid of instruments. (In fact, these irregularities add to the charm and the recognition of an engineering sketch as opposed to finished instrument drawings. A sketch will be viewed as a *preliminary* design, as should be the case.)
2. The engineer should focus upon the endpoints of the line while sketching. The *overall direction* of the line is more important than the elimination of irregularities in the line (although irregularities should be minimized in both number and extent).

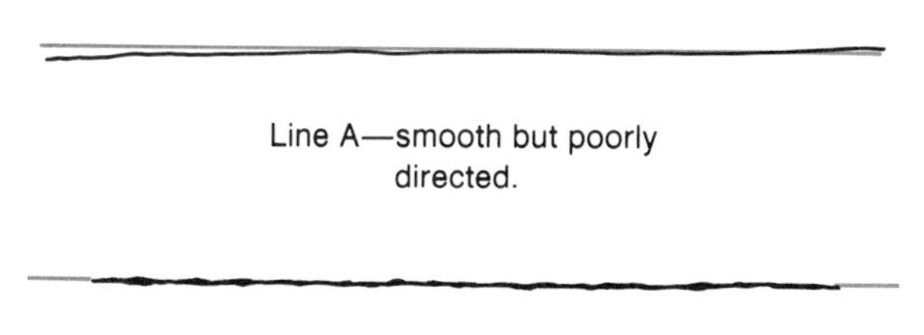

Figure 16.5 **A sketched straight line should be drawn along a linear path between specific endpoints. Line A is smoother than line B; however, it does not follow the specified linear path as well as line B. Despite its irregularity, line B is preferable to line A.**

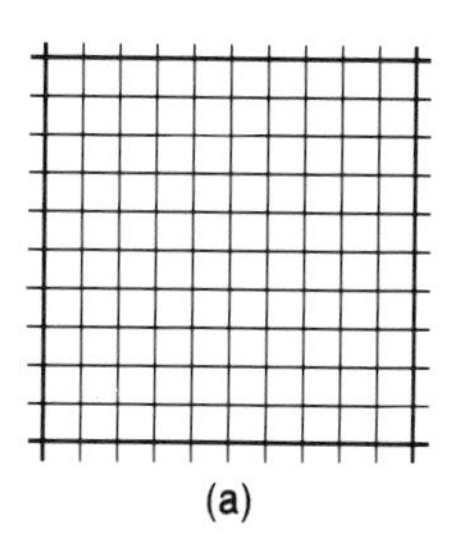

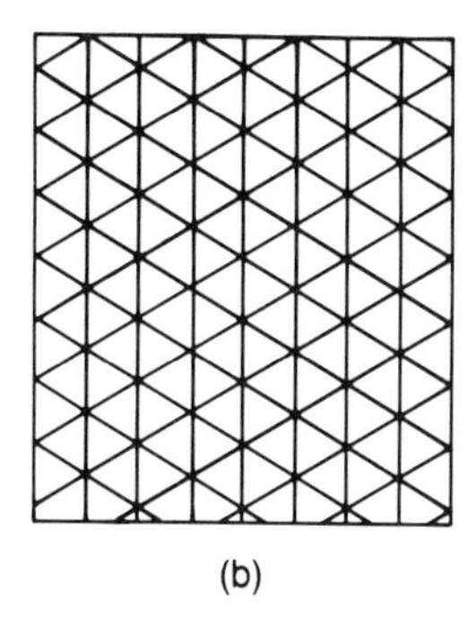

Figure 16.3 **Printed grid paper can be used to achieve consistency in direction and length of linework in sketches. A square pattern (a) and an isometric pattern (b) are shown here.**

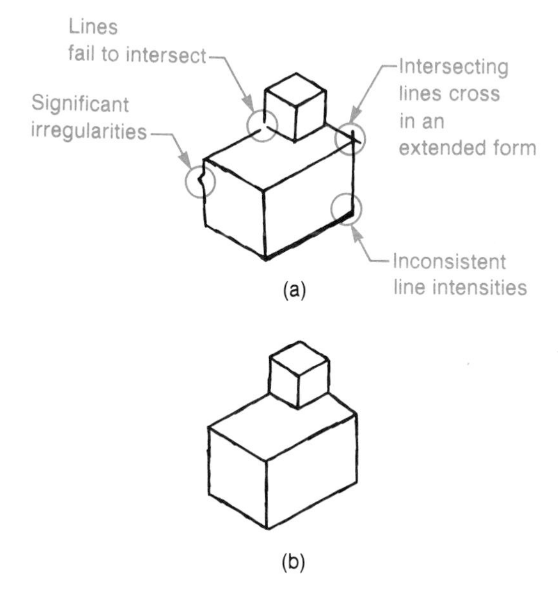

Figure 16.4 **A sketch must be carefully drawn. Compare the poorly drawn sketch with several errors (a) against the improved sketch with minimal errors (b).**

Common types of lines used in sketches (see Figure 16.6) include:

- **Visible lines,** which are dark and (relatively) thick.
- **Hidden lines,** which are of medium width and intensity.
- **Centerlines,** which are thin and sharply drawn.
- **Construction lines,** which are very thin and very light.

Instead of changing pencils in order to generate these different types of lines in a sketch, the engineer adjusts both the *pressure* applied to the pencil point and the *sharpness* of the pencil point. The sharpness is increased or decreased by rubbing the point on a piece of scratch paper. As Figure 16.6 indicates, a rounded point corresponds to a thicker and darker line; as the point is sharpened, it produces thinner and less intense lines.

16.3 Sketching Techniques

To create straight lines in a variety of directions, one can move, or rotate, the drawing paper to different orientations in order to maintain both comfort and accuracy. Remember, the user should concentrate on the endpoints and relative direction of each line as it is sketched. For example, as shown in Figure 16.7, a set of parallel lines at a particular angle is more easily drawn in either a horizontal or a vertical direction, so the paper should be rotated accordingly.

Right-handed individuals usually find that left-to-right motions are most effective for producing horizontal lines and downwardly directed strokes for vertical lines. Lines

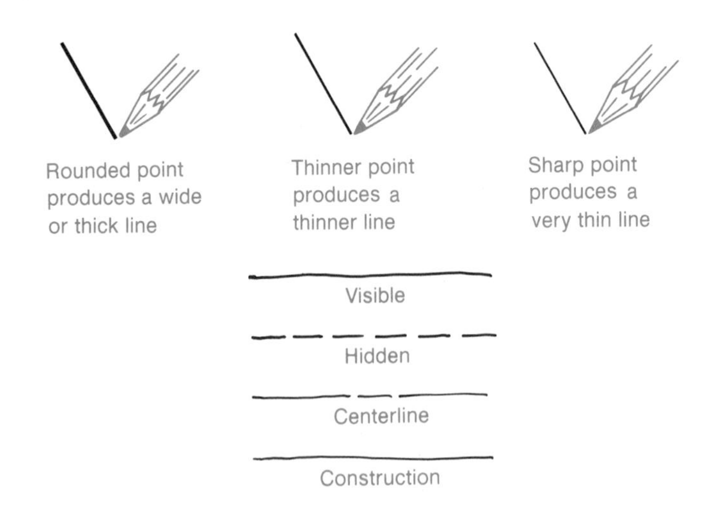

Figure 16.6 Four common types of lines used in engineering sketches. Lines are produced by modifying both the *pressure* applied to the pencil point and the *sharpness* of the point. A sharper point produces a thinner line.

drawn at particular angles (such as 45° or 60°) can be approximated by first drawing a set of perpendicular construction lines, then dividing the angle between them into equal segments (see Figure 16.8). This method provides an approximation of the desired angle of orientation for the linework without the need to measure any angles. (Remember, a freehand sketch should not require any instruments other than a pencil.)

Arcs can be similarly generated by first drawing a set of perpendicular construction centerlines, then marking the desired radius of the arc with respect to the centerpoint (Figure 16.9). Such radial distances are, of course, only approximations, not measured values. A light construction arc is then drawn, analyzed for accuracy and corrected; finally the finished arc is drawn with dark visible linework.

Circles are also created by using construction centerlines (Figure 16.9). A strip of scratch paper, along which is marked the approximate length for the diameter (or the radius) of the circle, may be used to increase accuracy in the size of the sketched circle. Construction lines are drawn, along which are marked eight (or more) points through which the circle should pass. A light construction circle is then drawn and checked for accuracy and corrected if necessary before the finished circle is drawn with dark visible linework.

Circles in isometric pictorials appear (distorted) as ellipses (see Figure 16.10). Construction centerlines should be used to specify the centerpoint of an isometric circle/ellipse. Radial distances are then marked off along these centerlines with respect to the centerpoint. (A piece of scratch paper on which the approximate radial distance is marked can be

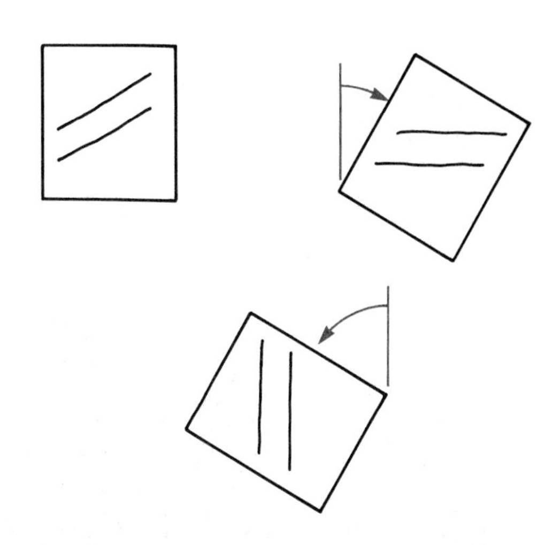

Figure 16.7 Parallel lines may be drawn at particular angles by rotating the paper to a comfortable position. Horizontal and vertical lines are usually more easily drawn than lines directed along other directions.

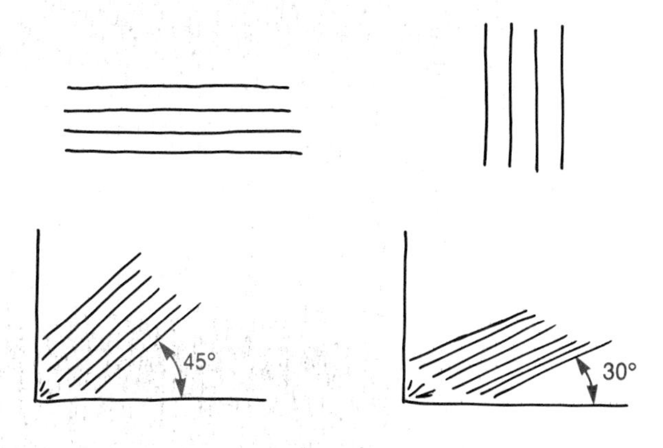

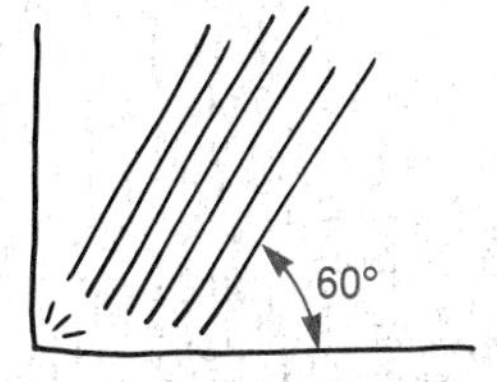

Figure 16.8 Lines may be drawn along particular directions by using a set of construction lines. The angle between these lines is then divided into equal segments (for examle, 15° increments) in order to obtain approximations of the desired angular measurements.

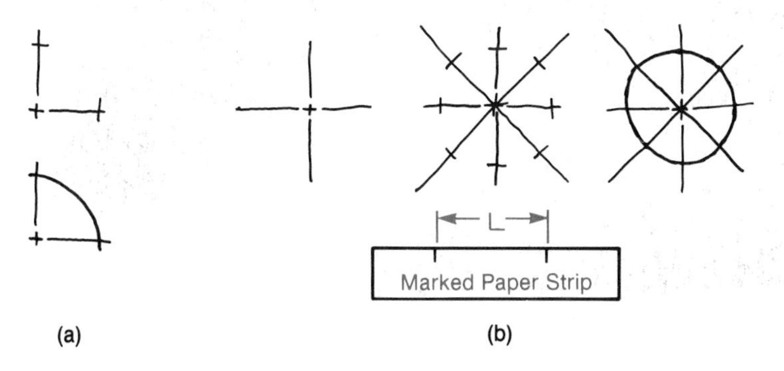

Figure 16.9 Construction centerlines may be used to develop arcs in sketches by first marking the radial distance along the construction centerlines and then drawing the finished arc (a). For circles, construction lines are used to form eight points that are equally distant from the centerpoint (a marked paper strip may be used to achieve consistency in the construction distances) and then drawing the finished circle through the construction points (b).

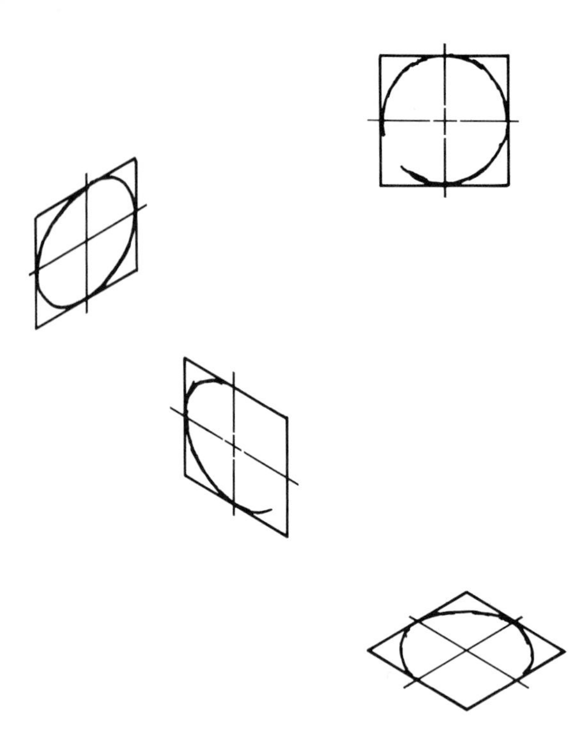

Figure 16.10 A circle may be drawn with the use of construction centerlines and the development of a rhombus.

used to maintain consistency in distances on the sketch.) A rhombus (a parallelogram with sides of equal length) is then formed by drawing straight lines—parallel to the axial directions associated with the isometric plane in which the circle is to be shown—through these radial construction points on the centerlines. These radial points are, in fact, points of tangency through which the circle/ellipse is to pass. The ellipse is sketched by drawing four construction arcs through these points of tangency. After checking the construction arcs for accuracy and correcting if necessary, the finished ellipse is drawn with dark finished linework. If a printed isometric grid is used, the centerlines and the rhombus can be identified without the use of construction lines along axial directions.

Finally, we should note that oblique pictorial sketching is often preferred over isometric sketching, since arcs and circles lying in planes perpendicular to the viewing direction can be quickly drawn as arcs and circles (see Section 15.5).

16.4 Cylinders in Isometric Sketches

The circular bases of cylinders (and cylindrical holes) appear distorted as ellipses in isometric sketches. In order to *orient* these ellipses correctly with respect to the axis of the cylinder, *always draw the major (largest) axis of the ellipse perpendicular to the axis of the cylinder* (see Figures 16.11 and 16.12). Thus, if printed isometric grid paper is not available,

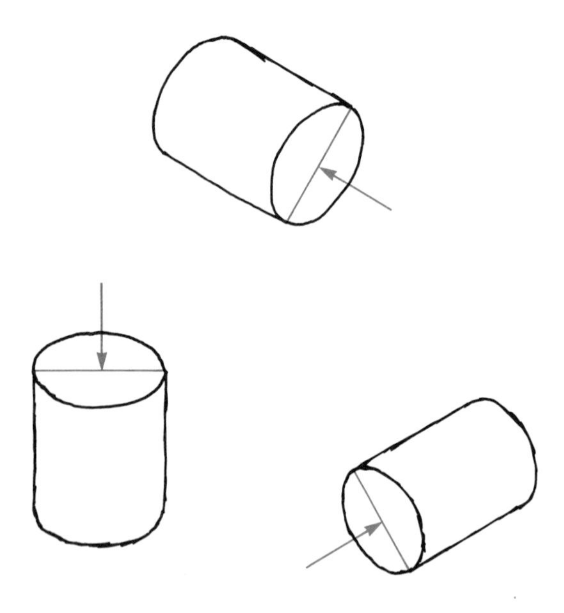

Figure 16.11 The major axis of the ellipse that represents the base of a cylinder in an isometric sketch should always be perpendicularly oriented to the cylindrical axis.

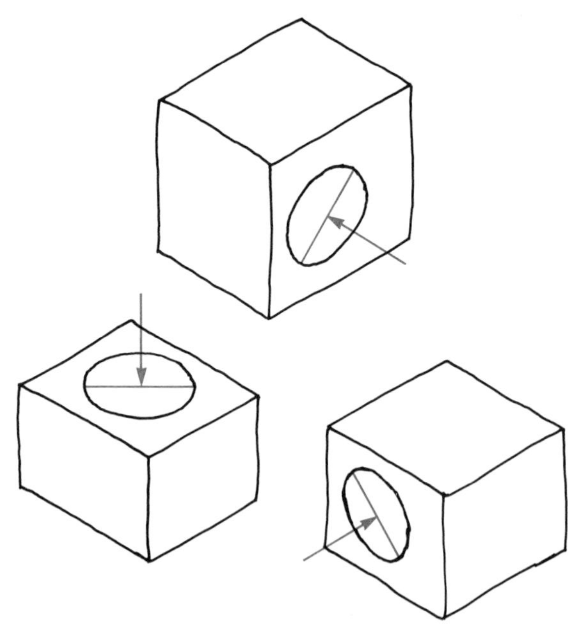

Figure 16.12 The circular boundary of a cylindrical hole, seen as an ellipse in an isometric sketch, should always be perpendicularly oriented with respect to the axis of the hole.

one need only identify the direction of the cylindrical axis on the sketch, then develop the cylindrical base so that the major axis of the ellipse is perpendicular to this cylindrical base.

16.5 Shading

Shading can enhance an engineering drawing by identifying different surfaces with patterns of varying intensity and direction. In addition, shading can specify surface texture. Light, upon striking an object, is more highly reflected from finished (machined or polished) surfaces than from rough or unfinished surfaces. As a result, shading patterns can be used to indicate the relative reflectivity—and the corresponding relative finishes— of different surfaces. Particular types of material can also be identified by shading; for example, wood can be identified by including grainlines in the sketch.

Boundary edges between surfaces are not truly distinct lines in space, but are in fact nonplanar surfaces. Figure 16.13(a) indicates the continuity between intersecting surfaces by showing them as areas of reflected light. For the sake of speed, however, such detailed considerations are usually neglected, and the intersection between planar surfaces is indicated by straight lines (b).

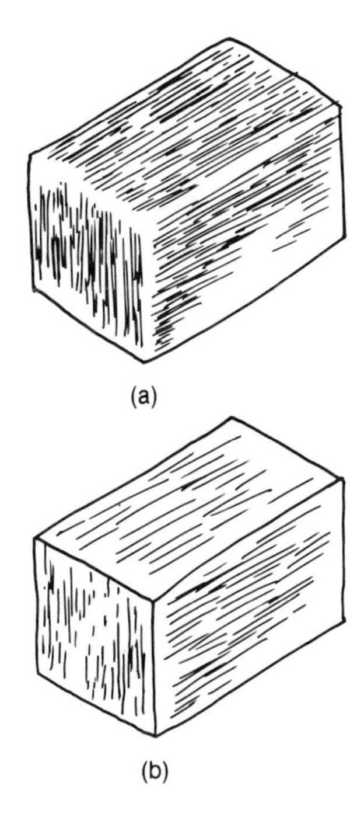

Figure 16.13 The intersection between planar surfaces is actually smooth and continuous (a), but for the sake of speed, such intersections are usually sketched as discrete lines (b).

16.6 Layout of a Sketch

Orthographic sketches and **pictorial sketches** can both be constructed through the use of the following general procedure (Figure 16.14):

1. *Lightly box in the height, width, and depth distances for each view* of the object in the sketch. (If orthographic views are to be sketched, allow space for dimensioning linework and separation of the views.)

2. *Add interior linework* (in-lines), including centerlines for arcs and circles.

3. *Darken the linework for surface boundaries.*

4. *Darken and complete* dimension lines (if any) and add any other data to be included in the sketch. Do not erase light construction work.

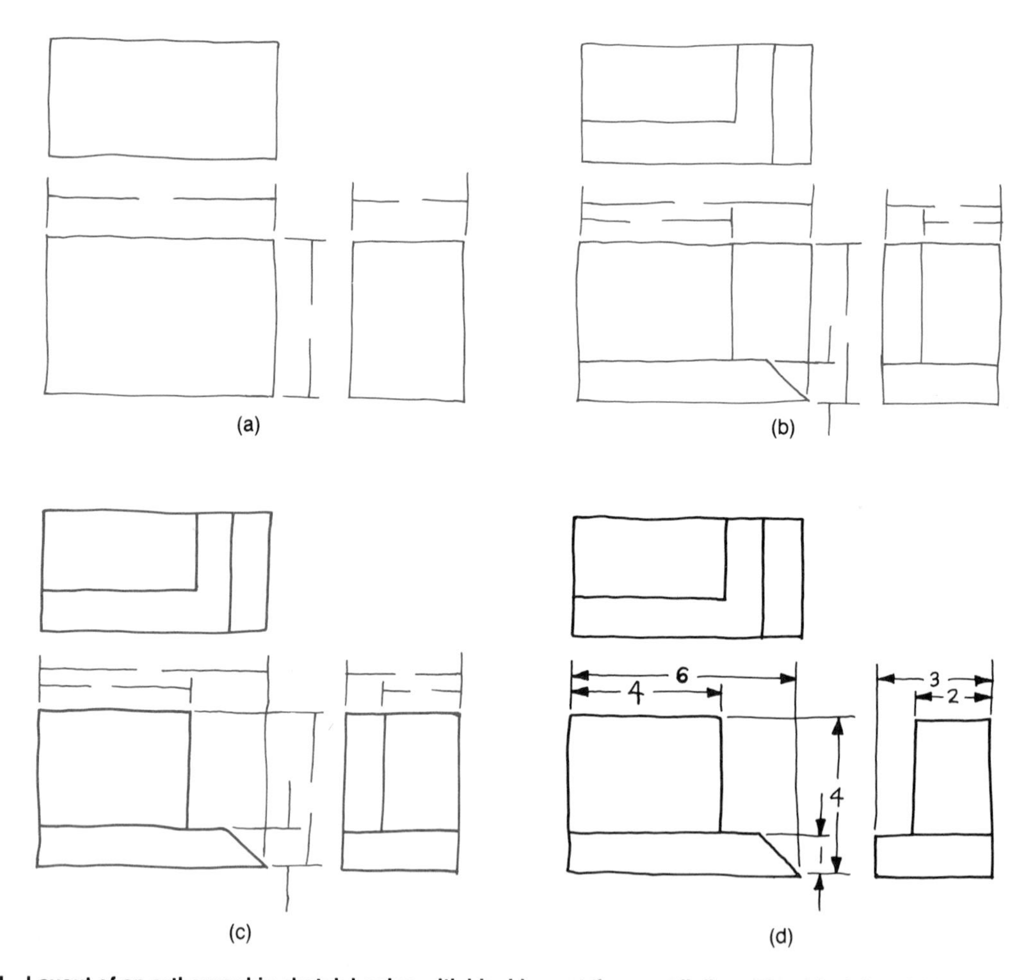

Figure 16.14 **Layout of an orthographic sketch begins with blocking out the overall dimensions (height, width, and depth) for each view (a), then adding interior linework with light construction lines (b), then darkening and finishing the surface boundaries (c), and finally adding dimensions and other details (d). The procedure is similar for pictorial sketches.**

In Retrospect

- Sketching (or freehand drawing) is a necessary skill for the engineer to graphically communicate ideas and design concepts with speed and without the aid of drawing instruments.
- Sketches should be carefully drawn. Intersections between lines should be clearly indicated. Finished linework should be consistent in line width and intensity.
- Printed grids can be helpful in maintaining consistency in the relative orientation of lines and the distances between lines in a sketch.
- Light construction linework should not be erased from a sketch unless it detracts from clarity or appearance.
- A straight line should be drawn by concentrating upon its endpoints. The relative direction of the line is more important in a sketch than any small irregularities contained within the line (although irregularities must be *minimized*).
- Common types of linework used in sketching include visible, hidden, center, and construction lines. Different types of lines are generated by modifying both the pressure applied to the pencil point and the sharpness of the pencil point.
- The drawing paper should be rotated to more easily produce inclined lines.
- Lines at specific angles (such as 45° or 60°) to the horizontal direction can be approximated by dividing the 90° angle between two perpendicular construction lines into appropriate segments.
- Arcs and circles can be created with the aid of construction centerlines and the use of a sheet of scratch paper on which is marked the (approximate) value for the radius of the arc or circle under consideration.
- Ellipses can be drawn to represent distorted circles in isometric sketches by using construction centerlines to form a rhombus from which the ellipse is then formed.
- The circular base of a cylinder (or cylindrical hole) will appear as an ellipse in an isometric sketch. The major axis of such an ellipse should be perpendicularly oriented with respect to the axial direction of the cylinder.
- Shading can be used in a sketch to enhance clarity and specify particular materials and surface textures.
- The procedure for creating orthographic and pictorial sketches includes (1) blocking out the general region of a view in accordance with the overall dimensions of the object, (2) drawing interior linework with construction lines, (3) darkening surface boundaries, and (4) adding dimensions and otherwise completing the finished sketch.

ENGINEERING IN ACTION

Built-in Locks Prevent Tool Loss

Design for dropproof hand tools is based on a labyrinth containing springs, plungers, balls, and detents.

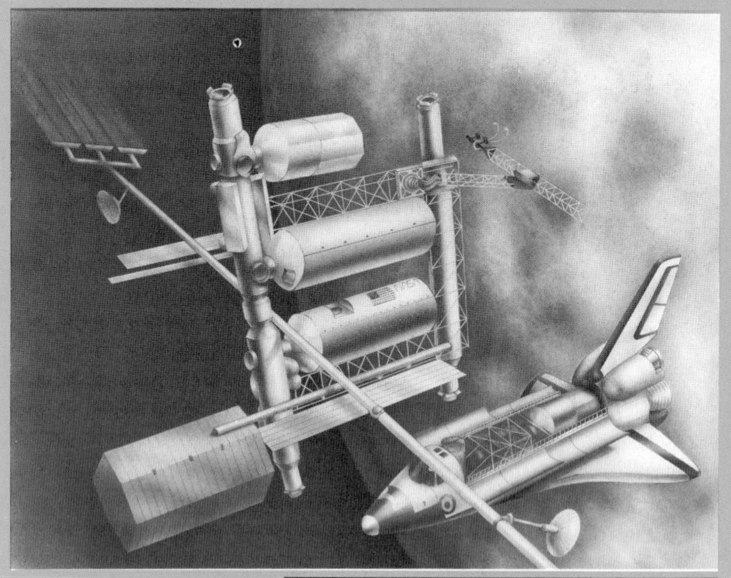

■ **Problem:** The loss of a component is inconvenient or hazardous when two-part hand tools are used on scaffolds, underwater, or in space and other high-risk work areas. In an explosive atmosphere, for example, a dropped part can generate sparks that ignite an explosion. Tethers attached to both parts prevent accidental dropping but interfere with relative motion between the parts.

■ **Solution:** Interlocking sets of spring-loaded plungers, spring-loaded balls, and ball detents require the continuous engagement of an untethered wrench socket. These elements are built into the handle, socket, and retaining-pin assemblies of a three-part wrench system. For example, the handle and socket can be joined or separated only when the retaining pin is inserted to depress a plunger in the socket housing. Otherwise, a ball in the housing is forced into its locked position. The retaining pin also contains a plunger-ball-detent assembly which must be actuated before the pin can be inserted in or removed from a drilled passageway in the socket mechanism. This arrangement permits testing of the handle and socket lock before disengaging the pin. The overall design requires a socket to be joined with either the handle or retaining pin at all times. Tethers attached to the handle and pin thus make the design dropproof and permit unencumbered motion of the socket. The wrench is an invention of Bruce McCandless of Johnson Space Center, Houston.

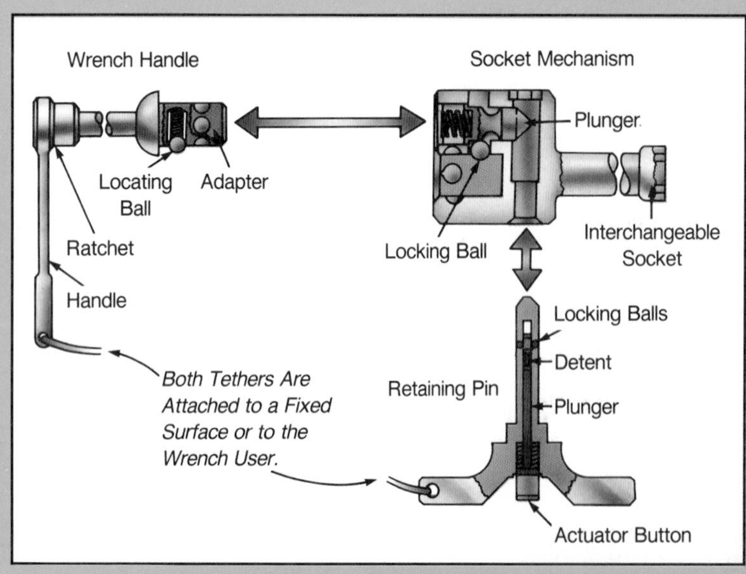

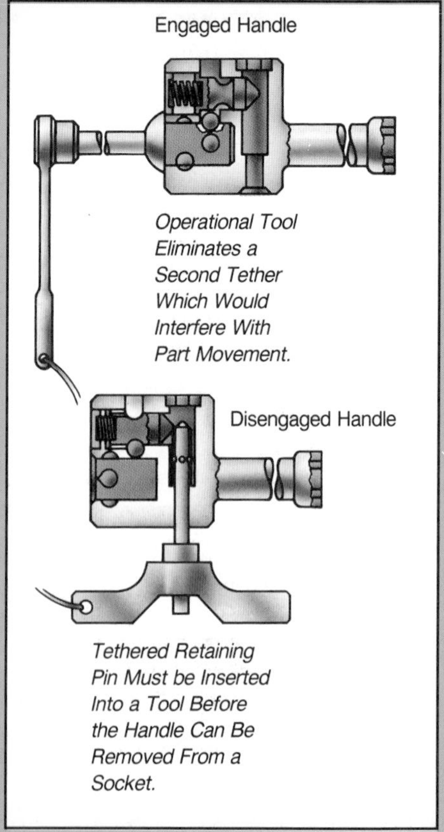

Source: Reprinted from MACHINE DESIGN, Sept. 22, 1983. Copyright 1983, by Penton/IPC., Inc., Cleveland, Ohio, p. 30.

Exercises

Your instructor will specify the appropriate scale for each exercise. Each exercise refers to a specific exercise in a preceding chapter. In each case develop a *sketch* of the object in accordance with the directions of your instructor, who will specify whether the sketch should consist of a pictorial description of the object, a fully dimensioned set of orthographic views, or so on.

EX16.1. See EX4.7.

EX16.2. See EX4.8.

EX16.3. See EX4.12.

EX16.4. See EX4.13.

EX16.5. See EX4.14.

EX16.6. See EX4.15.

EX16.7. See EX4.19.

EX16.8. See EX4.21.

EX16.9. See EX4.22.

EX16.10. See EX4.23.

EX16.11. See EX4.25.

EX16.12. See EX4.26.

EX16.13. See EX4.28.

EX16.14. See EX4.29.

EX16.15. See EX4.30.

EX16.16. See EX4.31.

EX16.17. See EX4.32.

EX16.18. See EX4.33.

EX16.19. See EX4.35.

EX16.20. See EX4.36.

EX16.21. See EX4.37.

EX16.22. See EX4.39.

EX16.23. See EX4.41.

EX16.24. See EX4.44.

EX16.25. See EX4.45.

EX16.26. See EX4.47.

EX16.27. See EX4.48.

EX16.28. See EX4.49.

EX16.29. See EX4.50.

EX16.30. See EX4.51.

EX16.31. See EX4.52.

EX16.32. See EX4.53.

EX16.33. See EX4.54.

EX16.34. See EX4.55.

EX16.35. See EX4.56.

EX16.36. See EX4.57.

EX16.37. See EX4.58.

EX16.38. See EX4.59.

EX16.39. See EX5.1.

EX16.40. See EX5.2.

EX16.41. See EX5.3.

EX16.42. See EX5.4.

EX16.43. See EX5.5.

EX16.44. See EX5.6.

EX16.45. See EX5.7.

EX16.46. See EX5.8.

EX16.47. See EX5.9.

EX16.48. See EX6.1.

EX16.49. See EX6.2.

EX16.50. See EX6.3.

EX16.51. See EX6.4.

EX16.52. See EX8.1.

EX16.53. See EX8.2.

EX16.54. See EX8.3.

EX16.55. See EX8.4.

EX16.56. See EX8.5.

EX16.57. See EX8.6.

EX16.58. See EX8.7.

EX16.59. See EX8.8.

EX16.60. See EX8.9.

EX16.61. See EX8.10.

EX16.62. See EX8.11.

EX16.63. See EX8.12.

17 Functional Graphs, Charts, and Diagrams

Preview Engineers and scientists very often represent the **functional relationships** between system parameters (such as temperature, velocity, force, torque, current, voltage, pressure, and volume flow rate) through the use of two- (or more) dimensional graphs. Such graphs are generated from experimental data that is used to construct a mathematical model (an equation or a set of equations) of the dependence of one system parameter upon another (independent and controlled) parameter. These functional graphs are a *critical* communication and analytical tool in both engineering and science.

This chapter initially focuses on the construction of functional graphs from given sets of experimental data, together with the development of appropriate equations describing the relationship between the variables appearing in such graphs. **Linear equations, power equations,** and **exponential equations** are the most common (simple) mathematical models of physical phenomena and are therefore discussed in detail. In addition, the use of **semilogarithmic** and **logarithmic** grid paper in the graphic analysis and interpretation of experimental data is explored. The application of functional graphs in engineering is both broad in scope and of great value; it should be mastered by the engineering student.

Engineers should also develop the ability to present a wide variety of data (for example, experimentally determined results, economic trends in the marketing of a product, characteristics of materials, and forces acting upon an object in motion) with *accuracy* and *clarity.* Such data may need to be presented to both technical and nontechnical audiences, so it must be in a form that can be quickly and easily interpreted. The focus of the presentation must be directed toward the most significant aspect(s) of the data, yet it must not be improperly manipulated to give a distorted view of the information. These considerations create a challenge in communication.

To meet the challenge, the engineer must be familiar with the wide variety of graphic formats that can be used to present information in clear, accurate, and easily understood terms. Different types of data should be presented in different formats in accordance with the mutual needs of the engineer and the audience. These graphic formats, or **charts,** are reviewed in this chapter.

Learning Objectives

Upon completion of this chapter, the reader should be able to:

- Represent the relationship between system parameters through the use of two-dimensional graphs with functional curves.
- Identify both the abscissa and the ordinate of a graph.
- Differentiate between multiple data sets that are included in a single graph by identification callout

letters and distinctive linework for curves representing such data sets.

- Judiciously choose the appropriate scales for a graph so that functional curves are not unnecessarily compressed and so that the relative amounts of given quantities in a graph are accurately represented.
- Insert lettering and numerical values in graphs in accordance with accepted practice (including title block information).
- Specify which curves in a graph are based upon experimental data points and which curves are generated from mathematical relationships between variables.
- Describe the behavior of physical phenomena through the use of functional curves and corresponding appropriate equations such as:

$$Y = b + mX \quad \text{(linear equation)}$$

$$Y = A(X)^J \quad \text{(power equation)}$$

and

$$Y = Ae^{JX} \quad \text{(exponential equation)}$$

- Apply both logarithmic and semilogarithmic paper in the graphic analysis and interpretation of experimental data.
- Use a wide variety of charting formats to effectively present data to both technical and nontechnical audiences so that it will be quickly and properly interpreted by members of both groups.
- Recognize that in the design and development of a chart one must evaluate: (1) the type of data to be presented, (2) the audience to be addressed, and (3) the relevant conclusions and relationships to be emphasized in the presentation.
- Apply line or curve charts to present the variation of one (dependent) parameter with that of another (independent) parameter.
- Differentiate between the various types of line charts that can be used to present information, including multiple-curve charts, multiple-amount scale charts, multiple-time charts, cumulative-curve charts, step or staircase charts, and histograms.
- Emphasize the size of the variation in a particular parameter through the judicious use of surface charts.
- Describe various types of column or vertical-bar charts, horizontal-bar charts (including the 100-percent sliding chart and the paired-bar chart), and the pie chart.
- Appreciate the danger associated with the removal of zero coordinate positions from a chart (although such a reduction in the scales of a chart may be attractive since it does conserve space).

17.1 Experimental Data and Graphs

The graphic presentation of data is a valuable communication tool for accurately and efficiently describing the behavior of physical phenomena. The functional dependence of a particular variable (such as temperature) on a system parameter (such as pressure or volume) may be described by a two-dimensional graph like the one shown in Figure 17.1. Each axis of the graph represents one of the quantities under consideration—one axis for the independent or controlled variable (denoted as X in the figure) and one axis for the dependent variable (denoted as Y). The horizontal axis (known as the **abscissa**) is commonly used to represent the independent variable (X), and the vertical axis (known as the **ordinate**) is commonly used for the dependent variable (Y).

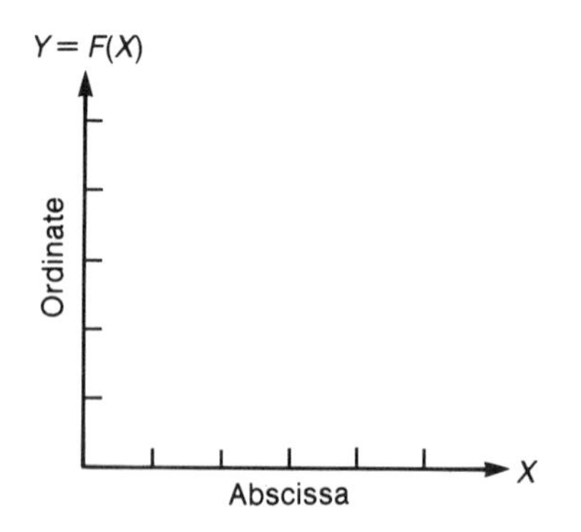

Figure 17.1 Axes of a functional (two-dimensional) graph. The independent variable is denoted as *X* and the dependent variable as *Y*. The abscissa (horizontal axis) is commonly used for the independent axis and the ordinate (vertical axis) for the dependent variable.

Empirical relationships between the variables of a system are often determined through a series of experiments in which different sets of data are generated. These data sets are frequently displayed in a single graph to emphasize similarities or differences among the experimental results. Such **multiple data sets** in a single graph must be clearly identified by **callout letters or callout numbers** (together with a table that keys these letters or numbers to the appropriate conditions under investigation) and/or by **distinctive linework** for each (such as solid lines for one curve and dashed lines for another; see Figure 17.2).

In summary then:

1. Two-dimensional graphs may be used to represent the functional dependence of a variable upon the value of another (controlled and independent) variable of a system.
2. The horizontal axis is known as the abscissa; the vertical axis is called the ordinate.
3. The abscissa is traditionally used to represent the independent variable.
4. Multiple data sets in a single graph should be identified by callout letters and distinctive linework for each.

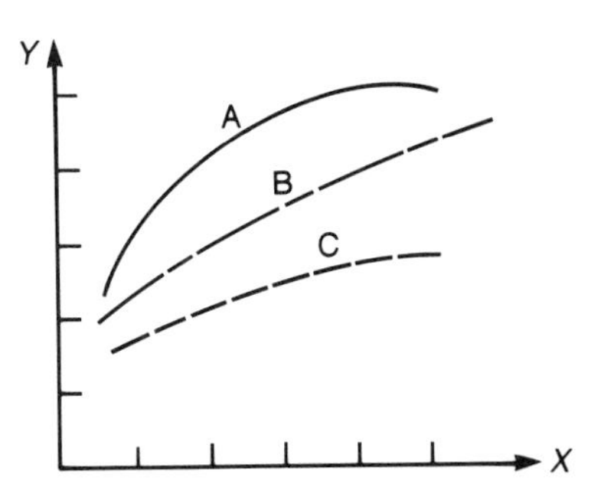

Figure 17.2 Multiple data sets (in this case curves) in a graph require distinct linework for each and/or identification notation by callout letters (as shown here) or callout numbers.

17.2 Scales, Lettering, and Title Blocks

The choice of **scales** is a critical decision in the development of a functional graph. A poor choice may lead the observer to totally incorrect conclusions about the data. One consideration in selecting scales is that they not result in unnecessarily compressed or extended functional curves. Figure 17.3 presents the preferable range for functional curves in a graph (a) and two linear (straight-line) graphs that have been unduly compressed by poor scaling practices.

Another consideration in selecting scales is to avoid confusing the observer about the *relative* values of any quantities presented in a graph. In Figure 17.4 the average annual salaries of employees in a fictitious company entitled AA Engineering Corporation are presented for comparison purposes. The first graph (a) fails to accurately represent the relative salaries for graduates in each of three engineering disciplines due to the absence of a **full scale**—the scale does not begin at zero. Because the second graph (b) includes a full scale—the scale begins at zero—the reader can see how the three salaries compare.

Horizontal scales traditionally increase from left to right, whereas vertical scales increase from bottom to top. Lettering and numerical values should be entered for reading from the bottom or from the right side (as in dimensioning of orthographic views; see Chapter 8). Scales are placed outside the graphic area for greatest clarity. In addition, the caption for a scale should include both the *unit of measurement* used and an unambiguous specification of the particular *variable* under consideration. The number of digits used in a scale

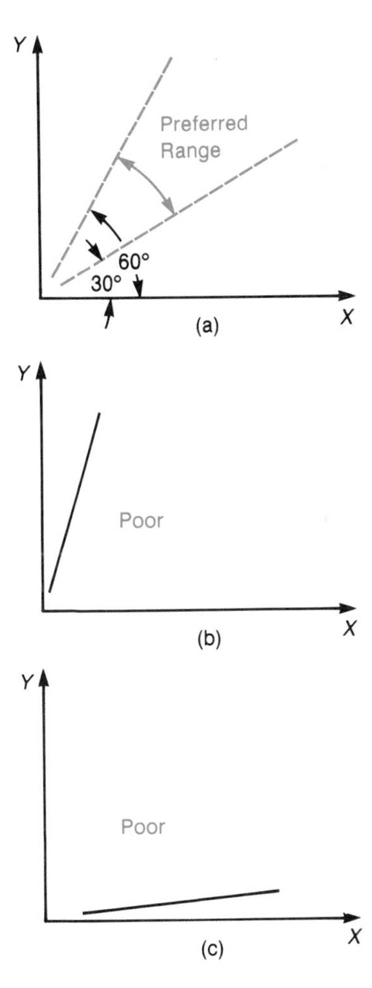

Figure 17.3 Scales should be chosen that allow functional relationships to be easily discerned. The preferred range (a) avoids undue compression or extension of functional curves (b and c).

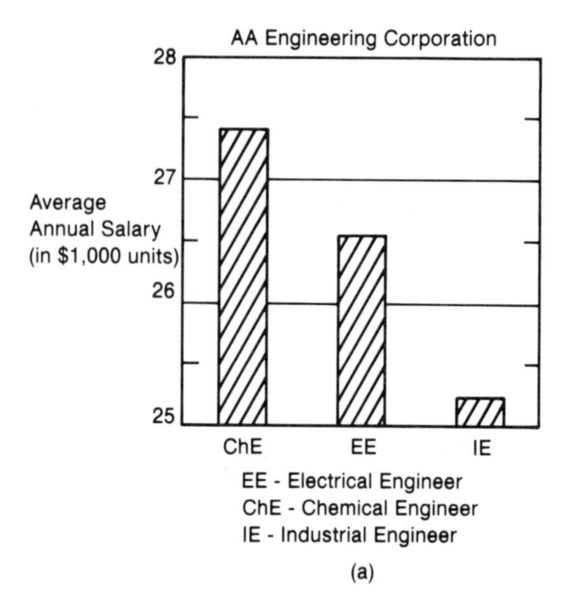

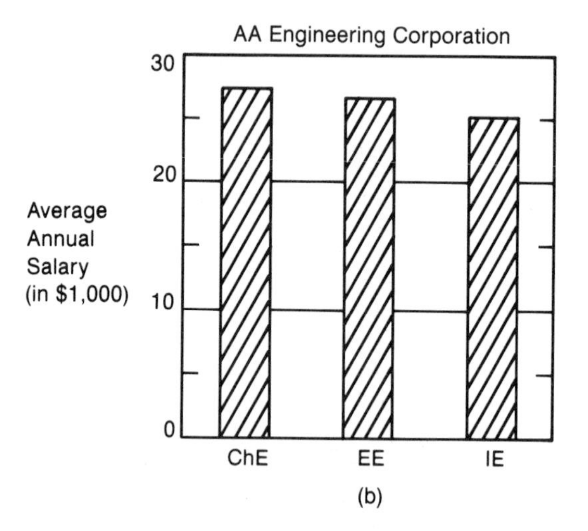

Figure 17.4 Accurate visual comparison of relative amounts requires a full scale. Since the first graph (a) does not begin at zero, its representation is inaccurate. The second graph (b) presents an accurate representation because it includes a full scale.

should be minimal; only significant digits are needed at coordinate rulings. A multiplication factor (for example, $x10^4$) should be introduced in the unit of measurement specification in order to minimize the number of digits used at coordinate rulings.

A **title block** should be included with a graph in which all additional information (such as source of data, date, and experimental specifications) is presented. A border should be placed around this information to clearly separate it from the other portions of the graph.

In summary, then:

1. Scales should be chosen so that functional curves are not unnecessarily compressed.
2. Full scales should be used to accurately represent relative amounts of given quantities.
3. Horizontal and vertical scale values increase from left to right and bottom to top, respectively, in a traditional layout.
4. Lettering and numerical values should be read from the bottom or right side of the graph.
5. Scales are normally placed outside the graphic region.
6. Scales should include the identification of both the units of measurement and the particular variables under consideration.
7. The number of digits entered at coordinate rulings in a graph should be minimized through the use of a multiplication factor.
8. A title block should include all the additional information needed to completely identify the graph, its data, and its source. A border should be used to separate the title block from the main body of the graph.

17.3 Experimentally Determined Data and Theoretical Curves

Functional graphs should clearly specify the basis for a curve. Experimentally determined data points should be denoted by circles, crosses, or other **point markers,** as shown in Figure 17.5(a). Theoretical curves (generated through the use of equations or mathematical relationships between the variables) are drawn without the inclusion of plotted data points (b).

17.4 Linear Graphs

Linear graphs, as the name implies, depict proportional (straight-line) dependence of one variable upon another. Figure 17.6 presents a sample linear graph in which the dependence of the variable Y is given as a function of X in accordance with the following **linear equation** (the most important equations will be numbered to facilitate classroom instruction):

$$Y = b + mX \qquad \text{EQ. 17-1}$$

where b represents the **intercept** of the equation and m denotes the **slope** of the line. The intercept represents the value of Y when X is set equal to zero (that is, b is—literally—the intersection of the straight line with the Y-axis, or ordinate). The slope of the line represents the relative vertical rise, ΔY, with respect to a corresponding change, ΔX, in the independent variable; that is:

$$m \equiv \frac{\Delta Y}{\Delta X} \qquad \text{EQ. 17-2}$$

where the **rise,** ΔY, is given by:

$$\Delta Y \equiv Y_2 - Y_1$$

and the **run** ΔX is defined by:

$$\Delta X \equiv X_2 - X_1$$

(where Y_1, Y_2, X_1, and X_2 are the coordinate values of specific data points on the straight line, as shown in Figure 17.6). Linear equations can be generated from experimental data points through the calculation of the rise and the run associated with these points (that is, under the assumption that the data do indeed behave in accordance with a linear equation).

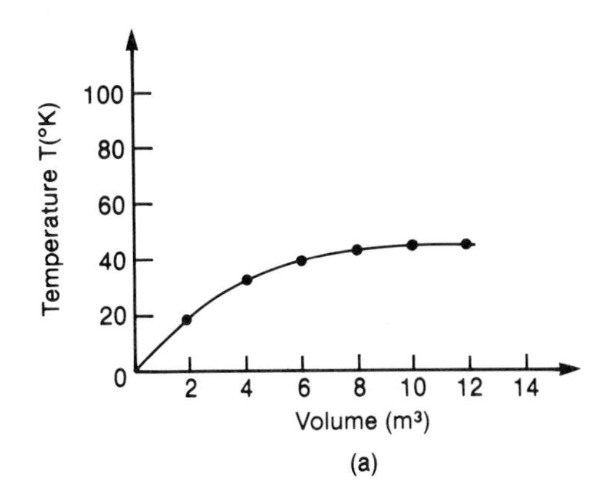

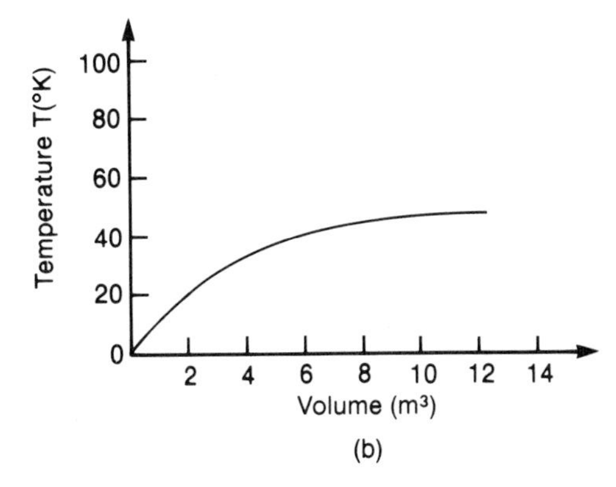

Figure 17.5 An experimentally determined curve is indicated by the inclusion of data points (a). A theoretical curve does not have data points (b).

17.5 Power Equation

The dependence of a variable, Y, upon another quantity, X, may not be simply linear; instead, the behavior may be represented by a **power equation** of the form

$$Y = A(X)^J \qquad \text{EQ. 17-3}$$

where A represents a constant coefficient and J denotes the specific value of the **power dependence** of Y upon X. Linear dependence corresponds to a value of unity for J ($J = 1$); parabolic dependence corresponds to $J = 2$, and so on. Figure 17.7 presents several power curves for various values of J. (As an example of a parabolic dependence of one variable upon another, consider the behavior of a falling body within a gravitational field: the vertical distance Y through which the body travels is given by

$$Y = \frac{1}{2}gt^2 + v_0 t \qquad \text{EQ. 17-4}$$

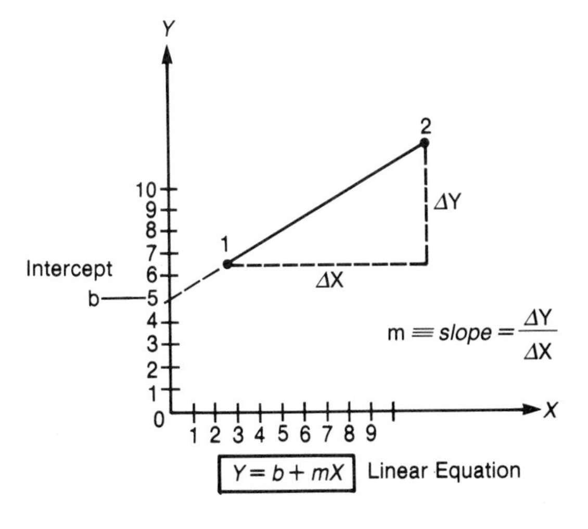

Figure 17.6 Linear (straight-line) behavior.

Highlight

Euclid, Descartes, and Analytic Geometry

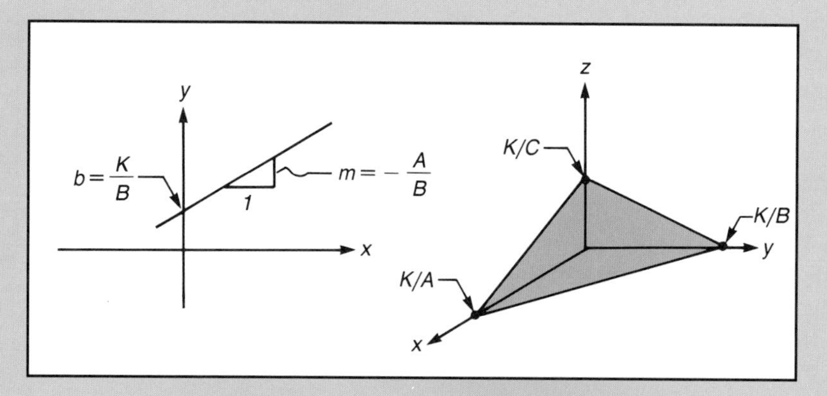

Euclid (325–270 B.C., approximately) provided a succinct presentation of axioms and postulates in his textbook, *Elements,* which formed the basis for Euclidean geometry. Euclid's geometry was used to describe objects with straight lines and circles. Conic sections (such as ellipses and parabolas) were known in Euclid's time, but not well understood until the work of René Descartes (1596–1650) combined geometry with algebra to form **analytic geometry.**

Descartes (known as the father of modern philosophy) recognized that one could use a coordinate system based on three mutually perpendicular planes to describe the location of a point in three-dimensional space. Furthermore, he realized that curves in space could then be represented algebraically in terms of the (x, y, z) coordinates of each point on the curve. (These x, y, z coordinates are known as **Cartesian** in honor of Descartes.) For example, the equation

$$Ax + By = K \quad \text{EQ. 1}$$

represents a **straight line** in space in terms of a set of specific constants A, B, and K. Rewriting this expression in the form

$$y = \frac{K}{B} - \frac{A}{B}x \quad \text{EQ. 2}$$

we see that it is indeed equivalent to the equation

$$y = b + mx \quad \text{EQ. 3}$$

if we identify the slope m and the intercept b for this line as

$$m = -\frac{A}{B} \quad \text{EQ. 4a}$$

and

$$b = \frac{K}{B} \quad \text{EQ. 4b}$$

Similarly, a **plane** in space is represented by the equation

$$Ax + By + Cz = K \quad \text{EQ. 5}$$

(where A, B, C, and K are specific constants). We know that three points in space can also be used to define a unique plane containing these points. For example, the three points of intersection between the plane represented by the equation

$$Ax + By + Cz = K$$

and the (x, y, z) axes can be used to define this plane. To find the intersection of this plane with the x-axis, simply set both y and z equal to zero in the above equation (thereby restricting our attention to the x-axis), so that

$$x = \frac{K}{A} \quad \text{EQ. 6}$$

The intersections of this plane with the y and z axes are similarly given by

$$y = \frac{K}{B} \quad \text{EQ. 7}$$

$$z = \frac{K}{C} \quad \text{EQ. 8}$$

Kepler showed that the orbital paths of planets about the sun are elliptical. In addition, Galileo showed that the path of a cannonball is parabolic. Descartes' work in "coordinate geometry" allowed the mathematical representation of ellipses, parabolas, and other conic sections (not allowed by Euclidean geometry), thereby providing the tools needed for the development of the calculus by Newton and Leibniz—and the subsequent formulation of many scientific principles upon which engineering is based. In our modern world, algebraic and geometric representations of curves are used in computer-aided design (CAD) systems to create, analyze, and fabricate engineering products. Descartes' contributions to mathematics and geometry—and thus to science and engineering—are enormous.

where v_0 represents the initial vertical velocity of the falling body, g is the local acceleration constant due to gravity, and t denotes the independent variable of time. In the absence of an initial velocity ($v_0 = 0$), this expression is identical to the general power equation when $J = 2$, and $A = \frac{1}{2}g$. In other words, there is a parabolic dependence of the distance Y upon the time t.)

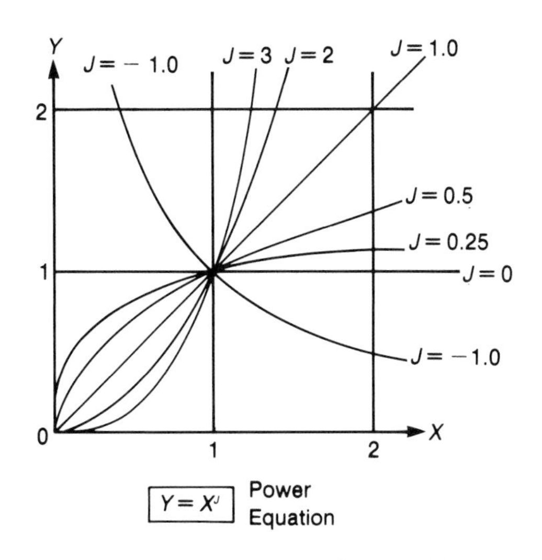

Figure 17.7 Behavior in accordance with power equations.

17.6 Rectification of a Power Equation

A power equation of the form shown in Figure 17.7 can be transformed (or **rectified**) into an equivalent linear expression through a calculation of the logarithms for each side of the equation. For example, the power equation

$$Y = A(X)^J \qquad \text{EQ. 17-5}$$

upon rectification, becomes

$$\log(Y) = \log(A) + J\log(X) \qquad \text{EQ. 17-6}$$

where the logarithm of a number (for the base 10) is defined by the relations

$$B = (10)^C \qquad \text{EQ. 17-7}$$

and

$$\log(B) = C \qquad \text{EQ. 17-8}$$

Equation 17-6 is a linear equation of the form

$$y = a + Jx \qquad \text{EQ. 17-9}$$

where

$$y \equiv \log(Y) \qquad \text{EQ. 17-10a}$$

$$a \equiv \log(A) \qquad \text{EQ. 17-10b}$$

and

$$x \equiv \log(X) \qquad \text{EQ. 17-10c}$$

Rectification of a power equation can be accomplished graphically through the use of logarithmic (log-log) paper. A power equation is rectified through the calculation of logarithmic values for points along the curve (Figure 17.8), which may then be plotted on a two-dimensional linear

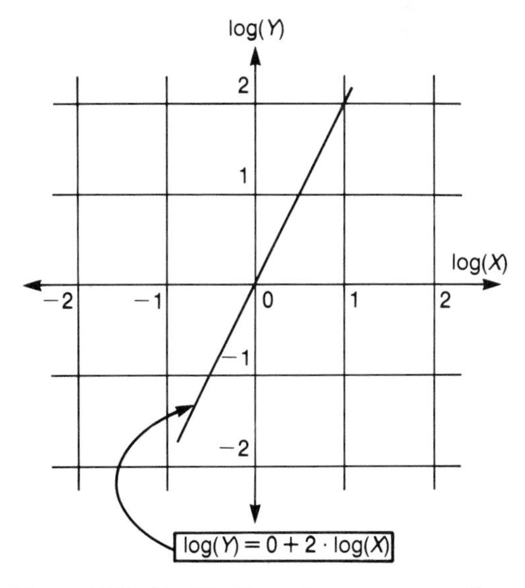

Figure 17.8 Rectification of a power equation produces linear relationships among logarithms.

graph (Figure 17.9); that is, one calculates and plots such values as

$$0.3 = \log(2)$$

$$0.601 = \log(4)$$

$$1 = \log(10)$$

$$1.477 = \log(30)$$

$$2 = \log(100)$$

The use of logarithmic paper (which is commercially available) eliminates the need to actually calculate the logarithmic values because the scales of such paper are given in terms of the arguments of the logarithms (see Figure 17.10). As a result, one may plot *directly* the logarithms of such values as 2, 4, 10, 30, and 100 on this type of paper.

Experimental data, if plotted on logarithmic paper, can be used to determine if the functional dependence between two variables should be expressed in the form of a power equation—that is, if the logarithmically plotted data points

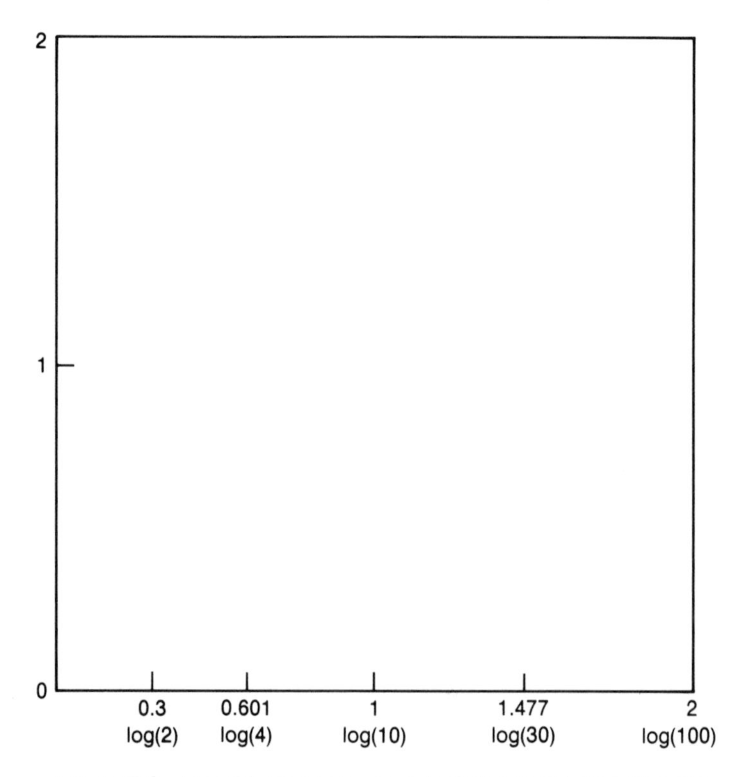

Figure 17.9 Logarithmic values are calculated and plotted along linear scales.

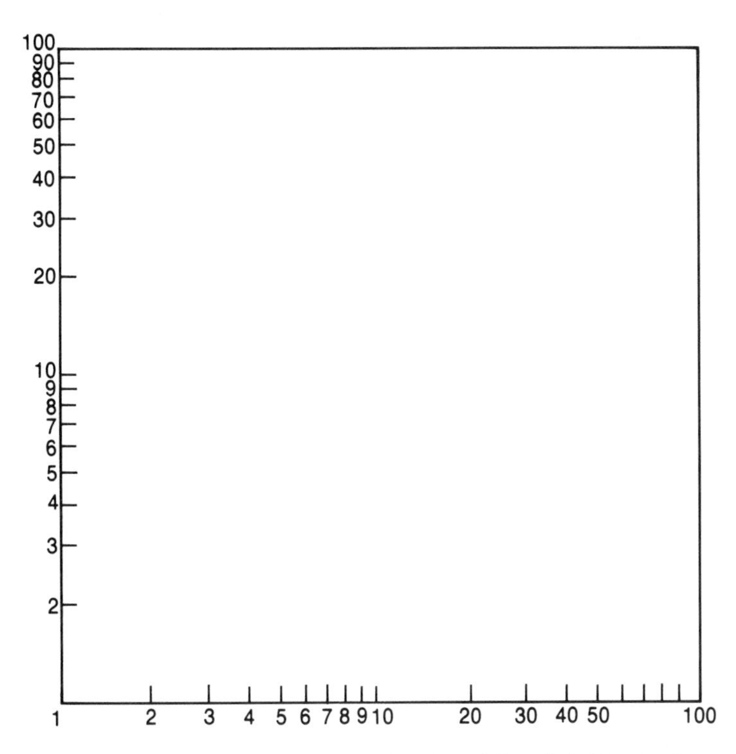

Figure 17.10 Logarithmic values may be plotted directly (without calculations) on standard log paper.

follow a straight-line path, the behavior can be represented by a power equation. In addition, the slope of a straight-line plot on logarithmic paper corresponds to the power J for the functional relation between the variables.

In summary, then:

1. Power equations may be rectified into equivalent linear forms through the use of logarithmic (log-log) paper or via calculation of the logarithms of sample data points.

2. The slope of a logarithmic plot of experimental data (which behaves in accordance with a power equation) corresponds to the value J for the power in the functional relationship between variables.

17.7 Exponential Equations

Functional dependence between variables Y and X may be exponential, such as

$$Y = Ae^{JX} \quad \text{EQ. 17-11}$$

or, alternatively,

$$X = Ae^{JY} \quad \text{EQ. 17-12}$$

Highlight

Analytic Geometry and Algebraic Manipulation

We know that algebraic expressions represent geometric elements; for example, the equation

$$Ax + By + Cz = K \qquad \text{EQ. 1}$$

(where A, B, C, and K are specific constants) describes a plane in terms of Cartesian coordinates x, y, and z (Figure 1).

Similarly, the *manipulation* of algebraic equations can be interpreted geometrically. For example, if we consider the plane represented by the expression

$$Dx + Ey + Cz = L \qquad \text{EQ. 2}$$

(where D, E, C, and L are constants—see Figure 2), one possible manipulation of Equations 1 and 2 involves the subtraction of Equation 2 from Equation 1, thereby giving

$$(A-D)x + (B-E)y = K-L \qquad \text{EQ. 3}$$

As seen in Figure 3, Equation 3 corresponds to the *projection* of the line of intersection between the two given planes onto the x, y coordinate plane. We eliminated the z-dependence of the intersection line via subtraction of Equations 1 and 2, thereby effectively projecting this line onto the x, y plane.

When working with algebraic expressions, we should remember that a geometric interpretation of our work is possible and may, in fact, clarify the meaning of our results.

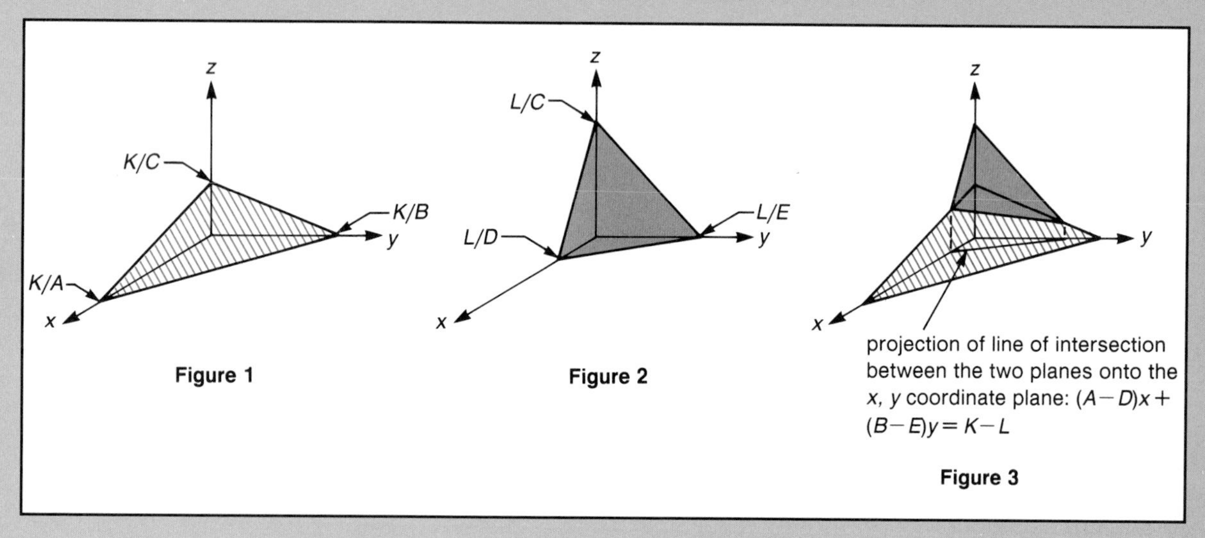

Figure 1

Figure 2

Figure 3

(see Figure 17.11). A general mathematical form corresponding to exponential behavior is

$$Y = A(B)^X \qquad \text{EQ. 17-13}$$

where the base B must be specified. If B equals 10, we then obtain

$$\log(Y) = \log(A) + X\log(10)$$

$$= \log(A) + X(1) \qquad \text{EQ. 17-14}$$

or equivalently,

$$y = a + X \qquad \text{EQ. 17-15}$$

where

$$y \equiv \log(Y)$$

and

$$a \equiv \log(A)$$

Equation 17-15 may be plotted directly on semilogarithmic paper, in which one scale is given in terms of logarithmic values (base 10) of Y with the other scale in units of the independent variable X (Figure 17.12). If B is equal to e^J, where J denotes the instantaneous percentage of growth in the dependent variable, and e represents Euler's number given by

$$e \equiv \text{Euler's number} \simeq 2.71828$$

the functional dependence of one variable upon another is given by Equation 7-11.

If we calculate the logarithms for each side of Equation 7-11, we obtain

$$\ln(Y) = \ln(A) + JX \ln(e) = \ln(A) + JX(1)$$

or

$$y = a + JX \qquad \text{EQ. 17-16}$$

where

$$y \equiv \ln(Y)$$

$$a \equiv \ln(A)$$

and

$$1 = \ln(e)$$

where the notation $\ln(Y)$ indicates that the logarithm *to the base, e,* is to be calculated. Equation 17-16 is a linear expression relating y to X; a linear plot may be constructed directly if one uses semilogarithmic paper (base e). Such a plot will then provide values for a (the intercept) and J (the slope), from which A may then be determined for the exponential

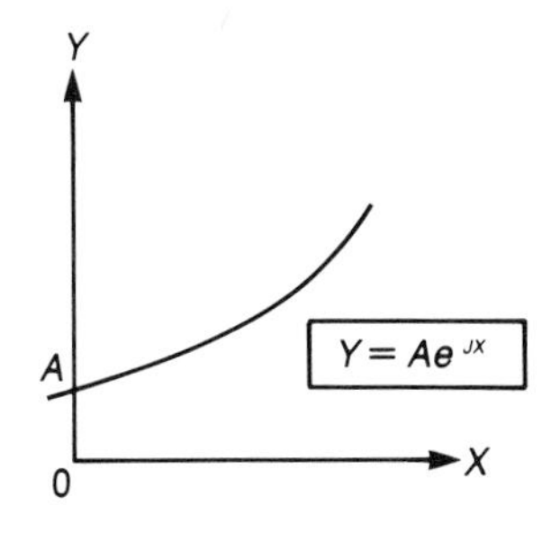

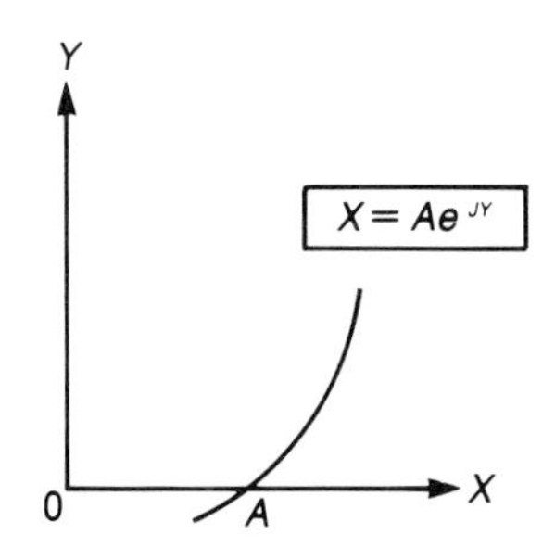

Figure 17.11 Behavior in accordance with exponential equations.

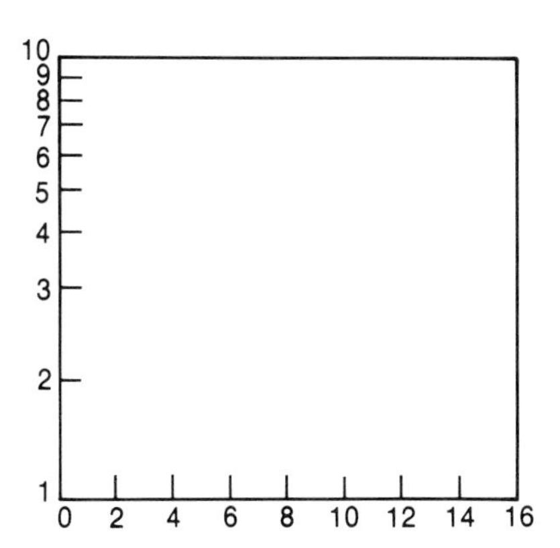

Figure 17.12 Exponential curves may be transformed graphically into equivalent linear functions with the use of semilogarithmic paper.

expression in Equation 17-11. A similar determination of the constants appearing in Equation 17-12 may be performed with semilogarithmic paper.

Notice that an expression of the form given in Equation 17-11 indicates that the functional curve intercepts the Y-axis at $Y = A$; a functional curve that is described by an expression of the form given in Equation 17-12 corresponds to an intercept of the X-axis at $X = A$.

In summary, then:

1. Exponential behavior may be represented by a equation of the form

$$Y = A(B)^X$$

If B is equal to e^J, where e is Euler's number, the functional behavior is described by

$$Y = A\, e^{JX}$$

If the exponential curve intercepts the X-axis at $X = A$, the appropriate equation is of the form

$$X = A\, e^{JY}$$

2. Semilogarithmic paper allows one to plot data points directly without calculating any logarithms. Such plots—if linear—may then be used to determine the constants (A and J) in the exponential expression that describes the functional relation between system variables.

17.8 Chart Design

Proper chart design requires that the engineer evaluate (1) the type of data to be presented in the chart, (2) the audience to be addressed, and (3) the relevant relationships and conclusions to be emphasized. *The message of a chart must be clearly and simply expressed.* Lettering on the chart should not distract the viewer—for example, by varying unnecessarily in style. The title should succinctly state the subject under consideration. Clarifying details (such as dates, and axes' identification labels and scales) should be presented in smaller type than that used for the title and should include sufficient information to avoid ambiguity.

17.9 Line (or Curve) Charts

The **line chart,** also called a **curve chart,** presents the variation of one (dependent) parameter with that of another (independent) parameter. Rectangular coordinate axes are used for the base (or scaling) lines of the parameters. Tic marks can be used to indicate specific values along these axes. Data points are plotted and then connected by solid lines to reveal any variation in the behavior of the system under consideration.

A variety of line charts are possible. A standard line chart, showing the variation of one parameter with respect to another parameter, is presented in Figure 17.13. A **multiple-curve chart** shows the relative variation of two or more parameters with respect to a third (Figure 17.14). **Multiple-amount scale charts** are similar to multiple-curve charts except that, as the name implies, multiple scales or axes are included to allow the presentation of curves differing significantly in size and/or units (Figure 17.15). (This type of chart can be misinterpreted by the casual viewer who fails to recognize the use of multiple scales; therefore, it should be used with caution.)

Multiple-time charts allow one to economically present data varying over different time periods of similar length. For example, twelve-month variations in data for several different yearly intervals can be included in a single chart for economy and ease in comparison (Figure 17.16).

A **cumulative-curve chart** emphasizes the accumulation in the value of a parameter during a given time interval (Figure 17.17). **Step, or staircase, charts** and **histograms** allow presentation of data that has been collected over distinct intervals of the independent parameter (which is usually time). As a result, any changes in the dependent variable's value are shown as distinct and discontinuous. A smooth-curve chart, such as that shown in Figure 17.13, cannot be used here because it would imply that any change is continuous and that it is based upon data collected at distinct values of the independent variable rather than over a finite period or interval, as is actually the case. A histogram includes distinct bound-

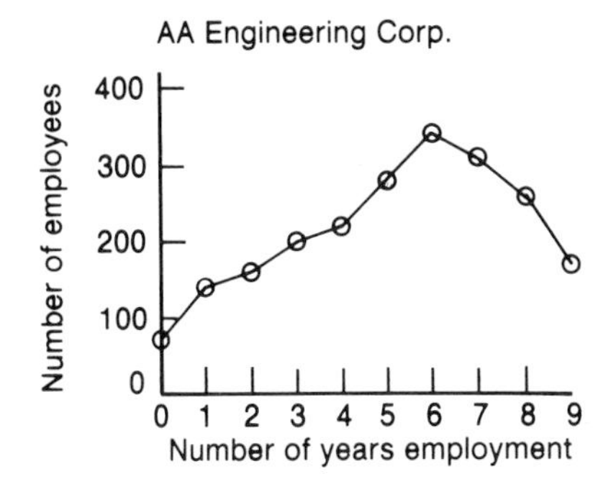

Figure 17.13 A standard line chart shows the variation of one parameter with respect to another.

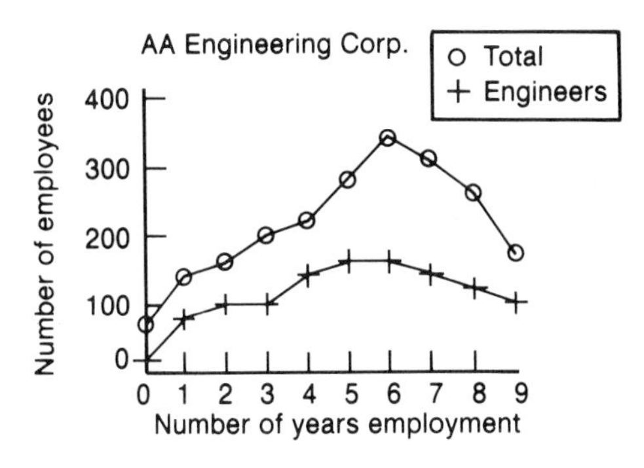

Figure 17.14 A multiple-curve chart shows the relative variation of two or more parameters with respect to a third.

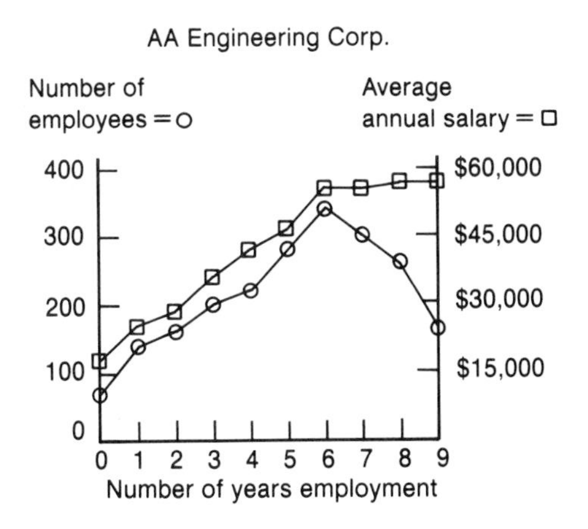

Figure 17.15 A multiple-amount scale chart allows the presentation of curves differing significantly in size and/or units.

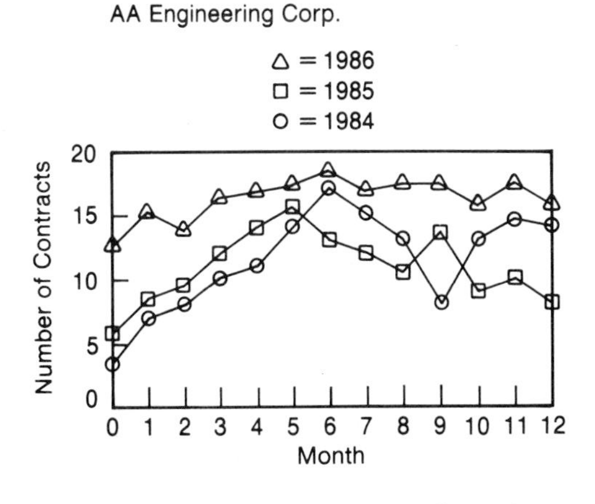

Figure 17.16 A multiple-time chart allows the economic presentation of data varying over different time periods of similar length.

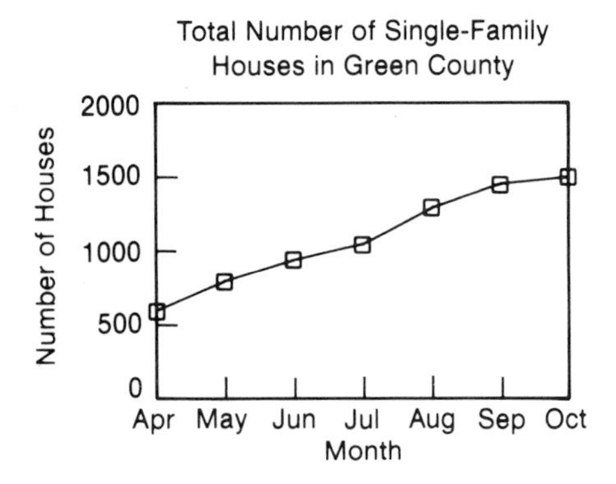

Figure 17.17 A cumulative-curve chart shows the accumulation in the value of a parameter during a given time interval.

ary lines between intervals, leading to a column format (see Figure 17.18).

As mentioned before, a common error in the use of charts is the removal of a zero coordinate position along one or both scales (recall the discussion in Section 17.2). In some instances, the absence of such zero coordinate positions does not adversely affect the reader's interpretation of the data and, in fact, conserves space in the charting region. In many cases, however, the absence of zero coordinate positions in a chart leads the viewer to an incorrect conclusion (refer again to Figure 17.4).

17.10 Surface Charts

A **surface chart** is similar to a line chart in its appearance and application; however, it emphasizes the size of the variation in a particular parameter by filling in the area below the curve with crosshatching, shading, photographs, and other highlighting patterns. A 100-percent surface chart is shown in Figure 17.19; notice how the relative sizes of the parameters are clearly and immediately apparent in this type of chart.

17.11 Column or Vertical-Bar Charts

A chart similar in appearance to the histogram and used to present numerical values of a parameter during particular intervals of the independent variable is the **column chart** or **vertical-bar chart.** Values of the dependent parameter are represented by relative heights of the column (or bar). The finite spacing between columns (or bars) enhances the clarity of the chart and emphasizes the intervals of the independent parameter during which data is obtained.

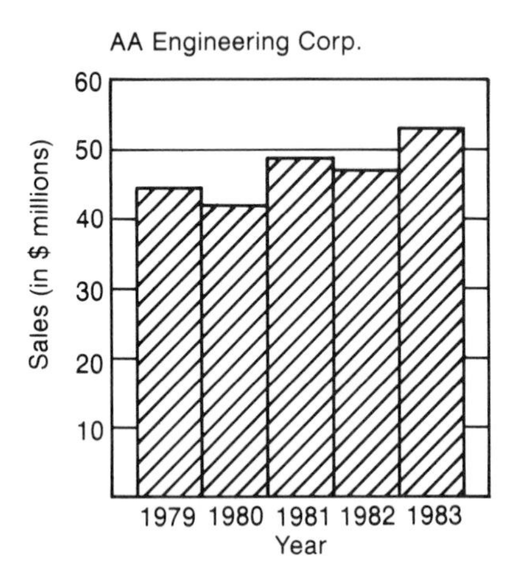

Figure 17.18 A histogram allows the presentation of data that has been collected over distinct intervals of the independent parameter (usually time).

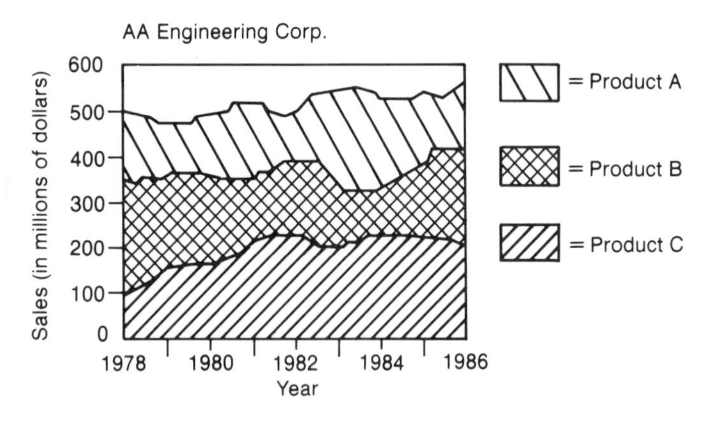

Figure 17.19 A surface chart emphasizes the size of the variation of a parameter by filling in the area below the curve with some type of highlighting pattern.

A standard column chart is shown in Figure 17.20. Another type is the **grouped-column chart,** in which comparisons between different sets of data are facilitated by distinctive identification patterns or colors for each data set, together with an identification legend for each pattern (Figure 17.21).

The **subdivided-column chart** clearly specifies the variation of different portions of a total data set with respect to an independent parameter. An identification pattern or color, together with an accompanying key or legend, is used to distinguish each portion of the whole under consideration (Figure 17.22).

17.12 Horizontal-Bar Charts

Horizontal-bar charts are used when the relative sizes of different data sets must be compared but one or more of the values is too large in comparison to the other values to be shown in a vertical-bar (or column) chart. (Available space in charts should be used as economically as possible without unnecessary or confusing concentration of information.)

The width between horizontal bars is usually one-half of the bar width. Spacing between bars should be chosen so that information is neither crowded nor difficult for the viewer to correlate.

A standard horizontal bar chart is shown in Figure 17.23. In a **grouped-bar chart,** which is a horizontal version of a grouped-column chart, the variations of two or three different sets of data are compared (Figure 17.24). In a **subdivided-bar chart,** which is a horizontal version of a subdivided-

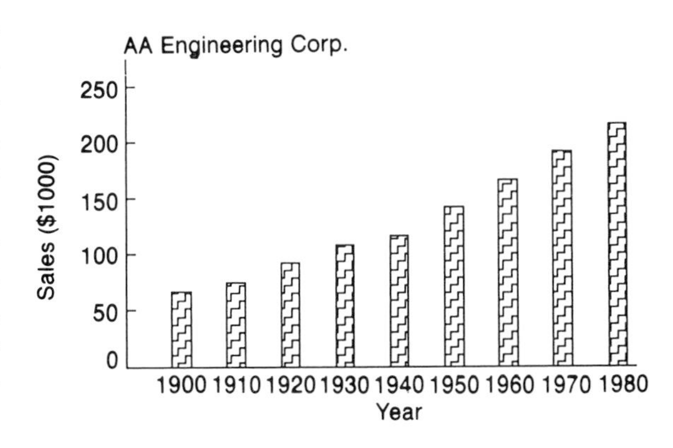

Figure 17.20 A basic column (or vertical-bar) chart shows numerical values of a parameter during particular intervals.

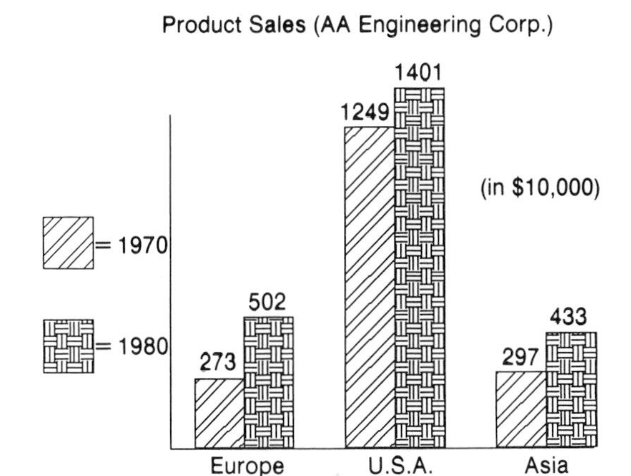

Figure 17.21 A grouped-column chart uses distinctive identification patterns of colors for each data set.

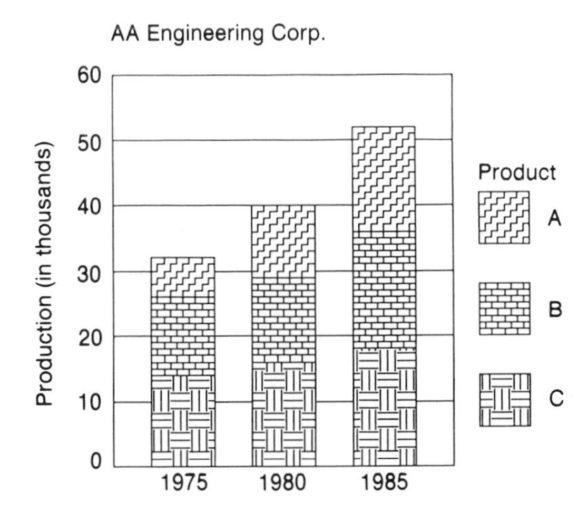

Figure 17.22 A subdivided-column chart uses distinctive identification patterns or colors for each portion of a data set.

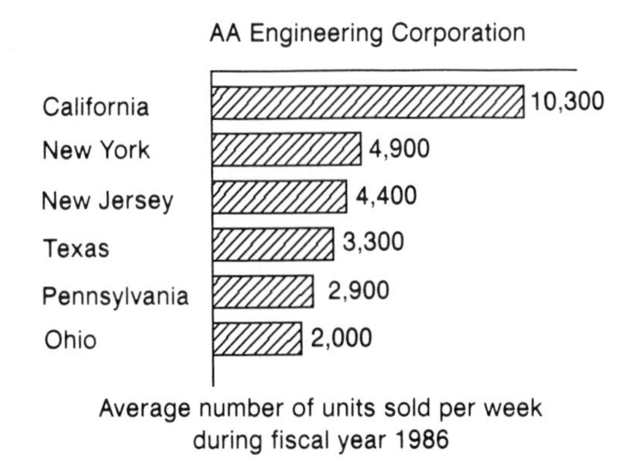

Figure 17.23 A horizontal-bar chart is used when one or more data sets to be compared is too large in comparison to the other values to be shown in a vertical-bar chart.

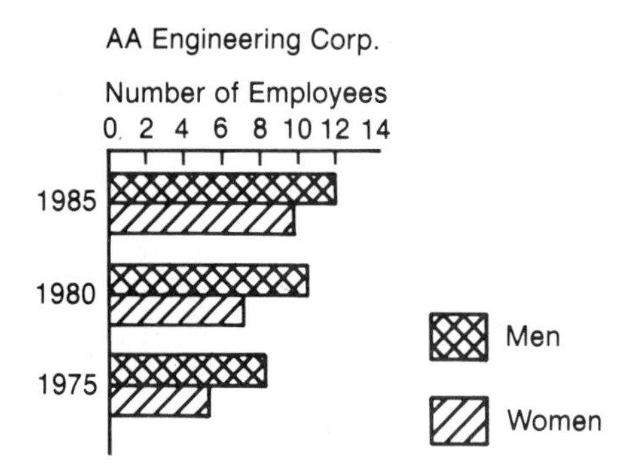

Figure 17.24 A grouped-bar chart is a horizontal version of a grouped-column chart (see Figure 17.21).

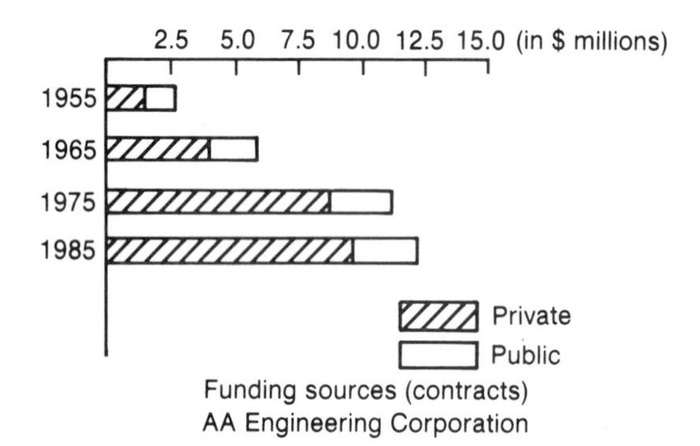

Figure 17.25 A subdivided-bar chart is a horizontal version of a subdivided-column chart (see Figure 17.22).

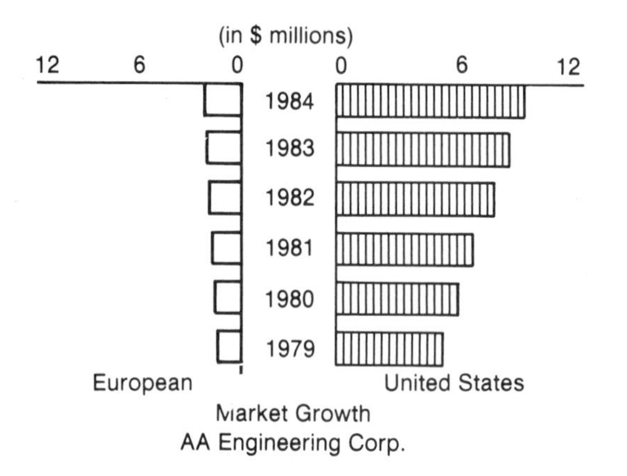

Figure 17.26 A paired-bar chart allows comparison of the variations of coupled sets of data with respect to two independent parameters.

column chart, the relative amounts of the components forming a whole are presented (Figure 17.25). In a **paired-bar chart,** coupled sets of data are compared in terms of their variation with respect to two independent parameters (Figure 17.26).

In the **100-percent sliding-bar chart,** the length of the bar corresponds to 100 percent of a quantity; the bar is slid horizontally with respect to a base line or reference position in order to indicate the relative portions of the quantity with respect to two independent parameters (Figure 17.27).

17.13 Pie Charts

Another popular chart that can be used to compare relative portions (percentages) of a given whole is the **pie chart** (Figure 17.28). It is composed of a circle divided into appropriate segments. The pie chart is an effective tool if the data are divided into a *small number* of segments. However, a bar chart is preferable if the number of components is large due to difficulties in labeling (and reading) small segments of a pie chart.

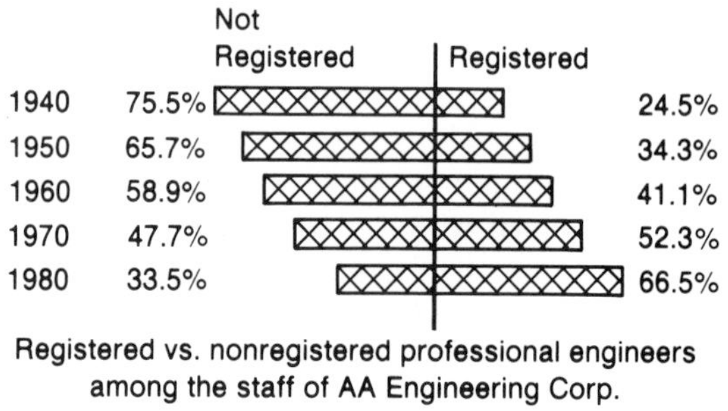

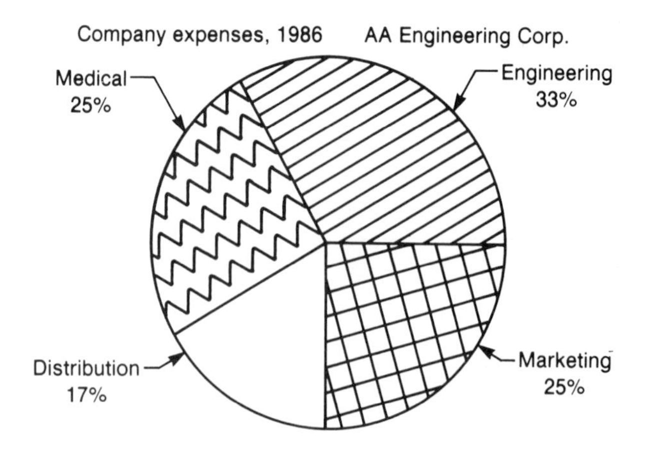

Figure 17.27 A 100-percent sliding-bar chart indicates the relative portions of a quantity with respect to two independent parameters.

Company expenses, 1986 AA Engineering Corp.

Medical 25%

Engineering 33%

Distribution 17%

Marketing 25%

Figure 17.28 A pie chart shows relative portions of a given whole. It is most effective for comparison of a small number rather than a large number of portions.

In Retrospect

- Two-dimensional graphs may be used to represent the functional dependence of a variable upon the value of another (controlled and independent) variable of a system.
- The horizontal axis of a graph is known as the abscissa; the vertical axis is called the ordinate.
- Multiple data sets in a single graph should be identified by callout letters and distinctive linework.
- Scales should be chosen so that functional curves are not unnecessarily compressed.
- Full scales should be used (if possible) to accurately represent relative amounts of given quantities.
- Horizontal and vertical scale values increase from left to right and from bottom to top, respectively, in traditional layouts.
- Lettering and numerical values should be read from the bottom or right side of a graph.
- Scales are normally placed outside the graphic region.
- Scales should include the identification of both the units of measurement and the variables under consideration.
- The number of significant digits at coordinate rulings in a graph should be minimized through the use of an appropriate multiplication factor.
- A title block should include all additional information needed for a complete identification of a graph, its data, and all sources of information. A border should be used to clearly separate the title block from the main body of the graph.
- Experimental data points should be specified on a graph. Theoretical curves should be drawn without the inclusion of plotted points.
- A linear relationship between two variables, Y and X, can be represented by an equation of the form

$$Y = b + mX \qquad \text{EQ. 17-1}$$

where b is the intercept and m represents the slope of the straight line corresponding to this equation.

- The behavior of many physical systems can be accurately represented by a power equation of the form

$$Y = A(X)^J \qquad \text{EQ. 17-3}$$

where Y is the dependent variable, X represents the independent quantity and where both A and J are appropriate constants.

- Power equations may be rectified into equivalent linear forms through the use of logarithmic paper or via calculation of the logarithms of sample data points.
- The slope of a logarithmic plot of experimental data (which behaves in accordance with a power equation) corresponds to the value J for the power in the functional relationship between variables.
- Exponential behavior may be represented by an equation of the form

$$Y = A(B)^X \qquad \text{EQ. 17-13}$$

If B is equal to e^J, where e is Euler's number, the functional behavior is described by

$$Y = Ae^{JX} \qquad \text{EQ. 17-11}$$

If the exponential curve intercepts the X-axis at $X = A$, the appropriate exponential equation is

$$X = Ae^{JY} \qquad \text{EQ. 17-12}$$

- Semilogarithmic paper may be used to plot data points directly without calculating any logarithms. Such a plot—if linear—may then be used to determine the constants (A and J) in the exponential expression that describes the functional relation between system variables.
- Engineers must be able to present a wide variety of data with accuracy and clarity. Charts allow the engineer to present such data in a way that will be quickly and properly interpreted by both technical and nontechnical audiences.
- Proper chart design requires that the engineer evaluate (1) the type of data to be presented, (2) the audience to be addressed, and (3) the relevant relationships and conclusions to be emphasized.
- The line (or curve) chart is used to present the variation of one (dependent) parameter with that of another (independent) parameter. In addition to the standard line chart, there are multiple-curve charts, multiple-amount scale charts, multiple-time charts, cumulative-curve charts, step (or staircase) charts, and histograms.

ENGINEERING IN ACTION

The Violin Goes Electronic

Can a Stradivarius be simulated?

In most design projects, the goal is to build some ideal end product. In violin design, the problem is reversed. Ideal products already exist in the form of a few high-quality instruments built by master craftsmen over 400 years ago. The problem of the modern instrument maker is to duplicate that sound.

Until recently, those probing the mysteries of violin design tried to duplicate the vibration system. They built instruments that copied, as exactly as possible, the shape, materials, and manufacturing processes used to produce the original instruments. Their work was inhibited by the fact that only a few violins (such as those made by Stradivari and Guarnieri) were considered by musicians to be first rate. And finding an owner of such an instrument who would allow it to be disassembled and tested was, and still is, extremely rare.

A new, revolutionary approach, made possible through improvements in miniature electronic sensors and advanced wave-form analysis, is to duplicate the sound, not the instrument. With this concept, only the barest essentials of the violin are needed: the strings, bridge, and something to anchor the string ends. Pickups under each string feed signals to filter arrays that alter the signals and feed them to speakers. Alterations in the signals are the principal determinants of sound quality.

One effort to electronically duplicate the sounds of a high-quality conventional violin is being carried out at the University of California/San Diego department of music. Experts in three areas are involved: Max Mathews, an acoustic expert who has been designing electronic violins for 10 years, is supplying instruments; Dr. Richard Moore, who has degrees in music and electrical engineering, is working on the analysis and production of the sound; and Janos Negyesy, a concert violinist, provides professional feedback to the other two. Goal of the designers is an affordable instrument that provides quality sound. It is estimated that the electronic violin could be manufactured for less than $1,000. Prices of high quality conventional violins begin around $100,000.

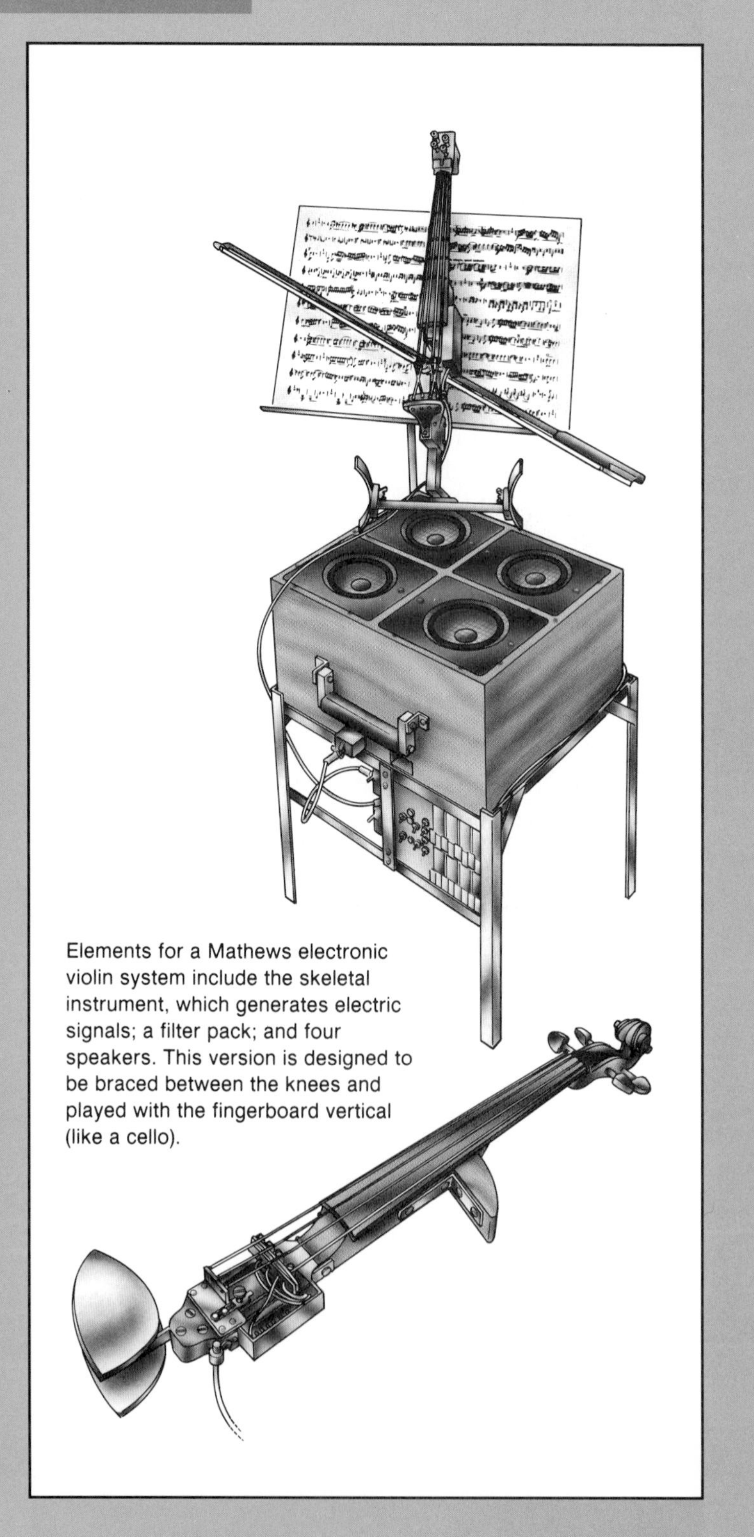

Elements for a Mathews electronic violin system include the skeletal instrument, which generates electric signals; a filter pack; and four speakers. This version is designed to be braced between the knees and played with the fingerboard vertical (like a cello).

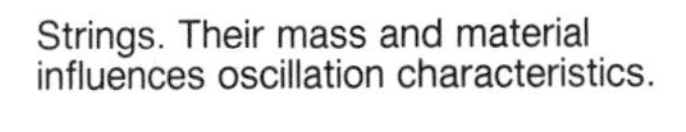
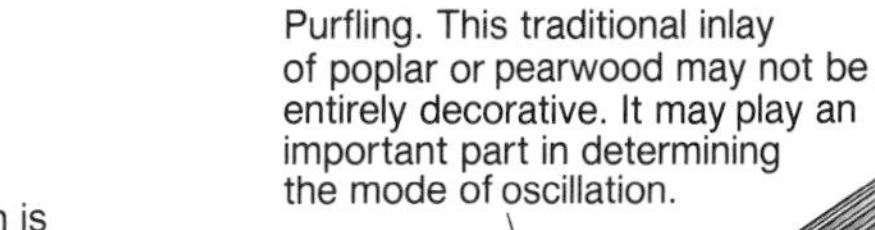

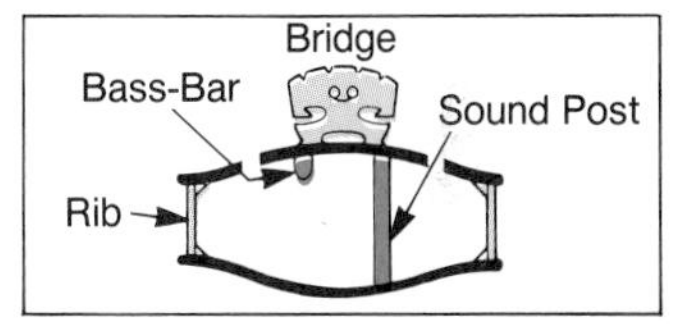

Forming a Classical Sound

Probing the Old Master's secrets:

The traditional technique for analyzing a violin's performance is to try to untangle the jungle of variables that go together to produce the instrument's sound. Although each factor listed can be evaluated separately, how they interact to produce sound has yet to be completely understood.

Four miniature contact microphones detect string motion and convert it to output signals. Four separate filter systems, amplifiers, and speakers are used to eliminate intermodulation distortion of the string signals. Filter circuits are adjusted to give each string's signal a particular set of gain and attenuation characteristics.

The sound the electronic violin produces is said to be comparable to a quality violin, particularly if the listener is away from the speaker's near field distortion pattern.

Mechanics of the electronic instrument, such as the very stiff bridge, introduce factors that may prevent an exact reproduction of a traditional quality violin. However, the present electronic instruments can offer the advantage of a full sound at any volume, something only the best traditional violins can provide.

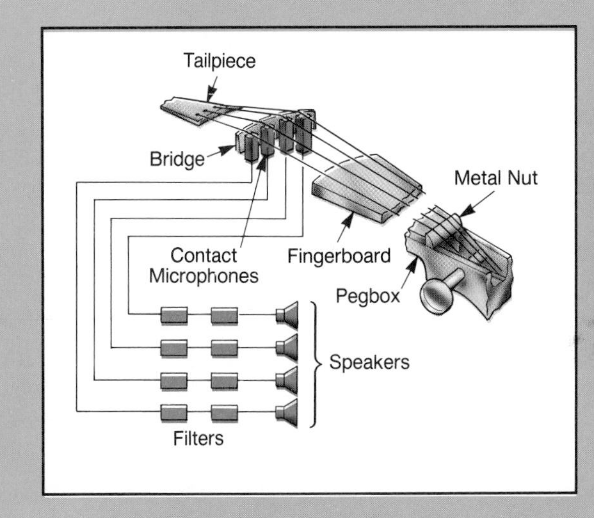

Source: Reprinted from MACHINE DESIGN, April 22, 1982. Copyright, 1982, by Penton/IPC., Inc., Cleveland, Ohio, pp. 80, 81.

- Surface charts emphasize the size of the variation in a particular parameter by filling in the area under the curve with a pattern.
- In vertical-bar (or column) charts, values of the dependent parameter are represented by relative heights of the column. The finite spacing between columns enhances the clarity of the chart and emphasizes the intervals of the independent parameter during which data is obtained.
- Horizontal-bar charts are used when the relative size of different data sets must be compared but one or more of the values is too large in comparison to the other values to allow convenient use of a vertical-bar (or column) format. Such charts include the grouped-bar chart, the subdivided-bar chart, the paired-bar chart, and the 100-percent sliding-bar chart.
- Another popular graphic format in which data may be effectively presented is the pie chart, which is most effective when the data have been divided into a small number rather than a large number of segments.

Exercises

EX17.1. Draw a 2-dimensional functional graph that relates the following system parameters. Draw a straight line through the data points.

Year *X* (independent parameter)	Annual Gross Income (in $1,000) *Y* (dependent variable)
1960	100
1965	250
1970	620
1975	1,560
1980	3,900
1985	9,750

Determine the *intercept, b,* and the *slope, m,* for the straight line that *approximately* describes the behavior of this system. Also, determine the gross income one would *expect* to be generated in 1990, based upon the above data and the *assumption* that this system will continue to operate (and grow in gross income level) according to the linear equation obtained from the preceding data.

EX17.2. Draw a multiple-curve chart for the following industrial data for the AA Engineering Corporation. Use distinctive linework and notation to distinguish between different data sets in the chart. Draw a curve through each set of data points. Label the axes properly.

Year	Annual Gross Income (*I*) (in $1,000)	Annual Costs (*C*) ($1,000)	Annual Profit (*I-C*) ($1,000)
1960	100	550	−450
1965	250	780	−530
1970	620	950	−330
1975	1,560	1,240	+320
1980	3,900	1,670	+2,230
1985	9,750	2,890	+6,860

EX17.3. Draw functional graphs for the following power equations. The abscissa should vary from $X=0$ to $X=10$. Use at least ten data points.

a. $Y=X^{-1}$
b. $Y=X^{-0.5}$
c. $Y=X^{0.5}$
d. $Y=X^{1}$
e. $Y=X^{2}$
f. $Y=X^{3}$

EX17.4. Rectify the power equation represented by the functional graph developed in each of the following problems:

a. EX17.3a.
b. EX17.3b.
c. EX17.3c.
d. EX17.3d.
e. EX17.3e.
f. EX17.3f.

Draw an appropriate functional graph for each of the rectified (linear) relations between X and Y in these problems. Write the equation that describes each rectified relationship between X and Y.

EX17.5. Draw functional graphs for the following equations. The abscissa should vary over an appropriate range of your choice; however, you should use at least ten data points for each graph.

a. $Y = 2e^{X}$
b. $Y = 0.5e^{2X}$
c. $Y = 3e^{-2X}$
d. $Y = 5e^{4X}$
e. $Y = 100e^{5X}$
f. $X = 2e^{3Y}$
g. $X = 3e^{2Y}$
h. $X = 1.5e^{-Y}$
i. $X = 5e^{-2Y}$
j. $X = 25e^{4Y}$

EX17.6. Use appropriate graph paper to draw linear representations of the relationships between the parameters X and Y in the exponential equations given in the following problems.

a. EX17.5a.
b. EX17.5b.
c. EX17.5c.
d. EX17.5d.
e. EX17.5e.
f. EX17.5f.
g. EX17.5g.
h. EX17.5h.
i. EX17.5i.
j. EX17.5j.

EX17.7. Draw a line chart for the relationship between the "Year" and the "Annual Costs" given for the AA Engineering Corporation in EX17.2.

The following data will be used in EX17.8 through 17.15.

BB Engineering Corp.

Year	Number of Employees	Annual Costs (in $1,000)	Annual Gross Income (in $1,000)	Number of Units Produced	(Total No.) Patents Held
1960	8	123	66	575	2
1965	23	2,100	2,056	15,500	12
1970	38	3,870	4,830	35,300	15
1975	106	12,400	17,200	120,000	28
1980	157	28,300	33,500	187,000	34
1985	145	34,100	40,200	163,000	39

EX17.8. Draw a cumulative-curve chart for the total number of patents issued to the BB Engineering Corporation between the years 1960 and 1985 (inclusive).

EX17.9. Draw a histogram for the number of employees of the BB Engineering Corporation between the years 1960 and 1985 (inclusive).

EX17.10. Draw a surface chart that includes data for the "Annual Costs" and the "Annual Gross Income" for the BB Engineering Corporation between the years 1960 and 1985 (inclusive).

EX17.11. Draw a column chart for the number of units (product units) produced by the BB Engineering Corporation between the years 1960 and 1985 (inclusive).

EX17.12. Draw a grouped-column chart that compares the number of employees at the BB Engineering Corporation between the years 1960 and 1985 (inclusive) with the number of patents held by the firm during these years.

EX17.13. Draw a bar chart for the "Annual Costs" of the BB Engineering Corporation between the years 1960 and 1985 (inclusive).

EX17.14. Draw a grouped-bar chart for the "Annual Costs" and the "Annual Gross Income" for the BB Engineering Corporation between the years 1960 and 1985 (inclusive).

EX17.15. Draw a paired-bar chart for the number of employees versus the number of units produced by the BB Engineering Corporation between the years 1960 and 1985 (inclusive).

EX17.16. Draw a pie chart for the following data:

Area of Costs/Expenditures	Percentage Total Costs
Research and Development	21.0
Engineering Design	11.0
Tooling/Equipment	15.0
Labor (general)	33.0
Marketing/Distribution	20.0

18 Engineering Design

Preview We continue our excursion into the world of engineering and technology by examining the engineering design process for developing solutions to technical problems. Such topics as computer-aided design and computer-aided manufacturing (CAD/CAM), problem definition, goals' and constraints' specifications, design analysis, materials and fabrication processes, patents, and design presentations are reviewed. Each of these elements is a part of engineering design, and each should be considered carefully by the engineer in developing a design solution to a problem.

Learning Objectives

Upon completion of this chapter, the reader should be able to:

- Define the design process as a step-by-step procedure that is applied in the development of a solution to an engineering problem.
- Explain why design engineers must be aware of the needs and probable reactions of those who will use the proposed design solutions to a problem.
- Identify the numerous factors that can influence a design effort, including physical or environmental considerations, economic conditions or constraints, material characteristics, standardization guidelines, storage and waste disposal requirements, legal restrictions, and the availability of materials and/or components.
- Recognize that an effective engineer must be familiar with a broad range of physical phenomena so that he or she can apply this information to the design of engineering solutions.
- Identify the general characteristics of metals, plastics, and wood, together with the fabrication processes commonly associated with these materials.
- Select the optimum concept or design solution from a set of alternative designs by applying a decision matrix or similar analytical tool.
- Identify the benefits provided by United States patents for design solutions, together with the three principle criteria used to determine if an applicant should be granted a patent.

18.1 Computer-Aided Design and Computer-Aided Manufacturing (CAD/CAM)

Engineering design is a creative, analytical, and structured activity in which solutions to difficult technical problems are developed. Before discussing the procedure for engineer-

Figure 18.1 The crosshead of the new No. 1 press shapes the workpiece at the No. 2 forge shop at Bethlehem Steel Corporation's plant in Bethlehem, Pennsylvania. The press is guided by press columns, weighing 92,000 pounds apiece. An operator in a computerized pulpit *(see insert)* controls the operations of the press. The pulpit and the press are part of a $19-million modernization program that was completed in 1982. The modernization program has enabled the company to develop previously untapped product lines and, at the same time, improve the quality of its existing products. (Courtesy Bethlehem Steel.)

ing design, we will briefly review the subjects of **computer-aided design (CAD)** and **computer-aided manufacturing (CAM).**

Computers are used in both CAD and CAM to assist the engineer in the development, analysis, and fabrication of engineering designs (Figure 18.1). CAD allows the engineer to create and modify alternative design solutions to a problem, thereby increasing both the speed and the accuracy with which these solutions are generated.

Many factories now use industrial robots (Figures 18.2 and 18.3) for automated fabrication and assembly operations in order to:

- Increase production levels.
- Improve product quality and consistency.
- Ease the workload of employees.
- Increase the safety of the workplace by reducing hazards.

CAM focuses upon the use of a computer to monitor and control such automated manufacturing facilities.

CAD provides the engineer with the capability to:

- Create a mathematical (usually geometric) model of a design, which can then be analyzed with respect to specific operating conditions and modified as necessary.
- Store (for later recall) the geometric and mass data that define a particular design within the computer's data base.

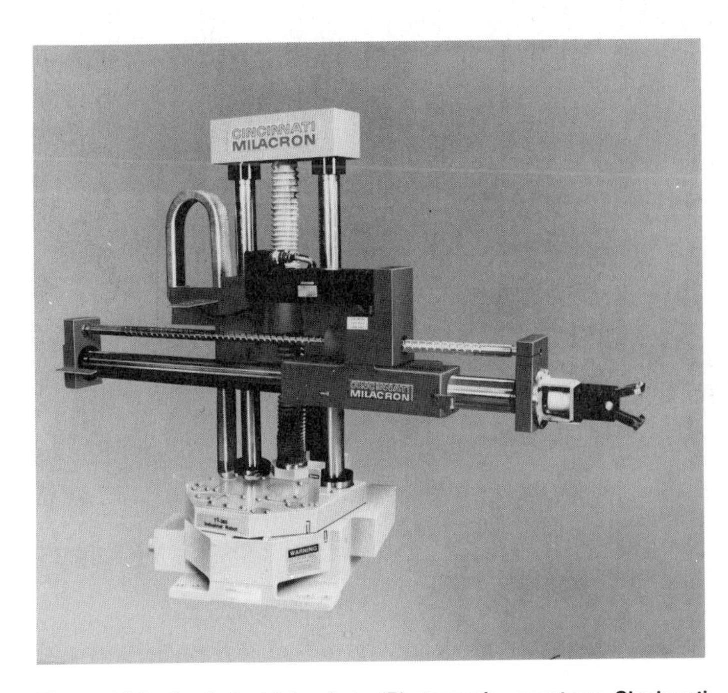

Figure 18.2 An industrial robot. (Photograph—courtesy Cincinnati Milacron.)

Figure 18.3 Industrial robots are used in automated fabrication and assembly operations. (Photograph—courtesy Cincinnati Milacron.)

- Analyze the behavior of a design within a specific environment and for a given set of operating conditions.
- Generate a complete set of working drawings (orthographic, pictorial, exploded assembly, and so forth) of the design as needed.
- Control the fabrication of a design part (via control of the fabrication machinery) in accordance with the geometric, mass, and other data used to define the design. (This type of activity is known as **numerical control (NC) machining;** see Figures 18.4 and 18.5.)

As a result, the computer is now used to assist the engineer from the initial stages of the design process—in which the problem to be solved is carefully defined in terms of specific goals and design constraints—through the creation of imaginative and feasible alternative solutions to the problem and, finally, to the manufacture of the final design.

Although the introduction of the computer into the design process has greatly expanded the engineer's ability to quickly create and fabricate solutions to challenging techni-

Figure 18.4 Numerically controlled (NC) machinery is used to fabricate engineering designs. (Photograph—courtesy Cincinnati Milacron.)

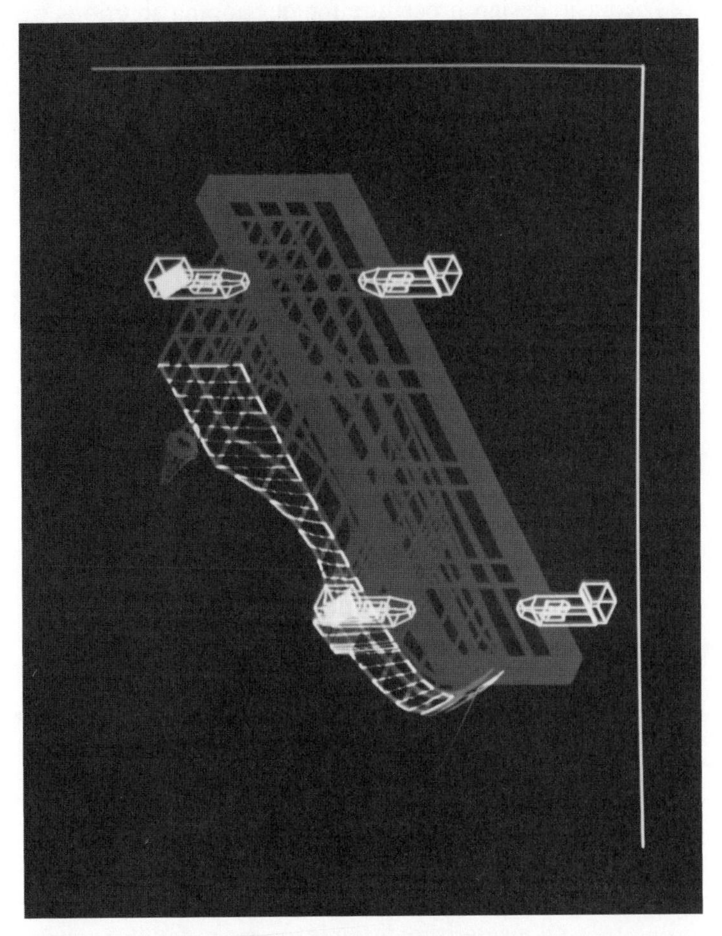

Figure 18.5 Once created and verified, a surface model is used to program numerical control (NC) toolpaths. Here a Computervision CDS 4000 system allows NC tools and toolpaths to be displayed dynamically. Dynamic display of the tool moving along the toolpath helps verify the programming and detect interference with fixtures. (Courtesy of Computervision Corporation.)

cal problems, it is still only another tool—it has no value beyond that which the engineer gives it by imaginative, skillful use. The development of creative and practical engineering designs continues to be primarily dependent upon the engineer rather than any tool, even a sophisticated computer.

18.2 Engineering Design

Engineering design is a creative process used to develop solutions to existing engineering problems that need to be solved for the benefit of humankind. To arrive at the best solution for a given problem, the design engineer must not only be imaginative but also willing to work with other professionals toward the common goal of a solution.

The basic design procedure for developing solutions to engineering problems is divided into specific phases and each phase is further divided into specific stages as follows:

- **Initial Phase**
  - Stage 1. Problem statement and justification.
  - Stage 2. Tasks/goals definition.
  - Stage 3. Tasks specifications (constraints).
- **Creative Phase**
  - Stage 4. Generation of ideas.
  - Stage 5. Conceptualization and development of solutions.
  - Stage 6. Selection of optimum concept/solution.
- **Verification Phase**
  - Stage 7. Theoretical design analysis.
  - Stage 8. Experimental testing.
- **Presentation Phase**
  - Stage 9. Final design report and oral presentation.
- **Industrial Phase**
  - Stage 10. Production design/working drawings.
  - Stage 11. Manufacturing.
- **Commercial Phase**
  - Stage 12. Distribution.
  - Stage 13. Consumption.

The final stage is followed by feedback from the consumer or user of the engineering design to the designers; as a result, a **second-generation design** may then be developed in response to the needs and concerns of the user. Figure 18.6 summarizes this design process.

Before we consider the design process in detail, we will first focus upon a variety of factors that affect our ability to apply this process in developing a solution to an engineering problem.

18.3 Personal and Societal Values in Design

It is important to recognize that our personal and societal values may affect our ability to design, develop, or choose the optimum solution to a particular engineering problem. Each of us must make decisions in our lives—both major (such as whether to earn a college degree, marry, move to another country, enter a particular profession) and minor (such as whether to purchase a newspaper or wear a hat). In each case we try to base our decision upon as much accurate information as possible; in each case we try to be as logical and objective as possible. Logical thought, objectivity, and accurate information are also necessary ingredients in engineering decision-making. As engineers or technicians, we must realize that our individual attitudes toward particular

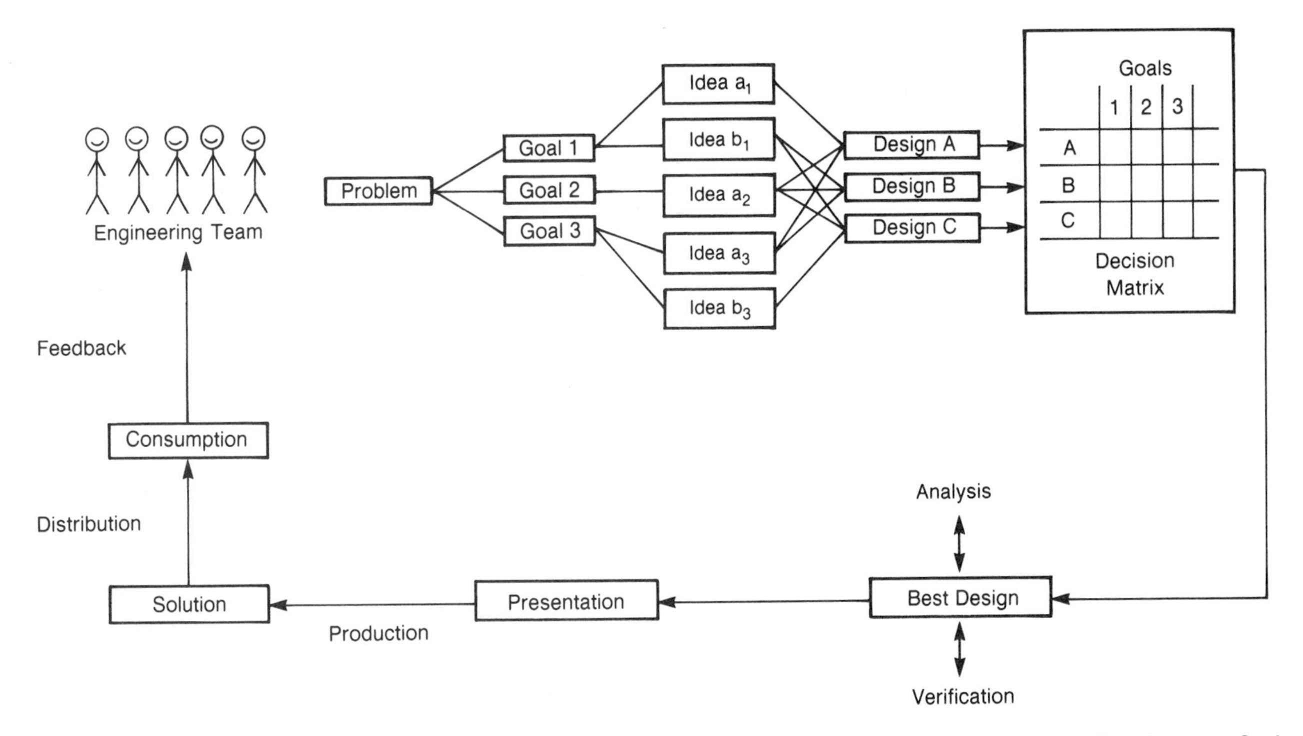

Figure 18.6 The design process can be modeled in terms of specific stages that form a feedback loop to the engineering team. Such feedback facilitates the development of improved second-generation (and third-generation, and so on) design solutions.

types of problems may affect our decisions. In addition, the reaction of the consumer to an engineering solution may be affected by that consumer's personal and societal values.

As an example, consider the following situation: a vagrant approaches two people and asks each person for some money. One person decides against giving the vagrant any money, basing his decision upon the premise that the vagrant would spend the money to become intoxicated. This person believes that he is helping the vagrant by *not* providing any money. The second person responds to the vagrant's request by giving one dollar to him; this decision is also based upon the belief that it will aid the vagrant (a conviction that no one should be refused assistance for which he has asked). In this example, we see that two entirely opposite paths have been chosen to meet the same objective, that of providing the vagrant with assistance. The path chosen by each person was determined by individual past experiences and individual systems of values and ethics.

Similarly, engineers seeking the same goal (the best solution to a given problem) may choose entirely different paths to that goal because they have different perspectives, or viewpoints, resulting from different sets of personal values. Engineering decisions may not reflect the effects of different value systems as starkly as our example concerning a vagrant, but the same principle operates. For example, an engineer may have to choose between two solutions, one safer but more expensive than the other. A compromise between the two goals of safety and minimum cost must be achieved, and such a compromise may depend greatly upon the engineer's personal values as well as his or her technical expertise, professional position, and experience.

Engineers should also know when the users of their design solutions are members of a country, profession, or cultural group that is different from that of the engineer. When societal or cultural attitudes of the users differ from those of the engineer, a design solution considered excellent by the engineer may be rejected by the users. An example of such a design would be a device that solves a problem confronting a disabled person, but that also constantly reminds that person of his or her disability. Such a device fails to assist the user in the *optimum,* or best, way.

In summary, then, design engineers should try to be aware not only of the users' needs but also of the probable reactions of the users to proposed design solutions. In addition, engineers should recognize that their personal values may affect their ability to determine the best solution to a given problem. We are *not* suggesting that engineers should—or even could—change their values but merely that they should be aware of how their values could affect their ability to develop and choose the optimum solution to a particular problem.

18.4 Human Factors in Design

The engineering area known as **human factors** concerns the effects of human behavior in the development of design solutions. The effects of a working environment upon employees, specific interactions between a device and its operator, and typical human-body dimensions are some of the concerns of the human factors design engineer as he or she strives to achieve the most effective and efficient solution to a particular problem.

Professionals from such diversified disciplines as industrial engineering, medicine, psychology, and sociology are involved in the design process. Measurements of the typical male and female hands, arms, legs, feet, and so forth, are crucial to designs that involve body movements or comfort, such as wheelchairs, automobile instrumentation, crutches, pens, and doorknobs. Proper design engineering strives to improve the working climate of individuals by minimizing fatigue, boredom, and unsafe practices. The effects of temperature, sound, light, and other stimuli upon human beings are recorded, analyzed, and applied to achieve optimum working conditions.

Color patterns for control panels are designed to be within the visual range of the typical human being and to be effective indicators of the system's current condition. Certain colors will be chosen for particular purposes, such as to indicate dangerous temperature levels or to remind the machine operator of a specific task to be performed. Sound may also be used to guide the worker through a series of tasks.

If many people are working within a system, the system must be designed to avoid crowding, provide proper lighting, and, in general, allow comfortable and efficient working conditions. Physiological and psychological data must be obtained in order to design a system with ample workspace, lighting, and other positive characteristics.

Design engineering depends upon many people (representing many disciplines) working together toward a common goal: the well-being of those who will operate or be affected by the design solution. Safety, comfort, the emotional and psychological reactions of the user *all* must be considered by the design engineer and his or her colleagues as they develop the best solution to a problem.

18.5 Additional Factors That May Influence Design Efforts

In addition to the human factors discussed in the previous section (such as human dimensions, motor activities, physical and psychological reactions to a particular working environment) there are other influences that directly affect design efforts.

Physical or environmental factors include atmospheric conditions, space allocations, reactions to chemical compounds and applied forces. Severe changes in temperature (naturally occurring or artificially produced by machinery) can cause deformation or cracking of material. High humidity levels can result in corrosion or electrical short circuits. Intense light or vibrations may create unsafe conditions.

Economic conditions can also affect design. A product may be directed toward a particular consumer group in a specific economic level; as a result, production costs must be limited to a certain range. Corporate interests usually require that a specific amount of profit be obtained for a product solution to a problem. In addition, pollution control requirements for a design may be so expensive that the design becomes financially infeasible.

Material factors may significantly influence a design. Deformation of material under stress is usually undesirable; therefore, materials must be chosen that are resistant to stress effects. Marketing of a product can be affected by its appearance (color, shape, texture), odor, or chemical interaction with the consumer (for example, jewelry that causes skin irritation). Durability is a goal common to most designs; materials should be chosen that will endure in a particular working environment—within the restrictions of cost and other constraints.

Standardization for production processes, components, and systems by professional, corporate or governmental groups affects design. The *availability of raw materials, safety conditions,* (of the production workers, distributors, and the consumer), *storage conditions,* (such as the shelf life of the product and the need for a specific storage environment in terms of temperature or humidity), and *waste disposal* can influence the acceptance or rejection of a particular design.

Learning Check

Why must the reactions and needs of many groups (such as government officials, corporate managers, distributors, environmental protection organizations, materials suppliers, and, particularly, the consumers) be anticipated—if possible—during the design of an engineering product or system?

Answer If the design team fails to recognize these reactions or needs, it may design a solution to a problem that fails to adequately serve the consumer or that is never manufactured due to legal or economic difficulties.

18.6 Formulating a Problem Statement

The initial step of the design process is to identify the specific problem to be solved. Marketing surveys may be used by a manufacturer to discover a *need* of the consumer. An existing product may require modifications if it is to effectively compete in the marketplace. Scientific discoveries or new technology may create the need for a particular engineering solution. The project may be initiated in any one of these (or other) ways.

After the problem is identified, it must be carefully defined. *The problem statement may be more difficult to develop than the design solution(s) to the problem.* The statement should not be vague; a *specific* problem must be defined if one is to develop *specific* solutions. However, the designer must avoid stating the problem in terms of a particular solution! Otherwise, one is restricted to a particular design path, and the danger exists that only designs similar to the particular solution under consideration will be devel-

Highlight

Thomas Edison—Inventor

Thomas Alva Edison (1847–1931) is considered by many to be history's greatest inventor. He patented more inventions than any other person. In addition, he pioneered the concept of an industrial research laboratory and also created the first consulting engineering firm. His inventions include the first commercially viable electric light, waxed paper, the mimeograph, improvements in both telegraphy and the telephone, the phonograph, an electric generating system (so that the electric light would be practical), and the use of a film strip to produce motion pictures (his firm produced the first story-telling motion picture, *The Great Train Robbery*, in 1903).

Edison's design for an electric light that could be produced and operated at sufficiently low cost and for sufficiently lengthy periods of time changed the world. This design (with United States patent number 222,898) first provided illumination for forty continuous hours, beginning October 21, 1879.

One statement of Edison's is particularly significant for anyone who is involved in engineering design:

> Genius is one percent inspiration and ninety-nine percent perspiration.

We would all do well to remember this observation.

Thomas Edison

oped. Creativity is thus constrained, and the best solution to the problem may not be found!

The problem statement should identify the specific *function* of the product or system to be developed. Such a function statement will assist the designer in understanding the problem under consideration. Again, however, one must beware of defining the problem in terms of a particular solution.

The reason why a solution to the problem is needed must be identified. Sometimes the reason may seem obvious, for example, a fire alarm for those who are both blind and deaf appears to be an obviously necessary product. Other times the reason stems from the desires of management; the design team may be assigned to a particular problem by an employer who wishes to enter a certain product market or who wishes to improve an existing product. Whatever the reason, it must be specified by the designer. In other words, the designer must justify—in specific and concrete terms—the effort and cost required to solve the problem. Engineering is the application of science and technology to solve practical problems that must be solved for the benefit of society. No engineering project should be frivolous or impractical. Many times, the design engineer must submit a formal proposal for the work planned, proving to those funding the efforts that the resulting design will be valuable.

Since a poorly conceived problem statement can greatly hinder the development of a feasible solution, the time and effort necessary to carefully define the problem is a wise investment that will produce dividends throughout the design process. Indeed, the designer may discover, as he or she develops the justification for a design solution to a problem, that the initial problem statement is too vague or ill-conceived to support continuation of the design effort.

18.7 Identifying Design Goals

After a problem has been identified and the need for a design solution justified, the problem must be defined in terms of specific **design goals** that must be achieved by any viable solution. To formulate these goals, the designer must gather information about the problem. A **patent search** (see Section 18.17.3) may be necessary in order to determine if there are any existing solutions to the problem. The designer may also contact other engineers who have worked to develop solutions for similar problems. Solutions that were rejected for some reason may have been developed by these other engineers, and knowing the reasons for their rejection may prevent the designer of a similar effort from treading a fruitless path already trod by others.

If the design is to be part of a larger system, the entire system must be defined (for example, the design of an automobile seat requires that the entire passenger compartment—and, in fact, the entire vehicle to some extent—be defined). The environment in which the solution must operate should be identified. The physical factors that may affect the design must be determined. It is particularly important that the specific users, operators, or consumers for whom the solution is to be developed be identified. The needs of the user cannot be met if the user is not identified. Also, it may be necessary to place specific requirements upon the solution in accordance with the specific abilities or needs of a particular user group.

Again, if the goals that must be achieved by the design solution are not carefully defined, then the development of the best solution to the problem will be more difficult or even impossible to achieve. During the entire design process, these project goals should be continuously reevaluated. Initial goals may be divided into increasingly more specific tasks. Additional goals will probably be identified. The list of goals will remain in a state of flux until the best solution to the problem has been developed.

The following *general* goals are usually associated with engineering problems, although this list is far from being all-inclusive.

1. **Use of standard parts.** Many standard parts are available in the marketplace from manufacturers and should be incorporated in a design where possible, since it is usually less costly (in terms of money, effort, and time) to purchase these existing components "off the shelf" than to machine the component from raw materials. In addition, an investment in expensive tooling equipment may be required for the fabrication of customized components. Such expenses may result in the rejection of a design solution as infeasible. The design engineer should be familiar with as many different types of components (and their manufacturers) as possible (for example, electrical relays, solenoids, various types of gearing, motors, piping, and leadscrews). Catalogs are available from most manufacturing firms that describe the components available for purchase. Familiarity with a wide variety of components facilitates the generation of more creative and feasible designs.

2. **Safety.** Obviously, a design solution should not threaten the safety of those who will produce, distribute, or operate it. The general goal of safety can usually be divided into specific goals such as electrical safety, mechanical safety, and environmental safety. Each of these categories may be subdivided, depending upon the problem under investigation.

3. **Durability.** Any product must be designed to resist wear and operate properly for a specified amount of time. The durability of the system's components may vary, but the design should include provisions for minimizing wear of certain key components, such as a provision for lubrication of moving parts or use of specific (wear-resistant) materials.

4. **Minimum maintenance.** Another general engineering goal, related to durability, is to minimize necessary maintenance work. Maintenance is required of virtually all engineering products, and it can be expensive in terms of materials and labor. For a successful and competitive product, maintenance should be minimized (and, if possible, eliminated).

5. **Public acceptance.** The consumer must respond positively to the product, or it will fail in the marketplace. To increase the likelihood of public acceptance, the designer should identify the user as precisely as possible in terms of mental and physical abilities, economic level, and specific *needs* to be satisfied by the design solution. Although marketing surveys may have been conducted by the engineering firm, the individual engineer should endeavor to learn more about the user group to be served. *Do not work in a vacuum; know and understand your customer!*

6. **Reliability.** The design solution should perform each and every time it is used (at least, this is the ideal behavior sought by the engineer).

7. **Performance.** The design must perform well in the given environmental and operating conditions. The engineering team must anticipate the reactions of the design to its environment (temperature, humidity, intensity of applied forces, and so forth) and its ability to operate according to specifications (time of operation, conservation of power, and so forth).

8. **Minimum cost.** Costs must be identified, estimated, and minimized. Usually costs can be categorized as follows: costs for development of the design solution, costs for production, costs for distribution, and costs for advertising and promotion.

9. **Operating conditions.** The design should be easy to understand and to operate by the user. Once again, it is imperative that the user be identified; different operating conditions should be developed for users with differing abilities and needs.

In addition to the preceding *general* engineering goals, each design must achieve *specific* goals that reflect the particular problem under consideration. As an example, consider the problem of restraining the driver and passengers of an automobile at the time of impact during a crash. One might be tempted to define the goals in terms of a specific solution—for example, improved seat belts or air bags. This would be a mistake since it would limit the imagination of the design engineer. Instead, the problem might be divided into time periods: before the crash, during the crash, and after the crash. Such a division would then suggest certain specific goals:

1. Before the crash the design must not severely restrict the movement of the user or cause discomfort.
2. During the crash the design must operate effectively and automatically since the user cannot be expected to initiate or control its operation under such conditions.
3. After the crash the design must not restrict the user from quickly escaping the vehicle.

Many other goals (regarding weight, size, shape, and other factors) may also be specific to the problem.

18.8 Task Specifications: Identifying Design Boundaries

The next stage of the design process (following the identification of problem goals) is the identification of **design boundaries,** which are the specifications, constraints, or restrictions associated with each goal and limiting design solutions. **Research** must be conducted by the design team in order to determine these boundaries. A list of limits, or constraints, must be generated to guide the designers as they develop possible solutions to the problem. These constraints may be physical, human, economic, or legal.

- **Physical** constraints include space allocation or dimension requirements, weight limits, material characteristics, energy or power interactions, and so forth.
- **Functional,** or operational, constraints include acceptable vibrational ranges, operating time, and so forth.
- **Environmental** constraints include moisture limits, dust levels, temperature ranges, intensity of light, noise levels, and so forth.
- **Human factors** include the strength, intelligence, and anatomical dimensions of the user.
- **Economic** constraints include production costs, depreciation, operating costs, service or maintenance requirements, and the existence of competitive solutions to the same problem in the marketplace.
- **Legal** constraints include safety requirements, environmental or pollution control codes and production standards.

It is the responsibility of the design engineer to identify—through exhaustive research—all of the constraints, or design boundaries, that are necessary to prevent development of a solution that is illegal, or physically or financially infeasible.

18.9 The Creative Phase

18.9.1 Formulating Ideas

After identifying the problem, its associated goals, and the restrictions that must be placed upon all solutions, the engineer enters the **creative phase** of the design process, in which specific ideas must be generated to achieve each design goal. The designer may employ knowledge of components to develop ideas or he or she may simply allow imagination to produce as many different ideas as possible.

One method of producing ideas is to consider each goal separately and try to identify as many different ways as possible to achieve that goal. **Individual creativity** can be stimulated by **group discussion,** in which the designer gathers other design team members together to develop ideas. In a group discussion it is best to refrain from categorizing each idea as acceptable or unfeasible. The group is merely trying to generate as many different ideas as possible. Some of these ideas will be excellent, some will be mediocre, and most will require *modification* during the design process.

Sometimes a person feels inhibited in a group discussion, afraid to suggest an idea that may sound silly. With modifications, however, a so-called silly idea may become the key to finding a solution to the problem.

All ideas and points of discussion should be carefully recorded in a group discussion. Usually the group will meet several times during this stage of the design process, and it is very easy to forget or incorrectly remember the details of a meeting that was held several days earlier if notes of that meeting are not available for review. This stage of the design process requires patience and discipline as well as creativity. Designers should not allow themselves to be distracted or interrupted during a creative period.

The list of goals developed before the group discussion may need to be changed as a result of the better understanding of the problem the designer will have after this creative phase. New goals will be targeted, and original goals may need to be divided into several distinct tasks.

Finally, it is important to realize that total solutions to the problem are *NOT* being developed during this stage; only individual ideas to achieve individual goals are being generated. The designer might list goals and ideas in tabular form; for example, if one is designing a new lamp, one goal could be:

Goal 1: Product must not tip over; that is, it must be stable.
Idea *a*—use a very large base to balance product.
Idea *b*—use a tripod.
Idea *c*—use a base composed of very heavy material.
Idea *d*—secure product via bolts, glue, cord, etc., to a table or supporting structure.

More ideas would probably follow, and a similar list of ideas would be generated for each additional goal.

In summary, then:

1. Develop ideas for achieving each design goal, one goal at a time.
2. Avoid any preliminary evaluation or criticism of ideas—that is, defer judgment.
3. *Record all ideas!*

18.9.2 Creativity

The design process is meant to produce alternate, feasible design solutions for the problem under consideration—yet how does one create these alternate designs? Philosophers, psychologists, sociologists, engineers, businesspeople, artists, and many others have devoted much effort in their attempts to understand and stimulate creativity in human beings. It is our contention that *every* person can be creative. *Creativity is a universal talent.*

We each bring to engineering our unique perspective, experience, education, and creativity. In fact, society does not require simply another mechanical engineer who can solve a problem in statics or an electrical engineer who can apply Kirchoff's laws in an analysis of a circuit—there are already tens of thousands of engineers who have the ability to solve problems in analysis. What society needs is creative engineers who can develop new applications of scientific principles in the form of innovative products.

One might then ask, can creativity be taught? The answer is that a person's innate ability to create can be developed through practice. *We can teach ourselves to be creative* through the application of the many techniques to stimulate creativity that have been developed over the years. These techniques include:

- **Bionics,** in which one searches within nature for design concepts (for example, radar is based upon the use of echoes for flight navigation by bats).
- **Brainstorming,** in which ideas are generated by a group (or an individual) without any evaluation of these ideas occurring during the creative idea-generating session. Ideas are later compared and evaluated to determine which concepts are most feasible for inclusion in a design solution.
- **Checklisting,** in which a series of key words, phrases, or questions is used to stimulate ideas about the problem under consideration.

For further details about these and other creativity techniques, the reader is referred to "Stimulating Creativity in Engineering Design Efforts" by G. Voland in *Engineering Design Graphics Journal* 49, no. 1 (Winter 1985): pp. 23–29, published by the American Association for Engineering Education (ASEE), Washington, D.C.

18.10 Knowledge of Physical Effects—A Key to Creative Design

One fact that is seldom emphasized in discussions of design is that *we can apply only that knowledge with which we are familiar.* The successful design engineer continuously seeks to increase his or her understanding of physical phenomena (and their theoretical interpretations), because (1) a thorough understanding of the scientific principles and empirical data that relate different physical quantities (or system parameters) is required for proper *analysis* of new designs, and (2) scientific principles and facts often provide the designer with innovative ideas for new designs, in other words, they can inspire creativity!

Consider the use of the physical phenomenon known as the Magnus effect in the development of Jacques Cousteau's experimental research vessel Calypso II (described in the "Engineering in Action" section near the end of this chapter). This is an example of truly innovative engineering design.

Learning Check

Why should we—as engineers—be familiar with scientific principles and phenomena?

Answer Scientific principles and phenomena provide us with a "source-pool" of ideas for accomplishing specific tasks, in addition to allowing us to analyze proposed design solutions.

18.11 Conceptualization: Formulating Preliminary Design Solutions

The ideas generated during the creative phase of the design process are now used to form **preliminary design solutions** which are meant to be total solutions to the problem rather than just a solution to one of the problem's goals. Preliminary designs are not, however, *fully developed* solutions to the problem under consideration; they only describe possible solution concepts that must be compared and analyzed.

To formulate a preliminary design solution, the engineer takes an idea, a_1, (from a set of ideas, a_1, b_1, c_1, and so forth) for achieving goal 1 and combines it with another idea, b_2, (from a set of ideas, a_2, b_2, c_2, and so forth) for achieving goal 2. These two goals are then combined with still another idea, a_3, for achieving goal 3, and so on.

Certain ideas will naturally combine well because of (1) the materials to be used for each idea, (2) the function to be achieved, or (3) some other reason. Other ideas will necessarily be mutually exclusive because of their characteristics. This procedure allows the designer to formulate as many different solutions to the problem as possible by combining individual ideas for achieving the specific problem goals.

Each total solution, or concept, should be described in a brief written summary, together with *sketches* of the concept, its components, and the relationship between the solution and the system within which it is to operate. Sketches improve the designer's ability to visualize and thus to evaluate the concept. Sketches also allow the designer to communicate ideas more easily and accurately to other members of the design team (see Chapter 16).

18.12 Materials and Fabrication Processes in Engineering Design

18.12.1 The Significance of Materials Selection and Fabrication Processes

Before an optimum concept is selected from the set of preliminary solutions generated during the conceptualization stage of the design process, manufacturing materials and processes should be considered in terms of such questions as:

- What materials are available?
- How will each material, if selected, aid or hinder efforts to achieve the design requirements and specifications (material strength, shape, electrical and thermal conductivities, corrosion resistance, weight, availability of materials, costs, elasticity, color, response to variations in temperature and humidity, surface finish, and so forth)?
- What manufacturing processes are available to the producer or supplier of the design?
- How will each manufacturing process affect the production of the design in terms of costs, accuracy, size restrictions, fabrication time, assembly, plant space and equipment, use of personnel, waste of raw materials, and so forth?

Some designers fail to answer such questions during the initial development, selection, and analysis of design alternatives because:

- They are not familiar with manufacturing materials and their characteristics.
- They are not familiar with manufacturing methods.
- They believe that someone else—perhaps an engineer in the company's production facility—is responsible for the selection of materials and processes.
- They do not realize the importance of materials selection in engineering design.
- They believe that the information about materials and manufacturing processes is not easily available (of course, many texts, handbooks, periodicals and manufacturers' catalogs are available among the literature—but this information must be *sought* by the designer).
- They believe that they are familiar with materials and processes when, in fact, their knowledge is limited and/or out-of-date.
- They believe that their firm will not invest in any manufacturing equipment not presently in use at the plant facility.

We now present a brief review of some common engineering materials and manufacturing processes. This review is far from all-inclusive; it is intended as a checklist for the reader, in order to demonstrate the variety of available materials and processes, together with the significance of materials and process selection in engineering design efforts.

18.12.2 Characteristics and Fabrication Processes of Metals

Common manufacturing metals include aluminum, chromium, copper, iron, lead, magnesium, molybdenum, nickel, steel, tin, titanium and zinc. In general, most metals are (1) good conductors of electricity and heat, (2) ductile, and (3) lustrous.

Particularly important characteristics of some of these metals are as follows:

- **Aluminum:** Lightweight, corrosion-resistant, good conductor of heat and electricity, soft, relatively weak. Often alloyed with copper, manganese, nickel, titanium, zinc, magnesium, and silicon in order to increase both strength and hardness.
- **Copper:** High electrical conductivity, ductile, corrosion-resistant. A major component in alloys—alloyed with tin to form bronze and with zinc to form brass.
- **Iron:** Historically, the most important of the basic metals. Commonly available in three forms:
  - Gray iron—the most common material for sand castings; hard, brittle, inexpensive.
  - Malleable iron—tougher, less brittle, more expensive than gray iron.
  - Wrought iron—ductile, corrosion-resistant, mostly replaced in applications today by steel and other materials; composed of very pure iron, through which is distributed fibrous particles of siliceous slag.
- **Magnesium:** Lightweight, relatively expensive, weak, corrodes unless protected (for example, with paint.).
- **Nickel:** Base metal of many nonferrous, corrosion-resistant alloys.
- **Steel:** Oxidation is used to reduce amounts of carbon, manganese, phosphorus, silicon, and sulfur in mixtures of steel scrap and pig iron, from which steel is produced. Heat treating is used to produce desirable characteristics of strength, hardness, and ductility in various combinations and degrees.
- **Titanium:** Corrosion-resistant, high ratio of strength to weight, relatively expensive, difficult to fabricate.
- **Zinc:** Used for galvanizing (making steel relatively corrosion-resistant) and as an element in brass (with copper) and other alloys.

Common fabrication processes involving metals include casting, extrusion, forging, machining, powder metallurgy (sintering) and stamping and forming (Figures 18.7 and 18.8). Details of the processes follow.

Casting In casting processes, molten metal is poured, or injected, into a mold, after which it is allowed to solidify into the desired form. (Some machining may be required to properly finish the work.) The different types of casting methods vary in:

- Materials used for the molds.
- Techniques used to place the molten metal in the mold.
- Molten metals to be cast.

The process for **sand casting** is as follows:

1. The **pattern** (from which the mold impression is formed) is made, usually of wood or metal, in the form of the desired product.

Figure 18.7 Molten steel rushes from a tilted electric furnace into a huge ladle at Bethlehem Steel Corporation's Bethlehem, Pa., plant. Both temperature and atmosphere can be closely controlled in an electric furnace, making it ideal for producing steel to exacting specifications: high-alloy steels, stainless steels or special steels requiring close metallurgical control. (Courtesy Bethlehem Steel.)

Figure 18.8 Steel sheet enters temper mill at Bethlehem Steel Corporation's Burns Harbor, Ind., plant where the metal is rolled through a single stand to restore proper stiffness lost by the metal during the annealing process. Tempering (or skin passing) also improves flatness and imparts desired surface finish to the product. Temper mill has a maximum production speed of 4,120 feet per minute. Pipelike device above coil is an electric-eye monitor. It is used to set coil in correct position so that the uncoiling steel sheet feeds properly into the mill. (Courtesy Bethlehem Steel.)

2. The **mold** is formed (usually in two parts) from a mixture of moist sand and clay.

3. The mold is packed around the pattern in a container (a wooden or metal box) known as a **flask.**

4. Separately-molded **cores** are placed in the mold prior to pouring the molten metal in order to form cavities within the finished piece. Openings through which the molten metal can be poured into the mold are known as **gates;** additional passages **(risers)** allow the metal to flow throughout the mold, (The excess metals, hardened in the gates and risers, are known as **sprues** and must be removed from the finished product).

5. The molten metal is allowed to harden, after which the mold is broken away, cores are removed, and sprues are cut away. The casting is then cleaned by sand-blasting or some other means.

6. Surfaces may need to be machined for close tolerances or smooth finishing.

The advantages of sand casting are that it uses inexpensive patterns; it can be used with all castable metals; intricate pieces can be cast; and there are virtually no restrictions on the size of the cast (ounces to tons). The disadvantages are that it produces a rough surface finish and is inaccurate.

The process for **shell-mold casting** is as follows:

1. A metal (usually alloyed steel, iron or bronze) pattern is constructed.

2. The pattern is mounted on a flat metal plate, after which the pattern and plate are heated to between 375°F and 600°F.

3. The heated plate is clamped to a box that contains a sand and thermosetting plastic resin.

4. Inverting the box for five to fifteen seconds allows the sand-resin mixture to form a shell (with a thickness of approximately 0.25 in.) around the heated pattern.

5. Excess sand is allowed to fall away by returning the box to its original orientation; the plate and shell are then heated for an additional fifteen seconds to three minutes for final curing.

6. The shell is removed from the pattern via ejector pins.

7. The shell sections are clamped or cemented together, thereby forming a mold; molten metal is then poured through the gate, allowed to harden, and cleaned and finished.

The advantages of shell-mold casting are that detailed work can be cast; it produces a smooth surface; it is accurate and casting is consistent. The disadvantage is that the molds are more expensive than the ones used in sand casting.

The process for **plaster-mold casting** (for small, nonferrous metal castings) is as follows:

1. A metal pattern is split, mounted on a plate, and covered with a special plaster that is allowed to harden.

2. The pattern is removed, and the plaster mold is baked for increased hardening.

3. Finally, the casting metal is poured, cooled, and cleaned.

The advantages of plaster-mold casting are that it produces a very fine finish; detailed work can be cast; it is accurate; and it eliminates machining and its associated costs. The disadvantages are slower cooling rates and higher costs for molds than in sand castings.

The process for **investment-mold casting (lost wax** or **precision** castings) is:

1. A die is constructed that is the reverse of the cast to be produced.

2. Wax or thermoplastic resin patterns are cast in the die under pressure.

3. The patterns are then dipped in a very fine powder of refractory material and placed in a flask.

4. A mixture of refractory investment material is poured about the pattern in the flask and allowed to harden.

5. The flask is heated at high temperatures in order to cure the investment material and melt out the pattern.

6. Molten metal is then poured into the hot mold, after which the cast is cooled and cleaned.

The advantages of investment-mold casting are that it is accurate; it produces a very fine finish; and small, intricate parts can be cast. The disadvantage is that castings are expensive (although machining costs can be saved).

The process for **die (pressure) casting** is as follows:

1. Pressure is used to introduce and hold molten metal in a die until hardening has occurred.

2. The mold is then opened, the cast is ejected, and the process is repeated.

The advantage of die (pressure) casting is that it is inexpensive. The disadvantages are that designs are restricted because the mold is not destroyed (the molds must allow easy removal of the casts), and dies are expensive.

Another casting method is **permanent-mold casting,** in which permanent metal (iron or steel) molds are used (usually for lower melting point alloys). The advantages of permanent-mold casting are that it is accurate and produces a very fine finish. The disadvantages are that designs are restricted because the mold is not destroyed (the molds must allow easy removal of the casts), and molds are expensive—therefore, this method should be used only for high-quantity production runs.

Finally, another casting method is **centrifugal-mold casting,** in which cylindrical casts can be made in whirling metal molds. Molten metal is pressed against the wall of the mold by centrifugal force, thereby creating a symmetrical hole in the center of the cast. The advantages of centrifugal-mold casting are that it produces a smooth finish and may save costs due to less material wastes. The disadvantage is that in some alloys, elements of greater weight may concentrate in the outer layers; therefore, the process is not advisable for parts where strength is needed on the inner surfaces (such as bearings).

Extrusion In the extrusion process, hydraulic pressure is used to push hot (soft but not liquid) metal through the opening of a die, thereby forcing the metal into the desired shape. Tubing, rods, moldings, and wires of copper, magnesium, aluminum, or brass can be produced via the extrusion process. (There is also a process known as **cold extrusion** in which pressure is applied to a cold metal in order to obtain the desired shape—one can form deep cans and cups in this way.) The advantages of extrusion are that it is relatively inexpensive for low-quantity production runs; it can produce complex shapes; and wall thickness can be controlled by it. The disadvantage is that the surface finish it produces is not very fine.

Forging In the forging process, heated (soft but not liquid) metal is formed into the desired shape via applied pressure or impact. One can obtain greater strength and toughness in many metals and alloys (such as steel, brass, copper, aluminum, and bronze) through forging than from the casting process; however, forging is usually more expensive than casting.

Common types of forging techniques include:

- **Open-die forging.** As the name of this technique implies, the die and the blank are exposed so that the metal can be hammered into the desired shape (as in blacksmithing). This is a useful process for producing blanks for dies, gears, and so forth. The advantages of open-die forging are that it is relatively inexpensive, and it allows easy shaping of the design. The disadvantages are that it is inaccurate, and that usually further machining is required for proper tolerances—which entails additional costs.
- **Drop forging.** A pair of heavy steel dies are hammered against a hot blank between them. Parts requiring strength with minimum bulk (such as automobile crankshafts and connecting rods) can be produced with this technique. The advantages of drop forging are that it is accurate, and it allows one to form relatively complex shapes. The disadvantage is that dies are expensive; therefore, this method should be used only for high-quantity production runs.
- **Press forging.** This process is quite similar to drop forging in that closed dies are used to form the hot metal; however, a squeezing action is used to form the part under high pressure (primarily used for aluminum, brass, or magnesium parts). The advantages are that large parts (such as aircraft frame sections) can be formed via this process, and greater density can be obtained. The disadvantages are that it is expensive, and large pressures are required.
- **Upset (machine) forging.** This is another process in which pressure is used to squeeze heated metal into the desired shape. It is useful in producing bolt-heads, flanges, and projections inexpensively.
- **Rolling.** Hot metal billets are rolled into sheets, bars, rods, and so forth. This is a common process in steel mills.

Powder Metallurgy (Sintering) The process of powder metallurgy (sintering) is a valuable manufacturing process—with multiple advantages—in which a fine powdered mixture of metals (and possibly ceramics) is pressed into the shape of the desired piece under pressures of 20 to 50 tons per square inch. The product is then heated in a sintering furnace in order to increase its strength.

In addition, a **coining** operation may be used in which the piece is further pressed (under 40 to 50 tons per square inch) in a die in order to improve its finish, minimize distortions, and produce a more consistent density. The density and toughness of the product may be further controlled through an **infiltration** process in which a metal (with a lower melting point than the powdered metal part, or **sintering**) is heated, together with the sintering, until it melts and is absorbed via capillary action into the piece.

Powder metallurgy is used for high-quantity production runs of parts for which machining costs can be saved. So-called oil-less bearings, cams, !ock cylinders, gears, and many other products can be manufactured (with a wide range of desirable characteristics) from different combinations of metals and other materials.

The advantages of powder metallurgy is that machining can be eliminated; it produces a fine finish; it is accurate; a variety of physical characteristics (such as weight, hardness, heat resistance, etc.) can be obtained with it; complex shapes can be formed with it; porosity can be controlled with it; and many combinations of materials are possible with it, some of which can be formed only by this process.

The disadvantages are that very large pressures are required, thereby limiting the size of the dies (and therefore of the parts); materials can be expensive; and dies are expensive; therefore, this process should be used only for high-quantity production runs.

Stamping and Forming Processes In stamping and forming processes, metal sheets or strips are punched or otherwise formed into the desired shape. Some of these methods are:

- **Stamping,** in which a punch presses the metal against a die block, thereby forming the desired finished piece. Cutting, shearing, perforating, and bending can be achieved in this way.
- **Drawing,** in which the metal is gradually shaped into a deep-drawn form (such as cups or cans) via several operations with dies.
- **Roll forming,** an inexpensive process in which rolls are used to gradually shape the sheet metal.
- **Spinning,** which allows one to form flat metal disks. A metal blank is connected to a pattern and then rotated in a lathe while being worked into the desired shape with the use of hand tools or rollers.

18.12.3 Machining Techniques

After a part has been roughly formed via casting, forging, stamping, and so forth, it often requires **machining,** (in

Figure 18.9 **Hot saw cuts finished rolled shapes to customer length after delivery from the 48-inch finishing mill at Bethlehem Steel Corporation's Bethlehem, Pa., plant (Courtesy Bethlehem Steel).**

which additional material is removed) in order to produce an accurate, finished piece. Drilling, tapping, sawing, boring, grinding, lathe working, reaming, milling, planing, and shaping are examples of common machining processes (Figures 18.9 and 18.10).

- **Drilling, boring, tapping.** A drill press is used to create holes in a part. Drill bits of varying diameters are inserted into the press and then rotated into the fixed piece. Multiple-spindle drill heads can be used to quickly produce multiple holes in a part during mass-production runs. Additional attachments allow screw threads to be cut in the drilled hole (tapping) or form a tapered (countersinking) or enlarged (counterboring) end of a hole for a screw head.
- **Grinding.** Abrasives are used (often in the form of grinding wheels or belts) to remove small amounts of surface material in order to produce a finer finish and greater accuracy.
- **Lathe work.** In this type of process, the workpiece (usually a blank) is rotated while a cutting edge forms the material into the desired shape. Many operations (such as facing, boring, tapering, and tapping) can be performed on a lathe, producing very accurate results.

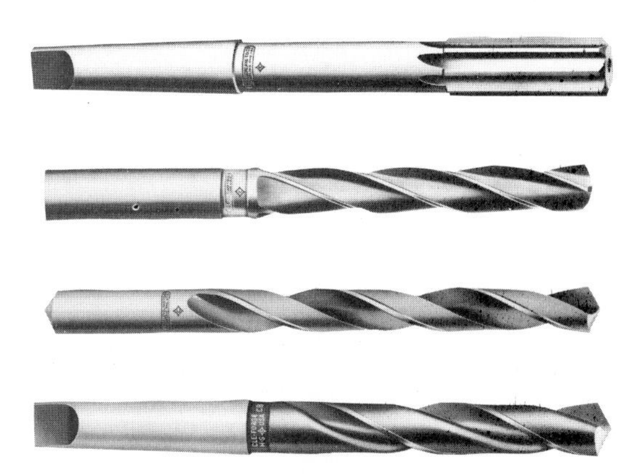

Figure 18.10 **Examples of a reamer and various drills used in machining operations. (Courtesy of Cleveland Twist Drill Company.)**

- **Milling.** Rotating multiple-tooth cutters on a milling machine allow a skilled operator to cut, groove, face, and so forth. Helical and spur gears, spiral reamers, keyways, splines, drills, and other relatively complex results can be produced via milling operations with various attachments.
- **Reaming.** The accuracy or finish of an existing hole can be improved with the use of a reamer, which cuts along the *sides* of the hole.
- **Sawing.** Metal-cutting band saws can be used to slice work into desired shapes. Steel, with a thickness of several inches, can be cut with these machines. Internal cuts can be obtained by cutting the saw blade, inserting it through an interior starter-hole in the material, and then welding the blade into a single piece once again.
- **Shaping and planing.** A shaper is a tool with a single cutting edge that is used to cut angular openings, splines, keyways, and so forth. One can use a planar to produce straight slots or grooves or to plane surfaces on larger pieces. These machines are usually used for small-quantity production runs or for tool and die work because the processes are (relatively) time-consuming.

18.12.4 Characteristics and Fabrication Processes of Plastics

Increasingly, plastics are being used to manufacture both new and established products (previously formed in metal, wood, or other nonplastics). Properties and characteristics shared by most plastics include:

- Plasticity—the property of being easily formed, under applied pressure or heat, into a desired shape, which is then retained upon the removal of the heat or pressure.
- Light weight.
- Low thermal conductivity compared to conductivity of metals, making plastic a good thermal insulator.
- Smooth surface finishes.
- Formation of the final part via a single operation (usually).
- Low electrical conductivities (in many cases).
- Wide range of colors (in many cases).
- Weak or low-impact strength compared to the strength of most metals (although some plastics, such as polycarbonate, high-density polyethylene, and cellulose proprionate, do have high-impact strength).

The unique combination of these properties is the basis for the demand for plastics in such industrial/commercial applications as (thermal and electrical) insulation, cushioning, and packaging or housing material.

Plastics are formed from such natural raw materials as petroleum, water, coal, lime, agricultural products, and fluorspar. These materials of nature are used to manufacture *chemical* raw materials such as phenol, formaldehyde, acetylene, hydrogen chloride, cellulose, urea and acetic acid. Finally, reactions involving these chemical raw materials produce such plastics as phenol formaldehyde, urea formaldehyde, cellulose acetate and polyvinyl chloride. The chemical reactions produce **monomers** (or simple chemical compounds), which can then be used to produce large molecules—often composed of thousands of atoms—known as **polymers.** The prefix *poly* is attached to the name of the *monomer* from which the polymer has been constructed. For example, polystyrene is polymerized from styrene, and polyvinyl chloride is formed from the monomer vinyl chloride, which in turn is the result of a reaction between the raw chemical materials acetylene and hydrogen chloride (formed from the natural raw materials water, coal, lime, and salt). Variations in polymeric mixtures provide the tremendous range of characteristics and properties exhibited by plastics.

There are two general categories or types of plastics: **thermoplastics** and **thermosets** (or thermosetting plastics). Thermoplastics can be repeatedly reformed upon heating, whereas thermosets undergo a chemical change during the curing process that makes them retain their final shape even when heated.

Thermoplastics include the styrenes, vinyls, cellulosics, acrylics, polyethylene, nylon, and teflon. Thermosets include the phenolics, polyester, epoxies, ureas, melamine, and silicone. Tables 18.1 and 18.2 present some thermoplastics and thermosets, together with examples of their applications.

Some of the fabrication processes used for plastics are blow molding, casting, coating, compression molding, extrusion, injection molding, laminating, rotational molding, thermoforming and transfer molding.

Blow Molding In the process of blow molding, air is blown into a molten thermoplastic balloon, which is contained within a mold, thereby forcing the plastic into the desired shape. The finished product is then cooled and removed from the mold. This process is useful in manufacturing hollow products such as squeezable bottles.

Casting In the process of casting plastics, soft thermoplastics or liquid thermosets are placed in a mold and then cured via heating or chemical action. Tubes, rods, film, and so forth are produced by this process.

Coating In the process of plastic coating, thermosets or thermoplastics are spread onto the object in need of coating. Spraying, dipping, roller coating, and other methods can be used.

Table 18.1 Examples of Thermoplastics, Their Properties and Applications

Type	Characteristics	Examples of Applications
Cellulose Acetate	Easily fabricated; wide color range; cannot withstand humidity or high temperatures.	Jewelry, novelties.
Fluorocarbons	Very low moisture absorption; withstands extremes in temperature; difficult to mold; good impact strength; hard (toxic fumes may be given off by burning Teflon).	Coatings on cooking pans, insulation, heating cables, nonlubricated bearings, gaskets.
Nylon	Withstands high (300–350°F) temperatures; durable; chemically resistant.	Combs, gears, bearings, stockings, clothing, washers.
Methyl Methacrylate (tradenames—Lucite, Plexiglass)	Very transparent; limited temperature range; expensive.	Lenses, signs, windows.
Polyethylene	Withstands low temperatures; flexible.	Squeeze bottles, bags for food products, housewares, pipes.
Polypropylene	Chemically resistant; very lightweight; heat-resistant.	Ropes, wire coatings, textiles, packaging, heat-sterilizable bottles.
Polystyrene	Low moisture absorption; transparent; moderate cost.	Coatings, storage boxes, lamp globes, tableware.
Vinyls	Chemically resistant; ages well; strong; low moisture absorption.	Phonograph records, curtains, garden hoses, boots, pipe, automobile roof coverings.

Table 18.2 Examples of Thermosets, Their Properties and Applications

Type	Characteristics	Examples of Applications
Alkyds	Withstands high temperatures; good for outdoor uses; surface hardness; wide coloring range.	Electrical insulation, lacquers, molding, coatings, buttons.
Epoxies	Adhesive qualities; low shrinkage.	Adhesives.
Melamine	Low water absorption; odorless; tasteless; hard.	Distributor heads, buttons, laminated surface.
Phenolics	Strong; hard; inexpensive; heat-resistant.	Television cabinets, handles, knobs, electrical sockets.
Polyesters	Wide coloring range; water-resistant.	Binders for low-pressure laminates for automobile bodies, boat hulls, coatings.

Compression Molding In the process of compression molding, thermosetting plastics are formed (this method is seldom applied to thermoplastics). The material is placed in a mold, heated, and then pressed by a ram that forces the plastic to flow into the desired shape of the mold. After curing, the formed product is removed from the mold (pressures usually range from 300 psi to 8,000 psi).

Extrusion Similar to the extrusion process used in metal fabrication, the process of extrusion for plastics allows one to form thermoplastics into tubes, rods, sheets, and other desired shapes. Granular material is heated in a chamber until a fluid is formed and then transferred (under pressure) through a die opening in the shape of the desired product. The plastic is then cooled and cut, if necessary. This is a high-quantity production process.

Injection Molding Usually used for thermoplastic fabrication—although thermosets can also be formed in this way—the process of injection molding forces molten material, under very high pressure (12,000–30,000 psi), through a nozzle into a cool, closed mold in which the plastic completes its formation. The process is rapid and suitable for high-quantity production runs.

Laminating In the process of laminating, sheets of wood, paper, fabric, or other material are coated with plastic resin, then stacked and pressed between two highly polished steel plates in a hydraulic press under heat and pressure until cured. Low-pressure (0–400 psi) laminating is used to produce automobile bodies, luggage, boat hulls, and so forth, whereas high-pressure (1,000–2,000 psi) laminating results in electrical insulation sheets and other products. Thermosets are commonly used in the formation of laminates.

Rotational Molding In the process of rotational molding, molding powders are enclosed within a two-piece mold which is then placed inside (and rotated within) a heating chamber. Centrifugal force results in the thermoplastic material forming against the walls of the mold. Hollow, one-piece products—such as dolls and street lamps—can be formed with this process.

Thermoforming Also known as **forming,** the process of thermoforming shapes—through the application of air or mechanical forces—heated thermoplastic sheets into the desired form of a mold. Variations of the process include vacuum forming, pressure forming, and slip forming, among others.

Transfer Molding The die used in the process of transfer molding is in two sections, one of which is an open die used as a loading chamber for preformed, preheated pellets. The loading chamber is heated after filling and closed with a plunger. The fluid formed as a result of this process is then forced through a small opening into the final mold. Intricate pieces, with fragile cores, can be produced in this way.

18.12.5 Conclusion on Materials and Fabrication Processes

The extensive variety of materials and fabrication processes (of which only a select sample have been described here) indicates that the design engineer must carefully match (1) the product's specifications and requirements, (2) his or her firm's manufacturing capabilities, and (3) available materials and processes in order to produce a successful design. It is imperative that engineering students (and practicing engineers) familiarize themselves with manufacturing materials and methods. Many references, handbooks, and data sources are readily available for this purpose.

18.13 Selecting the Optimum Concept: The Decision Matrix

Each total solution, or concept, must be evaluated in terms of its ability to achieve the design goals. During this evaluation process, each design should be modified (if possible) in order to improve its ability to achieve each goal.

No design is a perfect solution to a practical problem. Every engineering device in existence can be improved; most devices will eventually undergo modification and improvement. A design solution represents a *compromise* between goals. A design that achieves one goal may be totally ineffective in regard to another goal. The designer, therefore, must determine the relative importance of the design goals (for example, is durability more important than minimizing cost?) Such 'weighing' of goals can be very subjective.

As an attempt to increase objectivity in this evaluation process, designers may develop a **decision matrix** that includes:

- Ratings indicating the relative significance of each goal **(weighting factors).**
- Estimates of the ability of each design solution to achieve a particular goal **(rating factors).**
- Numerical scores that are the products of weighting and rating factors **(decision factors).**
- Numerical scores that equal the sum of all decision factors for a particular preliminary design **(total decision factors).**

These numerical scores are based upon the information about the problem that the design engineer has obtained, together with the designer's own experience and expertise. A decision matrix is used to estimate the *potential benefits or losses* that may result from the proposed design solutions.

Table 18.3 presents a sample decision matrix in which the columns are defined by the goals (general and specific) identified during the earlier stages of the design process, and the rows of the matrix represent particular concepts or total design solutions that have been developed by the engineering team. (We are using only ten goals here—eight general goals, two specific goals; you may need to identify many more goals for your problem solutions.)

The top row of the matrix in Table 18.3 is labeled "Final Weighting Factor"; the value given in each column represents the relative significance of each goal. In this example, goal 1 (safety) receives a factor of 0.149, whereas goal 2 (minimum maintenance) receives a factor of 0.015, meaning that goal 1 is *perceived* by the designer to be more important than goal 2 by a factor of ten. As shown in Table 18.4, these *final* weighting factors can be developed by listing all goals together with an *initial* weighting factor that is assigned to each goal. (Again, these initial factors are based upon the engineering team's understanding of the problem and the requirements for any solution to this problem.)

These initial weighting factors are then summed (the sum is 1.34 in Table 18.4). Each initial factor is multiplied by the inverse of the sum (1/1.34 in our example); the products of this multiplication are the final weighting factors which will sum to 1.00. In this way, we obtain final weighting factors that represent *percentages* of the total (100 percent) significance of all combined goals.

Each solution is then compared to the other proposed solutions in terms of its ability to achieve each goal. These comparisons are reflected by the rating factors appearing in Table 18.3 (based upon a scale of zero to 10, where zero represents total failure of the design to achieve a particular goal and 10 represents complete attainment of a goal by a solution). In our example, design B (with a rating factor equal to 9 for goal 1) is *expected* to achieve goal 1 much more effectively than design A or design C. Since it did not receive a rating of 10 for this goal, however, it is presumably unsafe under some (perhaps unlikely) operating conditions.

Finally, decision factors for each design are obtained, according to the relation

$$\text{Decision Factor} = (\text{Weighting Factor}) \times (\text{Rating Factor})$$

For example, design A receives a decision factor of 0.596 with respect to goal 1 ($0.149 \times 4 = 0.596$). These decision

Table 18.3 Sample Decision Matrix

Goals[a]	Safety	(Minimum) Maintenance	Cost	Durability	Standard Parts	Public Acceptance	Reliability	Perf.[b]	9[c]	10[c]	Total
Final weighting factor (WF)	0.149	0.015	0.060	0.075	0.060	0.104	0.119	0.119	0.15	0.13	1.00
Design A	4 (= RF)[d]	2	8	5	3	5	8	5	7	6	5.52
	0.596 (= DF)[e]	0.030	0.480	0.375	0.180	0.520	0.952	0.595	1.05	0.78	
Design B	9	7	6	6	8	7	6	9	7	8	7.33
	1.340	0.107	0.360	0.450	0.480	0.728	0.714	1.071	1.05	1.04	
Design C	3	5	4	5	4	3	2	3	5	6	4.48
	0.447	0.750	0.240	0.375	0.240	0.312	0.238	0.357	0.75	0.78	

[a]Performance
[b]Total number of goals may be much larger than the number used in this example.
[c]Goals 9 and 10 are specific to the problem under consideration; they are not identified here.
[d]RF = Rating factor
[e]DF = Decision factor
(RF x WF)

Table 18.4 Development of Final Weighting Factors

Goal	Initial Weighting Factor	x Scaling Factor	= Final Weighting Factor
1	0.20	1/1.34	0.149
2	0.02	1/1.34	0.015
3	0.08	1/1.34	0.060
4	0.10	1/1.34	0.075
5	0.08	1/1.34	0.060
6	0.14	1/1.34	0.104
7	0.16	1/1.34	0.119
8	0.16	1/1.34	0.119
9	0.20	1/1.34	0.149
10	0.18	1/1.34	0.130
TOTAL:	1.34	-----	1.000

factors for *each* design are then summed, providing a total decision factor for each design (shown in the last column of Table 18.3). Comparison of these total decision factors allows the designer to identify the solution that most effectively achieves all desired goals—that is, the solution offering the greatest value, as reflected by the largest total decision factor. If two or more designs receive total decision factors that are very similar, the designer should reevaluate these designs before concluding that one design is superior to another.

During the entire evaluation process, the designer should record the *reasons* for the weighting and rating factors attributed to each goal and each design, respectively. The designer should be able to justify every value appearing in the decision matrix.

A decision matrix allows one to predict, with some accuracy, the benefits and the costs associated with each design solution. Furthermore, the optimum design can be identified, together with its weaknesses or shortcomings. Nevertheless, the evaluation process is very *subjective.* It can be strongly influenced by the personal values of the engineer performing the evaluation, his or her expertise, familiarity with the problem, and the available information concerning the problem. If information concerning the problem goals, the proposed solution, and the relationships between these goals and solutions is incomplete, unavailable or unreliable, evaluation can be very difficult.

18.14 Analysis and CAD

After the optimum, or best, concept has been identified, the design must be analyzed in depth. The design engineer applies knowledge of physics, chemistry, computer programming, and his or her specific engineering discipline to evaluate the ability of the design to solve the problem. The design is studied in terms of such considerations as:

- Its behavior according to the laws of physics.
- Its reaction to its environment.
- The choice of appropriate construction materials.
- The availability of these materials and design components.
- The necessity of all design features.
- Costs.
- Marketing appeal.
- Its potential effects upon the users.

Depending upon the problem and the proposed solution, it may also be necessary to build a prototype or working model of the design solution for experimental analysis. Also, certain designs are sometimes produced in limited quantity for testing in the marketplace in order to obtain consumer reactions. Hence, a design may be analyzed *theoretically, experimentally* or through *market-sampling* procedure. If weaknesses or flaws in the design are discovered as a result of these analyses, the proposed solution is modified or rejected.

Today, CAD (computer-aided design) has become an extremely valuable analytical tool that can be used to compare and evaluate alternative design solutions (in ways somewhat similar to those of the decision matrix). In addition, such CAD systems allow the engineer to generate detailed written and graphic descriptions of a proposed design—including material and size specifications that are consistent with the goals and constraints associated with the problem to be solved and that are often developed through a detailed analysis of the design designated as being optimum. Computer technology, therefore, allows us to analyze designs with greater speed and accuracy than ever before—and then store the results or produce written and graphic documentation, as needed (Figures 18.11, 18.12, and 18.13).

18.15 Written and Oral Presentations

After the feasibility of the final design solution has been verified, the designer must present the design in a formal report. This report may include both written and oral presentations of (1) the chosen design, (2) the process employed during the development of this design, (3) the alternative designs that were rejected (and the reasons for rejection), and any other relevant data. Management will use the report to determine if the design should be produced, modified, or rejected; therefore, the report must be carefully prepared by the designer, with ideas and recommendations presented with *clarity.*

The **written presentation** should be comprehensive and detailed, yet concise. All important conclusions and recommendations should be stated succinctly and presented in such a way that the reader can easily identify the major

Figure 18.11 A Control Data Corporation ICEM Workstation with which a solid model of an engineering design may be developed. (Photo courtesy of Control Data Corporation.)

Figure 18.12 A fabricated part is compared to the solid model which is shown on a Control Data Corporation ICEM Workstation. (Photo courtesy of Control Data Corporation.)

findings of the report. The topics should be chosen and ordered so that the discussion flows easily from one section of the report to the next. In order to enhance clarity, graphic communication is usually employed via orthographic (detail and assembly) drawings, schematics, pictorials, technical notes and so forth. The technical level of the report should be consistent with the technical background of the audience.

A **materials list** (bill of materials) must also be provided specifying all *standard components*[1] by name (and identification or model number), specifications, quantity required for the design, material, and so forth. In addition, all *raw materials*[2] that are needed for the design and that can be purchased in bulk must be identified. Finally, a *cost breakdown* for materials, labor, overhead (such as electricity, heating, taxes, rent, and depreciation expenses), advertising, shipping, and other expenses must be provided.

References, sources of information and any other appropriate written material that informs the reader about the design solution and its development should also be included. The engineer should examine the written report to ensure that proper grammar and spelling have been used.

---

1. These components may include relays, solenoids, nuts, screws, bolts, motors, gears, lights, switches, and so forth.

2. Such materials include sheet metal, wood, cement, paint, and so forth.

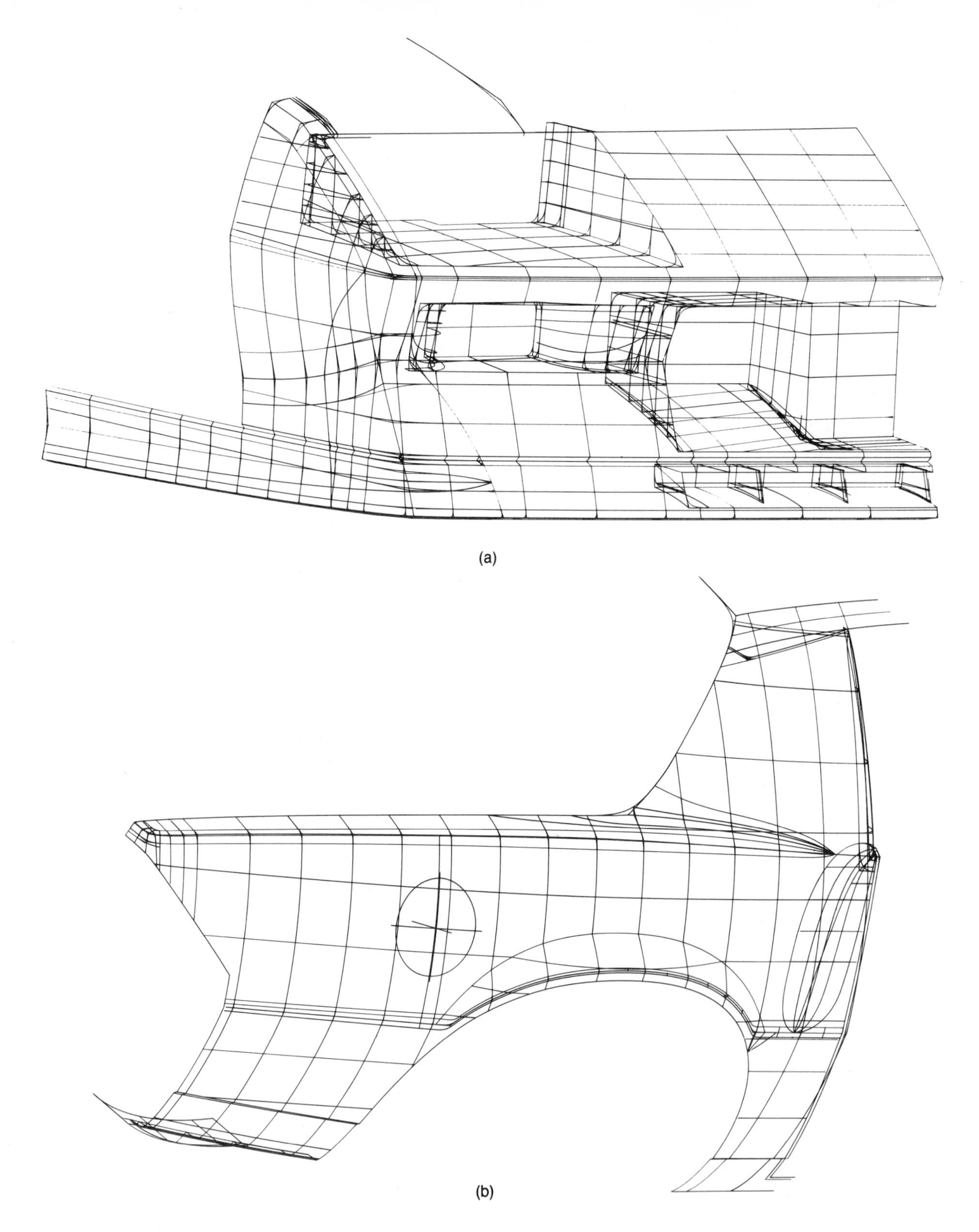

Figure 18.13 A computer may also be used to generate detailed perspective drawings of an object, based upon geometric and other design data. (a) Automobile front bumper and air dam. (b) Automobile quarter panel. (Courtesy of General Motors Corporation.)

A **draft** of the written report should be prepared before the final version is generated. The draft should be written quickly without undue concern about choice of words but simply concentrating upon the key points or ideas that form the basis for the report. Each statement should be supported with data (which may be included within an appendix). After the draft has been written, it should be carefully *reviewed and revised* for the particular audience to which it is addressed. The final written report is more likely to be clear, accurate, and well-organized if this procedure is followed.

The written report should be accompanied by a **letter of transmittal,** or cover letter, that states that the report is attached and that summarizes the significant conclusions or recommendations within the report). The report should include the following elements:

- Title page
- Table of contents
  - List of figures
  - List of tables
- Problem statement
- Methodology (a review of the procedure or method used to solve the problem)
- Main text or body of report (the details of the report are given in this section or sections; graphic descriptions should be included where appropriate.)
- Analysis of alternative solutions/designs
- Results/presentation of best design solution
- Recommendations, implementation of results, and any evaluation(s)
- Conclusion
- Bibliography
- Appendices

Finally, the report should be *bound* in an appropriate manner.

The **oral presentation** should be prepared for the same audience addressed by the written report. The designer thus has the opportunity to emphasize verbally the significant results of the work, answer questions from the audience, and defend any perceived weaknesses inherent in the final design.

Oral presentations must be prepared as carefully as the written report. (It is usually advisable to provide the members of the audience with copies of the written report—or at least summaries of the report—several days before the oral presentation is to be given. This will provide the audience members with the opportunity to familiarize themselves with the details of the report.) The designer should prepare the oral presentation with attention to seating arrangements for the audience, the room facilities (such as projection equipment, blackboard, chalk, microphones, water, and screen) and the visual aids (such as charts, slides, and illustrations) that are so vital to a truly effective presentation.

The designer should *rehearse* the oral presentation until he or she is completely familiar with its order, content, and required time. If there is a time limit for the presentation, it should be short enough to easily fit within the allotted time. The speaker should remember that in rehearsal he or she will not be distracted by the presence of the audience, whereas during the actual presentation some amount of time might be lost due to nervousness or unforeseen delays. The oral presentation should be planned to allow for such delays.

The designer should briefly review the problem for the audience, since people usually need a minute or two for their attention to become fully directed toward a topic of discussion. The oral report should be consistent with the written report in its order of topics, conclusions, and so forth, but it should not—indeed it cannot—include all of the information contained within the written report.

Finally, the designer's manner of presentation should be enthusiastic and friendly. He or she should be honest in identifying the major weaknesses of the design, together with justifications for including these weaknesses.

18.16 Production, Distribution, and Consumption of Design Solutions

After receiving final approval for production, the design must be completely described with detail and assembly drawings. These may include fully dimensioned orthographic views, pictorials, schematics, and so forth.

The final stages of the design process include the manufacturing of the product, its distribution to the marketplace, and consumption by the users. We will not discuss the details of these stages here, except to note that there may be feedback information about the product (and its performance) provided to the manufacturer by the consumers. This feedback information could identify unforeseen weaknesses in the design or suggest additional features that should be included in an improved version of the product.

Remember, the consumer is not only *a* member of the design team, but he or she is *the most important member* of the design team!

18.17 Patenting Engineering Designs[1]

18.17.1 Historical Review

As noted by Joenk (1979a), the United States patenting system is based upon the principle of *quid pro quo* (something

1. References are given at the end of this section.

for something). In exchange for publicizing his or her invention, a person receives the right to *exclude others* from making, selling, or using the invention in the United States for a period of seventeen years (nonrenewable).

The first *recorded* patent to be granted for an invention was awarded by the Republic of Florence in 1421 to the designer of a barge with hoisting gear for loading and unloading marble. Subsequent early patents were granted in Venice (1469), in Germany (1484), and in France (1543).

In 1641 Samuel Winslow was granted the first patent in colonial America (by the Massachusetts General Court) for a salt-making process. In 1783 the Continental Congress recommended the enactment (by each state) of copyright acts; then in 1789 the United States Constitution empowered the federal government

> To promote the Progress of Science and useful Arts by securing for limited Times to Authors and Inventors the exclusive Right to their respective Writings and Discoveries.

Finally, the first United States Patent Act (April 10, 1790) resulted in the granting of the first U.S. patent to Samuel Hopkins for "Making Pot and Pearl Ashes" on July 31, 1790.

18.17.2 Foremost Criteria

Patent examiners are particularly concerned with three criteria in determining if an applicant should be granted a patent: novelty, usefulness, and nonobviousness (see the Patent Law listed under 35 U.S.C. 101, 102, 103—the relevant sections of Title 35 of the United States Code; also see *Patent Laws,* which is available from the Superintendent of Documents, Government Printing Office, Washington, D.C. 20402).

The first criteria, **novelty,** means that the invention must be *demonstrably different* from "prior art." (All work in the field of the invention is "prior art"; an inventor is expected to be familiar with this prior art material.)

In the United States the person(s) who is (are) recognized as the **first to invent** (as opposed to the **first to file**) is (are) granted the patent for an invention. (All inventors of a single invention must apply together for a patent.) In order to receive such recognition, the inventor must demonstrate (1) the earliest date of conception of the invention and (2) diligence in reducing it to practice without any period of abandonment (Joenk, 1979a). Therefore, it is highly recommended that the inventor keep complete records during the period of development before filing the patent application. The following list provides hints on record-keeping (Franz and Child, 1979).

- Records should be written.
- Obtain competent witnesses of the recording:
    - Preferably two or more other persons; inventors cannot serve as witnesses to their own inventions.

- Each witness should date and sign each page of the inventor's notes and sketches as soon as possible after these records are generated.
- Tests and experiments should also be witnessed—with corresponding dated, signed entries in the record book.
- Witnesses must be available (easily located) in the future.
- Witnesses should have the technical training or knowledge to understand the invention and its use.
- Witnesses should not be related to the inventor.
- An inventor's protection that his or her witnesses will not steal his/her work is their signatures as witnesses.

- Records should be kept in a bound laboratory notebook with numbered pages.
- Entries should be in ink and dated; any delays in conducting the work should be explained in writing.
- The advantages and uses of the invention should be identified, together with complete descriptions of the design components and specifications.
- All related papers (correspondence, sales slips, components' sales literature, and so forth) should be saved.
- If a correction is necessary, line through the error and initial and date the change.
- Later additions should be entered in an ink of a different color, then initialed and dated.
- Finally, sign each page as the "inventor" and enter the date.

The invention must be **reduced to practice,** or completed. Such a reduction-to-practice can be **actual** (that is, a prototype can be built) or **constructive** (that is, a complete patent application is filed that satisfies the requirements of the *United States Patent and Trademark Office,* abbreviated here as PTO). Although expensive, an actual reduction can lead to improvements and needed modifications in the design.

The inventor must avoid (1) publishing a description of the invention, (2) offering a product in which the invention is incorporated, and (3) allowing anyone else to use the invention (other than on an experimental basis). In other words, disclosure before filing a patent application must be avoided.

The second criteria, **usefulness,** means that a desired objective (or objectives) must be achieved by the invention. That is, there must be some practical utility associated with the invention. Immoral, fraudulent, frivolous, and antipublic policy uses are not patentable (Joenk, 1979a).

The third criteria, **nonobviousness,** means that the invention must be deemed to have required more than (1) ordinary skill to design or (2) the mere addition or duplication of components of an existing (patented) design. However, if a new result is achieved through the new application of an old design, then that new application may be patentable.

18.17.3 Additional Guidelines

Before applying for a patent, one should consider the advantages of obtaining professional assistance. The more than 9,000 patent agents and attorneys who are registered to represent inventors may do so before the United States PTO. An annual listing of these agents and attorneys is available from the PTO *(Attorneys and Agents Registered to Practice Before the U.S. Patent and Trademark Office).* Patent attorneys are knowledgeable about the patent laws and technically competent as well. Chester Carlson was a patent lawyer at the time he invented the xerographic copying process (Vanderbilt, 1979). A key goal of the agent or attorney is the maximization of the coverage and protection that will be provided by the patent by carefully writing the claims listed in the application (Nussbaumer, 1979).

A **patent search** must be conducted. The public file of the PTO in Arlington, Virginia (two identical files are maintained at the PTO—one for use by the public and one for use by the employees of the PTO, particularly patent examiners), as well as the partial U.S. patent collections located in many libraries throughout the nation, can be examined in order to determine if the inventor has designed a truly new invention. The U.S. Patent Classification System contains approximately 350 distinct classes. The scheme is as follows: a class is listed, followed by a series of subclasses (**mainline,** or primary, divisions)—each of which is further divided into additional subcategories (there are about 100,000 subdivisions within the system). The format is as follows:

CLASS
- MAINLINE (Primary) division
  - Secondary division
  - Secondary division
    - Subcategory
    - Subcategory
      - ----
      - ----
      - ----
- MAINLINE division
  - ----
  - ----
  - ----

Each patent is classified within the system according to the *most comprehensive claim* contained within the patent. A classification label is given to each patent in the form:

CLASS NUMBER—SUBCLASS NUMBER

where the subclass refers to the most distinct category to which the patent belongs. The PTO offers such publications as the *Manual of Classification* and the *Index to U.S. Patent Classification.* The reader is also referred to the paper by Dood (1979).

A **disclosure,** which describes the field of technology in which the invention belongs, the components of the design (and their functions)—with sketches, and the advantages/differences of the new invention relative to the prior art of the field, will aid the searcher (Joenk, 1979a).

The search is crucial since its results will determine if an application should be submitted. If the decision is made to submit an application, the following sections should be included (Joenk, 1979a):

- **Specification** (or text portion), which identifies
  - Any appropriate previous patents or applications of the inventor that are related to the current application.
  - The technical field to which the invention belongs.
  - The prior art of the invention—in order to expedite the search and examination of the application by the PTO.
  - The problem and its solution by the invention (including advantages over the prior art).
  - Any figures or drawings as necessary.
  - The invention—via a sufficiently detailed and complete description of the invention and its use(s).
  - The claims to be protected by the patent.
- An **abstract,** which summarizes the disclosure and claims.
- An **oath** or **declaration** in which the applicant states
  - One's belief that he or she is the original and first inventor of the described invention.
  - His or her citizenship.

Note that the **claims** determine the legal coverage to be provided by the patent. Each claim is written as a single sentence, beginning with the phrase "I (We) claim" or "What is claimed is." This phrase is used only once, even if multiple claims are made. All claims after the first one are written as if this phrase were prefixed to them. In addition, claims are ordered from the most general to the most specific. Any element of the invention that is mentioned as a claim should have been described earlier in the specification portion of the application (Landis, 1974).

18.17.4 Patent Drawings

As required by 35 U.S.C. 113, "When the nature of the case admits, the applicant shall furnish a drawing." A **patent drawing** may consist of several figures. These figures must show every feature of the invention, although *conventional* characteristics can be represented by a *standard* (or labeled) **graphic symbol.** See *Guide for Patent Draftsmen* and Joenk (1979b) for graphic symbols.

Any "improved" portion of an old device should be shown, both disconnected from and in connection with the main apparatus. Also, one view of the invention should be representative and suitable to serve as the only illustration to be published on the title page of the patent and in the announcement of the patent in the *Official Gazette* of the PTO.

The standards for the drawing(s) are as follows:

- **Paper/ink:** White paper of a thickness equivalent to that of two-ply or three-ply bristleboard with an erasurable surface. India ink should be used for black linework.
- **Size/margins:** Each sheet should be 8½ in. by 14 in. (21.6 cm. by 35.6 cm) with a top margin (along one of the edges of 8½ in. in length) of 2 in. (5.1 cm.) and bottom/side margins of 0.25 in. (6.4 mm.). The working space will then be 8 in. by 11¾ in.
- **Identification block:** The identification of the drawing should be contained within a rectangle 2¾ in. (7.0 cm.) in width and ¹¹⁄₁₆ in. (17.5 mm.) in height, located at the upper edge of the paper. This identification may consist of the name of the inventor and the case number of the patent attorney's name and the docket number; in addition, the sheet number and total number of sheets can be included (for example, "sheet 2 of 3").
- **Linework:** Performed with drafting instruments (or by other professional means) for high-quality reproduction. Spacing between parallel hatchwork lines should be at least 0.05 in. (1.3 mm). Crowding should be avoided; if necessary, several sheets may be used to describe the invention.
- **Reference characters, symbols, and legends:** Reference numerals should be at least ⅛ in. (3.22 mm.) in height and should not be encircled; lines are used to connect the numerals to the parts of the design used to identify a particular component in all views in which that component appears. Standard mechanical and electrical (graphic) symbols can be used to identify conventional components within the design; legends may also be used where needed.

18.17.5 Conclusion on Patenting Engineering Designs

Patent rights can be traded, licensed, sold, or assigned—as can other forms of personal property (Wolber, 1979). As Hill (1970) notes, the United States Patent System has provided inventors with the necessary protection, opportunity and hope of reward to encourage the growth of new technology and industrial products.

References for This Section

Dood, K. J., "The U.S. Patent Classification System", in *IEEE Transactions on Professional Communication,* Volume PC-22, June 1979, pp. 95–100.

Franz, W. L., and Child, J. S., Jr., "Good Habits before Filing a Patent Application", in *IEEE Transactions on Professional Communication,* Volume PC-22, June 1979, pp. 86–88.

Hill, P. H., *The Science of Engineering Design,* Holt, Rinehart and Winston, Inc., New York, NY, 1970.

Joenk, R. J., "Patents: Incentive to Innovate and Communication—An Introduction", *IEEE Transactions on Professional Communication,* Volume PC-22, June 1979a, pp. 46–58.

Joenk, R. J., "Guide for Patent Drawings", in *IEEE Transactions on Professional Communication,* Volume PC-22, June 1979b, pp. 109–111.

Landis, J. L., *Mechanics of Patent Claim Drafting,* Practicing Law Institute, New York, NY, G1-0633, 1974.

Nussbaumer, H. J., "Patents and the Engineer", in *IEEE Transactions on Professional Communication,* Volume PC-22, June 1979, pp. 68–72.

Skolnik, H., "Historical Aspects of Patent Systems", in *IEEE Transactions on Professional Communication,* Volume PC-22, June 1979, pp. 59–62.

Vanderbilt, B. M., "The Plight of the Independent Inventor", in *IEEE Transactions on Professional Communication,* Volume PC-22, June 1979, pp. 63–67.

Wolber, W. G., "The Business Value of Patents", in *IEEE Transactions on Professional Communication,* Volume PC-22, June 1979, pp. 73–76.

Attorneys and Agents Registered to Practice Before the Patent and Trademark Office, Superintendent of Documents, Government Printing Office, Washington, D.C., 20402.

Guide for Patent Draftsmen, U.S. Department of Commerce Patent and Trademark Office, available from the Superintendent of Documents, Washington, D.C. 20402.

Index to United States Patent Classification, U.S. Department of Commerce Patent and Trademark Office, available from the Superintendent of Documents, Washington, D.C., 20402.

Manual of Classification, U.S. Department of Commerce Patent and Trademark Office, available from the Superintendent of Documents, Washington, D.C. 20402.

Official Gazette, U.S. Department of Commerce Patent and Trademark Office, available from the Superintendent of Documents, Washington, D.C., 20402.

Patent Laws, available from the Superintendent of Documents, Government Printing Office, Washington, D.C. 20402.

In Retrospect

- The design process is a step-by-step procedure that is applied in the development of a solution to an engineering problem.
- Design engineers must be aware of the user's needs and, if possible, the probable reactions of the users to proposed design solutions. Engineers must recognize that personal and societal values can affect their ability to design, develop, and evaluate alternative solutions to a problem.
- Design engineering considers human factors in the development of design solutions. Human-machine interactions, the environment of an operating design system, and typical human body dimensions are some of the concerns of the design engineer.
- Additional factors that may influence design efforts include physical or environmental considerations, economic conditions, material characteris-

tics, standardization guidelines, storage and waste disposal requirements, and the availability of materials and/or components.

- The initial step of the design process is to determine the specific problem to be solved. Marketing surveys and other tools can be used to identify a need.
- After the problem has been identified and the need for a solution has been justified, the problem must be defined in terms of *general* engineering goals, which are common to virtually all engineering problems, and *specific* goals, which are unique to the particular problem under investigation. General goals include the use of standard parts, safety, durability, maintenance, public acceptance, reliability, performance, cost, and operating conditions. Each general goal may be divided into a set of specific goals that directly relate to the particular problem under consideration. Precisely and fully defined goals will assist the designer in obtaining an objective perspective of the problem to be solved and the characteristics of the solutions to be developed.
- The next stage of the design process involves the identification of the specifications, or constraints, associated with each goal that restricts the design solutions possible. These constraints may be physical, functional, environmental, economic, or legal.
- Numerous ideas should be generated for the achievement of each design goal. A complete set of ideas (one for each goal) can then be formed to produce a design solution to the problem during the conceptualization stage of the design process. Several complete, innovative, and feasible design solutions can thereby be created.
- *To be effective, an engineer must be familiar with a broad range of physical phenomena* so that he or she can apply this information in the design of engineering solutions. Scientific principles, empirical data, and available design components form a source-pool of ideas, which can help create innovative solutions to challenging problems.
- The optimum concept must be selected from the set of proposed design solutions to a problem. One technique for aiding the analysis and comparison of alternative solutions is the decision matrix, with which designs are compared in terms of their relative ability to achieve the project goals.
- After the optimum, or best, solution has been identified, it must be analyzed in terms of the following (in addition to other considerations).
  - Its behavior according to the laws of physics.
  - Its reaction to its operating environment.
  - The choice of appropriate construction materials.
  - The availability of these materials and design components.
  - The necessity of all design features.
  - Costs.
  - Marketing.
  - Its potential effects upon users.
- Patent rights can be traded, licensed, sold, or assigned. The United States patent system has provided inventors with the necessary protection, opportunity, and hope of reward to encourage the growth of new technology and industrial products. Patent examiners are particularly concerned with three criteria in determining if an applicant should be granted a patent: novelty, usefulness, and nonobviousness. Patent records also provide another source-pool of ideas for achieving specific engineering goals; these ideas can be modified or applied in new and different ways to solve existing problems. (And, of course, any design engineer should determine whether any patented solutions exist to the problem under consideration in order to avoid fruitless effort.)
- Engineers must be familiar with a variety of materials and fabrication processes, some of which were reviewed in this chapter.

ENGINEERING IN ACTION

Tube Sail Drives Cousteau's Experimental Boat

A Cleaner Calypso

Capt. Jacques Cousteau
President, Cousteau Society, Norfolk, VA

"The technology that permitted heavier-than-air flight will help us perfect sailing without sails. Our experimental catamaran symbolizes our hope that. . . we will soon cease to squander the limited resources of our planet."

When Jacques Cousteau decided he needed a replacement for his famous research vessel, Calypso, he began looking for a design that was in keeping with his clean-environment philosophy. High on his list of possibilities was sail power.

A team of French scientists—headed by Dr. Lucien Malavard, who has a special interest in the fluid mechanics of sailboats—was charged with considering the possibilities. Conventional sails were eliminated because they are too large, require a skilled crew, and need frequent maintenance. An idea that showed promise was a sail based on the Magnus-effect, a phenomenon in which lift forces are generated by a cylinder when it is rotated in a fluid.

In 1924, a ship was built using such tubular sails for propulsion; it operated successfully for several years. Two sails gave the ship a top speed of 11.5 mph. Rotation speed was 150 rpm. Although the French designers liked the basic idea, they wanted to eliminate the cost, complexity, and hazards of a rotating cylinder. This was done by using vents and a fan to create the same forces that occur when a cylinder rotates. The thrust force can be controlled by turning the tube relative to the wind or by changing the point at which air is drawn into the tubes. This is an important advantage over the rotating-tube sale, which has to change direction of rotation in order to change thrust direction.

Wind tunnel tests showed that an 800-ton ship with two cylinders, and having an area of 150 m², would have a top speed of 15 knots in a 24-knot wind. A 150-hp fan engine would be required.

With optimum flow, the tube sail can provide 90% of the energy needed to maintain a given speed. An equivalent motor ship would need a 1,500-hp motor and a conventional-sail ship would need a sail area of 750 m².

The design that resulted from the tests is a 44-ft-tall, 4.5-ft.-diameter tube sail. It is used as the chief propulsion source on a 65-ft. catamaran, the *Moulin à Vent* (Windmill). This test vessel has been operating in the Mediterranean.

If the design proves to be adequate, a single-hull *Calypso II,* powered by two tube sails, will be built.

Basic Operation

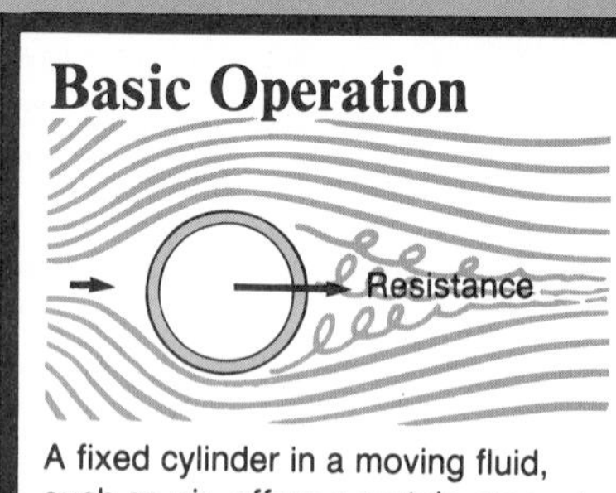

A fixed cylinder in a moving fluid, such as air, offers a certain amount of resistance to the fluid, causing turbulence in the downstream flow.

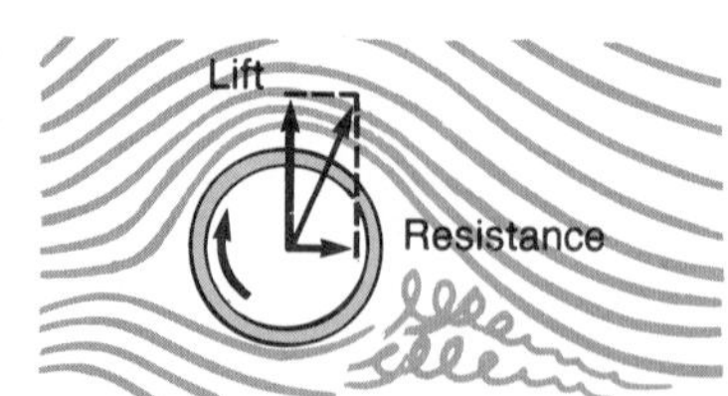

When the same cylinder is rotated in a moving fluid, a lifting force is created perpendicular to the flow direction. If the cylinder is placed on a ship, the lift force can be used as a means of propulsion.

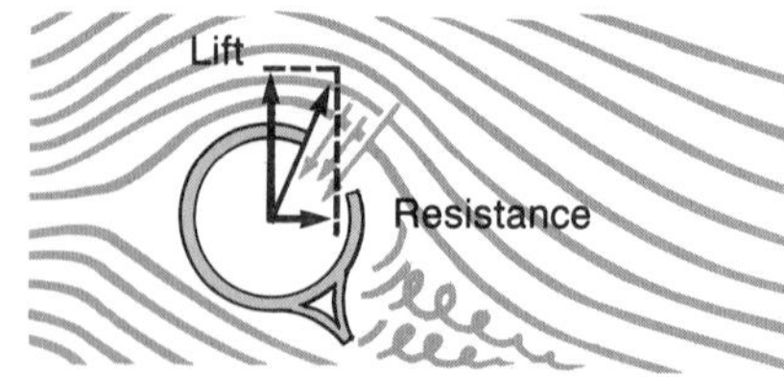

The lift effect can be achieved by adding a movable flap and drawing air into the cylinder through appropriately placed slots. The lift force achieved through this method is approximately the same as with the rotating tube, but the complex cylinder-rotation mechanism is eliminated.

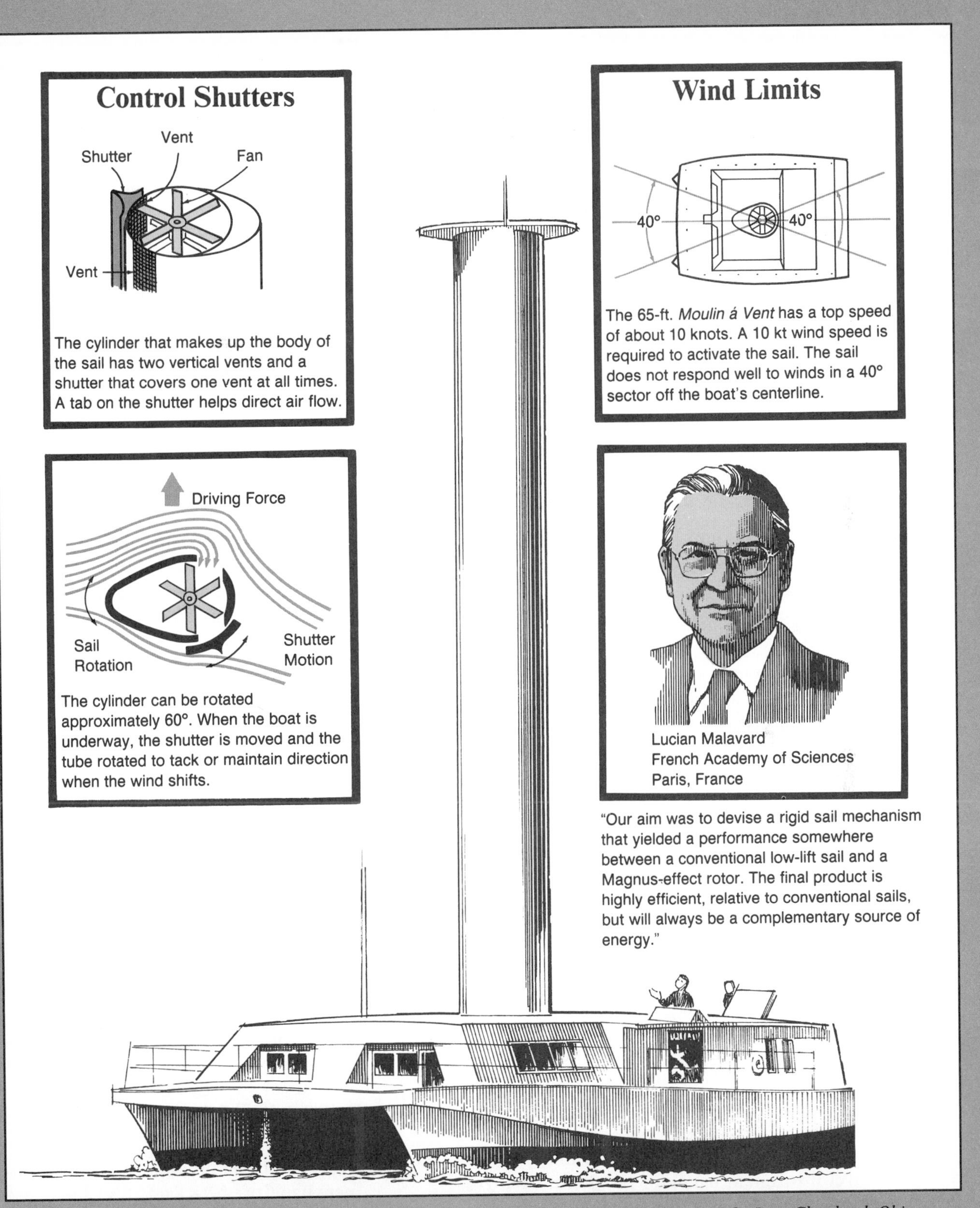

Source: Reprinted from MACHINE DESIGN, October 6, 1983, Copyright, 1983, by Penton/IPC., Inc., Cleveland, Ohio, pp. 102, 103.

Exercises

EX18.1. Choose a product or device and identify the functions of each component (for example, if a surface is covered by a plastic coating, explain why this coating was included in the design). Sketch the product or device and identify each component and its function/purpose.

EX18.2. Why is it imperative that a design problem not be defined in terms of a proposed or existing solution?

EX18.3 Identify three engineering products or designs that you believe to be innovative and unique. Explain how each design can be used to serve humankind, and justify your decision that the design is innovative.

EX18.4. Most preliminary designs are ultimately rejected. Why, then, should sketches of these designs be prepared during the design process? List as many reasons as possible.

EX18.5. Compare the following two approaches to design: the individual working alone and the design team working together. What are the positive and negative aspects of each approach? In what circumstances would you recommend that the individual work as a member of a team?

EX18.6. What is the difference between the *problem statement* and the *list of goals* that must be achieved by a solution? Be as specific as possible in stating this difference.

EX18.7. Why must a design engineer be thoroughly familiar with the goals that must be achieved by his or her solution(s) to a problem? Why is this *goal list* continually updated during the design effort? Why must the design engineer communicate with the potential user(s) of his or her design?

EX18.8. Using your own body dimensions as a guide, design an automobile seat and control instrument panel for the driver and passengers. Such factors as comfort, level of vision, arm length, steering wheel position, and the position of pedal controls will influence your design. Include a complete list of goals for this problem. Sketch your design solution. Identify and explain each feature of the design.

EX18.9. Review relevant literature that describes automobile design features. Determine if your automobile seat and instrument panel design of EX18.8 corresponds to the specifications described in this literature. Explain in detail which design features of your solution conform to industry specifications and which do not. Also, state whether the nonconforming specifications need to be modified to meet industrial specifications, and if so, describe these modifications in detail and with sketches.

EX18.10. An ice cream scooping device must be strong, durable, and comfortable for the user. Design a hand-held ice cream scooping device that satisfies these goals. (Consider that a worker in an ice cream parlor might need to scoop hundreds of servings during a single eight-hour shift. Your device should allow the worker to accomplish this task in an efficient and painless manner.)

EX18.11. Perform a patent search for ice cream scooping devices (see EX18.10) Identify the significant differences between all patented designs that you discover during this search. Include all relevant information used to identify the patent(s).

EX18.12. Modify the industrial hand-held ice cream scooping device described in EX18.10 so that it meets the needs for home use. (Answer the following question before you begin to modify an industrial design: What are the different goals that must be satisfied by a device designed for home use instead of industrial or commercial use?)

EX18.13. Redesign the automobile seat and instrument panel of EX18.8 and EX18.9 for use by a disabled person. Identify the specific abilities that a user would need to use your modified design. Identify and explain the function of each component in your design. Sketch the design and its components.

EX18.14. Prepare a brief report upon the engineering discipline of your choice. Identify (with examples) the types of engineering problems that are solved by those involved in this discipline.

EX18.15. Consider the engineering design process: identify the stages of this process in which each member (for example, engineer, craftsperson, stylist, accountant, lawyer) of the design team is actively involved. Develop a brief chart that describes this involvement in the process.

EX18.16. Consider a product of your choice (or perhaps one of the designs described in the "Engineering in Action" sections near the end of many chapters in this text). Identify as many specific goals for each design as possible; in addition, try to weight these goals relative to one another. (Remember to consider the user population of each design.)

EX18.17. Identify the problem specifications, or constraints, that must be placed upon a design of your choice (for example, one of the designs in the "Engineering in Action" sections throughout this text). Determine the type of boundary (physical, economic, legal, and so forth) of each specification/constraint.

Develop three different solutions to one or more of the following engineering problems. Include a list of specific goals, problem specifications, or constraints, ideas for achieving each design goal, sketches for each design, and so on—that is, be as detailed and conscientious in your effort as possible. Perform a patent search, if time permits. *Be both innovative and practical!*

EX18.18. A typing or computer terminal stand/table.

EX18.19. A garage door.

EX18.20. A kitchen appliance of your choice.

EX18.21. A device to automatically scramble eggs.

EX18.22. A collapsible crutch.

EX18.23. A fruit juicer.

EX18.24. An automobile jack.

EX18.25. An educational toy.

EX18.26. A hook or other fastening device which is unique.

EX18.27. A hand-held hammer.

EX18.28. A piece of sporting equipment of your choice (such as tennis racquet, baseball, basketball hoop, protective clothing).

EX18.29. A new device for lawn care (not necessarily a lawnmower).

EX18.30. A storage unit for records or cassettes.

EX18.31. An eating utensil that can be used to effectively manipulate food sometimes served in difficult form (such as spaghetti, grapefruit, pizza).

EX18.32. A container for liquids.

EX18.33. A container for crystals or powders.

EX18.34. An irrigation system for agricultural areas with little or no rainfall.

EX18.35. A briefcase.

EX18.36. Sleeping accommodations for several people in a small shelter.

EX18.37. A doghouse.

EX18.38. A tool for planing or otherwise shaping wood.

EX18.39. A flashlight.

EX18.40. A motor-driven conveyor belt system (you will need to become familiar with existing systems).

EX18.41. A stapler.

EX18.42. A tool or piece of equipment for exercising.

EX18.43. A problem of your choice.

EX18.44. One of the engineering design problems presented in the "Engineering in Action" sections within this text.

Extended Design Projects

The following problems may be used as term projects *in accordance with the directions of your instructor.* Each problem focuses upon a current need in occupational safety and health. It is recommended that student design *teams* be formed to develop creative and feasible solutions to each problem. Each problem is very challenging, as one would expect of any real-life engineering project. You may wish to define each problem in more specific or narrow terms so that a feasible design can be developed within your time constraints.

These problems have been developed by the National Institute for Occupational Safety and Health (NIOSH), United States Department of Health and Human Services for the American Society for Engineering Education (ASEE)—NIOSH Engineering Design Competition in Health and Safety. The author is grateful to Mr. John Talty, Project Coordinator, and Mr. William McKinnery, both of NIOSH (Figure 18.14).

(Source: Used by permission of the National Institute for Occupational Safety and Health.)

A. Asbestos Fibers

1. *Title:* Engineering control of asbestos fiber hazards.

2. *Problem:* Asbestos fibers have been found to cause cancer and respiratory disease (asbestosis) in workers. The basic hazards caused by worker exposure to asbestos were first reported in the United States in 1938. The permissible exposure limit (PEL) was first legally established at 12 fibers/cc of air in 1960. It was reduced to 5 fibers/cc in 1971 and 2 fibers/cc in 1976. NIOSH is currently recommending

Figure 18.14 The award-winning design project of the 1985 ASEE/NIOSH Engineering Design Competition in Health and Safety is represented by Jan Simon and Darren Martin (standing), two members of the Student Design Team from Texas A&M University. Kneeling are: John T. Talty, NIOSH; William M. McKinnery, NIOSH; Professor Gerard Voland, Northeastern University; and Dr. Melvin First, Harvard University. (Used by permission of the National Institute for Occupational Safety and Health.)

a PEL of 0.1 fibers/cc. A primary current use of asbestos is in the manufacture of products such as brake shoes, thermal insulation, floor tile, and cement pipe. In these uses of asbestos, one of the most difficult operations to control is dust generated during bag delivery, bag emptying (debagging) into hoppers for subsequent mixing, and proper disposal of empty delivery bags.

A typical process flow description is as follows:

a. Asbestos is received in bales that have been compressed and wrapped in plastic.
b. The asbestos bales are stored on pallets for use as needed.
c. Forklift trucks transport pallets to a bag emptying work station.
d. A worker transfers bales from pallets to a conveyor.
e. The conveyor feeds bales into bag opening equipment.
f. The bag opening equipment separates the bale from the wrapping which is discarded.
g. The asbestos is processed and transported to a hopper or mixing unit.

Workers who operate bag emptying equipment may be exposed to asbestos from the following primary sources:

a. Bales of asbestos received with, or developing, tears or holes.
b. Reentrainment of settled asbestos fibers from the floor, unused bags, and equipment.
c. Bag opening, emptying, and disposal unit operations.

3. *Solution needed:* Engineering control measures are essential to prevent worker exposure to asbestos fibers during debagging and bag disposal operations. Possible approaches worthy of investigation include:

a. Use of fully automated bag opening and empty bag compacting machinery which prevents the release of asbestos fibers.
b. Use of local exhaust ventilated, manual bag opening station with integral disposal of empty bags.
c. Other innovative process design improvements to prevent worker exposure to asbestos fibers.

4. *References:*

a. Bel-Tyne, "Fully Automatic Bag Slitter" (sales brochure, Bel-Tyne Co. Ltd., Stockport, England).
b. Environmental Protection Agency, "Control Techniques for Asbestos Air Pollutants," U.S. EPA (OAQPS) Report No. AP-117 (1973).
c. First, M. W., and Love, D., "Engineering Control of Asbestos," *American Industrial Hygiene Association Journal* 43, no. 9 (1982).
d. Heitbrink, W. A., "In-depth Survey Report—Control Technology for Richard Klinger, Inc." Unpublished report of U.S. DHHS (NIOSH) (1984).

e. National Institute for Occupational Safety and Health, "Criteria for a Recommended Standard—Occupational Exposure to Asbestos," U.S. DHHS (NIOSH) Publ. No. 72–10267 (1972).
f. National Institute for Occupational Safety and Health, "Revised Recommended Asbestos Standard," U.S. DHHS (NIOSH) Publ. No. 77–169 (1976).
g. National Institute for Occupational Safety and Health, and Occupational Safety and Health Administration, Asbestos Work Group, "Workplace Exposure to Asbestos—Review and Recommendations," U.S. DHHS (NIOSH) Publ. No. 81-103 (1980).
h. Occupational Safety and Health Administration, "General Industry Standards, Subpart Z—Toxic and Hazardous Substances," Sec. 1910.1001—Asbestos, U.S. DOL (OSHA) Publ. No. 2206 (1983).

B. Grain Dust

1. *Title:* Engineering control of grain dust explosions and fires.

2. *Problem:* There are approximately 15,000 grain-handling and grain-processing facilities in the United States. These facilities include grain elevators, feed mills, and other grain-processing plants.

Fires and explosions in these facilities have been reported in this country and abroad for almost 200 years. This danger is present in the industry because of the physical characteristics of organic dust that is generated while handling and processing grains.

Grain elevators are establishments which provide storage space and serve as collection and transfer points for grain and beans. Auxiliary operations such as sampling, weighing, blending, drying, cleaning, and fumigating may be performed. Feed mills are establishments engaged in the manufacture of feeds for animals.

The threat of dust fires and explosions and the corresponding severity of injuries and damage prompts the greatest safety concern in grain-handling and grain-processing facilities. Of all the industrial dust explosions in the United States, those in grain elevators are the most frequent and cause the most injuries and property damage. According to the National Fire Protection Association, about 48 percent of the total number of dust explosions in the United States (1900–1956) occurred in industries handling grain, feed, and flour.

A recent U.S. Department of Agriculture (USDA) compilation includes 434 explosions in U.S. grain-handling facilities in the 25-year period from 1958 through 1982 which resulted in 776 injuries and 209 deaths. The USDA also reported fire experience for the period from 1958 through

1975 on the basis of data provided by the National Fire Protection Association. The number of fires in the grain-handling industry during this 18-year period averaged about 2,700 incidents per year. Fires in grain elevators and feed mills result in the loss of millions of dollars in both direct expenses and lost time.

For a grain dust explosion to occur, the following conditions must be met:

- Grain dust must be present.
- An ignition source must be present.
- Oxygen must be present in a concentration to sustain rapid combustion.
- The grain dust must be well mixed with the oxygen at a concentration above the lower explosive limit.
- Ignition must occur in an enclosed space.

Grain breakage occurs initially at harvest and continues through each subsequent handling. Particles range in size from respirable (typically < 10 microns in aerodynamic diameter) to particles of 120 microns or more. The dust may be suspended in the air, settled onto horizontal surfaces, or adhered to vertical surfaces.

The explosibility of a particular dust is determined by its concentration in air and influenced by factors such as chemical composition and particle size. The presence of noncombustibles, such as mineral matter or moisture, decreases the explosibility. Increases in particle size also decrease explosibility.

The presence of layered dust is a significant problem. Dust settles not only on floors, ledges, and other horizontal surfaces, but also to some extent on vertical surfaces and ceilings. If agitated, layered dust may lead to explosive airborne concentrations. Burning or smoldering dust which is settled may also ignite airborne dust concentrations or become airborne itself. Dust on warm surfaces such as machinery, motors, bearings, or lighting fixtures tends to dry out and becomes susceptible to ignition at temperatures as low as 200°C (392°F). The layered dust is acknowledged to be the source of immensely damaging secondary explosions. The primary explosion resulting from ignition of airborne dust may be relatively small; however, pressure waves and structural vibrations dislodge layered dust which, in turn, explodes and dislodges more dust, propagating the explosion through the entire facility.

3. *Solution needed:* Dust control is essential as a worker protection measure to prevent explosions in grain-handling facilities. Possible approaches worthy of investigation include:

a. Use of grain additives that will reduce dust emissions and not adversely affect the grain quality.

b. Improvement in the design of bucket elevators to reduce dust emissions.

c. Other innovative process design improvements to prevent dust emissions and accumulations.

4. *References:*

a. United States Department of Agriculture, "Prevention of Dust Explosions in Grain Elevators—An Achievable Goal," Washington, D.C. (1980).

b. National Academy of Sciences, "Prevention of Grain Elevator and Mill Explosions," *Report of the Panel on Causes and Prevention of Grain Elevator Explosions,* NMAB 367–2 (Washington, D.C.: National Academy Press, 1982).

c. Veradke, M., and Chiotti, P., "A Bibliography of Topics Related to the Study of Grain-Dust Fire and Explosion with Key Word Indexes," Energy and Mineral Resources Research Institute, Iowa State University, Ames, Iowa. Project 400-25-04 (1978).

d. Theimer, O. F., "Cause and Prevention of Dust Explosions in Grain Elevators and Flour Mills," *Powder Technology* 8 (September–October 1973), pp. 137–147.

e. Wolanski, P. (Technical University of Warsaw), "Explosion Hazard of Agricultural Dusts," *Proceedings of the International Symposium on Grain Dust,* October 2–4, 1979, Kansas State University, Manhattan, Kansas, pp. 422–446.

f. Spencer, M. R., "Grain Mill Products," *Industrial Fire Hazards Handbook,* NFPA No. SPP–57, 1st ed. (Boston, Mass.: National Fire Protection Association), pp. 83–104, 639–654.

g. Jacobson, M.; Nagy, J.; Cooper, A.; and Ball, F. J., "Explosibility of Agricultural Dusts," *Report of Investigations 5753,* Washington, D.C., U.S. Department of the Interior, Bureau of Mines (1961).

h. Tait, S. R., and Tou, J. C., "The Effects of Fumigants on Grain Dust Explosions," *Journal of Hazardous Materials* 4, no. 2 (1980).

i. Aldis, David F., and Lai, Fang S., "Review of Literature Related to Engineering Aspects of Grain Dust Explosions," Washington, D.C., U.S. Department of Agriculture, Science and Education Administration (August 1979).

j. Inrie, C. D., "General Safety Practices: A Practical Guide to Elevator Design," Chapter 15, *Proceedings of the Elevator Design Conference,* September 27–28, 1979, National Grain and Feed Association, Kansas City, Missouri, pp. 330–343.

k. American Insurance Association, "Grain Elevator Dust Explosions," *Fire Protection Bulletin No. 79–02,* New York, New York (April 1978).

l. Factory Mutual Engineering Corporation, "Grain Storage and Milling," *Loss Prevention Data 7–75,* Norwood, Mass. (August 1975).

m. National Fire Protection Association, "Standard for the Prevention of Fires and Explosions in Grain Elevators and Facilities Handling Bulk Raw Agricultural Commodities," NFPA 61B–1980, Quincy, Mass.
n. National Academy of Sciences, "Pneumatic Dust Control in Grain Elevators: Guidelines for Design, Operation, and Maintenance," *Report of the Panel on Causes and Prevention of Grain Elevator Explosions,* NMAB 367–3 (Washington, D.C.: National Academy Press, 1982).
o. Winsett, W. G., "Dust Control in Grain Elevators," *Proceedings of the International Symposium on Grain Elevator Explosions* (Washington, D.C.: National Academy of Sciences, 1978), pp. 1–17.
p. Nelson, G. S., "Grain Dust Explosions Can Be Prevented," *Grain Age* (April 1979), pp. 34, 36, 42–44.
q. Anderson, R. J., "Report on the Grain Elevator Design Conference of the National Grain and Feed Association, Kansas City, Missouri, September 28–29, 1979; *Proceedings of the International Symposium on Grain Dust,* October 2–4, 1979, Kansas State University, Manhatten, Kansas, pp. 455–462.
r. Robinson, B. K., and Houston, O., "Preventive Maintenance: A Practical Guide to Elevator Design," Chapter 15, *Proceedings of the Elevator Design Conference,* National Grain and Feed Association, September 27–28, 1979, Kansas City, Missouri, pp. 348–382.
s. Sargent, L. M., "General Layout and Structural Design: A Practical Guide to Elevator Design, Chapter 3, *Proceedings of the Elevator Design Conference,* National Grain and Feed Association, September 27–28, 1979, Kansas City, Missouri, pp. 14–38.
t. Johnston, J. A., "Grain Elevator Monitoring Systems: A Practical Guide to Elevator Design, Chapter 11, *Proceedings of the Elevator Design Conference,* National Grain and Feed Association; September 27–28, 1979, Kansas City, Missouri, pp. 384–388.
u. Biorn, D. R., "Design, Operation, and Safety of Bucket Elevators: A Practical Guide to Elevator Design, Chapter 7, *Proceedings of the Elevator Design Conference,* National Grain and Feed Association, September 27–28, 1979, Kansas City, Missouri, pp. 114–163.
v. Gillis, J. P., "Explosion Venting and Suppression of Bucket Elevators," Washington, D.C., National Grain and Feed Association.
w. National Academy of Sciences, "Guidelines for the Investigation of Grain Dust Explosions," *Report of the Panel on Causes and Prevention of Grain Elevator Explosions,* NMAB 367–4. (Washington, D.C.: National Academy Press, 1983).
x. National Institute of Occupational Safety and Health,

"Occupational Safety in Grain Elevators and Feed Mills," DHHS (NIOSH) Publ. No. 83–126 (September 1983).

y. Cocker, J., and Getchel, N., "Controlling Dust in Agricultural Products with Additives," *Cereal Foods World* 23, no. 9 (1978).

z. Schofield, C., "Dust Generation and Control in Material Handling," *Bulk Solids Handling* 1, no. 3 (1981).

C. Tacky Aerosols

1. *Title:* Engineering control of tacky aerosol hazards.

2. *Problem:* The safe removal of tacky aerosols from the workplace is a major problem in the rubber and plastics industries. Recent studies in tire manufacturing plants have shown the presence of tacky solids within exhaust ventilation ducts. These tacky aerosols appear to collect on the walls of the ducting. The resultant clogging of ventilation ducts can result in reduced airflow at hoods which are intended to prevent worker exposure to air contaminants. In addition, several fires have occured within ventilation ducts causing significant physical damage to buildings and equipment.

3. *Solution needed:* Control of tacky aerosols in exhaust ventilation ducts is essential to prevent worker exposure to air contaminants and to prevent fires within ducts. Possible approaches worthy of investigation include:

a. Use of non-stick linings on ventilation duct interior surfaces.

b. Use of ventilation duct clean-out mechanism to periodically remove deposited materials from interior duct surfaces.

c. Other innovative process design improvements to prevent the fallout and deposition of tacky aerosols on interior duct surfaces.

4. *References:*

a. Liu, B. Y. H., and Agarwal, J. K., "Experimental Observation of Aerosol Deposition in Turbulent Flow, *Aerosol Science* 5, no. 2 (1974).

b. McKinnery, W., and Heitbrink, W., "Control of Air Contaminants in Tire Manufacturing." Unpublished report of U.S. DHHS (NIOSH) (1984).

D. General References

1. Burgess, W. A., *Recognition of Health Hazards in Industry: A Review of Materials and Processes* (New York: Wiley and Sons, 1980).

2. Committee on Industrial Ventilation, *Industrial Ventilation—A Manual of Recommended Practice* (Lansing,

Michigan: American Conference of Governmental Industrial Hygienists, 1982).

3. McElroy, F. E., ed., *Accident Prevention Manual for Industrial Operations,* 2 vols. (Chicago: National Safety Council, 1980).

4. National Institute of Occupational Safety and Health, "The Industrial Environment—Its Evaluation and Control," U.S. DHHS (NIOSH) Publ. No. 74–117.

5. Olishifski, J. B., ed., *Fundamentals of Industrial Hygiene* (Chicago: National Safety Council, 1979).

6. Parmeggiani, L., ed., *Encyclopedia of Occupational Safety and Health.* (Geneva: International Labour Office, 1983).

7. Soule, R. D., "Industrial Hygiene Engineering Controls," in Patty's *Industrial Hygiene and Toxicology,* Vol. 1. (New York: Wiley and Sons, 1978).

Appendixes

A. Prefixes, Units, and Conversions

- A.1. SI Unit Prefixes
- A.2. Common SI Units
- A.3. Decimal Equivalents of Inch Fractions
- A.4. Metric Equivalents

B. Preferred Limits and Fits for Cylindrical Parts

- B.1. Running and Sliding Fits
- B.2. Locational Clearance Fits
- B.3. Locational Transition Fits
- B.4. Locational Interference Fits
- B.5. Force and Shrink Fits

C. Preferred Metric Limits and Fits

- C.1. Preferred Hole Basis Clearance Fits
- C.2. Preferred Hole Basis Transition and Interference Fits
- C.3. Preferred Shaft Basis Clearance Fits
- C.4. Preferred Shaft Basis Transition and Interference Fits

D. Standard Thread Series

- D.1. ANSI Standard Series Threads (UN/UNR)
- D.2. Metric Screw Thread Series

E. Numbered and Lettered Twist Drill Sizes

F. American National Standard Square and Hex Bolts and Nuts

- F.1. Dimensions of Square Bolts
- F.2. Dimensions of Hex Bolts
- F.3. Dimensions of Square Nuts
- F.4. Dimensions of Hex Nuts and Hex Jam Nuts

G. American Standard Round Head Bolts

- G.1. Dimensions of Round Head Bolts
- G.2. Dimensions of Round Head Square Neck Bolts
- G.3. Dimensions of Round Head Ribbed Neck Bolts
- G.4. Dimensions of Round Head Fin Neck Bolts
- G.5. Dimensions of Step Bolts

H. American Standard Slotted Head Cap Screws

- H.1. Dimensions of Slotted Flat Countersunk Head Cap Screws
- H.2. Dimensions of Slotted Round Head Cap Screws
- H.3. Dimensions of Slotted Fillister Head Cap Screws

I. American Standard Machine Screws and Machine Screw Nuts

- I.1. Dimensions of Slotted Flat Countersunk Head Machine Screws
- I.2. Dimensions of Slotted Oval Countersunk Head Machine Screws
- I.3. Dimensions of Slotted Pan Head Machine Screws
- I.4. Dimensions of Slotted Fillister Head Machine Screws
- I.5. Dimensions of Slotted Truss Head Machine Screws
- I.6. Dimensions of Square and Hex Machine Screw Nuts

J. American Standard Head Set Screws

K. Woodruff Keys and Keyseats

- K.1. Woodruff Keys
- K.2. Keyseat Dimensions

L. American Standard Washers

- L.1. Dimensions of Regular Helical Spring Lock Washers
- L.2. Dimensions of Metric Plain Washers (General Purpose)

M. Clevis and Cotter Pins

- M.1. Dimensions of Clevis Pins
- M.2. Dimensions of Cotter Pins

Appendix A: Prefixes, Units, and Conversions

Table A.1: SI Unit Prefixes

femto	10^{-15}
pico	10^{-12}
nano	10^{-9}
micro	10^{-6}
milli	10^{-3}
centi	10^{-2}
deci	10^{-1}
deka	10^{1}
hecto	10^{2}
kilo	10^{3}
mega	10^{6}
giga	10^{8}
tera	10^{12}
peta	10^{15}

Table A.2: Common SI Units

Length	meter (m)
Plane angle	degree (°)
Area	square meter (m^2)
Volume	cubic meter (m^3)
Speed or Velocity	meters per second (m/s)
Acceleration	meters per second squared (m/s^2)
Mass	gram (g)
Density	kilograms per cubic meter (kg/m^3)
Force	newton (N)
Moment of force	newton-meter (N-m)
Pressure	Pascal (Pa)
Energy/work	joule (J)
Power	watt (W)
Temperature	kelvin (°K)
Eletric current	ampere (A)
Electromotive force	volt (V)
Electric resistance	ohm (Φ)
Electric charge	coulomb (C)

Table A.3: Decimal Equivalents of Inch Fractions

Inch Fraction	Decimal Equivalent
1/64	0.015625
1/32	0.031250
1/16	0.062500
1/8	0.125000
1/4	0.250000
1/2	0.500000
1	1.000000

Table A.4: Metric Equivalents

Inch (in.)[a]	Millimeter (mm.)
0.0394	1
0.0787	2
0.1181	3
0.1575	4
0.1969	5
0.2362	6
0.2756	7
0.3150	8
0.3543	9
0.3937	10

[a] To nearest fourth decimal place.

Appendix B: Preferred Limits and Fits for Cylindrical Parts

Table B.1: Running and Sliding Fits

Limits are in thousandths of an inch.

Limits for hole and shaft are applied algebraically to the basic size to obtain the limits of size for the parts.

Data in bold face are in accordance with ABC agreements.

Symbols H5, g5, etc., are Hole and Shaft designations used in ABC System.

Nominal Size Range Inches		Class RC 1			Class RC 2			Class RC 3			Class RC 4		
		Limits of Clearance	Standard Limits		Limits of Clearance	Standard Limits		Limits of Clearance	Standard Limits		Limits of Clearance	Standard Limits	
Over	To		Hole H5	Shaft g4		Hole H6	Shaft g5		Hole H7	Shaft f6		Hole H8	Shaft f7
0	− 0.12	0.1 0.45	+ 0.2 0	− 0.1 − 0.25	0.1 0.55	+ 0.25 0	− 0.1 − 0.3	0.3 0.95	+ 0.4 0	− 0.3 − 0.55	0.3 1.3	+ 0.6 0	− 0.3 − 0.7
0.12	− 0.24	0.15 0.5	+ 0.2 0	− 0.15 − 0.3	0.15 0.65	+ 0.3 0	− 0.15 − 0.35	0.4 1.12	+ 0.5 0	− 0.4 − 0.7	0.4 1.6	+ 0.7 0	− 0.4 − 0.9
0.24	− 0.40	0.2 0.6	0.25 0	− 0.2 − 0.35	0.2 0.85	+ 0.4 0	− 0.2 − 0.45	0.5 1.5	+ 0.6 0	− 0.5 − 0.9	0.5 2.0	+ 0.9 0	− 0.5 − 1.1
0.40	− 0.71	0.25 0.75	+ 0.3 0	− 0.25 − 0.45	0.25 0.95	+ 0.4 0	− 0.25 − 0.55	0.6 1.7	+ 0.7 0	− 0.6 − 1.0	0.6 2.3	+ 1.0 0	− 0.6 − 1.3
0.71	− 1.19	0.3 0.95	+ 0.4 0	− 0.3 − 0.55	0.3 1.2	+ 0.5 0	− 0.3 − 0.7	0.8 2.1	+ 0.8 0	− 0.8 − 1.3	0.8 2.8	+ 1.2 0	− 0.8 − 1.6
1.19	− 1.97	0.4 1.1	+ 0.4 0	− 0.4 − 0.7	0.4 1.4	+ 0.6 0	− 0.4 − 0.8	1.0 2.6	+ 1.0 0	− 1.0 − 1.6	1.0 3.6	+ 1.6 0	− 1.0 − 2.0
1.97	− 3.15	0.4 1.2	+ 0.5 0	− 0.4 − 0.7	0.4 1.6	+ 0.7 0	− 0.4 − 0.9	1.2 3.1	+ 1.2 0	− 1.2 − 1.9	1.2 4.2	+ 1.8 0	− 1.2 − 2.4
3.15	− 4.73	0.5 1.5	+ 0.6 0	− 0.5 − 0.9	0.5 2.0	+ 0.9 0	− 0.5 − 1.1	1.4 3.7	+ 1.4 0	− 1.4 − 2.3	1.4 5.0	+ 2.2 0	− 1.4 − 2.8
4.73	− 7.09	0.6 1.8	+ 0.7 0	− 0.6 − 1.1	0.6 2.3	+ 1.0 0	− 0.6 − 1.3	1.6 4.2	+ 1.6 0	− 1.6 − 2.6	1.6 5.7	+ 2.5 0	− 1.6 − 3.2
7.09	− 9.85	0.6 2.0	+ 0.8 0	− 0.6 − 1.2	0.6 2.6	+ 1.2 0	− 0.6 − 1.4	2.0 5.0	+ 1.8 0	− 2.0 − 3.2	2.0 6.6	+ 2.8 0	− 2.0 − 3.8
9.85	−12.41	0.8 2.3	+ 0.9 0	− 0.8 − 1.4	0.8 2.9	+ 1.2 0	− 0.8 − 1.7	2.5 5.7	+ 2.0 0	− 2.5 − 3.7	2.5 7.5	+ 3.0 0	− 2.5 − 4.5
12.41	−15.75	1.0 2.7	+ 1.0 0	− 1.0 − 1.7	1.0 3.4	+ 1.4 0	− 1.0 − 2.0	3.0 6.6	+ 0	− 3.0 − 4.4	3.0 8.7	+ 3.5 0	− 3.0 − 5.2
15.75	−19.69	1.2 3.0	+ 1.0 0	− 1.2 − 2.0	1.2 3.8	+ 1.6 0	− 1.2 − 2.2	4.0 8.1	+ 1.6 0	− 4.0 − 5.6	4.0 10.5	+ 4.0 0	− 4.0 − 6.5
19.69	−30.09	1.6 3.7	+ 1.2 0	− 1.6 − 2.5	1.6 4.8	+ 2.0 0	− 1.6 − 2.8	5.0 10.0	+ 3.0 0	− 5.0 − 7.0	5.0 13.0	+ 5.0 0	− 5.0 − 8.0
30.09	−41.49	2.0 4.6	+ 1.6 0	− 2.0 − 3.0	2.0 6.1	+ 2.5 0	− 2.0 − 3.6	6.0 12.5	+ 4.0 0	− 6.0 − 8.5	6.0 16.0	+ 6.0 0	− 6.0 −10.0
41.49	−56.19	2.5 5.7	+ 2.0 0	− 2.5 − 3.7	2.5 7.5	+ 3.0 0	− 2.5 − 4.5	8.0 16.0	+ 5.0 0	− 8.0 −11.0	8.0 21.0	+ 8.0 0	− 8.0 −13.0
56.19	−76.39	3.0 7.1	+ 2.5 0	− 3.0 − 4.6	3.0 9.5	+ 4.0 0	− 3.0 − 5.5	10.0 20.0	+ 6.0 0	−10.0 −14.0	10.0 26.0	+10.0 0	−10.0 −16.0
76.39	−100.9	4.0 9.0	+ 3.0 0	− 4.0 − 6.0	4.0 12.0	+ 5.0 0	− 4.0 − 7.0	12.0 25.0	+ 8.0 0	−12.0 −17.0	12.0 32.0	+12.0 0	−12.0 −20.0
100.9	−131.9	5.0 11.5	+ 4.0 0	− 5.0 − 7.5	5.0 15.0	+ 6.0 0	− 5.0 − 9.0	16.0 32.0	+10.0 0	−16.0 −22.0	16.0 36.0	+16.0 0	−16.0 −26.0
131.9	−171.9	6.0 14.0	+ 5.0 0	− 6.0 − 9.0	6.0 19.0	+ 8.0 0	− 6.0 −11.0	18.0 38.0	+ 8.0 0	−18.0 −26.0	18.0 50.0	+20.0 0	−18.0 −30.0
171.9	−200	8.0 18.0	+ 6.0 0	− 8.0 −12.0	8.0 22.0	+10.0 0	− 8.0 −12.0	22.0 48.0	+16.0 0	−22.0 −32.0	22.0 63.0	+25.0 0	−22.0 −38.0

Source: ANSI B4.1-1967 (1974), AMERICAN NATIONAL STANDARD PREFERRED LIMITS AND FITS FOR CYLINDRICAL PARTS.

Table B.1—*Continued*

Limits are in thousandths of an inch.

Limits for hole and shaft are applied algebraically to the basic size to obtain the limits of size for the parts

Data in bold face are in accordance with ABC agreements

Symbols H8, e7, etc., are Hole and Shaft designations used in ABC System.

Class RC 5			Class RC 6			Class RC 7			Class RC 8			Class RC 9			Nominal Size Range Inches	
Limits of Clearance	Standard Limits Hole H8	Standard Limits Shaft e7	Limits of Clearance	Standard Limits Hole H9	Standard Limits Shaft e8	Limits of Clearance	Standard Limits Hole H9	Standard Limits Shaft d8	Limits of Clearance	Standard Limits Hole H10	Standard Limits Shaft c9	Limits of Clearance	Standard Limits Hole H11	Standard Limits Shaft	Over	To
0.6 1.6	+ 0.6 − 0	− 0.6 − 1.0	0.6 2.2	+ 1.0 − 0	− 0.6 − 1.2	1.0 2.6	+ 1.0 0	− 1.0 − 1.6	2.5 5.1	+ 1.6 0	− 2.5 − 3.5	4.0 8.1	+ 2.5 0	− 4.0 − 5.6	0	0.12
0.8 2.0	+ 0.7 − 0	− 0.8 − 1.3	0.8 2.7	+ 1.2 − 0	− 0.8 − 1.5	1.2 3.1	+ 1.2 0	− 1.2 − 1.9	2.8 5.8	+ 1.8 0	− 2.8 − 4.0	4.5 9.0	+ 3.0 0	− 4.5 − 6.0	0.12	0.24
1.0 2.5	+ 0.9 − 0	− 1.0 − 1.6	1.0 3.3	+ 1.4 − 0	− 1.0 − 1.9	1.6 3.9	+ 1.4 0	− 1.6 − 2.5	3.0 6.6	+ 2.2 0	− 3.0 − 4.4	5.0 10.7	+ 3.5 0	− 5.0 − 7.2	0.24	0.40
1.2 2.9	+ 1.0 − 0	− 1.2 − 1.9	1.2 3.8	+ 1.6 − 0	− 1.2 − 2.2	2.0 4.6	+ 1.6 0	− 2.0 − 3.0	3.5 7.9	+ 2.8 0	− 3.5 − 5.1	6.0 12.8	+ 4.0 − 0	− 6.0 − 8.8	0.40	0.71
1.6 3.6	+ 1.2 − 0	− 1.6 − 2.4	1.6 4.8	+ 2.0 − 0	− 1.6 − 2.8	2.5 5.7	+ 2.0 0	− 2.5 − 3.7	4.5 10.0	+ 3.5 0	− 4.5 − 6.5	7.0 15.5	+ 5.0 0	− 7.0 − 10.5	0.71	1.19
2.0 4.6	+ 1.6 − 0	− 2.0 − 3.0	2.0 6.1	+ 2.5 − 0	− 2.0 − 3.6	3.0 7.1	+ 2.5 0	− 3.0 − 4.6	5.0 11.5	+ 4.0 0	− 5.0 − 7.5	8.0 18.0	+ 6.0 0	− 8.0 − 12.0	1.19	1.97
2.5 5.5	+ 1.8 − 0	− 2.5 − 3.7	2.5 7.3	+ 3.0 − 0	− 2.5 − 4.3	4.0 8.8	+ 3.0 0	− 4.0 − 5.8	6.0 13.5	+ 4.5 0	− 6.0 − 9.0	9.0 20.5	+ 7.0 0	− 9.0 − 13.5	1.97	3.15
3.0 6.6	+ 2.2 − 0	− 3.0 − 4.4	3.0 8.7	+ 3.5 − 0	− 3.0 − 5.2	5.0 10.7	+ 3.5 0	− 5.0 − 7.2	7.0 15.5	+ 5.0 0	− 7.0 − 10.5	10.0 24.0	+ 9.0 0	− 10.0 − 15.0	3.15	4.73
3.5 7.6	+ 2.5 − 0	− 3.5 − 5.1	3.5 10.0	+ 4.0 − 0	− 3.5 − 6.0	6.0 12.5	+ 4.0 0	− 6.0 − 8.5	8.0 18.0	+ 6.0 0	− 8.0 − 12.0	12.0 28.0	+ 10.0 0	− 12.0 − 18.0	4.73	7.09
4.0 8.6	+ 2.8 − 0	− 4.0 − 5.8	4.0 11.3	+ 4.5 0	− 4.0 − 6.8	7.0 14.3	+ 4.5 0	− 7.0 − 9.8	10.0 21.5	+ 7.0 0	− 10.0 − 14.5	15.0 34.0	+ 12.0 0	− 15.0 − 22.0	7.09	9.85
5.0 10.0	+ 3.0 0	− 5.0 − 7.0	5.0 13.0	+ 5.0 0	− 5.0 − 8.0	8.0 16.0	+ 5.0 0	− 8.0 − 11.0	12.0 25.0	+ 8.0 0	− 12.0 − 17.0	18.0 38.0	+ 12.0 0	− 18.0 − 26.0	9.85	12.41
6.0 11.7	+ 3.5 0	− 6.0 − 8.2	6.0 15.5	+ 6.0 0	− 6.0 − 9.5	10.0 19.5	+ 6.0 0	− 10.0 − 13.5	14.0 29.0	+ 9.0 0	− 14.0 − 20.0	22.0 45.0	+ 14.0 0	− 22.0 − 31.0	12.41	15.75
8.0 14.5	+ 4.0 0	− 8.0 −10.5	8.0 18.0	+ 6.0 0	− 8.0 −12.0	12.0 22.0	+ 6.0 0	− 12.0 − 16.0	16.0 32.0	+10.0 0	− 16.0 − 22.0	25.0 51.0	+ 16.0 0	− 25.0 − 35.0	15.75	19.69
10.0 18.0	+ 5.0 0	−10.0 −13.0	10.0 23.0	+ 8.0 0	−10.0 −15.0	16.0 29.0	+ 8.0 0	− 16.0 − 21.0	20.0 40.0	+12.0 0	− 20.0 − 28.0	30.0 62.0	+ 20.0 0	− 30.0 − 42.0	19.69	30.09
12.0 22.0	+ 6.0 0	−12.0 −16.0	12.0 28.0	+10.0 0	−12.0 −18.0	20.0 36.0	+10.0 0	− 20.0 − 26.0	25.0 51.0	+16.0 0	− 25.0 − 35.0	40.0 81.0	+ 25.0 0	− 40.0 − 56.0	30.09	41.49
16.0 29.0	+ 8.0 0	−16.0 −21.0	16.0 36.0	+12.0 0	−16.0 −24.0	25.0 45.0	+12.0 0	− 25.0 − 33.0	30.0 62.0	+20.0 0	− 30.0 − 42.0	50.0 100	+ 30.0 0	− 50.0 − 70.0	41.49	56.19
20.0 36.0	+10.0 0	−20.0 −26.0	20.0 46.0	+16.0 0	−20.0 −30.0	30.0 56.0	+16.0 0	− 30.0 − 40.0	40.0 81.0	+25.0 0	− 40.0 − 56.0	60.0 125	+ 40.0 0	− 60.0 − 85.0	56.19	76.39
25.0 45.0	+12.0 0	−25.0 −33.0	25.0 57.0	+20.0 0	−25.0 −37.0	40.0 72.0	+20.0 0	− 40.0 − 52.0	50.0 100	+30.0 0	− 50.0 − 70.0	80.0 160	+ 50.0 0	− 80.0 −110	76.39	100.9
30.0 56.0	+16.0 0	−30.0 −40.0	30.0 71.0	+25.0 0	−30.0 −46.0	50.0 91.0	+25.0 0	− 50.0 − 66.0	60.0 125	+40.0 0	− 60.0 − 85.0	100 200	+ 60.0 0	−100 −140	100.9	131.9
35.0 67.0	+20.0 0	−35.0 −47.0	35.0 85.0	+30.0 0	−35.0 −55.0	60.0 110.0	+30.0 0	− 60.0 − 80.0	80.0 160	+50.0 0	− 80.0 −110	130 260	+ 80.0 0	−130 −180	131.9	171.9
45.0 86.0	+25.0 0	−45.0 −61.0	45.0 110.0	+40.0 0	−45.0 −70.0	80.0 145.0	+40.0 0	− 80.0 −105.0	100 200	+60.0 0	−100 −140	150 310	+100 0	−150 −210	171.9	200

Source: ANSI B4.1-1967 (1974), AMERICAN NATIONAL STANDARD PREFERRED LIMITS AND FITS FOR CYLINDRICAL PARTS.

Table B.2: Locational Clearance Fits

Limits are in thousandths of an inch.

Limits for hole and shaft are applied algebraically to the basic size to obtain the limits of size for the parts.

Data in bold face are in accordance with ABC agreements.

Symbols H6, h5, etc., are Hole and Shaft designations used in ABC System.

Nominal Size Range Inches		Class LC 1			Class LC 2			Class LC 3			Class LC 4			Class LC 5		
		Limits of Clearance	Standard Limits		Limits of Clearance	Standard Limits		Limits of Clearance	Standard Limits		Limits of Clearance	Standard Limits		Limits of Clearance	Standard Limits	
Over	To		Hole H6	Shaft h5		Hole H7	Shaft h6		Hole H8	Shaft h7		Hole H10	Shaft h9		Hole H 7	Shaft g6
0 –	0.12	0 0.45	+ 0.25 – 0	+ 0 –0.2	0 0.65	+ 0.4 – 0	+ 0 –0.25	0 1	+ 0.6 – 0	+ 0 – 0.4	0 2.6	+ 1.6 – 0	+ 0 – 1.0	0.1 0.75	+ 0.4 – 0	– 0.1 – 0.35
0.12–	0.24	0 0.5	+ 0.3 – 0	+ 0 –0.2	0 0.8	+ 0.5 – 0	+ 0 – 0.3	0 1.2	+ 0.7 – 0	+ 0 – 0.5	0 3.0	+ 1.8 – 0	+ 0 – 1.2	0.15 0.95	+ 0.5 – 0	– 0.15 – 0.45
0.24–	0.40	0 0.65	+ 0.4 – 0	+ 0 –0.25	0 1.0	+ 0.6 – 0	+ 0 – 0.4	0 1.5	+ 0.9 – 0	+ 0 – 0.6	0 3.6	+ 2.2 – 0	+ 0 – 1.4	0.2 1.2	+ 0.6 – 0	– 0.2 – 0.6
0.40–	0.71	0 0.7	+ 0.4 – 0	+ 0 –0.3	0 1.1	+ 0.7 – 0	+ 0 – 0.4	0 1.7	+ 1.0 – 0	+ 0 – 0.7	0 4.4	+ 2.8 – 0	+ 0 – 1.6	0.25 1.35	+ 0.7 – 0	– 0.25 – 0.65
0.71–	1.19	0 0.9	+ 0.5 – 0	+ 0 –0.4	0 1.3	+ 0.8 – 0	+ 0 – 0.5	0 2	+ 1.2 – 0	+ 0 – 0.8	0 5.5	+ 3.5 – 0	+ 0 – 2.0	0.3 1.6	+ 0.8 – 0	– 0.3 – 0.8
1.19–	1.97	0 1.0	+ 0.6 – 0	+ 0 –0.4	0 1.6	+ 1.0 – 0	+ 0 – 0.6	0 2.6	+ 1.6 – 0	+ 0 – 1	0 6.5	+ 4.0 – 0	+ 0 – 2.5	0.4 2.0	+ 1.0 – 0	– 0.4 – 1.0
1.97–	3.15	0 1.2	+ 0.7 – 0	+ 0 –0.5	0 1.9	+ 1.2 – 0	+ 0 – 0.7	0 3	+ 1.8 – 0	+ 0 – 1.2	0 7.5	+ 4.5 – 0	+ 0 – 3	0.4 2.3	+ 1.2 – 0	– 0.4 – 1.1
3.15–	4.73	0 1.5	+ 0.9 – 0	+ 0 –0.6	0 2.3	+ 1.4 – 0	+ 0 – 0.9	0 3.6	+ 2.2 – 0	+ 0 – 1.4	0 8.5	+ 5.0 – 0	+ 0 – 3.5	0.5 2.8	+ 1.4 – 0	– 0.5 – 1.4
4.73–	7.09	0 1.7	+ 1.0 – 0	+ 0 –0.7	0 2.6	+ 1.6 – 0	+ 0 – 1.0	0 4.1	+ 2.5 – 0	+ 0 – 1.6	0 10	+ 6.0 – 0	+ 0 – 4	0.6 3.2	+ 1.6 – 0	– 0.6 – 1.6
7.09–	9.85	0 2.0	+ 1.2 – 0	+ 0 –0.8	0 3.0	+ 1.8 – 0	+ 0 – 1.2	0 4.6	+ 2.8 – 0	+ 0 – 1.8	0 11.5	+ 7.0 – 0	+ 0 – 4.5	0.6 3.6	+ 1.8 – 0	– 0.6 – 1.8
9.85–	12.41	0 2.1	+ 1.2 – 0	+ 0 –0.9	0 3.2	+ 2.0 – 0	+ 0 – 1.2	0 5	+ 3.0 – 0	+ 0 – 2.0	0 13	+ 8.0 – 0	+ 0 – 5	0.7 3.9	+ 2.0 – 0	– 0.7 – 1.9
12.41–	15.75	0 2.4	+ 1.4 – 0	+ 0 –1.0	0 3.6	+ 2.2 – 0	+ 0 – 1.4	0 5.7	+ 3.5 – 0	+ 0 – 2.2	0 15	+ 9.0 – 0	+ 0 – 6	0.7 4.3	+ 2.2 – 0	– 0.7 – 2.1
15.75–	19.69	0 2.6	+ 1.6 – 0	+ 0 –1.0	0 4.1	+ 2.5 – 0	+ 0 – 1.6	0 6.5	+ 4 – 0	+ 0 – 2.5	0 16	+ 10.0 – 0	+ 0 – 6	0.8 4.9	+ 2.5 – 0	– 0.8 – 2.4
19.69–	30.09	0 3.2	+ 2.0 – 0	+ 0 –1.2	0 5.0	+ 3 – 0	+ 0 – 2	0 8	+ 5 – 0	+ 0 – 3	0 20	+12.0 – 0	+ 0 – 8	0.9 5.9	+ 3.0 – 0	– 0.9 – 2.9
30.09–	41.49	0 4.1	+ 2.5 – 0	+ 0 –1.6	0 6.5	+ 4 – 0	+ 0 – 2.5	0 10	+ 6 – 0	+ 0 – 4	0 26	+16.0 – 0	+ 0 –10	1.0 7.5	+ 4.0 – 0	– 1.0 – 3.5
41.49–	56.19	0 5.0	+ 3.0 – 0	+ 0 –2.0	0 8.0	+ 5 – 0	+ 0 – 3	0 13	+ 8 – 0	+ 0 – 5	0 32	+20.0 – 0	+ 0 –12	1.2 9.2	+ 5.0 – 0	– 1.2 – 4.2
56.19–	76.39	0 6.5	+ 4.0 – 0	+ 0 –2.5	0 10	+ 6 – 0	+ 0 – 4	0 16	+10 – 0	+ 0 – 6	0 41	+25.0 – 0	+ 0 –16	1.2 11.2	+ 6.0 – 0	– 1.2 – 5.2
76.39–	100.9	0 8.0	+ 5.0 – 0	+ 0 –3.0	0 13	+ 8 – 0	+ 0 – 5	0 20	+12 – 0	+ 0 – 8	0 50	+30.0 – 0	+ 0 –20	1.4 14.4	+ 8.0 – 0	– 1.4 – 6.4
100.9 –	131.9	0 10.0	+ 6.0 – 0	+ 0 –4.0	0 16	+10 – 0	+ 0 – 6	0 26	+16 – 0	+ 0 –10	0 65	+40.0 – 0	+ 0 –25	1.6 17.6	+10.0 – 0	– 1.6 – 7.6
131.9 –	171.9	0 13.0	+ 8.0 – 0	+ 0 –5.0	0 20	+12 – 0	+ 0 – 8	0 32	+20 – 0	+ 0 –12	0 8	+ 50.0 – 0	+ 0 –30	1.8 21.8	+12.0 – 0	– 1.8 – 9.8
171.9 –	200	0 16.0	+10.0 – 0	+ 0 –6.0	0 26	+16 – 0	+ 0 –10	0 41	+25 – 0	+ 0 –16	0 100	+60.0 – 0	+ 0 –40	1.8 27.8	+16.0 – 0	– 1.8 –11.8

Source: ANSI B4.1-1967 (1974), AMERICAN NATIONAL STANDARD PREFERRED LIMITS AND FITS FOR CYLINDRICAL PARTS.

Limits are in thousandths of an inch.

Limits for hole and shaft are applied algebraically to the basic size to obtain the limits of size for the parts.

Data in bold face are in accordance with ABC agreements.

Symbols H9, f8, etc., are Hole and Shaft designations used in ABC System.

Class LC 6			Class LC 7			Class LC 8			Class LC 9			Class LC 10			Class LC 11			Nominal Size Range Inches
Limits of Clearance	Standard Limits		Limits of Clearance	Standard Limits		Limits of Clearance	Standard Limits		Limits of Clearance	Standard Limits		Limits of Clearance	Standard Limits		Limits of Clearance	Standard Limits		
	Hole H9	Shaft f8		Hole H10	Shaft e9		Hole H10	Shaft d9		Hole H11	Shaft c10		Hole H12	Shaft		Hole H13	Shaft	Over To
0.3 1.9	+ 1.0 0	− 0.3 − 0.9	0.6 3.2	+ 1.6 0	− 0.6 − 1.6	1.0 3.6	+ 1.6 − 0	− 1.0 − 2.0	2.5 6.6	+ 2.5 − 0	− 2.5 − 4.1	4 12	+ 4 − 0	− 4 − 8	5 17	+ 6 − 0	− 5 − 11	0 – 0.12
0.4 2.3	+ 1.2 0	− 0.4 − 1.1	0.8 3.8	+ 1.8 0	− 0.8 − 2.0	1.2 4.2	+ 1.8 − 0	− 1.2 − 2.4	2.8 7.6	+ 3.0 − 0	− 2.8 − 4.6	4.5 14.5	+ 5 − 0	− 4.5 − 9.5	6 20	+ 7 − 0	− 6 − 13	0.12 – 0.24
0.5 2.8	+ 1.4 0	− 0.5 − 1.4	1.0 4.6	+ 2.2 0	− 1.0 − 2.4	1.6 5.2	+ 2.2 − 0	− 1.6 − 3.0	3.0 8.7	+ 3.5 − 0	− 3.0 − 5.2	5 17	+ 6 − 0	− 5 − 11	7 25	+ 9 − 0	− 7 − 16	0.24 – 0.40
0.6 3.2	+ 1.6 0	− 0.6 − 1.6	1.2 5.6	+ 2.8 0	− 1.2 − 2.8	2.0 6.4	+ 2.8 − 0	− 2.0 − 3.6	3.5 10.3	+ 4.0 − 0	− 3.5 − 6.3	6 20	+ 7 − 0	− 6 − 13	8 28	+ 10 − 0	− 8 − 18	0.40 – 0.71
0.8 4.0	+ 2.0 0	− 0.8 − 2.0	1.6 7.1	+ 3.5 0	− 1.6 − 3.6	2.5 8.0	+ 3.5 − 0	− 2.5 − 4.5	4.5 13.0	+ 5.0 − 0	− 4.5 − 8.0	7 23	+ 8 − 0	− 7 − 15	10 34	+ 12 − 0	− 10 − 22	0.71 – 1.19
1.0 5.1	+ 2.5 0	− 1.0 − 2.6	2.0 8.5	+ 4.0 0	− 2.0 − 4.5	3.0 9.5	+ 4.0 − 0	− 3.0 − 5.5	5 15	+ 6 − 0	− 5 − 9	8 28	+ 10 − 0	− 8 − 18	12 44	+ 16 − 0	− 12 − 28	1.19 – 1.97
1.2 6.0	+ 3.0 0	− 1.2 − 3.0	2.5 10.0	+ 4.5 0	− 2.5 − 5.5	4.0 11.5	+ 4.5 − 0	− 4.0 − 7.0	6 17.5	+ 7 − 0	− 6 − 10.5	10 34	+ 12 − 0	− 10 − 22	14 50	+ 18 − 0	− 14 − 32	1.97 – 3.15
1.4 7.1	+ 3.5 0	− 1.4 − 3.6	3.0 11.5	+ 5.0 0	− 3.0 − 6.5	5.0 13.5	+ 5.0 − 0	− 5.0 − 8.5	7 21	+ 9 − 0	− 7 − 12	11 39	+ 14 − 0	− 11 − 25	16 60	+ 22 − 0	− 16 − 38	3.15 – 4.73
1.6 8.1	+ 4.0 0	− 1.6 − 4.1	3.5 13.5	+ 6.0 0	− 3.5 − 7.5	6 16	+ 6 − 0	− 6 −10	8 24	+ 10 − 0	− 8 − 14	12 44	+ 16 − 0	− 12 − 28	18 68	+ 25 − 0	− 18 − 43	4.73 – 7.09
2.0 9.3	+ 4.5 0	− 2.0 − 4.8	4.0 15.5	+ 7.0 0	− 4.0 − 8.5	7 18.5	+ 7 − 0	− 7 −11.5	10 29	+ 12 − 0	− 10 − 17	16 52	+ 18 − 0	− 16 − 34	22 78	+ 28 − 0	− 22 − 50	7.09 – 9.85
2.2 10.2	+ 5.0 0	− 2.2 − 5.2	4.5 17.5	+ 8.0 0	− 4.5 − 9.5	7 20	+ 8 − 0	− 7 −12	12 32	+ 12 − 0	− 12 − 20	20 60	+ 20 − 0	− 20 − 40	28 88	+ 30 − 0	− 28 − 58	9.85 – 12.41
2.5 12.0	+ 6.0 0	− 2.5 − 6.0	5.0 20.0	+ 9.0 0	− 5 −11	8 23	+ 9 − 0	− 8 −14	14 37	+ 14 − 0	− 14 − 23	22 66	+ 22 − 0	− 22 − 44	30 100	+ 35 − 0	− 30 − 65	12.41 – 15.75
2.8 12.8	+ 6.0 0	− 2.8 − 6.8	5.0 21.0	+ 10.0 0	− 5 −11	9 25	+10 − 0	− 9 −15	16 42	+ 16 − 0	− 16 − 26	25 75	+ 25 − 0	− 25 − 50	35 115	+ 40 − 0	− 35 − 75	15.75 – 19.69
3.0 16.0	+ 8.0 0	− 3.0 − 8.0	6.0 26.0	+12.0 − 0	− 6 −14	10 30	+12 − 0	−10 −18	18 50	+ 20 − 0	− 18 − 30	28 88	+ 30 − 0	− 28 − 58	40 140	+ 50 − 0	− 40 − 90	19.69 – 30.09
3.5 19.5	+10.0 0	− 3.5 − 9.5	7.0 33.0	+16.0 − 0	− 7 −17	12 38	+16 − 0	−12 −22	20 61	+ 25 − 0	− 20 − 36	30 110	+ 40 − 0	− 30 − 70	45 165	+ 60 − 0	− 45 −105	30.09 – 41.49
4.0 24.0	+12.0 0	− 4.0 −12.0	8.0 40.0	+20.0 − 0	− 8 −20	14 46	+20 − 0	−14 −26	25 75	+ 30 − 0	− 25 − 45	40 140	+ 50 − 0	− 40 − 90	60 220	+ 80 − 0	− 60 −140	41.49 – 56.19
4.5 30.5	+16.0 0	− 4.5 −14.5	9.0 50.0	+25.0 − 0	− 9 −25	16 57	+25 − 0	−16 −32	30 95	+ 40 − 0	− 30 − 55	50 170	+ 60 − 0	− 50 110	70 270	+100 − 0	− 70 −170	56.19 – 76.39
5.0 37.0	+20.0 0	− 5 −17	10.0 60.0	+30.0 − 0	−10 −30	18 68	+30 − 0	−18 −38	35 115	+ 50 − 0	− 35 − 65	50 210	+ 80 − 0	− 50 −130	80 330	+125 − 0	− 80 −205	76.39 – 100.9
6.0 47.0	+25.0 0	− 6 −22	12.0 67.0	+40.0 − 0	−12 −27	20 85	+40 − 0	−20 −45	40 140	+ 60 − 0	− 40 − 80	60 260	+100 − 0	− 60 −160	90 410	+160 − 0	− 90 −250	100.9 – 131.9
7.0 57.0	+30.0 0	− 7 −27	14.0 94.0	+50.0 − 0	−14 −44	25 105	+50 − 0	−25 −55	50 180	+ 80 − 0	− 50 − 100	80 330	+125 − 0	− 80 −205	100 500	+200 − 0	−100 −300	131.9 – 171.9
7.0 72.0	+40.0 0	− 7 −32	14.0 114.0	+60.0 − 0	−14 −54	25 125	+60 − 0	−25 −65	50 210	+100 − 0	− 50 − 110	90 410	+160 − 0	− 90 −250	125 625	+250 − 0	−125 −375	171.9 – 200

Source: ANSI B4.1-1967 (1974), AMERICAN NATIONAL STANDARD PREFERRED LIMITS AND FITS FOR CYLINDRICAL PARTS.

Table B.3: Locational Transition Fits

Limits are in thousandths of an inch.

Limits for hole and shaft are applied algebraically to the basic size to obtain the limits of size for the mating parts.

Data in bold face are in accordance with ABC agreements.

"Fit" represents the maximum interference (minus values) and the maximum clearance (plus values).

Symbols H7, js6, etc., are Hole and Shaft designations used in ABC System.

Nominal Size Range Inches	Class LT 1			Class LT 2			Class LT 3			Class LT 4			Class LT 5			Class LT 6		
	Fit	Standard Limits		Fit	Standard Limits		Fit	Standard Limits		Fit	Standard Limits		Fit	Standard Limits		Fit	Standard Limits	
Over To		Hole H7	Shaft js6		Hole H8	Shaft js7		Hole H7	Shaft k6		Hole H8	Shaft k7		Hole H7	Shaft n6		Hole H7	Shaft n7
0 – 0.12	-0.10 +0.50	+0.4 -0	+0.10 -0.10	-0.2 +0.8	+0.6 -0	+0.2 -0.2							-0.5 +0.15	+0.4 -0	+0.5 +0.25	-0.65 +0.15	+0.4 -0	-0.65 +0.25
0.12 – 0.24	-0.15 +0.65	+0.5 -0	+0.15 -0.15	-0.25 +0.95	+0.7 -0	+0.25 -0.25							-0.6 +0.2	+0.5 -0	+0.6 +0.3	-0.8 +0.2	+0.5 -0	+0.8 +0.3
0.24 – 0.40	-0.2 +0.8	+0.6 -0	+0.2 -0.2	-0.3 +1.2	+0.9 -0	+0.3 -0.3	-0.5 +0.5	+0.6 -0	+0.5 +0.1	-0.7 +0.8	+0.9 -0	+0.7 +0.1	-0.8 +0.2	+0.6 -0	+0.8 +0.4	-1.0 +0.2	+0.6 -0	+1.0 +0.4
0.40 – 0.71	-0.2 +0.9	+0.7 -0	+0.2 -0.2	-0.35 +1.35	+1.0 -0	+0.35 -0.35	-0.5 +0.6	+0.7 -0	+0.5 +0.1	-0.8 +0.9	+1.0 -0	+0.8 +0.1	-0.9 +0.2	+0.7 -0	+0.9 +0.5	-1.2 +0.2	+0.7 -0	+1.2 +0.5
0.71 – 1.19	-0.25 +1.05	+0.8 -0	+0.25 -0.25	-0.4 +1.6	+1.2 -0	+0.4 -0.4	-0.6 +0.7	+0.8 -0	+0.6 +0.1	-0.9 +1.1	+1.2 -0	+0.9 +0.1	-1.1 +0.2	+0.8 -0	+1.1 +0.6	-1.4 +0.2	+0.8 -0	+1.4 +0.6
1.19 – 1.97	-0.3 +1.3	+1.0 -0	+0.3 -0.3	-0.5 +2.1	+1.6 -0	+0.5 -0.5	-0.7 +0.9	+1.0 -0	+0.7 +0.1	-1.1 +1.5	+1.6 -0	+1.1 +0.1	-1.3 +0.3	+1.0 -0	+1.3 +0.7	-1.7 +0.3	+1.0 -0	+1.7 +0.7
1.97 – 3.15	-0.3 +1.5	+1.2 -0	+0.3 -0.3	-0.6 +2.4	+1.8 -0	+0.6 -0.6	-0.8 +1.1	+1.2 -0	+0.8 +0.1	-1.3 +1.7	+1.8 -0	+1.3 +0.1	-1.5 +0.4	+1.2 -0	+1.5 +0.8	-2.0 +0.4	+1.2 -0	+2.0 +0.8
3.15 – 4.73	-0.4 +1.8	+1.4 -0	+0.4 -0.4	-0.7 +2.9	+2.2 -0	+0.7 -0.7	-1.0 +1.3	+1.4 -0	+1.0 +0.1	-1.5 +2.1	+2.2 -0	+1.5 +0.1	-1.9 +0.4	+1.4 -0	+1.9 +1.0	-2.4 +0.4	+1.4 -0	+2.4 +1.0
4.73 – 7.09	-0.5 +2.1	+1.6 -0	+0.5 -0.5	-0.8 +3.3	+2.5 -0	+0.8 -0.8	-1.1 +1.5	+1.6 -0	+1.1 +0.1	-1.7 +2.4	+2.5 -0	+1.7 +0.1	-2.2 +0.4	+1.6 -0	+2.2 +1.2	-2.8 +0.4	+1.6 -0	+2.8 +1.2
7.09 – 9.85	-0.6 +2.4	+1.8 -0	+0.6 -0.6	-0.9 +3.7	+2.8 -0	+0.9 -0.9	-1.4 +1.6	+1.8 -0	+1.4 +0.2	-2.0 +2.6	+2.8 -0	+2.0 +0.2	-2.6 +0.4	+1.8 -0	+2.6 +1.4	-3.2 +0.4	+1.8 -0	+3.2 +1.4
9.85 – 12.41	-0.6 +2.6	+2.0 -0	+0.6 -0.6	-1.0 +4.0	+3.0 -0	+1.0 -1.0	-1.4 +1.8	+2.0 -0	+1.4 +0.2	-2.2 +2.8	+3.0 -0	+2.2 +0.2	-2.6 +0.6	+2.0 -0	+2.6 +1.4	-3.4 +0.6	+2.0 -0	+3.4 +1.4
12.41 – 15.75	-0.7 +2.9	+2.2 -0	+0.7 -0.7	-1.0 +4.5	+3.5 -0	+1.0 -1.0	-1.6 +2.0	+2.2 -0	+1.6 +0.2	-2.4 +3.3	+3.5 -0	+2.4 +0.2	-3.0 +0.6	+2.2 -0	+3.0 +1.6	-3.8 +0.6	+2.2 -0	+3.8 +1.6
15.75 – 19.69	-0.8 +3.3	+2.5 -0	+0.8 -0.8	-1.2 +5.2	+4.0 -0	+1.2 -1.2	-1.8 +2.3	+2.5 -0	+1.8 +0.2	-2.7 +3.8	+4.0 -0	+2.7 +0.2	-3.4 +0.7	+2.5 -0	+3.4 +1.8	-4.3 +0.7	+2.5 -0	+4.3 +1.8

Source: ANSI B4.1-1967 (1974), AMERICAN NATIONAL STANDARD PREFERRED LIMITS AND FITS FOR CYLINDRICAL PARTS.

Table B.4: Locational Interference Fits

Limits are in thousandths of an inch.
Limits for hole and shaft are applied algebraically to the basic size to obtain the limits of size for the parts.
Data in bold face are in accordance with ABC agreements,
Symbols H7, p6, etc., are Hole and Shaft designations used in ABC System.

Nominal Size Range Inches		Class LN 1			Class LN 2			Class LN 3		
		Limits of Interference	Standard Limits		Limits of Interference	Standard Limits		Limits of Interference	Standard Limits	
Over	To		Hole H6	Shaft n5		Hole H7	Shaft p6		Hole H7	Shaft r6
0	0.12	0 0.45	+ 0.25 − 0	+0.45 +0.25	0 0.65	+ 0.4 − 0	+ 0.65 + 0.4	0.1 0.75	+ 0.4 − 0	+ 0.75 + 0.5
0.12	0.24	0 0.5	+ 0.3 − 0	+0.5 +0.3	0 0.8	+ 0.5 − 0	+ 0.8 + 0.5	0.1 0.9	+ 0.5 0	+ 0.9 + 0.6
0.24	0.40	0 0.65	+ 0.4 − 0	+0.65 +0.4	0 1.0	+ 0.6 − 0	+ 1.0 + 0.6	0.2 1.2	+ 0.6 − 0	+ 1.2 + 0.8
0.40	0.71	0 0.8	+ 0.4 − 0	+0.8 +0.4	0 1.1	+ 0.7 − 0	+ 1.1 + 0.7	0.3 1.4	+ 0.7 − 0	+ 1.4 + 1.0
0.71	1.19	0 1.0	+ 0.5 − 0	+1.0 +0.5	0 1.3	+ 0.8 − 0	+ 1.3 + 0.8	0.4 1.7	+ 0.8 − 0	+ 1.7 + 1.2
1.19	1.97	0 1.1	+ 0.6 − 0	+1.1 +0.6	0 1.6	+ 1.0 − 0	+ 1.6 + 1.0	0.4 2.0	+ 1.0 − 0	+ 2.0 + 1.4
1.97	3.15	0.1 1.3	+ 0.7 − 0	+1.3 +0.7	0.2 2.1	+ 1.2 − 0	+ 2.1 + 1.4	0.4 2.3	+ 1.2 − 0	+ 2.3 + 1.6
3.15	4.73	0.1 1.6	+ 0.9 − 0	+1.6 +1.0	0.2 2.5	+ 1.4 − 0	+ 2.5 + 1.6	0.6 2.9	+ 1.4 − 0	+ 2.9 + 2.0
4.73	7.09	0.2 1.9	+ 1.0 − 0	+1.9 +1.2	0.2 2.8	+ 1.6 − 0	+ 2.8 + 1.8	0.9 3.5	+ 1.6 − 0	+ 3.5 + 2.5
7.09	9.85	0.2 2.2	+ 1.2 − 0	+2.2 +1.4	0.2 3.2	+ 1.8 − 0	+ 3.2 + 2.0	1.2 4.2	+ 1.8 − 0	+ 4.2 + 3.0
9.85	12.41	0.2 2.3	+ 1.2 − 0	+2.3 +1.4	0.2 3.4	+ 2.0 − 0	+ 3.4 + 2.2	1.5 4.7	+ 2.0 − 0	+ 4.7 + 3.5
12.41	15.75	0.2 2.6	+ 1.4 − 0	+2.6 +1.6	0.3 3.9	+ 2.2 − 0	+ 3.9 + 2.5	2.3 5.9	+ 2.2 − 0	+ 5.9 + 4.5
15.75	19.69	0.2 2.8	+ 1.6 − 0	+2.8 +1.8	0.3 4.4	+ 2.5 − 0	+ 4.4 + 2.8	2.5 6.6	+ 2.5 − 0	+ 6.6 + 5.0
19.69	30.09		+ 2.0 − 0		0.5 5.5	+ 3 − 0	+ 5.5 + 3.5	4 9	+ 3 − 0	+ 9 + 7
30.09	41.49		+ 2.5 − 0		0.5 7.0	+ 4 − 0	+ 7.0 + 4.5	5 11.5	+ 4 − 0	+11.5 + 9
41.49	56.19		+ 3.0 − 0		1 9	+ 5 − 0	+ 9 + 6	7 15	+ 5 − 0	+15 +12
56.19	76.39		+ 4.0 − 0		1 11	+ 6 − 0	+11 + 7	10 20	+ 6 − 0	+20 +16
76.39	100.9		+ 5.0 − 0		1 14	+ 8 − 0	+14 + 9	12 25	+ 8 − 0	+25 +20
100.9	131.9		+ 6.0 − 0		2 18	+10 − 0	+18 +12	15 31	+10 − 0	+31 +25
131.9	171.9		+ 8.0 − 0		4 24	+12 − 0	+24 +16	18 38	+12 − 0	+38 +30
171.9	200		+10.0 − 0		4 30	+16 − 0	+30 +20	24 50	+16 − 0	+50 +40

Source: ANSI B4.1-1967 (1974), AMERICAN NATIONAL STANDARD PREFERRED LIMITS AND FITS FOR CYLINDRICAL PARTS.

Table B.5: Force and Shrink Fits

Limits are in thousandths of an inch.

Limits for hole and shaft are applied algebraically to the basic size to obtain the limits of size for the parts.

Data in bold face are in accordance with ABC agreements.

Symbols H7, s6, etc., are Hole and Shaft designations used in ABC System.

Nominal Size Range Inches	Class FN 1			Class FN 2			Class FN 3			Class FN 4			Class FN 5		
Over – To	Limits of Interference	Standard Limits Hole H6	Standard Limits Shaft	Limits of Interference	Standard Limits Hole H7	Standard Limits Shaft s6	Limits of Interference	Standard Limits Hole H7	Standard Limits Shaft t6	Limits of Interference	Standard Limits Hole H7	Standard Limits Shaft u6	Limits of Interference	Standard Limits Hole H8	Standard Limits Shaft x7
0 – 0.12	0.05 0.5	+0.25 −0	+0.5 +0.3	0.2 0.85	+0.4 −0	+0.85 +0.6				0.3 0.95	+0.4 −0	+0.95 +0.7	0.3 1.3	+0.6 −0	+1.3 +0.9
0.12 – 0.24	0.1 0.6	+0.3 −0	+0.6 +0.4	0.2 1.0	+0.5 −0	+1.0 +0.7				0.4 1.2	+0.5 −0	+1.2 +0.9	0.5 1.7	+0.7 −0	+1.7 +1.2
0.24 – 0.40	0.1 0.75	+0.4 −0	+0.75 +0.5	0.4 1.4	+0.6 −0	+1.4 +1.0				0.6 1.6	+0.6 −0	+1.6 +1.2	0.5 2.0	+0.9 −0	+2.0 +1.4
0.40 – 0.56	0.1 0.8	−0.4 −0	+0.8 +0.5	0.5 1.6	+0.7 −0	+1.6 +1.2				0.7 1.8	+0.7 −0	+1.8 +1.4	0.6 2.3	+1.0 −0	+2.3 +1.6
0.56 – 0.71	0.2 0.9	+0.4 −0	+0.9 +0.6	0.5 1.6	+0.7 −0	+1.6 +1.2				0.7 1.8	+0.7 −0	+1.8 +1.4	0.8 2.5	+1.0 −0	+2.5 +1.8
0.71 – 0.95	0.2 1.1	+0.5 −0	+1.1 +0.7	0.6 1.9	+0.8 −0	+1.9 +1.4				0.8 2.1	+0.8 −0	+2.1 +1.6	1.0 3.0	+1.2 −0	+3.0 +2.2
0.95 – 1.19	0.3 1.2	+0.5 −0	+1.2 +0.8	0.6 1.9	+0.8 −0	+1.9 +1.4	0.8 2.1	+0.8 −0	+2.1 +1.6	1.0 2.3	+0.8 −0	+2.3 +1.8	1.3 3.3	+1.2 −0	+3.3 +2.5
1.19 – 1.58	0.3 1.3	+0.6 −0	+1.3 +0.9	0.8 2.4	+1.0 −0	+2.4 +1.8	1.0 2.6	+1.0 −0	+2.6 +2.0	1.5 3.1	+1.0 −0	+3.1 +2.5	1.4 4.0	+1.6 −0	+4.0 +3.0
1.58 – 1.97	0.4 1.4	+0.6 −0	+1.4 +1.0	0.8 2.4	+1.0 −0	+2.4 +1.8	1.2 2.8	+1.0 −0	+2.8 +2.2	1.8 3.4	+1.0 −0	+3.4 +2.8	2.4 5.0	+1.6 −0	+5.0 +4.0
1.97 – 2.56	0.6 1.8	+0.7 −0	+1.8 +1.3	0.8 2.7	+1.2 −0	+2.7 +2.0	1.3 3.2	+1.2 −0	+3.2 +2.5	2.3 4.2	+1.2 −0	+4.2 +3.5	3.2 6.2	+1.8 −0	+6.2 +5.0
2.56 – 3.15	0.7 1.9	+0.7 −0	+1.9 +1.4	1.0 2.9	+1.2 −0	+2.9 +2.2	1.8 3.7	+1.2 −0	+3.7 +3.0	2.8 4.7	+1.2 −0	+4.7 +4.0	4.2 7.2	+1.8 −0	+7.2 +6.0
3.15 – 3.94	0.9 2.4	+0.9 −0	+2.4 +1.8	1.4 3.7	+1.4 −0	+3.7 +2.8	2.1 4.4	+1.4 −0	+4.4 +3.5	3.6 5.9	+1.4 −0	+5.9 +5.0	4.8 8.4	+2.2 −0	+8.4 +7.0
3.94 – 4.73	1.1 2.6	+0.9 −0	+2.6 +2.0	1.6 3.9	+1.4 −0	+3.9 +3.0	2.6 4.9	+1.4 −0	+4.9 +4.0	4.6 6.9	+1.4 −0	+6.9 +6.0	5.8 9.4	+2.2 −0	+9.4 +8.0
4.73 – 5.52	1.2 2.9	+1.0 −0	+2.9 +2.2	1.9 4.5	+1.6 −0	+4.5 +3.5	3.4 6.0	+1.6 −0	+6.0 +5.0	5.4 8.0	+1.6 −0	+8.0 +7.0	7.5 11.6	+2.5 −0	+11.6 +10.0
5.52 – 6.30	1.5 3.2	+1.0 −0	+3.2 +2.5	2.4 5.0	+1.6 −0	+5.0 +4.0	3.4 6.0	+1.6 −0	+6.0 +5.0	5.4 8.0	+1.6 −0	+8.0 +7.0	9.5 13.6	+2.5 −0	+13.6 +12.0
6.30 – 7.09	1.8 3.5	+1.0 −0	+3.5 +2.8	2.9 5.5	+1.6 −0	+5.5 +4.5	4.4 7.0	+1.6 −0	+7.0 +6.0	6.4 9.0	+1.6 −0	+9.0 +8.0	9.5 13.6	+2.5 −0	+13.6 +12.0
7.09 – 7.88	1.8 3.8	+1.2 −0	+3.8 +3.0	3.2 6.2	+1.8 −0	+6.2 +5.0	5.2 8.2	+1.8 −0	+8.2 +7.0	7.2 10.2	+1.8 −0	+10.2 +9.0	11.2 15.8	+2.8 −0	+15.8 +14.0
7.88 – 8.86	2.3 4.3	+1.2 −0	+4.3 +3.5	3.2 6.2	+1.8 −0	+6.2 +5.0	5.2 8.2	+1.8 −0	+8.2 +7.0	8.2 11.2	+1.8 −0	+11.2 +10.0	13.2 17.8	+2.8 −0	+17.8 +16.0
8.86 – 9.85	2.3 4.3	+1.2 −0	+4.3 +3.5	4.2 7.2	+1.8 −0	+7.2 +6.0	6.2 9.2	+1.8 −0	+9.2 +8.0	10.2 13.2	+1.8 −0	+13.2 +12.0	13.2 17.8	+2.8 −0	+17.8 +16.0
9.85 – 11.03	2.8 4.9	+1.2 −0	+4.9 +4.0	4.0 7.2	+2.0 −0	+7.2 +6.0	7.0 10.2	+2.0 −0	+10.2 +9.0	10.0 13.2	+2.0 −0	+13.2 +12.0	15.0 20.0	+3.0 −0	+20.0 +18.0
11.03 – 12.41	2.8 4.9	+1.2 −0	+4.9 +4.0	5.0 8.2	+2.0 −0	+8.2 +7.0	7.0 10.2	+2.0 −0	+10.2 +9.0	12.0 15.2	+2.0 −0	+15.2 +14.0	17.0 22.0	+3.0 −0	+22.0 +20.0
12.41 – 13.98	3.1 5.5	+1.4 −0	+5.5 +4.5	5.8 9.4	+2.2 −0	+9.4 +8.0	7.8 11.4	+2.2 −0	+11.4 +10.0	13.8 17.4	+2.2 −0	+17.4 +16.0	18.5 24.2	+3.5 +0	+24.2 +22.0
13.98 – 15.75	3.6 6.1	+1.4 −0	+6.1 +5.0	5.8 9.4	+2.2 −0	+9.4 +8.0	9.8 13.4	+2.2 −0	+13.4 +12.0	15.8 19.4	+2.2 −0	+19.4 +18.0	21.5 27.2	+3.5 −0	+27.2 +25.0
15.75 – 17.72	4.4 7.0	+1.6 −0	+7.0 +6.0	6.5 10.6	+2.5 −0	+10.6 +9.0	9.5 13.6	+2.5 −0	+13.6 +12.0	17.5 21.6	+2.5 −0	+21.6 +20.0	24.0 30.5	+4.0 −0	+30.5 +28.0
17.72 – 19.69	4.4 7.0	+1.6 −0	+7.0 +6.0	7.5 11.6	+2.5 −0	+11.6 +10.0	11.5 15.6	+2.5 −0	+15.6 +14.0	19.5 23.6	+2.5 −0	+23.6 +22.0	26.0 32.5	+4.0 −0	+32.5 +30.0

Source: ANSI B4.1-1967 (1974), AMERICAN NATIONAL STANDARD PREFERRED LIMITS AND FITS FOR CYLINDRICAL PARTS.

Limits are in thousandths of an inch.

Limits for hole and shaft are applied algebraically to the basic size to obtain the limits of size for the parts.

Data in bold face are in accordance with ABC agreements.

Symbols H7, s6, etc., are Hole and Shaft designations used in ABC System.

Nominal Size Range Inches Over – To	Class FN 1			Class FN 2			Class FN 3			Class FN 4			Class FN 5		
	Limits of Interference	Standard Limits		Limits of Interference	Standard Limits		Limits of Interference	Standard Limits		Limits of Interference	Standard Limits		Limits of Interference	Standard Limits	
		Hole H6	Shaft		Hole 17	Shaft s6		Hole H7	Shaft t6		Hole H7	Shaft u6		Hole H8	Shaft x7
19.69 – 24.34	6.0 9.2	+ 2.0 – 0	+ 9.2 + 8.0	9.0 14.0	+ 3.0 – 0	+ 14.0 + 12.0	15.0 20.0	+ 3.0 – 0	+ 20.0 + 18.0	22.0 27.0	+ 3.0 – 0	+ 27.0 + 25.0	30.0 38.0	+ 5.0 – 0	+ 38.0 + 35.0
24.34 – 30.09	7.0 10.2	+ 2.0 – 0	+10.2 + 9.0	11.0 16.0	+ 3.0 – 0	+ 16.0 + 14.0	17.0 22.0	+ 3.0 – 0	+ 22.0 + 20.0	27.0 32.0	+ 3.0 – 0	+ 32.0 + 30.0	35.0 43.0	+ 5.0 – 0	+ 43.0 + 40.0
30.09 – 35.47	7.5 11.6	+ 2.5 – 0	+11.6 +10.0	14.0 20.5	+ 4.0 – 0	+ 20.5 + 18.0	21.0 27.5	+ 4.0 – 0	+ 27.5 + 25.0	31.0 37.5	+ 4.0 – 0	+ 37.5 + 35.0	44.0 54.0	+ 6.0 – 0	+ 54.0 + 50.0
35.47 – 41.49	9.5 13.6	+ 2.5 – 0	+13.6 +12.0	16.0 22.5	+ 4.0 – 0	+ 22.5 + 20.0	24.0 30.5	+ 4.0 – 0	+ 30.5 + 28.0	36.0 43.5	+ 4.0 – 0	+ 43.5 + 40.0	54.0 64.0	+ 6.0 – 0	+ 64.0 + 60.0
41.49 – 48.28	11.0 16.0	+ 3.0 – 0	+16.0 +14.0	17.0 25.0	+ 5.0 – 0	+ 25.0 + 22.0	30.0 38.0	+ 5.0 – 0	+ 38.0 + 35.0	45.0 53.0	+ 5.0 – 0	+ 53.0 + 50.0	62.0 75.0	+ 8.0 – 0	+ 75.0 + 70.0
48.28 – 56.19	13.0 18.0	+ 3.0 – 0	+18.0 +16.0	20.0 28.0	+ 5.0 – 0	+ 28.0 + 25.0	35.0 43.0	+ 5.0 – 0	+ 43.0 + 40.0	55.0 63.0	+ 5.0 – 0	+ 63.0 + 60.0	.72.0 85.0	+ 8.0 – 0	+ 85.0 + 80.0
56.19 – 65.54	14.0 20.5	+ 4.0 – 0	+20.5 +18.0	24.0 34.0	+ 6.0 – 0	+ 34.0 + 30.0	39.0 49.0	+ 6.0 – 0	+ 49.0 + 45.0	64.0 74.0	+ 6.0 – 0	+ 74.0 + 70.0	90.0 106	+10.0 – 0	+106 +100
65.54 – 76.39	18.0 24.5	+ 4.0 – 0	+24.5 +22.0	29.0 39.0	+ 6.0 – 0	+ 39.0 + 35.0	44.0 54.0	+ 6.0 – 0	+ 54.0 + 50.0	74.0 84.0	+ 6.0 – 0	+ 84.0 + 80.0	110 126	+10.0 – 0	+126 +120
76.39 – 87.79	20.0 28.0	+ 5.0 – 0	+28.0 +25.0	32.0 45.0	+ 8.0 – 0	+ 45.0 + 40.0	52.0 65.0	+ 8.0 – 0	+ 65.0 + 60.0	82.0 95.0	+ 8.0 – 0	+ 95.0 + 90.0	128 148	+12.0 – 0	+148 +140
87.79 – 100.9	23.0 31.0	+ 5.0 – 0	+31.0 +28.0	37.0 50.0	+ 8.0 – 0	+ 50.0 + 45.0	62.0 75.0	+ 8.0 – 0	+ 75.0 + 70.0	92.0 105	+ 8.0 – 0	+105 +100	148 168	+12.0 – 0	+168 +160
100.9 – 115.3	24.0 34.0	+ 6.0 – 0	+34.0 +30.0	40.0 56.0	+10.0 – 0	+ 56.0 + 50.0	70.0 86.0	+10.0 – 0	+ 86.0 + 80.0	110 126	+10.0 – 0	+126 +120	164 190	+16.0 – 0	+190 +180
115.3 – 131.9	29.0 39.0	+ 6.0 – 0	+39.0 +35.0	50.0 66.0	+10.0 – 0	+ 66.0 + 60.0	80.0 96.0	+10.0 – 0	+ 96.0 + 90.0	130 146	+10.0 – 0	+146 +140	184 210	+16.0 – 0	+210 +200
131.9 – 152.2	37.0 50.0	+ 8.0 – 0	+50.0 +45.0	58.0 78.0	+12.0 – 0	+ 78.0 + 70.0	88.0 108	+12.0 – 0	+108 +100	148 168	+12.0 – 0	+168 +160	200 232	+20.0 – 0	+232 +220
152.2 – 171.9	42.0 55.0	+ 8.0 – 0	+55.0 +50.0	68.0 88.0	+12.0 – 0	+ 88.0 + 80.0	108 128	+12.0 – 0	+128 +120	168 188	+12.0 – 0	+188 +170	230 262	+20.0 – 0	+262 +250
171.9 – 200	50.0 66.0	+10.0 – 0	+66.0 +60.0	74.0 100	+16.0 – 0	+100 + 90	124 150	+16.0 – 0	+150 +140	184 210	+16.0 – 0	+210 +200	275 316	+ 2.5 – 0	+316 +300

Source: ANSI B4.1-1967 (1974), AMERICAN NATIONAL STANDARD PREFERRED LIMITS AND FITS FOR CYLINDRICAL PARTS.

Appendix C: Preferred Metric Limits and Fits

Table C.1: Preferred Hole Basis Clearance Fits

Dimensions in mm.

BASIC SIZE		LOOSE RUNNING Hole H11	LOOSE RUNNING Shaft c11	LOOSE RUNNING Fit	FREE RUNNING Hole H9	FREE RUNNING Shaft d9	FREE RUNNING Fit	CLOSE RUNNING Hole H8	CLOSE RUNNING Shaft f7	CLOSE RUNNING Fit	SLIDING Hole H7	SLIDING Shaft g6	SLIDING Fit	LOCATIONAL CLEARANCE Hole H7	LOCATIONAL CLEARANCE Shat h6	LOCATIONAL CLEARANCE Fit
1	MAX	1.060	0.940	0.180	1.025	0.980	0.070	1.014	0.994	0.030	1.010	0.998	0.018	1.010	1.000	0.016
	MIN	1.000	0.880	0.060	1.000	0.955	0.020	1.000	0.984	0.006	1.000	0.992	0.002	1.000	0.994	0.000
1.2	MAX	1.260	1.140	0.180	1.225	1.180	0.070	1.214	1.194	0.030	1.210	1.198	0.018	1.210	1.200	0.016
	MIN	1.200	1.080	0.060	1.200	1.155	0.020	1.200	1.184	0.006	1.200	1.192	0.002	1.200	1.194	0.000
1.6	MAX	1.660	1.540	0.180	1.625	1.580	0.070	1.614	1.594	0.030	1.610	1.598	0.018	1.610	1.600	0.016
	MIN	1.600	1.480	0.060	1.600	1.555	0.020	1.600	1.584	0.006	1.600	1.592	0.002	1.600	1.594	0.000
2	MAX	2.060	1.940	0.180	2.025	1.980	0.070	2.014	1.994	0.030	2.010	1.998	0.018	2.010	2.000	0.016
	MIN	2.000	1.880	0.060	2.000	1.955	0.020	2.000	1.984	0.006	2.000	1.992	0.002	2.000	1.994	0.000
2.5	MAX	2.560	2.440	0.180	2.525	2.480	0.070	2.514	2.494	0.030	2.510	2.498	0.018	2.510	2.500	0.016
	MIN	2.500	2.380	0.060	2.500	2.455	0.020	2.500	2.484	0.006	2.500	2.492	0.002	2.500	2.494	0.000
3	MAX	3.060	2.940	0.180	3.025	2.980	0.070	3.014	2.994	0.030	3.010	2.998	0.018	3.010	3.000	0.016
	MIN	3.000	2.880	0.060	3.000	2.955	0.020	3.000	2.984	0.006	3.000	2.992	0.002	3.000	2.994	0.000
4	MAX	4.075	3.930	0.220	4.030	3.970	0.090	4.018	3.990	0.040	4.012	3.996	0.024	4.012	4.000	0.020
	MIN	4.000	3.855	0.070	4.000	3.940	0.030	4.000	3.978	0.010	4.000	3.988	0.004	4.000	3.992	0.000
5	MAX	5.075	4.930	0.220	5.030	4.970	0.090	5.018	4.990	0.040	5.012	4.996	0.024	5.012	5.000	0.020
	MIN	5.000	4.855	0.070	5.000	4.940	0.030	5.000	4.978	0.010	5.000	4.988	0.004	5.000	4.992	0.000
6	MAX	6.075	5.930	0.220	6.030	5.970	0.090	6.018	5.990	0.040	6.012	5.996	0.024	6.012	6.000	0.020
	MIN	6.000	5.855	0.070	6.000	5.940	0.030	6.000	5.978	0.010	6.000	5.988	0.004	6.000	5.992	0.000
8	MAX	8.090	7.920	0.260	8.036	7.960	0.112	8.022	7.987	0.050	8.015	7.995	0.029	8.015	8.000	0.024
	MIN	8.000	7.830	0.080	8.000	7.924	0.040	8.000	7.972	0.013	8.000	7.986	0.005	8.000	7.991	0.000
10	MAX	10.090	9.920	0.260	10.036	9.960	0.112	10.022	9.987	0.050	10.015	9.995	0.029	10.015	10.000	0.024
	MIN	10.000	9.830	0.080	10.000	9.924	0.040	10.000	9.972	0.013	10.000	9.986	0.005	10.000	9.991	0.000
12	MAX	12.110	11.905	0.315	12.043	11.950	0.136	12.027	11.984	0.061	12.018	11.994	0.035	12.018	12.000	0.029
	MIN	12.000	11.795	0.095	12.000	11.907	0.050	12.000	11.966	0.016	12.000	11.983	0.006	12.000	11.989	0.000
16	MAX	16.110	15.905	0.315	16.043	15.950	0.136	16.027	15.984	0.061	16.018	15.994	0.035	16.018	16.000	0.029
	MIN	16.000	15.795	0.095	16.000	15.907	0.050	16.000	15.966	0.016	16.000	15.983	0.006	16.000	15.989	0.000
20	MAX	20.130	19.890	0.370	20.052	19.935	0.169	20.033	19.980	0.074	20.021	19.993	0.041	20.021	20.000	0.034
	MIN	20.000	19.760	0.110	20.000	19.883	0.065	20.000	19.959	0.020	20.000	19.980	0.007	20.000	19.987	0.000
25	MAX	25.130	24.890	0.370	25.052	24.935	0.169	25.033	24.980	0.074	25.021	24.993	0.041	25.021	25.000	0.034
	MIN	25.000	24.760	0.110	25.000	24.883	0.065	25.000	24.959	0.020	25.000	24.980	0.007	25.000	24.987	0.000
30	MAX	30.130	29.890	0.370	30.052	29.935	0.169	30.033	29.980	0.074	30.021	29.993	0.041	30.021	30.000	0.034
	MIN	30.000	29.760	0.110	30.000	29.883	0.065	30.000	29.959	0.020	30.000	29.980	0.007	30.000	29.987	0.000

Source: ANSI B4.2-1978, AMERICAN NATIONAL STANDARD PREFERRED METRIC LIMITS AND FITS.

Table C.1—*Continued*

Dimensions in mm.

BASIC SIZE		LOOSE RUNNING Hole H11	Shaft c11	Fit	FREE RUNNING Hole H9	Shaft d9	Fit	CLOSE RUNNING Hole H8	Shaft f7	Fit	SLIDING Hole H7	Shaft g6	Fit	LOCATIONAL CLEARANCE Hole H7	Shat h6	Fit
40	MAX	40.160	39.880	0.440	40.062	39.920	0.204	40.039	39.975	0.089	40.025	39.991	0.050	40.025	40.000	0.041
	MIN	40.000	39.720	0.120	40.000	39.858	0.080	40.000	39.950	0.025	40.000	39.975	0.009	40.000	39.984	0.000
50	MAX	50.160	49.870	0.450	50.062	49.920	0.204	50.039	49.975	0.089	50.025	49.991	0.050	50.025	50.000	0.041
	MIN	50.000	49.710	0.130	50.000	49.858	0.080	50.000	49.950	0.025	50.000	49.975	0.009	50.000	49.984	0.000
60	MAX	60.190	59.860	0.520	60.074	59.900	0.248	60.046	59.970	0.106	60.030	59.990	0.059	60.030	60.000	0.049
	MIN	60.000	59.670	0.140	60.000	59.826	0.100	60.000	59.940	0.030	60.000	59.971	0.010	60.000	59.981	0.000
80	MAX	80.190	79.850	0.530	80.074	79.900	0.248	80.046	79.970	0.106	80.030	79.990	0.059	80.030	80.000	0.049
	MIN	80.000	79.660	0.150	80.000	79.826	0.100	80.000	79.940	0.030	80.000	79.971	0.010	80.000	79.981	0.000
100	MAX	100.220	99.830	0.610	100.087	99.880	0.294	100.054	99.964	0.125	100.035	99.988	0.069	100.035	100.000	0.057
	MIN	100.000	99.610	0.170	100.000	99.793	0.120	100.000	99.929	0.036	100.000	99.966	0.012	100.000	99.978	0.000
120	MAX	120.220	119.820	0.620	120.087	119.880	0.294	120.054	119.964	0.125	120.035	119.988	0.069	120.035	120.000	0.057
	MIN	120.000	119.600	0.180	120.000	119.793	0.120	120.000	119.929	0.036	120.000	119.966	0.012	120.000	119.978	0.000
160	MAX	160.250	159.790	0.710	160.100	159.855	0.345	160.063	159.957	0.146	160.040	159.986	0.079	160.040	160.000	0.065
	MIN	160.000	159.540	0.210	160.000	159.755	0.145	160.000	159.917	0.043	160.000	159.961	0.014	160.000	159.975	0.000
200	MAX	200.290	199.760	0.820	200.115	199.830	0.400	200.072	199.950	0.168	200.046	199.985	0.090	200.046	200.000	0.075
	MIN	200.000	199.470	0.240	200.000	199.715	0.170	200.000	199.904	0.050	200.000	199.956	0.015	200.000	199.971	0.000
250	MAX	250.290	249.720	0.860	250.115	249.830	0.400	250.072	249.950	0.168	250.046	249.985	0.090	250.046	250.000	0.075
	MIN	250.000	249.430	0.280	250.000	249.715	0.170	250.000	249.904	0.050	250.000	249.956	0.015	250.000	249.971	0.000
300	MAX	300.320	299.670	0.970	300.130	299.810	0.450	300.081	299.944	0.189	300.052	299.983	0.101	300.052	300.000	0.084
	MIN	300.000	299.350	0.330	300.000	299.680	0.190	300.000	299.892	0.056	300.000	299.951	0.017	300.000	299.968	0.000
400	MAX	400.360	399.600	1.120	400.140	399.790	0.490	400.089	399.938	0.208	400.057	399.982	0.111	400.057	400.000	0.093
	MIN	400.000	399.240	0.400	400.000	399.650	0.210	400.000	399.881	0.062	400.000	399.946	0.018	400.000	399.964	0.000
500	MAX	500.400	499.520	1.280	500.155	499.770	0.540	500.097	499.932	0.228	500.063	499.980	0.123	500.063	500.000	0.103
	MIN	500.000	499.120	0.480	500.000	499.615	0.230	500.000	499.869	0.068	500.000	499.940	0.020	500.000	499.960	0.000

Source: ANSI B4.2-1978, AMERICAN NATIONAL STANDARD PREFERRED METRIC LIMITS AND FITS.

Table C.2: Preferred Hole Basis Transition and Interference Fits

Dimensions in mm.

BASIC SIZE		LOCATIONAL TRANSN. Hole H7	Shaft k6	Fit	LOCATIONAL TRANSN. Hole H7	Shaft n6	Fit	LOCATIONAL INTERF. Hole H7	Shaft p6	Fit	MEDIUM DRIVE Hole H7	Shaft s6	Fit	FORCE Hole H7	Shaft u6	Fit
1	MAX	1.010	1.006	0.010	1.010	1.010	0.006	1.010	1.012	0.004	1.010	1.020	-0.004	1.010	1.024	-0.008
	MIN	1.000	1.000	-0.006	1.000	1.004	-0.010	1.000	1.006	-0.012	1.000	1.014	-0.020	1.000	1.018	-0.024
1.2	MAX	1.210	1.206	0.010	1.210	1.210	0.006	1.210	1.212	0.004	1.210	1.220	-0.004	1.210	1.224	-0.008
	MIN	1.200	1.200	-0.006	1.200	1.204	-0.010	1.200	1.206	-0.012	1.200	1.214	-0.020	1.200	1.218	-0.024
1.6	MAX	1.610	1.606	0.010	1.610	1.610	0.006	1.610	1.612	0.004	1.610	1.620	-0.004	1.610	1.624	-0.008
	MIN	1.600	1.600	-0.006	1.600	1.604	-0.010	1.600	1.606	-0.012	1.600	1.614	-0.020	1.600	1.618	-0.024
2	MAX	2.010	2.006	0.010	2.010	2.010	0.006	2.010	2.012	0.004	2.010	2.020	-0.004	2.010	2.024	-0.008
	MIN	2.000	2.000	-0.006	2.000	2.004	-0.010	2.000	2.006	-0.012	2.000	2.014	-0.020	2.000	2.018	-0.024
2.5	MAX	2.510	2.506	0.010	2.510	2.510	0.006	2.510	2.512	0.004	2.510	2.520	-0.004	2.510	2.524	-0.008
	MIN	2.500	2.500	-0.006	2.500	2.504	-0.010	2.500	2.506	-0.012	2.500	2.514	-0.020	2.500	2.518	-0.024
3	MAX	3.010	3.006	0.010	3.010	3.010	0.006	3.010	3.012	0.004	3.010	3.020	-0.004	3.010	3.024	-0.008
	MIN	3.000	3.000	-0.006	3.000	3.004	-0.010	3.000	3.006	-0.012	3.000	3.014	-0.020	3.000	3.018	-0.024
4	MAX	4.012	4.009	0.011	4.012	4.016	0.004	4.012	4.020	0.000	4.012	4.027	-0.007	4.012	4.031	-0.011
	MIN	4.000	4.001	-0.009	4.000	4.008	-0.016	4.000	4.012	-0.020	4.000	4.019	-0.027	4.000	4.023	-0.031
5	MAX	5.012	5.009	0.011	5.012	5.016	0.004	5.012	5.020	0.000	5.012	5.027	-0.007	5.012	5.031	-0.011
	MIN	5.000	5.001	-0.009	5.000	5.008	-0.016	5.000	5.012	-0.020	5.000	5.019	-0.027	5.000	5.023	-0.031
6	MAX	6.012	6.009	0.011	6.012	6.016	0.004	6.012	6.020	0.000	6.012	6.027	-0.007	6.012	6.031	-0.011
	MIN	6.000	6.001	-0.009	6.000	6.008	-0.016	6.000	6.012	-0.020	6.000	6.019	-0.027	6.000	6.023	-0.031
8	MAX	8.015	8.010	0.014	8.015	8.019	0.005	8.015	8.024	0.000	8.015	8.032	-0.008	8.015	8.037	-0.013
	MIN	8.000	8.001	-0.010	8.000	8.010	-0.019	8.000	8.015	-0.024	8.000	8.023	-0.032	8.000	8.028	-0.037
10	MAX	10.015	10.010	0.014	10.015	10.019	0.005	10.015	10.024	0.000	10.015	10.032	-0.008	10.015	10.037	-0.013
	MIN	10.000	10.001	-0.010	10.000	10.010	-0.019	10.000	10.015	-0.024	10.000	10.023	-0.032	10.000	10.028	-0.037
12	MAX	12.018	12.012	0.017	12.018	12.023	0.006	12.018	12.029	0.000	12.018	12.039	-0.010	12.018	12.044	-0.015
	MIN	12.000	12.001	-0.012	12.000	12.012	-0.023	12.000	12.018	-0.029	12.000	12.028	-0.039	12.000	12.033	-0.044
16	MAX	16.018	16.012	0.017	16.018	16.023	0.006	16.018	16.029	0.000	16.018	16.039	-0.010	16.018	16.044	-0.015
	MIN	16.000	16.001	-0.012	16.000	16.012	-0.023	16.000	16.018	-0.029	16.000	16.028	-0.039	16.000	16.033	-0.044
20	MAX	20.021	20.015	0.019	20.021	20.028	0.006	20.021	20.035	-0.001	20.021	20.048	-0.014	20.021	20.054	-0.020
	MIN	20.000	20.002	-0.015	20.000	20.015	-0.028	20.000	20.022	-0.035	20.000	20.035	-0.048	20.000	20.041	-0.054
25	MAX	25.021	25.015	0.019	25.021	25.028	0.006	25.021	25.035	-0.001	25.021	25.048	-0.014	25.021	25.061	-0.027
	MIN	25.000	25.002	-0.015	25.000	25.015	-0.028	25.000	25.022	-0.035	25.000	25.035	-0.048	25.000	25.048	-0.061
30	MAX	30.021	30.015	0.019	30.021	30.028	0.006	30.021	30.035	-0.001	30.021	30.048	-0.014	30.021	30.061	-0.027
	MIN	30.000	30.002	-0.015	30.000	30.015	-0.028	30.000	30.022	-0.035	30.000	30.035	-0.048	30.000	30.048	-0.061

Source: ANSI B4.2-1978, AMERICAN NATIONAL STANDARD PREFERRED METRIC LIMITS AND FITS.

Table C.2—*Continued*

Dimensions in mm.

BASIC SIZE		LOCATIONAL TRANSN. Hole H7	Shaft k6	Fit	LOCATIONAL TRANSN. Hole H7	Shaft n6	Fit	LOCATIONAL INTERF. Hole H7	Shaft p6	Fit	MEDIUM DRIVE Hole H7	Shaft s6	Fit	FORCE Hole H7	Shaft u6	Fit
40	MAX	40.025	40.018	0.023	40.025	40.033	0.008	40.025	40.042	-0.001	40.025	40.059	-0.018	40.025	40.076	-0.035
	MIN	40.000	40.002	-0.018	40.000	40.017	-0.033	40.000	40.026	-0.042	40.000	40.043	-0.059	40.000	40.060	-0.076
50	MAX	50.025	50.018	0.023	50.025	50.033	0.008	50.025	50.042	-0.001	50.025	50.059	-0.018	50.025	50.086	-0.045
	MIN	50.000	50.002	-0.018	50.000	50.017	-0.033	50.000	50.026	-0.042	50.000	50.043	-0.059	50.000	50.070	-0.086
60	MAX	60.030	60.021	0.028	60.030	60.039	0.010	60.030	60.051	-0.002	60.030	60.072	-0.023	60.030	60.106	-0.057
	MIN	60.000	60.002	-0.021	60.000	60.020	-0.039	60.000	60.032	-0.051	60.000	60.053	-0.072	60.000	60.087	-0.106
80	MAX	80.030	80.021	0.028	80.030	80.039	0.010	80.030	80.051	-0.002	80.030	80.078	-0.029	80.030	80.121	-0.072
	MIN	80.000	80.002	-0.021	80.000	80.020	-0.039	80.000	80.032	-0.051	80.000	80.059	-0.078	80.000	80.102	-0.121
100	MAX	100.035	100.025	0.032	100.035	100.045	0.012	100.035	100.059	-0.002	100.035	100.093	-0.036	100.035	100.146	-0.089
	MIN	100.000	100.003	-0.025	100.000	100.023	-0.045	100.000	100.037	-0.059	100.000	100.071	-0.093	100.000	100.124	-0.146
120	MAX	120.035	120.025	0.032	120.035	120.045	0.012	120.035	120.059	-0.002	120.035	120.101	-0.044	120.035	120.166	-0.109
	MIN	120.000	120.003	-0.025	120.000	120.023	-0.045	120.000	120.037	-0.059	120.000	120.079	-0.101	120.000	120.144	-0.166
160	MAX	160.040	160.028	0.037	160.040	160.052	0.013	160.040	160.068	-0.003	160.040	160.125	-0.060	160.040	160.215	-0.150
	MIN	160.000	160.003	-0.028	160.000	160.027	-0.052	160.000	160.043	-0.068	160.000	160.100	-0.125	160.000	160.190	-0.215
200	MAX	200.046	200.033	0.042	200.046	200.060	0.015	200.046	200.079	-0.004	200.046	200.151	-0.076	200.046	200.265	-0.190
	MIN	200.000	200.004	-0.033	200.000	200.031	-0.060	200.000	200.050	-0.079	200.000	200.122	-0.151	200.000	200.236	-0.265
250	MAX	250.046	250.033	0.042	250.046	250.060	0.015	250.046	250.079	-0.004	250.046	250.169	-0.094	250.046	250.313	-0.238
	MIN	250.000	250.004	-0.033	250.000	250.031	-0.060	250.000	250.050	-0.079	250.000	250.140	-0.169	250.000	250.284	-0.313
300	MAX	300.052	300.036	0.048	300.052	300.066	0.018	300.052	300.088	-0.004	300.052	300.202	-0.118	300.052	300.382	-0.298
	MIN	300.000	300.004	-0.036	300.000	300.034	-0.066	300.000	300.056	-0.088	300.000	300.170	-0.202	300.000	300.350	-0.382
400	MAX	400.057	400.040	0.053	400.057	400.073	0.020	400.057	400.098	-0.005	400.057	400.244	-0.151	400.057	400.471	-0.378
	MIN	400.000	400.004	-0.040	400.000	400.037	-0.073	400.000	400.062	-0.098	400.000	400.208	-0.244	400.000	400.435	-0.471
500	MAX	500.063	500.045	0.058	500.063	500.080	0.023	500.063	500.108	-0.005	500.063	500.292	-0.189	500.063	500.580	-0.477
	MIN	500.000	500.005	-0.045	500.000	500.040	-0.080	500.000	500.068	-0.108	500.000	500.252	-0.292	500.000	500.540	-0.580

Source: ANSI B4.2-1978, AMERICAN NATIONAL STANDARD PREFERRED METRIC LIMITS AND FITS.

Table C.3: Preferred Shaft Basis Clearance Fits

Dimensions in mm.

BASIC SIZE		LOOSE RUNNING Hole C11	LOOSE RUNNING Shaft h11	LOOSE RUNNING Fit	FREE RUNNING Hole D9	FREE RUNNING Shaft h9	FREE RUNNING Fit	CLOSE RUNNING Hole F8	CLOSE RUNNING Shaft h7	CLOSE RUNNING Fit	SLIDING Hole G7	SLIDING Shaft h6	SLIDING Fit	LOCATIONAL CLEARANCE Hole H7	LOCATIONAL CLEARANCE Shaft h6	LOCATIONAL CLEARANCE Fit
1	MAX	1.120	1.000	0.180	1.045	1.000	0.070	1.020	1.000	0.030	1.012	1.000	0.018	1.010	1.000	0.016
	MIN	1.060	0.940	0.060	1.020	0.975	0.020	1.006	0.990	0.006	1.002	0.994	0.002	1.000	0.994	0.000
1.2	MAX	1.320	1.200	0.180	1.245	1.200	0.070	1.220	1.200	0.030	1.212	1.200	0.018	1.210	1.200	0.016
	MIN	1.260	1.140	0.060	1.220	1.175	0.020	1.206	1.190	0.006	1.202	1.194	0.002	1.200	1.194	0.000
1.6	MAX	1.720	1.600	0.180	1.645	1.600	0.070	1.620	1.600	0.030	1.612	1.600	0.018	1.610	1.600	0.016
	MIN	1.660	1.540	0.060	1.620	1.575	0.020	1.606	1.590	0.006	1.602	1.594	0.002	1.600	1.594	0.000
2	MAX	2.120	2.000	0.180	2.045	2.000	0.070	2.020	2.000	0.030	2.012	2.000	0.018	2.010	2.000	0.016
	MIN	2.060	1.940	0.060	2.020	1.975	0.020	2.006	1.990	0.006	2.002	1.994	0.002	2.000	1.994	0.000
2.5	MAX	2.620	2.500	0.180	2.545	2.500	0.070	2.520	2.500	0.030	2.512	2.500	0.018	2.510	2.500	0.016
	MIN	2.560	2.440	0.060	2.520	2.475	0.020	2.506	2.490	0.006	2.502	2.494	0.002	2.500	2.494	0.000
3	MAX	3.120	3.000	0.180	3.045	3.000	0.070	3.020	3.000	0.030	3.012	3.000	0.018	3.010	3.000	0.016
	MIN	3.060	2.940	0.060	3.020	2.975	0.020	3.006	2.990	0.006	3.002	2.994	0.002	3.000	2.994	0.000
4	MAX	4.145	4.000	0.220	4.060	4.000	0.090	4.028	4.000	0.040	4.016	4.000	0.024	4.012	4.000	0.020
	MIN	4.070	3.925	0.070	4.030	3.970	0.030	4.010	3.988	0.010	4.004	3.992	0.004	4.000	3.992	0.000
5	MAX	5.145	5.000	0.220	5.060	5.000	0.090	5.028	5.000	0.040	5.016	5.000	0.024	5.012	5.000	0.020
	MIN	5.070	4.925	0.070	5.030	4.970	0.030	5.010	4.988	0.010	5.004	4.992	0.004	5.000	4.992	0.000
6	MAX	6.145	6.000	0.220	6.060	6.000	0.090	6.028	6.000	0.040	6.016	6.000	0.024	6.012	6.000	0.020
	MIN	6.070	5.925	0.070	6.030	5.970	0.030	6.010	5.988	0.010	6.004	5.992	0.004	6.000	5.992	0.000
8	MAX	8.170	8.000	0.260	8.076	8.000	0.112	8.035	8.000	0.050	8.020	8.000	0.029	8.015	8.000	0.024
	MIN	8.080	7.910	0.080	8.040	7.964	0.040	8.013	7.985	0.013	8.005	7.991	0.005	8.000	7.991	0.000
10	MAX	10.170	10.000	0.260	10.076	10.000	0.112	10.035	10.000	0.050	10.020	10.000	0.029	10.015	10.000	0.024
	MIN	10.080	9.910	0.080	10.040	9.964	0.040	10.013	9.985	0.013	10.005	9.991	0.005	10.000	9.991	0.000
12	MAX	12.205	12.000	0.315	12.093	12.000	0.136	12.043	12.000	0.061	12.024	12.000	0.035	12.018	12.000	0.029
	MIN	12.095	11.890	0.095	12.050	11.957	0.050	12.016	11.982	0.016	12.006	11.989	0.006	12.000	11.989	0.000
16	MAX	16.205	16.000	0.315	16.093	16.000	0.136	16.043	16.000	0.061	16.024	16.000	0.035	16.018	16.000	0.029
	MIN	16.095	15.890	0.095	16.050	15.957	0.050	16.016	15.982	0.016	16.006	15.989	0.006	16.000	15.989	0.000
20	MAX	20.240	20.000	0.370	20.117	20.000	0.169	20.053	20.000	0.074	20.028	20.000	0.041	20.021	20.000	0.034
	MIN	20.110	19.870	0.110	20.065	19.948	0.065	20.020	19.979	0.020	20.007	19.987	0.007	20.000	19.987	0.000
25	MAX	25.240	25.000	0.370	25.117	25.000	0.169	25.053	25.000	0.074	25.028	25.000	0.041	25.021	25.000	0.034
	MIN	25.110	24.870	0.110	25.065	24.948	0.065	25.020	24.979	0.020	25.007	24.987	0.007	25.000	24.987	0.000
30	MAX	30.240	30.000	0.370	30.117	30.000	0.169	30.053	30.000	0.074	30.028	30.000	0.041	30.021	30.000	0.034
	MIN	30.110	29.870	0.110	30.065	29.948	0.065	30.020	29.979	0.020	30.007	29.987	0.007	30.000	29.987	0.000

Source: ANSI B4.2-1978, AMERICAN NATIONAL STANDARD PREFERRED METRIC LIMITS AND FITS.

Table C.3—*Continued*

Dimensions in mm.

BASIC SIZE		LOOSE RUNNING Hole C11	Shaft h11	Fit	FREE RUNNING Hole D9	Shaft h9	Fit	CLOSE RUNNING Hole F8	Shaft h7	Fit	SLIDING Hole G7	Shaft h6	Fit	LOCATIONAL CLEARANCE Hole H7	Shaft h6	Fit
40	MAX	40.280	40.000	0.440	40.142	40.000	0.204	40.064	40.000	0.089	40.034	40.000	0.050	40.025	40.000	0.041
	MIN	40.120	39.840	0.120	40.080	39.938	0.080	40.025	39.975	0.025	40.009	39.984	0.009	40.000	39.984	0.000
50	MAX	50.290	50.000	0.450	50.142	50.000	0.204	50.064	50.000	0.089	50.034	50.000	0.050	50.025	50.000	0.041
	MIN	50.130	49.840	0.130	50.080	49.938	0.080	50.025	49.975	0.025	50.009	49.984	0.009	50.000	49.984	0.000
60	MAX	60.330	60.000	0.520	60.174	60.000	0.248	60.076	60.000	0.106	60.040	60.000	0.059	60.030	60.000	0.049
	MIN	60.140	59.810	0.140	60.100	59.926	0.100	60.030	59.970	0.030	60.010	59.981	0.010	60.000	59.981	0.000
80	MAX	80.340	80.000	0.530	80.174	80.000	0.248	80.076	80.000	0.106	80.040	80.000	0.059	80.030	80.000	0.049
	MIN	80.150	79.810	0.150	80.100	79.926	0.100	80.030	79.970	0.030	80.010	79.981	0.010	80.000	79.981	0.000
100	MAX	100.390	100.000	0.610	100.207	100.000	0.294	100.090	100.000	0.125	100.047	100.000	0.069	100.035	100.000	0.057
	MIN	100.170	99.780	0.170	100.120	99.913	0.120	100.036	99.965	0.036	100.012	99.978	0.012	100.000	99.978	0.000
120	MAX	120.400	120.000	0.620	120.207	120.000	0.294	120.090	120.000	0.125	120.047	120.000	0.069	120.035	120.000	0.057
	MIN	120.180	119.780	0.180	120.120	119.913	0.120	120.036	119.965	0.036	120.012	119.978	0.012	120.000	119.978	0.000
160	MAX	160.460	160.000	0.710	160.245	160.000	0.345	160.106	160.000	0.146	160.054	160.000	0.079	160.040	160.000	0.065
	MIN	160.210	159.750	0.210	160.145	159.900	0.145	160.043	159.960	0.043	160.014	159.975	0.014	160.000	159.975	0.000
200	MAX	200.530	200.000	0.820	200.285	200.000	0.400	200.122	200.000	0.168	200.061	200.000	0.090	200.046	200.000	0.075
	MIN	200.240	199.710	0.240	200.170	199.885	0.170	200.050	199.954	0.050	200.015	199.971	0.015	200.000	199.971	0.000
250	MAX	250.570	250.000	0.860	250.285	250.000	0.400	250.122	250.000	0.168	250.061	250.000	0.090	250.046	250.000	0.075
	MIN	250.280	249.710	0.280	250.170	249.885	0.170	250.050	249.954	0.050	250.015	249.971	0.015	250.000	249.971	0.000
300	MAX	300.650	300.000	0.970	300.320	300.000	0.450	300.137	300.000	0.189	300.069	300.000	0.101	300.052	300.000	0.084
	MIN	300.330	299.680	0.330	300.190	299.870	0.190	300.056	299.948	0.056	300.017	299.968	0.017	300.000	299.968	0.000
400	MAX	400.760	400.000	1.120	400.350	400.000	0.490	400.151	400.000	0.208	400.075	400.000	0.111	400.057	400.000	0.093
	MIN	400.400	399.640	0.400	400.210	399.860	0.210	400.062	399.943	0.062	400.018	399.964	0.018	400.000	399.964	0.000
500	MAX	500.880	500.000	1.280	500.385	500.000	0.540	500.165	500.000	0.228	500.083	500.000	0.123	500.063	500.000	0.103
	MIN	500.480	499.600	0.480	500.230	499.845	0.230	500.068	499.937	0.068	500.020	499.960	0.020	500.000	499.960	0.000

Source: ANSI B4.2-1978, AMERICAN NATIONAL STANDARD PREFERRED METRIC LIMITS AND FITS.

Table C.4: Preferred Shaft Basis Transition and Interference Fits

Dimensions in mm.

BASIC SIZE		LOCATIONAL TRANSN. Hole K7	Shaft h6	Fit	LOCATIONAL TRANSN. Hole N7	Shaft h6	Fit	LOCATIONAL INTERF. Hole P7	Shaft h6	Fit	MEDIUM DRIVE Hole S7	Shaft h6	Fit	FORCE Hole U7	Shaft h6	Fit
1	MAX	1.000	1.000	0.006	0.996	1.000	0.002	0.994	1.000	0.000	0.986	1.000	-0.008	0.982	1.000	-0.012
	MIN	0.990	0.994	-0.010	0.986	0.994	-0.014	0.984	0.994	-0.016	0.976	0.994	-0.024	0.972	0.994	-0.028
1.2	MAX	1.200	1.200	0.006	1.196	1.200	0.002	1.194	1.200	0.000	1.186	1.200	-0.008	1.182	1.200	-0.012
	MIN	1.190	1.194	-0.010	1.186	1.194	-0.014	1.184	1.194	-0.016	1.176	1.194	-0.024	1.172	1.194	-0.028
1.6	MAX	1.600	1.600	0.006	1.596	1.600	0.002	1.594	1.600	0.000	1.586	1.600	-0.008	1.582	1.600	-0.012
	MIN	1.590	1.594	-0.010	1.586	1.594	-0.014	1.584	1.594	-0.016	1.576	1.594	-0.024	1.572	1.594	-0.028
2	MAX	2.000	2.000	0.006	1.996	2.000	0.002	1.994	2.000	0.000	1.986	2.000	-0.008	1.982	2.000	-0.012
	MIN	1.990	1.994	-0.010	1.986	1.994	-0.014	1.984	1.994	-0.016	1.976	1.994	-0.024	1.972	1.994	-0.028
2.5	MAX	2.500	2.500	0.006	2.496	2.500	0.002	2.494	2.500	0.000	2.486	2.500	-0.008	2.482	2.500	-0.012
	MIN	2.490	2.494	-0.010	2.486	2.494	-0.014	2.484	2.494	-0.016	2.476	2.494	-0.024	2.472	2.494	-0.028
3	MAX	3.000	3.000	0.006	2.996	3.000	0.002	2.994	3.000	0.000	2.986	3.000	-0.008	2.982	3.000	-0.012
	MIN	2.990	2.994	-0.010	2.986	2.994	-0.014	2.984	2.994	-0.016	2.976	2.994	-0.024	2.972	2.994	-0.028
4	MAX	4.003	4.000	0.011	3.996	4.000	0.004	3.992	4.000	0.000	3.985	4.000	-0.007	3.981	4.000	-0.011
	MIN	3.991	3.992	-0.009	3.984	3.992	-0.016	3.980	3.992	-0.020	3.973	3.992	-0.027	3.969	3.992	-0.031
5	MAX	5.003	5.000	0.011	4.996	5.000	0.004	4.992	5.000	0.000	4.985	5.000	-0.007	4.981	5.000	-0.011
	MIN	4.991	4.992	-0.009	4.984	4.992	-0.016	4.980	4.992	-0.020	4.973	4.992	-0.027	4.969	4.992	-0.031
6	MAX	6.003	6.000	0.011	5.996	6.000	0.004	5.992	6.000	0.000	5.985	6.000	-0.007	5.981	6.000	-0.011
	MIN	5.991	5.992	-0.009	5.984	5.992	-0.016	5.980	5.992	-0.020	5.973	5.992	-0.027	5.969	5.992	-0.031
8	MAX	8.005	8.000	0.014	7.996	8.000	0.005	7.991	8.000	0.000	7.983	8.000	-0.008	7.978	8.000	-0.013
	MIN	7.990	7.991	-0.010	7.981	7.991	-0.019	7.976	7.991	-0.024	7.968	7.991	-0.032	7.963	7.991	-0.037
10	MAX	10.005	10.000	0.014	9.996	10.000	0.005	9.991	10.000	0.000	9.983	10.000	-0.008	9.978	10.000	-0.013
	MIN	9.990	9.991	-0.010	9.981	9.991	-0.019	9.976	9.991	-0.024	9.968	9.991	-0.032	9.963	9.991	-0.037
12	MAX	12.006	12.000	0.017	11.995	12.000	0.006	11.989	12.000	0.000	11.979	12.000	-0.010	11.974	12.000	-0.015
	MIN	11.988	11.989	-0.012	11.977	11.989	-0.023	11.971	11.989	-0.029	11.961	11.989	-0.039	11.956	11.989	-0.044
16	MAX	16.006	16.000	0.017	15.995	16.000	0.006	15.989	16.000	0.000	15.979	16.000	-0.010	15.974	16.000	-0.015
	MIN	15.988	15.989	-0.012	15.977	15.989	-0.023	15.971	15.989	-0.029	15.961	15.989	-0.039	15.956	15.989	-0.044
20	MAX	20.006	20.000	0.019	19.993	20.000	0.006	19.986	20.000	-0.001	19.973	20.000	-0.014	19.967	20.000	-0.020
	MIN	19.985	19.987	-0.015	19.972	19.987	-0.028	19.965	19.987	-0.035	19.952	19.987	-0.048	19.946	19.987	-0.054
25	MAX	25.006	25.000	0.019	24.993	25.000	0.006	24.986	25.000	-0.001	24.973	25.000	-0.014	24.960	25.000	-0.027
	MIN	24.985	24.987	-0.015	24.972	24.987	-0.028	24.965	24.987	-0.035	24.952	24.987	-0.048	24.939	24.987	-0.061
30	MAX	30.006	30.000	0.019	29.993	30.000	0.006	29.986	30.000	-0.001	29.973	30.000	-0.014	29.960	30.000	-0.027
	MIN	29.985	29.987	-0.015	29.972	29.987	-0.028	29.965	29.987	-0.035	29.952	29.987	-0.048	29.939	29.987	-0.061

Source: ANSI B4.2-1978, AMERICAN NATIONAL STANDARD PREFERRED METRIC LIMITS AND FITS.

Table C.4—*Continued*

Dimensions in mm.

BASIC SIZE		LOCATIONAL TRANSN. Hole K7	Shaft h6	Fit	LOCATIONAL TRANSN. Hole N7	Shaft h6	Fit	LOCATIONAL INTERF. Hole P7	Shaft h6	Fit	MEDIUM DRIVE Hole S7	Shaft h6	Fit	FORCE Hole U7	Shaft h6	Fit
40	MAX	40.007	40.000	0.023	39.992	40.000	0.008	39.983	40.000	-0.001	39.966	40.000	-0.018	39.949	40.000	-0.035
	MIN	39.982	39.984	-0.018	39.967	39.984	-0.033	39.958	39.984	-0.042	39.941	39.984	-0.059	39.924	39.984	-0.076
50	MAX	50.007	50.000	0.023	49.992	50.000	0.008	49.983	50.000	-0.001	49.966	50.000	-0.018	49.939	50.000	-0.045
	MIN	49.982	49.984	-0.018	49.967	49.984	-0.033	49.958	49.984	-0.042	49.941	49.984	-0.059	49.914	49.984	-0.086
60	MAX	60.009	60.000	0.028	59.991	60.000	0.010	59.979	60.000	-0.002	59.958	60.000	-0.023	59.924	60.000	-0.057
	MIN	59.979	59.981	-0.021	59.961	59.981	-0.039	59.949	59.981	-0.051	59.928	59.981	-0.072	59.894	59.981	-0.106
80	MAX	80.009	80.000	0.028	79.991	80.000	0.010	79.979	80.000	-0.002	79.952	80.000	-0.029	79.909	80.000	-0.072
	MIN	79.979	79.981	-0.021	79.961	79.981	-0.039	79.949	79.981	-0.051	79.922	79.981	-0.078	79.879	79.981	-0.121
100	MAX	100.010	100.000	0.032	99.990	100.000	0.012	99.976	100.000	-0.002	99.942	100.000	-0.036	99.889	100.000	-0.089
	MIN	99.975	99.978	-0.025	99.955	99.978	-0.045	99.941	99.978	-0.059	99.907	99.978	-0.093	99.854	99.978	-0.146
120	MAX	120.010	120.000	0.032	119.990	120.000	0.012	119.976	120.000	-0.002	119.934	120.000	-0.044	119.869	120.000	-0.109
	MIN	119.975	119.978	-0.025	119.955	119.978	-0.045	119.941	119.978	-0.059	119.899	119.978	-0.101	119.834	119.978	-0.166
160	MAX	160.012	160.000	0.037	159.988	160.000	0.013	159.972	160.000	-0.003	159.915	160.000	-0.060	159.825	160.000	-0.150
	MIN	159.972	159.975	-0.028	159.948	159.975	-0.052	159.932	159.975	-0.068	159.875	159.975	-0.125	159.785	159.975	-0.215
200	MAX	200.013	200.000	0.042	199.986	200.000	0.015	199.967	200.000	-0.004	199.895	200.000	-0.076	199.781	200.000	-0.190
	MIN	199.967	199.971	-0.033	199.940	199.971	-0.060	199.921	199.971	-0.079	199.849	199.971	-0.151	199.735	199.971	-0.265
250	MAX	250.013	250.000	0.042	249.986	250.000	0.015	249.967	250.000	-0.004	249.877	250.000	-0.094	249.733	250.000	-0.238
	MIN	249.967	249.971	-0.033	249.940	249.971	-0.060	249.921	249.971	-0.079	249.831	249.971	-0.169	249.687	249.971	-0.313
300	MAX	300.016	300.000	0.048	299.986	300.000	0.018	299.964	300.000	-0.004	299.850	300.000	-0.118	299.670	300.000	-0.298
	MIN	299.964	299.968	-0.036	299.934	299.968	-0.066	299.912	299.968	-0.088	299.798	299.968	-0.202	299.618	299.968	-0.382
400	MAX	400.017	400.000	0.053	399.984	400.000	0.020	399.959	400.000	-0.005	399.813	400.000	-0.151	399.586	400.000	-0.378
	MIN	399.960	399.964	-0.040	399.927	399.964	-0.073	399.902	399.964	-0.098	399.756	399.964	-0.244	399.529	399.964	-0.471
500	MAX	500.018	500.000	0.058	499.983	500.000	0.023	499.955	500.000	-0.005	499.771	500.000	-0.189	499.483	500.000	-0.477
	MIN	499.955	499.960	-0.045	499.920	499.960	-0.080	499.892	499.960	-0.108	499.708	499.960	-0.292	499.420	499.960	-0.580

Source: ANSI B4.2-1978, AMERICAN NATIONAL STANDARD PREFERRED METRIC LIMITS AND FITS.

Appendix D: Standard Thread Series

Table D.1: Standard Series Threads (UN/UNR)[1]

Nominal Size		Basic Major Diameter	Threads/in.											Nominal Size
			Series With Graded Pitches			Series With Constant Pitches								
Primary	Secondary		Coarse UNC	Fine UNF	Extra-Fine UNEF	4UN	6UN	8UN	12UN	16UN	20UN	28UN	32UN	
0		0.0600	—	80	—	—	—	—	—	—	—	—	—	0
	1	0.0730	64	72	—	—	—	—	—	—	—	—	—	1
2		0.0860	56	64	—	—	—	—	—	—	—	—	—	2
	3	0.0990	48	56	—	—	—	—	—	—	—	—	—	3
4		0.1120	40	48	—	—	—	—	—	—	—	—	—	4
5		0.1250	40	44	—	—	—	—	—	—	—	—	—	5
6		0.1380	32	40	—	—	—	—	—	—	—	—	UNC	6
8		0.1640	32	36	—	—	—	—	—	—	—	—	UNC	8
10		0.1900	24	32	—	—	—	—	—	—	—	—	UNF	10
	12	0.2160	24	28	32	—	—	—	—	—	—	UNF	UNEF	12
1/4		0.2500	20	28	32	—	—	—	—	—	UNC	UNF	UNEF	1/4
5/16		0.3125	18	24	32	—	—	—	—	—	20	28	UNEF	5/16
3/8		0.3750	16	24	32	—	—	—	—	UNC	20	28	UNEF	3/8
7/16		0.4375	14	20	28	—	—	—	—	16	UNF	UNEF	32	7/16
1/2		0.5000	13	20	28	—	—	—	—	16	UNF	UNEF	32	1/2
9/16		0.5625	12	18	24	—	—	—	UNC	16	20	28	32	9/16
5/8		0.6250	11	18	24	—	—	—	12	16	20	28	32	5/8
	11/16	0.6875	—	—	24	—	—	—	12	16	20	28	32	11/16
3/4		0.7500	10	16	20	—	—	—	12	UNF	UNEF	28	32	3/4
	13/16	0.8125	—	—	20	—	—	—	12	16	UNEF	28	32	13/16
7/8		0.8750	9	14	20	—	—	—	12	16	UNEF	28	32	7/8
	15/16	0.9375	—	—	20	—	—	—	12	16	UNEF	28	32	15/16
1		1.0000	8	12	20	—	—	UNC	UNF	16	UNEF	28	32	1
	1 1/16	1.0625	—	—	18	—	—	8	12	16	20	28	—	1 1/16
1 1/8		1.1250	7	12	18	—	—	8	UNF	16	20	28	—	1 1/8
	1 3/16	1.1875	—	—	18	—	—	8	12	16	20	28	—	1 3/16
1 1/4		1.2500	7	12	18	—	—	8	UNF	16	20	28	—	1 1/4
	1 5/16	1.3125	—	—	18	—	—	8	12	16	20	28	—	1 5/16
1 3/8		1.3750	6	12	18	—	UNC	8	UNF	16	20	28	—	1 3/8
	1 7/16	1.4375	—	—	18	—	6	8	12	16	20	28	—	1 7/16
1 1/2		1.5000	6	12	18	—	UNC	8	UNF	16	20	28	—	1 1/2
	1 9/16	1.5625	—	—	18	—	6	8	12	16	20	—	—	1 9/16
1 5/8		1.6250	—	—	18	—	6	8	12	16	20	—	—	1 5/8
	1 11/16	1.6875	—	—	18	—	6	8	12	16	20	—	—	1 11/16
1 3/4		1.7500	5	—	—	—	6	8	12	16	20	—	—	1 3/4
	1 13/16	1.8125	—	—	—	—	6	8	12	16	20	—	—	1 13/16
1 7/8		1.8750	—	—	—	—	6	8	12	16	20	—	—	1 7/8
	1 15/16	1.9375	—	—	—	—	6	8	12	16	20	—	—	1 15/16
2		2.0000	4 1/2	—	—	—	6	8	12	16	20	—	—	2
	2 1/8	2.1250	—	—	—	—	6	8	12	16	20	—	—	2 1/8
2 1/4		2.2500	4 1/2	—	—	—	6	8	12	16	20	—	—	2 1/4
	2 3/8	2.3750	—	—	—	—	6	8	12	16	20	—	—	2 3/8
2 1/2		2.5000	4	—	—	UNC	6	8	12	16	20	—	—	2 1/2
	2 5/8	2.6250	—	—	—	4	6	8	12	16	20	—	—	2 5/8
2 3/4		2.7500	4	—	—	UNC	6	8	12	16	20	—	—	2 3/4
	2 7/8	2.8750	—	—	—	4	6	8	12	16	20	—	—	2 7/8
3		3.0000	4	—	—	UNC	6	8	12	16	20	—	—	3
	3 1/8	3.1250	—	—	—	4	6	8	12	16	—	—	—	3 1/8
3 1/4		3.2500	4	—	—	UNC	6	8	12	16	—	—	—	3 1/4
	3 3/8	3.3750	—	—	—	4	6	8	12	16	—	—	—	3 3/8
3 1/2		3.5000	4	—	—	UNC	6	8	12	16	—	—	—	3 1/2
	3 5/8	3.6250	—	—	—	4	6	8	12	16	—	—	—	3 5/8
3 3/4		3.7500	4	—	—	UNC	6	8	12	16	—	—	—	3 3/4
	3 7/8	3.8750	—	—	—	4	6	8	12	16	—	—	—	3 7/8
4		4.0000	4	—	—	UNC	6	8	12	16	—	—	—	4
	4 1/8	4.1250	—	—	—	4	6	8	12	16	—	—	—	4 1/8
4 1/4		4.2500	—	—	—	4	6	8	12	16	—	—	—	4 1/4
	4 3/8	4.3750	—	—	—	4	6	8	12	16	—	—	—	4 3/8
4 1/2		4.5000	—	—	—	4	6	8	12	16	—	—	—	4 1/2
	4 5/8	4.6250	—	—	—	4	6	8	12	16	—	—	—	4 5/8
4 3/4		4.7500	—	—	—	4	6	8	12	16	—	—	—	4 3/4
	4 7/8	4.8750	—	—	—	4	6	8	12	16	—	—	—	4 7/8
5		5.0000	—	—	—	4	6	8	12	16	—	—	—	5
	5 1/8	5.1250	—	—	—	4	6	8	12	16	—	—	—	5 1/8
5 1/4		5.2500	—	—	—	4	6	8	12	16	—	—	—	5 1/4
	5 3/8	5.3750	—	—	—	4	6	8	12	16	—	—	—	5 3/8
5 1/2		5.5000	—	—	—	4	6	8	12	16	—	—	—	5 1/2
	5 5/8	5.6250	—	—	—	4	6	8	12	16	—	—	—	5 5/8
5 3/4		5.7500	—	—	—	4	6	8	12	16	—	—	—	5 3/4
	5 7/8	5.8750	—	—	—	4	6	8	12	16	—	—	—	5 7/8
6		6.0000	—	—	—	4	6	8	12	16	—	—	—	6

NOTE:
(1) Series designation shown indicates the UN thread form; however, the UNR thread form may be specified by substituting UNR in place of UN in all designations for external use only.

Source: ANSI B1.1-1982, AMERICAN NATIONAL STANDARD UNIFIED INCH SCREW THREADS (UN AND UNR THREAD FORM).

Table D.2: Metric Screw Threads

Nominal Diameter (mm.)	Pitch (mm.)	
	Coarse	Fine
1.6	0.35	
2	0.4	
2.5	0.45	
3	0.5	
3.5	0.6	
4	0.7	
5	0.8	
6	1	
8	1.25	1
10	1.5	1.25
12	1.75	1.25
14	2	1.5
16	2	1.5
18	2.5	1.5
20	2.5	1.5
22	2.5	1.5
24	3	2
27	3	2
30	3.5	2
33	3.5	2
36	4	2
39	4	2
42	4.5	2
45	4.5	1.5
48	5	2
56	5.5	2
60	5.5	1.5
64	6	2
72	6	2
80	6	1.5
90	6	2
100	6	2
110		2
120		2
130		2
140		2
150		2
160		3
170		3
180		3
190		3
200		3

Source: ANSI B1.13M-1979, which is based upon ISO 261-1973; in addition to these standard M profile diameter/pitch combinations for many common threads, these standards list numerous other threads.

Appendix E: Numbered and Lettered Twist Drill Sizes

Number	Diameter Inch	Diameter Millimeter	Number	Diameter Inch	Diameter Millimeter	Letter	Diameter Inch	Diameter Millimeter
1	0.2280	5.7912	41	0.0960	2.4384	A	0.234	5.944
2	0.2210	5.6134	42	0.0935	2.3622	B	0.238	6.045
3	0.2130	5.4102	43	0.0890	2.2606	C	0.242	6.147
4	0.2090	5.3086	44	0.0860	2.1844	D	0.246	6.248
5	0.2055	5.2197	45	0.0820	2.0828	E	0.250	6.350
6	0.2040	5.1816	46	0.0810	2.0574	F	0.257	6.528
7	0.2010	5.1054	47	0.0785	1.9812	G	0.261	6.629
8	0.1990	5.0800	48	0.0760	1.9304	H	0.266	6.756
9	0.1960	4.9784	49	0.0730	1.8542	I	0.272	6.909
10	0.1935	4.9149	50	0.0700	1.7780	J	0.277	7.036
11	0.1910	4.8514	51	0.0670	1.7018	K	0.281	7.137
12	0.1890	4.8006	52	0.0635	1.6129	L	0.290	7.366
13	0.1850	4.6990	53	0.0595	1.5113	M	0.295	7.493
14	0.1820	4.6228	54	0.0550	1.3970	N	0.302	7.601
15	0.1800	4.5720	55	0.0520	1.3208	O	0.316	8.026
16	0.1770	4.4958	56	0.0465	1.1684	P	0.323	8.204
17	0.1730	4.3942	57	0.0430	1.0922	Q	0.332	8.433
18	0.1695	4.3053	58	0.0420	1.0668	R	0.339	8.611
19	0.1660	4.2164	59	0.0410	1.0414	S	0.348	8.839
20	0.1610	4.0894	60	0.0400	1.0160	T	0.358	9.093
21	0.1590	4.0386	61	0.0390	0.9906	U	0.368	9.347
22	0.1570	3.9878	62	0.0380	0.9652	V	0.377	9.576
23	0.1540	3.9116	63	0.0370	0.9398	W	0.386	9.804
24	0.1520	3.8608	64	0.0360	0.9144	X	0.397	10.084
25	0.1495	3.7973	65	0.0350	0.8890	Y	0.404	10.262
26	0.1470	3.7338	66	0.0330	0.8382	Z	0.413	10.490
27	0.1440	3.6576	67	0.0320	0.8128			
28	0.1405	3.5560	68	0.0310	0.7874			
29	0.1360	3.4544	69	0.0292	0.7417			
30	0.1285	3.2639	70	0.0280	0.7112			
31	0.1200	3.0480	71	0.0260	0.6604			
32	0.1160	2.9464	72	0.0250	0.6350			
33	0.1130	2.8702	73	0.0240	0.6096			
34	0.1110	2.8194	74	0.0225	0.5715			
35	0.1100	2.7940	75	0.0210	0.5334			
36	0.1065	2.7051	76	0.0200	0.5080			
37	0.1040	2.6416	77	0.0180	0.4572			
38	0.1015	2.5781	78	0.0160	0.4064			
39	0.0995	2.5273	79	0.0145	0.3638			
40	0.0980	2.4892	80	0.0135	0.3429			

Appendix F: American National Standard Square and Hex Bolts and Nuts

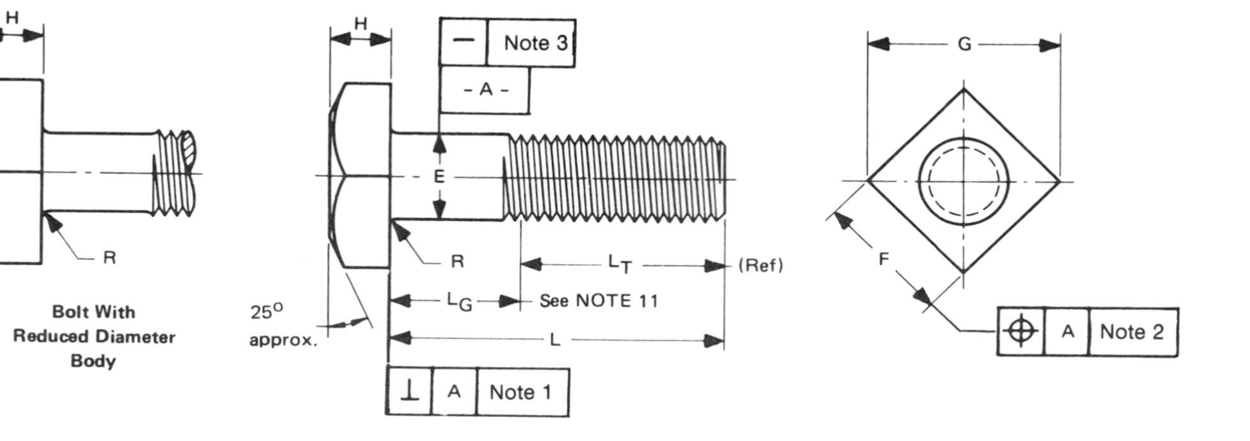

Table F.1: Dimensions of Square Bolts

Nominal Size or Basic Product Dia (17)		E	F			G		H			R		L_T	
		Body Dia (7), (14)	Width Across Flats (4)			Width Across Corners		Height			Radius of Fillet		Thread Length For Bolt Lengths (11)	
													6 in. and shorter	over 6 in.
		Max	Basic	Max	Min	Max	Min	Basic	Max	Min	Max	Min	Basic	Basic
1/4	0.2500	0.260	3/8	0.375	0.362	0.530	0.498	11/64	0.188	0.156	0.03	0.01	0.750	1.000
5/16	0.3125	0.324	1/2	0.500	0.484	0.707	0.665	13/64	0.220	0.186	0.03	0.01	0.875	1.125
3/8	0.3750	0.388	9/16	0.562	0.544	0.795	0.747	1/4	0.268	0.232	0.03	0.01	1.000	1.250
7/16	0.4375	0.452	5/8	0.625	0.603	0.884	0.828	19/64	0.316	0.278	0.03	0.01	1.125	1.375
1/2	0.5000	0.515	3/4	0.750	0.725	1.061	0.995	21/64	0.348	0.308	0.03	0.01	1.250	1.500
5/8	0.6250	0.642	15/16	0.938	0.906	1.326	1.244	27/64	0.444	0.400	0.06	0.02	1.500	1.750
3/4	0.7500	0.768	1 1/8	1.125	1.088	1.591	1.494	1/2	0.524	0.476	0.06	0.02	1.750	2.000
7/8	0.8750	0.895	1 5/16	1.312	1.269	1.856	1.742	19/32	0.620	0.568	0.06	0.02	2.000	2.250
1	1.0000	1.022	1 1/2	1.500	1.450	2.121	1.991	21/32	0.684	0.628	0.09	0.03	2.250	2.500
1 1/8	1.1250	1.149	1 11/16	1.688	1.631	2.386	2.239	3/4	0.780	0.720	0.09	0.03	2.500	2.750
1 1/4	1.2500	1.277	1 7/8	1.875	1.812	2.652	2.489	27/32	0.876	0.812	0.09	0.03	2.750	3.000
1 3/8	1.3750	1.404	2 1/16	2.062	1.994	2.917	2.738	29/32	0.940	0.872	0.09	0.03	3.000	3.250
1 1/2	1.5000	1.531	2 1/4	2.250	2.175	3.182	2.986	1	1.036	0.964	0.09	0.03	3.250	3.500

NOTES: ***1. Bearing Surface.*** *A die seam across the bearing surface is permissible. Bearing surface shall be perpendicular to the axis of the body within a tolerance of 3 deg. for 1 in. size and smaller, and 2 deg. for sizes larger than 1 in. Angularity measurement shall be taken at a location to avoid interference from a die seam.*

2. True Position of Head. *The axis of the head shall be located at true position with respect to the axis of the body (determined over a distance under the head equal to one diameter) within a tolerance zone having a diameter equivalent to 6% of the maximum width across flats, regardless of feature size.*

3. Straightness. *Shanks of bolts shall be straight within the following limits: for bolts within nominal lengths to and including 12 in., the maximum camber shall be 0.006 in. per inch of bolt length. Bolts with nominal lengths over 12 in. to and including 24 in., the maximum camber shall be 0.008 in. per inch of length.*

Source: ANSI B18.2.1-1981, AMERICAN NATIONAL STANDARD SQUARE AND HEX BOLTS AND SCREWS - INCH SERIES.

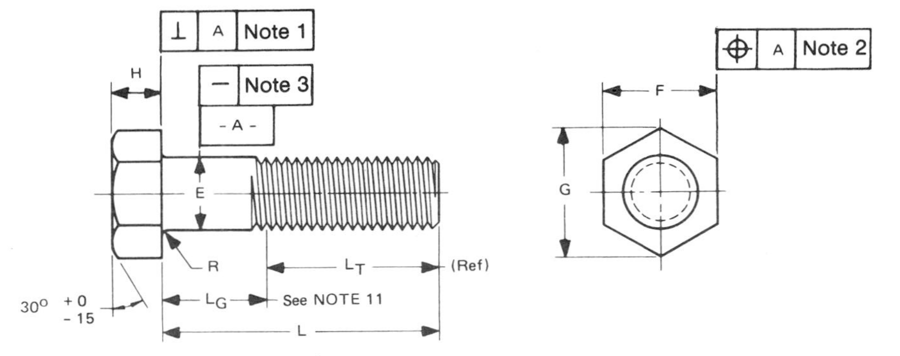

Table F.2: Dimensions of Hex Bolts

Nominal Size or Basic Product Dia (17)		E: Body Dia (7)	F: Width Across Flats (4)			G: Width Across Corners		H: Height			R: Radius of Fillet		L_T: Thread Length For Bolt Lengths (11), 6 in. and shorter	L_T: Thread Length For Bolt Lengths (11), over 6 in.
		Max	Basic	Max	Min	Max	Min	Basic	Max	Min	Max	Min	Basic	Basic
1/4	0.2500	0.260	7/16	0.438	0.425	0.505	0.484	11/64	0.188	0.150	0.03	0.01	0.750	1.000
5/16	0.3125	0.324	1/2	0.500	0.484	0.577	0.552	7/32	0.235	0.195	0.03	0.01	0.875	1.125
3/8	0.3750	0.388	9/16	0.562	0.544	0.650	0.620	1/4	0.268	0.226	0.03	0.01	1.000	1.250
7/16	0.4375	0.452	5/8	0.625	0.603	0.722	0.687	19/64	0.316	0.272	0.03	0.01	1.125	1.375
1/2	0.5000	0.515	3/4	0.750	0.725	0.866	0.826	11/32	0.364	0.302	0.03	0.01	1.250	1.500
5/8	0.6250	0.642	15/16	0.928	0.906	1.083	1.033	27/64	0.444	0.378	0.06	0.02	1.500	1.750
3/4	0.7500	0.768	1 1/8	1.125	1.088	1.299	1.240	1/2	0.524	0.455	0.06	0.02	1.750	2.000
7/8	0.8750	0.895	1 5/16	1.312	1.269	1.516	1.447	37/64	0.604	0.531	0.06	0.02	2.000	2.250
1	1.0000	1.022	1 1/2	1.500	1.450	1.732	1.653	43/64	0.700	0.591	0.09	0.03	2.250	2.500
1 1/8	1.1250	1.149	1 11/16	1.688	1.631	1.949	1.859	3/4	0.780	0.658	0.09	0.03	2.500	2.750
1 1/4	1.2500	1.277	1 7/8	1.875	1.812	2.165	2.066	27/32	0.876	0.749	0.09	0.03	2.750	3.000
1 3/8	1.3750	1.404	2 1/16	2.062	1.994	2.382	2.273	29/32	0.940	0.810	0.09	0.03	3.000	3.250
1 1/2	1.5000	1.531	2 1/4	2.250	2.175	2.598	2.480	1	1.036	0.902	0.09	0.03	3.250	3.500
1 3/4	1.7500	1.785	2 5/8	2.625	2.538	3.031	2.893	1 5/32	1.196	1.054	0.12	0.04	3.750	4.000
2	2.0000	2.039	3	3.000	2.900	3.464	3.306	1 11/32	1.388	1.175	0.12	0.04	4.250	4.500
2 1/4	2.2500	2.305	3 3/8	3.375	3.262	3.897	3.719	1 1/2	1.548	1.327	0.19	0.06	4.750	5.000
2 1/2	2.5000	2.559	3 3/4	3.750	3.625	4.330	4.133	1 21/32	1.708	1.479	0.19	0.06	5.250	5.500
2 3/4	2.7500	2.827	4 1/8	4.125	3.988	4.763	4.546	1 13/16	1.869	1.632	0.19	0.06	5.750	6.000
3	3.0000	3.081	4 1/2	4.500	4.350	5.196	4.959	2	2.060	1.815	0.19	0.06	6.250	6.500
3 1/4	3.2500	3.335	4 7/8	4.875	4.712	5.629	5.372	2 3/16	2.251	1.936	0.19	0.06	6.750	7.000
3 1/2	3.5000	3.589	5 1/4	5.250	5.075	6.062	5.786	2 5/16	2.380	2.057	0.19	0.06	7.250	7.500
3 3/4	3.7500	3.858	5 5/8	5.625	5.437	6.495	6.198	2 1/2	2.572	2.241	0.19	0.06	7.750	8.000
4	4.0000	4.111	6	6.000	5.800	6.928	6.612	2 11/16	2.764	2.424	0.19	0.06	8.250	8.500

*NOTES: **1. Bearing Surface.** A die seam across the bearing surface is permissible. Bearing surface shall be perpendicular to the axis of the body within a tolerance of 3 deg. for 1 in. size and smaller, and 2 deg. for sizes larger than 1 in. Angularity measurement shall be taken at a location to avoid interference from a die seam.*

***2. True Position of Head.** The axis of the head shall be located at true position with respect to the axis of the body (determined over a distance under the head equal to one diameter) within a tolerance zone having a diameter equivalent to 6% of the maximum width across flats, regardless of feature size.*

***3. Straightness.** Shanks of bolts shall be straight within the following limits: for bolts within nominal lengths to and including 12 in., the maximum camber shall be 0.006 in. per inch of bolt length. Bolts with nominal lengths over 12 in. to and including 24 in., the maximum camber shall be 0.008 in. per inch of length.*

Source: ANSI B18.2.1-1981, AMERICAN NATIONAL STANDARD SQUARE AND HEX BOLTS AND SCREWS - INCH SERIES.

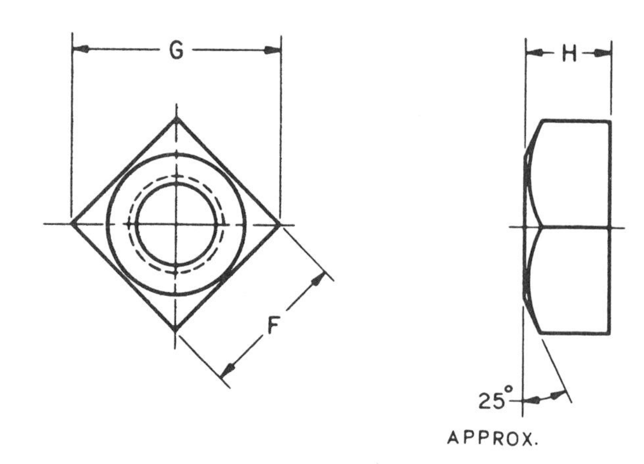

Table F.3: Dimensions of Square Nuts

Nominal Size or Basic Major Dia of Thread		F Width Across Flats			G Width Across Corners		H Thickness		
		Basic	Max	Min	Max	Min	Basic	Max	Min
1/4	0.2500	7/16	0.438	0.425	0.619	0.584	7/32	0.235	0.203
5/16	0.3125	9/16	0.562	0.547	0.795	0.751	17/64	0.283	0.249
3/8	0.3750	5/8	0.625	0.606	0.884	0.832	21/64	0.346	0.310
7/16	0.4375	3/4	0.750	0.728	1.061	1.000	3/8	0.394	0.356
1/2	0.5000	13/16	0.812	0.788	1.149	1.082	7/16	0.458	0.418
5/8	0.6250	1	1.000	0.969	1.414	1.330	35/64	0.569	0.525
3/4	0.7500	1 1/8	1.125	1.088	1.591	1.494	21/32	0.680	0.632
7/8	0.8750	1 5/16	1.312	1.269	1.856	1.742	49/64	0.792	0.740
1	1.0000	1 1/2	1.500	1.450	2.121	1.991	7/8	0.903	0.847
1 1/8	1.1250	1 11/16	1.688	1.631	2.386	2.239	1	1.030	0.970
1 1/4	1.2500	1 7/8	1.875	1.812	2.652	2.489	1 3/32	1.126	1.062
1 3/8	1.3750	2 1/16	2.062	1.994	2.917	2.738	1 13/64	1.237	1.169
1 1/2	1.5000	2 1/4	2.250	2.175	3.182	2.986	1 5/16	1.348	1.276

Source: ANSI B18.2.2-1972, AMERICAN NATIONAL STANDARD SQUARE AND HEX NUTS.

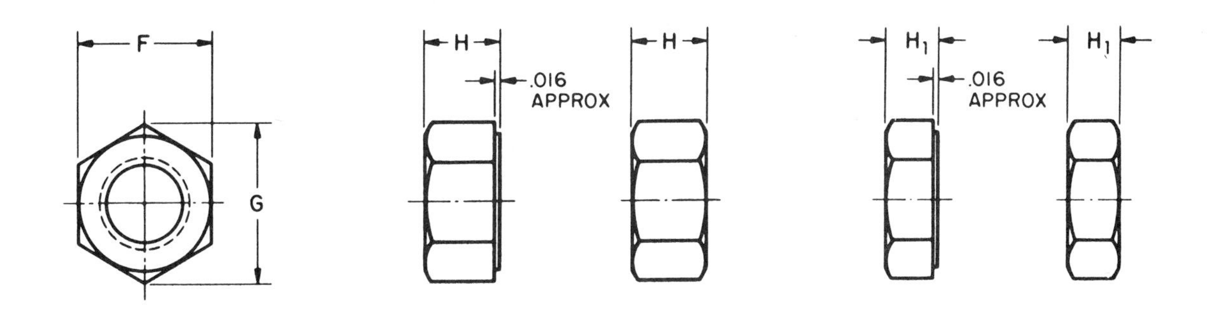

Table F.4: Dimensions of Hex Nuts and Hex Jam Nuts

Nominal Size or Basic Major Dia of Thread		F			G		H			H_1		
		Width Across Flats			Width Across Corners		Thickness Hex Nuts			Thickness Hex Jam Nuts		
		Basic	Max	Min	Max	Min	Basic	Max	Min	Basic	Max	Min
1/4	0.2500	7/16	0.438	0.428	0.505	0.488	7/32	0.226	0.212	5/32	0.163	0.150
5/16	0.3125	1/2	0.500	0.489	0.577	0.557	17/64	0.273	0.258	3/16	0.195	0.180
3/8	0.3750	9/16	0.562	0.551	0.650	0.628	21/64	0.337	0.320	7/32	0.227	0.210
7/16	0.4375	11/16	0.688	0.675	0.794	0.768	3/8	0.385	0.365	1/4	0.260	0.240
1/2	0.5000	3/4	0.750	0.736	0.866	0.840	7/16	0.448	0.427	5/16	0.323	0.302
9/16	0.5625	7/8	0.875	0.861	1.010	0.982	31/64	0.496	0.473	5/16	0.324	0.301
5/8	0.6250	15/16	0.938	0.922	1.083	1.051	35/64	0.559	0.535	3/8	0.387	0.363
3/4	0.7500	1 1/8	1.125	1.088	1.299	1.240	41/64	0.665	0.617	27/64	0.446	0.398
7/8	0.8750	1 5/16	1.312	1.269	1.516	1.447	3/4	0.776	0.724	31/64	0.510	0.458
1	1.0000	1 1/2	1.500	1.450	1.732	1.653	55/64	0.887	0.831	35/64	0.575	0.519
1 1/8	1.1250	1 11/16	1.688	1.631	1.949	1.859	31/32	0.999	0.939	39/64	0.639	0.579
1 1/4	1.2500	1 7/8	1.875	1.812	2.165	2.066	1 1/16	1.094	1.030	23/32	0.751	0.687
1 3/8	1.3750	2 1/16	2.062	1.994	2.382	2.273	1 11/64	1.206	1.138	25/32	0.815	0.747
1 1/2	1.5000	2 1/4	2.250	2.175	2.598	2.480	1 9/32	1.317	1.245	27/32	0.880	0.808

Source: ANSI B18.2.2-1972, AMERICAN NATIONAL STANDARD SQUARE AND HEX NUTS.

Appendix G: American Standard Round Head Bolts

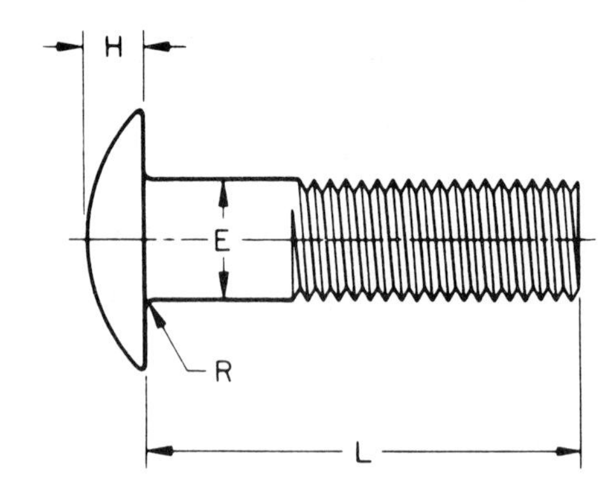

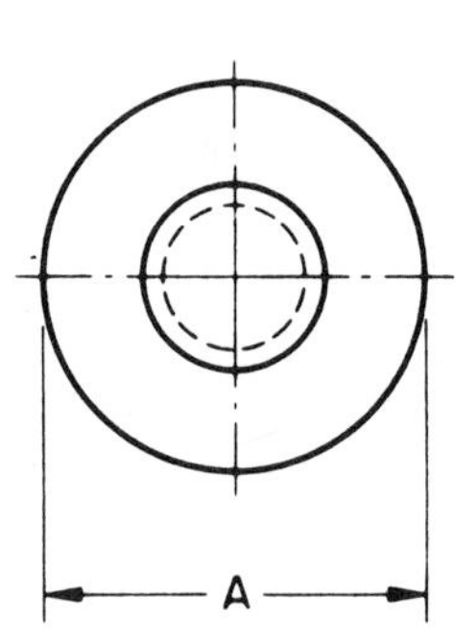

Table G.1: Dimensions of Round Head Bolts

Nominal Size[1] or Basic Bolt Diameter		E Body Diameter		A Head Diameter		H Head Height		R Fillet Radius
		Max	Min	Max	Min	Max	Min	Max[2]
No. 10	0.1900	0.199	0.182	0.469	0.438	0.114	0.094	0.031
1/4	0.2500	0.260	0.237	0.594	0.563	0.145	0.125	0.031
5/16	0.3125	0.324	0.298	0.719	0.688	0.176	0.156	0.031
3/8	0.3750	0.388	0.360	0.844	0.782	0.208	0.188	0.031
7/16	0.4375	0.452	0.421	0.969	0.907	0.239	0.219	0.031
1/2	0.5000	0.515	0.483	1.094	1.032	0.270	0.250	0.031
5/8	0.6250	0.642	0.605	1.344	1.219	0.344	0.313	0.062
3/4	0.7500	0.768	0.729	1.594	1.469	0.406	0.375	0.062
7/8	0.8750	0.895	0.852	1.844	1.719	0.469	0.438	0.062
1	1.0000	1.022	0.976	2.094	1.969	0.531	0.500	0.062

[1] Where specifying nominal size in decimals, zeros preceding decimal and in the fourth decimal place shall be omitted.

[2] The minimum radius is one half of the value shown.

Source: ANSI B18.5-1978, AMERICAN NATIONAL STANDARD ROUND HEAD BOLTS.

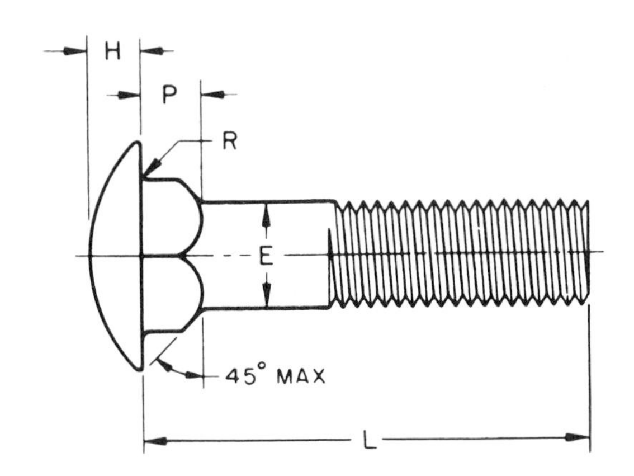

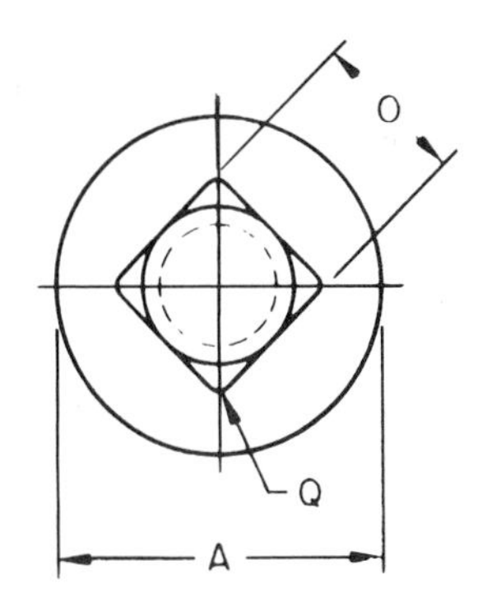

Table G.2: Dimensions of Round Head Square Neck Bolts

Nominal Size[1] or Basic Bolt Diameter		E Body Diameter		A Head Diameter		H Head Height		O Square Width		P Square Depth		Q Corner Radius on Square	R Fillet Radius
		Max	Min	Max	Min	Max	Min	Max	Min	Max	Min	Max	Max[2]
No. 10	0.1900	0.199	0.182	0.469	0.438	0.114	0.094	0.199	0.185	0.125	0.094	0.031	0.031
1/4	0.2500	0.260	0.237	0.594	0.563	0.145	0.125	0.260	0.245	0.156	0.125	0.031	0.031
5/16	0.3125	0.324	0.298	0.719	0.688	0.176	0.156	0.324	0.307	0.187	0.156	0.031	0.031
3/8	0.3750	0.388	0.360	0.844	0.782	0.208	0.188	0.388	0.368	0.219	0.188	0.047	0.031
7/16	0.4375	0.452	0.421	0.969	0.907	0.239	0.219	0.452	0.431	0.250	0.219	0.047	0.031
1/2	0.5000	0.515	0.483	1.094	1.032	0.270	0.250	0.515	0.492	0.281	0.250	0.047	0.031
5/8	0.6250	0.642	0.605	1.344	1.219	0.344	0.313	0.642	0.616	0.344	0.313	0.078	0.062
3/4	0.7500	0.768	0.729	1.594	1.469	0.406	0.375	0.768	0.741	0.406	0.375	0.078	0.062
7/8	0.8750	0.895	0.852	1.844	1.719	0.469	0.438	0.895	0.865	0.469	0.438	0.094	0.062
1	1.0000	1.022	0.976	2.094	1.969	0.531	0.500	1.022	0.990	0.531	0.500	0.094	0.062

[1] Where specifying nominal size in decimals, zeros preceding decimal and in the fourth decimal place shall be omitted.

[2] The minimum radius is one half of the value shown.

Source: ANSI B18.5-1978, AMERICAN NATIONAL STANDARD ROUND HEAD BOLTS.

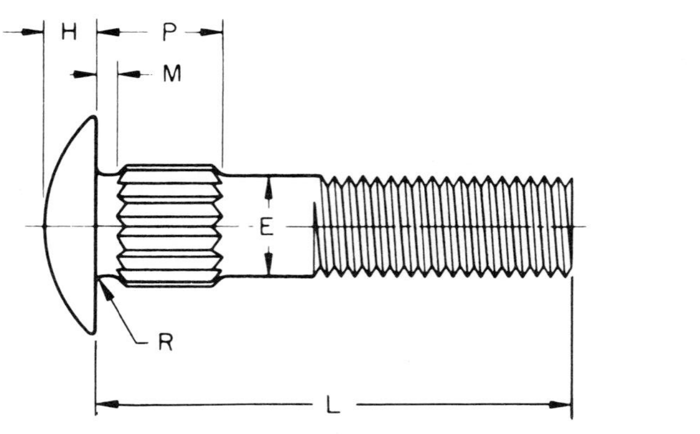

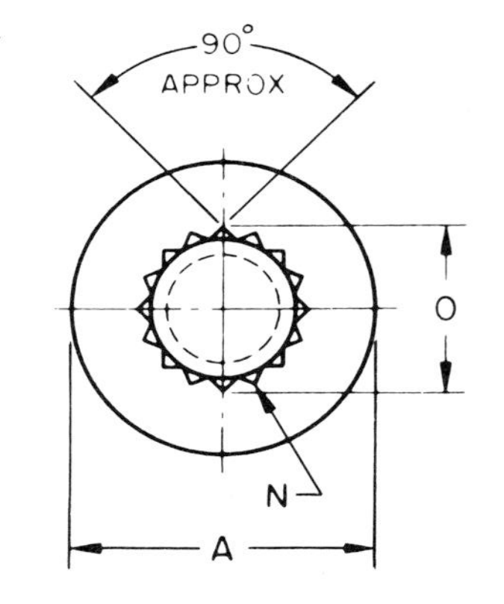

Table G.3: Dimensions of Round Head Ribbed Neck Bolts

Nominal Size[1] or Basic Bolt Diameter		E		A		H		M		N	O	P			R
		Body Diameter		Head Diameter		Head Height		Head to Ribs		Number of Ribs	Diameter Over Ribs	Depth Over Ribs			Fillet Radius
								For Lengths of				For Lengths of			
								7/8 and Shorter	1 in. and Longer			7/8 and Shorter	1 in and 1 1/8	1/4 and Longer	
		Max	Min	Max	Min	Max	Min	±0.031*		Approx	Min	±0.031			Max[2]
No. 10	0.1900	0.199	0.182	0.469	0.438	0.114	0.094	0.031*	0.063	9	0.210	0.250	0.407	0.594	0.031
1/4	0.2500	0.260	0.237	0.594	0.563	0.145	0.125	0.031*	0.063	10	0.274	0.250	0.407	0.594	0.031
5/16	0.3125	0.324	0.298	0.719	0.688	0.176	0.156	0.031*	0.063	12	0.340	0.250	0.407	0.594	0.031
3/8	0.3750	0.388	0.360	0.844	0.782	0.208	0.188	0.031*	0.063	12	0.405	0.250	0.407	0.594	0.031
7/16	0.4375	0.452	0.421	0.969	0.907	0.239	0.219	0.031*	0.063	14	0.470	0.250	0.407	0.594	0.031
1/2	0.5000	0.515	0.483	1.094	1.032	0.270	0.250	0.031*	0.063	16	0.534	0.250	0.407	0.594	0.031
5/8	0.6250	0.642	0.605	1.344	1.219	0.344	0.313	0.094	0.094	19	0.660	0.313	0.438	0.625	0.062
3/4	0.7500	0.768	0.729	1.594	1.469	0.406	0.375	0.094	0.094	22	0.785	0.313	0.438	0.625	0.062

[1] Where specifying nominal size in decimals, zeros preceding decimal and in the fourth decimal place shall be omitted.

[2] The minimum radius is one half of the value shown.

*Tolerance on the No. 10 through 1/2 in. sizes for nominal lengths 7/8 in. and shorter shall be +0.031 and −0.000.

Source: ANSI B18.5-1978, AMERICAN NATIONAL STANDARD ROUND HEAD BOLTS.

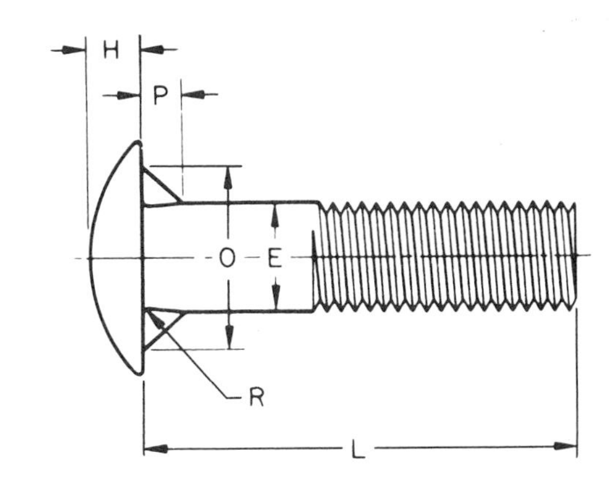

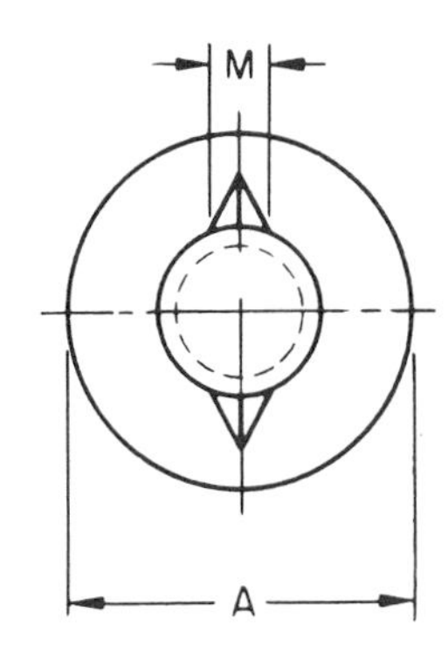

Table G.4: Dimensions of Round Head Fin Neck Bolts

Nominal Size[1] or Basic Bolt Diameter		E Body Diameter		A Head Diameter		H Head Height		M Fin Thickness		O Distance Across Fins		P Fin Depth		R Fillet Radius
		Max	Min	Max	Min	Max	Min	Max	Min	Max	Min	Max	Min	Max[2]
No. 10	0.1900	0.199	0.182	0.469	0.438	0.114	0.094	0.098	0.078	0.395	0.375	0.088	0.078	0.031
1/4	0.2500	0.260	0.237	0.594	0.563	0.145	0.125	0.114	0.094	0.458	0.438	0.104	0.094	0.031
5/16	0.3125	0.324	0.298	0.719	0.688	0.176	0.156	0.145	0.125	0.551	0.531	0.135	0.125	0.031
3/8	0.3750	0.388	0.360	0.844	0.782	0.208	0.188	0.161	0.141	0.645	0.625	0.151	0.141	0.031
7/16	0.4375	0.452	0.421	0.969	0.907	0.239	0.219	0.192	0.172	0.739	0.719	0.182	0.172	0.031
1/2	0.5000	0.515	0.483	1.094	1.032	0.270	0.250	0.208	0.188	0.833	0.813	0.198	0.188	0.031

[1] Where specifying nominal size in decimals, zeros preceding decimal and in the fourth decimal place shall be omitted.

[2] The minimum radius is one half of the value shown.

Source: ANSI B18.5-1978, AMERICAN NATIONAL STANDARD ROUND HEAD BOLTS.

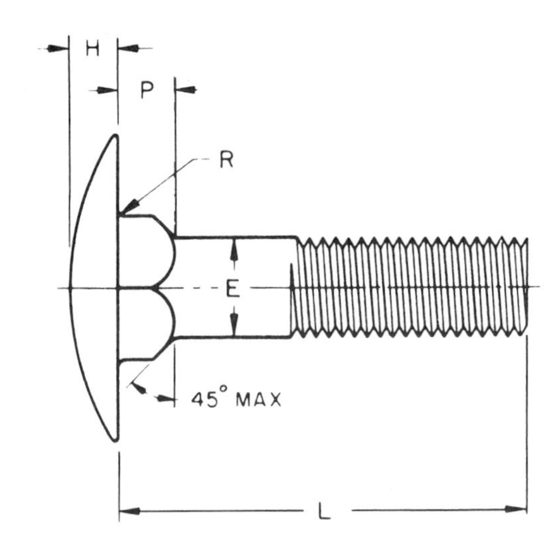

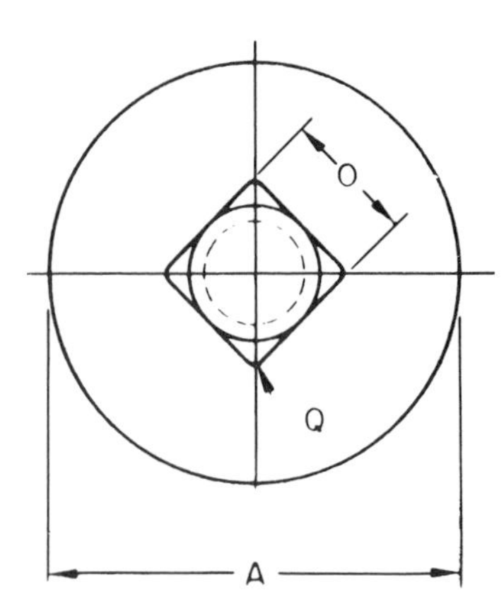

Table G.5: Dimensions of Step Bolts

Nominal Size[1] or Basic Bolt Diameter	E		A		H		O		P		Q	R
	Body Diameter		Head Diameter		Head Height		Square Width		Square Depth		Center Radius on Square	Fillet Radius
	Max	Min	Max	Min	Max	Min	Max	Min	Max	Min	Max	Max[2]
No. 10 0.1900	0.199	0.182	0.656	0.625	0.114	0.094	0.199	0.185	0.125	0.094	0.031	0.031
1/4 0.2500	0.260	0.237	0.844	0.813	0.145	0.125	0.260	0.245	0.156	0.125	0.031	0.031
5/16 0.3125	0.324	0.298	1.031	1.000	0.176	0.156	0.324	0.307	0.187	0.156	0.031	0.031
3/8 0.3750	0.388	0.360	1.219	1.188	0.208	0.188	0.388	0.368	0.219	0.188	0.047	0.031
7/16 0.4375	0.452	0.421	1.406	1.375	0.239	0.219	0.452	0.431	0.250	0.219	0.047	0.031
1/2 0.5000	0.515	0.483	1.594	1.563	0.270	0.250	0.515	0.492	0.281	0.250	0.047	0.031

[1] Where specifying nominal size in decimals, zeros preceding decimal and in the fourth decimal place shall be omitted.

[2] The minimum radius is one half of the value shown.

Source: ANSI B18.5-1978, AMERICAN NATIONAL STANDARD ROUND HEAD BOLTS.

Appendix H: American Standard Slotted Head Cap Screws

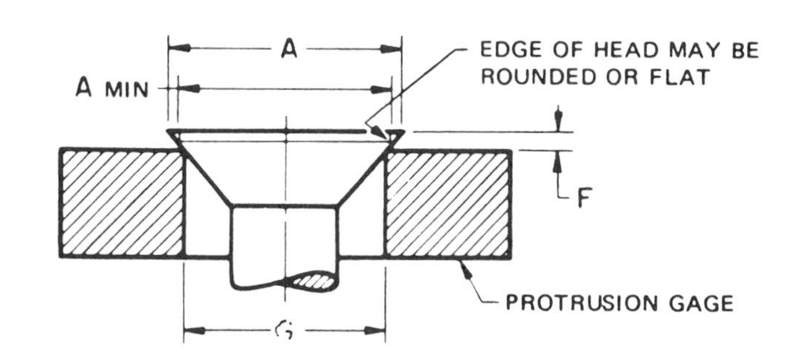

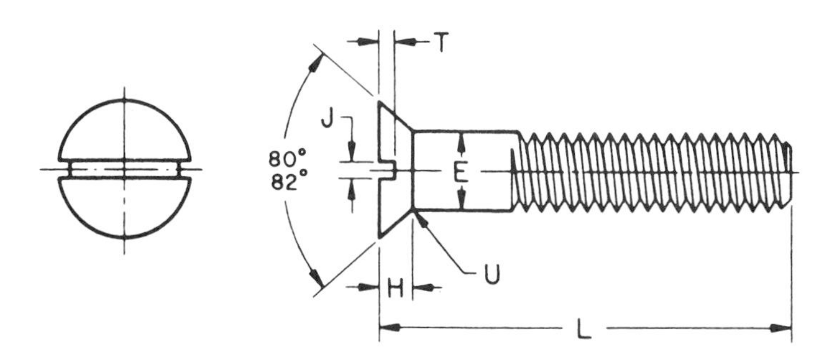

Table H.1: Dimensions of Slotted Flat Countersunk Head Cap Screws

Nominal Size[1] or Basic Screw Diameter	E		A		H[2]	J		T		U	F[3]		G[3]
	Body Diameter		Head Diameter		Head Height	Slot Width		Slot Depth		Fillet Radius	Protrusion Above Gaging Diameter		Gaging Diameter
	Max	Min	Max, Edge Sharp	Min, Edge Rounded or Flat	Ref	Max	Min	Max	Min	Max	Max	Min	
1/4 0.2500	0.2500	0.2450	0.500	0.452	0.140	0.075	0.064	0.068	0.045	0.100	0.046	0.030	0.424
5/16 0.3125	0.3125	0.3070	0.625	0.567	0.177	0.084	0.072	0.086	0.057	0.125	0.053	0.035	0.538
3/8 0.3750	0.3750	0.3690	0.750	0.682	0.210	0.094	0.081	0.103	0.068	0.150	0.060	0.040	0.651
7/16 0.4375	0.4375	0.4310	0.812	0.736	0.210	0.094	0.081	0.103	0.068	0.175	0.065	0.044	0.703
1/2 0.5000	0.5000	0.4930	0.875	0.791	0.210	0.106	0.091	0.103	0.068	0.200	0.071	0.049	0.756
9/16 0.5625	0.5625	0.5550	1.000	0.906	0.244	0.118	0.102	0.120	0.080	0.225	0.078	0.054	0.869
5/8 0.6250	0.6250	0.6170	1.125	1.020	0.281	0.133	0.116	0.137	0.091	0.250	0.085	0.058	0.982
3/4 0.7500	0.7500	0.7420	1.375	1.251	0.352	0.149	0.131	0.171	0.115	0.300	0.099	0.068	1.208
7/8 0.8750	0.8750	0.8660	1.625	1.480	0.423	0.167	0.147	0.206	0.138	0.350	0.113	0.077	1.435
1 1.0000	1.0000	0.9900	1.875	1.711	0.494	0.188	0.166	0.240	0.162	0.400	0.127	0.087	1.661
1 1/8 1.1250	1.1250	1.1140	2.062	1.880	0.529	0.196	0.178	0.257	0.173	0.450	0.141	0.096	1.826
1 1/4 1.2500	1.2500	1.2390	2.312	2.110	0.600	0.211	0.193	0.291	0.197	0.500	0.155	0.105	2.052
1 3/8 1.3750	1.3750	1.3630	2.562	2.340	0.665	0.226	0.208	0.326	0.220	0.550	0.169	0.115	2.279
1 1/2 1.5000	1.5000	1.4880	2.812	2.570	0.742	0.258	0.240	0.360	0.244	0.600	0.183	0.124	2.505

1 Where specifying nominal size in decimals, zeros preceding decimal and in the fourth decimal place shall be omitted.
2 Tabulated values determined from formula for maximum H, Appendix III.
3 No tolerance for gaging diameter is given. If the gaging diameter of the gage used differs from tabulated value, the protrusion will be affected accordingly and the proper protrusion values must be recalculated using the formulas shown in Appendix II.

Source: ANSI B18.6.2-1972, AMERICAN NATIONAL STANDARD SLOTTED HEAD CAP SCREWS, SQUARE HEAD SET SCREWS, AND SLOTTED HEADLESS SET SCREWS.

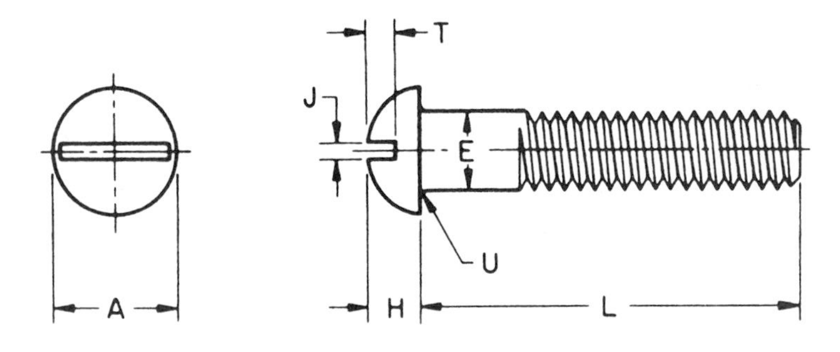

Table H.2: Dimensions of Slotted Round Head Cap Screws

Nominal Size[1] or Basic Screw Diameter		E Body Diameter		A Head Diameter		H Head Height		J Slot Width		T Slot Depth		U Fillet Radius	
		Max	Min	Max	Min	Max	Min	Max	Min	Max	Min	Max	Min
1/4	0.2500	0.2500	0.2450	0.437	0.418	0.191	0.175	0.075	0.064	0.117	0.097	0.031	0.016
5/16	0.3125	0.3125	0.3070	0.562	0.540	0.245	0.226	0.084	0.072	0.151	0.126	0.031	0.016
3/8	0.3750	0.3750	0.3690	0.625	0.603	0.273	0.252	0.094	0.081	0.168	0.138	0.031	0.016
7/16	0.4375	0.4375	0.4310	0.750	0.725	0.328	0.302	0.094	0.081	0.202	0.167	0.047	0.016
1/2	0.5000	0.5000	0.4930	0.812	0.786	0.354	0.327	0.106	0.091	0.218	0.178	0.047	0.016
9/16	0.5625	0.5625	0.5550	0.937	0.909	0.409	0.378	0.118	0.102	0.252	0.207	0.047	0.016
5/8	0.6250	0.6250	0.6170	1.000	0.970	0.437	0.405	0.133	0.116	0.270	0.220	0.062	0.031
3/4	0.7500	0.7500	0.7420	1.250	1.215	0.546	0.507	0.149	0.131	0.338	0.278	0.062	0.031

[1] Where specifying nominal size in decimals, zeros preceding decimal and in the fourth decimal place shall be omitted.

Source: ANSI B18.6.2-1972, AMERICAN NATIONAL STANDARD SLOTTED HEAD CAP SCREWS, SQUARE HEAD SET SCREWS, AND SLOTTED HEADLESS SET SCREWS.

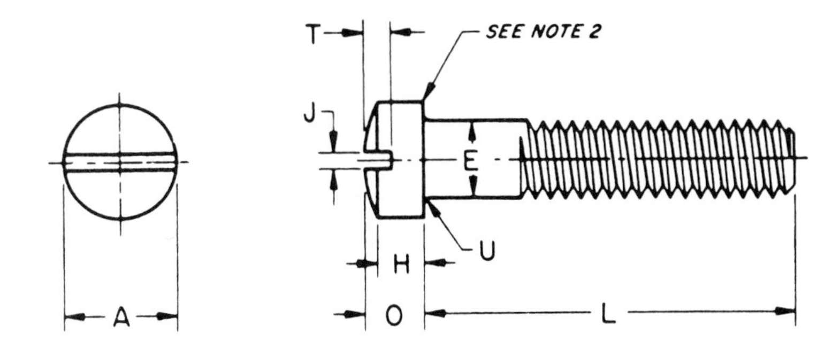

Table H.3: Dimensions of Slotted Fillister Head Cap Screws

Nominal Size[1] or Basic Screw Diameter	E		A		H		O		J		T		U	
	Body Diameter		Head Diameter		Head Side Height		Total Head Height		Slot Width		Slot Depth		Fillet Radius	
	Max	Min	Max	Min	Max	Min	Max	Min	Max	Min	Max	Min	Max	Min
1/4 0.2500	0.2500	0.2450	0.375	0.363	0.172	0.157	0.216	0.194	0.075	0.064	0.097	0.077	0.031	0.016
5/16 0.3125	0.3125	0.3070	0.437	0.424	0.203	0.186	0.253	0.230	0.084	0.072	0.115	0.090	0.031	0.016
3/8 0.3750	0.3750	0.3690	0.562	0.547	0.250	0.229	0.314	0.284	0.094	0.081	0.142	0.112	0.031	0.016
7/16 0.4375	0.4375	0.4310	0.625	0.608	0.297	0.274	0.368	0.336	0.094	0.081	0.168	0.133	0.047	0.016
1/2 0.5000	0.5000	0.4930	0.750	0.731	0.328	0.301	0.413	0.376	0.106	0.091	0.193	0.153	0.047	0.016
9/16 0.5625	0.5625	0.5550	0.812	0.792	0.375	0.346	0.467	0.427	0.118	0.102	0.213	0.168	0.047	0.016
5/8 0.6250	0.6250	0.6170	0.875	0.853	0.422	0.391	0.521	0.478	0.133	0.116	0.239	0.189	0.062	0.031
3/4 0.7500	0.7500	0.7420	1.000	0.976	0.500	0.466	0.612	0.566	0.149	0.131	0.283	0.223	0.062	0.031
7/8 0.8750	0.8750	0.8660	1.125	1.098	0.594	0.556	0.720	0.668	0.167	0.147	0.334	0.264	0.062	0.031
1 1.0000	1.0000	0.9900	1.312	1.282	0.656	0.612	0.803	0.743	0.188	0.166	0.371	0.291	0.062	0.031

[1] Where specifying nominal size in decimals, zeros preceding decimal and in the fourth decimal place shall be omitted.

[2] A slight rounding of the edges at periphery of head shall be permissible provided the diameter of the bearing circle is equal to no less than 90 per cent of the specified minimum head diameter.

Source: ANSI B18.6.2-1972, AMERICAN NATIONAL STANDARD SLOTTED HEAD CAP SCREWS, SQUARE HEAD SET SCREWS, AND SLOTTED HEADLESS SET SCREWS.

Appendix I: American Standard Machine Screws and Machine Screw Nuts

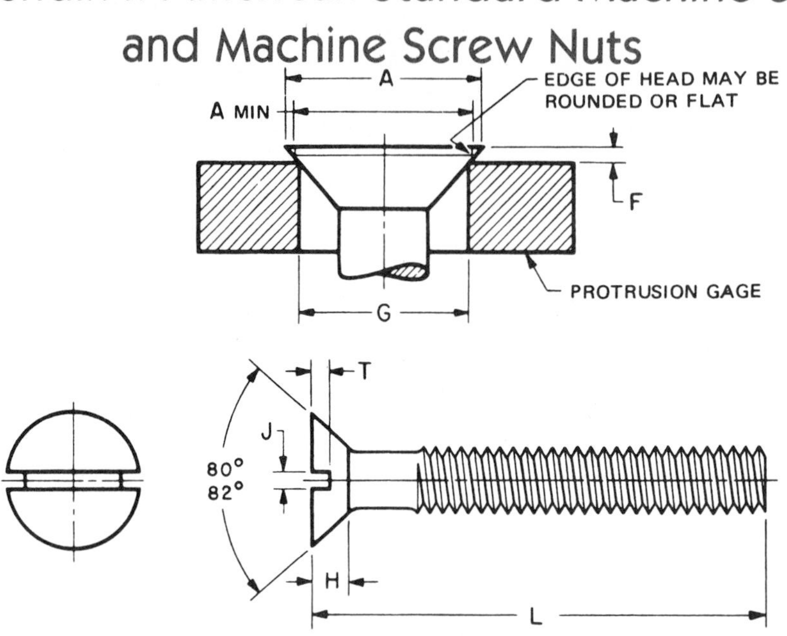

Table I.1: Dimensions of Slotted Flat Countersunk Head Machine Screws

Nominal Size[1] or Basic Screw Diameter		L	A		H	J		T		F		G
		These Lengths or Shorter are Undercut.	Head Diameter		Head Height	Slot Width		Slot Depth		Protrusion Above Gaging Diameter		Gaging Diameter
			Max, Edge Sharp	Min, Edge Rounded or Flat	Ref	Max	Min	Max	Min	Max	Min	
0000	0.0210	—	0.043	0.037	0.011	0.008	0.004	0.007	0.003	*	*	*
000	0.0340	—	0.064	0.058	0.016	0.011	0.007	0.009	0.005	*	*	*
00	0.0470	—	0.093	0.085	0.028	0.017	0.010	0.014	0.009	*	*	*
0	0.0600	1/8	0.119	0.099	0.035	0.023	0.016	0.015	0.010	0.026	0.016	0.078
1	0.0730	1/8	0.146	0.123	0.043	0.026	0.019	0.019	0.012	0.028	0.016	0.101
2	0.0860	1/8	0.172	0.147	0.051	0.031	0.023	0.023	0.015	0.029	0.017	0.124
3	0.0990	1/8	0.199	0.171	0.059	0.035	0.027	0.027	0.017	0.031	0.018	0.148
4	0.1120	3/16	0.225	0.195	0.067	0.039	0.031	0.030	0.020	0.032	0.019	0.172
5	0.1250	3/16	0.252	0.220	0.075	0.043	0.035	0.034	0.022	0.034	0.020	0.196
6	0.1380	3/16	0.279	0.244	0.083	0.048	0.039	0.038	0.024	0.036	0.021	0.220
8	0.1640	1/4	0.332	0.292	0.100	0.054	0.045	0.045	0.029	0.039	0.023	0.267
10	0.1900	5/16	0.385	0.340	0.116	0.060	0.050	0.053	0.034	0.042	0.025	0.313
12	0.2160	3/8	0.438	0.389	0.132	0.067	0.056	0.060	0.039	0.045	0.027	0.362
1/4	0.2500	7/16	0.507	0.452	0.153	0.075	0.064	0.070	0.046	0.050	0.029	0.424
5/16	0.3125	1/2	0.635	0.568	0.191	0.084	0.072	0.088	0.058	0.057	0.034	0.539
3/8	0.3750	9/16	0.762	0.685	0.230	0.094	0.081	0.106	0.070	0.065	0.039	0.653
7/16	0.4375	5/8	0.812	0.723	0.223	0.094	0.081	0.103	0.066	0.073	0.044	0.690
1/2	0.5000	3/4	0.875	0.775	0.223	0.106	0.091	0.103	0.065	0.081	0.049	0.739
9/16	0.5625	—	1.000	0.889	0.260	0.118	0.102	0.120	0.077	0.089	0.053	0.851
5/8	0.6250	—	1.125	1.002	0.298	0.133	0.116	0.137	0.088	0.097	0.058	0.962
3/4	0.7500	—	1.375	1.230	0.372	0.149	0.131	0.171	0.111	0.112	0.067	1.186

[1] Where specifying nominal size in decimals, zeros preceding decimal and in the fourth decimal place shall be omitted.

Source: ANSI B18.6.3-1972, AMERICAN NATIONAL STANDARD MACHINE SCREWS AND MACHINE SCREW NUTS.

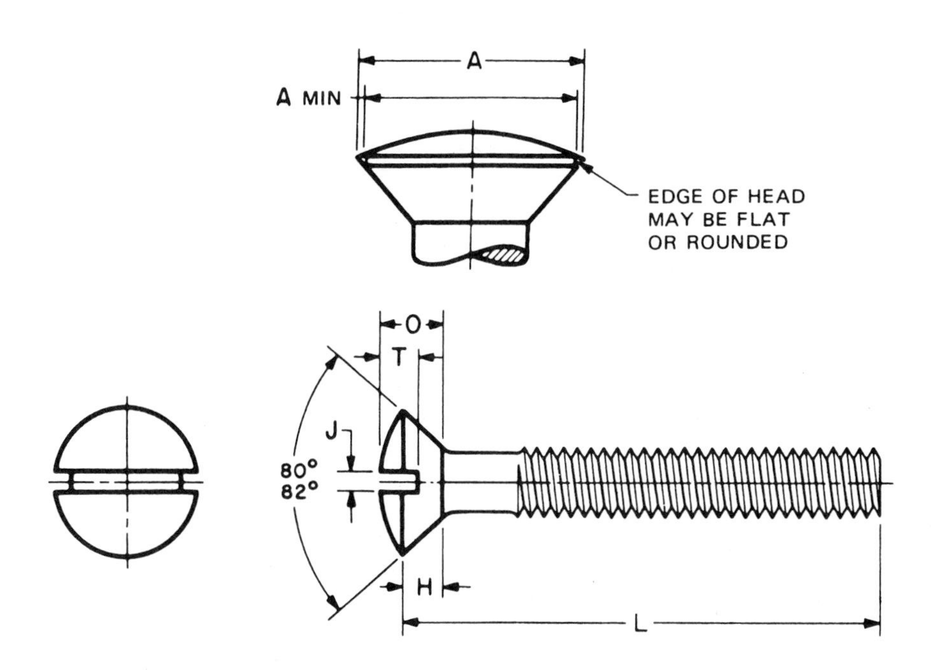

Table I.2: Dimensions of Slotted Oval Countersunk Head Machine Screws

Nominal Size[1] or Basic Screw Diameter		L	A		H	O		J		T	
		These Lengths or Shorter are Undercut	Head Diameter		Head Side Height	Total Head Height		Slot Width		Slot Depth	
			Max, Edge Sharp	Min, Edge Rounded or Flat	Ref	Max	Min	Max	Min	Max	Min
00	0.0470	—	0.093	0.085	0.028	0.042	0.034	0.017	0.010	0.023	0.016
0	0.0600	1/8	0.119	0.099	0.035	0.056	0.041	0.023	0.016	0.030	0.025
1	0.0730	1/8	0.146	0.123	0.043	0.068	0.052	0.026	0.019	0.038	0.031
2	0.0860	1/8	0.172	0.147	0.051	0.080	0.063	0.031	0.023	0.045	0.037
3	0.0990	1/8	0.199	0.171	0.059	0.092	0.073	0.035	0.027	0.052	0.043
4	0.1120	3/16	0.225	0.195	0.067	0.104	0.084	0.039	0.031	0.059	0.049
5	0.1250	3/16	0.252	0.220	0.075	0.116	0.095	0.043	0.035	0.067	0.055
6	0.1380	3/16	0.279	0.244	0.083	0.128	0.105	0.048	0.039	0.074	0.060
8	0.1640	1/4	0.332	0.292	0.100	0.152	0.126	0.054	0.045	0.088	0.072
10	0.1900	5/16	0.385	0.340	0.116	0.176	0.148	0.060	0.050	0.103	0.084
12	0.2160	3/8	0.438	0.389	0.132	0.200	0.169	0.067	0.056	0.117	0.096
1/4	0.2500	7/16	0.507	0.452	0.153	0.232	0.197	0.075	0.064	0.136	0.112
5/16	0.3125	1/2	0.635	0.568	0.191	0.290	0.249	0.084	0.072	0.171	0.141
3/8	0.3750	9/16	0.762	0.685	0.230	0.347	0.300	0.094	0.081	0.206	0.170
7/16	0.4375	5/8	0.812	0.723	0.223	0.345	0.295	0.094	0.081	0.210	0.174
1/2	0.5000	3/4	0.875	0.775	0.223	0.354	0.299	0.106	0.091	0.216	0.176
9/16	0.5625	—	1.000	0.889	0.260	0.410	0.350	0.118	0.102	0.250	0.207
5/8	0.6250	—	1.125	1.002	0.298	0.467	0.399	0.133	0.116	0.285	0.235
3/4	0.7500	—	1.375	1.230	0.372	0.578	0.497	0.149	0.131	0.353	0.293

[1]**Where specifying nominal size in decimals, zeros preceding decimal and in the fourth decimal place shall be omitted.**

Source: ANSI B18.6.3-1972, AMERICAN NATIONAL STANDARD MACHINE SCREWS AND MACHINE SCREW NUTS.

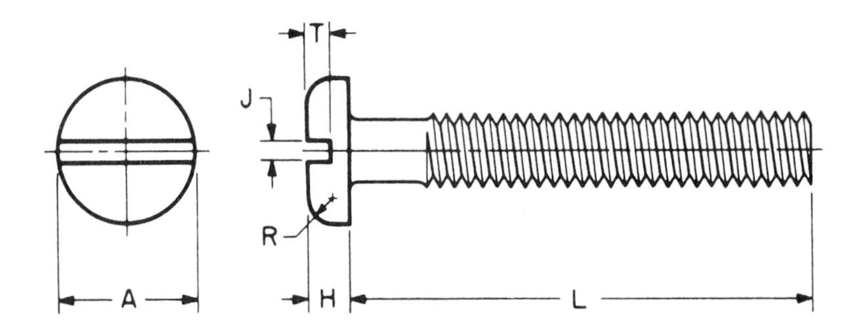

Table I.3: Dimensions of Slotted Pan Head Machine Screws

Nominal Size[1] or Basic Screw Diameter		A Head Diameter		H Head Height		R Head Radius	J Slot Width		T Slot Depth	
		Max	Min	Max	Min	Max	Max	Min	Max	Min
0000	0.0210	0.042	0.036	0.016	0.010	0.007	0.008	0.004	0.008	0.004
000	0.0340	0.066	0.060	0.023	0.017	0.010	0.012	0.008	0.012	0.008
00	0.0470	0.090	0.082	0.032	0.025	0.015	0.017	0.010	0.016	0.010
0	0.0600	0.116	0.104	0.039	0.031	0.020	0.023	0.016	0.022	0.014
1	0.0730	0.142	0.130	0.046	0.038	0.025	0.026	0.019	0.027	0.018
2	0.0860	0.167	0.155	0.053	0.045	0.035	0.031	0.023	0.031	0.022
3	0.0990	0.193	0.180	0.060	0.051	0.037	0.035	0.027	0.036	0.026
4	0.1120	0.219	0.205	0.068	0.058	0.042	0.039	0.031	0.040	0.030
5	0.1250	0.245	0.231	0.075	0.065	0.044	0.043	0.035	0.045	0.034
6	0.1380	0.270	0.256	0.082	0.072	0.046	0.048	0.039	0.050	0.037
8	0.1640	0.322	0.306	0.096	0.085	0.052	0.054	0.045	0.058	0.045
10	0.1900	0.373	0.357	0.110	0.099	0.061	0.060	0.050	0.068	0.053
12	0.2160	0.425	0.407	0.125	0.112	0.078	0.067	0.056	0.077	0.061
1/4	0.2500	0.492	0.473	0.144	0.130	0.087	0.075	0.064	0.087	0.070
5/16	0.3125	0.615	0.594	0.178	0.162	0.099	0.084	0.072	0.106	0.085
3/8	0.3750	0.740	0.716	0.212	0.195	0.143	0.094	0.081	0.124	0.100
7/16	0.4375	0.863	0.837	0.247	0.228	0.153	0.094	0.081	0.142	0.116
1/2	0.5000	0.987	0.958	0.281	0.260	0.175	0.106	0.091	0.161	0.131
9/16	0.5625	1.041	1.000	0.315	0.293	0.197	0.118	0.102	0.179	0.146
5/8	0.6250	1.172	1.125	0.350	0.325	0.219	0.133	0.116	0.197	0.162
3/4	0.7500	1.435	1.375	0.419	0.390	0.263	0.149	0.131	0.234	0.192

1 Where specifying nominal size in decimals, zeros preceding decimal and in the fourth decimal place shall be omitted.

Source: ANSI B18.6.3-1972, AMERICAN NATIONAL STANDARD MACHINE SCREWS AND MACHINE SCREW NUTS.

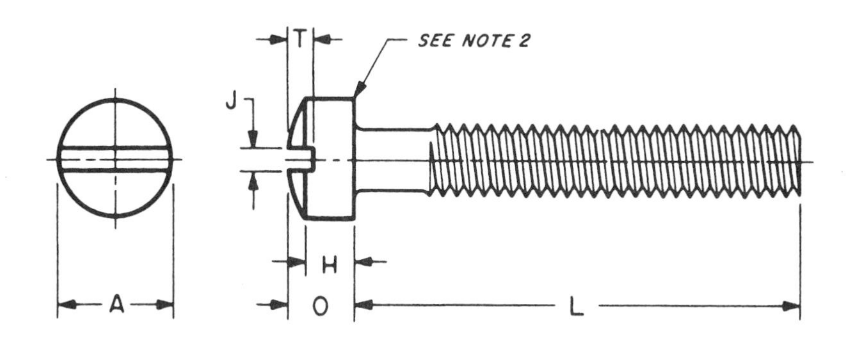

Table I.4: Dimensions of Slotted Fillister Head Machine Screws

Nominal Size[1] or Basic Screw Diameter	A Head Diameter		H Head Side Height		O Total Head Height		J Slot Width		T Slot Depth	
	Max	Min	Max	Min	Max	Min	Max	Min	Max	Min
0000 0.0210	0.038	0.032	0.019	0.011	0.025	0.015	0.008	0.004	0.012	0.006
000 0.0340	0.059	0.053	0.029	0.021	0.035	0.027	0.012	0.006	0.017	0.011
00 0.0470	0.082	0.072	0.037	0.028	0.047	0.039	0.017	0.010	0.022	0.015
0 0.0600	0.096	0.083	0.043	0.038	0.055	0.047	0.023	0.016	0.025	0.015
1 0.0730	0.118	0.104	0.053	0.045	0.066	0.058	0.026	0.019	0.031	0.020
2 0.0860	0.140	0.124	0.062	0.053	0.083	0.066	0.031	0.023	0.037	0.025
3 0.0990	0.161	0.145	0.070	0.061	0.095	0.077	0.035	0.027	0.043	0.030
4 0.1120	0.183	0.166	0.079	0.069	0.107	0.088	0.039	0.031	0.048	0.035
5 0.1250	0.205	0.187	0.088	0.078	0.120	0.100	0.043	0.035	0.054	0.040
6 0.1380	0.226	0.208	0.096	0.086	0.132	0.111	0.048	0.039	0.060	0.045
8 0.1640	0.270	0.250	0.113	0.102	0.156	0.133	0.054	0.045	0.071	0.054
10 0.1900	0.313	0.292	0.130	0.118	0.180	0.156	0.060	0.050	0.083	0.064
12 0.2160	0.357	0.334	0.148	0.134	0.205	0.178	0.067	0.056	0.094	0.074
1/4 0.2500	0.414	0.389	0.170	0.155	0.237	0.207	0.075	0.064	0.109	0.087
5/16 0.3125	0.518	0.490	0.211	0.194	0.295	0.262	0.084	0.072	0.137	0.110
3/8 0.3750	0.622	0.590	0.253	0.233	0.355	0.315	0.094	0.081	0.164	0.133
7/16 0.4375	0.625	0.589	0.265	0.242	0.368	0.321	0.094	0.081	0.170	0.135
1/2 0.5000	0.750	0.710	0.297	0.273	0.412	0.362	0.106	0.091	0.190	0.151
9/16 0.5625	0.812	0.768	0.336	0.308	0.466	0.410	0.118	0.102	0.214	0.172
5/8 0.6250	0.875	0.827	0.375	0.345	0.521	0.461	0.133	0.116	0.240	0.193
3/4 0.7500	1.000	0.945	0.441	0.406	0.612	0.542	0.149	0.131	0.281	0.226

[1] Where specifying nominal size in decimals, zeros preceding decimal and in the fourth decimal place shall be omitted.

[2] A slight rounding of the edges at periphery of head shall be permissible provided the diameter of the bearing circle is equal to no less than 90 per cent of the specified minimum head diameter.

Source: ANSI B18.6.3-1972, AMERICAN NATIONAL STANDARD MACHINE SCREWS AND MACHINE SCREW NUTS.

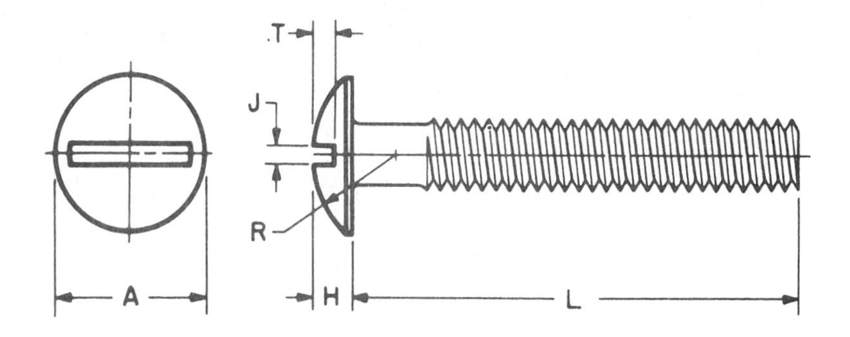

Table I.5: Dimensions of Slotted Truss Head Machine Screws

Nominal Size[1] or Basic Screw Diameter	A		H		R	J		T	
	Head Diameter		Head Height		Head Radius	Slot Width		Slot Depth	
	Max	Min	Max	Min	Max	Max	Min	Max	Min
0000 0.0210	0.049	0.043	0.014	0.010	0.032	0.009	0.005	0.009	0.005
000 0.0340	0.077	0.071	0.022	0.018	0.051	0.013	0.009	0.013	0.009
00 0.0470	0.106	0.098	0.030	0.024	0.070	0.017	0.010	0.018	0.012
0 0.0600	0.131	0.119	0.037	0.029	0.087	0.023	0.016	0.022	0.014
1 0.0730	0.164	0.149	0.045	0.037	0.107	0.026	0.019	0.027	0.018
2 0.0860	0.194	0.180	0.053	0.044	0.129	0.031	0.023	0.031	0.022
3 0.0990	0.226	0.211	0.061	0.051	0.151	0.035	0.027	0.036	0.026
4 0.1120	0.257	0.241	0.069	0.059	0.169	0.039	0.031	0.040	0.030
5 0.1250	0.289	0.272	0.078	0.066	0.191	0.043	0.035	0.045	0.034
6 0.1380	0.321	0.303	0.086	0.074	0.211	0.048	0.039	0.050	0.037
8 0.1640	0.384	0.364	0.102	0.088	0.254	0.054	0.045	0.058	0.045
10 0.1900	0.448	0.425	0.118	0.103	0.283	0.060	0.050	0.068	0.053
12 0.2160	0.511	0.487	0.134	0.118	0.336	0.067	0.056	0.077	0.061
1/4 0.2500	0.573	0.546	0.150	0.133	0.375	0.075	0.064	0.087	0.070
5/16 0.3125	0.698	0.666	0.183	0.162	0.457	0.084	0.072	0.106	0.085
3/8 0.3750	0.823	0.787	0.215	0.191	0.538	0.094	0.081	0.124	0.100
7/16 0.4375	0.948	0.907	0.248	0.221	0.619	0.094	0.081	0.142	0.116
1/2 0.5000	1.073	1.028	0.280	0.250	0.701	0.106	0.091	0.161	0.131
9/16 0.5625	1.198	1.149	0.312	0.279	0.783	0.118	0.102	0.179	0.146
5/8 0.6250	1.323	1.269	0.345	0.309	0.863	0.133	0.116	0.196	0.162
3/4 0.7500	1.573	1.511	0.410	0.368	1.024	0.149	0.131	0.234	0.182

[1] Where specifying nominal size in decimals, zeros preceding decimal and in the fourth decimal place shall be omitted.

Source: ANSI B18.6.3-1972, AMERICAN NATIONAL STANDARD MACHINE SCREWS AND MACHINE SCREW NUTS.

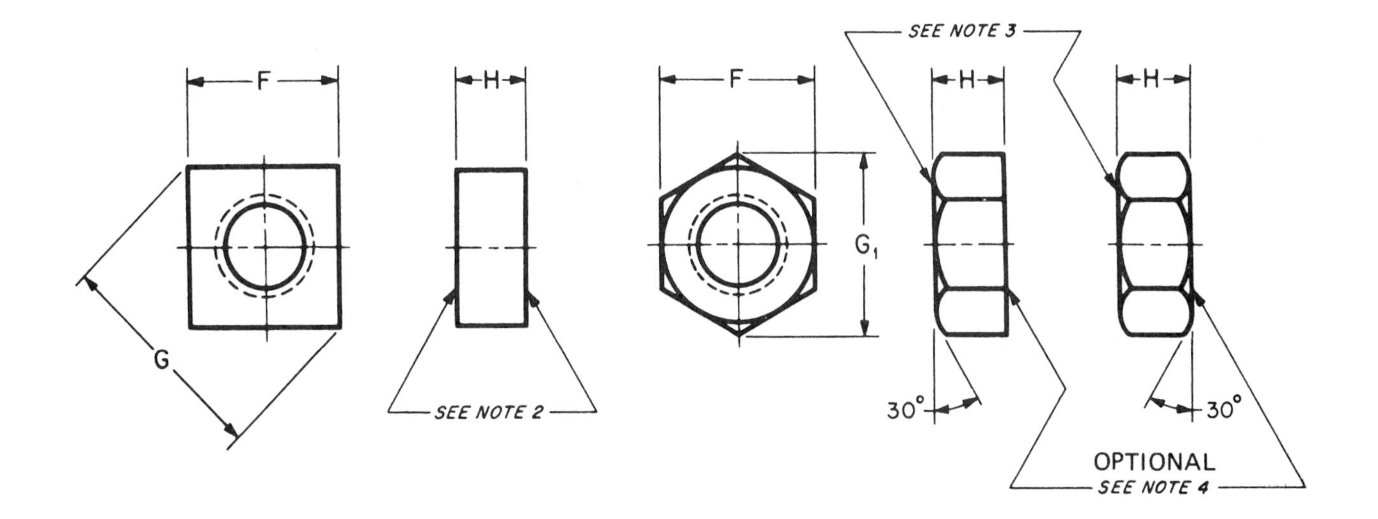

Table I.6: Dimensions of Square and Hex Machine Screws Nuts

Nominal Size[1] or Basic Thread Diameter	F			G		G_1		H	
	Width Across Flats			Width Across Corners				Thickness	
				Square		Hex			
	Basic	Max	Min	Max	Min	Max	Min	Max	Min
0 0.0600	5/32	0.156	0.150	0.221	0.206	0.180	0.171	0.050	0.043
1 0.0730	5/32	0.156	0.150	0.221	0.206	0.180	0.171	0.050	0.043
2 0.0860	3/16	0.188	0.180	0.265	0.247	0.217	0.205	0.066	0.057
3 0.0990	3/16	0.188	0.180	0.265	0.247	0.217	0.205	0.066	0.057
4 0.1120	1/4	0.250	0.241	0.354	0.331	0.289	0.275	0.098	0.087
5 0.1250	5/16	0.312	0.302	0.442	0.415	0.361	0.344	0.114	0.102
6 0.1380	5/16	0.312	0.302	0.442	0.415	0.361	0.344	0.114	0.102
8 0.1640	11/32	0.344	0.332	0.486	0.456	0.397	0.378	0.130	0.117
10 0.1900	3/8	0.375	0.362	0.530	0.497	0.433	0.413	0.130	0.117
12 0.2160	7/16	0.438	0.423	0.619	0.581	0.505	0.482	0.161	0.148
1/4 0.2500	7/16	0.438	0.423	0.619	0.581	0.505	0.482	0.193	0.178
5/16 0.3125	9/16	0.562	0.545	0.795	0.748	0.650	0.621	0.225	0.208
3/8 0.3750	5/8	0.625	0.607	0.884	0.833	0.722	0.692	0.257	0.239

[1] Where specifying nominal size in decimals, zeros preceding decimal and in the fourth decimal place shall be omitted.

[2] Square machine screw nuts shall have tops and bottoms flat, without chamfer. The bearing surface shall be perpendicular to the axis of the threaded hole pitch cylinder within a tolerance of 4 deg.

[3] Hexagon machine screw nuts shall have tops flat with chamfered corners. Diameter of top circle shall be equal to the maximum width across flats within a tolerance of minus 15 per cent. The bearing surface shall be perpendicular to the axis of the threaded hole pitch cylinder within a tolerance of 4 deg.

[4] Bottoms of hexagon machine screw nuts are normally flat, but for special purposes may be chamfered, if so specified by purchaser.

Source: ANSI B18.6.3-1972, AMERICAN NATIONAL STANDARD MACHINE SCREWS AND MACHINE SCREW NUTS.

Appendix J: American Standard Square Head Set Screws

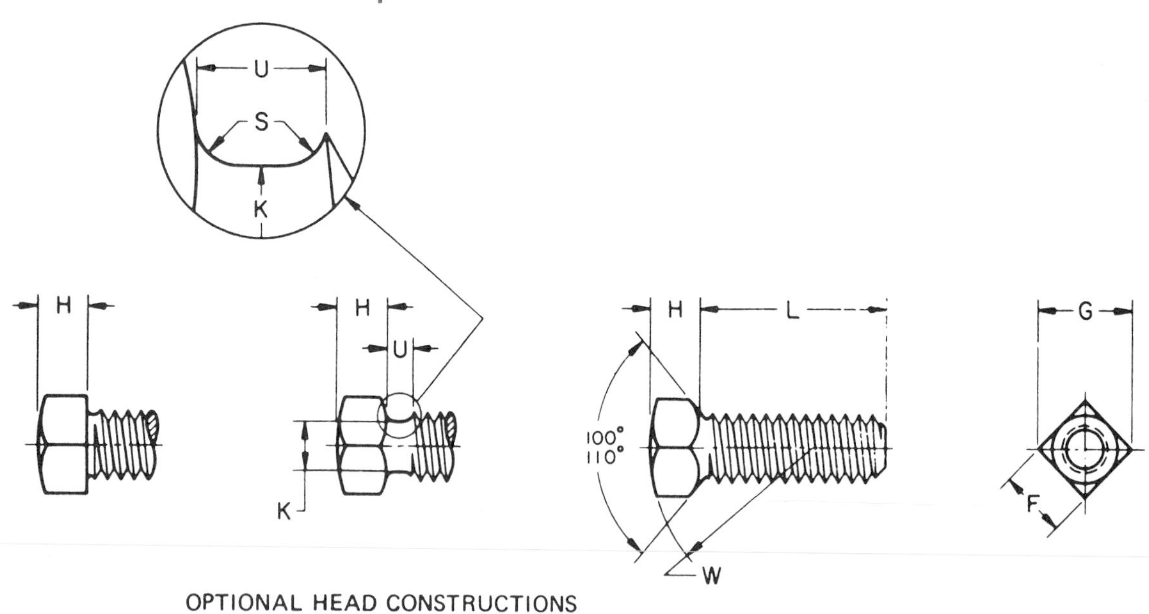

OPTIONAL HEAD CONSTRUCTIONS

Table J.1: Dimensions of Square Head Set Screws

Nominal Size[1] or Basic Screw Diameter	F Width Across Flats		G Width Across Corners		H Head Height		K Neck Relief Diameter		S Neck Relief Fillet Radius	U Neck Relief Width	W Head Radius
	Max	Min	Max	Min	Max	Min	Max	Min	Max	Min	Min
10 0.1900	0.188	0.180	0.265	0.247	0.148	0.134	0.145	0.140	0.027	0.083	0.48
1/4 0.2500	0.250	0.241	0.354	0.331	0.196	0.178	0.185	0.170	0.032	0.100	0.62
5/16 0.3125	0.312	0.302	0.442	0.415	0.245	0.224	0.240	0.225	0.036	0.111	0.78
3/8 0.3750	0.375	0.362	0.530	0.497	0.293	0.270	0.294	0.279	0.041	0.125	0.94
7/16 0.4375	0.438	0.423	0.619	0.581	0.341	0.315	0.345	0.330	0.046	0.143	1.09
1/2 0.5000	0.500	0.484	0.707	0.665	0.389	0.361	0.400	0.385	0.050	0.154	1.25
9/16 0.5625	0.562	0.545	0.795	0.748	0.437	0.407	0.454	0.439	0.054	0.167	1.41
5/8 0.6250	0.625	0.606	0.884	0.833	0.485	0.452	0.507	0.492	0.059	0.182	1.56
3/4 0.7500	0.750	0.729	1.060	1.001	0.582	0.544	0.620	0.605	0.065	0.200	1.88
7/8 0.8750	0.875	0.852	1.237	1.170	0.678	0.635	0.731	0.716	0.072	0.222	2.19
1 1.0000	1.000	0.974	1.414	1.337	0.774	0.726	0.838	0.823	0.081	0.250	2.50
1 1/8 1.1250	1.125	1.096	1.591	1.505	0.870	0.817	0.939	0.914	0.092	0.283	2.81
1 1/4 1.2500	1.250	1.219	1.768	1.674	0.966	0.908	1.064	1.039	0.092	0.283	3.12
1 3/8 1.3750	1.375	1.342	1.945	1.843	1.063	1.000	1.159	1.134	0.109	0.333	3.44
1 1/2 1.5000	1.500	1.464	2.121	2.010	1.159	1.091	1.284	1.259	0.109	0.333	3.75

[1] Where specifying nominal size in decimals, zeros preceding decimal and in the fourth decimal place shall be omitted.

Source: ANSI B18.6.2-1972, AMERICAN NATIONAL STANDARD SLOTTED HEAD CAP SCREWS, SQUARE HEAD SET SCREWS, AND SLOTTED HEADLESS SET SCREWS.

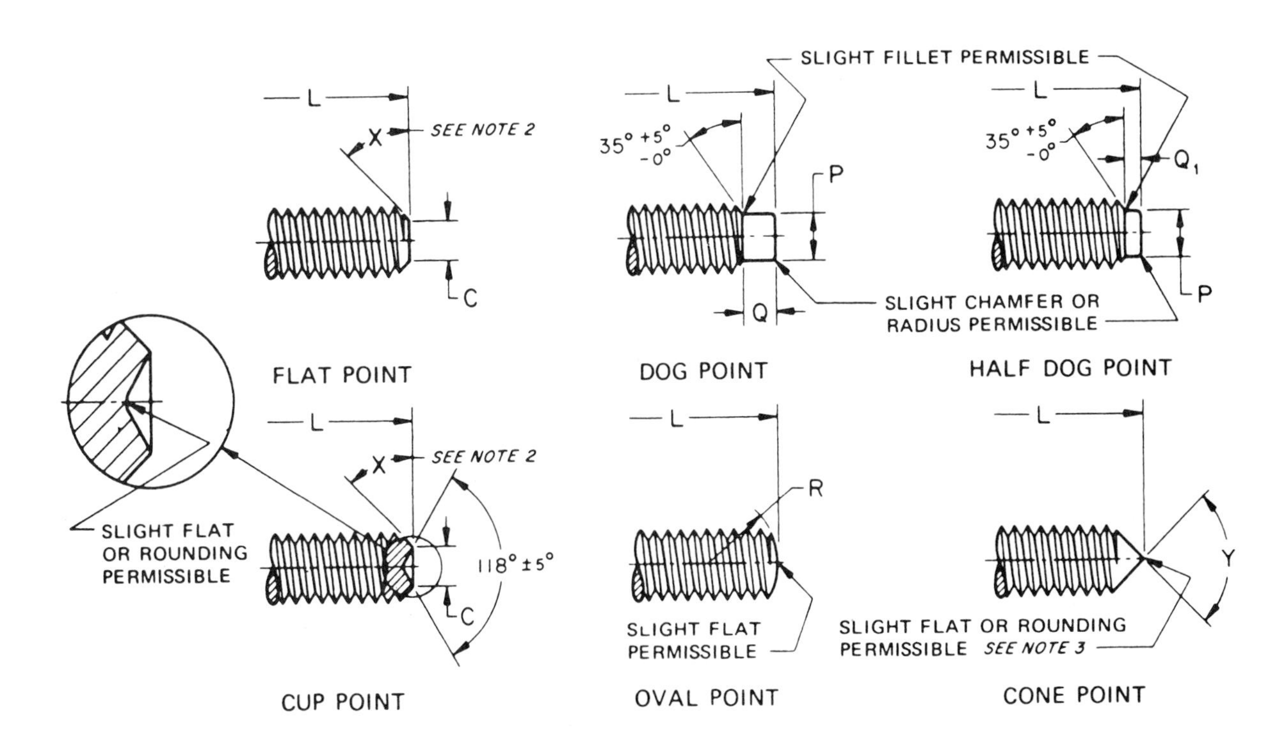

Table J.1—*Continued*

Nominal Size[1] or Basic Screw Diameter		C: Cup and Flat Point Diameters		P: Dog and Half Dog Point Diameters		Q: Point Length, Dog		Q_1: Point Length, Half Dog		R: Oval Point Radius	Y: Cone Point Angle 90° ±2° For These Nominal Lengths or Longer; 118° ±2° For Shorter Screws
		Max	Min	Max	Min	Max	Min	Max	Min	+0.031 -0.000	
10	0.1900	0.102	0.088	0.127	0.120	0.095	0.085	0.050	0.040	0.142	1/4
1/4	0.2500	0.132	0.118	0.156	0.149	0.130	0.120	0.068	0.058	0.188	5/16
5/16	0.3125	0.172	0.156	0.203	0.195	0.161	0.151	0.083	0.073	0.234	3/8
3/8	0.3750	0.212	0.194	0.250	0.241	0.193	0.183	0.099	0.089	0.281	7/16
7/16	0.4375	0.252	0.232	0.297	0.287	0.224	0.214	0.114	0.104	0.328	1/2
1/2	0.5000	0.291	0.270	0.344	0.334	0.255	0.245	0.130	0.120	0.375	9/16
9/16	0.5625	0.332	0.309	0.391	0.379	0.287	0.275	0.146	0.134	0.422	5/8
5/8	0.6250	0.371	0.347	0.469	0.456	0.321	0.305	0.164	0.148	0.469	3/4
3/4	0.7500	0.450	0.425	0.562	0.549	0.383	0.367	0.196	0.180	0.562	7/8
7/8	0.8750	0.530	0.502	0.656	0.642	0.446	0.430	0.227	0.211	0.656	1
1	1.0000	0.609	0.579	0.750	0.734	0.510	0.490	0.260	0.240	0.750	1 1/8
1 1/8	1.1250	0.689	0.655	0.844	0.826	0.572	0.552	0.291	0.271	0.844	1 1/4
1 1/4	1.2500	0.767	0.733	0.938	0.920	0.635	0.615	0.323	0.303	0.938	1 1/2
1 3/8	1.3750	0.848	0.808	1.031	1.011	0.698	0.678	0.354	0.334	1.031	1 5/8
1 1/2	1.5000	0.926	0.886	1.125	1.105	0.760	0.740	0.385	0.365	1.125	1 3/4

1 Where specifying nominal size in decimals, zeros preceding decimal and in the fourth decimal place shall be omitted.
2 Point angle X shall be 45° plus 5°, minus 0°, for screws of nominal lengths equal to or longer than those listed in Column Y, and 30° minimum for screws of shorter nominal lengths.
3 The extent of rounding or flat at apex of cone point shall not exceed an amount equivalent to 10 per cent of the basic screw diameter.

Source: ANSI B18.6.2-1972, AMERICAN NATIONAL STANDARD SLOTTED HEAD CAP SCREWS, SQUARE HEAD SET SCREWS, AND SLOTTED HEADLESS SET SCREWS.

Appendix K: Woodruff Keys and Keyseats

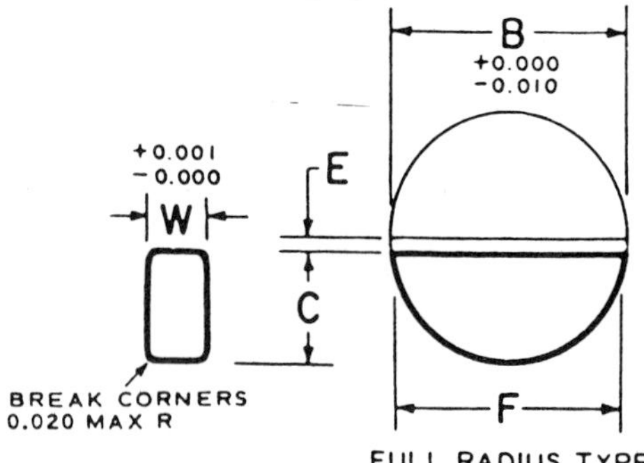

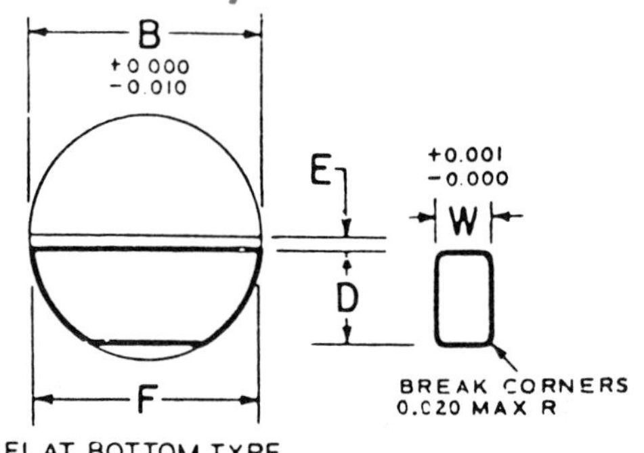

Table K.1: Woodruff Keys

Key No.	Nominal Key Size W × B	Actual Length F +0.000-0.010	Height of Key				Distance Below Center E
			C		D		
			Max	Min	Max	Min	
202	$\frac{1}{16} \times \frac{1}{4}$	0.248	0.109	0.104	0.109	0.104	$\frac{1}{64}$
202.5	$\frac{1}{16} \times \frac{5}{16}$	0.311	0.140	0.135	0.140	0.135	$\frac{1}{64}$
302.5	$\frac{3}{32} \times \frac{5}{16}$	0.311	0.140	0.135	0.140	0.135	$\frac{1}{64}$
203	$\frac{1}{16} \times \frac{3}{8}$	0.374	0.172	0.167	0.172	0.167	$\frac{1}{64}$
303	$\frac{3}{32} \times \frac{3}{8}$	0.374	0.172	0.167	0.172	0.167	$\frac{1}{64}$
403	$\frac{1}{8} \times \frac{3}{8}$	0.374	0.172	0.167	0.172	0.167	$\frac{1}{64}$
204	$\frac{1}{16} \times \frac{1}{2}$	0.491	0.203	0.198	0.194	0.188	$\frac{3}{64}$
304	$\frac{3}{32} \times \frac{1}{2}$	0.491	0.203	0.198	0.194	0.188	$\frac{3}{64}$
404	$\frac{1}{8} \times \frac{1}{2}$	0.491	0.203	0.198	0.194	0.188	$\frac{3}{64}$
305	$\frac{3}{32} \times \frac{5}{8}$	0.612	0.250	0.245	0.240	0.234	$\frac{1}{16}$
405	$\frac{1}{8} \times \frac{5}{8}$	0.612	0.250	0.245	0.240	0.234	$\frac{1}{16}$
505	$\frac{5}{32} \times \frac{5}{8}$	0.612	0.250	0.245	0.240	0.234	$\frac{1}{16}$
605	$\frac{3}{16} \times \frac{5}{8}$	0.612	0.250	0.245	0.240	0.234	$\frac{1}{16}$
406	$\frac{1}{8} \times \frac{3}{4}$	0.740	0.313	0.308	0.303	0.297	$\frac{1}{16}$
506	$\frac{5}{32} \times \frac{3}{4}$	0.740	0.313	0.308	0.303	0.297	$\frac{1}{16}$
606	$\frac{3}{16} \times \frac{3}{4}$	0.740	0.313	0.308	0.303	0.297	$\frac{1}{16}$
806	$\frac{1}{4} \times \frac{3}{4}$	0.740	0.313	0.308	0.303	0.297	$\frac{1}{16}$
507	$\frac{5}{32} \times \frac{7}{8}$	0.866	0.375	0.370	0.365	0.359	$\frac{1}{16}$
607	$\frac{3}{16} \times \frac{7}{8}$	0.866	0.375	0.370	0.365	0.359	$\frac{1}{16}$
707	$\frac{7}{32} \times \frac{7}{8}$	0.866	0.375	0.370	0.365	0.359	$\frac{1}{16}$
807	$\frac{1}{4} \times \frac{7}{8}$	0.866	0.375	0.370	0.365	0.359	$\frac{1}{16}$
608	$\frac{3}{16} \times 1$	0.992	0.438	0.433	0.428	0.422	$\frac{1}{16}$
708	$\frac{7}{32} \times 1$	0.992	0.438	0.433	0.428	0.422	$\frac{1}{16}$
808	$\frac{1}{4} \times 1$	0.992	0.438	0.433	0.428	0.422	$\frac{1}{16}$
1008	$\frac{5}{16} \times 1$	0.992	0.438	0.433	0.428	0.422	$\frac{1}{16}$
1208	$\frac{3}{8} \times 1$	0.992	0.438	0.433	0.428	0.422	$\frac{1}{16}$
609	$\frac{3}{16} \times 1\frac{1}{8}$	1.114	0.484	0.479	0.475	0.469	$\frac{5}{64}$
709	$\frac{7}{32} \times 1\frac{1}{8}$	1.114	0.484	0.479	0.475	0.469	$\frac{5}{64}$
809	$\frac{1}{4} \times 1\frac{1}{8}$	1.114	0.484	0.479	0.475	0.469	$\frac{5}{64}$
1009	$\frac{5}{16} \times 1\frac{1}{8}$	1.114	0.484	0.479	0.475	0.469	$\frac{5}{64}$

Source: USAS B17.2-1967.

Table K.1—*Continued*

Key No.	Nominal Key Size W × B	Actual Length F +0.000-0.010	Height of Key				Distance Below Center E
			C		D		
			Max	Min	Max	Min	
610	3/16 × 1 1/4	1.240	0.547	0.542	0.537	0.531	5/64
710	7/32 × 1 1/4	1.240	0.547	0.542	0.537	0.531	5/64
810	1/4 × 1 1/4	1.240	0.547	0.542	0.537	0.531	5/64
1010	5/16 × 1 1/4	1.240	0.547	0.542	0.537	0.531	5/64
1210	3/8 × 1 1/4	1.240	0.547	0.542	0.537	0.531	5/64
811	1/4 × 1 3/8	1.362	0.594	0.589	0.584	0.578	3/32
1011	5/16 × 1 3/8	1.362	0.594	0.589	0.584	0.578	3/32
1211	3/8 × 1 3/8	1.362	0.594	0.589	0.584	0.578	3/32
812	1/4 × 1 1/2	1.484	0.641	0.636	0.631	0.625	7/64
1012	5/16 × 1 1/2	1.484	0.641	0.636	0.631	0.625	7/64
1212	3/8 × 1 1/2	1.484	0.641	0.636	0.631	0.625	7/64

All dimensions given are in inches.

The key numbers indicate nominal key dimensions. The last two digits give the nominal diameter B in eighths of an inch and the digits preceding the last two give the nominal width W in thirty-seconds of an inch.

Example:

No. 204 indicates a key 2/32 × 4/8 or 1/16 × 1/2.
No. 808 indicates a key 8/32 × 8/8 or 1/4 × 1.
No. 1212 indicates a key 12/32 × 12/8 or 3/8 × 1 1/2.

Source: USAS B17.2-1967.

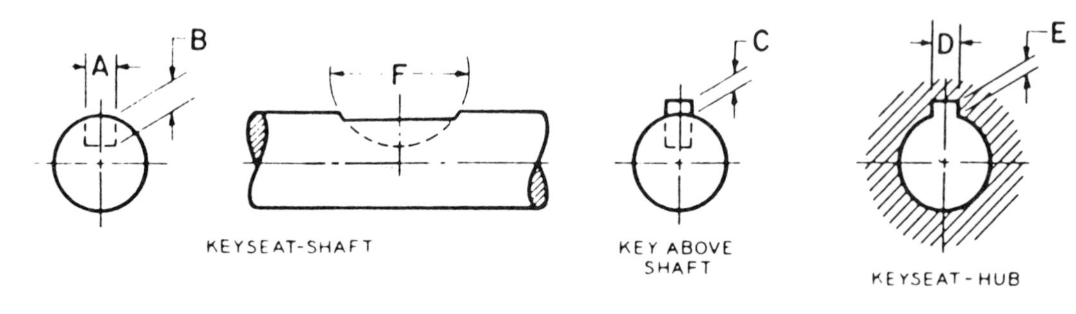

Table K.2 Keyseat Dimensions

Key Number	Nominal Size Key	Keyseat — Shaft					Key Above Shaft	Keyseat - Hub	
		Width A*		Depth B	Diameter F		Height C	Width D	Depth E
		Min	Max	+0.005 −0.000	Min	Max	+0.005 −0.005	+0.002 −0.000	+0.005 −0.000
202	1/16 x 1/4	0.0615	0.0630	0.0728	0.250	0.268	0.0312	0.0635	0.0372
202.5	1/16 x 5/16	0.0615	0.0630	0.1038	0.312	0.330	0.0312	0.0635	0.0372
302.5	3/32 x 5/16	0.0928	0.0943	0.0882	0.312	0.330	0.0469	0.0948	0.0529
203	1/16 x 3/8	0.0615	0.0630	0.1358	0.375	0.393	0.0312	0.0635	0.0372
303	3/32 x 3/8	0.0928	0.0914	0.1202	0.375	0.393	0.0469	0.0948	0.0529
403	1/8 x 3/8	0.1240	0.1255	0.1045	0.375	0.393	0.0625	0.1260	0.0685
204	1/16 x 1/2	0.0615	0.0630	0.1668	0.500	0.518	0.0312	0.0635	0.0372
304	3/32 x 1/2	0.0928	0.0943	0.1511	0.500	0.518	0.0469	0.0948	0.0529
404	1/8 x 1/2	0.1240	0.1255	0.1355	0.500	0.518	0.0625	0.1260	0.0685
305	3/32 x 5/8	0.0928	0.0943	0.1981	0.625	0.643	0.0469	0.0948	0.0529
405	1/8 x 5/8	0.1240	0.1255	0.1825	0.625	0.643	0.0625	0.1260	0.0685
505	5/32 x 5/8	0.1553	0.1568	0.1669	0.625	0.643	0.0781	0.1573	0.0841
605	3/16 x 5/8	0.1863	0.1880	0.1513	0.625	0.643	0.0937	0.1885	0.0997
406	1/8 x 3/4	0.1240	0.1255	0.2455	0.750	0.768	0.0625	0.1260	0.0685
506	5/32 x 3/4	0.1553	0.1568	0.2299	0.750	0.768	0-.0781	0.1573	0.0841
606	3/16 x 3/4	0.1863	0.1880	0.2143	0.750	0.768	0.0937	0.1885	0.0997
806	1/4 x 3/4	0.2487	0.2505	0.1830	0.750	0.768	0.1250	0.2510	0.1310
507	5/32 x 7/8	0.1553	0.1568	0.2919	0.875	0.895	0.0781	0.1573	0.0841
607	3/16 x 7/8	0.1863	0.1880	0.2763	0 875	0.895	0.0937	0.1885	0.0997
707	7/32 x 7/8	0.2175	0.2193	0.2607	0.875	0.895	0.1093	0.2198	0.1153
807	1/4 x 7/8	0.2487	0.2505	0.2450	0.875	0.895	0.1250	0.2510	0.1310
608	3/16 x 1	0.1863	0.1880	0.3393	1.000	1.020	0.0937	0.1885	0.0997
708	7/32 x 1	0.2175	0.2193	0.3237	1.000	1.020	0.1093	0.2198	0.1153
808	1/4 x 1	0.2487	0.2505	0.3080	1.000	1.020	0.1250	0.2510	0.1310
1008	5/16 x 1	0.3111	0.3130	0.2768	1.000	1.020	0.1562	0.3135	0.1622
1208	3/8 x 1	0.3735	0.3755	0.2455	1.000	1.020	0.1875	0.3760	0.1935
609	3/16 x 1 1/8	0.1863	0.1880	0.3853	1.125	1.145	0.0937	0.1885	0.0997
709	7/32 x 1 1/8	0.2175	0.2193	0.3697	1.125	1.145	0.1093	0.2198	0.1153
809	1/4 x 1 1/8	0.2487	0.2505	0.3540	1.125	1.145	0.1250	0.2510	0.1310
1009	5/16 x 1 1/8	0.3111	0.3130	0.3228	1.125	1.145	0.1562	0.3135	0.1622

Source: USAS B17.2-1967.

Appendix L: American Standard Washers

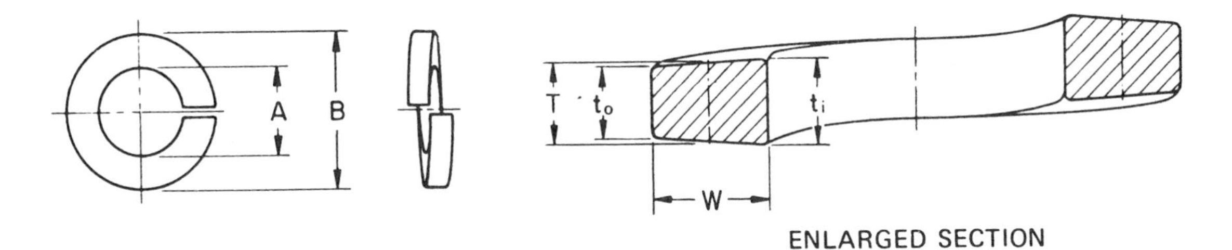

Table L.1: Dimensions of Regular Helical Spring Lock Washers[1]

Nominal Washer Size		A Inside Diameter		B Outside Diameter	T Mean Section Thickness $\left(\frac{t_i + t_o}{2}\right)$	W Section Width
		Max	Min	Max[2]	Min	Min
No. 2	0.086	0.094	0.088	0.172	0.020	0.035
No. 3	0.099	0.107	0.101	0.195	0.025	0.040
No. 4	0.112	0.120	0.114	0.209	0.025	0.040
No. 5	0.125	0.133	0.127	0.236	0.031	0.047
No. 6	0.138	0.148	0.141	0.250	0.031	0.047
No. 8	0.164	0.174	0.167	0.293	0.040	0.055
No. 10	0.190	0.200	0.193	0.334	0.047	0.062
No. 12	0.216	0.227	0.220	0.377	0.056	0.070
1/4	0.250	0.262	0.254	0.489	0.062	0.109
5/16	0.312	0.326	0.317	0.586	0.078	0.125
3/8	0.375	0.390	0.380	0.683	0.094	0.141
7/16	0.438	0.455	0.443	0.779	0.109	0.156
1/2	0.500	0.518	0.506	0.873	0.125	0.171
9/16	0.562	0.582	0.570	0.971	0.141	0.188
5/8	0.625	0.650	0.635	1.079	0.156	0.203
11/16	0.688	0.713	0.698	1.176	0.172	0.219
3/4	0.750	0.775	0.760	1.271	0.188	0.234
13/16	0.812	0.843	0.824	1.367	0.203	0.250
7/8	0.875	0.905	0.887	1.464	0.219	0.266
15/16	0.938	0.970	0.950	1.560	0.234	0.281
1	1.000	1.042	1.017	1.661	0.250	0.297
1 1/16	1.062	1.107	1.080	1.756	0.266	0.312
1 1/8	1.125	1.172	1.144	1.853	0.281	0.328
1 3/16	1.188	1.237	1.208	1.950	0.297	0.344
1 1/4	1.250	1.302	1.271	2.045	0.312	0.359
1 5/16	1.312	1.366	1.334	2.141	0.328	0.375
1 3/8	1.375	1.432	1.398	2.239	0.344	0.391
1 7/16	1.438	1.497	1.462	2.334	0.359	0.406
1 1/2	1.500	1.561	1.525	2.430	0.375	0.422

[1] Formerly designated Medium Helical Spring Lock Washers.
[2] The maximum outside diameters specified allow for the commercial tolerances on cold drawn wire.

Source: ANSI B18.21.1-1972, AMERICAN NATIONAL STANDARD LOCK WASHERS

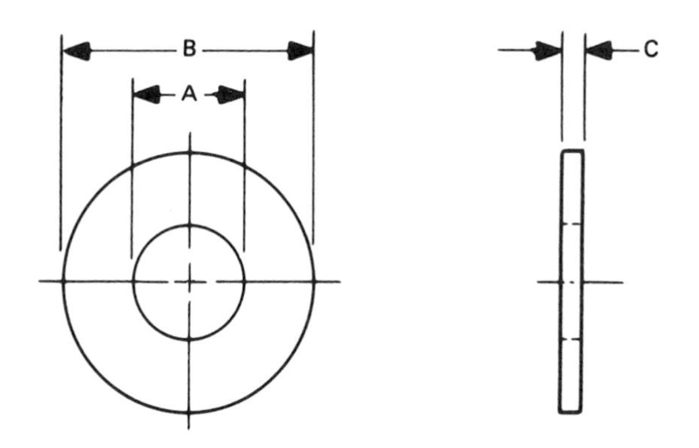

Table L.2: Dimensions of Metric Plain Washers (General Purpose)

Nominal Washer Size[1]	Washer Series	A Inside Diameter		B Outside Diameter		C Thickness	
		Max[2]	Min[3]	Max	Min	Max	Min
1.6	Narrow	2.09	1.95	4.00	3.70	0.70	0.50
	Regular	2.09	1.95	5.00	4.70	0.70	0.50
	Wide	2.09	1.95	6.00	5.70	0.90	0.60
2	Narrow	2.64	2.50	5.00	4.70	0.90	0.60
	Regular	2.64	2.50	6.00	5.70	0.90	0.60
	Wide	2.64	2.50	8.00	7.64	0.90	0.60
2.5	Narrow	3.14	3.00	6.00	5.70	0.90	0.60
	Regular	3.14	3.00	8.00	7.64	0.90	0.60
	Wide	3.14	3.00	10.00	9.64	1.20	0.80
3	Narrow	3.68	3.50	7.00	6.64	0.90	0.60
	Regular	3.68	3.50	10.00	9.64	1.20	0.80
	Wide	3.68	3.50	12.00	11.57	1.40	1.00
3.5	Narrow	4.18	4.00	9.00	8.64	1.20	0.80
	Regular	4.18	4.00	10.00	9.64	1.40	1.00
	Wide	4.18	4.00	15.00	14.57	1.75	1.20
4	Narrow	4.88	4.70	10.00	9.64	1.20	0.80
	Regular	4.88	4.70	12.00	11.57	1.40	1.00
	Wide	4.88	4.70	16.00	15.57	2.30	1.60
5	Narrow	5.78	5.50	11.00	10.57	1.40	1.00
	Regular	5.78	5.50	15.00	14.57	1.75	1.20
	Wide	5.78	5.50	20.00	19.48	2.30	1.60
6	Narrow	6.87	6.65	13.00	12.57	1.75	1.20
	Regular	6.87	6.65	18.80	18.37[4]	1.75	1.20
	Wide	6.87	6.65	25.40	24.88	2.30	1.60
8	Narrow	9.12	8.90	18.80	18.37	2.30	1.60
	Regular	9.12	8.90	25.40	24.48[4]	2.30	1.60
	Wide	9.12	8.90	32.00	31.38	2.80	2.00

Source: ANSI B18.22M-1981, AMERICAN NATIONAL STANDARD METRIC PLAIN WASHERS

Table L.2—*Continued*

Nominal Washer Size[1]	Washer Series	A		B		C	
		Inside Diameter		Outside Diameter		Thickness	
		Max[2]	Min[3]	Max	Min	Max	Min
10	Narrow	11.12	10.85	20.00	19.48	2.30	1.60
	Regular	11.12	10.85	28.00	27.48	2.80	2.00
	Wide	11.12	10.85	39.00	38.38	3.50	2.50
12	Narrow	13.57	13.30	25.40	24.88	2.80	2.00
	Regular	13.57	13.30	34.00	33.38	3.50	2.50
	Wide	13.57	13.30	44.00	43.38	3.50	2.50
14	Narrow	15.52	15.25	28.00	27.48	2.80	2.00
	Regular	15.52	15.25	39.00	38.38	3.50	2.50
	Wide	15.52	15.25	50.00	49.38	4.00	3.00
16	Narrow	17.52	17.25	32.00	31.38	3.50	2.50
	Regular	17.52	17.25	44.00	43.38	4.00	3.00
	Wide	17.52	17.25	56.00	54.80	4.60	3.50
20	Narrow	22.32	21.80	39.00	38.38	4.00	3.00
	Regular	22.32	21.80	50.00	49.38	4.60	3.50
	Wide	22.32	21.80	66.00	64.80	5.10	4.00
24	Narrow	26.12	25.60	44.00	43.38	4.60	3.50
	Regular	26.12	25.60	56.00	54.80	5.10	4.00
	Wide	26.12	25.60	72.00	70.80	5.60	4.50
30	Narrow	33.02	32.40	56.00	54.80	5.10	4.00
	Regular	33.02	32.40	72.00	70.80	5.60	4.50
	Wide	33.02	32.40	90.00	88.60	6.40	5.00
36	Narrow	38.92	38.30	66.00	64.80	5.60	4.50
	Regular	38.92	38.30	90.00	88.60	6.40	5.00
	Wide	38.92	38.30	110.00	108.60	8.50	7.00

NOTES:
(1) Nominal washer sizes are intended for use with comparable nominal screw or bolt sizes.
(2) See 4.3 for maximum permissible I.D. at the punch exit side.

Source: ANSI B18.22M-1981, AMERICAN NATIONAL STANDARD METRIC PLAIN WASHERS

Appendix M: Clevis and Cotter Pins

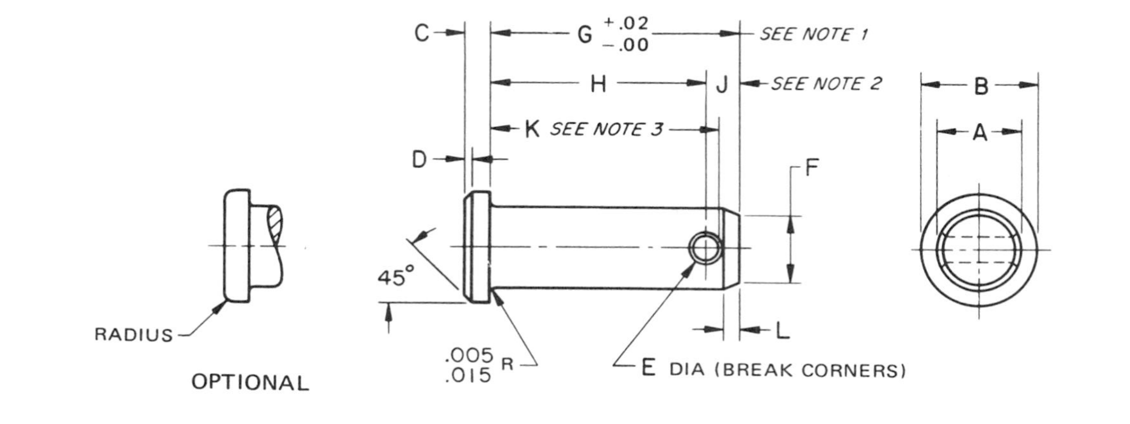

Table M.1: Dimensions of Clevis Pins

Nominal Size[1] or Basic Pin Diameter		A Shank Diameter		B Head Diameter		C Head Height		D Head Chamfer	E Hole Diameter		F Point Diameter		G[2] Pin Length	H Head to Center of Hole		J[3] End to Center Ref	K[4] Head to Edge of Hole Ref		L Point Length		Recommended Cotter Pin Nominal Size	
		Max	Min	Max	Min	Max	Min	±0.01	Max	Min	Max	Min	Basic	Max	Min	Basic	Max	Min	Max	Min		
3/16	0.188	0.186	0.181	0.32	0.30	0.07	0.05	0.02	0.088	0.073	0.15	0.14	0.58	0.504	0.484	0.09	0.548	0.520	0.055	0.035	1/16	0.062
1/4	0.250	0.248	0.243	0.38	0.36	0.10	0.08	0.03	0.088	0.073	0.21	0.20	0.77	0.692	0.672	0.09	0.736	0.708	0.055	0.035	1/16	0.062
5/16	0.312	0.311	0.306	0.44	0.42	0.10	0.08	0.03	0.119	0.104	0.26	0.25	0.94	0.832	0.812	0.12	0.892	0.864	0.071	0.049	3/32	0.093
3/8	0.375	0.373	0.368	0.51	0.49	0.13	0.11	0.03	0.119	0.104	0.33	0.32	1.06	0.958	0.938	0.12	1.018	0.990	0.071	0.049	3/32	0.093
7/16	0.438	0.436	0.431	0.57	0.55	0.16	0.14	0.04	0.119	0.104	0.39	0.38	1.19	1.082	1.062	0.12	1.142	1.114	0.071	0.049	3/32	0.093
1/2	0.500	0.496	0.491	0.63	0.61	0.16	0.14	0.04	0.151	0.136	0.44	0.43	1.36	1.223	1.203	0.15	1.298	1.271	0.089	0.063	1/8	0.125
5/8	0.625	0.621	0.616	0.82	0.80	0.21	0.19	0.06	0.151	0.136	0.56	0.55	1.61	1.473	1.453	0.15	1.548	1.521	0.089	0.063	1/8	0.125
3/4	0.750	0.746	0.741	0.94	0.92	0.26	0.24	0.07	0.182	0.167	0.68	0.67	1.91	1.739	1.719	0.18	1.830	1.802	0.110	0.076	5/32	0.156
7/8	0.875	0.871	0.866	1.04	1.02	0.32	0.30	0.09	0.182	0.167	0.80	0.79	2.16	1.989	1.969	0.18	2.080	2.052	0.110	0.076	5/32	0.156
1	1.000	0.996	0.991	1.19	1.17	0.35	0.33	0.10	0.182	0.167	0.93	0.92	2.41	2.239	2.219	0.18	2.330	2.302	0.110	0.076	5/32	0.156

[1] Where specifying nominal size in decimals, zeros preceding decimal shall be omitted.

[2] Lengths tabulated are intended for use with standard clevises, without spacers. When required, it is recommended that other pin lengths be limited wherever possible to nominal lengths in 0.06 in. increments.

[3] Basic "J" dimension (distance from centerline of hole to end of pin) is specified for calculating hole location from underside of head on pins of lengths not tabulated.

[4] Reference dimension provided for convenience in design layout and is not subject to inspection.

Source: ANSI B18.8.1-1972, AMERICAN NATIONAL STANDARD CLEVIS PINS AND COTTER PINS.

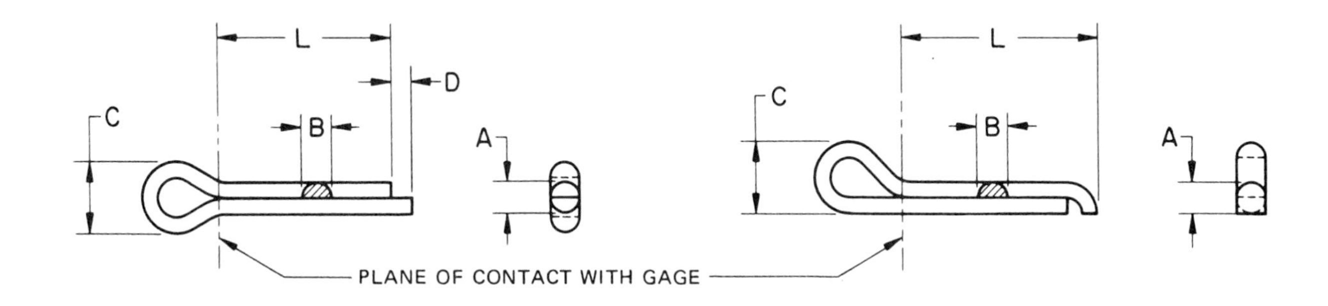

EXTENDED PRONG
SQUARE CUT TYPE

HAMMER LOCK TYPE

Table M.2: Dimensions of Cotter Pins

Nominal Size[1] or Basic Pin Diameter	A		B		C	D	Recommended Hole Size
	Total Shank Diameter		Wire Width		Head Diameter	Extended Prong Length	
	Max	Min	Max	Min	Min	Min	
1/32 0.031	0.032	0.028	0.032	0.022	0.06	0.01	0.047
3/64 0.047	0.048	0.044	0.048	0.035	0.09	0.02	0.062
1/16 0.062	0.060	0.056	0.060	0.044	0.12	0.03	0.078
5/64 0.078	0.076	0.072	0.076	0.057	0.16	0.04	0.094
3/32 0.094	0.090	0.086	0.090	0.069	0.19	0.04	0.109
7/64 0.109	0.104	0.100	0.104	0.080	0.22	0.05	0.125
1/8 0.125	0.120	0.116	0.120	0.093	0.25	0.06	0.141
9/64 0.141	0.134	0.130	0.134	0.104	0.28	0.06	0.156
5/32 0.156	0.150	0.146	0.150	0.116	0.31	0.07	0.172
3/16 0.188	0.176	0.172	0.176	0.137	0.38	0.09	0.203
7/32 0.219	0.207	0.202	0.207	0.161	0.44	0.10	0.234
1/4 0.250	0.225	0.220	0.225	0.176	0.50	0.11	0.266
5/16 0.312	0.280	0.275	0.280	0.220	0.62	0.14	0.312
3/8 0.375	0.335	0.329	0.335	0.263	0.75	0.16	0.375
7/16 0.438	0.406	0.400	0.406	0.320	0.88	0.20	0.438
1/2 0.500	0.473	0.467	0.473	0.373	1.00	0.23	0.500
5/8 0.625	0.598	0.590	0.598	0.472	1.25	0.30	0.625
3/4 0.750	0.723	0.715	0.723	0.572	1.50	0.36	0.750

[1]Where specifying nominal size in decimals, zeros preceding decimal shall be omitted.

Source: ANSI B18.8.1-1972, AMERICAN NATIONAL STANDARD CLEVIS PINS AND COTTER PINS.

Appendix N: Examples of Working Drawings

This appendix includes a set of working drawings for a 8¾-inch J44 rock bit cone used for oil drilling through hard rock. In addition, it includes a photograph of a 7⅞-inch J22 "Tri-cone" rock bit assembly. (The cones used in this assembly are similar, but not identical, to those described in the set of working drawings.)

Three distinct cone designs are required for a rock bit assembly. The first steel-toothed rock-drilling bit, designed and built by the Howard Hughes Tool Company, was used to sink an oil well at Goose Creek, Texas, in 1908. This bit allowed previously untapped oil deposits *below hard rock* to be drilled. Earlier wells had been sunk by applying the percussion method, that is, literally pounding soft rock repeatedly with a cutting tool. (See Figure N.3–N.5 for examples of professional engineering drawings of rock bit cones.)

Figures N.6 and N.7 are professional engineering drawings of a yoke and a lever, respectively.

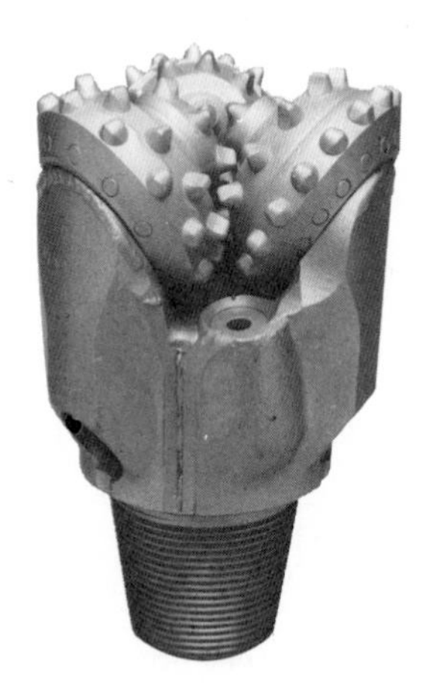

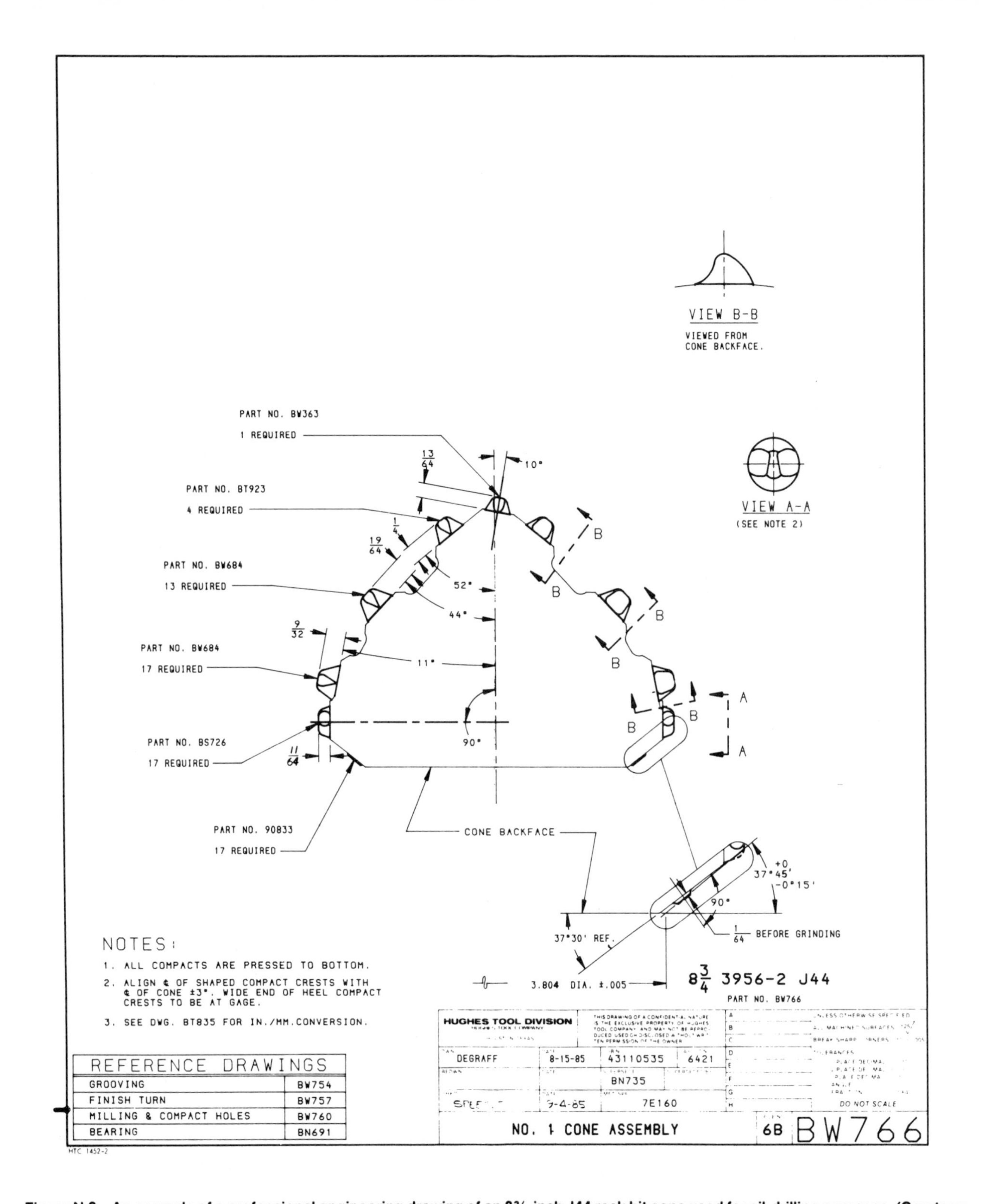

Figure N.2 An example of a professional engineering drawing of an 8¾-inch J44 rock bit cone used for oil-drilling purposes. (Courtesy of Hughes Tool Division, Houston, Texas.)

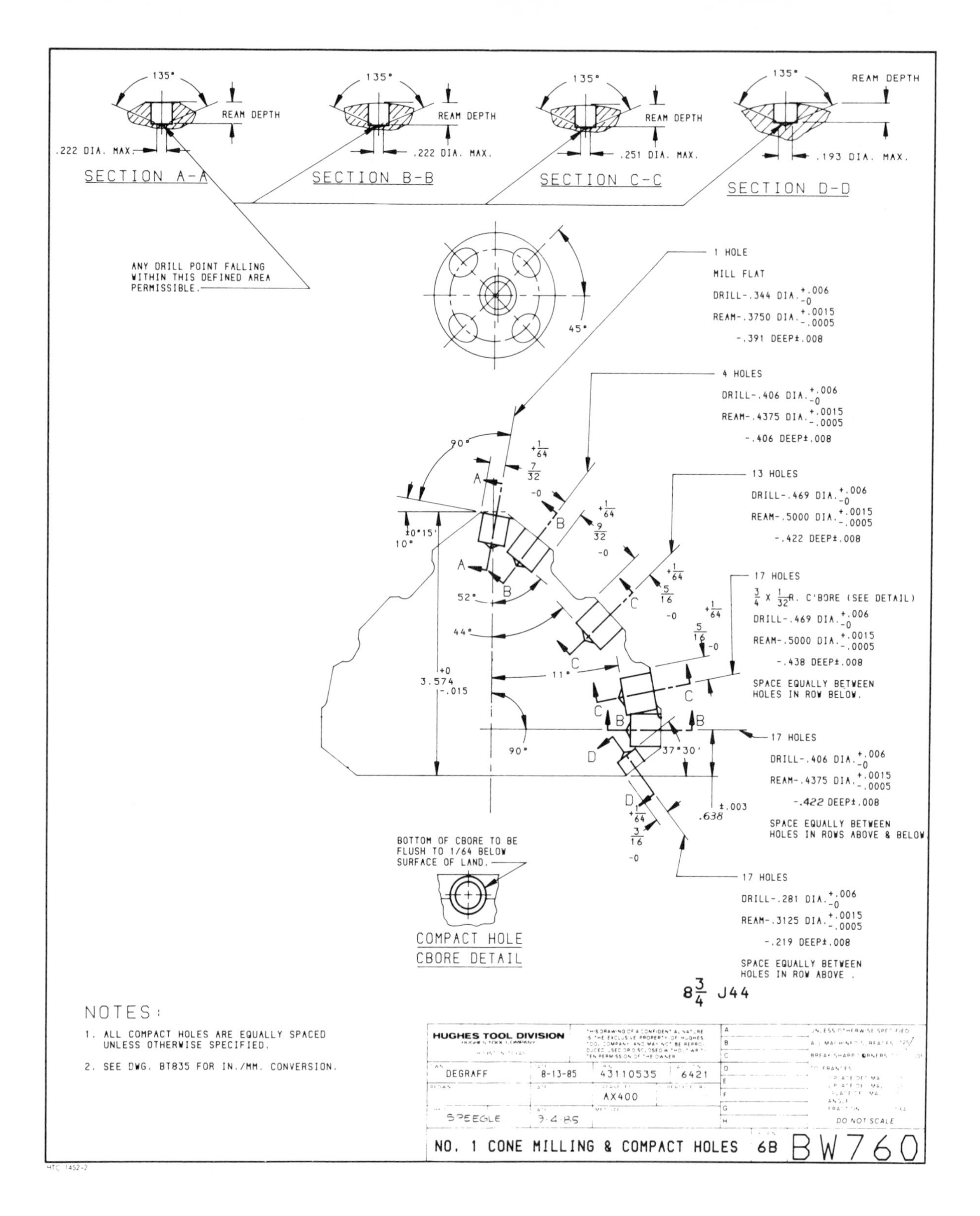

Figure N.3 An example of a professional engineering drawing in which the symmetry of the objects is used to present data in an efficient and compact form. (Courtesy of Hughes Tool Division, Houston, Texas.)

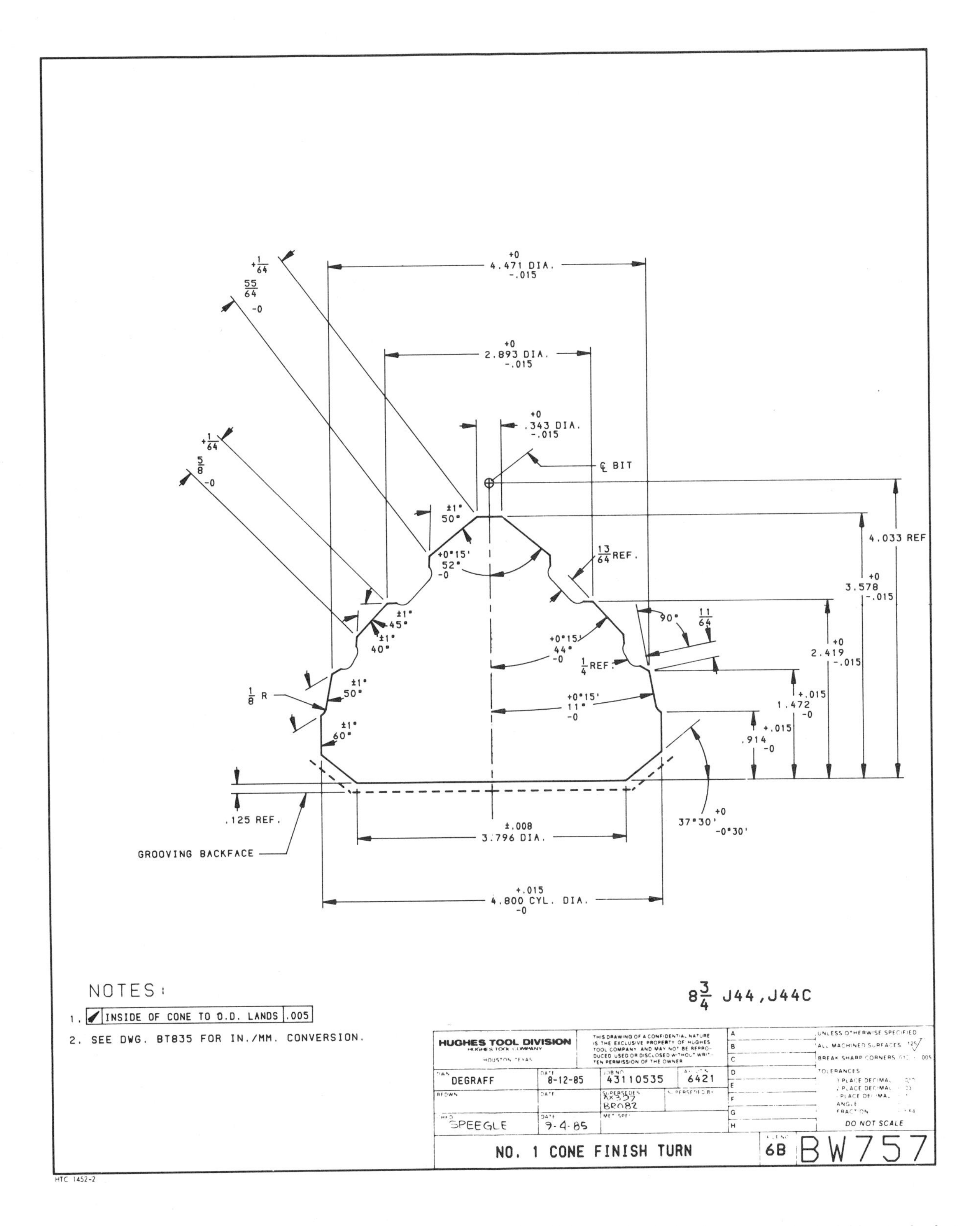

Figure N.4 An example of a professional engineering drawing in which additional detail for the fabrication of a rock bit cone is shown. (Courtesy of Hughes Tool Division, Houston, Texas.)

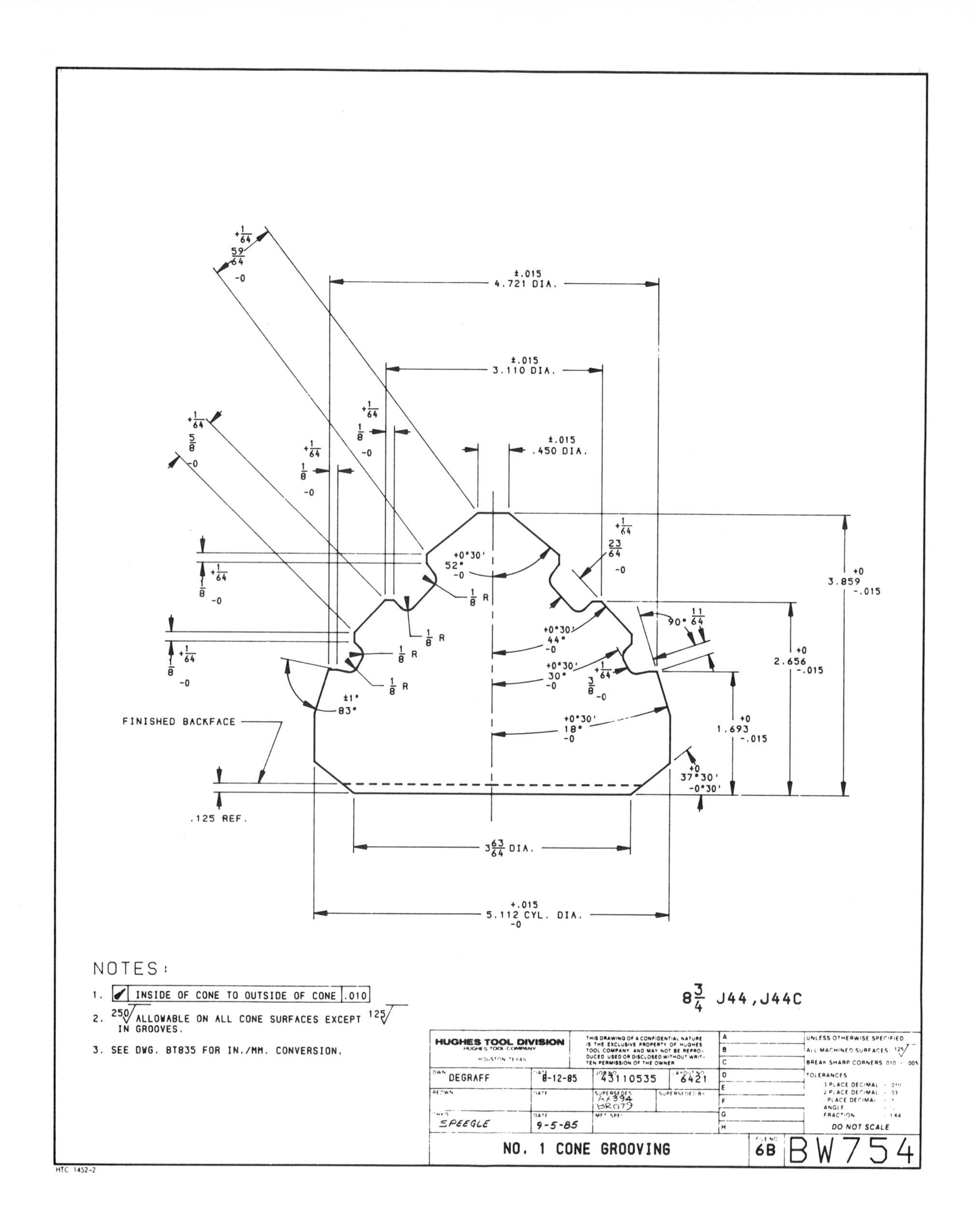

Figure N.5 Additional information for the fabrication of the rock bit cone is presented in this engineering drawing. (Courtesy of Hughes Tool Division, Houston, Texas.)

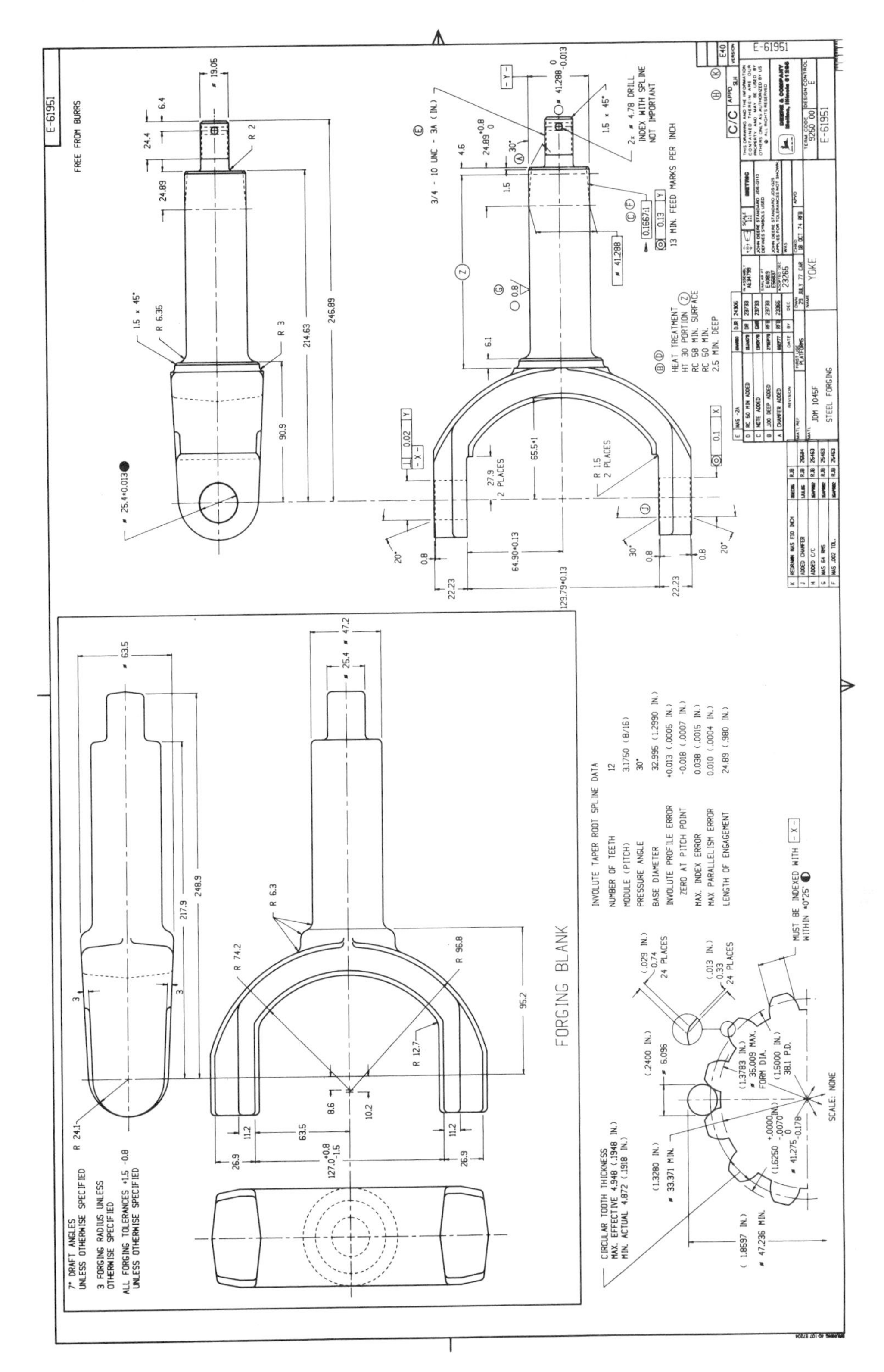

Figure N.6 An example of a professional working drawing. Note the detailed information that is provided for the manufacture of this part, including dimensioning and tolerancing data. (Courtesy of Deere & Company.) See section 10.12 on working drawings. This is Figure 10.16.

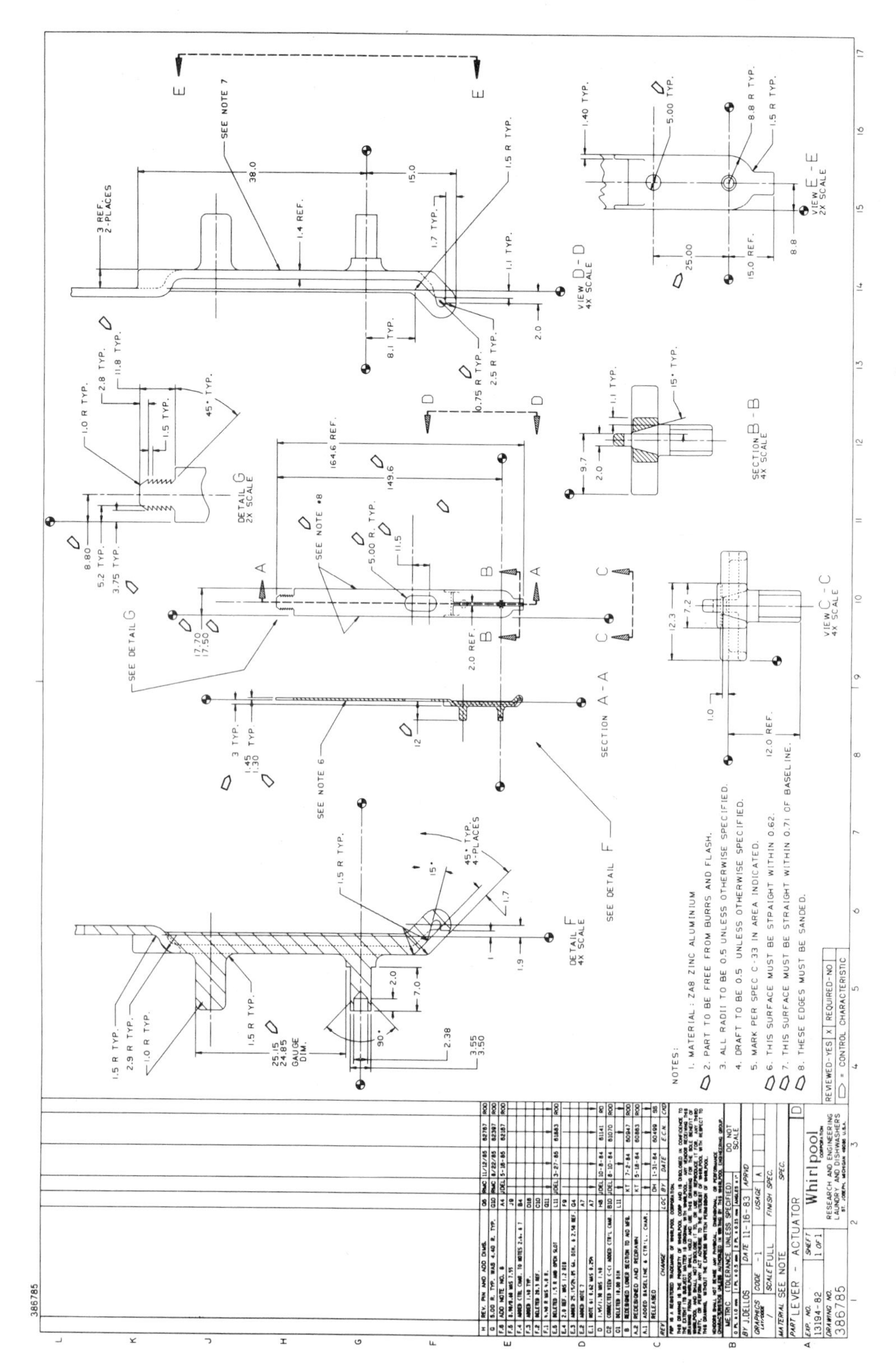

Figure N.7 A professional working drawing in which sectional and enlarged detail views are included for maximum clarity. Carefully read and coordinate the various views that are shown in this drawing. (Drawing courtesy of Whirlpool Corporation, Benton Harbor, MI.) See section 10.12 on working drawings. This is Figure 10.17.

Index